BIOCHEMISTRY

THIRD EDITION

GEOFFREY ZUBAY
COLUMBIA UNIVERSITY

WCB Wm. C. Brown Publishers
Dubuque, Iowa • Melbourne, Australia • Oxford, England

Book Team

Editor *Kevin Kane*
Developmental Editor *Megan Johnson*
Production Editor *Sherry Padden*
Designer *Mark Elliot Christianson*
Art Editor *Miriam J. Hoffman/Janice M. Roerig*
Photo Editor *Lori Gockel*
Permissions Editor *Vicki Krug*
Visuals/Design Developmental Consultant *Marilyn A. Phelps*

Wm. C. Brown Publishers
A Division of Wm. C. Brown Communications, Inc.

Vice President and General Manager *Beverly Kolz*
National Sales Manager *Vincent R. Di Blasi*
Assistant Vice President, Editor-in-Chief *Edward G. Jaffe*
Director of Marketing *John W. Calhoun*
Marketing Manager *Carol J. Mills*
Advertising Manager *Amy Schmitz*
Director of Production *Colleen A. Yonda*
Manager of Visuals and Design *Faye M. Schilling*
Design Manager *Jac Tilton*
Art Manager *Janice Roerig*
Publishing Services Manager *Karen J. Slaght*
Permissions/Records Manager *Connie Allendorf*

Wm. C. Brown Communications, Inc.

Chairman Emeritus *Wm. C. Brown*
Chairman and Chief Executive Officer *Mark C. Falb*
President and Chief Operating Officer *G. Franklin Lewis*
Corporate Vice President, President of WCB Manufacturing *Roger Meyer*

Biochemistry: © Orion/West Light
Artistic rendition of the double helix
Volume One: Energy, Cells, and Catalysis: Photograph by A. M. Lesk
Model for hemoglobin
Volume Two: Catabolism and Biosynthesis: Photograph by A. M. Lesk
Model for phosphofructokinase
Volume Three: Genetics and Physiology: Photograph by A. M. Lesk
Model for λ cro protein binding to DNA

Copyedited by *Emily Arulpragasam*

Freelance permissions editor *Karen Dorman*

The credits section for this book begins on page C-1 and is considered an extension of the copyright page.

Library of Congress Catalog Card Number: 91–76759

ISBN 0–697–14267–1 **Biochemistry** (Casebound)
0–697–14878–5 **Volume One: Energy, Cells, and Catalysis**
0–697–14879–3 **Volume Two: Catabolism and Biosynthesis**
0–697–14880–7 **Volume Three: Genetics and Physiology**
0–697–14877–7 **Biochemistry** (Boxed Set)

Printed in the United States of America by Wm. C. Brown Communications, Inc., 2460 Kerper Boulevard, Dubuque, IA 52001

10 9 8 7 6 5 4 3 2 1

Publisher's Note

Biochemistry, third edition, is available as a full-length casebound text, boxed set, or as three paperbound separates

Binding Option	Description	ISBN
Biochemistry (Casebound)	The full-length text, Chapters 1–36	0-697-14267-1
Biochemistry, Volume One (Paperbound)	Chapters 1–11	0-697-14878-5
Biochemistry, Volume Two (Paperbound)	Chapters 12–24	0-697-14879-3
Biochemistry, Volume Three (Paperbound)	Chapters 25–36	0-697-14880-7
Biochemistry, Boxed Set (Paperbound)	The full-length text in an attractive boxed set of all three paperback "splits".	0-697-14877-7

This book is dedicated to Ed Jaffe, the late Editor-in-Chief of the William C. Brown Publishing Company.

Ed passed away suddenly just before this text went to press. I am deeply saddened by his passing and that he will not see the text he backed so strongly. He was a man of wisdom and integrity who maintained an excellent rapport with the WCB staff and with its authors.

Brief Contents

Volume One

Volume Two

PART 4

Catabolism and the Generation of Chemical Energy 305

PART 5

Biosynthesis of the Building Blocks 473

Volume Three

PART 6

Storage and Utilization of Genetic Information 689

PART 7

Physiological Biochemistry 929

Contents

2

PART

Structure and Function of Major Components of the Cell 45

Chapter 3

The Building Blocks of Proteins: Amino Acids, Peptides, and Polypeptides 47

Geoffrey Zubay

Chapter 4

The Three-Dimensional Structure of Proteins 69

Geoffrey Zubay

Chapter 5

Functional Diversity of Proteins 106

Geoffrey Zubay and Irving Geis

Chapter 6

Carbohydrates, Glycoproteins, and Cell Walls 138

Pamela Stanley and Geoffrey Zubay

Chapter 14

The Tricarboxylic Acid Cycle 354

Daniel E. Atkinson and Geoffrey Zubay*

Chapter 15

Electron Transport and Oxidative Phosphorylation 379

William W. Parson

*Revised by Geoffrey Zubay from material in the preceding edition by Daniel E. Atkinson.

Chapter 16

Photosynthesis and Other Processes Involving Light 414

William W. Parson

Chapter 17

Metabolism of Fatty Acids 444

Dennis E. Vance

5 PART

Biosynthesis of the Building Blocks 473

Chapter 18

Biosynthesis of Amino Acids 475

H. Edwin Umbarger and Geoffrey Zubay

Chapter 21

Biosynthesis of Complex Carbohydrates 585

Pamela Stanley and Geoffrey Zubay

Chapter 22

Biosynthesis of Membrane Lipids and Related Substances 612

Dennis E. Vance

Chapter 23

Metabolism of Cholesterol 635

Dennis E. Vance

Chapter 24

Integration of Metabolism and Hormone Action 659

Richard Palmiter and Geoffrey Zubay

6 PART

Storage and Utilization of Genetic Information 689

Chapter 25

Structures of Nucleic Acids and Nucleoproteins 691

Geoffrey Zubay and Julius Marmur

Chapter 26

DNA Replication, Repair, and Recombination 721

Geoffrey Zubay and Julius Marmur

Chapter 27

DNA Manipulation and Its Applications 757

Geoffrey Zubay

Chapter 28

RNA Synthesis and Processing 791

Richard R. Burgess, Ann Baker Burgess, and Geoffrey Zubay

Chapter 29

Protein Synthesis, Targeting, and Turnover 830

James W. Bodley

Chapter 30

Regulation of Gene Expression in Prokaryotes 864

Geoffrey Zubay

Chapter 34

Carcinogenesis and Oncogenes 979

Geoffrey Zubay

Chapter 35

Neurotransmission 993

Gary R. Jacobson and Milton H. Saier, Jr.

Chapter 36

Vision 1012

William W. Parson

List of Supplementary Material

List of Contributors

Raymond L. Blakley
Chapter 20
Division of Biochemical and Clinical Pharmacology
St. Jude Children's Research Hospital
Memphis, TN 38101

James W. Bodley
Chapter 29
Biochemistry Department
University of Minnesota
Minneapolis, MN 55455

Ann Baker Burgess
Chapter 28
McArdle Laboratory for Cancer Research
University of Wisconsin
Madison, WI 53706

Richard R. Burgess
Chapter 28
McArdle Laboratory for Cancer Research
University of Wisconsin
Madison, WI 53706

Perry A. Frey
Chapter 11
Institute for Enzyme Research
University of Wisconsin
Madison, WI 53706

Irving Geis
Chapter 5
4700 Broadway
Apt. 4B
New York, NY 10040

Lloyd L. Ingram (Retired)
Chapter 2
Department of Biochemistry and Biophysics
University of California at Davis
Davis, CA 95616

Gary R. Jacobson
Chapters 7, 32, 35
Department of Biology
Boston University
Boston, MA 02215

Julius Marmur
Chapters 25, 26, 31
Department of Biochemistry
Albert Einstein College of Medicine
Bronx, NY 10461

Richard Palmiter
Chapter 24
Department of Biochemistry
University of Washington
Seattle, WA 98195

William W. Parson
Chapters 2, 8, 9, 10, 15, 16, 36
Department of Biochemistry
University of Washington
Seattle, WA 98195

Milton H. Saier, Jr.
Chapters 7, 32, 35
Department of Biology
University of California at San Diego
Lajolla, CA 92093

Pamela Stanley
Chapters 6, 21
Department of Cell Biology
Albert Einstein College of Medicine
Bronx, NY 10461

H. Edwin Umbarger
Chapters 18, 19
Department of Biological Sciences
Purdue University
West Lafayette, IN 47907

Dennis E. Vance
Chapters 7, 17, 22, 23
Lipid and Lipoprotein Group
Department of Biochemistry
University of Alberta
Edmonton, Alberta
Canada, T6G 2C2

Geoffrey Zubay
Chapters 1, 2, 3, 4, 5, 6, 12, 13, 14, 18, 19, 21, 24, 25, 26, 27, 28, 30, 31, 33, 34
Department of Biological Sciences
Columbia University
New York, NY 10027

Guided Tour through the Biochemistry Learning System

(A.)

CHAPTER OUTLINES

Each chapter begins with an outline. These will allow students to tell at a glance how the chapter is organized and what major topics have been included in the chapter.

2

CHAPTER

Thermodynamics in Biochemistry

(A.)

The primary usefulness of thermodynamics to biochemists lies in predicting whether particular chemical reactions could occur spontaneously. A simple illustration is to predict what compounds could possibly serve as energy sources for an organism. You are aware from everyday experience that oxidation of organic molecules by molecular oxygen releases energy. For example, wood or coal burns with a large output of heat. Similarly, organisms can obtain energy by oxidizing carbohydrates, fats, or proteins. Some organisms oxidize hydrocarbons, some oxidize reduced forms of sulfur, and others oxidize iron. But no organisms live by oxidizing molecular nitrogen, and the explanation lies in thermodynamics. The reaction cannot occur spontaneously. This example illustrates the importance of thermodynamics in controlling all life. Because organisms live by extracting chemical energy from their surroundings, thermodynamics is not an esoteric subject. It is a matter of life or death.

We say that thermodynamics determines whether a process "could" occur, because thermodynamics tells us only whether the process is possible, not whether it actually will occur in a finite period of time. The rate at which a thermodynamically possible reaction occurs depends on the detailed mechanism of the process. For a biochemical process to occur rapidly, appropriate enzymes must be available. The distinction between spontaneity and speed is more critical for biochemists than it is for chemists, because if a chemical reaction does not proceed rapidly a chemist can change the pressure or temperature or increase the concentration of the reactants. A living organism is under more rigid constraints: It must function at a fixed temperature and pressure and within a limited range of concentrations of reactants.

In this chapter we will elaborate on the thermodynamic quantities that we introduced in chapter 1. We will then expand the discussion to show how the concept of free energy is used in predicting biochemical pathways, and we will explore the central role of ATP in providing energy for biochemical reactions.

Thermodynamic Quantities

The properties of a substance can be classified as either intensive or extensive. Intensive properties, which include density, pressure, temperature, and concentration, do not depend on the amount of the material. Extensive properties, such as volume and weight, do depend on the amount. Most of the thermodynamic properties we will be discussing are extensive properties. These include energy (E), enthalpy (H), entropy (S), and free energy (G).

Energy, enthalpy, entropy, and free energy are all properties of the state of a substance. This means that they do not depend on how the substance was made or how it reached a particular state. In a chemical reaction, it is the difference between the initial and final states that is important; the pathway that is taken to get from the initial state to the final state has no bearing on whether the overall reaction releases or consumes energy (fig. 2.1).

Figure 6.16

Energetically favored conformations of ß(1,4)-linked D-glucose (*a*) and α(1,4)-linked D-glucose (*b*). Note that in the ß(1,4) configuration in (*a*), alternating residues are flipped 180° relative to one another so that long straight chains result. In the α(1,4) configuration (*b*), the chain has a natural curvature.

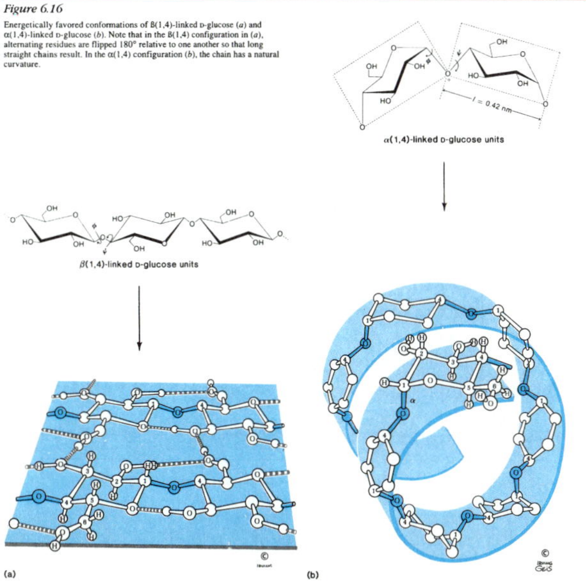

(B.)

to be discovered (1943) was the left-handed helix of amylose wound around molecules of iodine (fig. 6.17). This structure is responsible for the characteristic blue color of the amylose-iodine complex.

The extended-chain form of polyglucose has been exploited in nature for structural purposes, leaving by default the coiled form for use as an energy-storage macromolecule. Correlated with this functional difference is the omnipresence of degrading enzymes for glycogen and starch and the very limited phylogenetic distribution of comparable enzymes for cellulose. Cellulose is degraded in the gastrointestinal tract of herbivores, such as the cow, or in insects, such as termites, by a protozoan that synthesizes the enzyme cellulase. Humans do not possess this enzyme and hence cannot degrade cellulose.

148 Structure and Function of Major Components of the Cell

(B.)

DRAMATIC VISUALS PROGRAM

Colorful and informative photographs, illustrations, and tables enhance the learning program.

C.

D.

3A
BOX

Measurement of Ultraviolet Absorption in Solution

The general quantitative relationship that governs all absorption processes is called the Beer-Lambert law:

$$I = I_0 10^{-\epsilon cd}$$

where I_0 is the intensity of the incident radiation, I is the intensity of the radiation transmitted through a cell of thickness d (in centimeters) that contains a solution of concentration c (expressed either in moles per liter or in grams per 100 ml), and ϵ is the extinction coefficient, a characteristic of the substance being investigated (see fig. 3.7).

Light absorption is measured by a spectrophotometer as shown in the illustration. The spectrophotometer usually is capable of directly recording the absorbance A, which is related to I and I_0 by the equation

$$A = \log_{10} (I_0/I)$$

Hence $A = \epsilon cd$, and A is a direct measure of concentration. We can see from figure 3.7 that the ϵ values are largest for tryptophan and smallest for phenylalanine.

Since protein absorption maxima in the near-ultraviolet (240–300 nm) are determined by the content of the aromatic amino acids and their respective values, most proteins have absorption maxima in the 280-nm region. By contrast, absorption in the far-ultraviolet (around 190 nm) is shown by all polypeptides regardless of their aromatic amino acid content. The reason is that absorption in this region is due primarily to the peptide linkage.

Figure 1

Schematic diagram of a spectrophotometer for measuring light absorption. Laboratory instruments for making measurements are much more complex than this, but they all contain the same basic components: a light source, a monochromator, a sample, and a detector. λ is the wavelength of the light, I_0 and I are the incident light intensity and the transmitted light intensity, respectively, and d is the thickness of the absorbing solution.

Another convention for referring to configurations is called the R, S convention. As the R, S convention is not as popular for amino acids or sugars as it is for other types of biomolecules, such as lipids, we will not discuss this notation until chapter 11 (see box 11A).

Peptides and Polypeptides

Amino acids can link together by a covalent peptide bond between the α-carboxyl end of one amino acid and the α-amino end of another. Formally, this bond is formed by the loss of a water molecule, as shown in figure 3.9. The peptide bond has partial double-bond character owing to resonance effects; as a result, the C—N peptide linkage and all of the atoms directly connected to C and N lie in a planar configuration called the amide plane. In the following chapter we will see that this amide plane, by limiting the number of orientations available to the polypeptide chain, plays a major role in determining the three-dimensional structures of proteins.

Any number of amino acids can be joined by successive peptide linkages, forming a polypeptide chain. The polypeptide chain, like the dipeptide, has a directional sense. One end, called the N-terminal or amino-terminal end, has a free α-amino group, whereas the other end, the C-terminal or carboxyl-terminal end, has a free α-carboxyl group. The sequence of main-chain atoms from the N-terminal end to the C-terminal end is C_α—C—N—C_α, etc., and in the opposite direction it is C_α—N—C—C_α, etc. Short polypeptide chains, up to a length of about 20 amino acids, are called peptides or oligopeptides if they are fragments of whole polypeptide chains. A small protein molecule may contain a polypeptide chain of only 50 amino acids; a large protein may contain chains of 3,000 amino acids or more. One of the larger single polypeptide chains is that of the muscle protein myosin, which consists of approximately 1,750 amino acid residues. Figure 3.10 shows a section of a polypeptide chain as a linear array with α carbons and planar amides alternating as repeating units of the main chain. Different side chains are attached to each α carbon.

In addition to the covalent peptide bonds formed between adjacent amino acids within a polypeptide chain, covalent disulfide bonds can be formed within the same polypeptide chain or between different polypeptide chains (fig. 3.11). Such disulfide linkages have an important stabilizing influence on the structures formed by many proteins (see chapter 4).

54

Structure and Function of Major Components of the Cell

C.

SUPPLEMENTARY BOX MATERIAL

Throughout the text, supplemental information on experimental procedures, mechanisms, and methods of evaluation are presented in boxes

D.

UNDERLINED KEY CONCEPTS

Important terms and concepts are highlighted by underlining.

E.

END-OF-CHAPTER SUMMARIES

Each summary offers a concise review of the material covered in the chapter. Students can use the summary to review the chapter or to preview the important topics.

F.

SELECTED READINGS

Each chapter concludes with carefully selected references that contain further information on the topics covered in that chapter.

G.

END-OF-CHAPTER PROBLEMS

These problems will challenge students' mastery of the chapter's basic concepts. Odd-numbered problems are answered briefly in the back of the text.

E.

F.

G.

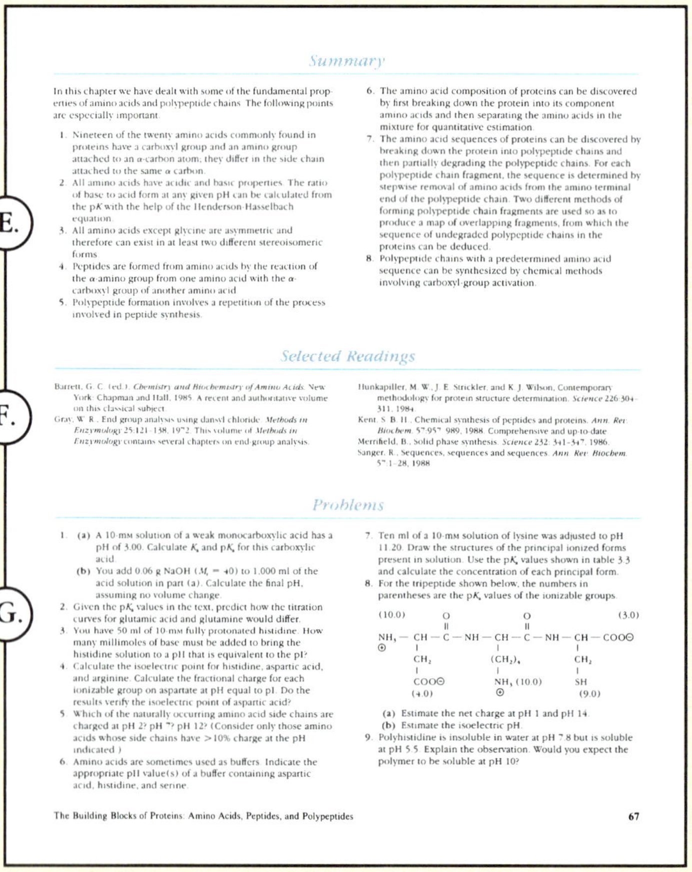
Summary

In this chapter we have dealt with some of the fundamental properties of amino acids and polypeptide chains. The following points are especially important.

1. Nineteen of the twenty amino acids commonly found in proteins have a carboxyl group and an amino group attached to an α-carbon atom; they differ in the side chain attached to the same α carbon.
2. All amino acids have acidic and basic properties. The ratio of base to acid form at any given pH can be calculated from the pK with the help of the Henderson-Hasselbach equation.
3. All amino acids except glycine are asymmetric and therefore can exist in at least two different stereoisomeric forms.
4. Peptides are formed from amino acids by the reaction of the α-amino group from one amino acid with the α-carboxyl group of another amino acid.
5. Polypeptide formation involves a repetition of the process involved in peptide synthesis.
6. The amino acid composition of proteins can be discovered by first breaking down the protein into its component amino acids and then separating the amino acids in the mixture for quantitative estimation.
7. The amino acid sequences of proteins can be discovered by breaking down the protein into polypeptide chains and then partially degrading the polypeptide chains. For each polypeptide chain fragment, the sequence is determined by stepwise removal of amino acids from the amino-terminal end of the polypeptide chain. Two different methods of forming polypeptide chain fragments are used so as to produce a map of overlapping fragments, from which the sequence of undegraded polypeptide chains in the proteins can be deduced.
8. Polypeptide chains with a predetermined amino acid sequence can be synthesized by chemical methods involving carboxyl-group activation.

Selected Readings

Barrett, G. C. (ed.). *Chemistry and Biochemistry of Amino Acids.* New York: Chapman and Hall, 1985. A recent and authoritative volume on this classical subject.

Gray, W. R., End group analysis using dansyl chloride. *Methods in Enzymology* 25:121–138, 1972. This volume of *Methods in Enzymology* contains several chapters on end-group analysis.

Hunkapiller, M. W., J. E. Strickler, and K. J. Wilson, Contemporary methodology for protein structure determination. *Science* 226:304–311, 1984.

Kent, S. B. H., Chemical synthesis of peptides and proteins. *Ann. Rev. Biochem.* 57:957–989, 1988. Comprehensive and up-to-date.

Merrifield, B., Solid phase synthesis. *Science* 232: 341–347, 1986.

Sanger, R., Sequences, sequences and sequences. *Ann. Rev. Biochem.* 57:1–28, 1988.

Problems

1. (a) A 10-mM solution of a weak monocarboxylic acid has a pH of 3.00. Calculate K_a and pK_a for this carboxylic acid.
   (b) You add 0.06 g NaOH (M_r = 40) to 1,000 ml of the acid solution in part (a). Calculate the final pH, assuming no volume change.
2. Given the pK_a values in the text, predict how the titration curves for glutamic acid and glutamine would differ.
3. You have 50 ml of 10-mM fully protonated histidine. How many millimoles of base must be added to bring the histidine solution to a pH that is equivalent to the pI?
4. Calculate the isoelectric point for histidine, aspartic acid, and arginine. Calculate the fractional charge for each ionizable group on aspartate at pH equal to pI. Do the results verify the isoelectric point of aspartic acid?
5. Which of the naturally occurring amino acid side chains are charged at pH 2? pH 7? pH 12? (Consider only those amino acids whose side chains have >10% charge at the pH indicated.)
6. Amino acids are sometimes used as buffers. Indicate the appropriate pH value(s) of a buffer containing aspartic acid, histidine, and serine.
7. Ten ml of a 10-mM solution of lysine was adjusted to pH 11.20. Draw the structures of the principal ionized forms present in solution. Use the pK_a values shown in table 3.3 and calculate the concentration of each principal form.
8. For the tripeptide shown below, the numbers in parentheses are the pK_a values of the ionizable groups.

   (a) Estimate the net charge at pH 1 and pH 14.
   (b) Estimate the isoelectric pH.
9. Polyhistidine is insoluble in water at pH 7.8 but is soluble at pH 5.5. Explain the observation. Would you expect the polymer to be soluble at pH 10?

The Building Blocks of Proteins: Amino Acids, Peptides, and Polypeptides

67

Preface

What makes biochemistry exciting? What makes biochemistry unique? What makes biochemistry so important? The answer to all three of these questions is virtually the same. Each and every reaction in biochemistry serves a function; that function is the maintenance and propagation of the living system. To understand biochemistry is to appreciate the place that each reaction occupies in the system.

Students of the previous generation discovered the principles that govern biochemistry, the basic reactions, and an immense body of facts consistent with these principles. Future biochemists to emerge from this generation of students will uncover the mysteries of developmental biology and physiology. They will find cures for cancers, cystic fibrosis, sickle cell anemia and many other diseases. My goal in putting this text together has been to convey my enthusiasm for this subject to this generation of students, to acquaint them with what is known and to prepare them for the discoveries of the future.

While the main principles of biochemistry are understood, the number of new facts that keep adding to the subject is immense. The field is becoming more and more subdivided as the volume of knowledge increases. No single person could directly cope with all the new information that is continuously appearing in the scientific literature. For the purpose of writing an authoritative up-to-date textbook on biochemistry, it would be ideal to involve a team of experts with hands-on experience. The contributing authors of this text are all research specialists who have made their mark in a specific area of biochemistry; they are teachers as well. Each contributing author could write an entire book on the chapter he/she has contributed to in this text. Recognizing that the primary goal of this text is to meet the needs of students facing the subject for the first time, each contributing author has had to pare down what he/she knows emphasizing principles, but not ignoring important facts. Each chapter has been carefully crafted with that goal in mind. My responsibility as coordinating author of this text has been to mold the contributions into a smooth reading, well integrated product.

This is the third edition of *Biochemistry.* Whereas I have had the same goal throughout this project, it is only with this edition, that the long-hoped-for product has finally emerged from our team approach. Reviewers teil us that this edition is not only consistently authoritative, it now reads like a cohesive, consistent single author textbook.

Each chapter emphasizes principles. Since the overriding principle is that every reaction serves a function, there is a strong emphasis on analysis of biochemical pathways and how different pathways are interrelated. In this pursuit balanced emphasis is given to chemistry and mechanisms, bioenergetics, metabolic regulation, and methods of biochemical analysis.

We have organized this text along conventional lines for teaching purposes. In Part 1 an overview is followed by an explanation of the importance of thermodynamics in biochemistry. In Part 2 we consider the structure and function of the major components of the cell. In Part 3 we discuss enzymes and coenzymes and the mechanisms of catalysis. Part 4 deals with energy producing catabolic processes, and Part 5 deals with energy consuming synthetic processes. In Part 6 we focus on nucleic acids and the way in which the biochemical information encoded in the chromosome is translated into the amino acid sequences found in proteins. Finally, in Part 7 we approach some of the applications of biochemistry to physiology.

Organizational Changes

The overall organization of the previous edition has been changed significantly:

- The chapter on methods for characterization and purification of proteins (formerly chapter 3) has been eliminated. Much of the material from this chapter has been disseminated into other chapters, appearing at the points where it is most relevant. The same goes for the discussion of methods in general, and there is a special appendix indicating where specific methods are discussed.

- The chapter on nucleic acid structure (formerly chapter 7) has been moved to a later part of the text (chapter 25). This has been done because nucleic acids are not discussed in any depth until Part 6. The general significance of nucleic acids is adequately discussed in chapter 1 so this should create no problem for the rare student who has not encountered this subject already.
- The chapter on origin of life has been eliminated. A few points about this subject are now made in chapters 1 and 28. This is a delightful subject and very close to my major research interests. However, it simply is not mainstream biochemistry and space would not allow us to keep this chapter.
- After many years of agonizing appraisal I have decided to place thermodynamics as the second chapter. Bioenergetics is so central to everything in biochemistry that I thought it should be discussed as soon as possible. At the same time an esoteric chapter on this subject could easily turn off many students. To avoid this we have considerably simplified this chapter to make it as palatable as possible without losing any of the significance of a more esoteric chapter.
- A new chapter on functional diversity of proteins (chapter 5) has been added. This gives us an opportunity to dwell on the very important subject of the relationship between protein structure and function. In addition we have used this chapter as a location to discuss the subject of protein purification.
- The chapters on lipids and membrane structures have been combined into one chapter. Students are more attentive to a description of lipids if they can see the important role they play in membrane structure.
- The chapters on lipid catabolism and synthesis have been combined in this edition (chapter 17). This makes it easier to discuss regulation of lipid metabolism.
- At the end of the treatment of intermediary metabolism (chapter 24) we have added a chapter on the integration of metabolism. In this chapter we have included most of the coverage of hormone action that was presented much later in the second edition. It is more relevant here as hormones are the key to the integration of metabolism in multicellular organisms.
- Two short chapters have been added to Part 7; chapter 33, on immunobiology, and chapter 34 on carcinogenesis. These are two hot topics that are of great interest to students. The chapters are kept short and simple for those who would choose to use them in a two term course. It would be best to consider them after the core chapters of Part 6 (i.e., chapters 25, 26, 28 and 29).

Content Features of This Edition

The coverage of biochemistry in this text is determined by the consensus feedback that we have received as well as the limitations of the present state of our knowledge. Coverage of the biochemistry of bacteria and mammals is emphasized over plant biochemistry.

Part 1 is entitled "An Overview of Biochemistry and Energy Considerations." The first chapter "Overview of Biochemistry," is very similar to the introductory chapter in the second edition. It presents the basic principles in chemistry and biology that relate to biochemistry. The second chapter, "Thermodynamics in Biochemistry," has been considerably rewritten and simplified over the corresponding chapter in the second edition, which had appeared as chapter 12. Chapter 2 elaborates on the thermodynamic quantities introduced in chapter 1, to show how the concept of free energy is used to predict the possibility that given reactions can occur. It also describes the central role of ATP in providing energy for biochemical reactions.

Part 2, "Structure and Function of Major Components of the Cell," contains five chapters describing the structure and function of proteins, carbohydrates, and lipids. The contents of Part 2 are similar to the contents of Part 1 in the previous edition, except that the discussion of nucleic acids has been moved to Part 6, and a new chapter 5, "Functional Diversity of Proteins," has been introduced.

Chapters 3, 4, and 5 present the basic structural and chemical properties of proteins. Chapter 3, "The Building Blocks of Proteins: Amino Acids, Peptides, and Polypeptides," is very similar to the comparable chapter in the second edition. Chapter 4, "The Three-Dimensional Structure of Proteins," describes the secondary, tertiary, and quaternary structures of proteins, the rules for which govern the folding of polypeptide chains and the ways in which these structures are determined. In this edition there is more coverage of the relationship between symmetry and quaternary structures than in the second edition. Chapter 5, "Functional Diversity of Proteins," focuses on the question of how protein structure relates to function and describes techniques used to isolate proteins.

Chapter 6, "Carbohydrates, Glycoproteins, and Cell Walls," describes the structural properties of sugars, oligosaccharides, and polysaccharides, along with the various roles that these compounds play as energy-storage molecules, components of glycoproteins, and in cell walls. This chapter's segment on glycoproteins has been considerably updated. Finally, in chapter 7, "Lipids and Membranes," two chapters from the second edition have been fused into one cohesive chapter. In it, the structural properties of lipids are discussed with reference to the roles lipids play as energy-storage molecules and in membrane structures.

Part 3, "Catalysis," is divided into four chapters, the first three of which have been completely rewritten, while the last has been simplified considerably, with discussions of lipid-soluble vitamins and the role of metals as cofactors added.

Chapter 8, "Enzyme Kinetics," features a description of kinetic methods for analyzing enzyme-catalyzed reactions. Chapter 9, "Mechanisms of Enzyme Catalysis," focuses on the way in which the structure of the enzyme active site is related to enzyme activity under different conditions. In this chapter a general discussion of factors determining enzyme specificity is followed by several detailed examples. Chapter 10, "Regulation of Enzyme Activities," describes enzymes regulated by structural modifications or the binding of specific factors at allosteric sites, and chapter 11, "Vitamins, Coenzymes, and Metal Cofactors," covers the function of small molecules acting in conjunction with enzymes as cocatalysts.

Part 4 "Catabolism and the Generation of Chemical Energy," is divided into six chapters. Since both Parts 4 and 5 are concerned with intermediary metabolism, it's not easy, or even appropriate, to separate the subject into artificially tidy discussions of catabolism and anabolism. Suffice it to say that the emphasis in Part 4 is on catabolism and the generation of chemical energy, while in Part 5 the emphasis is on biosynthesis. Chapter 12, "Metabolic Strategies," which initiates Part 4, relates strongly to both Parts 4 and 5, as does chapter 24, "Integration of Metabolism and Hormone Action," which concludes Part 5.

Chapter 12, "Metabolic Strategies," deals with the general ways in which metabolism is organized and the ways in which it is regulated. Chapters 13, 14, and 15 are devoted to carbohydrate metabolism in relationship to the generation of ATP. Chapter 13, "Glycolysis, Gluconeogenesis, and the Pentose Phosphate Pathway," describes anabolic as well as catabolic processes—namely, the synthesis of simple sugars and simple sugar polymers, as well as their breakdown. Both synthesis and breakdown are covered in the same chapter in order to facilitate meaningful discussion of the regulation of these two processes, which are intimately related.

Chapter 14, "The Tricarboxylic Acid Cycle," describes the further catabolism of sugars possible under aerobic conditions. Chapter 15, "Electron Transport and Oxidative Phosphorylation," complements chapter 14 because it shows how electrons released by the aerobic oxidation of sugars are channeled down the electron-transport system, ultimately to synthesize ATP. Chapter 16, "Photosynthesis and Other Processes Involving Light," appears immediately after the chapter on electron transport and the production of ATP because both chapters describe energy-generating metabolism. The complex light-powered processes are explained so that a minimal background in physics and chemistry is needed.

Part 4 concludes with Chapter 17, "Metabolism of Fatty Acids," which represents a fusion of two chapters from the second edition. The reasons for discussing both synthesis and breakdown of lipids in the same chapter are exactly the same as the reasons for doing this with sugars (as in chapter 13).

In many schools a one-semester treatment of biochemistry would end with chapter 17. Even though the first half of the text is tightly structured, instructors may still wish to seek additional material or skip around. Two recommendations involve chapter 25 and chapter 32. Chapter 25, "Structures of Nucleic Acids and Nucleoproteins," could be treated as a chapter in Part 2. If this is done, instructors should make sure students read the opening pages of chapter 20, which deal with the structure of nucleotides. Likewise, some instructors may wish to assign readings from chapter 32, which deals with mechanisms of membrane transport, after they have assigned chapter 7 on lipids and membranes.

Part 5, "Biosynthesis of the Building Blocks," contains seven chapters, mainly describing the biosynthesis and utilization of amino acids, nucleotides, complex carbohydrates, and lipids. Although there is a great deal of useful information in these chapters, most of Part 5, except for chapter 24, can be skipped if there is a strong desire to get to Part 6 as quickly as possible.

Chapter 18, "Biosynthesis of Amino Acids," describes the biosynthesis of amino acids. Because twenty amino acids found in proteins are made in bacteria and plants (whereas only a few amino acids are synthesized *de novo* in mammals), and because these pathways are better understood in bacteria, most of this chapter focuses on biosynthesis in bacteria. Nonprotein amino acids are discussed briefly. Finally, the problem of nitrogen fixation in biological systems is examined.

Chapter 19 is entitled "The Metabolic Fate of Amino Acids." It discusses the way in which amino acids, when in excess or when present as the only available carbon or energy source, are reduced to carbohydrates, which can serve as a source of energy or carbon skeletons for the synthesis of other molecules. A few products derived from amino acid metabolism are also discussed in detail, including porphyrins, biological active amines, glutathione, and peptide antibiotics.

Chapter 20, "Nucleotides," deals with the structures, synthesis, and breakdown of nucleotides, with considerable attention devoted to the role of nucleotide derivatives as inhibitors of nucleic acid synthesis and the important role they play in the medical field. The biosynthesis of nucleotide-containing coenzymes is also discussed in this chapter.

Chapter 21, "Biosynthesis of Complex Carbohydrates," describes the synthesis of complex carbohydrates that serve structural roles and the synthesis of oligosaccharides that exist in combination with proteins. A great deal of new information on this subject appears in this chapter, as does a detailed discussion of bacterial cell-wall synthesis.

Chapter 22, "Biosynthesis of Membrane Lipids and Related Substances," describes the metabolism of complex lipids. Phospholipids and sphingolipids are viewed in terms of their important structural roles in membranes, while eicosanoids are described in terms of functioning local hormones.

Chapter 23, "Metabolism of Cholesterol," describes the biosynthesis of cholesterol and some of its derivatives—bile acids and steroids. The chapter devotes considerable attention to the transport of these compounds, along with the problems associated with defects in cholesterol metabolism.

Chapter 24, "Integration of Metabolism and Hormone Action," concludes the two parts on intermediary metabolism. This chapter emphasizes the way in which intermediary metabolism is regulated in multicellular organisms.

Part 6, "Storage and Utilization of Genetic Information," features seven chapters describing the metabolism of nucleic acids and proteins. The entire section has been updated to keep pace with this rapidly changing area. Where useful, key genetic phenomena have also been explained.

This part begins with chapter 25, "Structures of Nucleic Acids and Nucleoproteins," which describes structures and functions of DNA and deoxyribonucleoproteins. It has intentionally been placed after structures of nucleotides have been discussed (in chapter 20) and before the structures of RNA and ribonucleoproteins are discussed (in chapter 28).

In chapter 26, "DNA Replication, Repair, and Recombination," the emphasis is on the metabolism of replication. The treatment is equally balanced between replication in prokaryotes and eukaryotes including animal viruses.

Chapter 27, "DNA Manipulation and Its Applications," begins with a discussion of the procedures for determining DNA sequence, proceeds with a discussion of the techniques for manipulating DNA sequences, and concludes by describing two important applications of DNA manipulation—the mapping of the globin gene family and the mapping of a gene responsible for the dreaded disease cystic fibrosis.

Chapter 28, "RNA Synthesis and Processing," which describes RNA synthesis and processing equally, presents a considerable amount of new information on transcription factors in animals cells and a variety of ways in which RNA is processed.

Chapter 29, "Protein Synthesis, Targeting, and Turnover," has been significantly expanded over the corresponding chapter in the second edition, particularly in the area of targeting, where many new discoveries have been made recently.

Chapter 30, "Regulation of Gene Expression in Prokaryotes," includes a discussion of several well-characterized gene systems found in *E coli*. It also features a treatment of the sequence of regulatory events involving bacteriophage λ infection and lysogeny.

Chapter 31, "Regulation of Gene Expression in Eukaryotes," begins by discussing several different gene systems in the intensively studied unicellular eukaryote *Saccharomyces cerevisiae*. Then it shifts to multicellular eukaryotes, focusing on the complex assemblages of protein factors involved in regulating various genes. Both this chapter and the previous one cover in considerable detail the structures of proteins that regulate gene expression and the ways in which they interact with DNA. The chapter concludes with a special section on the ways in which gene expression is regulated during early development.

Part 7, "Physiological Biochemistry," contains five chapters: "Mechanisms of Membrane Transport" (chapter 32), "Immunobiology" (chapter 33), "Carcinogenesis and Oncogenes" (chapter 34), "Neurotransmission" (chapter 35), and "Vision" (chapter 36). Except for the first one, all of these chapters are intentionally short, so that it will be convenient to assign them at different points in the course, depending on the taste of the instructor. The titles of the chapters are self-explanatory.

Learning Aids

We have included a number of in-text learning aids to help both the professor and the student navigate their way through the text.

The goal of this edition is to stress, not merely the facts, but especially the principles of biochemistry. In as many cases as is practical, explanations are given for how certain facts have been revealed by scientific investigations. Biochemists are sleuths always devising new methods to discover Nature's secrets. We felt it best to integrate discussion of the methods of biochemistry directly in the text at points where the methods are actually employed in biochemical investigations. To aid in location of a particular method description, we have placed a section entitled *A Student's Guide to Methods of Biochemical Analysis* in the appendices.

Great care has been taken to make explanations clear and to inject the maximum significance into topical headings. *Key terms* are always explained the first time they are introduced and *Key concepts* are underlined . Nevertheless, it is easy to forget the meaning of a particular term on occasion. To overcome this frustration we have included a *glossary* just before the index and a list of *common biochemical abbreviations,* we have placed a list just inside the back book cover.

As an additional learning aid, a *brief introduction* to each of the book's seven parts is included. Furthermore, each chapter begins with an *outline of contents* and an *introduction.*

All chapters have *summaries* at the end, and the longer chapters have periodic summaries within them. For those who would like to know more about any specific topic, *extensive references* are included at the end of each chapter.

The *exercise problems* at the end of the chapters are intended to reinforce what has been learned by the reading. *Brief answers* to selected problems are given at the end of the book, and instructors can obtain a *Solutions Manual* that provides complete solutions to all of the problems in the text (see the following section on ancillary materials).

Ancillary Material

In choosing the order of our presentation, we have been strongly influenced by extensive feedback from instructors of biochemistry. As always, however, some instructors would have preferred a different ordering of the subject matter. For their benefit we have suggested alternative possibilities for using the text in an accompanying instructor's manual. This manual also includes additional problems and solutions not offered in the text.

As an aid to course presentations, a set of 100 overhead color transparencies and 200 black-and-white transparency masters is available upon adoption of the text. Also available upon request to professors is a solutions manual, including worked-out solutions to all of the end-of-chapter problems. This manual is available for sale to students, but only at the instructor's request.

Acknowledgments

This preface is full of "I"s. That's because I, the coordinating author, wrote it. By contrast, the text itself is the brainchild of some twenty biochemists, of which I was only one. Nevertheless, because the manuscripts submitted by the contributing authors were modified by me, I must take the blame for any errors. If you should find errors, please notify me so that I can make corrections in the second printing. I would be most grateful for your input.

I would like to make a special note that Chapters 12, 13, and 14 were written by me based on Dr. Daniel Atkinson's material from the second edition of *Biochemistry.* So again, any errors should be considered my responsibility.

My fellow authors and I consulted many experts during the writing of each specific chapter, largely on an informal, "over the phone" basis. Although the list of their names is too long to include here, we are indebted to all of them. In addition, we had the benefit of formal reviews of chapters and whole sections of the text. I am deeply grateful to each of the reviewers, whose aid we will certainly wish to solicit for future editions. The reviewers included:

Teh-hui Kao *Penn State University*
Todd P. Silverstein *Willamette University*
Paul Melius *Auburn University*
Galen Mell *University of Montana*
Edward J. Miller *University of Alabama at Birmingham*
Rodney Cate *Midwestern State University*
Ricki Lewis *SUNY Albany*
Ronald A. MacQuarrie *University of Missouri-Kansas City*
Gerald W. Hart *Johns Hopkins University School of Medicine*
Larry Kirk *California State University at Chico*
Ralph C. Jacobson *California State University-San Luis Obispo*
Peter D. Jones *University of Tennessee*
C. Reynold Verret *Tulane University*
Arthur L. Haas, Ph.D *Medical College of Wisconsin*
Hans Gunderson *Northern Arizona University*
Ezio A. Moscatelli *University of Missouri at Columbia*
Gail Dinter-Gotleib *Drexel University*
David Gross *University of Massachusetts*
Edye E. Groseclose *Southeastern University of Health Sciences*
Maria O. Longas *Purdue University Calumet*
Bruce Howard Weber *California State University, Fullerton*
Donald R. Halenz *Pacific Union College*
Anthony T. Tu *Colorado State University*
Gary D. Small *University of South Dakota School of Medicine*
Celia Marshak *San Diego State University*
D. J. Davis *University of Arkansas*
Karl A. Wilson *SUNY at Binghamton*
Harry R. Matthews *University of California, Davis School of Medicine*
Richard E. Ebel *Virginia Polytechnic Institute and State University*
Malcolm Potts *Virginia Polytechnic Institute and State University*
Robert Lindquist *San Francisco State University*
Jeff Velten *New Mexico State University*

I am also very grateful to many workers at the Wm. C. Brown Publishing Company. First of all to Kevin Kane, who shared my vision and made a tremendous investment in our text. I am overwhelmed by the tremendous help received from Meg Johnson, who gave me advice on many aspects of the text, who obtained a multitude of top-notch reviews that were most useful in preparing the final manuscript, and who stuck by me during the hectic period of production. At times production problems became so great that it wasn't easy to keep going. I was very lucky to have a production coordinator, Sherry Padden, from whom I could get valuable advice on a day-by-day basis. She functioned well beyond the normal responsibilities of her job to help me and the other production workers to pull this project together and bring it to a successful conclusion.

Last but not least I once again had the opportunity to benefit from the superb copy editing of Emily Arulpragasam, who smoothed out bumps in the prose and helped me spot errors before they could be set in type. Her understanding of the subject was a major asset in making changes in the text that were always scientifically accurate.

Storage and Utilization of Genetic Information

6

PART

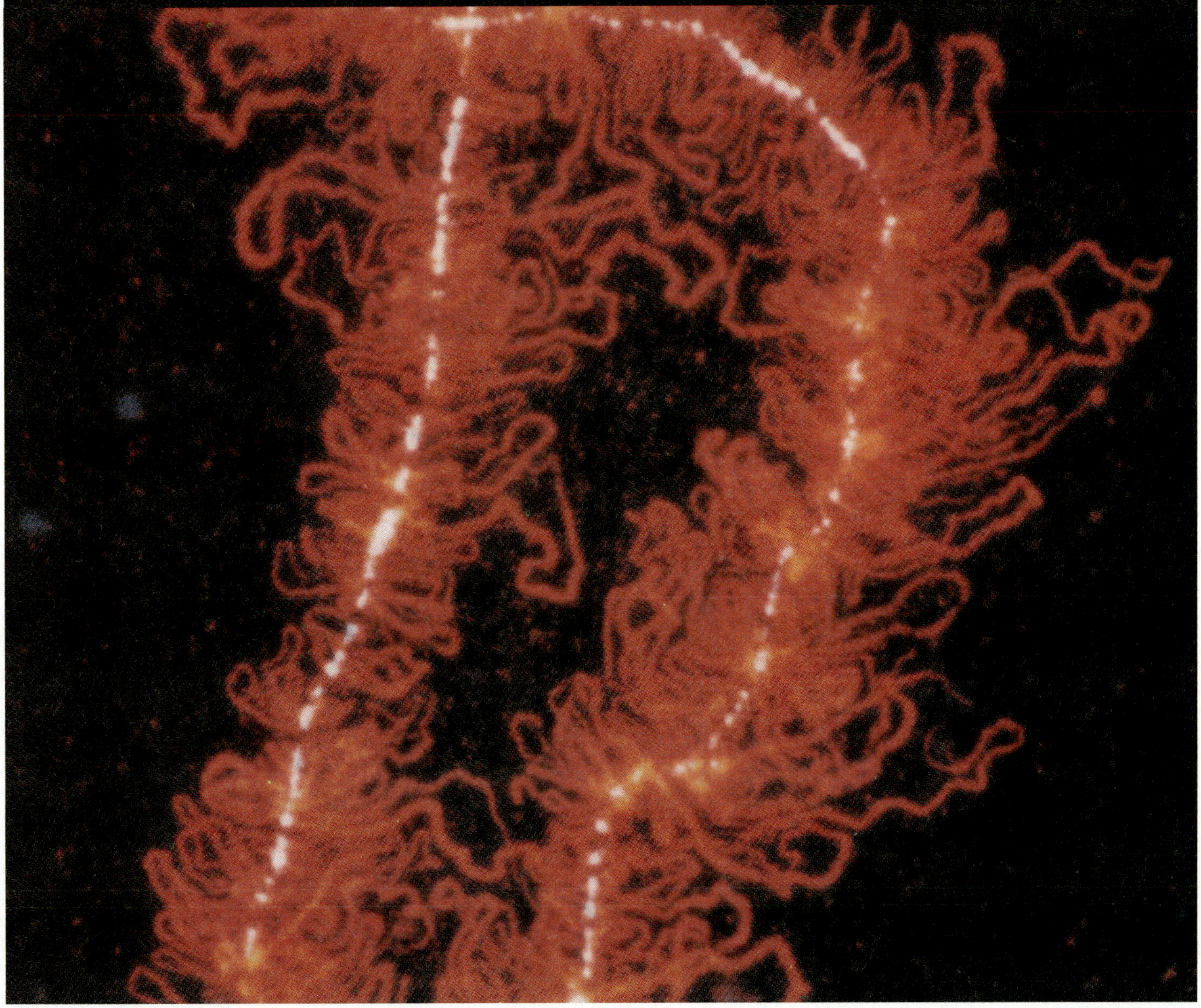

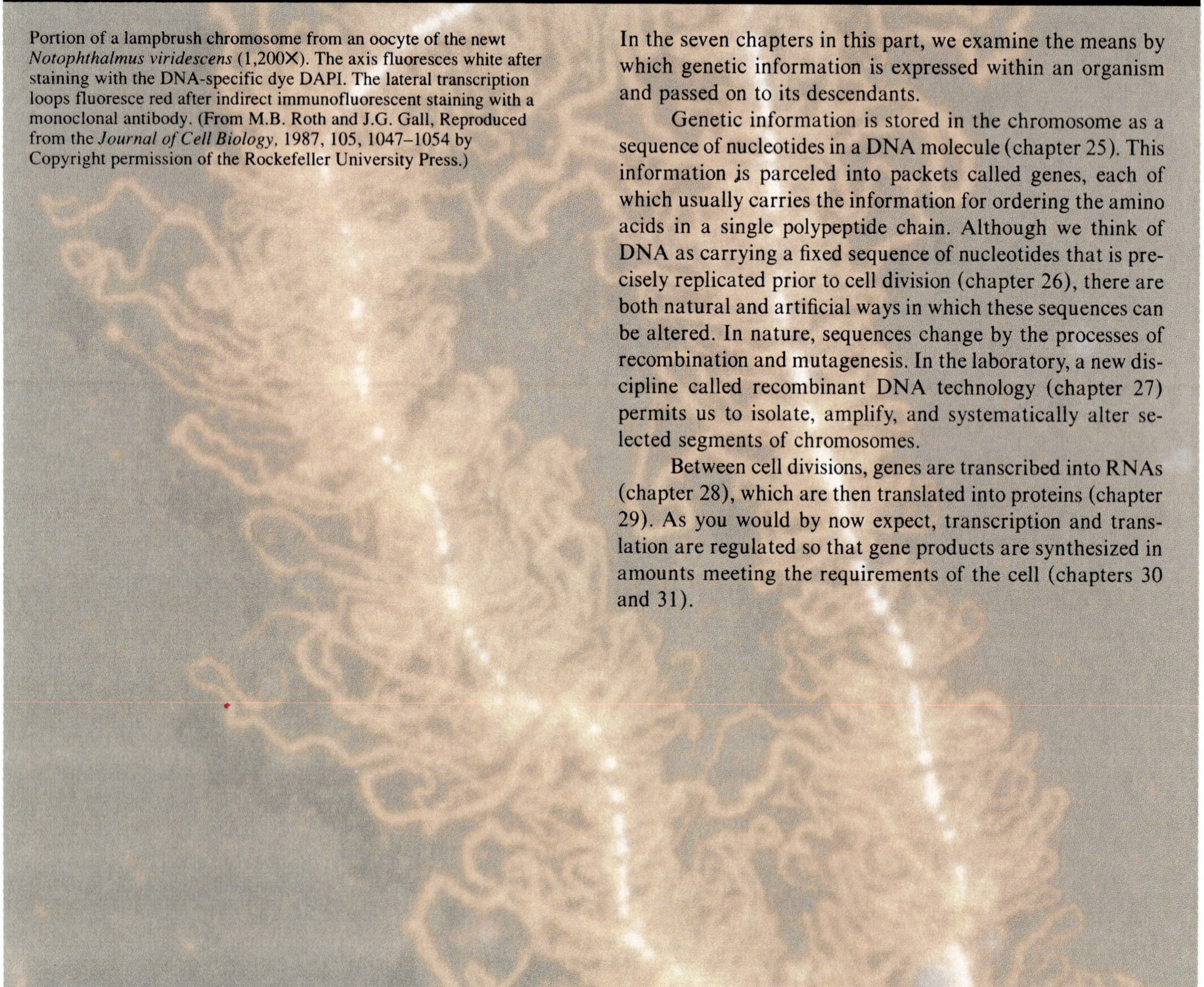

Portion of a lampbrush chromosome from an oocyte of the newt *Notophthalmus viridescens* (1,200×). The axis fluoresces white after staining with the DNA-specific dye DAPI. The lateral transcription loops fluoresce red after indirect immunofluorescent staining with a monoclonal antibody. (From M.B. Roth and J.G. Gall, Reproduced from the *Journal of Cell Biology*, 1987, 105, 1047–1054 by Copyright permission of the Rockefeller University Press.)

In the seven chapters in this part, we examine the means by which genetic information is expressed within an organism and passed on to its descendants.

Genetic information is stored in the chromosome as a sequence of nucleotides in a DNA molecule (chapter 25). This information is parceled into packets called genes, each of which usually carries the information for ordering the amino acids in a single polypeptide chain. Although we think of DNA as carrying a fixed sequence of nucleotides that is precisely replicated prior to cell division (chapter 26), there are both natural and artificial ways in which these sequences can be altered. In nature, sequences change by the processes of recombination and mutagenesis. In the laboratory, a new discipline called recombinant DNA technology (chapter 27) permits us to isolate, amplify, and systematically alter selected segments of chromosomes.

Between cell divisions, genes are transcribed into RNAs (chapter 28), which are then translated into proteins (chapter 29). As you would by now expect, transcription and translation are regulated so that gene products are synthesized in amounts meeting the requirements of the cell (chapters 30 and 31).

25

CHAPTER

Structures of Nucleic Acids and Nucleoproteins

Among the biopolymers, nucleic acids occupy a unique position. Not only are they involved in many important reactions, but they contain all biochemically useful information in the form of a genetic code. This information must be faithfully passed by replication from one cell generation to the next; it must be translated into proteins; and it should be capable of change by mutation to produce the variability on which evolutionary selection processes feed. In the following five chapters our focus will be on the biochemistry of DNA, RNA, and protein, with occasional reminders of the genetic significance of these processes.

Appropriately, we will begin this part of the text by considering certain experiments that demonstrated the genetic significance of nucleic acids. Following this, we will turn to the structural properties of nucleic acids (fig. 25.1). Next, we will discuss reactions involving DNA—first purely biochemical reactions (chapter 26), and then *in vitro* and *in vivo* reactions involving DNA manipulation (chapter 27). In chapter 28 we will deal with the process of information transfer from DNA to RNA, and finally, in chapter 29 we will consider the mechanism of information transfer from RNA to protein.

The Genetic Significance of Nucleic Acids

After the discovery, around the turn of the century, that genes are carried by chromosomes, a great deal of effort went into characterizing the sizes and shapes of chromosomes. But it was not until much later that significant progress was made in elucidating the chemical nature of the gene. As early as 1900, biologists were aware that chromosomes are composed of both nucleic acids and proteins. However, the seemingly simple chemical composition of nucleic acids misled early investigators into believing that nucleic acids were a purely structural component of the chromosome. They believed then that the arrangement of specific proteins along the chromosome accounted for gene specificity. This notion was dispelled in 1944 when Oswald T. Avery and his colleagues at Rockefeller Institute (now University) demonstrated that purified deoxyribonucleic acid (DNA) contains the genetic determinants of the bacterium *Diplococcus pneumoniae* (now Streptococcus). In this section we will discuss Avery's results, as well as some other historically important observations on the genetic significance of nucleic acids.

Figure 25.1

DNA is a long linear polymer of nucleotides. The nucleotides are very similar but they differ in several minor respects. All nucleotides are composed of a purine or a pyrimidine linked to a deoxyribose on one side. On the other side the deoxyribose is attached to a phosphate group. There are two purines, adenine and guanine, and two pyrimidines, cytosine and thymine. The polymeric structure is held together by phosphodiester linkages formed between the 5′ phosphate group of one nucleotide and the 3′OH group of the adjacent nucleotide. This gives the polynucleotide strands a directional sense (5′ to 3′). Two chains come together to form a duplex structure with hydrogen bonds formed between purines and pyrimidines on opposing chains (*a*). The two-chain structure twists to form a right-handed duplex structure (*b*). In eukaryotes the DNA wraps around a histone protein octamer, forming a hollow core (*c*). This structure forms a 200-nm-wide (2,000-Å) fiber by further coiling (*d*). The chromosome is a coiled-coil structure formed from very long 200-nm filaments that extend over the entire length of the chromosome (*e*).

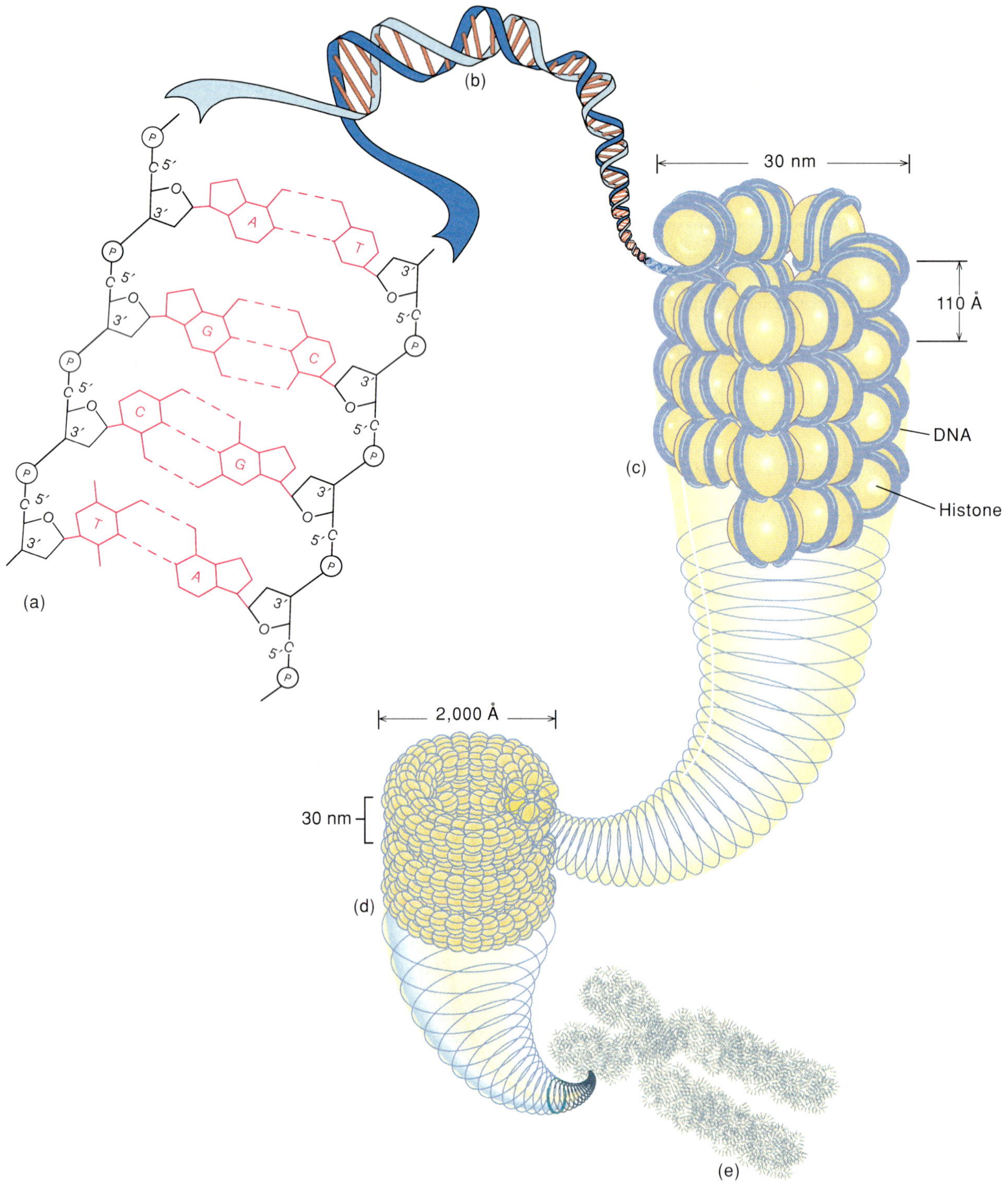

Figure 25.2

In vivo and *in vitro* evidence that DNA causes transformation of *D. pneumoniae*. (*a*) Transformation experiment by Griffith. R bacteria are nonvirulent. S bacteria are virulent. A mixture of R bacteria and heat-killed S bacteria is also virulent if transformation has occurred. (*b*) Transformation experiment by Avery and co-workers. When bacteria from a liquid culture are spread on a semisolid medium, each cell adheres to the medium at random. As time passes, the cells and their offspring grow and divide, leading to visible clones or colonies, each of which arose from a single cell. R and S cells each have a distinct clonal morphology. While R cells produce small rough clones, S colonies are smooth. R cells exposed to DNA from S cells produce a mixture of both types of clones: untransformed R colonies and transformed S colonies.

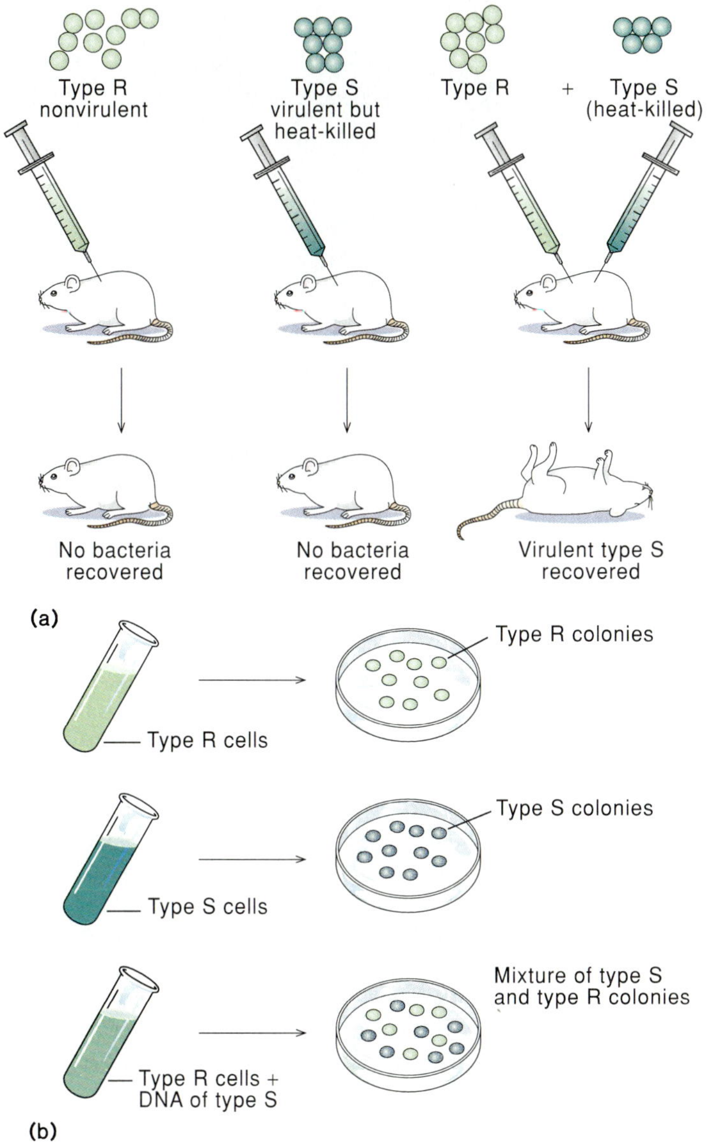

Transformation Is Mediated by DNA

In 1928 Fred Griffith was experimenting with two different strains of pneumococcus. Type S bacteria (S for smooth, from the appearance of bacterial colonies on agar plates) are encapsulated by polysaccharide. The capsules protect them from the host immune system, making the S bacteria pathogenic; even when small numbers of S bacteria are injected into mice, death results. By contrast, R bacteria (R for rough colonies) are nonencapsulated; they are readily attacked by the mouse's immune system and consequently are nonpathogenic. Although heat-killed S bacteria by themselves are nonvirulent, Griffith found that when heat-killed S bacteria were mixed with live R bacteria and introduced into a susceptible laboratory mouse, death of the animal frequently occurred (fig. 25.2*a*). S bacteria could then be detected in the blood. Apparently the genetic factor required for encapsulation was transferred from killed S cells to live R cells.

Griffith's result was duplicated *in vitro* by Dawson and Sia in 1930. The two bacterial strains, when grown in a liquid growth medium, can be distinguished by the different appearance of the R and S colonies on plates (see fig. 25.2*b*). Dawson and Sia found that, in parallel with the results that Griffith had obtained in the mouse, heat-treated extracts of S cells transformed R cells into S cells. Then Alloway, in Avery's laboratory, succeeded in isolating a stable, alcohol-precipitable transforming agent from killed smooth cultures. The *in vitro* results represented an important advance in experimental techniques. The ability to isolate the active agent (or "transforming principle," as it was then called) from S cells allowed investigators to use a quantitative approach similar to that used by biochemists to purify an enzyme.

More than ten years later, Avery provided convincing proof that the active agent in the S cell extracts was the cellular DNA. He and his co-workers did this by purifying the DNA and showing that it was extremely active in transforming R cells into S cells *in vitro* (see fig. 25.2*b*). The transforming activity in the purified extract was destroyed if the extract was first incubated with the enzyme DNase, an endodeoxyribonuclease that specifically degrades DNA. The activity was not affected by RNase, proteases, or enzymes that degrade capsular polysaccharides. Later studies showed that DNA could be used to transfer many other genetic traits between the appropriate pairs of donor and recipient bacterial strains. For example, resistance to the antibiotics streptomycin and penicillin could be transferred, with the DNA of resistant cells, to sensitive cells. In these studies, it appeared that the active genetic material being transferred faithfully reflected the mutational history and hence the genetic patterns of the donor strains. The alteration of the genetic composition of one cell strain as a result of exposure to DNA of another strain is called transformation.

Figure 25.3

Hershey-Chase experiment demonstrating that the DNA but not the protein of the T2 bacteriophage is passed from parent to progeny phage particles. (*a*) Viruses with ^{32}P-labeled DNA or ^{35}S-labeled protein were prepared by growing the viruses on bacteria in growth medium containing $^{32}P\text{-}PO_4^{3-}$ or $^{35}S\text{-}SO_4^{2-}$, respectively. (*b*) When labeled phage particles infect an unlabeled *E. coli* cell, only the ^{32}P label appears in the progeny phage particles, showing that the DNA structure but not the protein structure is preserved during phage replication. (*c*) T4 phage (closely related to phage T2) lysing an *E. coli* cell. (Courtesy of Lee D. Simon, Wakesman Institute for Microbiology, Rutgers University.)

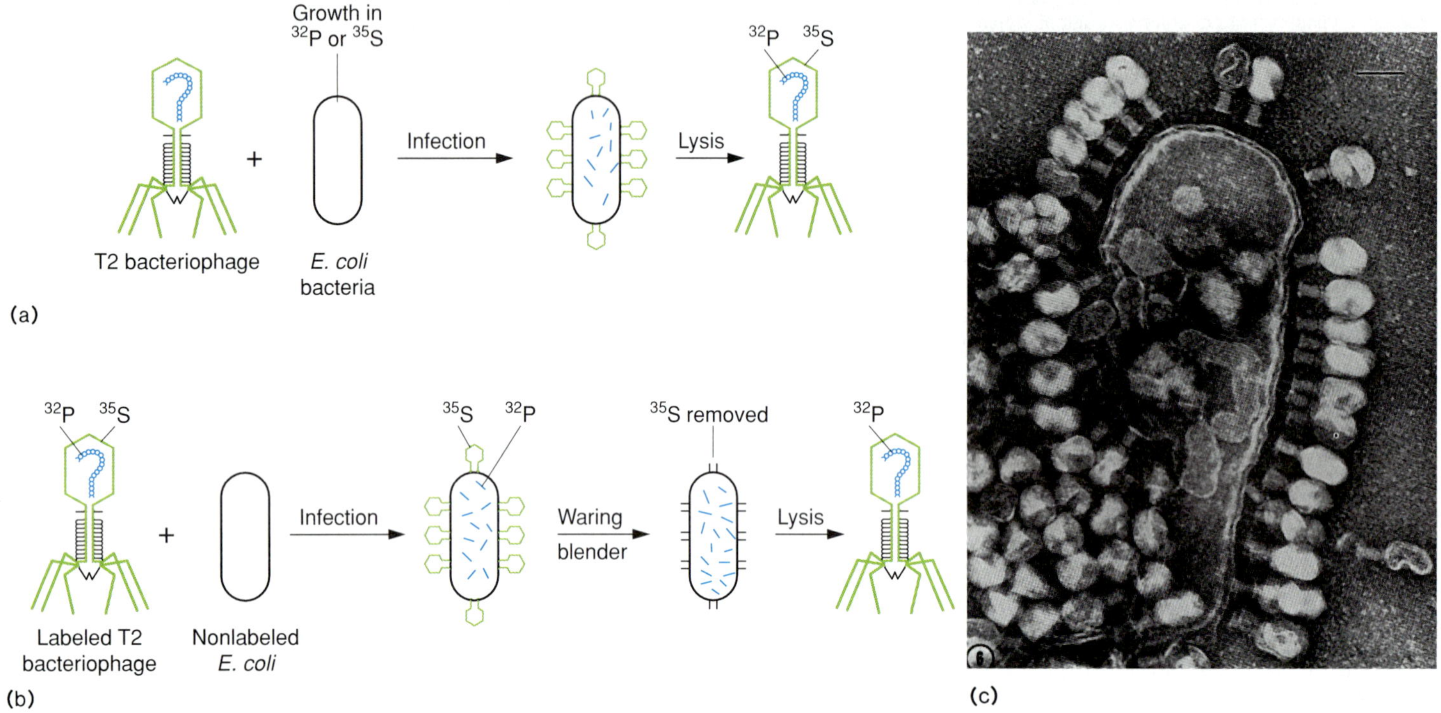

Transformation occurs naturally in only a small number of bacterial species other than *Streptococcus pneumoniae*. In organisms that are not transformed naturally, transformation can be induced by first making the cell envelope (e.g., of *E. coli*) partially permeable so that it can take up DNA. This technique is called artificial transformation. Conventional methods of inducing transformation include Ca^{2+} treatment and temperature shock, but for most organisms a more efficient method is electroporation, or electrical shock. For instance, by exposing yeast to electric field pulses it is possible to obtain transformation efficiencies that are 10^4 greater than that obtained by other methods. Transformation is an essential step in the cloning of genes. Transformation with self-replicating plasmids that may be carrying different genes has made it possible to select cells carrying the gene(s) of interest from large populations of cells.

Studies on Viruses Confirm the Genetic Nature of Nucleic Acid

For some time, doubts lingered about whether DNA, protein, or capsular polysaccharide was the transforming factor. During this period, other investigators performed some elegant experiments showing that nucleic acid was the genetic substance of bacterial viruses (bacteriophages) and plant viruses.

In 1952, A. D. Hershey and Martha Chase demonstrated the independent functions of viral protein and viral nucleic acid in the *E. coli* bacteriophage T2. This bacterial virus is composed of about equal weights of DNA and protein. The linear, duplex DNA is encompassed by a protein shell in a polyhedral "head" or capsid, which is connected to a protein tail that facilitates infection of cells. When a single bacteriophage particle infects a bacterium, it causes cell death and cell disruption (lysis), accompanied by the release of several hundred viruses within 30 min after infection. The fate of the DNA and the protein components of the infecting viruses can be followed by labeling with an appropriate radioactive isotope (fig. 25.3).

In an initial round of infection, Hershey and Chase used inorganic ^{32}P-labeled phosphate to label the DNA component of the virus and ^{35}S-labeled inorganic sulfate to label the protein component. The phosphorus became incorporated into the phosphoryl groups of the viral DNA and the sulfate into the sulfur-containing amino acids of the viral protein. In a second round of infection, they added the radioactively labeled bacteriophages to cells in an unlabeled medium. Shortly after the bacterial host was infected, they subjected the phage-bacterium complex to vigorous agitation in a Waring blender. This treatment sheared the phage from the attachment sites on the bacterial cell wall. The bacteria were then sedimented by centrifugation, leaving the DNA-free viral protein in the supernatant. Analysis for radioactivity showed that most of the ^{32}P-labeled viral DNA sedimented with the bacteria, whereas most of the ^{35}S-labeled viral protein remained in the supernatant. That result demonstrated that most of the viral DNA entered the cells (see fig. 25.3*b*) during infection. Quantitative isotopic analysis of the progeny phage particles indicated that they retained less than 1% of the original infecting viral protein but about two-thirds of the original infecting DNA. In other words, most of the original DNA that entered the cell on infection also became an integral part of the progeny phage, but the protein did not. Since the genetic material should survive replication, the results strongly implicated DNA as the genetic component of bacteriophage T2. Long after these experiments were performed, it was shown that purified, deproteinized phage DNA could itself produce phage after being taken up by host cells. Infection of bacteria with phage-derived DNA is referred to as transfection.

Early experiments on plant viruses provided even more convincing evidence identifying nucleic acid as the genetic substance of a virus. This work was done on tobacco mosaic virus (TMV), the first virus to be crystallized. TMV is a plant virus with a molecular weight of 40×10^6. The virion contains one molecule of single-stranded RNA with a molecular weight of 2×10^6 and 2,130 identical protein subunits, each with a molecular weight of 18,000. In 1956 Gierer and Schramm showed that the deproteinized viral RNA produced lesions on tobacco plant leaves similar in appearance to the lesions produced by the intact virus.

Shortly thereafter (1957) Fraenkel-Conrat and Singer obtained similar evidence that RNA was the genetic substance of TMV (fig. 25.4). Using two different strains of the virus, HR and TMV, each with readily identifiable protein components, they generated reconstituted viruses *in vitro* in all possible combinations:

HR-protein + HR-RNA
HR-protein + TMV-RNA
TMV-protein + HR-RNA
TMV-protein + TMV-RNA

The reconstituted virus made from purified RNA and protein components had infectious activities comparable to those of the normal viruses. After infection the protein components were examined. For all combinations the type of viral protein found in progeny virus was determined by the type of RNA in the reconstituted particles, a very elegant proof that the nucleic acid carries the genetic determinants of the viral protein.

The studies that we have described in this section show that the genetic information required for replication of a cell (e.g., pneumococcus) or a virus (T2, TMV) is carried by nucleic acids. In the case of a cell the genetic information is always carried by DNA. In the case of viruses the genetic information can be carried by either single- or double-stranded DNA or by RNA, depending on the virus. Sometimes the bacterial or viral DNAs are linear overall; sometimes they are circular. The only case in which circular RNAs have been encountered is in viroids—low-molecular-weight, single-stranded RNAs that cause several important diseases in cultivated plants but are not encapsidated. Viroids are the smallest known agents that cause infectious disease.

Figure 25.4

Reconstitution experiment with tobacco mosaic virus. The RNA and protein components of TMV may be separated and reconstituted. If two strains are used and if the RNA and protein parts from different viruses are used in the reconstitution experiment, the progeny virus will always reflect the parental type from which the RNA was taken.

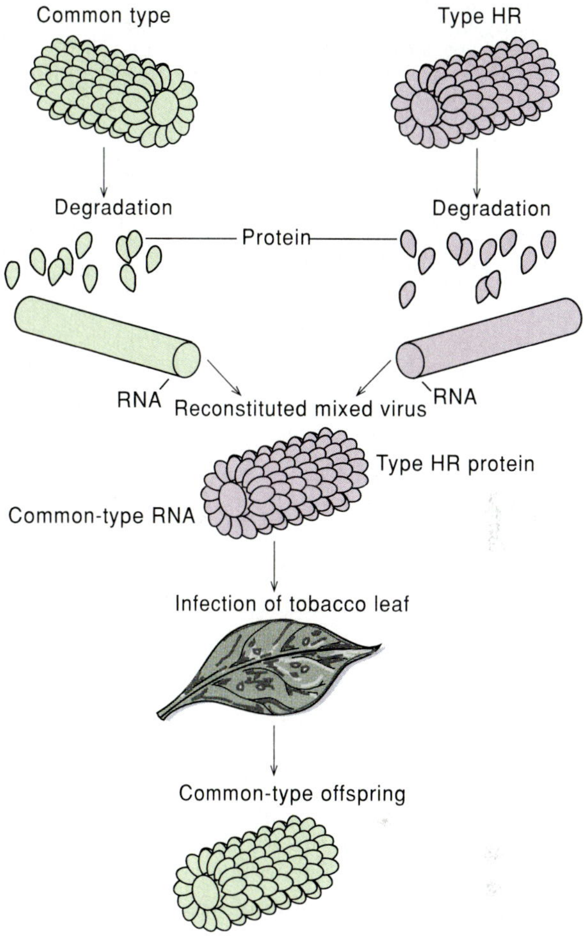

Structural Properties of DNA

The early work equating the genetic material with certain nucleic acids plunged genetics into an entirely new vocabulary of chemical terms. The genetic consequences of these early studies could be understood, but the chemistry was new. The remainder of this chapter focuses on the chemical and structural properties of nucleic acids.

As we have seen, genes can be composed of either DNA or RNA. However, RNA is the genetic component only in some viruses. Otherwise, DNA provides the genetic information, and

Figure 25.5

Genome size in different cells, viruses, and plasmids. Plasmids are small circular DNA molecules that replicate autonomously in cells harboring them. Unlike viruses, they do not form any complex nucleoprotein structures. In the case of plasmids, most viruses, and bacteria, the genome size is equivalent to the size of the chromosomal DNA because there is only one chromosome. For all of the remaining eukaryotic organisms that are listed, the genome is subdivided into two or more chromosomes. Some organisms contain over one hundred chromosomes. All chromosomes are believed to contain a single DNA molecule.

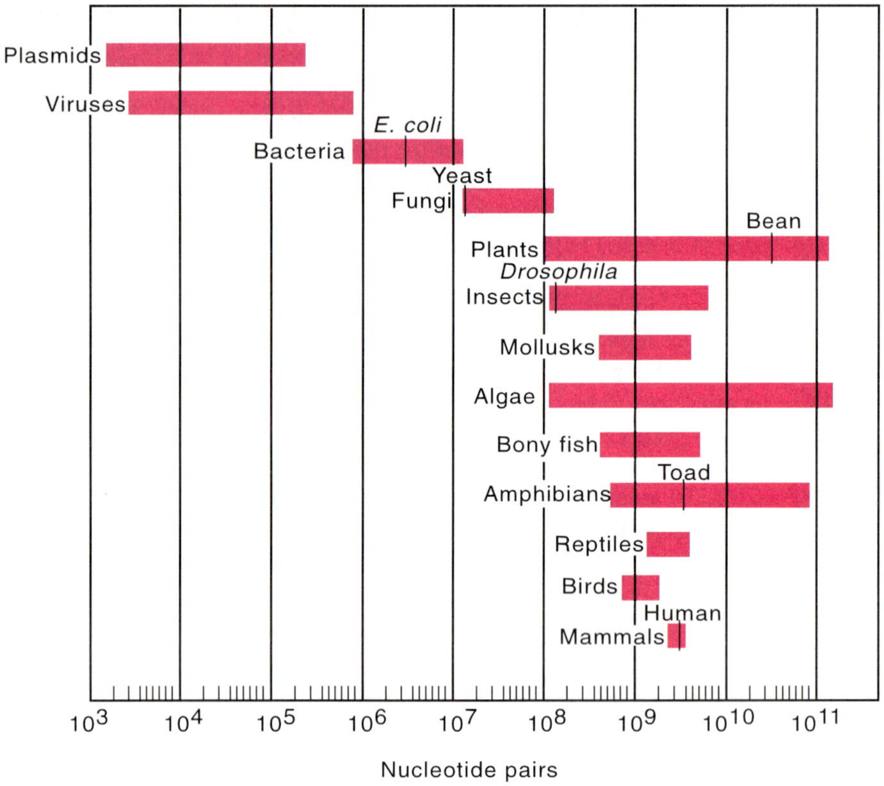

this is transcribed into RNA, which, in most cases, is subsequently used to direct protein synthesis. The different functions of the two cellular nucleic acids are reflected by their different locations in the cell; DNA is found almost exclusively in the cell nucleus, whereas the mature forms of RNA are located mainly in the cytoplasm, where protein synthesis takes place.

The amount of DNA per cell differs widely among different organisms (fig. 25.5). Mammalian cells contain about 1,000 times as much DNA as bacterial cells. Bacterial viruses such as the T-type bacteriophages that infect *E. coli* contain one-tenth to one-twentieth as much DNA as the bacterial host chromosome. The DNA of the smallest viruses is, in turn, about one-tenth of the amount found in the smallest T-phage DNA, containing barely enough genetic material to accommodate about ten genes. This finding is consistent with the fact that viruses do not contain sufficient genetic information for independent growth, but can only grow parasitically in the host cells they infect. On the other hand, the amount of DNA per cell is not always directly proportional to the amount of genetic information that an organism carries. This is because complex eukaryotes contain in their chromosomes a great deal of noninformational DNA, whose function is frequently unclear.

In addition to the main body of DNA associated with the cell or the virus, there is informational DNA found in special organelles such as chloroplasts and mitochondria. This DNA carries genes for some ribosomal RNAs and transfer RNAs and for proteins associated with the respective organelles. The simplest type of system using informational DNA is the plasmid. Plasmids vary in size and exist as naked DNA molecules that carry anywhere from 5 to 100 genes. They are very widespread in bacteria. They are active in both prokaryotes and eukaryotes and in the past ten years they have proved to be extremely useful for DNA manipulation (see chapter 27).

The Polynucleotide Chain Contains Mononucleotides Linked by Phosphodiester Bonds

Nucleotides are the building blocks of nucleic acids; we discussed nucleotide structures and biochemistry in chapter 20, which you may wish to review at this point or as the need for relevant details arises. When a 5′-phosphomononucleotide is joined by a phosphodiester bond to the 3′-OH group of another mononucleotide, a dinucleotide is formed. The 3′-5′-linked phosphodiester internucleotide structure of nucleic acids was firmly established in 1951 by Alexander Todd, who in 1957 was awarded a Nobel Prize for Chemistry as a result. Repetition of the phosphodiester linkage leads to the formation of polydeoxyribonucleotides in DNA or polyribonucleotides in RNA. The structure of a short polydeoxyribonucleotide is shown in figure 25.6. The polymeric structure consists of a sugar phosphate diester backbone with bases attached as distinctive side chains to the sugars.

Figure 25.6

The structure of a deoxyribonucleotide. Drawn in abbreviated form at lower left. The illustrated structure is written pTpApCpG.

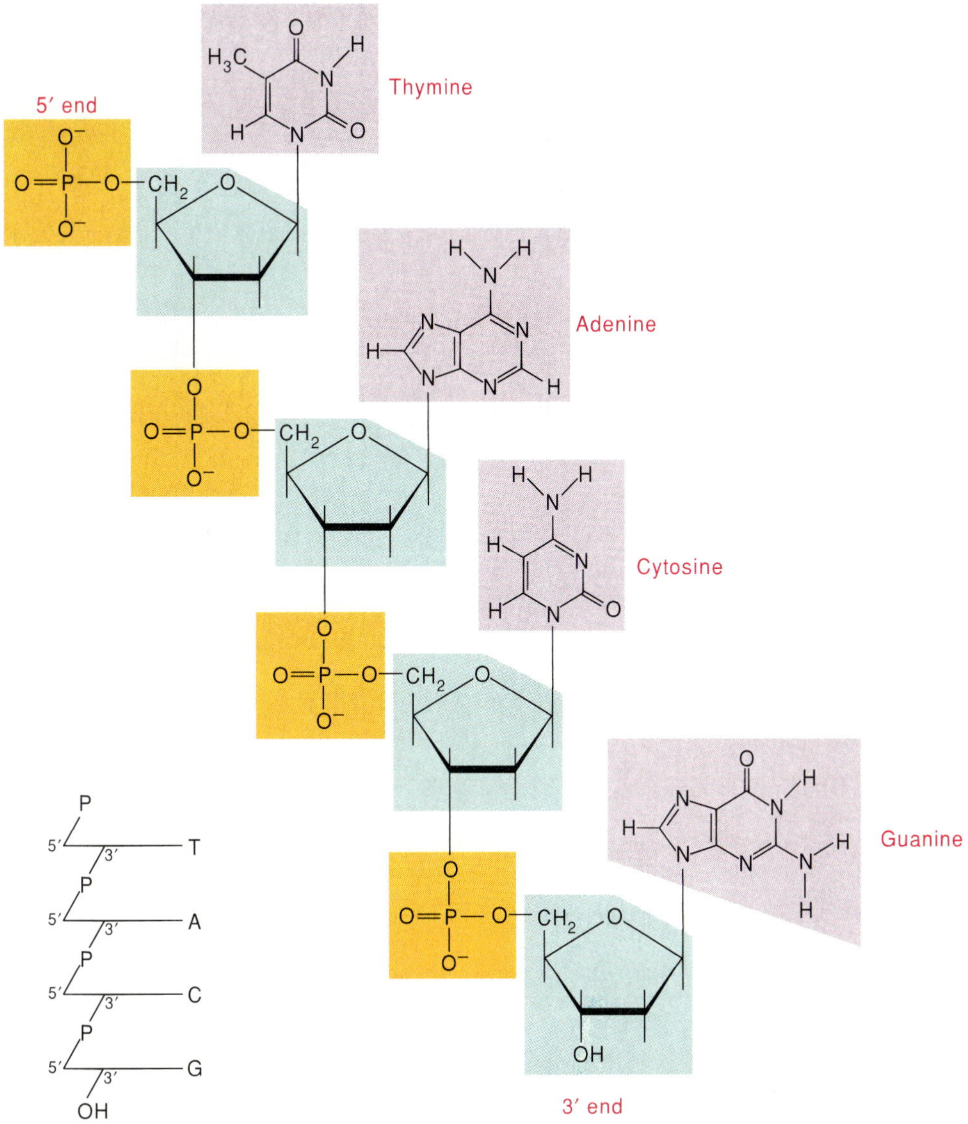

The polynucleotide chain has a directional sense with a 5′ and a 3′ end. Either of these ends may contain a free hydroxyl group or a phosphorylated hydroxyl group. The structure shown in figure 24.6 contains a phosphate group on the 5′ end but none on the 3′ end. By convention, we write a nucleic acid sequence from the 5′ to the 3′ end. Thus we would write the structure in figure 24.6 pTpApCpG. If it had no phosphate on the 5′ end, we would write the structure TpApCpG; alternatively, if the terminal phosphate were on the 3′ end rather than the 5′ end, we would write the structure TpApCpGp. When the phosphates are not indicated, we write the oligonucleotide TACG. If the oligonucleotide is part of a larger polynucleotide, we indicate this fact by dashes on either end: -TACG-. The letter "d" or "r" sometimes precedes the capital letter of the nucleotide to indicate a deoxyribo- or a ribo-derivative.

Most DNAs Exist as Double-Helix (Duplex) Structures

Like most other types of biological macromolecules, nucleic acids adopt highly organized three-dimensional structures. The dominant factors that determine nucleic acid conformation are the limitations imposed by the stereochemistry of the polynucleotide chains, the high negative charge resulting from the regularly repeating phosphate groups, and the noncovalent affinities between purine and pyrimidine bases.

A body of chemical information that proved vital to understanding DNA structure came from Erwin Chargaff's analyses of the nucleotide composition of duplex DNAs from various sources (table 25.1). Although the base compositions varied over a wide range, Chargaff found that within the DNA

Table 25.1
Base Composition of DNAs from Different Sources

	(A) Adenine	(G) Guanine	(C) Cytosine	(5-MC) 5-Methylcytosine	(T) Thymine	$\frac{A + T}{G + C + 5\text{-MC}}$
Human	30.4	19.6	19.9	0.7	30.1	1.53
Sheep	29.3	21.1	20.9	1.0	28.7	1.38
Ox	29.0	21.2	21.2	1.3	28.7	1.36
Rat	28.6	21.4	20.4	1.1	28.4	1.33
Hen	28.0	22.0	21.6		28.4	1.29
Turtle	28.7	22.0	21.3		27.9	1.31
Trout	29.7	22.2	20.5		27.5	1.34
Salmon	28.9	22.4	21.6		27.1	1.27
Locust	29.3	20.5	20.7	0.2	29.3	1.41
Sea urchin	28.4	19.5	19.3		32.8	1.58
Carrot	26.7	23.1	17.3	5.9	26.9	1.16
Clover	29.9	21.0	15.6	4.8	28.6	1.41
Neurospora crassa	23.0	27.1	26.6		23.3	0.86
Escherichia coli	24.7	26.0	25.7		23.6	0.93
T4 bacteriophage	32.3	17.6		16.7[a]	33.4	1.91

[a]In T bacteriophage all of the cytosine exists in the 5-hydroxymethyl form 5-HMC.

Figure 25.7

Dimensions and hydrogen bonding of (*a*) thymine to adenine and (*b*) cytosine to guanine. Note that there are two hydrogen bonds formed in the A-T base pair and three in the G-C base pair. The overall dimensions of the base pairs are the same. Consequently they will fit at any position in an otherwise regular polymeric structure (Source: S. Arnott et al., "Fourier synthesis studies of lithium DNA, part III: Hoogsteen models" in *Journal of Molecular Biology* 11:391, 1965. Copyright © 1965 Academic Press Ltd., London, England.)

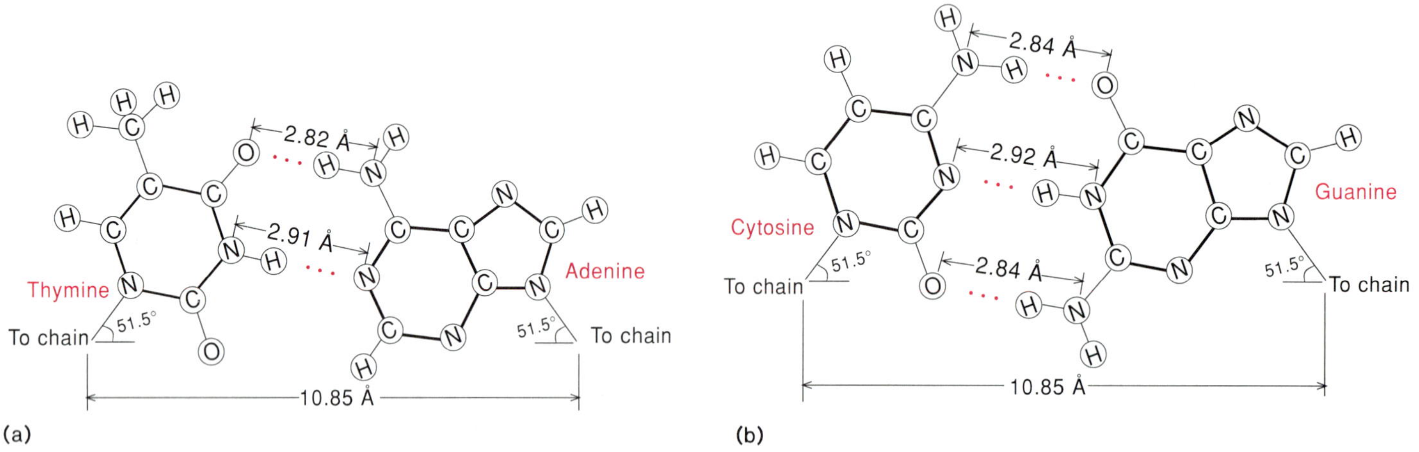

of each source that he examined, the amount of adenine (A) was very nearly equal to the amount of thymine (T), and the amount of guanine (G) was very nearly equal to the amount of cytosine (C). The cytosine is present as both unmodified cytosine and, to a lesser extent, 5-methylcytosine, which results from postreplicative modification. The two equalities were the first indication that regular complexes occur between A and T and between G and C in DNA.

While searching for the meaning of these equalities, James Watson discovered, with the help of molecular models, that between A and T and between G and C it was possible to form hydrogen-bonded base-paired structures that have the same overall dimensions (fig. 25.7). Two hydrogen bonds are formed in the A-T base-paired structure and three in the G-C base-paired structure. The hydrogen-bonded pairs are formed between bases of opposing strands and can arise only if the directional senses of the two interacting chains are opposite or antiparallel (fig. 25.8). Watson brought this information to the attention of his crystallographer colleague Francis Crick, who realized that the x-ray diffraction pattern produced by DNA could be interpreted in terms of a helix (see box 25A) composed of two polynucleotide strands. In this structure the planes of the base pairs are perpendicular to the helix axis, and the distance between adjacent pairs along the helix axis is 3.4 Å, bringing them into close contact (fig. 25.9). The structure repeats itself after about ten residues, or once every 34 Å along the helix axis; the repeating distance is referred to as the pitch length or just the pitch. An average-size bacterial gene is about 1,000 base pairs (bp) in length, equivalent to 100 helical turns. The complementary structure of duplex DNA suggests how the genetic material is faithfully replicated as well as how it is expressed.

An important feature of the helical structure is the grooved nature of the surface resulting from the helical twist. Alternating wide (major) and narrow (minor) grooves are displayed in a side view of the helix structure (see fig. 25.9*a* and *b*). Different sections of the purine and pyrimidine bases are exposed in these two grooves as indicated in figure 25.9*c*. Many different proteins interact with DNA; most of the interactions occur with the phosphoryl groups on the outer surface of the structure and with the purine and pyrimidine bases in the wide groove because of its greater accessibility. We will consider specific instances of DNA–protein interactions later (chapters 28, 30, and 31).

Hydrogen Bonds and Stacking Forces Stabilize the Double Helix

Several factors account for the stability of the double-helix structure. The negatively charged phosphoryl groups are all located on the outer surface, where they have a minimum effect on one another. The repulsive electrostatic interactions generated by these charged groups are often partly neutralized by interaction with cations such as Mg^{2+}, basic polyamines (such as putrescine and spermidine), and the positively charged side chains of chromosomal proteins. The core of the helix is composed of the base pairs, held together by the specific hydrogen bonds and also by favorable stacking interactions between the planes of adjacent base pairs. These stacking interactions are complex, involving dipole–dipole interactions and van der Waals forces; they result in a stacking energy comparable in magnitude to the stabilizing energy generated by the hydrogen bonds between the base pairs. The result is that stacking is maximized in most nucleic acid structures. In this connection, it is noteworthy that two fully extended polynucleotide strands can form a hydrogen-bonded base-paired complex that leads to a stepladderlike structure (fig. 25.10*b*). In this structure the chains do not form a helix but lie straight, with a distance of 6.8 Å between identical residues in the direction of the long axis. This 6.8-Å distance between adjacent base pairs produces a gap that would presumably have to be filled by water. Such a conformation does not result in an energetically favorable structure, however, since the planes of the bases would prefer to be in closer contact with each other than with water. The stepladder structure can be converted into the helix structure by a simple right-handed twist (see fig. 25.10*a*). When this is done, the distance between base pairs decreases until they are in close contact, with a spacing of 3.4 Å.

Figure 25.8

Segment of DNA, drawn to emphasize the hydrogen bonds formed between opposing chains. Each type of base is represented by a different color, with the sugar-phosphate backbones in black. Note the three hydrogen bonds in the G-C pairs and the two in the A-T pairs (A, red; T, green; G, yellow; C, blue). The two strands are antiparallel: one strand (left side) runs 5′ to 3′ from top to bottom and the other strand (right side) runs 5′ to 3′ from bottom to top. The planes of the base pairs are turned 90° to show the hydrogen bonds between the base pairs. (Adapted from "The Synthesis of DNA" by Arthur Kornberg. Copyright © 1968 by Scientific American, Inc.)

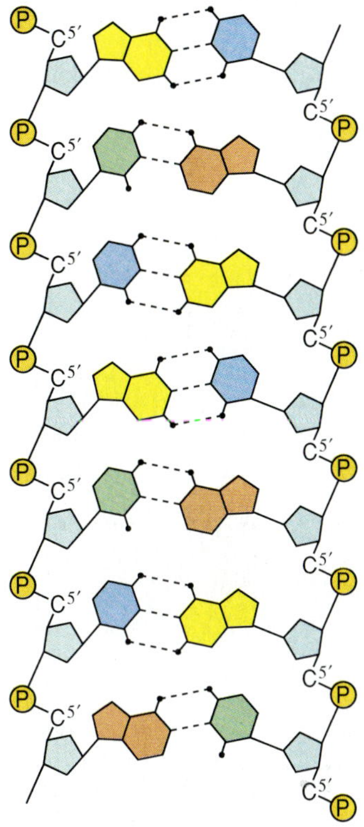

25A
BOX

X-Ray Diffraction of DNA

X-ray diffraction of DNA is performed in a manner similar to x-ray diffraction of fibrous proteins (see box 4A), except that the stretched fiber in this case contains many DNA molecules (see figure 1). The diffraction pattern recorded on the film is shown in figure 2. Note the strong 3.4-Å and 34-Å spacings and the central crosslike pattern, which reflects a helix structure. Watson and Crick were the first to appreciate the significance of these features. They concluded that the diffraction pattern was consistent with a hydrogen-bonded antiparallel double-helix structure—the model that is now accepted as a correct description of the structure of DNA.

Figure 1

Diffraction pattern of a fibrous sample of DNA. © M. H. F. Wilkins.

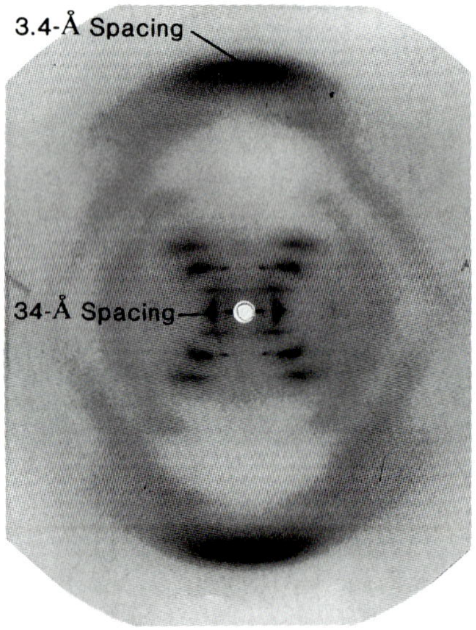

Figure 2

Camera setup for obtaining DNA diffraction pattern.

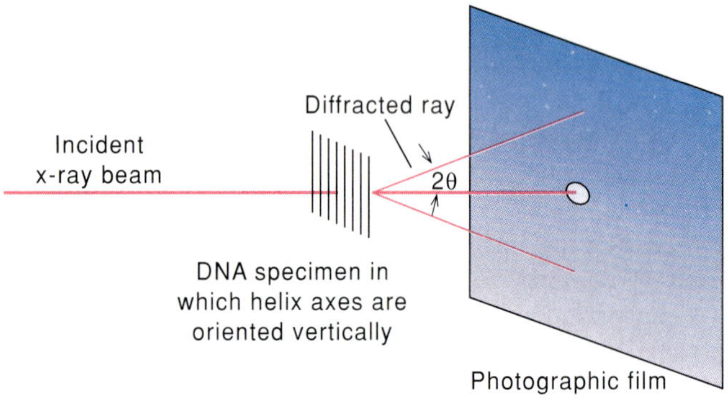

Conformational Variants of the Double-Helix Structure

The same base-pairing arrangement is found in all naturally occurring double-helix structures. However, the inherent flexibility in the furanose ring of the sugar and the degrees of freedom generated by several rotatable single bonds per residue—six in the sugar phosphate backbone and one in the C-1′ N-glycosidic linkage (fig. 25.11)—lead to considerable variation in the conformations adopted by double-helix structures. Four puckered conformations for the sugar, with small differences in stability, are shown in figure 25.12. In the double-helix structure shown in figure 25.8, the furanose rings are in the C-2′-endo conformation. This is believed to be the major conformation adopted by the sugars in DNA, both when free in aqueous solution and in chromosomes. It is known as the B form of DNA. When some of the water is removed from the hydrated DNA fibers, the double helices push even closer together and the structure changes to the so-called A form, which has about eleven bases per turn and base pairs that are tilted about 20° with respect to the helix axis (fig. 25.13*b*). In the A form, the furanose rings have changed their pucker to the C-3′-endo conformation. The furanose rings in RNA have a stronger preference for the C-3′-endo conformation, with the result that RNA double helices adopt a structure similar to the DNA A form, even at high degrees of hydration.

The most striking conformational variant observed for a DNA double helix with Watson-Crick base pairing is referred to as the Z form. This structure was first detected by

Figure 25.9

The most common form of the double-helix DNA. The base-paired structure shown in figure 25.8 forms the helix structure shown in (*a*) and (*b*) by a right-handed twist. The two strands are antiparallel as indicated by the curved arrows in (*a*). In (*b*) a space-filling model depicts the sugar-phosphate backbones as strings of mostly gray, red, white, and yellow spheres, while the base pairs are rendered as horizontal flat plates composed of dark blue spheres. In (*c*) the orientation of the groups in the base pairs with respect to the wide and narrow grooves is indicated. The wide and narrow grooves formed in the right-handed duplex structure are sometimes referred to as the major and minor grooves.

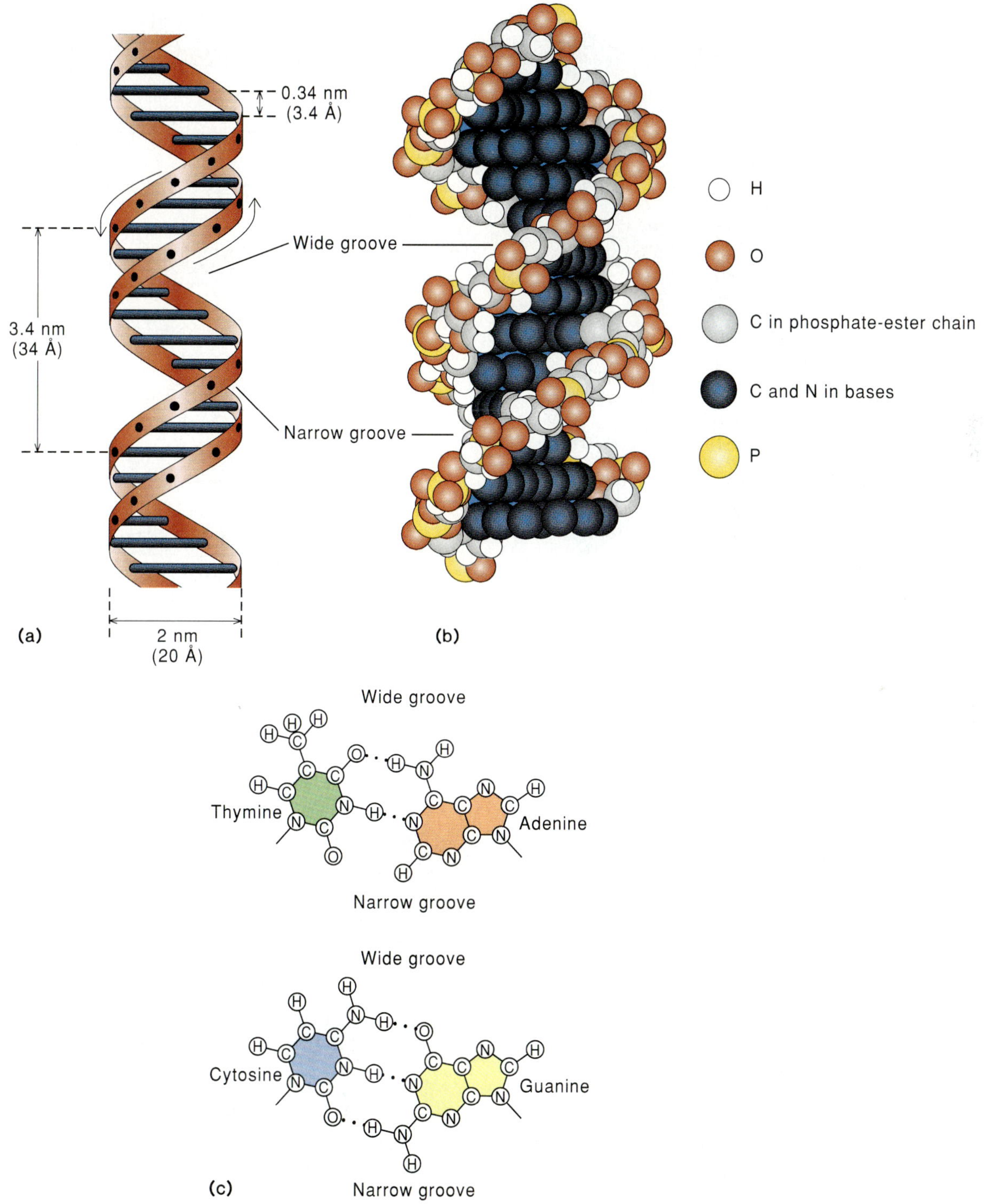

Figure 25.10

Different conformations of base-paired DNA: (*a*) the normal spiral ladder, (*b*) the untwisted straight ladder. The second structure is unstable; it can be converted into a spiral ladder by a right-handed twist, a change that permits the planes of the base pairs to come into close contact.

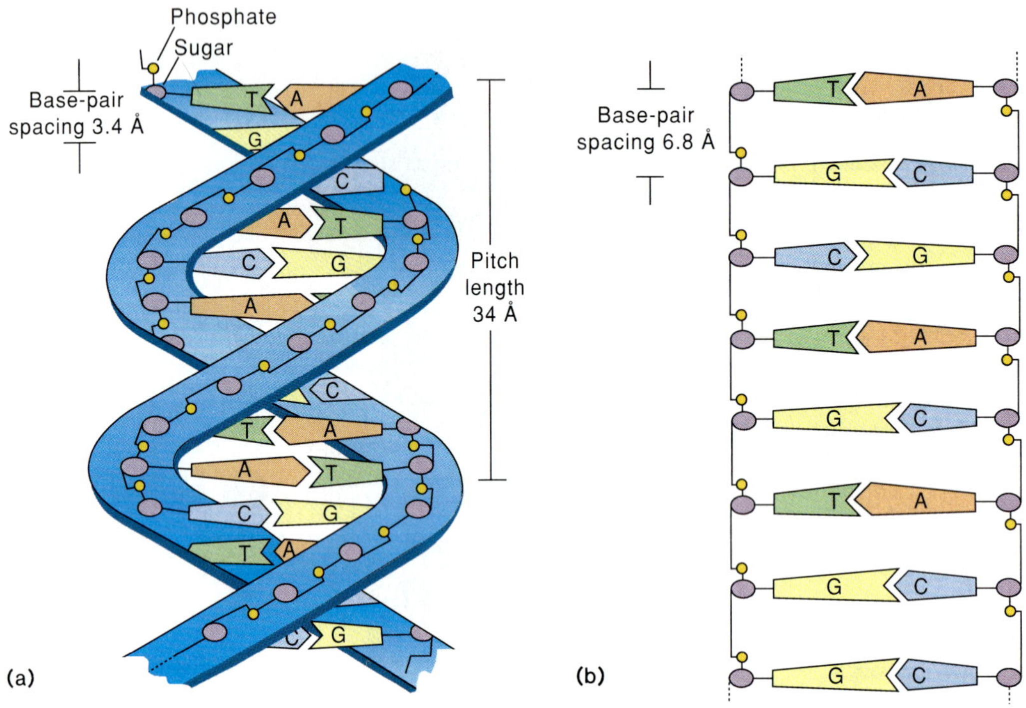

Figure 25.11

A segment of polynucleotide chain with rotatable bonds indicated by curved arrows. The presence of so many rotatable bonds permits the backbone of a polynucleotide chain to adopt many different conformations. (Source: W. K. Olson and P. J. Flory, "Different conformations of deoxyribose in DNA" in *Biopolymers*, 11:1, 1972. Copyright © 1972 John Wiley & Sons, Inc., New York, N.Y.)

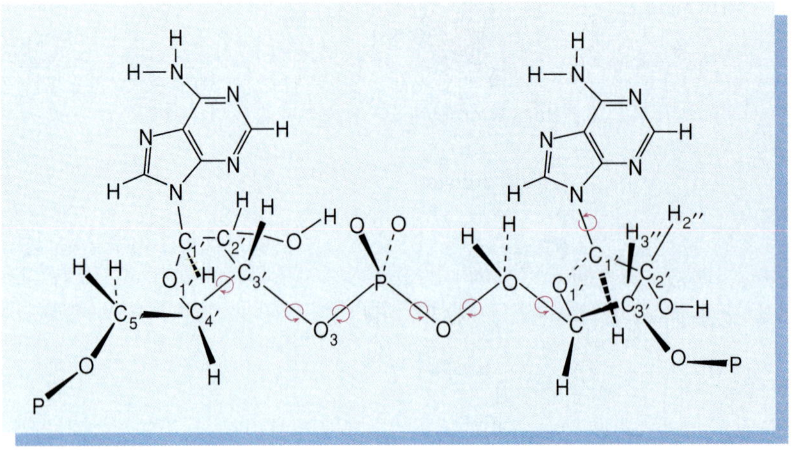

Figure 25.12

Four pucker conformations of the furanose rings of ribose and deoxyribose that are deemed energetically feasible. The flexibility of these furanose rings contributes to the possible variations in the conformation of the double helix.

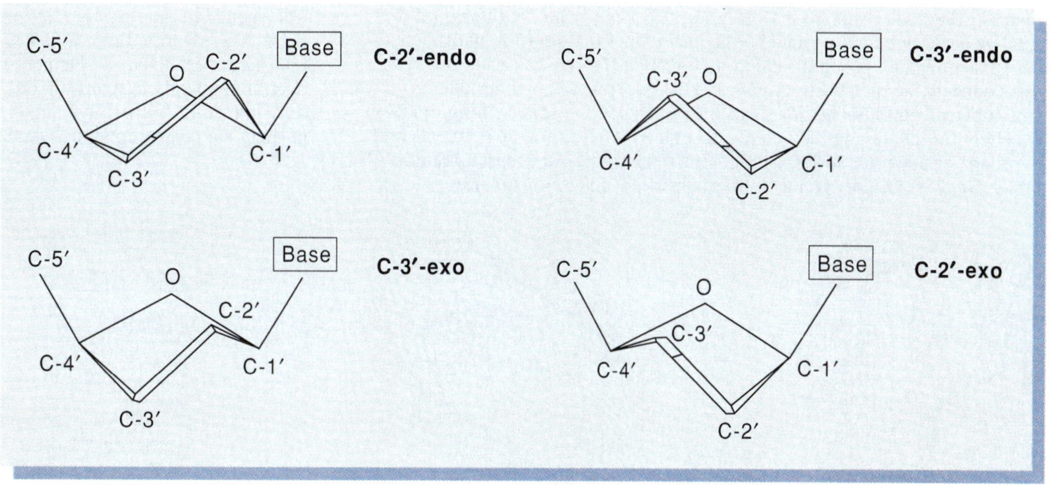

Figure 25.13

The A and B forms of DNA. (*Top*) A view perpendicular to the helix axis. (*Bottom*) A view of two adjacent base pairs, looking down the helix axis. A single turning is shown for each duplex. DNA usually is found in the B form. The A form has been observed in microcrystals of DNA from which a significant amount of the water has been removed. Dimensions of the two structures are compared in table 25.2.

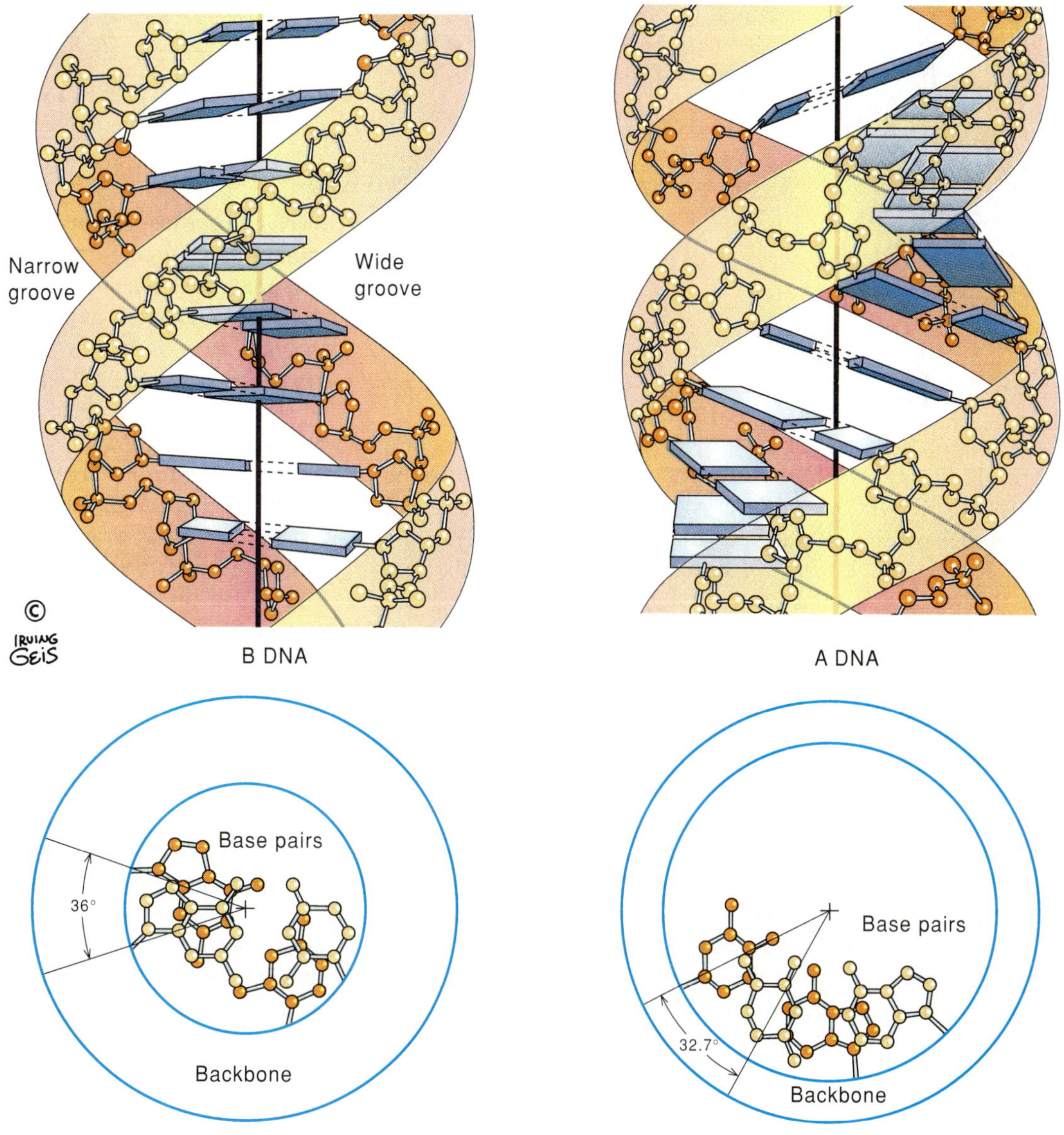

Figure 25.14

Space-filling models of (*a*) Z DNA and (*b*) B DNA. The irregularity of the Z DNA backbone is illustrated by the heavy lines that go from phosphate to phosphate residues along the chain. In contrast, B DNA has a smooth line that connects the phosphate groups and the two grooves, neither one of which extends into the helix axis of the molecule. The space-filling model is excellent for displaying the volume occupied by molecular constituents and the shape of the outer surface. By contrast, the framework representations of DNA (fig. 25.13) show connectivities and allow one to look inside.

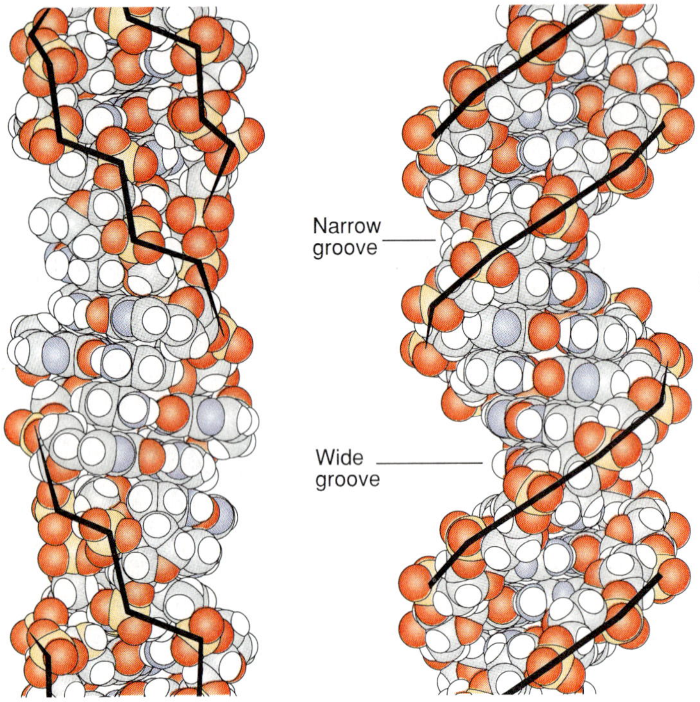

(a) Z DNA (b) B DNA

Figure 25.15

The change in topological relationship if a four-base-pair segment of B DNA is converted into Z DNA. Such a conversion could be accomplished by rotation of the bases relative to those in B DNA. This rotation is shown diagrammatically by coloring one surface of the bases. All of the colored areas are at the bottom in B DNA. In the segment of Z DNA, however, four of them are turned upward. The turning is indicated by the curved arrows.

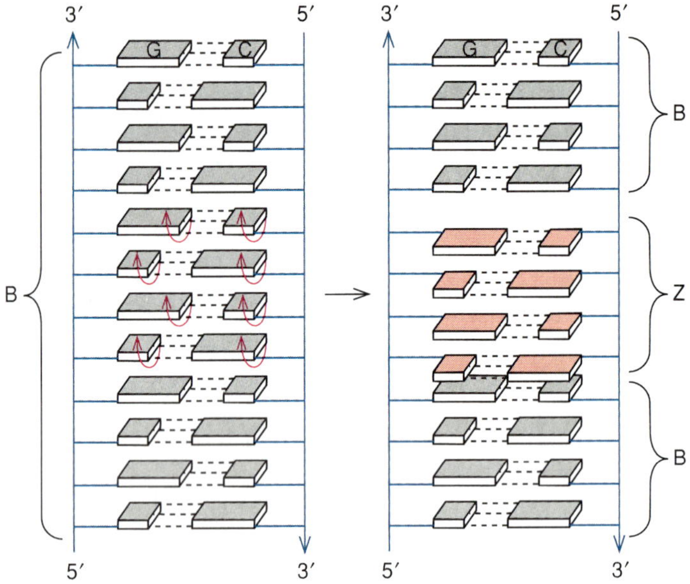

Alex Rich and his co-workers for the deoxyoligonucleotide d(CpGpCpGpCpG), which crystallizes into an antiparallel double helix with a left-handed rather than a right-handed twist (fig. 25.14). The Z form is a considerably slimmer helix than the B form and contains twelve base pairs per turn rather than ten. In the Z form, the planes of the base pairs are rotated approximately 180° with respect to the helix axis from their orientation in the B form (fig. 25.15). The flipping of the base pairs involves different conformational changes in the G and C residues in the alternating GC structure. In the case of the G residues the base is rotated by 180° about the glycosidic bond, a change resulting in a transition from the anti conformation found in B DNA to the syn conformation (fig. 25.16). Model-building studies indicate that the anti conformation of the nucleotide (found in B DNA) has less steric crowding than the syn conformation but that it is far easier for a purine nucleotide to adopt the syn conformation than for a pyrimidine nucleotide. In fact, cytidine remains in the anti conformation in Z DNA. The flipping of the cytidine base in going from the B to Z conformation involves rotation of the entire cytidine residue while maintaining the anti conformation. The effect is to make the sugar-phosphate backbone follow a zigzag course (see fig. 25.14). Thus the name Z DNA is an appropriate descriptive designation for this structure. A number of the structural parameters associated with the A, B, and Z helices are summarized in table 25.2.

Because of the different orientations of the G and C residues in Z DNA, this DNA conformation requires that the sequence of purine and pyrimidine bases be strictly alternating. Many other arrangements that involve alternating purine and pyrimidine residues can adopt the Z conformation. For example, a duplex containing alternating T and G residues in one strand and the complementary A and C residues in the other strand can adopt a Z conformation. An alternating A-and-T DNA sequence has never been observed in a Z conformation. The reason is believed to be the way that water molecules orient around an A-T base pair in the Z helix. The arrangement is not so satisfactory as that observed for the G-C base pair, creating a less stable structure. Other factors that favor the stability of Z DNA are methylation of the 5 position of C residues and negative supercoiling of the DNA. We will discuss negative supercoiling in the next section.

The biological significance of Z DNA is currently unclear. However, several cellular proteins that bind specifically to Z DNA have been isolated from the nuclei of *Drosophila* fruit flies. The mere existence of such proteins suggests that they may function in some specific role when they encounter stretches

Figure 25.16

The syn and anti conformations of deoxyadenosine and deoxycytidine. Purine nucleosides can readily adapt to either conformation by a simple rotation about the C-1—N-9 glycosidic bond. Pyrimidine nucleosides are considerably less stable in the syn conformation because of steric hindrance between the sugar and the C-2 carbonyl group.

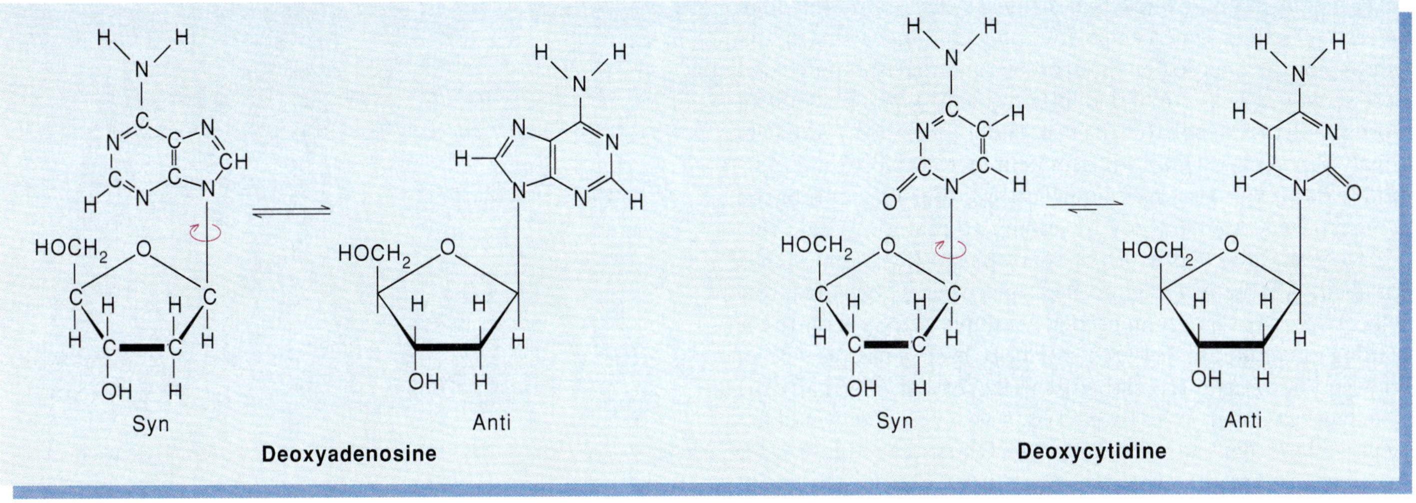

Table 25.2 Some Structural Parameters of A, B, and Z DNA			
	A DNA	**B DNA**	**Z DNA**
Helix sense	Right-handed	Right-handed	Left-handed
Residues per turn	11	10	12 (6 dimers)
Rise per residue	2.55 Å	3.4 Å	3.7 Å
Helix pitch	28 Å	34 Å	45 Å
Base pair tilt	20°	6°	7°
Rotation per residue	33°	36°	−60° (per dimer)
Glycosidic conformation			
Deoxycytidine	Anti	Anti	Anti
Deoxyguanosine	Anti	Anti	Syn
Sugar pucker			
Deoxycytidine	C-3′-endo	C-2′-endo	C-2′-endo
Deoxyguanosine	C-3′-endo	C-2′-endo	C-3′-endo

of DNA that can adopt a Z conformation. A mutation-genetic analysis of such Z-DNA-binding proteins would be helpful to ascertain their role(s). Current speculation on the biological significance of the Z conformation centers on the general notion that the conformation plays a regulatory role in gene expression and possible genetic recombination.

In addition to the A, B, and Z helices, many other conformations have been observed for DNA. All, however, preserve Watson-Crick hydrogen bonding. Synthetic polyribonucleotide complexes have been made that simulate the normal duplex structure. For example, poly(A) and poly(U) in 1:1 base-pair ratios form a duplex resembling the A form of DNA. In 1:2 ratios of one A chain to two U chains they form a triplex that displays additional kinds of hydrogen bonding. Two chains of A make a double helix and four chains of G can form a quadruple helix. A great variety of structures and hydrogen-bonding patterns is possible with polymers that have special sequences. The unusual hydrogen-bonded structures are difficult to detect in nature. Nevertheless, triple helices have been detected *in vivo*. To this, we would add that unusual pairings occur in transfer RNA (see chapter 28).

Duplex Structures Can Form Supercoils

The detection of different conformations of DNA underscores the inherent flexibility built into the DNA duplex. All the conformations discussed thus far involve regular linear duplexes. Energetically favorable interactions with other molecules, particularly proteins, can induce additional conformations that do not result in major changes in either base pairing or stacking. Several conformations are believed to play important roles in different situations. Bends are known to be important in structures formed by chromosomes (discussed later in this chapter). Cruciforms, in which a single chain folds back on itself into a hairpinlike duplex, are important as intermediates in DNA and RNA synthesis (chapters 26 and 28). Supercoiled DNA is a very common type of tertiary structure in which the double-helix segments twist around each other. Supercoiled DNA is topologically constrained by being covalently closed and circular, or by being complexed to proteins so that the ends of the DNA cannot rotate freely.

DNA can form right-handed (negatively supercoiled) or left-handed (positively supercoiled) supercoils (fig. 25.17). Negative supercoiling imparts a torsional stress to the DNA that favors unwinding, whereas positive supercoiling favors tighter winding of the double helix. Supercoiling imparts a more compact structure to a circular duplex, which makes it sediment more rapidly in a centrifuge. Thus, either a positively or a negatively supercoiled DNA will sediment more rapidly than a circular duplex with no supercoiling. Jerome Vinograd demonstrated this tendency by adding ethidium bromide (fig. 25.18) to circular DNA of the small phage PM2. This DNA has about 40 negatively supercoiled turns when it is isolated from cells. Addition of ethidium bromide to duplex DNA leads to the binding of ethidium between adjacent base pairs, a type of binding known as intercalation (see fig. 25.18). Normally the base pairs are nearly closely packed, so such binding would not be possible without an alteration in the DNA structure. In order to accommodate the ethidium, the duplex unwinds by about $-27°$ per base pair. This unwinding reduces the negative supercoiling in naturally occurring circular duplexes such as PM2 DNA. As more ethidium is added, enough unwinding takes place to eliminate all the supercoiling. At this point the DNA is in the most extended state and thus has its slowest sedimentation rate (fig. 25.19). Addition of more ethidium causes the DNA to adopt a positively supercoiled form that again increases its sedimentation rate.

Supercoiling of circular duplex DNA is quantitatively considered in terms of the linking number (L), an integer that specifies the number of complete turns made by one strand around the other. The linking number can change only if a covalent linkage in the DNA backbone is broken. If a molecule of DNA is projected onto a two-dimensional surface, the linking number is defined as the excess of right-handed over left-handed crossings of one strand over the other. Linear duplex DNA with free ends adopts a conformation in solution close to the B form, with about ten base pairs per turn. Therefore a closed circular duplex with this extent of twist is presumed to be under no torsional strain and is said to be relaxed. Because B DNA is a right-handed helix, the linking number is normally positive by the sign convention we have adopted. The values of the linking number of relaxed DNA, $L°$, will be distributed over a narrow range of integral values centered around 1 per 10 base pairs. DNA with a mean linking number smaller than this is termed negatively supercoiled, or underwound; DNA with a larger linking number is termed positively supercoiled or overwound. The deviation of the linking number from its relaxed value, $\Delta L = L - L°$, can be partitioned between twist (altered double helix coiling) and supercoiling:

$$L = \text{twist } (T) + \text{supercoiling } (S)^*$$

*It is common to use writhe (W) instead of supercoiling (S) in this equation. Writhe includes conformations other than supercoiled that are topologically equivalent as far as the linking number is concerned.

Figure 25.17

Topology of negative and positive supercoil.

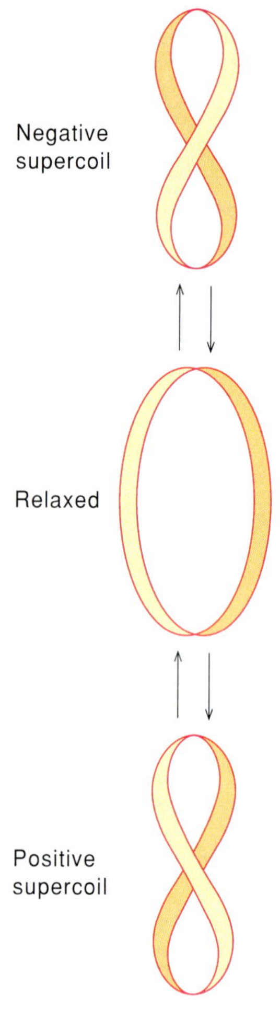

At the present time we do not know precisely how L will partition between twist and supercoiling for helices under torsional stress. For example, consider the hypothetical situation illustrated in figure 25.20. A 360-base-pair structure in the circular relaxed form ($L = +36$, $T = +36$, $S = 0$) is indicated to the left. Exposure to the bacterial enzyme DNA gyrase will introduce negative supercoils into such a structure in a reaction that requires ATP (see chapter 26). If four negative supertwists are introduced, L will be reduced to $+32$. Barring other changes, T will remain fixed and S will become -4. In fact, the torsional strain introduced by the four negative supertwists will tend to reduce T, causing a partial unwinding of the duplex. At one extreme, this effect could lead to the unwinding of four helical turns, in which case the supercoiling would disappear entirely ($L = +32$, $T = +32$, $S = 0$). The actual situation would probably lead to a reduction in the negative value of S and a concomitant reduction in T that would be spread over the entire duplex without any localized total unwinding of the double helix as pictured. Note that the linkage number L changes only when covalent bonds are broken, as in the case of gyrase treatment.

Figure 25.18

Ethidium bromide and DNA intercalated with ethidium bromide. Molecules such as ethidium bromide can intercalate DNA because they are flat rings of the same thickness as DNA base pairs. In order to accommodate the ethidium bromide molecule the duplex must untwist in the region of intercalation. This increases the separation between the planes of the base pairs. We saw the effect of complete untwisting in figure 25.10.

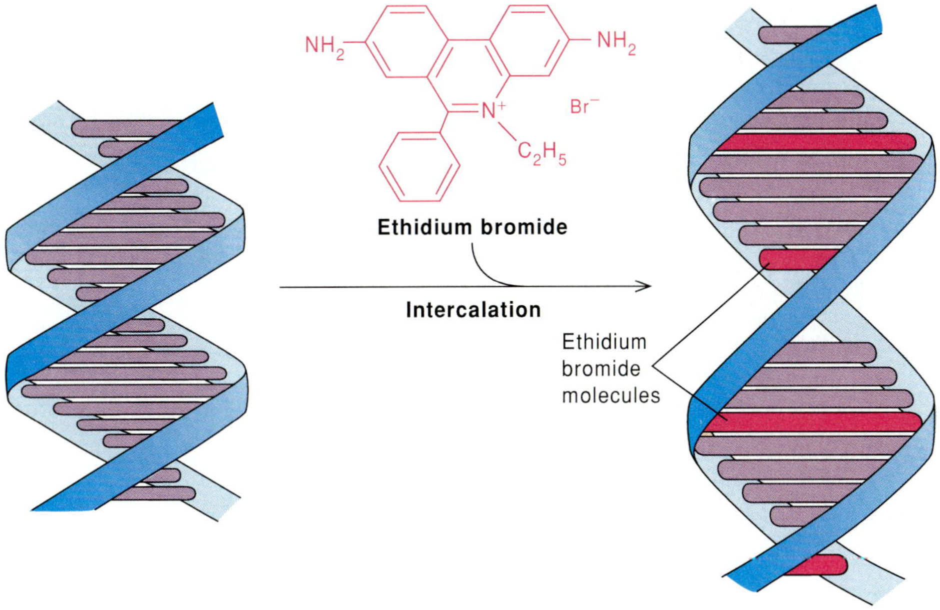

For DNA with a molecular weight of less than 10^7, agarose gel electrophoresis is a most effective method for assessing the extent of supercoiling. (For analysis of higher-molecular-weight DNAs, see boxes 25B and 25C). DNA isomers differing by 1 in linking number form separate bands in the gel (fig. 25.21). The more highly supercoiled molecule will migrate more rapidly through the gel as a result of its more compact structure.

In chapter 26 we will discuss enzymes called topoisomerases, which relax positively or negatively supercoiled DNA, as well as topoisomerases such as DNA gyrase (mentioned earlier), which only generate negatively supercoiled DNA from relaxed DNA. Gyrases that catalyze the formation of negatively supercoiled DNA have been found only in bacteria. Consistent with this fact, all double-helix DNA found in bacteria that is topologically constrained by not having free ends, such as circular DNA, is negatively supercoiled. Circular DNA isolated from virus-infected eukaryotic cells (for example, simian virus 40, or SV40 DNA) is frequently found to be negatively supercoiled, but only after deproteinization. In such instances, the DNA is not supercoiled in cells. Instead, the supercoiling results from removal of the proteins that normally are bound to the DNA and cause it to be underwound in its native state.

The biological importance of supercoiling has been clearly established only for DNA in bacteria. Drugs, such as novobiocin, that specifically inhibit DNA gyrase have a lethal effect except in strains that have a novobiocin-resistant DNA gyrase. On the other hand, mutants containing an altered topoisomerase, relaxing enzyme, which relaxes negatively supercoiled DNA, have been isolated from *E. coli*. Careful analysis of such mutants has shown that they are double mutants;

Figure 25.19

The sedimentation coefficients of closed circular phage PM2 DNA in 2.85-M CsCl containing varying amounts of ethidium bromide. The more compact the structure, the faster it will sediment. Supercoiled DNA is more compact than relaxed circular DNA. Thus when ethidium bromide is added to negatively supercoiled DNA the sedimentation constant decreases, indicating that the DNA is becoming more relaxed. A minimum sedimentation constant is reached when the fully relaxed structure is obtained. Addition of ethidium bromide beyond this point results in an increase in the sedimentation constant again, as the DNA becomes positively supercoiled. (Source: B. M. J. Revet et al., "Direct determination of the superhelix density of closed circular DNA by isometric titration" in *Nature (New Biol.)* 229:10, 1971. Copyright © 1971 Macmillan Magazines Ltd., London, England.)

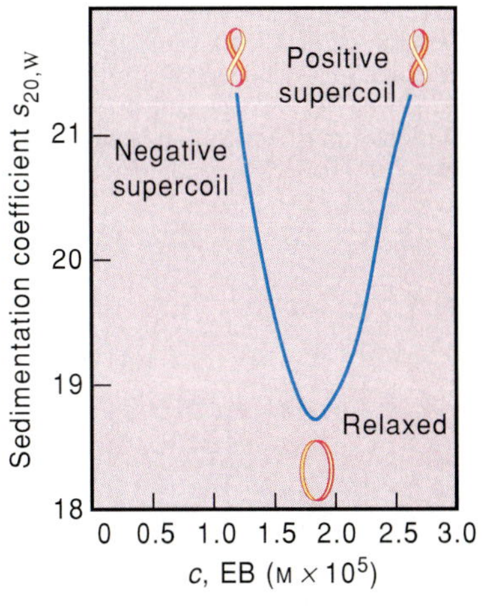

Figure 25.20

A circular duplex molecule in different topological states. (Adapted from a diagram supplied by M. Gellert.) The linking number can be changed only by breakage and re-formation of the phosphodiester linkages, as shown in the conversion of the relaxed circular form (1) to the strained negatively supercoiled form (2). The strain in the negatively supercoiled form can be partitioned in different ways between twist (*T*) and supercoiling (*S*) as shown in the interconversion between (2) and (3). No phosphodiester linkages are broken in making this interconversion and consequently there is no change in the linking number (*L*) as indicated.

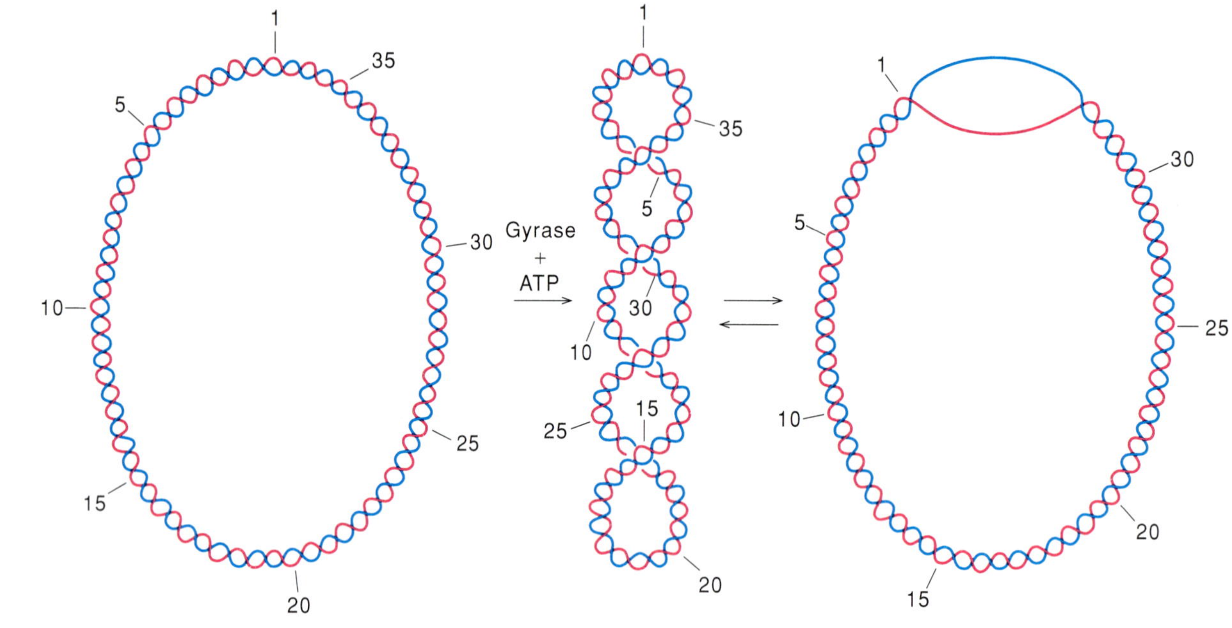

		(1) Relaxed	(2) Strained: supertwisted	(3) Strained: disrupted base pairs
Base pairs	(*bp*):	360	360	360
Linking number	(*L*):	36	32	32
Twist	(*T*):	36	36	32
Supercoiling	(*S*):	0	−4	0

Figure 25.21

Electrophoretic patterns of highly supercoiled or partially supercoiled DNA. Strip A represents a sample of circular duplex DNA obtained by deproteinization of the animal virus SV40. In strips B and C the DNA has been exposed for increasing times to an enzyme (topoisomerase) that catalyzes relaxation. Adjacent bands differ by 1 in linking number. (From W. Keller, Characterization of purified DNA-relaxing enzyme from human tissue culture cells, *Proc. Nat. Acad. Sci.* 72:2553, 1975.)

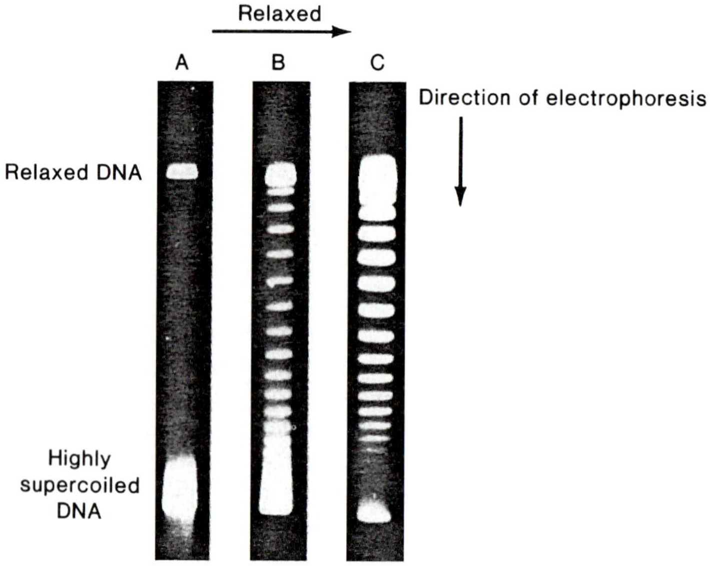

25B
BOX

Equilibrium-Density-Gradient Centrifugation

In equilibrium-density-gradient centrifugation, macromolecules such as nucleic acids or proteins are dissolved in a concentration of CsCl whose density is similar to that of the macromolecule. At high speeds, a nearly linear gradient of the salt is established. The macromolecules will form a narrow band (whose width is inversely related to their molecular weight) at a position where their density is equal to the density of the solution. If a mixture of DNA, RNA, and protein is centrifuged, the protein will band at a buoyant density of about 1.25 g/cc; DNA (depending on the percent G + C) will band at 1.710 g/cc; and RNA will settle to the bottom of the gradient, since its density is unusually high, about 1.9 g/cc. The greater the G + C content of DNA, the higher its buoyant density; the density in CsCl can in fact be used to determine the G + C content of native DNA. In contrast to velocity sedimentation, position of DNA in a density gradient is independent of linear duplex DNA size; the band width, however, is reciprocally related to size.

CsCl gradients can also be used to distinguish ^{15}N-labeled DNA from ^{14}N- or $^{14}N/^{15}N$-labeled DNA, or DNA containing 5-bromouracil from unsubstituted DNA. This feature has proved useful to establish the semiconservative replication of DNA and to separate replicated from unreplicated DNA (see chapter 26).

Finally, density-gradient centrifugation is very useful in the preparative separation of covalently closed circular (e.g., plasmid) DNA from linear (e.g., fragmented genomic) DNA in the presence of intercalating agents such as ethidium bromide. Because of its topological constraints, covalently closed circular DNA binds less of the dye. Since binding makes DNA lighter, its density is greater than that of linear (or nicked circular DNA) bound to ethidium bromide.

---

that is, they have an altered DNA gyrase that is less active, in addition to the defective relaxing enzyme. This observation suggests that in a normal bacterium the extent of negative supercoiling is carefully adjusted by the relative activities of the gyrase and relaxing enzymes. Negative supercoiling imparts a torsional stress or tension to the DNA structure that can be relieved to some extent by unwinding of the double helix. Both the initiation of DNA synthesis (chapter 26) and the transcription of certain DNA sequences or genes (chapter 28) are strongly dependent on negative supercoiling, probably because of the necessity to unwind at least one turn of the double helix during the initiation process. This same torsional stress explains why Z helix formation is encouraged by negative supercoiling; since in Z DNA the twists are left-handed, the transition from the B to Z form will tend to relax negatively supercoiled DNA.

DNA Denaturation Involves Separation of the Two Strands

The process of separating the polynucleotide strands of duplex nucleic acid structures is called denaturation. Denaturation disrupts the secondary binding forces that hold the strands together. Recall that the secondary binding forces include the edge-to-edge hydrogen bonds between the base pairs of opposing strands and the face-to-face stacking forces between the planes of adjacent base pairs. Individually, these secondary forces are weak, but when they act cooperatively they give rise to a DNA duplex that is highly stable in aqueous solution. The conditions required to denature DNA provide us with a measure of the strength of these interactions.

One of the simplest ways to denature DNA is by heating. The extent of denaturation at any temperature can be readily measured by the change in ultraviolet absorbance of a solution of DNA. A substantial rise in absorbance (hyperchromic shift) accompanies the transition from the native to the denatured state. When a solution of native DNA is slowly heated, the absorption at the ultraviolet maximum (at 260 nm) remains constant until an elevated temperature is reached, at which point the absorbance increases by about 40% over a narrow temperature range. This rise in absorbance coincides with the disruption of the regular base-paired structure, and the two polynucleotide strands separate from one another. The sharpness of the disruption of the regularly hydrogen-bonded base-paired native structure may be likened to the melting of a pure organic compound. It is customary to refer to this ultraviolet absorption temperature profile as a melting curve (curve *a* in figure 25.22).

The melting temperature, T_m, of DNA is defined as the temperature at the midpoint of the absorption increase. This is about 85° C for the example shown in figure 25.22. Rapid cooling of the denatured DNA solution leads to re-formation of intrastrand hydrogen bonds, but in a nonspecific, irregular manner. The absorbance decreases, but only by about three-fourths of the total original increase, and the decrease occurs over a much broader range of temperatures, as shown by curve *b* in figure 25.22. On subsequent reheating and cooling, the absorbance follows this cooling curve in the appropriate direction, a result indicating that denaturation produces an irreversible change.

25C
BOX

Application of Gel Electrophoresis for Chromosome Analysis

A variety of methods, both absolute (light scattering) and empirical (sedimentation, viscosity, gel electrophoresis), exist to determine the size of nucleic acids. The most accurate method is to determine the base sequence of the nucleic acid. Then, sequenced restriction fragments can be used as standards for comparison to samples whose molecular weight is being studied.

To analyze heterogeneous DNA populations, gel electrophoresis offers the best approach. However, very large DNA molecules—in the range of 150 kb or greater—tend to migrate with size-independent mobilities. (The abbreviation kb stands for "kilobase pair," or 1,000 base pairs.) A greatly improved technique can resolve large DNA molecules, the size of the *E. coli* genome and intact, individual yeast chromosomal DNA, on agarose gels. The gently isolated DNA samples are subjected alternately to two approximately orthogonal electric fields. First one field is applied and turned off. Then the other field is applied and turned off. This process is repeated with a pulse time of many seconds over a period of several hours. The precise pulse time depends on the size of the molecules being separated, being longer for larger molecules.

When large DNA molecules enter a gel in response to an electric field, the molecule must elongate parallel to the field. When the field is shut off and a new field is applied, perpendicular to the long axis of the DNA, the molecule must reorient. The reorientation time should be quite sensitive to the size or molecular weight of the DNA. Since the length of the stretched-out DNA molecule is generally larger than the pores in an agarose gel, this reorientation is absolutely essential if the DNA is to undergo any net migration in response to the new field. Therefore, the smaller molecules have a mobility advantage over the larger molecules, which results in a size-based separation. (See figure.)

Another development is the electrophoretic analysis of partially denatured DNA, a technique that is very useful in the detection of single base substitutions of cloned DNA. Fragments of DNA that are wild type or that contain a single base mismatch migrate into a polyacrylamide gel containing an ascending denaturing gradient of urea and formamide in the same direction as the electrical field. At a critical depth the mobility of partially denatured DNA slows down abruptly. The distance moved is a measure of the stability of the sequence, rather than the overall base composition or size of the remaining unmelted duplex. Differences in the partial denaturation of wild-type and mutant DNA molecules allow one to separate and detect them. This method, developed by L. Lerman, provides a quick method to detect differences, even a single base-pair change, in cloned homologous fragments from wild-type and mutant sources.

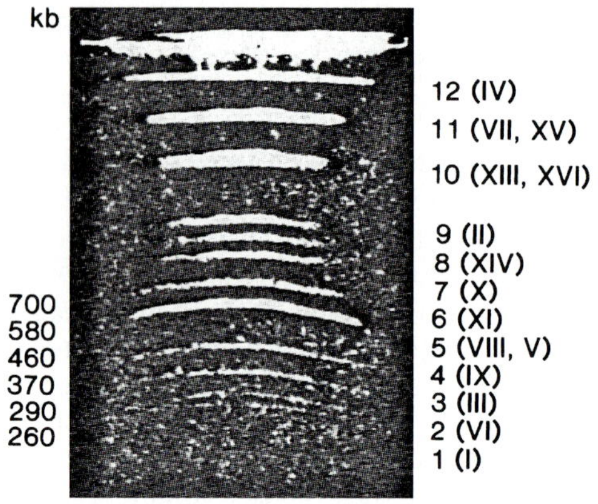

Figure 1

Separation of intact DNA from the yeast chromosome. The chromosome assignments are on the right; size markers, in kb, are on the left. The DNA bands were detected by staining with ethidium bromide. (From G. F. Carle and M. V. Olson, "An electrophoretic karyotype for yeast," *Proc. Natl. Acad. Sci.* 82, 3756-3760, 1985.)

Some DNAs do not occur naturally in the double-helix form. The melting curve is an excellent tool for detecting DNAs in the single-stranded conformation. For example, in certain bacteriophages that infect *E. coli,* such as ϕX174, fd, or M13, the DNA exists as a single, circular strand. The ultraviolet absorption temperature curve for the DNA of these phages is similar in shape to that observed for denatured DNA (curve *b* in figure 25.22). Broad melting curves also are characteristic of most RNAs, which rarely have regions of regular base pairing that extend for more than ten or twenty residues.

Melting curves also have provided evidence that the stability of the double-helix structure is a function of its base composition. The midpoint of denaturation (T_m) of naturally occurring DNAs is precisely correlated with the average base composition of the DNA: the higher the mole percent of G-C base pairs, the higher the T_m (fig. 25.23). This seems reasonable, since the G-C base pair contains three hydrogen bonds, whereas the A-T base pair contains only two (see fig. 25.7); thus we would expect DNA with a greater G-C content to be more stable. As we indicated earlier, base stacking is also believed to contribute to the stability of the duplex structure. In general, the interaction energy gained by stacking between adjacent G-C base pairs is greater than that gained by interaction between A-T base pairs. An excellent source of natural DNAs that differ in their content of G-C base pairs is bacteria. Members of the genus *Clostridium* have only 25% G-C in their DNA,

Figure 25.22

Effect of temperature on the relative absorbance of native, renatured, and denatured DNA. When native DNA is heated in aqueous solution, its absorbance does not change until a temperature of about 80° C is reached, after which the absorbance rises sharply, by about 40% (curve *a*). On cooling the absorbance falls, but along a different curve, and it does not return to its original value (curve *b*). Renatured DNA, in which the two strands have been brought back into perfect register, shows a sharp melting curve similar to native DNA (curve *c*). Renatured DNA is prepared from denatured DNA by holding the temperature at about 25° C below the denaturation temperature for an extended time. This subject is discussed in detail later in the text. The temperature at which the native DNA is half denatured is labeled T_m.

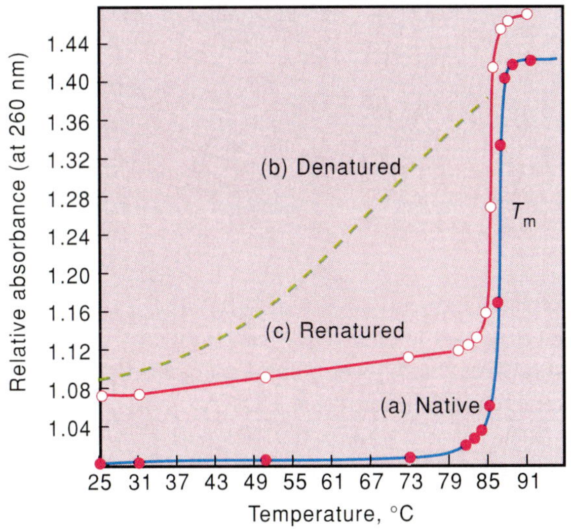

Figure 25.23

Dependence of the temperature midpoint (T_m) of DNA on the content of guanine and cytosine. As the percentage of G + C increases, the T_m increases. Two curves are shown to illustrate the point that the denaturation temperature is shifted to lower values when the ionic strength is lowered.

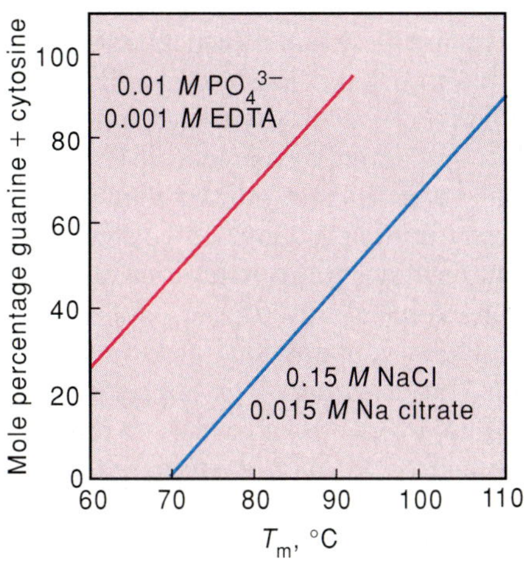

whereas some *Micrococcus* species have DNA with 72% G-C. *E. coli* DNA is about 50% G-C. Whereas the base composition of the DNA of bacteria along the genome is uniform, reflected in a narrow, symmetrical peak in a CsCl density gradient, the DNA of higher eukaryotic cells is not so uniform, with a minor fraction of DNA displaying a G-C content different from the major fraction of DNA. This minor fraction can be detected in a CsCl density gradient as a small band separate from the main band (see box 25B). The smaller band is termed satellite DNA.

Other factors present in aqueous solution can affect the stability of the double-helix structure in a positive or a negative way. For example, salt has a stabilizing effect, which is mainly due to the repulsive electrostatic interactions between the negatively charged phosphate groups. Salt shields this charge interaction and therefore stabilizes the duplex structure. Thus DNA in 0.15-M NaCl denatures at a T_m about 20° C higher than DNA in 0.01-M phosphate (see fig. 25.23). In pure water (no salt present) DNA denatures at room temperature. Extremes of pH also have a destabilizing effect on the double-helix structure. When the pH is above 11.5 or below 2.3, there is extensive deprotonization or protonization, respectively, of the hydrogen-bonding groups of the bases, which in turn disrupts the hydrogen-bonded structure. Alkali is an excellent DNA denaturant; it permits rapid separation of the strands without degradation. Alkali both denatures and degrades RNA to 2′(3′) mononucleotides.

Many solutes that can form hydrogen bonds also lower the melting temperature (decrease the stability) of double-helix structures. The organic compounds formamide and urea are frequently used to lower the denaturation temperature as well as to prevent reaggregation of denatured strands on subsequent cooling. This is important in DNA manipulations, where the wish is to avoid nonspecific aggregation. Reagents that increase the solubility of the DNA bases (e.g., methanol) or disrupt the water shell around them (e.g., trifluoracetate) reduce the hydrophobic interactions between the bases and lower the T_m. Most proteins that bind to DNA inhibit denaturation. However, some DNA-binding proteins destabilize the native state. Proteins of this class usually bind preferentially to single-stranded DNA, thereby favoring separation of the double strands. For example, gene 32 of bacteriophage T4 encodes a DNA-binding protein that is essential for bacteriophage T4 DNA replication. This protein induces the local unwinding of the DNA duplex, a process that facilitates its replication (chapter 26). Addition of appropriate amounts of the gene-32-encoded protein leads to complete denaturation of individual DNA molecules. This complete denaturation occurs because the gene-32 protein exhibits cooperative binding; that is, the binding of one gene-32 protein molecule to DNA facilitates binding of a second gene-32-encoded molecule at a neighboring site and continued binding at adjacent sites. A number of DNA-binding proteins similar to the gene-32 protein have been found in both prokaryotic and eukaryotic cells. They are believed to play an important role in DNA replication, recombination, and repair.

DNA Renaturation Involves Duplex Formation from Single Strands

We have pointed out that when a solution of heat-denatured DNA is allowed to cool rapidly, the regularly hydrogen-bonded structure does not re-form. However, reassembly of the two separated polynucleotide strands into the native structure, called renaturation, is possible under certain specialized conditions.

The first indication that renaturation was possible came from studies of Julius Marmur and Paul Doty, using transforming DNA in a biological assay. The basis of the assay was that denatured pneumococcus DNA is inactive in DNA-mediated transformation, whereas native DNA is active in transferring genetic traits, such as streptomycin resistance, from a donor to a recipient strain. When transforming DNA was heated and rapidly cooled it was biologically inactive; however, when denatured DNA was slowly cooled, a large fraction of the initial transforming activity was recovered.

Further experiments showed that the optimum temperature for renaturation is about 25° below the T_m. As is the case with denaturation, the renaturation process can be followed spectrophotometrically. If the temperature of a denatured DNA solution is maintained at $T_m - 25°$ C for a long period of time, the absorbance of the solution gradually decreases until it approaches a value close to that of native DNA (see curve *c* in figure 24.22). The optimum temperature for renaturation is frequently referred to as the annealing temperature. At such a temperature, irregularly hydrogen-bonded structures are unstable but regularly hydrogen-bonded structures are stable. Consequently, prolonged exposure of denatured DNA at the annealing temperature allows the DNA bases to explore various configurations until complementary regions of pairing are formed between otherwise separated strands.

Spectrophotometric methods and biological assays are not the only means of monitoring renaturation. Other methods are sometimes used, especially when it is desirable to separate the denatured and renatured fractions or when small, radioactive amounts of nucleic acids are being analyzed. One of these methods involves digestion of the nucleic acid with enzymes that preferentially degrade single-stranded DNAs. For this purpose, the S1 nuclease derived from the mold *Aspergillus* is commonly used. The amount of DNA resistant to digestion by this nuclease gives an accurate measure of the amount of duplex structure in a sample of partially renatured DNA. A more commonly used method involves binding to a column made from a calcium phosphate gel known as hydroxyapatite. For reasons that are unclear, duplex DNA binds to hydroxyapatite at salt concentrations where single-stranded DNA does not bind. This method must be used with caution as a quantitative measure of renaturation, because a DNA molecule that is partially duplex and partially single-stranded will also bind to hydroxyapatite. Hydroxyapatite chromatography is also useful as a preparative technique for separating rapidly reassociating fragments from slowly reassociating fragments. The significance of the rate of reassociation is explained later.

Renaturation Rate Measures Sequence Complexity

Marmur and Doty and their co-workers' discovery of the phenomenon of renaturation met with considerable interest. Many applications have grown out of the technique of renaturation or annealing (see box 25D).

Figure 25.24

Steps in denaturation and renaturation of a DNA duplex. In step 1 the temperature is raised to the point where the two strands of the duplex separate. If denatured DNA is slowly cooled, the events depicted as steps 2 and 3 follow. In step 2 there is a second-order reaction in which two complementary strands of DNA must collide and form interstrand hydrogen bonds over a limited region. Step 3 is a first-order reaction in which additional hydrogen bonds form between the complementary strands that are partially hydrogen-bonded (zippering). Once complementary strands are partially bonded, the zippering reaction will occur rapidly. In the overall process, step 2 is rate-limiting.

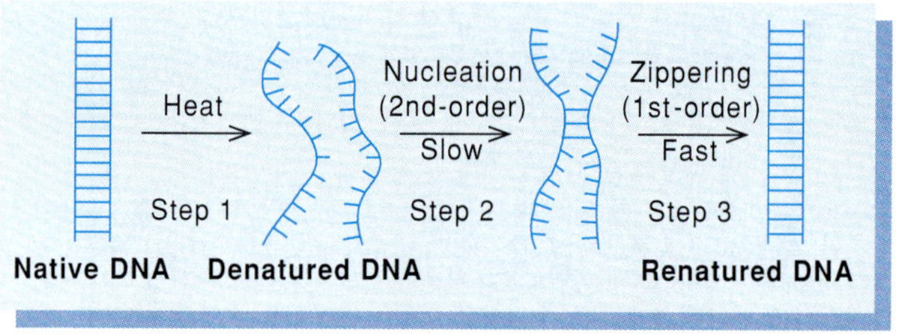

The renaturation rate of DNA has been used extensively to compare the base sequence complexity of DNAs from different sources. For a given weight concentration of DNA, the more homogeneous in sequence a DNA sample, the more rapidly it will renature. Thus a 0.01% solution of denatured T4 bacteriophage DNA renatures much more rapidly than does a 0.01% solution of *E. coli* DNA. In turn, bacterial DNA renatures much more rapidly than mammalian DNA. The steps involved in denaturation and renaturation are indicated in figure 25.24. Renaturation is a two-step process. In the first step, called nucleation, contact is made between two complementary regions of DNA. Nucleation is followed by a relatively rapid zippering up of adjoining base residues into a duplex structure. Thus nucleation is the rate-limiting step, and it is this step that must be examined to determine the time course of renaturation. Because nucleation involves interaction between two molecules, it should occur at a rate proportional to the square of the concentration of single strands; that is, it is a second-order reaction.

If *c* is the concentration of single-stranded DNA at time *t*, then the second-order rate equation for loss of single-stranded DNA is

$$-\frac{dc}{dt} = k_2c^2$$

where k_2 is the second-order rate constant. Given a starting concentration c_0 of completely denatured DNA, the amount of single-stranded DNA left after renaturation for time *t* is given by

$$\frac{c}{c_0} = \frac{1}{1 + k_2c_0t}$$

At time $t_{1/2}$, when half of the DNA is renatured, $c/c_0 = 0.5$ and $t = t_{1/2}$, from which it follows that

$$c_0t_{1/2} = \frac{1}{k_2}$$

25D
BOX

Techniques Using Nucleic Acid Renaturation

Nucleic acid hybridization takes place between any two complementary single-stranded nucleic acids that are annealed at $T_m - 25°$ C for prolonged periods of time. Since the rate of duplex formation depends on the concentration of the interacting complementary strands, the method can be used to measure the abundance of a specific nucleic acid in a mixture. Thus the number of *Drosophila* ribosomal RNA genes can be estimated by titrating a fixed amount of denatured genomic DNA with increasing amounts of rRNA. Specific DNA-RNA hybrids are detected by first immobilizing denatured DNA onto nitrocellulose filters, then by adding labeled RNA to serve as a probe. The labeled RNA will be retained by the nitrocellulose only if it is hybridized by complementarity to the DNA. The method can be modified for use in isolating specific mRNA sequences; RNA hybridized to immobilized, denatured DNA can be eluted following denaturation of the hybrids.

Hybridization can be used to detect specific sequences in the nucleus or chromosomes using a labeled probe and cells fixed to a microscope slide. Such *in situ* experiments serve to locate genes on chromosomes and were useful in demonstrating that highly repeated sequences in eukaryotic DNA are found in heterochromatin. In the Southern blotting technique, individual DNA fragments generated by restriction endonuclease digestion are first separated by gel electrophoresis; they are denatured and then transferred to a nitrocellulose support. The DNA fragments can then be probed for specific sequences using a suitable labeled DNA or RNA probe. In Northern blot hybridization, immobilized RNA separated by agarose gel electrophoresis under denaturing conditions is transferred to nitrocellulose and is detected by a labeled, complementary single-strand probe. These blotting techniques can be used to determine the size and abundance of the immobilized nucleic acid. The blotting and filter hybridization techniques are very useful in studying the level of expression of various genes and the effect of inducers, of repression, and of mutations on the level of expression.

Hybridization has also been exploited to detect specific sequences in recombinant clones. A labeled probe is applied to bacterial colonies (or plaques) that have been immobilized and the DNA denatured. Hybridization allows a worker to detect transformants (or transfectants) harboring recombinant vectors that carry the sequence of interest.

As a rule, c/c_0 is plotted as a function of c_0t. The resulting curve is referred to as a "cot" curve (fig. 25.25). Data are shown for DNAs and RNAs with varying nucleotide complexity N, which is defined as the number of nucleotides in a nonrepeating sequence. If there are no repeating sequences in the cellular DNA, then N is equal to the number of nucleotides in the genome. As you can see, $c_0t_{1/2}$ is proportional to N for these samples. This family of curves has been used to calibrate more complex situations.

For a given nucleotide complexity, the renaturation rate is also a function of the length L of the nucleic acid, i.e., the actual number of bases per single-stranded nucleic acid molecule present in a renaturation mixture. It can be shown that

$$k_2 = \frac{L^{0.5}}{N}$$

and at $c_0t_{1/2}$ should be proportional to $N/L^{0.5}$. When DNA samples of different initial lengths are compared, the effect of length is usually eliminated by shearing the DNAs of the different samples to a more or less uniform length, so that $c_0t_{1/2}$ can be used directly as a measure of N. This is usually the parameter of interest in a renaturation rate experiment.

Most prokaryotic DNAs yield simple monophasic cot curves with one inflection point (see fig. 25.25). In such instances, N is directly proportional to the amount of DNA in the chromosome. However, for the total DNA from a complex eukaryotic organism such as a human, the curves are more complex, containing more than one inflection point. A precise interpretation of the cot curve for human DNA, shown in figure 25.26, is not possible, but the following explanation is considered most likely. About 2% of the DNA (the foldback portion) renatures very rapidly. This behavior is due mostly to single-stranded regions of DNA that fold back on themselves, thereby forming hairpin duplex (cruciform) structures. Structures giving rise to cruciforms must have possessed inverted repeating sequences called palindromes. The next class of sequences to renature (designated the fast segment in the figure) accounts for about 5% of the total DNA of higher eukaryotes and has a $c_0t_{1/2}$ value of about 10^{-2}. The small amount of DNA, together with the rapid reassociation kinetics, suggests that about 5% of the nuclear DNA (fast) is present in a very large number of copies per genome (highly repetitive DNA). In the same figure, a region labeled intermediate can be seen that accounts for an additional 20% of the DNA (middle repetitive DNA). Finally, about 70% of the DNA has reassociation kinetics in the range expected for single-copy or unique DNA, i.e., DNA whose sequence is present as a single copy per haploid genome.

This type of renaturation kinetic analysis has been used for the gross characterization of DNA in many cell types as shown in figure 25.27. These data show that bacterial DNA

Figure 25.25

Reassociation of double-stranded nucleic acids from various sources. The genome size is indicated by arrows near the upper nomographic scale. Over a factor of 10^9, this value is proportional to the c_0t (the "cot") required for half-reaction. All DNAs were sheared so that they have approximately the same fragment size (about 400 nucleotides, single-stranded). Correction has been made to give the rate that would be observed at 0.18 M sodium ion concentration. No correction for temperature has been applied, since it was approximately optimum in all cases. The labels for the different DNAs should not concern the average reader. MS2 is the RNA sequence obtained from a bacterial virus. Mouse satellite and calf (nonrepetitive fraction) are fractions of the genome obtained from the indicated animals. (Source: R. J. Britten and D. E. Kohne, "Repeated sequences in DNA" in *Science,* 161:529, 1968. Copyright © 1968 American Association for the Advancement of Science, Washington, D.C.)

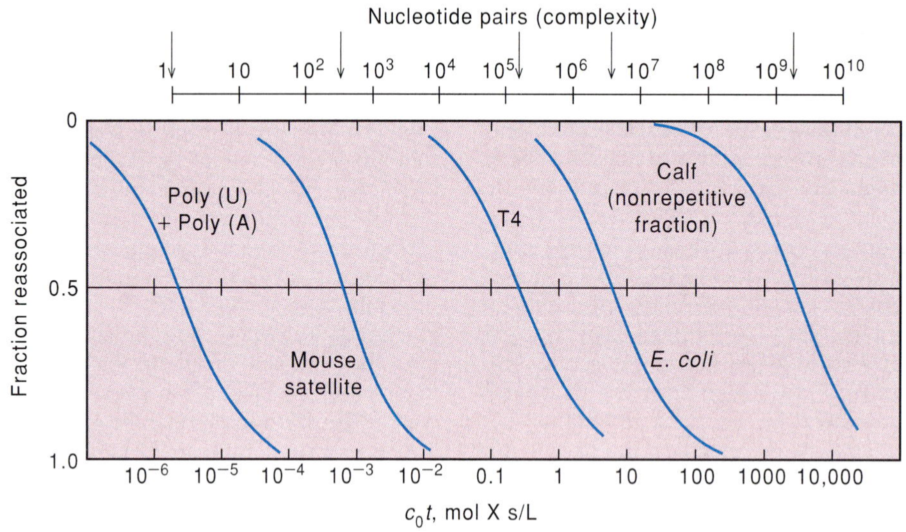

Figure 25.26

The cot curves for total human nuclear DNA. The total human DNA (and the DNA from other complex eukaryotic organisms) does not give a simple monophasic plot, but shows more than a single inflection point. Labels indicate the type of DNA that renatures at different cot values. (Source: Unpublished data of A. R. Mitchell.)

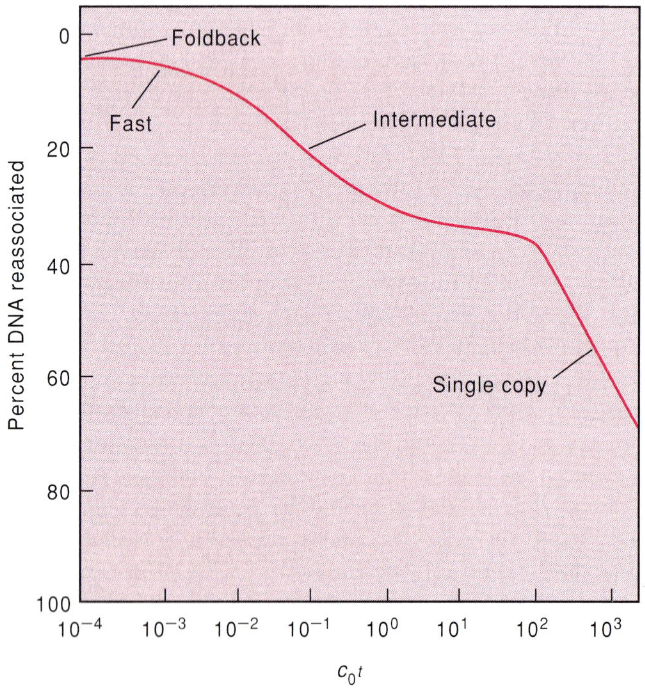

Figure 25.27

Distribution of single-copy DNA and repetitive-sequence DNA in various organisms. The width of bands and the number below the bands indicate the fraction of total cellular DNA in each class. For example, in calf about 55% of the DNA sequences are single copy, about 38% are present in somewhat less than 10 copies, and about 3% are present in somewhat less than 10 copies. (Adapted from "Repeated segments of DNA" by Roy J. Britten and David E. Kohne. Copyright © 1970 by Scientific American, Inc. All rights reserved.)

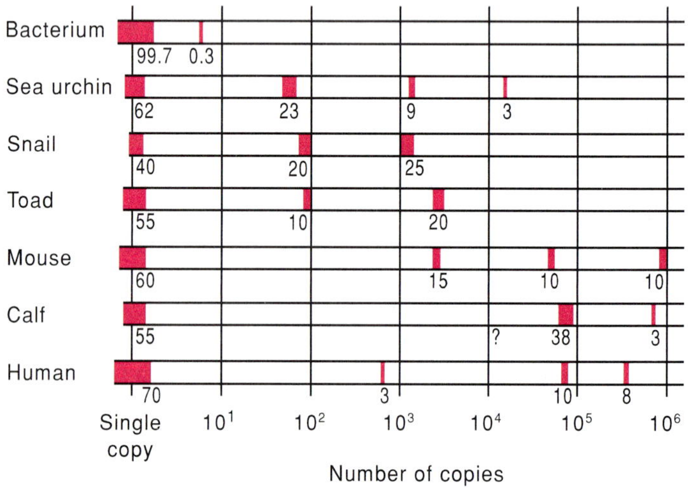

(*E. coli*) contains very little repetitive DNA (0.3%). The amount that is found is mainly accounted for by the eight genes for *E. coli* ribosomal RNA that have nearly identical sequences. In eukaryotes, single-copy DNA (i.e., nonrepetitive DNA) accounts for 40 to 70% of the DNA, most of the remainder being roughly divided between middle repetitive ($<10^4$ copies per genome) and highly repetitive ($>5 \times 10^4$ copies per genome). Further analyses of eukaryotic gene structure by a variety of other techniques have shown that most genes encode proteins that belong to the unique (single-copy) class. The middle repetitive class includes transfer RNA genes and ribosomal RNA genes that are involved in protein synthesis (chapter 29) as well as the genes encoding the nuclear proteins called histones. Some highly repetitive sequences occur in tandem; and still other repetitive DNA elements are distributed at random throughout the genome. It has also been argued that some families of repetitive DNA, referred to as "selfish DNA," represent "parasitic" sequences that replicate together with the genome without conferring any positive or negative characteristics to the host cells that harbor these sequences. Satellite DNA detected in CsCl density-gradient centrifugation is usually enriched in highly repetitive DNA.

Chromosome Structure

All types of nucleic acids interact with proteins. Chromosomal DNA forms stable nonspecific complexes with structural proteins that stabilize their tertiary structure; it also forms transient complexes with enzymes and regulatory proteins that are involved in DNA and RNA metabolism.

We have discussed nucleoprotein complexes at various points in this text where they are relevant to metabolism. Let us now look at the structure of DNA in *E. coli* and in eukaryotic nucleohistone and chromatin.

Physical Structure of the Bacterial Chromosome

The single chromosome of *E. coli* contains about 2×10^9 daltons or about 3×10^6 base pairs of DNA. If all of this DNA were in a duplex structure stretched end-to-end it would be 1 mm long, which is about 50 cell diameters. In fact, the chromosome is circular, centrally located in the cell, and highly folded, so that it is only about 2 μm across. By contrast to what happens in eukaryotes, no dramatic change in chromosome morphology is seen prior to cell division. Clearly, the degree of compaction observed throughout the cell cycle does not interfere with transcription to any great extent, since all the genes in *E. coli* are readily expressible.

Electron micrographs of the *E. coli* chromosome suggest a folded circular structure containing about 40 to 100 supercoiled loops (diagrammatically indicated in figure 25.28). It is believed that the folded structure is held together by an RNA-protein core, although the manner in which this is done is not well understood. The structure is further stabilized because the core forms a complex with positively charged polyamines and

Figure 25.28

The *E. coli* chromosome exists as a circular, folded, supercoiled duplex (*a*). This can be converted to a partially unfolded structure by brief treatment with RNase (*c*). There are 50 to 100 loops in the structure; supercoiling may be selectively eliminated from individual loops by single-strand nicking of the DNA within the loop (*b*).

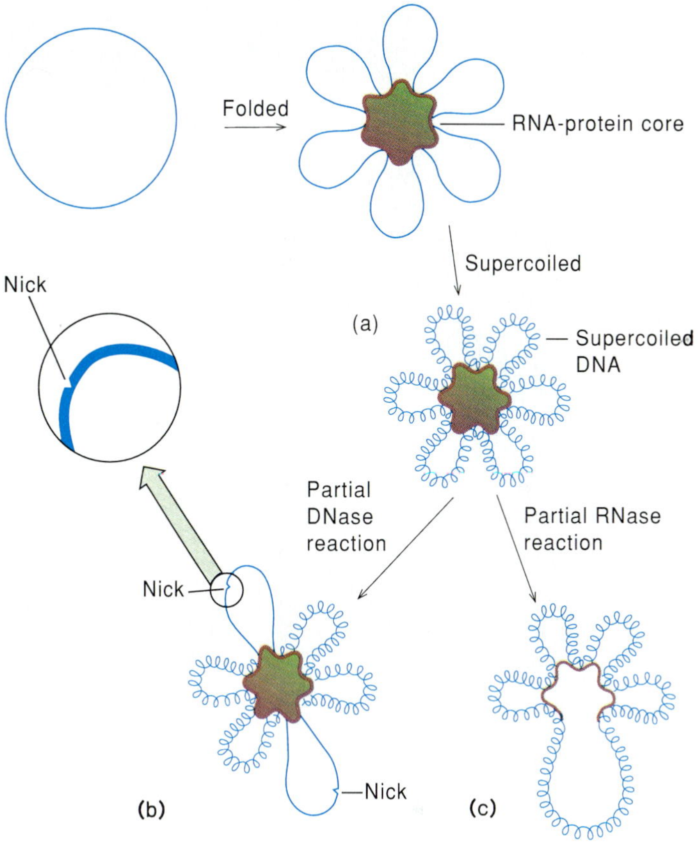

certain basic proteins. Pettijohn and his co-workers have provided evidence of such a core. They first showed that the individual supercoiled loops maintain their supercoiling independently of one another. Thus if a single nick is introduced into one of the loops by limited DNase action, that loop adopts an expanded relaxed conformation, but supercoiling in the other loops is maintained (see fig. 25.28). Limited RNase or protease treatment causes the partial breakdown of the looped structures without interfering with the supercoiling (see fig. 25.28). These results have led to the proposal that each of the loops is a domain whose lateral motion is restricted by an RNA-protein core complex.

The Genetic Map of E. coli

The circular *E. coli* chromosome contains enough base pairs to make about 3,000 average-size genes (1,000 base pairs each). The relative positions of over half of these genes are known. In regions where our understanding is reasonably complete, we gain the impression of an efficiently organized genome. Coding regions are interspersed with regulatory regions; there is no evidence for significant stretches of DNA with no function.

Figure 25.29

Swollen fibers of chromatin from the nucleus of the chicken red blood cell. The electron micrograph is enlarged about 325,000X and negatively stained with uranyl acetate. (Micrograph courtesy of A. L. Olins and D. E. Olins.)

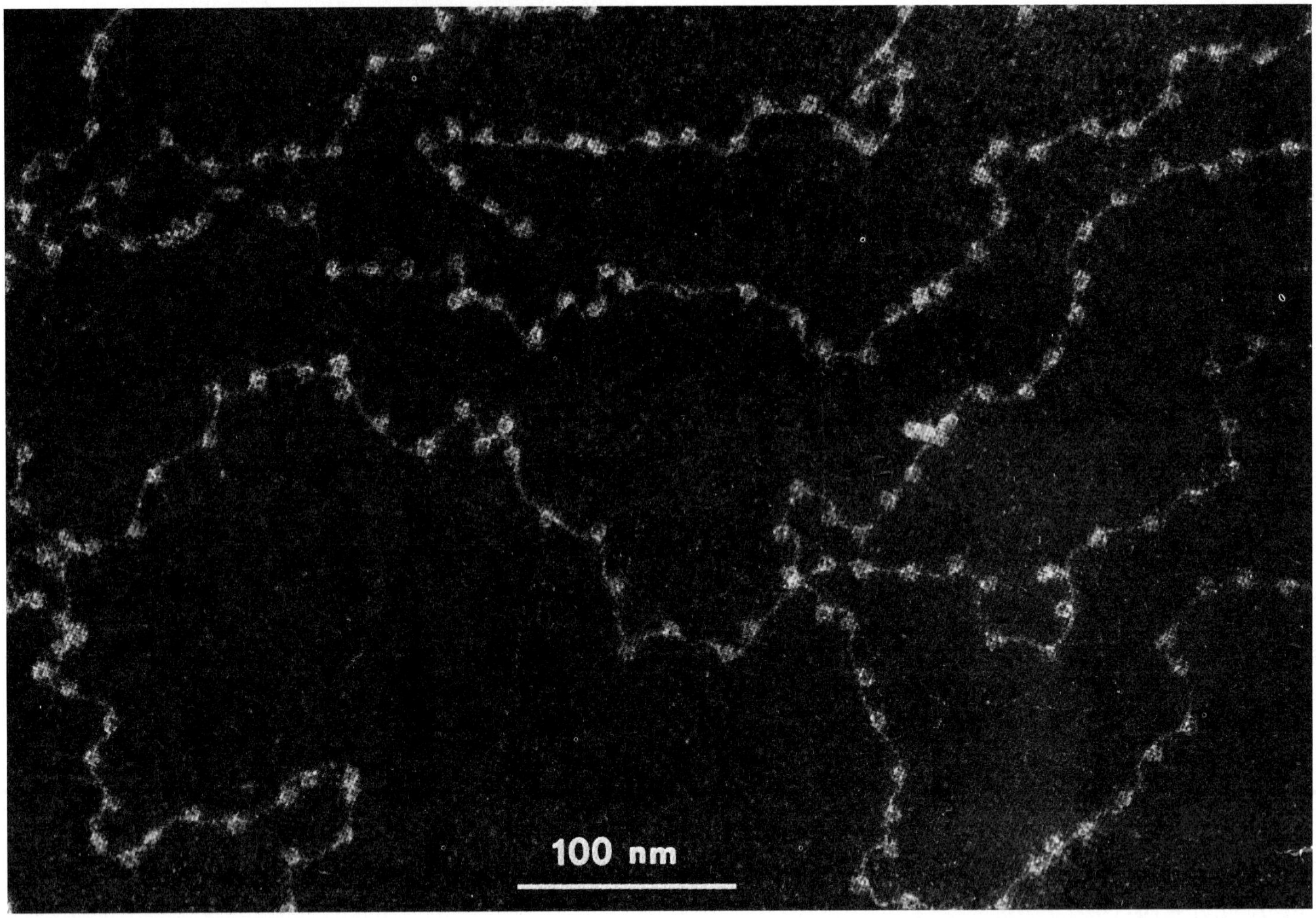

Frequently genes with a related function are tightly clustered. These clustered genes are usually transcribed into single expression units (messenger RNAs) containing the information for the synthesis of several functionally related proteins.

Eukaryotic DNA Is Complexed with Histones

DNA in eukaryotic chromosomes exists in a highly compacted form known as nucleohistone, a complex of DNA with approximately equal weights of five proteins known collectively as histones. Since the same kind of compaction occurs with duplex DNA of almost any sequence, it is presumed that a repeating aspect of the structure is recognized by the protein.

Electron microscopic and x-ray diffraction studies on compacted chromatin suggest that chromatin is a coiled coil. Data are scant, and some of the x-ray spacings observed still require explanation. A breakthrough came when D. E. Olins and A. L. Olins observed that chromatin viewed after sudden swelling in water showed a beaded structure (fig. 25.29). The beads, called nucleosomes, contain most of the histone; they are about 10 nm in diameter, and the spacing between the beads is about 14 nm. Brief enzymatic digestion of chromatin with micrococcal nuclease or some other endonucleases fragments this structure. The DNA from this partial digestion gives rise to a banded pattern on agarose gel electrophoresis that indicates nucleoprotein structures containing 200 base pairs of DNA or multiples thereof (400, 600, 800 bp, etc.). When the products of partial micrococcal nuclease digestion were fractionated by ultracentrifugation, the fractions examined in the electron microscope showed a direct correlation between the size of the DNA estimated on gels and the number of nucleosomes. Thus the most rapidly moving DNA band seen on gels was derived from a structure containing one nucleosome, and the second-fastest migrating species contained nucleosome dimers, etc. Evidently, the brief treatment with endonuclease preferentially cleaves DNA in the internucleosomal region, where the DNA

Table 25.3
Characteristics of Histones

Name	Ratio of Lysine to Arginine	M_r	Copies per Nucleosome
Histone H1[a]	20	21,000	1 (not in bead)
Histone H2a	1.2	14,500	2 (in bead)
Histone H2b	2.5	13,700	2 (in bead)
Histone H3	0.7	15,300	2 (in bead)
Histone H4	0.8	11,300	2 (in bead)

[a]Not found in lower eukaryotes such as yeast.

is least likely to be protected from enzyme attack. More exhaustive nuclease treatment gives rise to a single band on gels that contain a single nucleosome and 140 base pairs of DNA. The effect of the more exhaustive nuclease treatment is to degrade all the DNA that is not in direct contact with the nucleosome. Combined electrophoresis and electron microscopy results have led to a model for the beaded structure in which the nucleosome core particles contain clumps of histone complexed to 140 base pairs of DNA duplex, with about 60 base pairs of duplex serving as linkers between the core particles.

The histones present in chromatin have been extensively characterized, and their sequences are known. The classes of histones and their molecular weights are listed in table 25.3. The lysine-rich histone H1 is not present in the nucleosome core particle, as evidenced by its release on extensive nuclease treatment. Consistent with this conclusion is the finding that H1 is the only histone that readily exchanges between free and chromatin-bound histone. The other eight histones, two each from the other four histone classes, form a tightly complexed core particle that is conserved when chromosomes duplicate.

Many fine points about the structure of chromatin remain to be determined. An illustration of the probable DNA coiled-coil structure of the nucleosome core particle is presented in figure 25.30. There are 140 base pairs of DNA in a nuclease-resistant nucleosome core that make about 1.75 superhelical turns around a histone octamer. An additional 60 base pairs of spacer DNA connect the core particles. Electron microscopic observations indicate that the amount of linker DNA is actually different in different species and in different tissues of the same species, varying from about 20 to 95 base pairs.

Nuclear magnetic resonance spectroscopy reveals that nucleosomal DNA has a secondary structure quite similar to that of B DNA. The DNA is wound around the histone complex. Treatment with pancreatic DNase I (instead of with micrococcal nuclease) produces single-stranded fragments that are multiples of about ten nucleotides. The current interpretation of this observation is that the digestion by DNase I is confined to the exposed side of the DNA in the nucleosome core (see fig. 25.30*a*), the approach of the enzyme to the other side being hindered by the presence of the histones.

Salt bridges between positively charged basic amino acid side chains of histones and the DNA phosphates play an important role in stabilizing the DNA-histone complex.

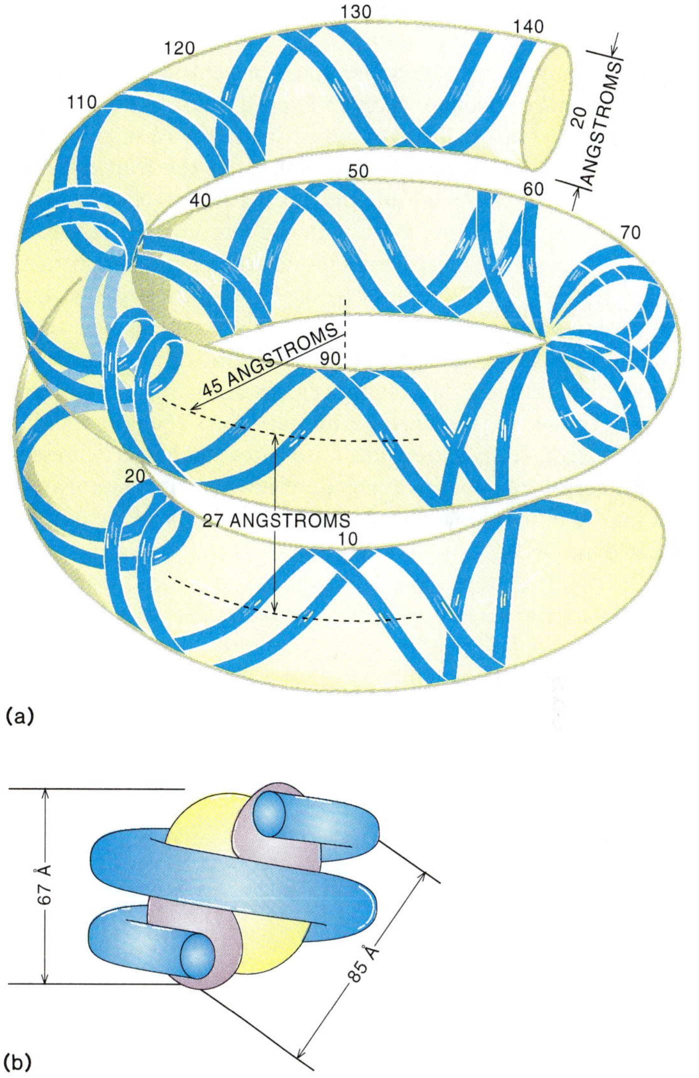

Figure 25.30

(*a*) Path of DNA that can account for the bipartite structure of the nucleosome core is a super helix with an external diameter of 110 Å and a pitch of 27 Å; the turns of the 20-Å-wide DNA helix are nearly in contact. There are about 80 nucleotide pairs of DNA per turn; the nucleosome core, an enzymatically reduced form of the nucleosome consisting of some 140 nucleotide pairs, has about one and three-quarter turns wrapped on it. The histone octamer complex, containing two each of histones H2a, H2b, H3, and H4, is packed on the inside of the DNA coiled-coil structure. (From "The Nucleosome" by Roger D. Kornberg and Aaron Klug. Copyright © 1981 Scientific American Inc. All rights reserved.) In (*b*) we see this histone octamer inserted into the nucleosome core. The H3-H4 tetramer is shown in yellow and an H2a-H2b dimer is shown at each end in purple.

Higher-order structures beyond that of the nucleosome itself have been inferred from light, x-ray, and neutron-scattering microscopy and visualized by electron microscopy. Thus by electron microscopy two types of fibers have been seen: a 10-nm fiber and a 30-nm fiber. The nucleosome can be arranged in a zigzag fashion to produce a fibril that is 10 nm wide, in which the nucleosomes are packed edge to edge rather than face to face. When the ionic strength is raised, the fibrils reversibly condense into an irregular supercoiled fiber about 30 nm in diameter. The nucleosome particles in this fiber are thought to have their cylindrical axes approximately perpendicular to the long axis of the fiber with six to seven nucleosomes per turn (fig. 25.31). Further coiling of these structures is necessary to explain the structures seen in mitotic chromosomes (e.g., see fig. 25.1), but evidence on the precise nature of these structures is lacking.

Organization of Genes within the Eukaryotic Chromosome

The organization of genes within the eukaryotic chromosome is far more complex and less well understood than in prokaryotes. It is clear that a much lower percentage of the DNA is used for coding in complex eukaryotes than in prokaryotes. *E. coli* contains about 3,000 genes; the human genome is 1,000 times larger than the *E. coli* genome, but it is very unlikely to contain 1,000 times the number of functional genes present in *E. coli*. Most estimates hover around 50,000. Even if there were 200,000 human genes, this still leaves approximately 90% of the DNA with no coding function. Measurements on the fruit fly *Drosophila melanogaster* indicate that coding regions of some genes probably account for no more than one-tenth of the base pairs within the gene. The discovery that noncoding regions (introns) occur between coding regions (exons) goes a long way toward explaining the excessive amounts of DNA that appear to be present in eukaryotes (see chapter 28), although it is probably not the whole story. Noncoding regions serving as control loci may be larger on the average in eukaryotes. Nonfunctioning genes that have lost their initiation sites for being expressed (promoters) and repetitive genetic elements with no apparent coding function also help to account for the large amount of noncoding DNA.

Not all repetitive DNA in eukaryotes is noncoding. Thus histone genes, for example, are typically reiterated many times. The five different histone genes are usually clustered, and this cluster is then tandemly repeated many (up to 100 or more) times. Ribosomal RNA genes are also tandemly clustered. Other nonidentical but functionally related genes that show clustering include the globin genes and the immunoglobulin genes.

Figure 25.31

Helical superstructures might be formed with increasing salt concentration (*bottom to top*) as is suggested here. The zigzag pattern of nucleosome (1, 2, 3, 4) closes up, eventually to form a solenoid, a helix with about six nucleosomes per turn. (The helix is probably more irregular than it is in this drawing.) Cross-linking data indicate that H1 molecules on adjacent nucleosomes make contact. Extrapolation from the zigzag form to the solenoid suggests (but does not prove) that the aggregation of H1 at higher ionic strengths gives rise to a helical H1 polymer (not shown) running down the center of the solenoid. In the absence of H1 (*bottom*) no ordered structures are formed. The details of H1 associations are not known at this time; the drawing is meant to indicate only that H1 molecules contact one another and linker DNA. (From "The Nucleosome" by Roger D. Kornberg and Aaron Klug. Copyright © 1981 Scientific American Inc. All rights reserved.)

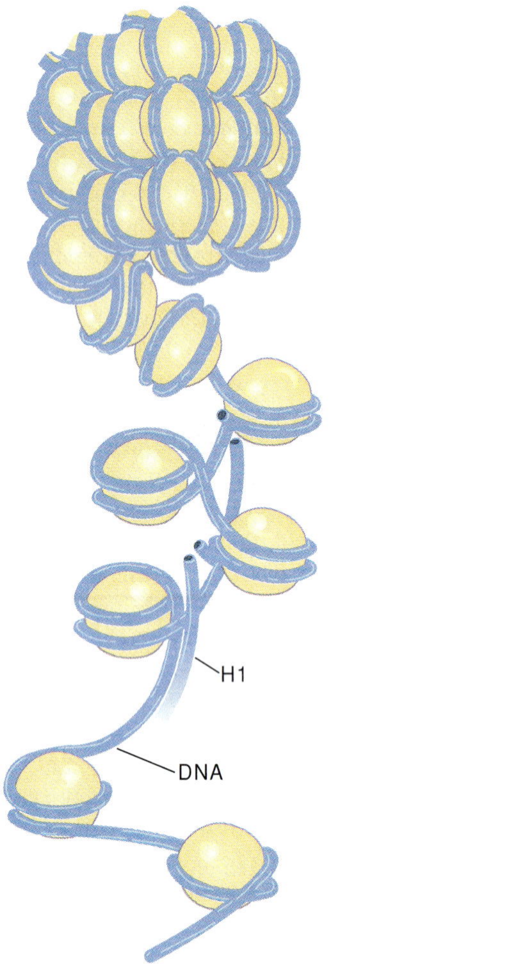

Summary

In this chapter we have been concerned with the nature and structure of DNA, RNA, and associated nucleoproteins. The highlights of the discussion are as follows.

1. The genetic material of cells and viruses consists of DNA or RNA. That DNA bears genetic information was first shown when the heritable transfer of various traits from one bacterial strain to another was found to be mediated by purified DNA.
2. All nucleic acids consist of covalently linked nucleotides. Each nucleotide has three characteristic components: a purine or pyrimidine base, a pentose, and a phosphate group. The purine or pyrimidine bases are linked to the C-1′ carbon of a deoxyribose sugar in DNA or a ribose sugar in RNA. The phosphate groups are linked to the sugar at the C-5′ and C-3′ positions. The purine bases in both DNA and RNA are always adenine (A) and guanine (G). The pyrimidine bases in DNA are thymine (T) and cytosine (C); in RNA they are uracil (U) and cytosine. The bases may be postreplicatively or posttranscriptionally modified by methylation or other reactions in certain circumstances.
3. DNA exists most typically as a double-stranded molecule, but in rare instances it exists (in some phages and viruses) in a single-stranded form. The continuity of the strands is maintained by repeating 3′,5′-phosphodiester linkages formed between the sugar and the phosphate groups; they constitute the covalent backbone of the macromolecule.

The side chains of the covalent backbone consist of the purine or pyrimidine bases. In double-stranded, or duplex, DNA the two chains are held together in an antiparallel arrangement.

4. The base composition of DNA varies characteristically from one species to another in the range of 25 to 75% guanine plus cytosine. Specific pairing occurs between bases on one strand and bases on the other strand. The complementary base pairs are either A and T, which can form two hydrogen bonds, or G and C, which can form three hydrogen bonds. The duplex is stabilized by the edge-to-edge hydrogen bonds formed between these planar base pairs and face-to-face interactions (stacking) between adjacent base pairs. Twisting of the duplex structure into a helix makes stacking interactions possible.
5. The right-handed helical structure, known as B DNA, is the most commonly occurring conformation of linear duplex DNA in nature. In this structure, the distance between stacked base pairs is 3.4 Å, with approximately ten base pairs per helical turn. The inherent flexibility of the structure, however, makes a variety of conformations possible under different conditions. In some instances, nucleotide sequence and degree of hydration dictate which conformations will be favored. DNA interacts with a variety of proteins inside the cell, and these proteins can also have a significant influence on its secondary and tertiary structure.
6. Circular DNA molecules, which are topologically confined so that their ends are not free to rotate, can form supercoils that are either right-handed (negative) or left-handed (positive). Negative supercoiling exerts a torsional tension favoring the untwisting of the primary right-handed double helix, whereas positive supercoiling has the opposite effect. Negatively supercoiled DNAs are most commonly observed in prokaryotes, which contain an enzyme that generates the supercoiled structure.
7. When duplex DNA (or RNA) is heated, it dissociates (denatures) into single strands. The temperature at which denaturation occurs (the melting temperature) is a measure of the stability of the duplex and is a function of the G-C content of the DNA.
8. A preparation of denatured DNA may be renatured by maintaining the temperature about 25° C below the melting temperature. The rate of renaturation is a measure of the sequence complexity of the DNA. In prokaryotes, which consist predominantly of unique sequences, the complexity (the number of base pairs) is approximately equal to the genome size. However, complex eukaryotic cells contain DNAs of varying sequence complexity that renature at quite different rates. The fastest-renaturing fractions are present in many copies per nucleus (usually detected as satellites in CsCl density gradients), whereas the slowest-renaturing fractions are present in single copies. Analyses by other techniques have shown that some of the repetitive DNA sequences exist as tandemly repeated structures, while other types of repetitive sequences are dispersed throughout the genome.
9. In the chromatin of eukaryotic cells, DNA forms a coiled-coil structure with approximately equal weights of five basic proteins known as histones. Four of these histones in pairs form an octamer around which the DNA duplex is wound in a left-handed helix. The DNA octamer complex is called a nucleosome.
10. Each nucleosome contains about 140 base pairs of DNA in a nuclease-resistant nucleosome core and approximately 60 base pairs of spacer between core particles. Histone H1 binds to the chromatin independently of the octamer and is the first histone to dissociate from the chromatin when the ionic strength is raised. Beyond the nucleosome, the higher-order structure of the chromosome involves coiled-coil structures with varying degrees of regularity.

Selected Readings

Avery, O. T., C. M. MacLeod, and C. McCarthy, Studies on the chemical nature of the substance inducing transformation of pneumococcal types. *J. Exp. Med.* 79:137–158, 1944.

Cantor, C. R., C. L. Smith, and M. K. Mathew, Pulsed-field gel electrophoresis of very large molecules. *Ann. Rev. Biophys. Chem.* 17:287–304, 1988.

Dickerson, R. E., The DNA helix and how it is read. *Sci. Am.* 249(6)d:94–111, 1983.

Felsenfeld, G., DNA. *Sci. Am.* 253(4):58–66, 1985.

Hershey, A. D., and M. Chase, Independent functions of viral proteins and nucleic acid in growth of bacteriophage. *J. Gen. Physiol.* 36:39–56, 1952.

Hillary, C. M., J. T. Finch, B. F. Luisi, and A. Klug, The structure of an oligo(dA),oligo(dT) tract and its biological implications. *Nature* 330:221–236, 1987.

Kim, S. H., Three-dimensional structure of transfer RNA. *Prog. Nuc. Acid Res. Mol. Biol.* 17:181–216, 1973.

Kornberg, R. D., and A. Klug, The nucleosome. *Sci. Am.* 244(2):52–64, 1981.

Lerman, L. S., S. G. Fischer, I. Hurley, K. Silverstein, and N. Lumelsky, Sequence-determined DNA separations. *Ann Rev. Biophys. Bioeng.* 13:399–423, 1983.

Morse, R. H., and R. T. Simpson, DNA in the nucleosome. *Cell* 54: 285–287, 1988.

Nadeau, J. G., and D. M. Crothers, Structural basis for DNA bending. *Proc. Natl. Acad. Sci.* 86:2622–2626, 1989.

Noller, H. F., Structure of ribosomal RNA. *Ann. Rev. Biochem.* 53: 119–162, 1984.

Rich, A., A. Nordheim, and A. H.-J. Wang, The chemistry and biology of left-handed Z DNA. *Ann. Rev. Biochem.* 53:791–846, 1984.

Richmond, T. J., J. T. Finch, B. Rushton, D. Rhoades, and A. Klug, Structure of the nucleosome core particle at 7 Å resolution. *Nature* 311:532–537, 1984.

Saenger, W., *Principles of Nucleic Acid Structure.* New York: Springer-Verlag, 1984.

Schmid, M. B. Structure and function of the bacterial chromosome. *Trends Biochem. Sci.* 13:131–135, 1988.

Schwartz, D. C., and C. R. Cantor, Separation of yeast chromosome-sized DNAs by pulsed field gradient gel electrophoresis. *Cell* 37:67–75, 1984.

Strobel, S. A., L. A. Doucette-Stamm, L. Riba, D. E. Housman, P. B. Dervan. Site-specific cleavage of human chromosome 4 mediated by triple-helix formation. *Science* 254: 1639–1642, 1991.

Structures of DNA, *Cold Spring Harbor Symp. Quant. Biol.* 47, 1983.
van Holde, K. E., *Chromatin.* New York: Springer-Verlag, 1988.
Watson, J. D., and F. H. C. Crick, Molecular structure of nucleic acids. *Nature* 171:737–738, 1953.
Wells, R. D., D. A. Collier, J. C. Hanvey, M. Shimizu, and F. Wohlrab, The chemistry and biology of unusual DNA structures adopted by oligopurine · oligopyrimidine sequences. *Faseb J.* 2:2939–2949, 1988.

Problems

1. Briefly describe how Avery was able to show that DNA is the genetic material in cells.
2. Summarize the evidence that RNA is the genetic material in tobacco mosaic virus (TMV).
3. Describe two physical methods that could be used to estimate the base composition of DNA. What would the data look like with two DNA samples, one with high G-C content and another with high A-T content? (Assume that the concentration of the samples is equal.)
4. Why is DNA denatured at either low pH (pH 2) or high pH (pH 11) and why is DNA stable at pH 7? (Hint: See p*K* values in table 20.2.)
5. Standard conditions for hydrolyzing RNA to nucleotides are 0.3 *N* NaOH, 37°, for 16 h. Draw a chemical reaction mechanism for this hydrolysis. Why is DNA not hydrolyzed under these conditions?
6. What effect would the following reagents have on the T_m of duplex DNA: 7-M urea, 90% formamide, higher concentrations of NaCl, pure water, and T4 gene 32-encoded protein? Explain how these chemicals act to affect DNA duplex stability.
7. Why can't RNA duplexes or RNA-DNA hybrids adopt the B conformation?
8. There are DNA-binding proteins that specifically bind to Z DNA. How could these proteins help stabilize DNA in the Z configuration? (Hint: How do single-stranded DNA-binding proteins destabilize duplex structures?)
9. Linear duplex DNA can bind more ethidium bromide than covalently closed circular DNA of the same molecular weight. Why? How could ethidium bromide be used to separate these two forms of DNA on CsCl gradients? (Hint: The ethidium cation has a density less than that of water.)
10. What is the structure of the nucleic acid called A, given the clues listed below?
    (a) Nucleic acid A sediments as a single species in a neutral sucrose density gradient.
    (b) In an alkaline sucrose gradient, one-half (structure B) of the mass of A sediments down the tube, the other half remains at the meniscus (top of tube).
    (c) The buoyant density of A is much higher than that of duplex DNA of the same G + C content.
    (d) The thermal transition profile of B is broad when A is melted.
    (e) When A is heated and rapidly cooled it gives rise to B and C. The buoyant density of C is heavier than that of B.
    (Hint: Single-stranded DNA is more dense than duplex DNA, and RNA is more dense than DNA.)
11. Renaturation of randomly sheared denatured DNA can be measured by the hypochromic shift at 260 nm, by S1-nuclease resistance, or by retention on hydroxyapatite. Which method is likely to give an overestimate for the extent of renaturation?
12. Give the relative times for 50% renaturation of the following pairs of denatured DNAs, starting with the same initial DNA concentrations.
    (a) T4 DNA and *E. coli* DNA, each sheared to an average single-strand length of 400 nucleotides.
    (b) Unsheared T4 DNA and sheared T4 DNA.
13. When histone proteins are isolated from chromatin their mass is equal to the DNA, and the ratio of four of the histones is 1:1:1:1 (H2a:H2b:H3:H4), while H1 is found in half the yield (0.5). Discuss whether or not these data fit the bead-and-string model for nucleosomes.
14. You are given a sample of nucleic acid. How would you determine (a) whether it is DNA or RNA and (b) whether it is single- or double-stranded?

26

CHAPTER

DNA Replication, Repair, and Recombination

Since the genetic information is contained in the sequence of bases in the DNA molecule, it is crucial that DNA be replicated with a very low error frequency. Otherwise the information transfer system would be destabilized. DNA must also resist damage and, if damage occurs, it must have a repair system that restores the original sequence. An elaborate enzymatic machinery has evolved to do just that. Recombination of segments of DNA between different chromosomes is sometimes used to facilitate repair and sometimes is used to facilitate orderly change. Many enzymes are involved in recombination processes. Our focus in this chapter is on the biochemistry of replication. We will also consider critical features of the repair and recombination processes (fig. 26.1).

The Semiconservative Replication of Duplex DNA

Double-helix DNA is relatively inert. Unwinding the double helix is essential to replication and transcription. When the DNA duplex replicates, each chain serves as a template for the synthesis of a complementary chain (fig. 26.2). The rules of Watson-Crick base pairing are strictly obeyed in the polymerization of monomers binding to a DNA template. Although semiconservative replication seems like a predictable mode of replication for a DNA molecule, it required a great deal of ingenuity to devise an experiment to demonstrate it the first time.

Matthew Meselson and Franklin Stahl conceived of a way of demonstrating the semiconservative mode of replication of the entire *E. coli* genome. Their approach made use of isotopes that would result in DNAs of altered densities after replication. For this purpose they grew *E. coli* cells for several generations on a defined medium in which all the nitrogen was the heavy ^{15}N isotope (normal nitrogen is ^{14}N). As a result, the DNA in the progeny cells had a greater than normal density, since the ^{15}N became incorporated into the bases of the DNA during replication. Next, they transferred the bacteria to growth medium containing normal ^{14}N-nitrogen, and allowed the cells

Figure 26.1

Major phases of metabolism associated with DNA: replication, repair, and recombination. *Replication* occurs by a semiconservative mechanism at specified origins of replication. Synthesis of new chains involves absorption of complementary nucleotides at the growth points, followed by covalent bond formation to the growing chain. Growth is always in the 5′ → 3′ direction (arrowhead in the figure represents the 3′ ends of the polymers). Since the two chains in a mature DNA duplex are oriented in an antiparallel manner, continuous growth is possible on one chain only. Growth on the opposing chain occurs in short spurts in a direction opposite to the general direction of replication. Replication is semiconservative; that is, after duplication each duplex contains one old chain from the "parent" duplex and one newly formed chain.

Repair is divided into two phases: recognition of an imperfection followed by removal of the imperfection, and repair synthesis. Some imperfections occur during the replication process; these imperfections are usually corrected immediately by the polymerase, which carries the necessary proofreading functions. Other imperfections occur at postreplication stages and are corrected by an auxiliary repair system that recognizes imperfections in the already formed DNA. There are a wide range of degradative enzymes that serve different functions; some of these nucleases operate in conjunction with repair or recombination systems and some function merely to break down the DNA. Nucleases can be broadly classified into exonucleases, which attack the DNA at free ends (3′ or 5′), and endonucleases, which attack the DNA somewhere in the interior.

Recombination processes exist for various purposes. The most general recombination system serves to recombine segments of DNA from homologous chromosomes. This type of recombination involves breakage, rejoining, and repair.

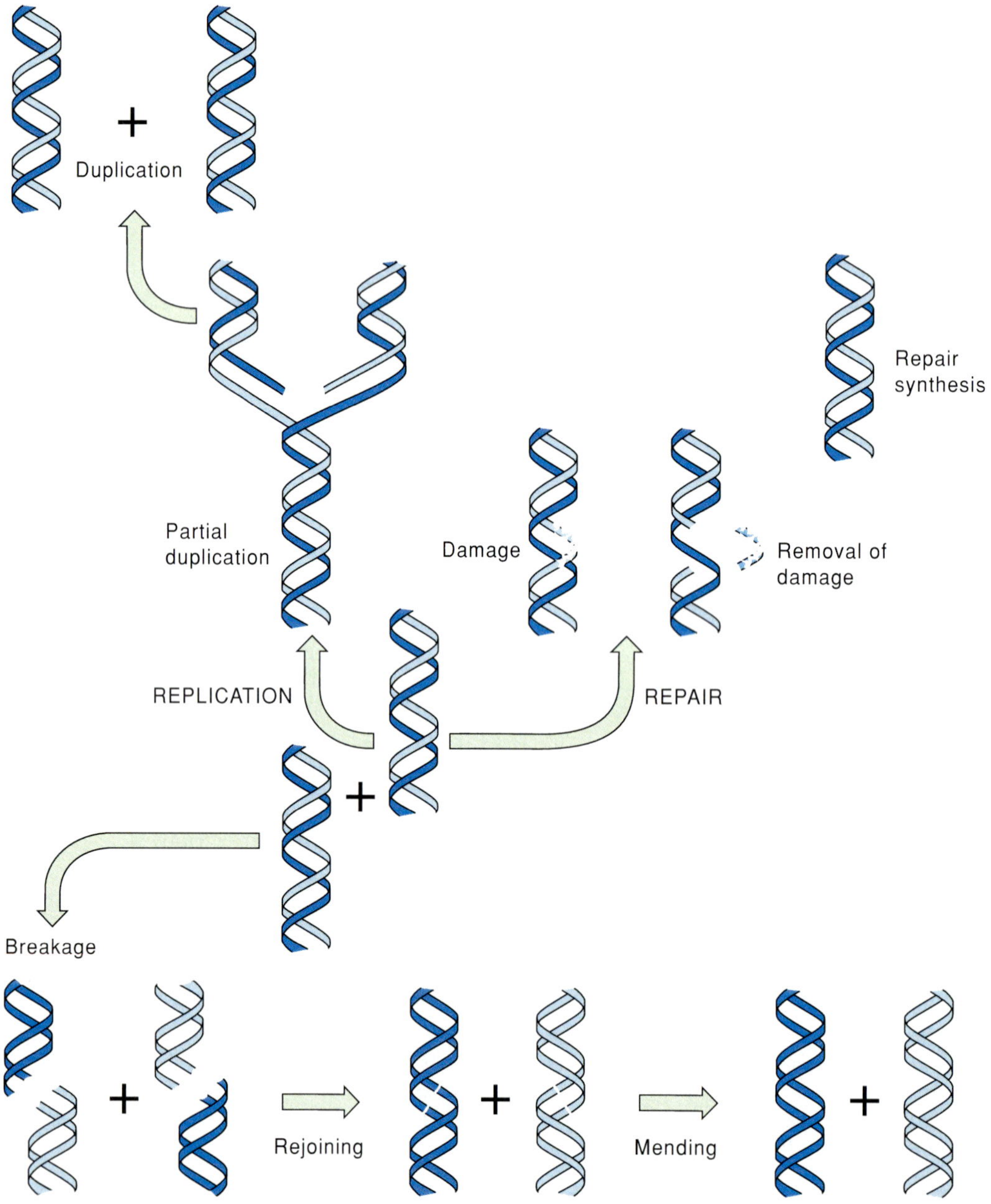

Figure 26.2

Watson-Crick model for DNA replication. The double helix unwinds at one end. New strand synthesis begins by absorption of mononucleotides to complementary bases on the old strands. These ordered nucleotides are then covalently linked into a polynucleotide chain, a process resulting ultimately in two daughter DNA duplexes.

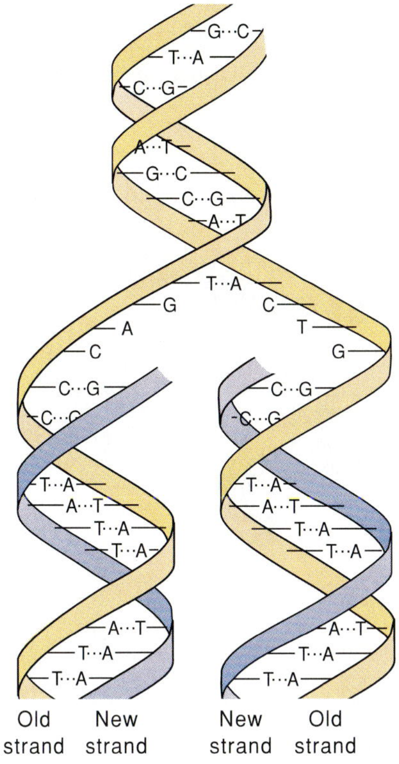

Figure 26.3

The Meselson-Stahl experiment demonstrating semiconservative replication for *E. coli* chromosomal DNA. CsCl density-gradient centrifugation is used to discriminate between DNAs of different densities. *E. coli* DNA has different densities when cells are grown in ^{14}N or ^{15}N medium (frames 1 and 2). When cells containing pure heavy DNA (^{15}N-^{15}N DNA) are grown in ^{14}N medium for one generation, all of the DNA is of intermediate density (^{14}N-^{15}N). After two generations of growth in ^{14}N medium, the cells contain equal amounts of light (^{14}N-^{14}N) DNA and intermediate density DNA (frame 5). In subsequent generations the hybrid DNA reappears in constant amounts, but the amount of light DNA increases. These results support the model of a semiconservative mode of DNA replication.

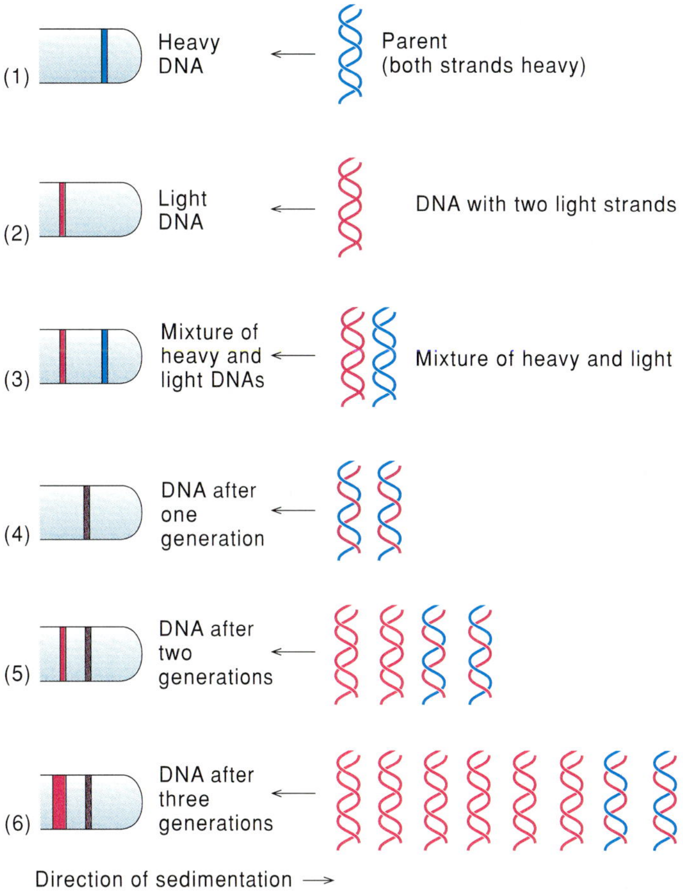

to go through one or more doublings. Then they isolated the DNA from these cells and analyzed it by CsCl density-gradient centrifugation (see box 25B). Pure ^{15}N-DNA produces a single band of DNA (fig. 26.3, frame 1). The same is true for ^{14}N-DNA (fig. 26.3, frame 2). The only difference is that denser DNA produces a band farther down the centrifuge tube. Thus the location of the DNA in the tube made it possible to monitor the density of the DNA. When cells containing pure ^{15}N-DNA were allowed to grow in ^{14}N medium for precisely one generation time, the only band visible in the isolated DNA was that corresponding to ^{15}N-^{14}N hybrid DNA (fig. 26.3, frame 4). This observation argued in favor of a semiconservative mode of replication, and the results in subsequent generations agreed with this view. Thus the next generation grown in light label showed equal amounts of hybrid-density and light-density DNA (fig. 26.3, frame 5), and the following generation showed a ratio of 3:1 of light to hybrid (fig. 26.3, frame 6). Besides demonstrating that *E. coli* DNA was replicated semiconservatively, the density transfer experiments also showed that all the chromosomal DNA replicated once, before it initiated a new round of DNA synthesis.

Similar experiments have been performed on mammalian cells grown in tissue culture, using bromouracil as a density label. Bromouracil contains a bromine atom instead of a methyl group on the 5 position of thymine. Bromouracil, when incorporated into DNA, can substitute for most of the thymidine, leading to DNA with a substantially higher than normal density. The results with eukaryotic DNA were found to parallel those in *E. coli*. Thus it appears that semiconservative replication of cellular DNA is a general phenomenon.

Taylor, Woods, and Hughes devised another method for labeling DNA to follow its replication. They selectively labeled chromosomal DNA with tritiated (^{3}H-labeled) thymidine and then visualized the distribution of the radioactive label directly in autoradiographs of chromosomes (fig. 26.4*a*); the same pattern of labeling was visible after the cells subsequently replicated in the absence of tritiated thymidine (fig. 26.4*b*). Bean seedlings were used for those experiments because cell division is very rapid in their growth tips. A sufficient period of time was allowed for some of the cells in the seedlings to undergo one round of DNA duplication (incorporating the radioactive label)

Figure 26.4

Autoradiographs of *Vicia faba* chromosomes labeled with [^{3}H]thymidine. The labeled thymidine becomes incorporated into the chromosomal DNA. A suitably labeled preparation is flattened and subjected to film exposure. Small dots indicate radioactive disintegration in the exposed film. (*a*) The first metaphase after replication in the presence of [^{3}H]thymidine. (*b*) The second metaphase after an additional replication in nonradioactive medium. (*c*) A diagrammatic interpretation of the results shown in (*a*) and (*b*). Radioactive single strands of DNA are shown in color. Radioactive chromatids at metaphase are also indicated in color. Colchicine has been used to inhibit spindle fiber formation and thus the anaphase separation of sister chromatids. Under these "C-metaphase" conditions, separation of sister chromatids is delayed. In (*a*) both sister chromatids are labeled uniformly. In (*b*) the sister chromatids are not labeled uniformly. The large chromosome at the top has one chromatid labeled and one virtually unlabeled. The homolog to its right has two exchanges (a labeled segment moved into the lower chromatid). The two small chromosomes to the lower left of it are lying one on top of the other. Both have one sister chromatid exchange. The small chromosome to the upper left has one sister chromatid exchange and the one at the lower left is lightly labeled but probably has two exchanges. (Autoradiographs courtesy of J. H. Taylor.)

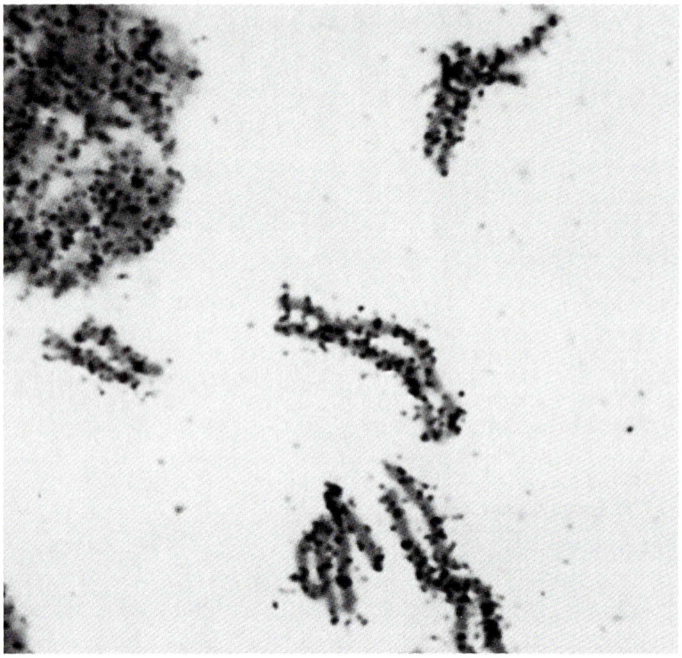

(a)

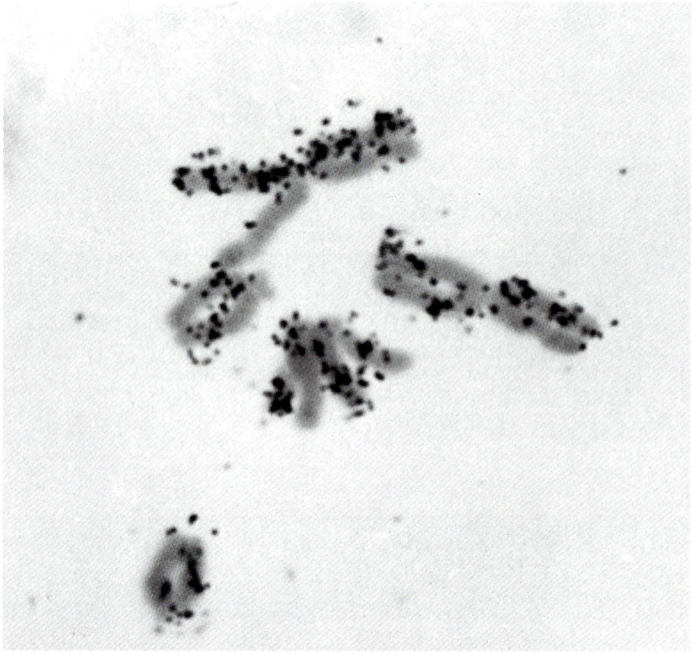

(b)

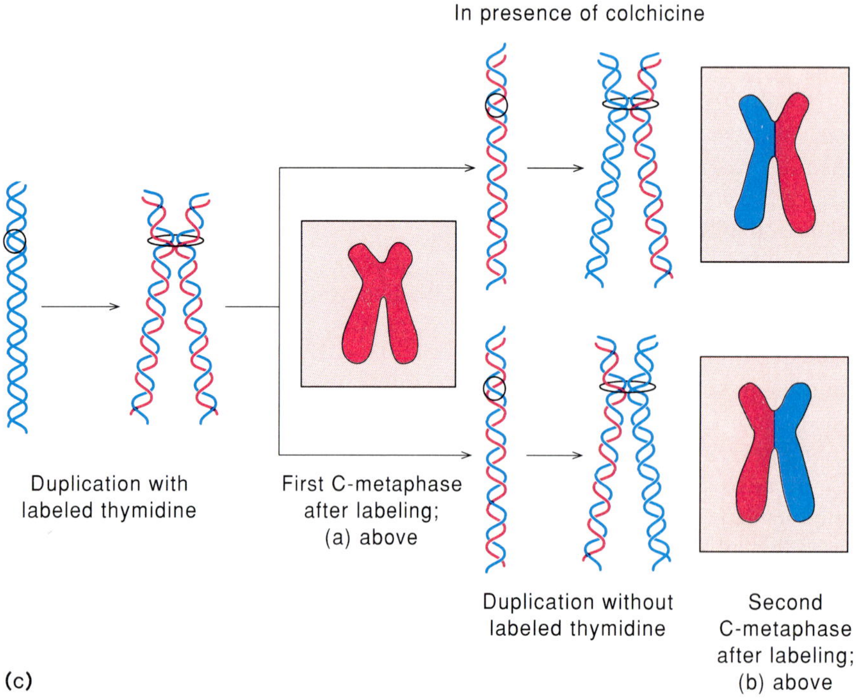

(c)

and cell division. After this, the seedlings were transferred to a fresh solution containing colchicine with the ^3H label. Colchicine is a plant alkaloid that inhibits normal mitosis by interacting with microtubule proteins necessary for mitotic spindle formation; it does not inhibit DNA synthesis nor the replication of chromosomes but delays the formation of daughter cells by inhibiting chromatid segregation. The advantage of colchicine treatment in this experiment is that in tissue so treated many cells become arrested in the state where the sister chromatids are paired. Cells that were examined shortly after the transfer from the [^3H]thymidine medium were examined at mitosis, and all the chromosomes appeared to be labeled uniformly (see fig. 26.4*a*). When the cells were allowed to duplicate their chromosomes in unlabeled medium, only one of the two chromatids in each chromosome pair was labeled (see fig. 26.4*b*). That is exactly the result that would be expected if each chromatid is composed of a single linear duplex of DNA that replicates semiconservatively (see fig. 26.4*c*). Similar results have been obtained for other eukaryotic cell types.

The universality of the semiconservative mode of DNA replication follows from the complementary nature of the DNA duplex. As the duplex unwinds, it presents two templates for the binding of complementary nucleotides. Subsequently, polymerization results in two duplexes, each with one old strand and one new strand. Thus the genetic information contained in the base sequence is directly transferred from one generation of DNA to the next by the capacity of DNA single strands to serve as templates for the assembly of complementary mononucleotides.

DNA Synthesis in Prokaryotes

More is known about the synthesis of genomic DNA in *E. coli* than in any other cell. The *E. coli* bacterium contains a single circular chromosome with about 4.5×10^6 base pairs. Most other bacteria have chromosomes of about the same size and, in some cases, the chromosomes have been found to be circular as well. The number of genes required for DNA synthesis is probably between 20 and 50. In figure 26.5 the locations of some of these genes are indicated, together with the locations of unique initiation (*oriC*) and termination points for DNA replication. Under optimal growth conditions the *E. coli* chromosome is replicated in 20–30 min.

The Circular Bacterial Chromosome Replicates Bidirectionally

Replication of the *E. coli* chromosome can be visualized by autoradiography of intact ^{13}H-labeled chromosomes, using a gentle isolation technique developed by Cairns and Davern. After completing one round of replication in labeled medium, chromosomes appear as uniformly labeled circles (fig. 26.6, top). Initiation of a second round of replication leads to the formation of a replication eye (fig. 26.6, middle). As synthesis proceeds in labeled medium, the size of the replication eye increases; the replicating chromosome at this stage is referred to as a theta structure because it resembles the Greek letter θ (fig. 26.6, bottom). Semiconservative replication is consistent with the density of the autoradiographic tracks observed on parts of the chromosome after one and two rounds of replication in [^3H]thymidine.

Figure 26.5

A diagram of *E. coli* chromosome. The origin and approximate region of termination of replication are indicated. Locations of some genes involved in DNA replication are also indicated. The functions of many of these genes are described in tables 26.1 and 26.2.

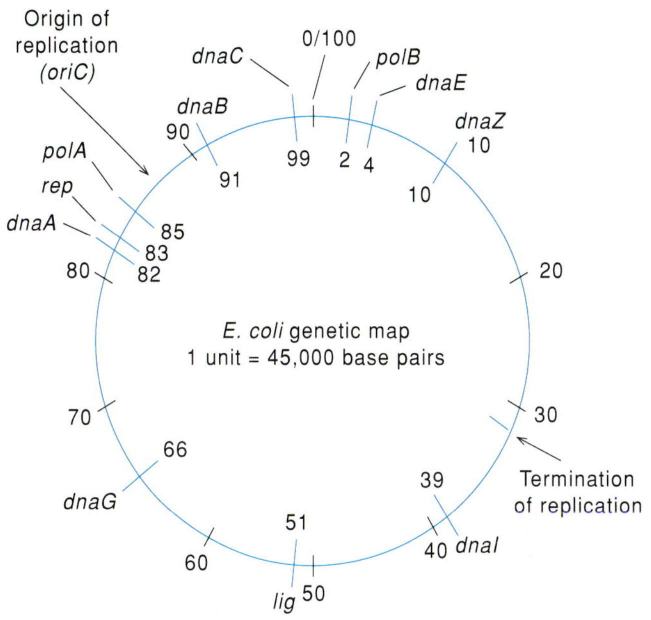

It is reasonable to conclude that the replication eye contains two partially separated parental DNA strands that are base-paired with strands of newly synthesized DNA. Not resolved by this type of observation is the question of whether replication occurs in one direction or both directions about the origin of replication. If growth is unidirectional, we would expect one growth point (fig. 26.7*a*), called a growth fork (or replication fork); if growth is bidirectional, we would expect two growth points or growth forks (fig. 26.7*b*). Although examples of both types of replication are observed, it has been shown in several cases that bacterial chromosomes, including that of *E. coli*, replicate bidirectionally.

Convincing evidence of bidirectional replication of *E. coli* was obtained by measuring gene frequency during replication. Shortly after initiation, those genes near the origin of replication that have been duplicated must be present at twice the number of copies as those genes far removed from the origin.

Figure 26.6

Simulated autoradiographs of the *E. coli* chromosome after one or more replications in the presence of [^{3}H]thymidine. After one round of replication the autoradiograph shows a circular structure that is uniformly labeled. The second round of replication begins with the formation of a replication eye. One branch in the replication eye is twice as strongly labeled as the remainder of the chromosome, indicating that this branch contains two labeled strands. This structure is consistent with semiconservative replication for the *E. coli* chromosome.

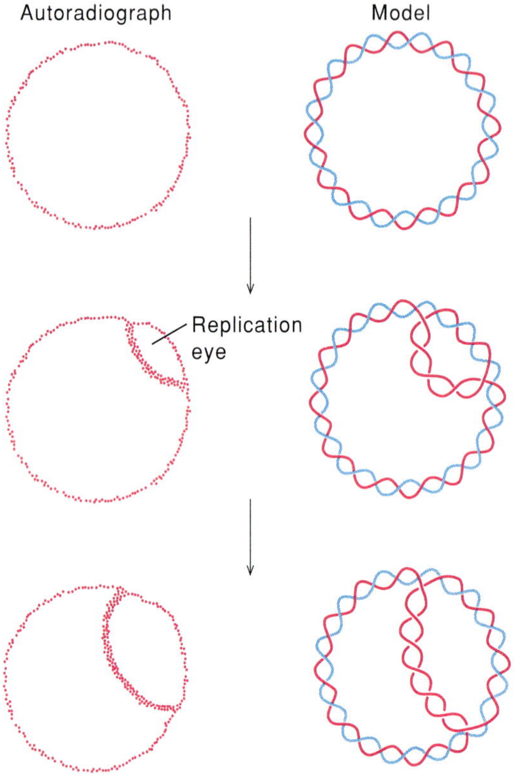

Figure 26.7

Schematic diagrams of two different modes of DNA synthesis at the growth fork(s). In unidirectional replication (*a*) there is one growth fork; in bidirectional replication (*b*) there are two. Color indicates regions containing newly synthesized DNA. Measurements of gene frequency make it possible to distinguish between unidirectional and bidirectional synthesis. (*c*) In unidirectional replication, the gene frequencies should be highest for genes located on one side of the origin. (*d*) In bidirectional replication, the gene frequencies should be equally high for regions symmetrically disposed about the origin.

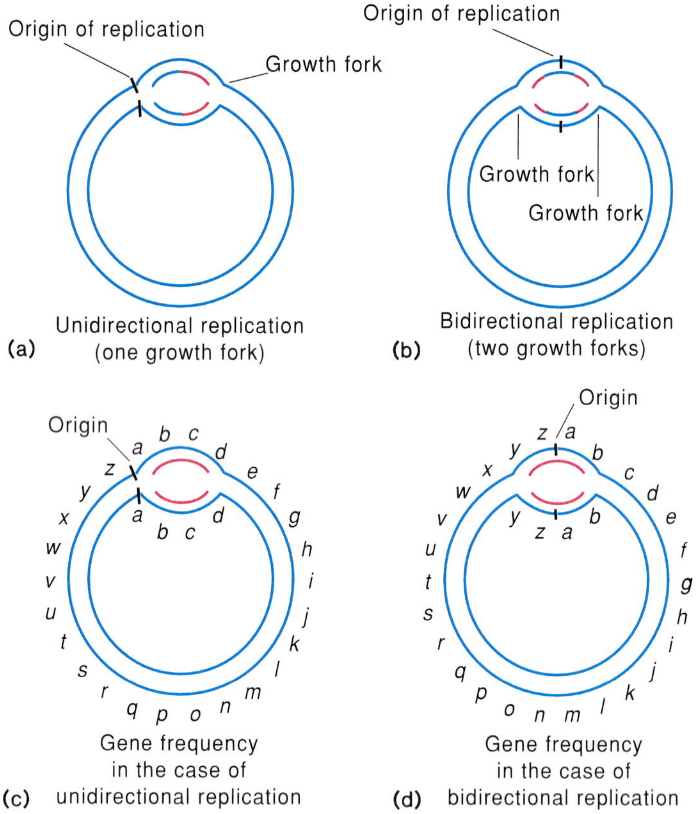

The numbers of genes can be assayed biologically by transformation (chapter 25) or transduction. If replication is unidirectional, a steady increase in gene number should occur in one direction along the circular chromosome (see fig. 26.7*c*). On the other hand, if replication is bidirectional, the increase in gene number should proceed in both directions from a common point, the origin (see fig. 26.7*d*). Detailed measurements indicated that the latter situation is true for the chromosomes of *E. coli* and *Bacillus subtilis*. These measurements also made it possible to locate the initiation and termination points on the chromosomes (see fig. 26.5).

The genetic evidence for bidirectional replication of the *E. coli* chromosome was supported by direct cytologic evidence. Bacteria were grown for a very short time in the presence of radioactive thymidine, and replicating chromosomes were examined by autoradiography; both of the forks in the replicating structures were intensely labeled (as in the colored portions of figure 26.7*b*). This result shows that both forks must be active during replication, a finding consistent with bidirectional growth. Replication was also shown to be bidirectional in bacteriophage λ DNA, through electron microscopy. By analyzing circular λ molecules at different stages of their replication, it was found that both ends of the replication eye moved relative to a single, unique cut made by a restriction endonuclease (we will consider the properties of restriction endonucleases later in this chapter and more fully in chapter 27). The experiments on bidirectionality of replication of the *E. coli* genome have been possible because there is a unique origin of replication. Once initiated, each replication fork travels about the same distance, each fork terminating at a point 180° removed from *oriC*.

Growth in Each Direction Is Discontinuous on at Least One Strand

Continuous synthesis on both strands of a replication fork would require synthesis in the 5′→3′ direction on one strand and in the 3′→5′ direction on the other strand because of the antiparallel nature of duplex DNA (fig. 26.8*a*). That possibility seems unlikely because the only known enzymes that catalyze DNA synthesis do so in the 5′→3′ direction. For this reason replication was postulated to be discontinuous on one of the branches at the replication fork (fig. 26.8*b*). Careful electron microscopic examination of replication forks in bacterial viruses has in fact

Figure 26.8

Models for synthesis at the replication fork. (*a*) Continuous synthesis on both strands. Note that both growth arrows are pointing in the same direction, which would require growth in the 5′ → 3′ direction on one strand and in the 3′ → 5′ direction on the other strand. If growth occurs only in the 5′ → 3′ direction, synthesis would have to be discontinuous on one strand, as in (*b*). Alternatively, it could be discontinuous on both strands (*c*).

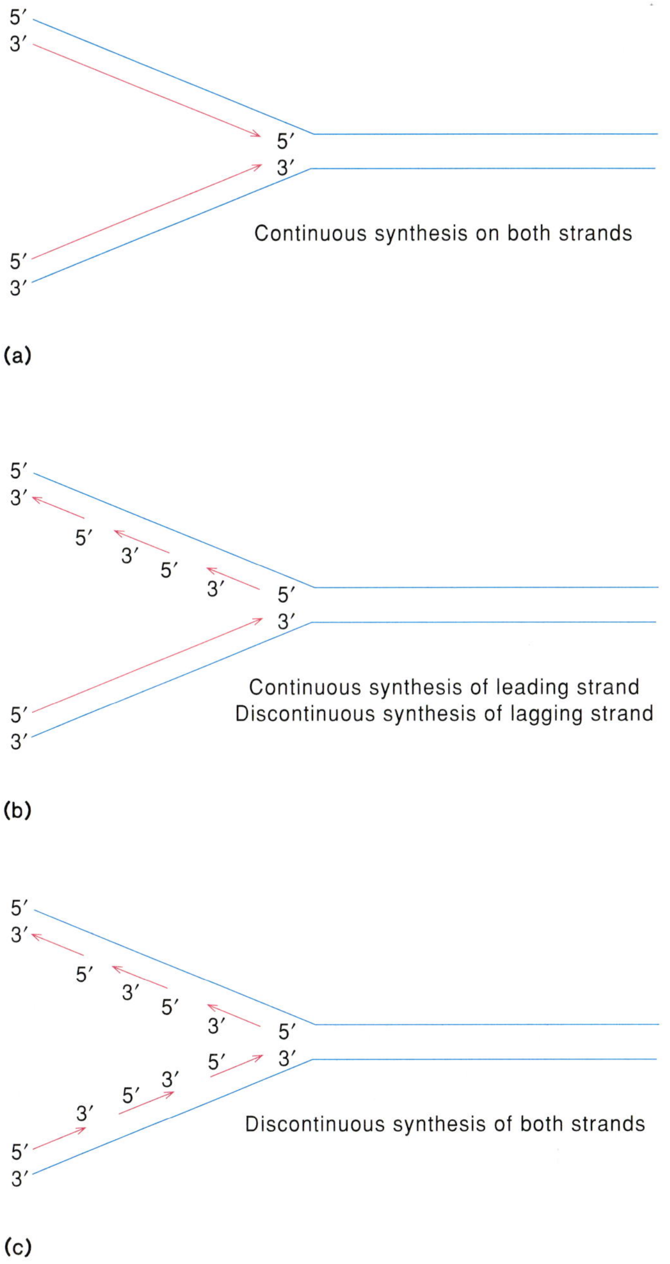

shown that transient gaps sometimes are apparent on one of the daughter DNA strands, close to the replication fork. Observations such as this led to the notion of a leading strand and a lagging strand (fig. 26.8*c*). In this model, 5′→3′ synthesis of the leading strand could occur continuously in the same direction as the unwinding of the replication fork. Synthesis of the lagging strand in the 5′→3′ direction could occur in discontinuous spurts in a direction opposite to DNA unwinding.

Figure 26.9

Sedimentation analysis of *E. coli* DNA from cells labeled with [³H]thymidine for different lengths of time. Sedimentation analysis is done on an alkaline sucrose gradient. The alkali denatures the DNA so that it becomes single-stranded. (*a*) Short-term labeling (2 to 10 s) preferentially labels the most slowly sedimenting DNA, in the size range of 1,000–2,000 base pairs. In long-term labeling (1 to 2 min), most of the labeled DNA is of much higher molecular weight. (*b*) In a pulse labeling experiment, the short-term labeling is done in the same way as in (*a*). The long-term labeling includes the addition of a large amount of nonradioactive substrate after a short time. This added substrate is called a chase, as it lowers the specific radioactivity of the labeled substrate so that no further incorporation of radioactivity is detectable. Long-term labeling is done in order to follow the fate of the short-term-labeled DNA.

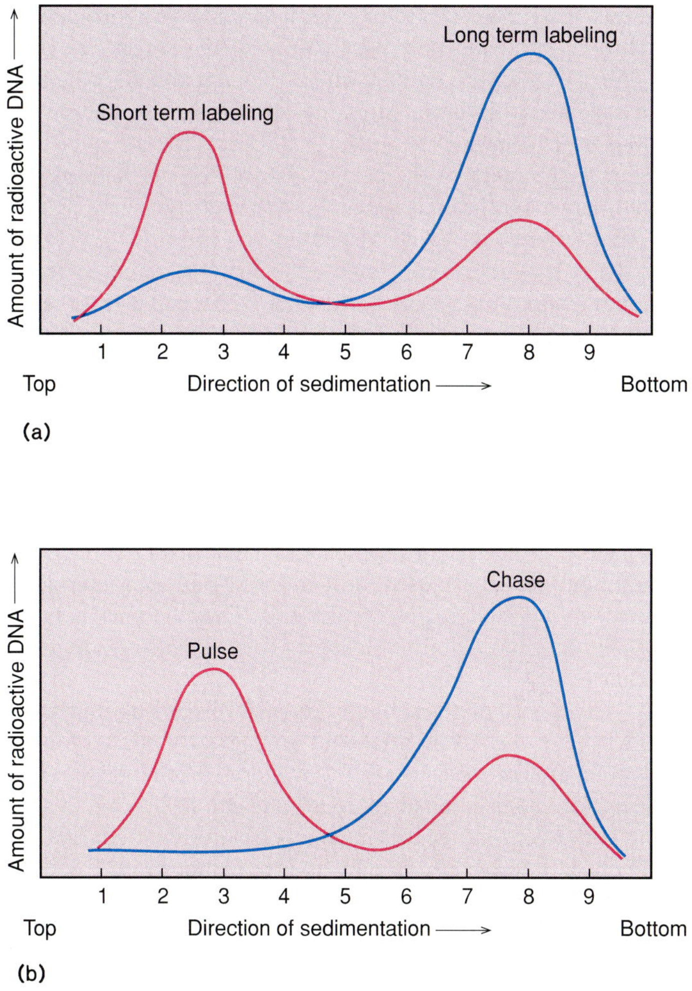

This concept of discontinuous synthesis was supported by Okazaki, who found that at least half the newly synthesized DNA is first made as small pieces (Okazaki fragments) that later become incorporated into large segments of DNA. Small replication fragments were detected by exposing growing cells to tritiated thymidine for a very short time (2 to 10 s) (a pulse), followed by rapid isolation of the radioactively labeled DNA. If replication was allowed to continue in unlabeled medium for several minutes (the chase period) most of the labeled DNA was found in much larger pieces of DNA (fig. 26.9). Sedimentation analysis in alkali (the alkali denatures the duplex DNA into single strands) provided an estimated length of 1,000–2,000 bases for the bacterial Okazaki fragments. The researchers concluded that the replicating polymerase must operate by synthesizing short fragments on the lagging strand as new points for the initiation of replication are presented by the progressive unwinding of the double helix at the growth fork.

A closer examination of the Okazaki fragments revealed short stretches of ribonucleotides at their 5′ ends. From this and many other observations made with different *in vitro* systems, it appears that a new DNA chain can be initiated only by attaching the first deoxynucleotide through its 5′ phosphate to the 3′-OH of a short RNA chain. An RNA oligonucleotide that functions in this capacity is called a primer. Primers are synthesized *de novo* at various secondary priming sites along the chromosome, and they base-pair with the single-stranded template DNA in the regions where they are found. Whether these secondary initiation sites are specific or are chosen at random is not known.

A summary of the currently accepted mechanism for discontinuous synthesis is given in figure 26.10. First, RNA primers are made on the single-strand region of the template; then DNA is synthesized. Finally, the RNA is removed from the fragments, the gaps are filled by a DNA polymerase, and the nicks are ligated. In *E. coli* the *dnaG* gene encodes a primase that generates primers from ribonucleotide precursors.

An Assemblage of Proteins Is Involved in the Replication Process

Many proteins are required for DNA replication in addition to the polymerizing enzyme. The process of discovering these enzymes has combined many approaches. Historically, three general methods have been used for identifying and characterizing the proteins involved in DNA replication. They are purification, reconstitution, and mutation. Ideally, all three methods are used together in suitable organisms.

The first method, purification, involves isolating proteins with enzymatic activities that are logically related to the replication process, such as DNA polymerases and ligases. This is the classical biochemical approach and can be applied to any biological system. After an enzyme is isolated and characterized, several approaches may be used to demonstrate that the purified enzyme is active in the replication process *in vivo*. Sometimes this can be done by using inhibitors that act on both the purified protein extracts and at the cellular level. The concentration of inhibitor required to inhibit the purified enzyme *in vivo* should be approximately the same as that required to inhibit the purified enzyme *in vitro*. Inhibitors are particularly useful in comparing and characterizing various polymerases that are present in the same cell and in comparing DNA polymerases extracted and purified from different organisms by monitoring their relatedness. Thus yeast DNA polymerase I is probably related to mammalian DNA polymerase α, since these are both resistant to the inhibitor butylphenyl-dGTP. In prokaryotes, mutations in specific genes have been very useful for confirming the functions of isolated proteins in the replication process. The induction of a new enzyme activity associated with a biological process, such as virus infection or cell proliferation, also provides useful evidence for the involvement of the enzyme.

A second method used to identify proteins needed for replication is reconstitution. Whole-cell lysates containing all of the components necessary for replication are fractionated and the DNA replication system is then reconstituted with various combinations of the purified or partially purified proteins. Components of the replication system are recognized on the basis of their ability to restore overall activity *in vitro*. This procedure can be applied to any organism, even when relevant genetic mutants are not available. However, mutant studies are usually required for final confirmation that a protein carries out a particular function *in vivo*. We will see an example of such confirmation a little later, when we discuss the role of the DNA polymerase I.

Figure 26.10

A model for discontinuous DNA synthesis. Synthesis occurs in a region that has been partially single-stranded. First, RNA primers are formed at various points on the single-stranded region (1). The DNA synthesis starts at the 3′ ends of the primers (2). The primers are removed (3). Gaps between DNA fragments are filled in by further DNA synthesis (4). The fragments are ligated to make one long continuous piece of DNA (5). Newly synthesized RNA and DNA are indicated in red.

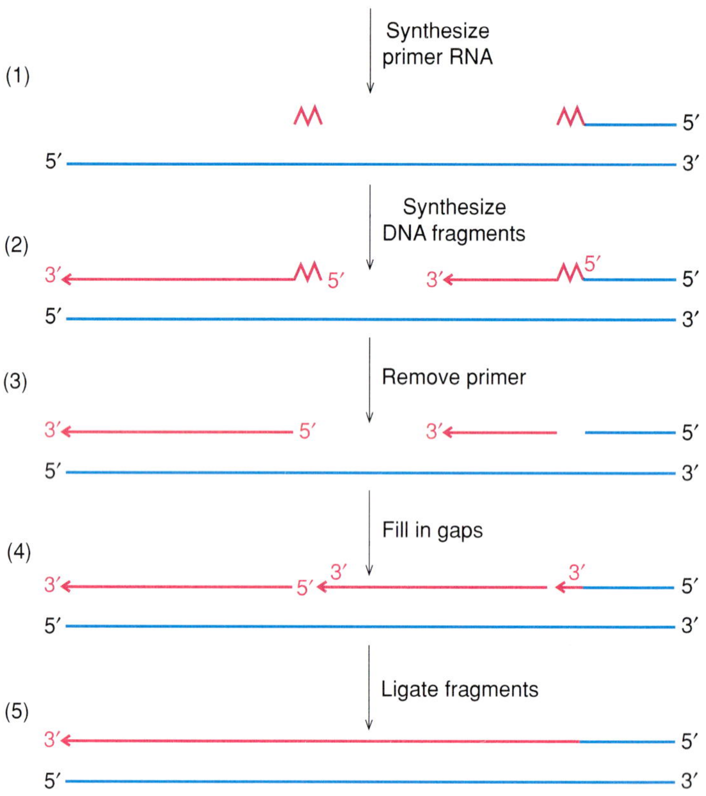

The third method of analysis uses mutants as the primary tool. This approach is most effective with organisms having well-characterized means of carrying out genetic analysis and exchange. It requires the isolation of conditionally lethal mutants, i.e., mutants that grow normally under one set of conditions but fail to do so under another set of conditions. Most commonly, temperature-sensitive DNA replication mutants are used. Such mutants have been isolated for *E. coli;* they grow normally at a low temperature (25–30° C), called the permissive temperature, but poorly or not at all at a high temperature (41° C), referred to as the nonpermissive temperature. Preliminary analysis of the temperature-sensitive step provides clues to the stage of replication affected. For example, the length of time required for DNA synthesis to stop, after cells have been shifted from permissive to nonpermissive temperatures, can indicate whether the mutation occurs in a protein involved in the initiation or the elongation reactions of DNA synthesis. If the mutation is in the gene for a protein required for elongation, DNA synthesis will stop almost immediately at the nonpermissive temperature. If the mutation affects a protein required only for initiation of replication, DNA synthesis will continue for some time and stop when the round of replication in progress is completed.

Once the stage of replication has been identified, *in vitro* assays are used to aid in purifying the corresponding proteins from fully competent wild-type cells. Extracts from cells with the temperature-sensitive defect will not synthesize DNA at the elevated temperature, but activity can be restored by adding the corresponding protein from wild-type cells. This complementation test can be used as an assay for the purification of particular replication proteins. To prove that the correct protein has been purified from wild-type cells, the proteins from the temperature-sensitive mutant also must be purified and shown to be abnormal. Frequently they exhibit unusual thermolability.

A Polymerase with the Expected Characteristics Is Discovered by Kornberg

The first DNA-polymerizing enzyme to be discovered was found in cell-free extracts of *E. coli.* We will examine this polymerase in some detail as an example of how biochemical and genetic studies can be used to determine the properties of an enzyme and its normal *in vivo* role. We will then consider the other major classes of proteins involved in *E. coli* DNA replication.

The Watson-Crick proposal of a complementary duplex structure for DNA stimulated a search for a DNA polymerase enzyme with certain implied properties. The enzyme should require an intact DNA chain to serve as a template for the binding of complementary bases, and the *de novo* synthesized DNA should be complementary to the template DNA chain. Kornberg and his co-workers isolated from cells of *E. coli* a DNA-synthesizing enzyme that satisfied these requirements. They gave it the name DNA polymerase; it is now known as DNA polymerase I (Pol I) or the Kornberg enzyme.

In its purified state, the Kornberg enzyme requires a DNA template, the four commonly occurring deoxynucleoside triphosphates, and Mg^{2+} ions for synthesizing DNA. The enzyme catalyzes the addition of mononucleotides to the 3′-OH end of a growing chain (fig. 26.11*a*). At the same time, the linkage between the α and β phosphates is broken, releasing an inorganic pyrophosphate group (PP_i). In fact, the energy provided by this cleavage drives the linking of the mononucleotide to the growing DNA chain. The bases added to the growing chain are determined by the sequence of bases in the DNA template. As nucleotides complementary to those on the template are added, the single-stranded DNA template is progressively converted to a double helix.

Physiochemical studies have shown that the enzyme has a complex surface, with specific attachment sites for the template chain, the growing chain, and monomer nucleoside triphosphate. The enzyme is highly selective, since it links to the growing chain only those nucleotides that form Watson-Crick base pairs with the template strand. As the new chain is lengthened by synthesis, the enzyme moves along the template one base at a time. That the newly synthesized DNA is a replica of the template strand has been shown in a number of ways—by gross base-composition analysis, by melting-curve profile of the hybrid formed between the template strand and the newly synthesized strand (see chapter 25 for an explanation of nucleic acid melting curves), and finally, through a series of manipulations, by the *de novo* synthesis of genetically active DNA. A unique property of DNA and RNA polymerases is that they require a template to exhibit their activity.

Phosphodiester bond formation probably occurs as a nucleophilic attack, by the 3′ hydroxyl group of the terminal mononucleotide residue at the growing end of the chain, on the α phosphorus atom of the entering nucleoside-5′-triphosphate. The attack causes displacement of the pyrophosphate group and formation of the internucleotide linkage (fig. 26.11*b*).

In addition to catalyzing the characteristic template-directed polymerization, DNA polymerase I is associated with an activity that can remove mononucleotides from the 3′ end of a polynucleotide strand. This second function is an exonuclease activity; it leads to hydrolytic cleavage of the bond between the 5′ phosphate of the terminal residue and the 3′-OH group of the penultimate residue. Since this is the same linkage made during synthesis, the degradation reaction may be thought of as a reversal of polymerization. For net chain propagation to occur beyond the 3′-OH end of the DNA strand, the polymerization rate must exceed the depolymerization rate. Polymerization is much faster than depolymerization if the correct base-paired nucleotide is inserted in the growing chain. If a mismatched base is accidentally inserted, the opposite is true and the mismatched base is usually removed. The combination of polymerization-depolymerization reactions serves as an "editing" or "proofreading" error-reducing mechanism in DNA synthesis. If the editing function is defective because of a mutation, a generalized increase in mutations occurs in cells harboring the altered nuclease.

Figure 26.11

Template and growing strands of DNA. (*a*) Nucleotides are added one at a time to the 3′-OH end of the growing chain. Only residues that form Watson-Crick H-bonded base pairs with the template strand are added. (*b*) Covalent bond formation between the 3′-OH end of the growing chain and the 5′ phosphate of the mononucleotide is accompanied by removal of the two terminal phosphates from the substrate nucleoside triphosphate.

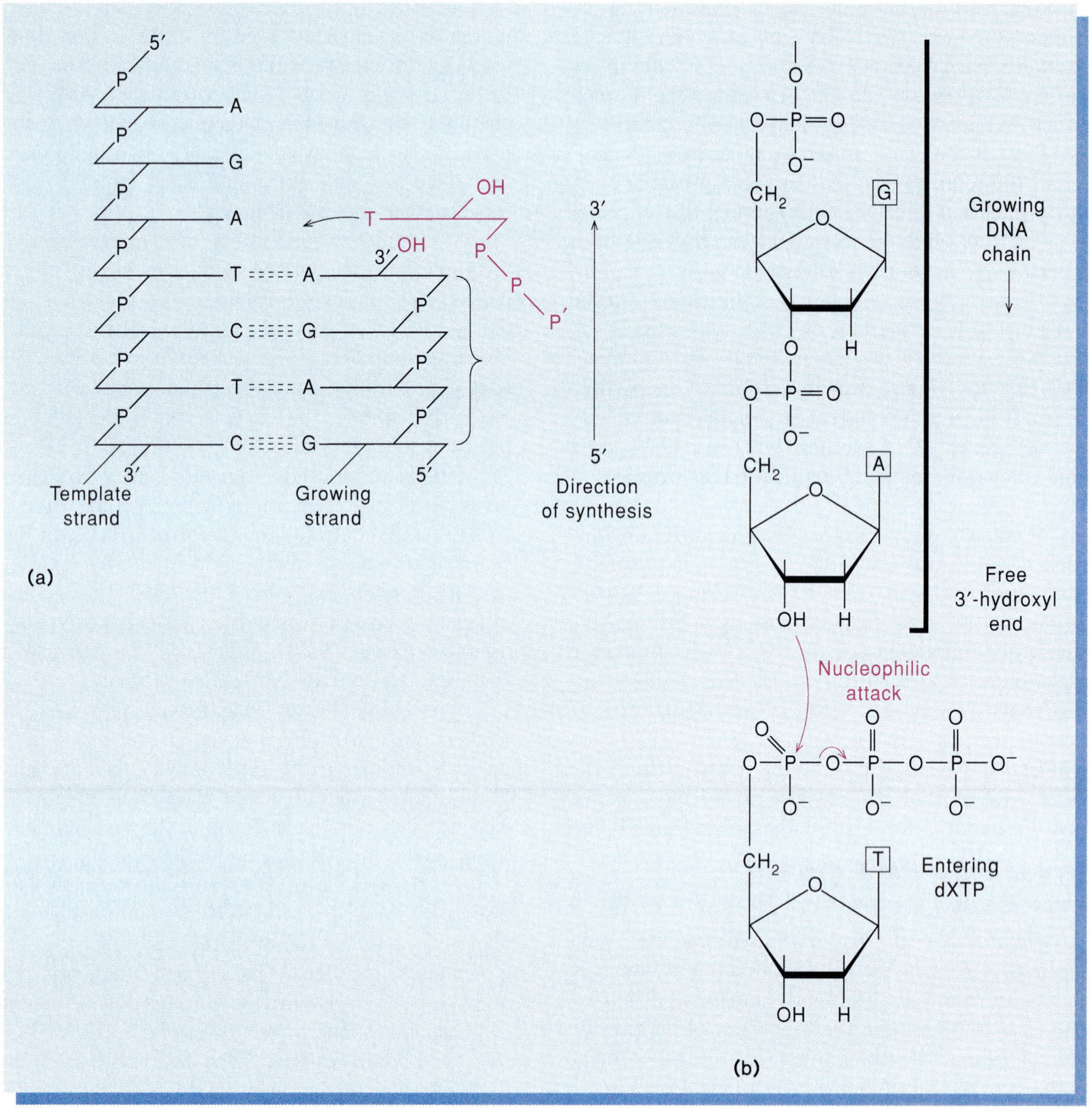

Pol I has another associated activity, catalyzing the 5′→3′ degradation of DNA. Whereas the 3′→5′ nuclease activity of the enzyme is much more effective on unpaired or mispaired bases, the 5′→3′ activity preferentially cleaves base-paired regions. The ability of DNA Pol I to degrade DNA (or RNA) in the 5′→3′ direction, as well as to carry out a polymerization reaction, suggests a possible function in removing the RNA primer prior to gap filling (see fig. 26.10). In fact, RNA primer removal and gap filling are probably catalyzed in rapid succession by the same DNA polymerase I molecule.

Crystallography Has Combined with Genetics to Produce a Detailed Picture of DNA Pol I Function

The next question that arises is whether the three activities of Pol I just described all originate from a single active site or more than one active site. Cleavage of Pol I by the protease subtilisin leads to a small fragment (M_r = 30,000) with 5′→3′ nuclease activity and a large fragment (M_r = 70,000), called the Klenow fragment, exhibiting the polymerization and 3′→5′ depolymerization activities.

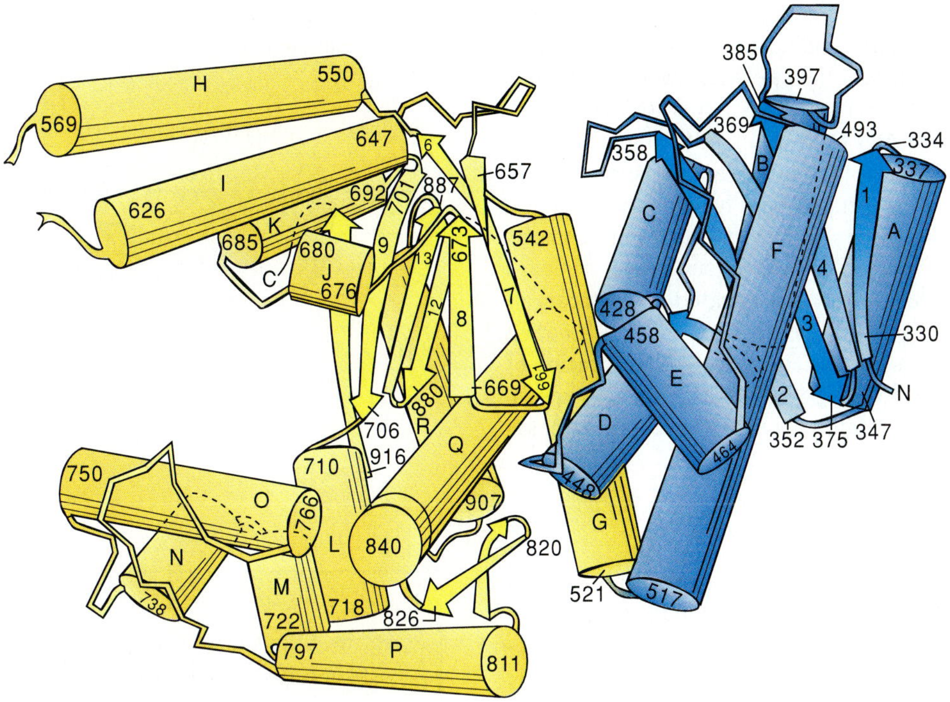

Figure 26.12

Schematic drawing of the large proteolytic (Klenow) fragment of *E. coli* DNA polymerase I. The domain catalyzing the 3′→5′ exonuclease reaction includes residues 324-520, while the polymerase domain includes residues 521 to the C terminus and contains a large cleft. (From D. L. Ollis, P. Brick, R. Hamlin, N. G. Xuong, and T. A. Steitz, Structure of large fragment of *Escherichia coli* DNA polymerase I complexed with dTMP, *Nature* 313:762-766, 1958.)

A series of investigations by Tom Steitz and his coworkers has led to a detailed understanding of the molecular anatomy of Klenow peptide fragment. Crystallography shows that the 605-amino-acid polypeptide is folded into two structural domains containing about 200 and 400 amino acids (fig. 26.12). The N-terminal third of the Klenow fragment (residues 324–517 in the Pol I sequence) forms the smaller of the two domains. This domain has a core of β-pleated sheet with α helices on both sides. In the crystal, this domain can bind two divalent metal ions and a molecule of dNMP. The larger domain (residues 521–928) forms a structure with a deep cleft, about 20–24 Å wide and 25–35 Å deep. A six-stranded antiparallel β sheet forms the bottom of the cleft and large protrusions of α helices form its sides. Steitz points out that this structure is similar in shape to a right hand grasping a rod. Thus one side of the cleft forms a wall, 50 Å long, that can be compared with the curled fingers of a right hand. The other side of the cleft is formed primarily by two long α helices, H and I, projecting from the protein like a thumb. Located at the tip of the thumblike protrusion are 50 amino acid residues connecting helices H and I that are partially disordered, indicating that this region is flexibly attached to the rest of the molecule.

The cleft in the large domain can accommodate double-stranded B DNA. The J and K α helices are placed partially into a major groove and may function like the thread of a nut to fix the exact position of the DNA major groove relative to the protein. Thus during DNA synthesis the polymerase would follow a spiral path along the DNA. This Klenow fragment-DNA complex is consistent with calculations that indicate that virtually all of the positive electrostatic charge of the Klenow fragment lies within the cleft. Some experimental evidence for the assignment of the DNA binding site is provided by genetics. The mutation polA6, which weakens the enzyme's interaction with DNA, is an Arg → His change within helix K.

When duplex DNA is bound, the flexible 50-amino-acid subdomain connecting helices H and I could fold across the top of the binding cleft, allowing the protein to surround the DNA template completely. This firm grip on the template should encourage polymerase to stay bound to the template so that it could catalyze the formation of many phosphodiester linkages in swift succession. The alternative to an enzyme that operates in this way (processively) would be a polymerase that dissociates every time it catalyzes the formation of a single phosphodiester linkage. It seems likely that a processive enzyme would be much more efficient, since it would not have to go through the time-consuming events of dissociation and reassociation with the template every time a nucleotide is added to the growing polymer. All DNA polymerases are processive enzymes, some more than others, as we will see when we discuss the mammalian DNA polymerases.

It is known that deoxynucleoside monophosphates (dNMPs) inhibit the 3′,5′-exonuclease reaction but not the polymerase reaction. Thus it is likely that the dNMP binding site corresponds to the position of the DNA 3′ terminus in the exonuclease active site. The dNMP phosphate would therefore mark the position of the phosphodiester bond to be cleaved in the exonuclease reaction. Crystal structure studies show that dNMP binds to the small domain of Klenow fragment. Furthermore, mutant studies show that discrete amino acid changes that eliminate the binding of dNMP also eliminate the 3′, 5′-exonuclease activity without perturbing the structure of the Klenow fragment. Such studies indicate that the small domain contains the 3′,5′-exonuclease active site. Moreover, the absence of an effect of these mutations on the polymerase activity indicates that the two active sites must be separate to some degree.

Although the precise structure for the smaller fragment cleaved in the subtilisin experiment has not been studied, it appears highly likely that the structure of intact Pol I consists of three domains with the approximate layout shown in figure 26.13. The 5′,3′-exonuclease activity is located downstream of the other two activities, a position consistent with its function of removing RNA primers on nascent strands. The location of the 3′,5′ activity relative to the polymerase activity is known more precisely; the two enzyme sites are located eight bases apart.

Two More DNA Polymerases Are Discovered in E. coli

Kornberg's early work was done on DNA polymerase in cell-free systems. The fact that the enzyme had so many of the properties expected for a DNA-replicating enzyme led most observers to believe that it was indeed the DNA-replicating enzyme of *E. coli.* But to obtain final proof that an enzyme functions in a given capacity *in vivo,* it is essential to isolate mutants in which the enzyme's behavior has come about by correlating *in vitro* biochemical behavior with the behavior of mutants in which DNA replication and repair are affected.

From the time of the discovery of DNA polymerase I, about twenty years were needed to clarify its physiological function. A major step in that direction was taken by Cairns and De Lucia, who laboriously scanned several thousand colonies of heavily mutagenized *E. coli* and found one that contained almost no DNA polymerase I polymerizing activity (1 to 2% of normal). That mutant grew well under normal conditions, a fact suggesting that the DNA-polymerizing activity of the Kornberg enzyme was not needed for replication. However, the mutant was more sensitive to ultraviolet irradiation than its wild-type parent. Ultraviolet irradiation was known to damage DNA. Therefore the increased sensitivity of the mutant suggested that, although DNA polymerase I was not the chromosome-replicating polymerase, it might be involved in repairing chromosome damage caused by ultraviolet light.

In cell-free extracts from a mutant *E. coli* strain that did not contain DNA polymerase I polymerizing activity, two additional DNA-polymerizing enzymes were subsequently detected. These were named DNA polymerases II and III. The behavior of conditional lethal mutants of DNA polymerase III led to the conclusion that DNA polymerase III is the main replication enzyme. Some of the properties of the three enzymes are summarized in table 26.1.

Figure 26.13

A schematic drawing of the apparent domain structure of *E. coli* DNA polymerase I. The solid line represents the experimentally determined Klenow fragment structure; the dashed lines indicate a possible location for the small fragment produced by proteolytic cleavage of Pol I. The orientation of the nicked DNA substrate is such that the 5′→3′ exonuclease domain would be able to interact with the DNA 5′ downstream of the primer terminus. (From C. M. Joyce and T. A. Steitz, "DNA polymerase I: From crystal structure to function via genetics," *Trends Biochem. Sci.* 12:288-292, 1987.)

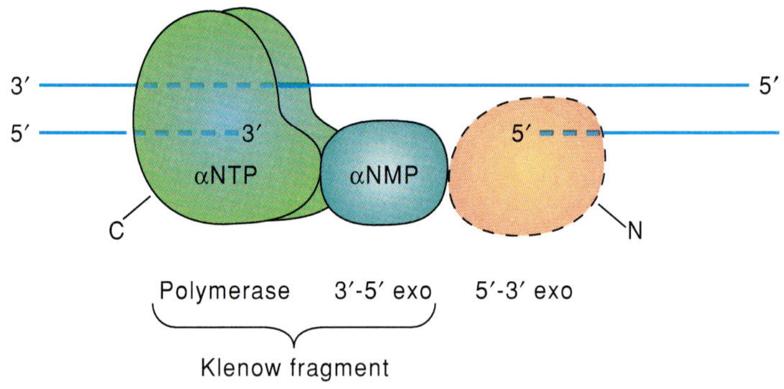

Table 26.1
Properties of Polymerases I, II, and III of *E. coli*

	Pol I	Pol II	Pol III
Molecules per cell	400	—	15
Turnover number[a]	600	30	9,000
Structural gene[b]	*polA*	*polB*	*polC* (*dnaE*)
5′→3′ polymerizing activity	+	+	+
3′→5′ exonuclease activity	+	+	+
5′→3′ exonuclease activity	+	−	−
Mutant loci	*polA*	*polB*	*polC, dnaN, dnaX, dnaZ, dnaQ*
Lethality	Viability reduced only when 5′→3′ exonuclease affected	No effect	Conditional lethality

[a]Nucleotides polymerized/min/molecule of enzyme at 37° C.
[b]Only the structural gene for the largest protein subunit in the enzyme is recorded.

For a few years following the observations of Cairns and De Lucia, it was assumed that DNA polymerase I was not important in replication but only in repair. However, further genetic and biochemical studies proved otherwise. The original mutated DNA polymerase I was defective only in its polymerizing function; its two degradation functions were intact. Subsequently, workers isolated conditional lethal mutants in which the 5′→3′ degradation function of Pol I was affected; the mutants cannot elongate DNA under nonpermissive conditions. Thus it was demonstrated that the degradation activity of DNA polymerase I is indispensable for chromosome replication.

DNA polymerase III, which functions as the main replicating enzyme, is a complex of at least seven different proteins (table 26.2). Genetic loci for some of the protein subunits found in the polymerase III enzyme have been identified (see table 26.2). Mutants in which θ is affected have not yet been isolated, so the physiological significance of θ remains unknown.

Table 26.2
Subunits and Subassemblies of Pol III Holoenzyme

Subunit	kDa	Gene	Subassembly			
α	130	*dnaE*				
ϵ	27.5	*dnaQ*	pol III (core)			
θ	10			pol III′		
					pol III*	
τ	71	*dnaX*				
γ	47.5	*dnaX*				holoenzyme
δ	35					
δ'	33		γ complex			
χ	15					
ψ	12					
β	40.6	*dnaN*				

Polynucleotide Ligase Links Long Chains Together

Discontinuous DNA synthesis requires an enzyme for joining the newly synthesized segments (see fig. 26.10). Initially, two enzymes of this type were discovered, and they have been given the general name polynucleotide ligase. In the ligase-catalyzed reaction, a single phosphodiester bond forms between long runs of discontinuous chains held in proper juxtaposition by a template chain (fig. 26.14). Since the joining reaction is vital to DNA replication, it is not surprising to find that ligases are ubiquitous in living cells. They differ in the source of activation energy used for phosphodiester bond formation. The *E. coli* bacterial ligase (encoded by the *lig* gene, table 26.3) uses nicotinamide adenine dinucleotide (NAD) as the source of energy: the T4 bacteriophage and mammalian DNA ligases use ATP. Ligase is important in recombinant DNA techniques because it can seal fragments that have single-stranded complementary termini. Such fragments can be generated by restriction endonucleases. If the concentration of DNA fragments is high enough, ligase will also covalently link blunt-ended fragments that lack single-stranded complementary termini.

Topoisomerases Catalyze Supercoiling

Supercoiled helices are frequently found in closed circular DNA molecules or in otherwise constrained DNA segments (see chapter 25). With the help of supercoiled DNA and advances in the technology of DNA analysis by agarose gel electrophoresis (see fig. 25.21), researchers have discovered several kinds

Figure 26.14

Steps in the sealing of a DNA nick, catalyzed by DNA ligase. The bacterial ligase uses NAD⁺ to make an enzyme-AMP intermediate. Mammalian DNA ligases and bacteriophage T4 ligase use ATP for the same purpose.

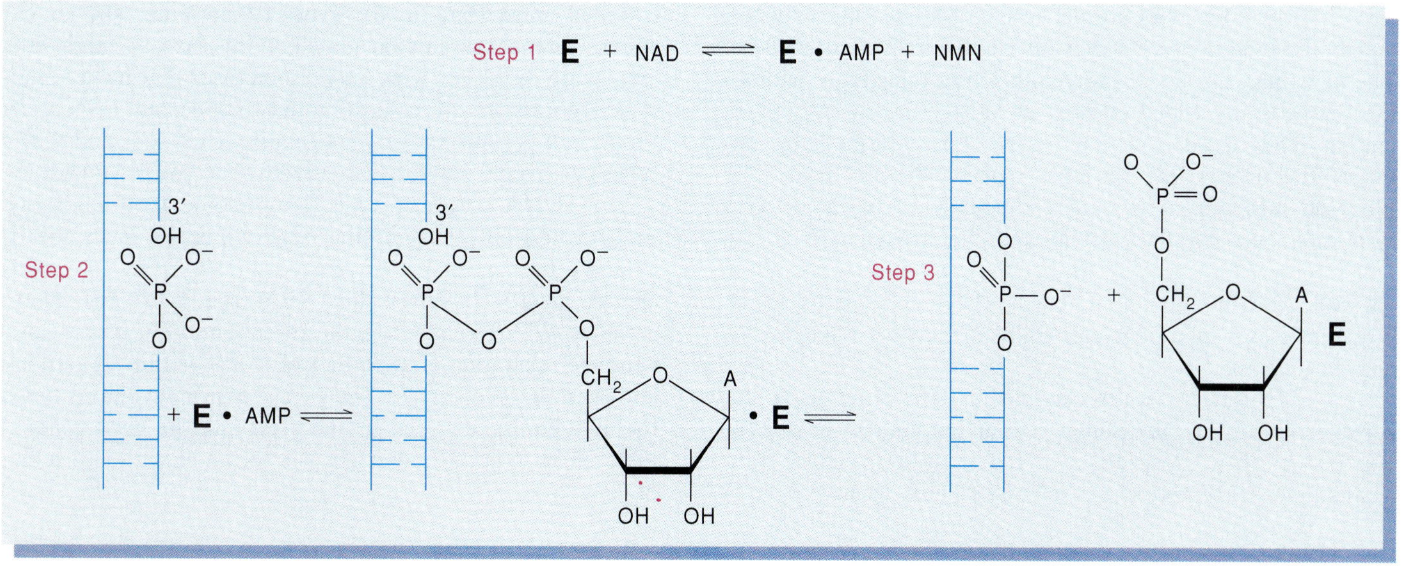

Table 26.3
Other Proteins Involved in DNA Synthesis in *E. coli*

Gene	Function of Encoded Protein	Size (Daltons × 10^3)
polA	In DNA polymerase I, extending Okazaki fragments and removing primers	109
dnaB	Interacts with primase (functions as hexamer)	50 monomer
dnaC	Stimulates priming	25 monomer
dnaG	Primase (acts as dimer)	60 monomer
dnaY	—	—
lig	Ligase	75
rep	Helicase, ATP-dependent	66
ssb	Single-strand DNA-binding protein	74
GyrA	Subunit of topoisomerase II encoding GyrA (sensitive to nalidixic acid)	100 monomer
GyrB	Subunit of topoisomerase II encoding GyrB (sensitive to coumermycin and novobiocin)	95 monomer

The collection of the *dnaB*-, *dnaC*-, and *dnaG*-encoded proteins, together with i, n, n″ and Y, constitute the primosome. Factor Y first recognizes a primosome assembly site, stimulating its DNA-dependent ATPase. The n protein then binds to the factor Y:DNA complex, with the subsequent addition of protein i, n″, and *dnaB*- and *dnaC*-encoded proteins, which together constitute the *prepriming complex*. It can move along DNA in the 5′→3′ direction when primase is bound (forming the *primosome*); primers are synthesized at various locations on the DNA template.

of enzymes that can either reduce or increase the linking number (L) of a supercoiled helix. Those enzymes, called topoisomerases, were first identified as activities capable of relaxing negatively supercoiled DNA in *E. coli* and in mouse embryo cells.

Topoisomerases are believed to be essential for DNA metabolism and they are widely distributed in living organisms. The enzymes can break and rejoin DNA repeatedly without the need for any added cofactor to supply the energy for rejoining. Thus it appears that the reaction intermediate conserves the energy of the DNA phosphodiester bond. The proposed mechanism for supercoiling involves covalent-bond formation between the protein and the broken end of the DNA, followed by resealing of the DNA and dissociation of the enzyme, after the linking number has been changed by one or more units. A 3′-phosphotyrosine linkage has been detected as an intermediate.

Topoisomerases are classified as type I or type II, according to whether they change the linking number in steps of one or steps of two, respectively. Type I enzymes produce transient single-strand breaks in the double helix, whereas type II enzymes produce transient double-strand breaks (fig. 26.15). Type I topoisomerase of *E. coli,* also known as omega protein, is the best-understood enzyme of this class. Its molecular weight is 110,000 and the native protein exists as a monomer in solution. The enzyme shows a preference for highly negatively supercoiled DNA and is inactive on positively supercoiled DNA. This enzyme can also catalyze the catenation (interlocking) of double-stranded circular DNA, or the separation of catenanes into simple circles provided at least one circle contains a single-stranded break (fig. 26.16). Type II topoisomerases can carry out the catenation reaction as well; in this case, no single-strand breaks are required. Type I topoisomerases that have been isolated from eukaryotic cells differ in two important respects from the *E. coli* protein: they do not require Mg^{2+} ions and they can relax positively as well as negatively supercoiled DNA.

DNA gyrase, a type II topoisomerase found only in prokaryotes, differs from other topoisomerases in its ability to catalyze the conversion of relaxed duplex DNA into a high-energy, negatively superhelical form (see fig. 26.15). That reaction is ATP-dependent. In the absence of ATP, gyrase can still catalyze relaxation of superhelical DNA or the catenation reactions described earlier. The gyrase enzyme contains two different subunits, encoded by the *gyrA* and the *gyrB* genes (see table 26.3). Normally it exists as a tetramer with two subunits of each type.

Figure 26.15

Type I and type II topoisomerases relax negatively supercoiled DNA in steps of one and steps of two, respectively. Type II topoisomerases can also add additional negative supercoils, as indicated by the double reaction arrow. The latter reaction requires energy input, which is encoded by ATP cleavage.

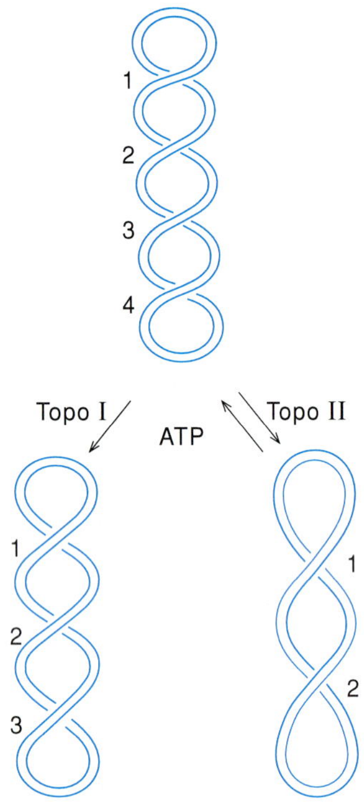

Figure 26.16

Catenation by topoisomerases. (*a*) Two circular DNAs can be catenated by type I topoisomerase only if one of the DNAs is nicked. This is not necessary when using a type II topoisomerase. (*b*) Electron micrographs of catenated DNA before (i) and after (ii) incubation with DNA gyrase. The catenane contains one large circular DNA and one small circular pBNP66 plasmid DNA. (Source: M. Gellert et al., "DNA gyrase: Site-specific interactions and transient double-strand breakage of DNA" in *Cold Spring Harbor Symposium on Quantitative Biology,* 45:301, 1981. Copyright © 1981 Cold Spring Harbor Laboratory Press, Cold Spring Harbor, N.Y.)

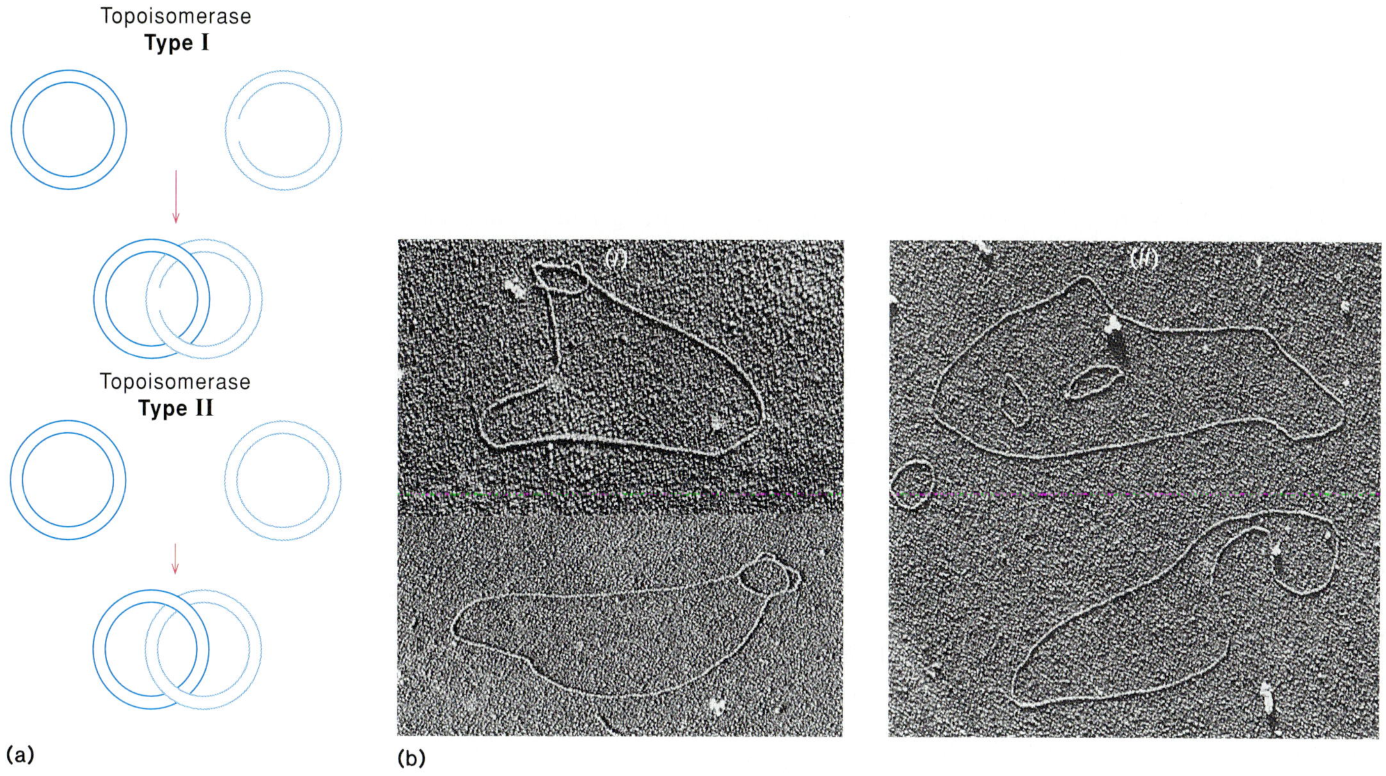

Topoisomerases must serve vital functions, since mutants carrying defective topoisomerases grow very poorly. We do not know all the functions of topoisomerases, but probably they include the elimination of tangled DNA by decatenation reactions, as well as the elimination of the extreme supercoiling that would otherwise result from the unwinding of the duplex structure during DNA replication (see the following subsection). Topoisomerase is likely to act in the final stages of the separation of replicating circular DNA.

Helicases Catalyze Duplex Unwinding

The parental double helix must unwind at the growth fork so that it can present additional single-stranded regions to serve as templates for continued replication. Enzymes that catalyze unwinding are called helicases. The rep protein of *E. coli* is believed to be a helicase, since DNA elongation is slower when the rep protein is not present. During the unwinding of duplex DNA by helicase, ATP is consumed, indicative that the reaction requires energy. Helicase is one of the components of the prepriming *E. coli* replication complex, probably facilitating both priming and replication fork movement.

The unwinding of a circular duplex should tend to generate positively supercoiled DNA. As unwinding progresses, the increasing degree of positive supercoiling would lead to considerable strain in the duplex structure. It seems likely that those tensions are relieved in bacterial systems by the counteracting tendency of gyrase to cause negative supercoiling. Type I bacterial topoisomerases cannot relax positively supercoiled DNA and therefore would be of little value in this regard. By contrast, in eukaryotes containing no gyraselike enzyme, the type I topoisomerases are competent in relaxing positively supercoiled DNA.

Single-Strand Binding Protein Stabilizes Single Strands

Single-strand DNA-binding protein (SSB) is found in abundance in *E. coli.* It has several functions, including the presentation of the template to the polymerase and productive interactions with several replication proteins. This and other evidence (to be described later) indicate that SSB plays a role in DNA replication, and probably in repair and recombination as well. It is believed to act in concert with the rep protein to facilitate unwinding at the growth fork. Since it binds preferentially to single-stranded DNA, it probably facilitates unwinding by inhibiting rewinding. Possibly, also, it protects the single-stranded DNA from attack by endogenous nucleases.

Single-Stranded DNA Viruses Replicate through a Duplex Intermediate

Our understanding of prokaryotic DNA replication has proceeded most rapidly through investigations on viruses because of their relative simplicity. Bacteriophages that infect *E. coli* vary considerably in size and structure. Each type replicates its chromosome in a unique manner and relies on the host enzymatic machinery in a way that suits its needs. In general, there is an inverse relationship between the size of the virus chromosome and the degree of dependency on host enzymes. Large viruses, such as bacteriophage T4, encode several genome-replicating functions that have properties analogous to those of closely related host protein, but that for various reasons are preferred by the virus (sometimes only because they are made in larger amounts).

The bacteriophage ϕX174 has a small, single-stranded circular DNA chromosome (5,386 bases) that depends on many *E. coli* proteins for replication. During replication, ϕX174 DNA passes from a single-stranded to a double-stranded form and then back to a single-stranded form. The replication process is particularly well understood, and findings with ϕX174 not only reveal the replication process of the viral DNA but also give insights concerning certain host proteins used in host chromosome replication. The ease with which ϕX174 DNA can be isolated intact and assayed biologically, through its ability to infect *E. coli* spheroplasts, makes it an ideal template for the study of the fidelity of its replication *in vitro.*

When ϕX174 infects *E. coli* it loses its protein coat, so that only the circular viral DNA enters the cytoplasm. During the first phase of the replication process, the single-stranded viral DNA must be converted to a double-stranded form. This double-stranded form serves two purposes, as an intermediate in replication (replicative form) and as a template for synthesis of viral RNA. Since no viral RNA or proteins can be made until after the viral DNA is converted to the double-stranded form, it follows that the single-to-double-strand transition must be carried out by preexisting host cellular proteins. This point has been verified by experiments showing that the double-stranded form of the virus is made in the presence of chloramphenicol, which inhibits protein synthesis, whereas further steps in viral DNA replication cannot occur in the presence of this antibiotic.

Kornberg, his co-workers, and Hurwitz and his colleagues have constructed a cell-free system containing ϕX174 DNA and purified proteins that can carry out many of the steps believed to be involved in ϕX174 DNA replication *in vivo.* Before DNA synthesis can be initiated on the single-stranded circular DNA, a primer must be synthesized by the host primase. Two types of complexes have been made in which primase is active. The first complex is formed by adding primase to a preformed complex of DNA and dnaB-encoded protein. Primer formation occurs in this system at random sites on the DNA. In the second complex, which is sequence-specific, several additional proteins are required before dnaB protein or primase can bind (fig. 26.17). These include n, n′, n″, dnaC, and single-strand binding (SSB) protein. When the DNA is uniformly coated with SSB, random primer formation is prevented because the dnaB protein cannot displace SSB from the DNA to form a complex. Under these conditions n′ makes an initial complex at a specific site on the DNA (step I in fig. 26.17). This site is adjacent to a 44-base region on the single-stranded DNA that forms a hairpin double helix to which SSB does not bind. Other proteins—n, n″—join the initial complex, resulting in the displacement of some SSB. Finally, the dnaB, dnaC, and i proteins become part of the complex, as more SSB is displaced; the resulting multiprotein complex is known as the preprimosome. Addition of primase to this complex completes assembly of the primosome (step II, fig. 26.17). The primosome can migrate on the single-stranded DNA in the 5′→3′ direction. In this process, still more SSB is removed from the single-stranded DNA (step III, fig. 26.17).

This movement of the primosome is an energy-dependent reaction requiring ATP. At various points during its journey the primosome pauses to synthesize a short stretch of primer (step IV, fig. 26.17). DNA polymerase III holoenzyme uses these primers as initiation sites for DNA synthesis (step V, fig. 26.17). Although primase initiates limited stretches of RNA synthesis at several locations on the template, the necessary primosome can be assembled only at a unique site around base 2,300 (step I, fig. 26.17). The primosome complex is very stable and can survive many rounds of replication. The Okazaki fragments formed by primase and DNA polymerase III (step V, fig. 26.17) require DNA polymerase I and ligase for completion (step VI, fig. 26.17). DNA polymerase I closes the gaps as it removes the RNA on the 5′ ends of the fragments. Ligase joins the nicked fragments of DNA together. The circular duplex so formed is referred to as replicative form DNA (RF). If gyrase is present, this molecule will become negatively supercoiled. The RF is used as a template to synthesize more RF and viral messenger RNA, which, as we will see, is transcribed entirely from the minus strand of the RF. The viral messenger RNA serves in turn as the template for the synthesis of viral proteins that are essential in subsequent steps in viral DNA replication.

After single-stranded viral DNA (SS DNA) has been converted to the replicative form duplex (SS→RF), the parental RF is duplicated to produce multiple copies of RF (RF→RF). In this reaction the viral DNA, known as the plus strand, and the complementary DNA, known as the minus strand, are replicated by distinct mechanisms. Synthesis of the viral plus strand begins when the viral gene A protein induces the cleavage of the viral plus strand in the RF at position 4,305 (fig. 26.18). The bifunctional A protein not only cleaves the plus strand at this point, but also becomes covalently attached to the 5′ end of the interrupted strand (see fig. 26.18). More copies of the replicative duplex form are initiated from this point. Nucleotides are added continuously to the 3′-OH end of the plus strand. The new plus strand displaces the old plus strand during this phase of the reaction. After one viral single-stranded plus

Figure 26.17

Proposed mechanism for the conversion of single-stranded ϕX174 DNA to the duplex replicative form (SS → RF reaction). (Source: K. Arai et al., "Mechanism of *dnaB* protein action V. Association of *dnaB* protein, protein *n′* and other prepriming proteins in the primosome of DNA replication" in *Journal of Biological Chemistry,* 256:5280, 1981. Copyright © 1981 American Society for Biochemistry and Molecular Biology, Inc., Bethesda, Md.)

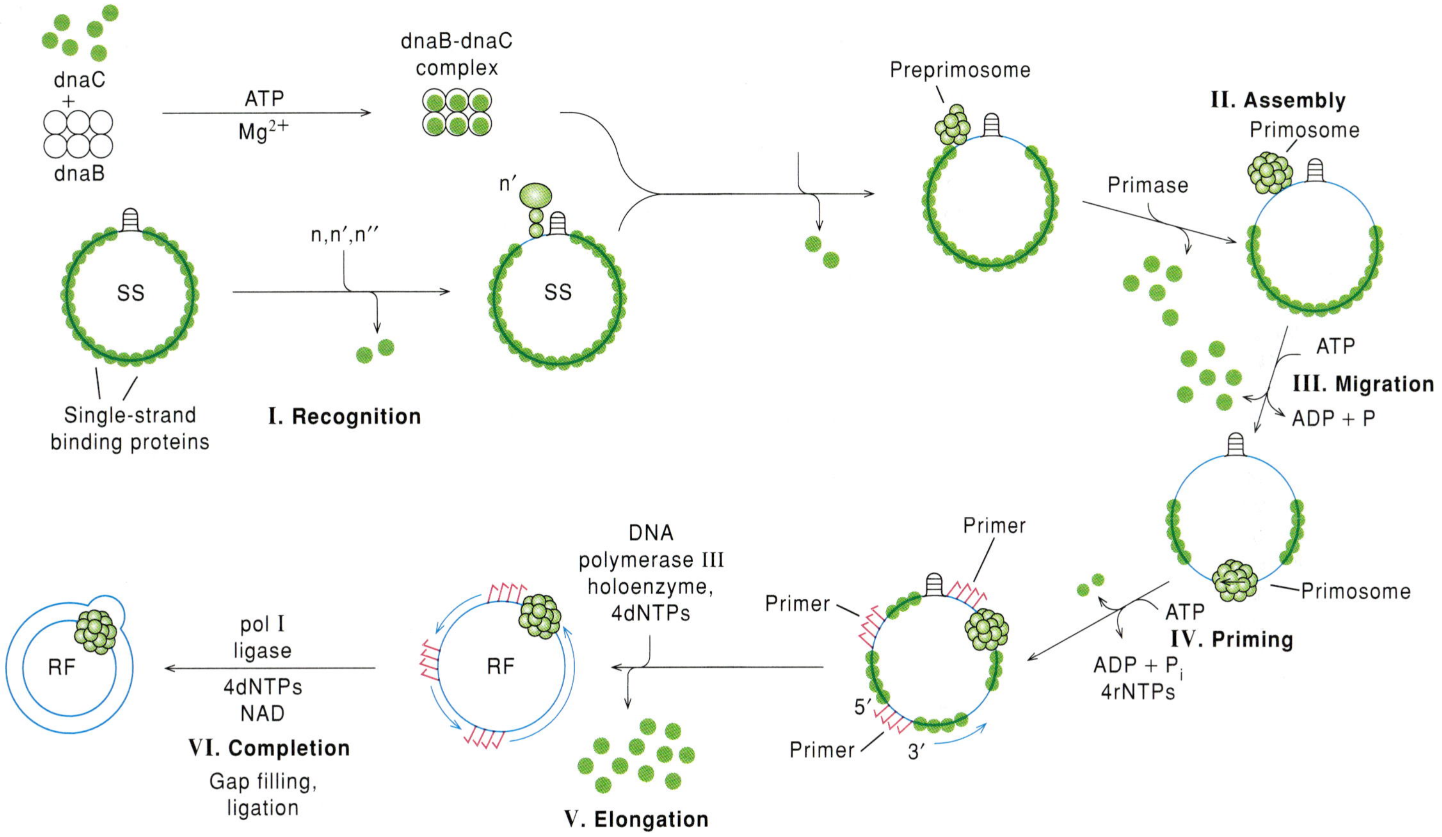

Figure 26.18

Scheme for ϕX174 RF replication in two stages. Continuous replication initiated by gene A protein cleavage generates viral (+) circles, and discontinuous replication of the viral circles by the SS → RF system produces RF. In the presence of phage-encoded maturation and capsid proteins, viral circles (RF → SS) are encapsulated rather than replicated. (Source: A. Kornberg, *DNA Replication.* Copyright © 1980, W. H. Freeman, New York, N.Y.)

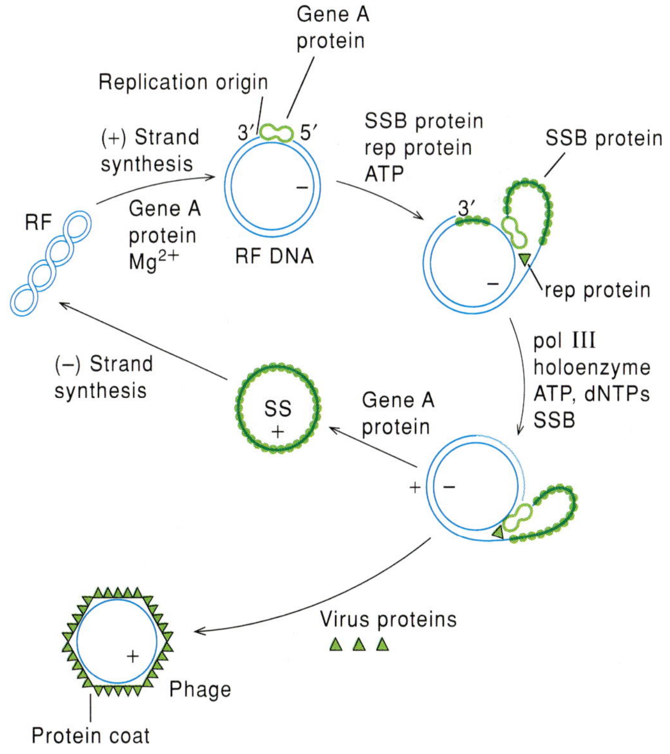

strand has been displaced, it becomes covalently closed through the action of the viral gene A protein (acting as a specific ligase) and can either serve as the template for synthesis of more RF or become a component (in the genome) of the mature virus. During the synthesis of plus strand it can be seen that the minus strand is being used as the template in a continuous fashion: its action could be likened to that of a rolling circle. The rolling-circle mode of replication is used by a number of bacterial viruses (e.g., bacteriophage λ) at some stage during their replication cycle.

Host proteins required during plus-strand synthesis include the host rep protein, or helicase, as well as DNA polymerase III and single-strand binding protein (SSB). As we have noted, the function of helicase is to facilitate unwinding of the duplex. Helicase is not absolutely required for DNA synthesis using a duplex template, but it greatly accelerates the process. SSB acts together with helicase by complexing with single-stranded regions, thus inhibiting rewinding. The viral gene A protein is crucial in maintaining the integrity of the displaced virus plus strand during synthesis. It does this by holding the otherwise free end and ensuring that the strand closes at the proper point without releasing itself from the template.

What we know about additional steps in the replication process comes primarily from an analysis of the events occurring *in vivo*. After about 60 copies of the replicative form have been made (20 min after infection), the infected cell switches to making single-stranded circles exclusively. Presumably some regulatory device is involved to ensure that further single-stranded circles do not get converted into duplex forms at this stage. The protein encoded by the phage genes *C* and *J* are thought to play a key role in diverting the single-stranded DNA into mature viral particles. In that process, the plus-strand circles get packaged into phage protein shells as they are being synthesized, in preparation for making completed viruses. The packaging step concludes the replication cycle for ϕX174 DNA.

Initiation of Chromosomal Replication in E. coli *Occurs at a Unique Location*

Replication of the *E. coli* chromosome has been more difficult to study than replication of phage DNA, in part because of its large size. As we stated earlier, genetic evidence has indicated that host proteins involved in ϕX174 replication might also be involved in similar ways in *E. coli* DNA replication. In particular, the movement of the primosome complex suggests a mechanism for discontinuous DNA synthesis on the lagging strand of the replication fork. The primosome could migrate on the lagging strand as the template unwinds, stopping periodically to allow primase to form primer (fig. 26.19). The elongation steps, so clearly elucidated in ϕX174 minus-strand synthesis, could account for lagging-strand synthesis at the *E. coli* replication fork.

That leaves two aspects of *E. coli* chromosome replication to be accounted for: initiation at the origin of replication (location indicated in figure 26.5) and termination. Study of termination awaits the reconstitution of an *in vitro* system containing all the proteins necessary for replication. To explore the

Figure 26.19

Role of the primosome at the bacterial replication fork. The primosome, assembled at or near the DNA replication origin, migrates continuously on the lagging strand as the replication fork moves. (Source: K. Arai et al., "Movement and site selection for priming by the primosome in phage ϕX174 DNA replication," in *Proceedings of the National Academy of Sciences USA* 78:711, 1981. Copyright © 1981 National Academy of Sciences, Washington, D.C.)

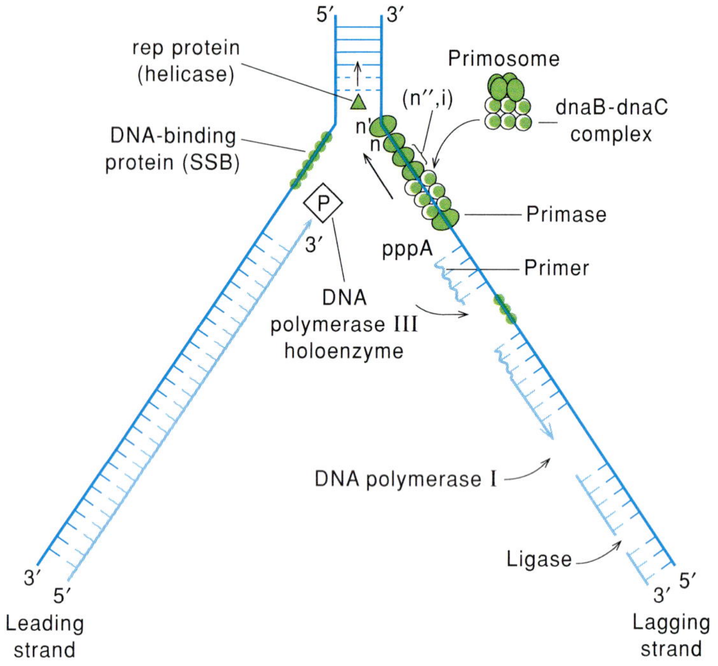

mechanism of initiation, Kornberg and his co-workers developed a cell-free system for bidirectional replication using small circular DNA molecules, known as hybrid plasmids, that contain the origin of the *E. coli* chromosome (*oriC*) together with purified proteins from *E. coli.* These ongoing studies are beginning to show how bacterial chromosome synthesis is initiated.

The use of plasmids is crucial in such studies because they greatly facilitate detection of initiation on a much smaller and more tractable piece of DNA. Plasmids contain about 245 base pairs (bp) of the minimal origin sequence from the *oriC* region of the *E. coli* chromosome. The 245-bp *oriC* region has a considerable number of GATC sites in which the cytosine residues are specifically methylated by an *E. coli* enzyme encoded by the *dam* gene. Methylation may be important in the control of initiation of replication at *oriC*.

The initiation of replication at *oriC* can be divided into three steps:

1. Generation of a template (requiring DNA gyrase, topoisomerase I, and a histonelike DNA-binding protein called HU).
2. Transcription and priming (requiring RNA polymerase and the dnaA and dnaB proteins).
3. Primer processing (requiring RNase H, an enzyme that degrades RNA when it is base-paired to DNA or RNA, and some yet unidentified factor).

Figure 26.20

Hypothetical asymmetric dimeric structure suggested by Kornberg for the DNA polymerase III holoenzyme of *E. coli.* The holoenzyme appears to be organized as an asymmetric dimer; a pair of core subassemblies, each with a potential for polymerase action, has an asymmetric distribution of auxiliary subunits that may endow each one with different properties, one suited to the continuous synthesis of one strand and the other to the discontinuous synthesis of the other strand. The α subunit contains the polymerase activity. The ε subunit contains the 3′ to 5′ exonuclease proofreading activity, which is active only on attachment to the α subunit. Several of the subunits are essential to processivity. Among them, the ß subunit is held loosely in the holoenzyme and dissociates during the cycling of the polymerase from one template to another. (Source: A. Kornberg, "DNA replication" in *Journal of Biological Chemistry* 263:1-4, 1988. Copyright © 1988 American Society for Biochemistry and Molecular Biology Inc., Bethesda, Md.)

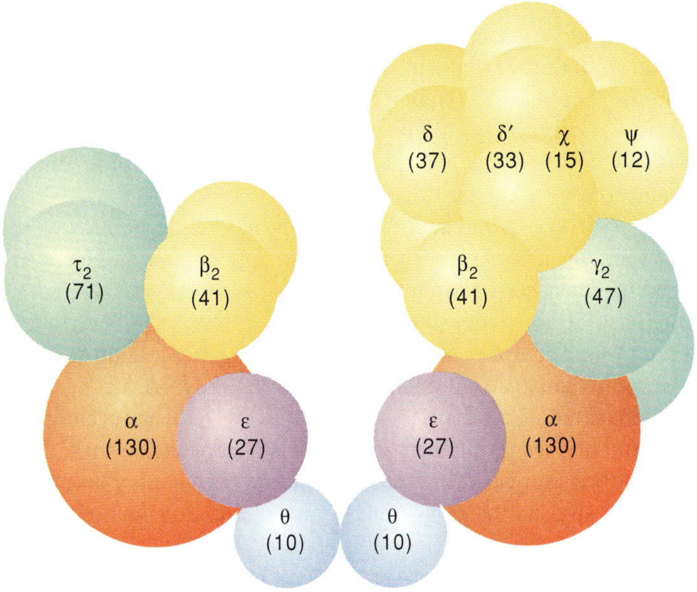

In addition to the primosomal proteins and the Pol III holoenzyme, other, as yet unidentified proteins are required for *oriC*-dependent DNA replication. The requirements for RNA polymerase and DNA gyrase have been shown by their specific sensitivities to the inhibitors rifampicin and nalidixic acid (and novobiocin), respectively. Rifampicin is known specifically to inhibit RNA polymerase (see chapter 28), and nalidixic acid and novobiocin specifically inhibit the action of the subunits of DNA gyrase (see chapter 25). The characterization of the proteins required in this system is under intensive investigation. It is noteworthy that RNA polymerase catalyzes primer synthesis at *oriC* but that primase is used at other sites along the *E. coli* chromosome.

Synthesis May Take Place Concurrently on Both Strands

Concurrent replication of both strands, rather than the jerky sequence of synthesis of one strand and then the other, might be achieved if priming of nascent fragments of the lagging strand were integrated with continuous synthesis of the leading strand. Concurrent replication would also require a more complex holoenzyme possessing primase and twin active sites for polymerization (fig. 26.20). Evidence for such a structure includes (1) twin polymerase subassemblies in the Pol III holoenzyme and (2) complexing of primase by some polymerases. In this hypothetical scheme for concurrent replication, looping of the lagging strand by 180° gives it the same orientation as the leading strand at the replication fork (fig. 26.21). A primer generated by primase is extended by polymerase as the lagging strand template is drawn past it. When synthesis reaches the 5′ end of the previous nascent fragment, the lagging strand template is released. Helicase action and continuous synthesis of the leading strand periodically expose lengths of template for priming of nascent fragments.

Figure 26.21

Hypothetical scheme for concurrent replication of leading and lagging strands by an asymmetric, dimeric polymerase associated with a primosome and a helicase in a replisome. In this hypothetical scheme for concurrent replication, looping of the lagging strand by 180° gives it the same orientation as the leading strand at the fork. A primer generated by primase is extended by polymerase as the lagging strand template is drawn past it. When synthesis reaches the 5′ end of the previous nascent fragment, the lagging strand template is released and unlooped. Helicase action and continuous synthesis of the leading strand periodically expose lengths of template for priming of nascent fragments. (Source: A. Kornberg, "DNA replication" in *Journal of Biological Chemistry,* 263:1-4, 1988. Copyright © 1988 American Society for Biochemistry and Molecular Biology Inc., Bethesda, Md.)

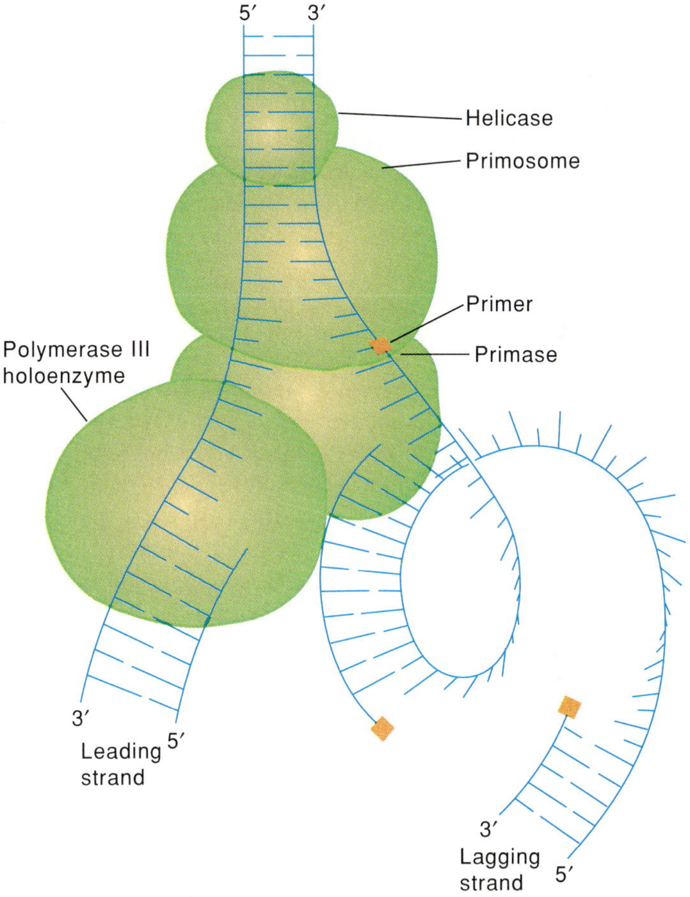

DNA Synthesis and Chromosomal Replication in Eukaryotes

The most productive approach to understanding DNA replication in eukaryotes has been to isolate proteins that exhibit activities logically related to DNA replication and to isolate replicative intermediates. Different forms of DNA polymerase, DNA ligases that require ATP, topoisomerases, single-stranded DNA-binding proteins, unwinding enzymes, and degradation enzymes have all been isolated from eukaryotic cells. We can presume that many of these have vital functions *in vivo;* some of the functions have been verified in yeast with the help of mutants.

Two major systems are currently being exploited in studying eukaryotic DNA synthesis. They are animal viruses and yeast. The yeast cell system and its extensively known genetics makes it possible to isolate mutants by conventional means. The use of yeast also makes it possible to exploit reverse genetics—that is, purified proteins can be used to clone the genes encoding them, then the cloned genes can be used to disrupt, by homologous integrative transformation, their genomically located counterparts.

Eukaryotic Cells Also Have Several DNA Polymerases

Four nuclear species of DNA polymerases have been characterized in eukaryotes, designated α, β, δ and ϵ. An additional polymerase γ is located in mitochondria. The amount of DNA polymerase α is relatively high in cells actively engaged in DNA synthesis. This and its key role in the replication of SV40 DNA (see below) implicate the α polymerase in host chromosomal DNA synthesis. The δ enzyme has an important function in the DNA replication of certain animal viruses including SV40; it seems likely that it has a comparable function in host chromosomal DNA synthesis.

DNA polymerases β and γ are present in eukaryotic cells in resting and differentiated states. The levels of β and γ enzymes do not change markedly in proliferating cells. The function of the β enzyme is unclear, although its localization in the nucleus has led to the speculation that it may be involved in repair or replication of nuclear DNA. DNA polymerase γ is usually localized in mitochondria, and is believed to be involved in replication of the organelle's DNA.

Eukaryotic Chromosomes Contain Several Origins of Replication

Although the general features of DNA replication in eukaryotes are thought to be similar to those in prokaryotes, there are some interesting differences. The chromosomes of higher eukaryotic organisms are quite large, in some cases a thousand times larger than their bacterial counterparts. In order for such large DNA molecules to replicate in a reasonable length of time, they must have multiple origins of replication. The simultaneous synthesis of DNA at several points along the chromosome has been demonstrated by incorporating radioactive nucleotides into the replicating chromosomes for a short period and then observing the distribution of radioactive DNA by autoradiography (fig. 26.22). Multiple regions (or replicons) of incorporated label are observed, with replication proceeding bidirectionally from each of these regions. Termination of replication occurs at the point where the growth forks from two adjacent replication units meet (see fig. 26.22). DNA on the lagging strand of a fork is made discontinuously. The Okazaki fragments are much shorter than those found in prokaryotes, averaging between 100 and 200 nucleotides in length. Synthesis of these DNA fragments is initiated on RNA primers that are found covalently attached to the 5′ ends of newly synthesized fragments. Chromosomal replication occurs only during the S phase of the cell cycle.

Are eukaryotic replication origins located at unique sites? This is the case in animal viruses such as SV40 and polyoma DNA. It is also true for the yeast *S. cerevisiae;* in this eukaryotic microorganism the origins are termed ARS (autonomously replicating sequences). ARS sequences can be shown to act as origins by providing this function to yeast plasmids that lack them and are the site of replication bubbles when monitored by electron microscopy.

DNA in eukaryotic chromosomes is associated with histones in complexes called nucleosomes (the nucleosome structure is described in chapter 25). These nucleoprotein complexes serve to condense exceedingly long DNA molecules into much more compact structures, but during replication the nucleosomes must be disassembled so that the DNA strands can be separated. The disassembly of the DNA-histone complex presumably occurs directly in front of the replication fork. Two observations suggest that nucleosome disassembly may be a rate-limiting step in the migration of the replication fork through chromatin:

1. The rate of migration of replication forks is slower in eukaryotes (with nucleosomes) than in prokaryotes (lacking nucleosomes).
2. The length of the Okazaki replication fragments is similar to the length of DNA between adjacent nucleosomes (about 200 base pairs).

Before the newly replicated DNA is reassembled into nucleosomes, the RNA primers must be removed by an as yet unknown enzyme, the gaps must be filled in by DNA polymerase, and the replication fragments must be linked together by DNA ligase. Replication origins are activated in a specific temporal order, the overall number of origins depending on the time required to replicate the cell's genome. Those genes that are actively transcribed are replicated in early S phase; if an actively transcribed gene is transposed to a late-replicating region, it is still replicated early in S phase.

The yeast genome has 17 chromosomes, with a combined size of 1.35×10^4 kb. There are about 400 ARS sequences, where DNA replication is believed to be initiated. DNA-binding proteins that bind specifically to ARS sequences have been identified; they could be analogous to the SV40 viral T antigen, to be described later. Many mutant genes that affect

Figure 26.22

Multiple-origin model for eukaryotic chromosomal DNA replication. (*a*) Autoradiograph of short-term labeling of a eukaryotic chromosome during replication and its interpretation. (*b*) Overall replication scheme for a eukaryotic chromosome. Only a short region of the chromosome is shown. It is believed that replication origins are relatively free of proteins.

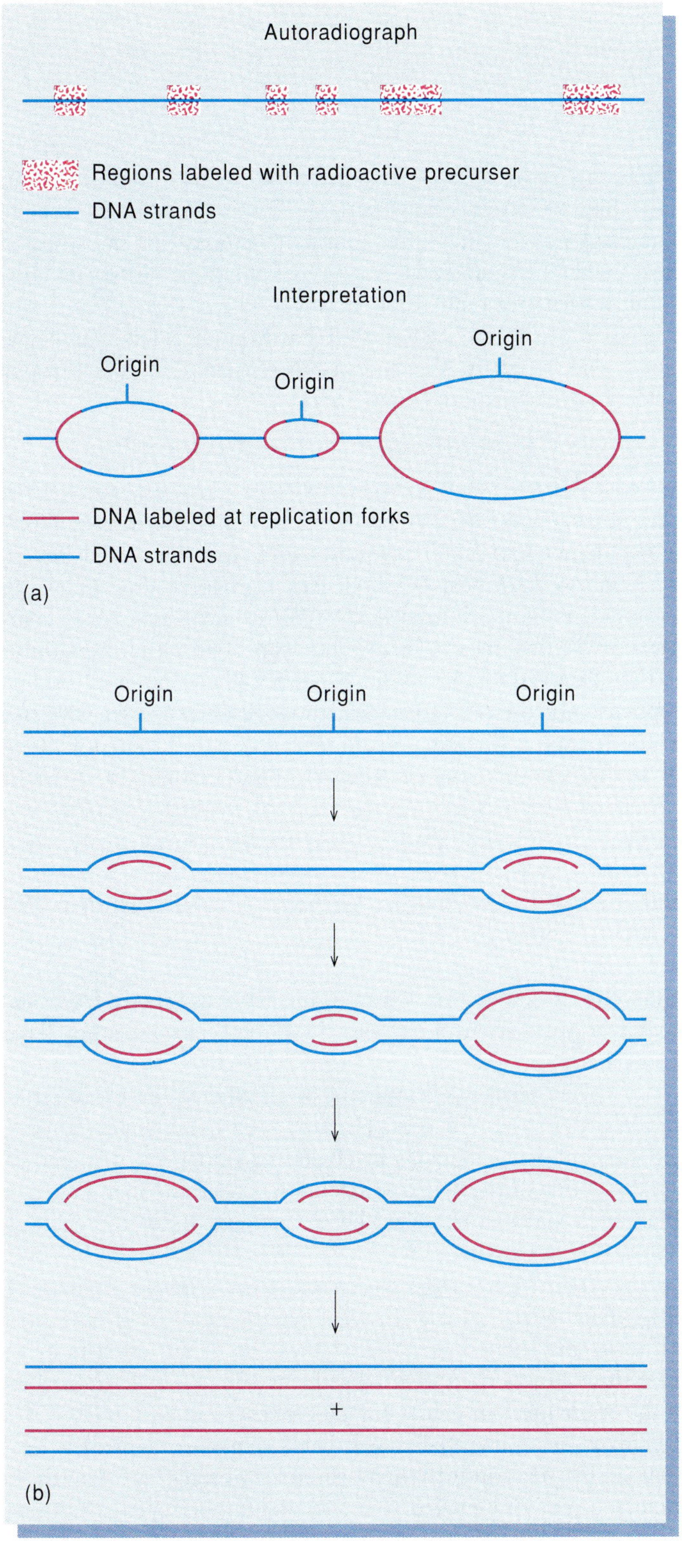

replication in yeast have been mapped and their effects have been characterized. The products of some of these genes are known. Most notably it is known that yeast DNA polymerase I, encoded by the CDC 17 gene, is similar to mammalian DNA polymerase α while yeast polymerase III, encoded by the CDC 2 gene, is most like the mammalian DNA polymerase δ.

Telomerase Facilitates Replication at the Ends of a Eukaryotic Chromosome

The linear structure of the eukaryotic chromosome creates a problem in replication as DNA chains always start from a primer. If an RNA primer was to form as a complement on the 3′ end of a DNA strand there would be no obvious way of replacing the primer with DNA. As a result, a segment of DNA from the ends of the chromosomal DNA would be lost every time it goes through a round of replication. This problem is overcome by a special DNA structure at the chromosome ends and a unique enzyme, DNA telomerase.

The ends of linear eukaryotic chromosomes are terminated in a loop or hairpin by a covalent linkage formed between the 3′ and 5′ ends of opposing chains. Short sequences tandemly repeated many times (telomeres) adjoin the chromosome ends. In humans the repeating sequence is TTAGGG at the 3′ ends and complementary sequences at the opposing 5′ ends. Prior to replication, the hairpin end is scissored and the protruding 3′ end is extended by a unique reverse transcriptase that is partly RNA and partly protein. The RNA of this enzyme associates via hydrogen bonds with complementary sequences on the 3′ ends of the DNA and serves as a template for extension of the 3′ end of the DNA. This 3′ extension of the DNA provides the repetitive sequences occurring in the mature chromosome as well as a region on which an RNA primer can be formed, which need not be replaced by DNA during the chromosome replication process.

Mitochondrial DNA Replicates Continuously on Both Strands

Mitochondria in mammalian cells contain circular, supercoiled DNA (mtDNA) about 5 μ in size ($M_r = 10 \times 10^6$; about 15,000 bp). It appears to be replicated by DNA polymerase γ, since this is the only DNA polymerase present in mammalian mitochondria. Furthermore, mtDNA replicates even in cells where DNA polymerase α is absent or inhibited by aphidicolin, a highly specific inhibitor of α polymerase (fig. 26.23).

The replication of animal mtDNA has been studied most extensively in mouse L cells (fig. 26.24), which have two origins of replication, one for each strand (referred to as H and L). DNA replication is initiated at the first site on the H strand (O_H) to form a displacement loop (D loop). It is a triple-stranded structure that includes the newly synthesized short daughter H strand. Replication continues unidirectionally until completion. When H strand synthesis is two-thirds complete, L strand synthesis is initiated (O_L) and elongated in the direction opposite to that of H strand synthesis. The daughter molecules segregate and synthesis proceeds to completion prior to closure with the

Figure 26.23

Structure of the DNA polymerase α inhibitor aphidicolin.

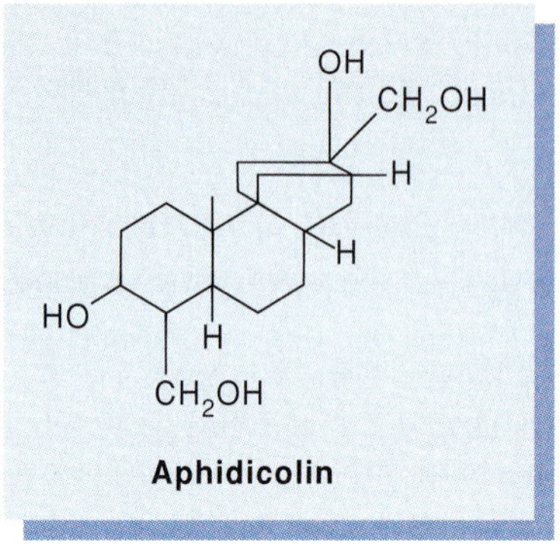

Figure 26.24

Replication model for mouse mitochondrial DNA. (Blue solid line, parental heavy (H) strands. Black solid lines, parental light (L) strands. Blue dashed lines, daughter H strands. Black dashed lines, daughter L strands.) The order of replication is clockwise, starting at D mtDNA. O_H and O_L are the origins of H- and L-strand synthesis, respectively. The double arrows reflect the metabolic instability of D-loop strands and consequent equilibrium between D mtDNA and C mtDNA. Expanded D-loop replicative intermediates are termed Exp-D prior to initiation of L-strand synthesis and Exp-D(l) after initiation of L-strand synthesis. The caret marks the interruption of at least one phosphodiester bond in the H strand of the daughter molecule. (ß Gpc = gapped circular daughter molecule.) Each replicative form is discussed in order in the text.

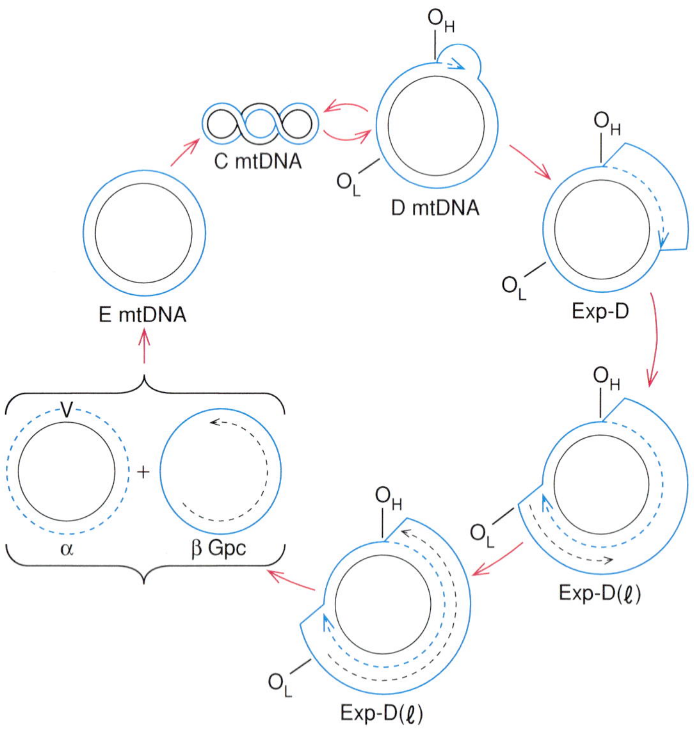

introduction of about 100 negative superhelical turns. The overall rate of mtDNA synthesis is about 270 nucleotides/min, about 0.5% the rate at which *E. coli* DNA replicates.

An interesting observation is that mammalian mtDNA contains a small percentage of ribonucleotides, which have been monitored by measuring susceptibility to alkali and to ribonuclease H. It has been found that mouse L-cell mtDNA has ribosubstitutions in the two replication origin regions. Their function is unknown.

Replication of Animal Viruses

There are two major reasons for studying animal viruses: (1) They are often serious pathogens and (2) as in the case of bacterial viruses, they offer many advantages for the study of host-cell DNA replication. Most types of animal viruses undergo semiconservative replication; however, two of them do not. Retroviruses are RNA viruses that go through a DNA intermediate, and hepatitis B virus is a DNA virus that goes through an RNA intermediate (table 26.4).

SV40 Is Most Similar to the Host in Its Mode of Replication

The viral genome of SV40 consists of a circular duplex DNA molecule of about 5,000 base pairs with one origin of DNA replication. As with most DNA viruses, replication takes place in the nucleus of the host cell. The SV40 viral genome is complexed with histones to form a nucleoprotein structure similar to that observed for chromatin. Since SV40 encodes only a single replication protein (T antigen), the virus makes extensive use of the cellular replication machinery. As a result there are likely to be many similarities between viral and cellular DNA replication. In both cases initiation of DNA synthesis involves two nascent strands. The leading strand grows continuously while the other, lagging strand, grows discontinuously by joining together small (about 200 bp) segments of DNA that are independently initiated with RNA primers. Completion of replication occurs when two oppositely moving forks meet. In linear cellular chromosomes the two merging forks originate from adjacent origins, while in circular SV40 chromosomes they have a single origin.

The recent development of an efficient cell-free replication system has greatly accelerated progress in understanding the molecular mechanisms involved. An important dividend of the dissection of the cell-free system has been the identification and functional characterization of components of the cellular replication apparatus.

The SV40 origin of replication is a 64-bp segment of viral genome (fig. 26.25). Careful genetic analysis of base substitution mutations has revealed that the origin consists of at least three functionally distinct sequence domains. At the center of the origin are four copies of a pentameric sequence (GAGGC) organized as an inverted repeat. This sequence element is recognized by the viral initiation protein, T antigen; the T antigen molecule binds to each of the four pentamer repeats. On one

Table 26.4
Animal Viruses Requiring DNA Synthesis

Family	Examples	Genome Size (kb)	Virion
1. dsDNA Viruses Papovavirus	Polyoma, SV40, wart viruses	5–8	Naked, icosahedral 
Adenovirus	Many human and animal adenoviruses	35–40	Naked, icosahedral 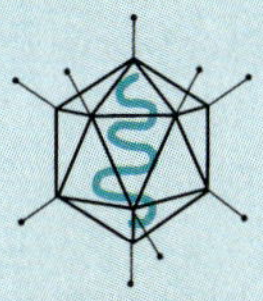
Herpesvirus	Herpes simplex I, II, varicella-zoster (chickenpox), Epstein-Barr virus	120–200	Enveloped, icosahedral 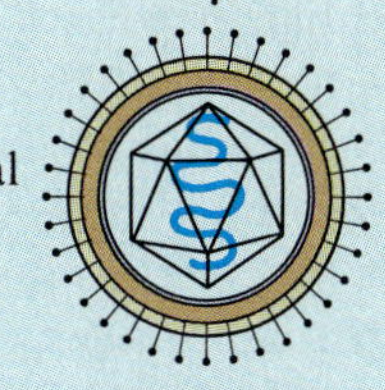
2. ssDNA Viruses Parvovirus	Adeno-associated virus, minute virus of mice, canine, feline, and human parvovirus	4–5	Naked, icosahedral 
3. RNA-DNA Viruses Retrovirus	Rous sarcoma, avian, feline, and murine leukemia, mouse mammary tumor virus, HTLV	7–10 (diploid)	Enveloped, icosahedral 
4. DNA-RNA Viruses Hepadnavirus	Hepatitis B virus of humans, birds, and rodents	3	Enveloped, icosahedral 

Figure 26.25

The SV40 origin of replication. The origin consists of at least three functionally distinct sequence domains. At the center of the origin are four copies of a pentameric sequence (GAAGC) organized as an inverted repeat. This sequence element is recognized by the viral initiation protein, T antigen. On one side of the T antigen-binding site is a 17-bp segment containing A-T base pairs. On the other side of the T antigen-binding site is a 15-bp imperfect palindrome of unknown function. All three sequence domains of the origin are required for SV40 DNA replication.

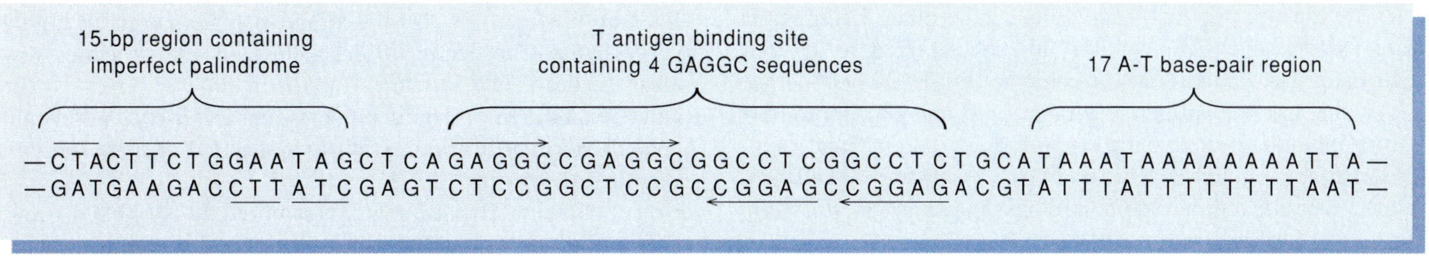

side of the T antigen-binding site is a 17-bp segment containing A-T base pairs. It is suspected that this is the initial site of strand opening during initiation of SV40 DNA replication. On the other side of the T antigen-binding site is a 15-bp imperfect palindrome of unknown function. All three sequence domains of the origin are required for SV40 DNA replication.

In addition to its specific binding activity, the δ2 kDa T antigen has an intrinsic helicase activity. Once it is bound to the origin, T antigen is capable of catalyzing the ATP-dependent unwinding of the two DNA strands. Unwinding appears to be a critical step that establishes the replication forks and generates the substrate that is required for the priming and elongation of nascent strands.

In addition to specific nucleotide sequence elements, the T antigen-mediated unwinding reaction requires accessory proteins contributed by the host cell. For example, a single-stranded DNA-binding protein is required to prevent reassociation of the single strands exposed during unwinding. It seems likely that this function is normally fulfilled by a cellular protein designated replication protein RP-A. This protein is absolutely required for SV40 DNA replication in the reconstituted cell-free system. The protein consists of three tightly associated subunits of 70 kDa, 32 kDa, and 14 kDa. The largest subunit binds specifically to single-stranded DNA. Heterologous single-stranded DNA binding proteins, such as *E. coli* SSB, will substitute for RP-A in the unwinding reaction; however, *E. coli* SSB cannot replace RP-A in the complete DNA replication reaction, a fact indicating that RP-A must play other roles in the replication process.

Of the four distinguishable nuclear DNA polymerase activities α, β, δ, and ϵ, it appears that α and δ are required for SV40 DNA replication. DNA polymerase α has long been considered to be the major replicative polymerase in animal cells. The enzyme is composed of four distinct subunits. The largest subunit (180 kDa) contains the polymerase active site while the smallest subunit (50 kDa) contains a primase capable of synthesizing short RNA transcripts that can serve as primers for subsequent DNA chain elongation by the catalytic subunit. The mammalian DNA polymerase α is not a highly processive enzyme, as fewer than 100 nucleotides are polymerized per binding event under the usual assay conditions.

There is evidence that DNA polymerase α can form a specific complex with the SV40 T antigen. Thus antibodies that bind to T antigen (or DNA polymerase α) will usually coprecipitate the two proteins from crude cell extracts. T antigen is also associated with a cellular function referred to as p53. Murine p53 has been shown to compete with DNA polymerase α for binding to T antigen (see table 34.3) and appears to be a strong inhibitor of the replication both *in vivo* and *in vitro.* These observations coupled with the finding that transformed cells frequently contain a mutated p53 protein suggest a regulatory role for p53 in host chromosomal DNA synthesis.

DNA polymerase δ has no primase activity. A 37-kDa protein called proliferating cell nuclear antigen (PCNA) greatly augments the activity of DNA polymerase δ with template/primers containing long single-stranded regions. In the presence

Figure 26.26

Hypothetical scheme for the concurrent replication of leading and lagging strands at the replication fork on the SV40 chromosome. The scheme follows the same general notion as the one proposed by Kornberg for *E. coli* chromosomal replication (see fig. 26.21). In this scheme polymerase δ functions on the leading strand and polymerase α functions in conjunction with primase on the lagging strand. The 37-kDa protein PCNA greatly augments the activity of polymerase δ but has no effect on polymerase α. (Source: B. Stillman, "Initiation of eukaryotic DNA replication *in vitro*" in *BioEssays* 9:56, 1988.)

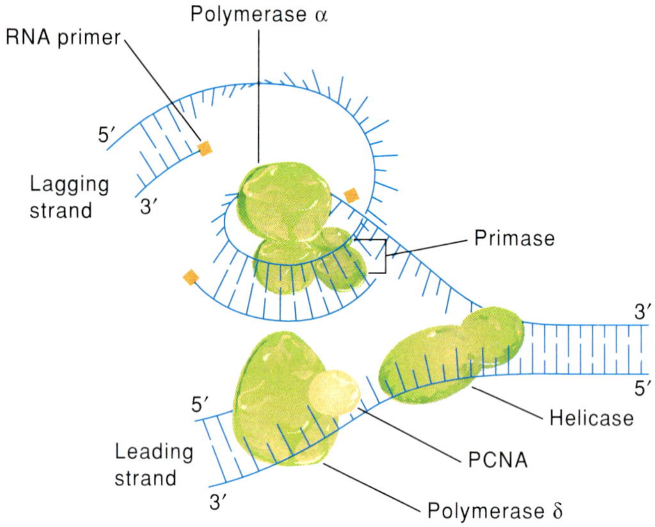

of PCNA, DNA polymerase δ is a highly processive enzyme capable of catalyzing the polymerization of at least 1,000 nucleotides per binding event. PCNA has no effect on the activity or processivity of DNA polymerase α. It has been demonstrated by direct reconstitution that PCNA is required for efficient SV40 DNA replication in cell-free systems and is probably involved in DNA chain elongation. In the absence of PCNA, initiation of DNA synthesis at the origin occurs, but only short nascent strands, containing a maximum of a few hundred nucleotides, are synthesized.

It has been suggested that DNA polymerase δ serves as the leading-strand polymerase and DNA polymerase α serves as the lagging-strand polymerase. Such a model is consistent with the known biochemical properties of the two enzymes. We would expect the leading-strand polymerase to be highly processive and to derive little benefit from an associated primase activity. The lagging-strand polymerase, on the other hand, would require only moderate processivity, but would benefit enormously from a tightly associated primase activity (fig. 26.26).

Topoisomerase activity is required for SV40 DNA replication. It has two distinct roles. One role is to act as a swivel to relieve superhelical tension that would otherwise hinder the unwinding of the parental strands as the replication forks advance. Either of the two known mammalian DNA topoisomerases, topoisomerase I or topoisomerase II, could provide this

Figure 26.27

(*a*) A model of synthesis of adenoviral DNA in which synthesis of a new strand displaces the parental strand. If the sequences of the inverted terminal repetition base-pair form a panhandle intermediate, then the double-stranded panhandle has the same terminal structure as parental viral DNA and will presumably be recognized by the enzyme complex responsible for initiation of viral DNA synthesis. Although not shown here, *r* strands displaced during the first step in synthesis could obviously form an analogous panhandle intermediate, with the same terminal double-strand sequence. (*b*) A similar displacement mechanism in which the possible role of the 5′ terminal protein in initiation of adenoviral DNA synthesis is illustrated. (From J. Tooze, *The Molecular Biology of DNA Tumor Viruses,* Cold Spring Harbor Press, New York, 1980. Reprinted with permission.)

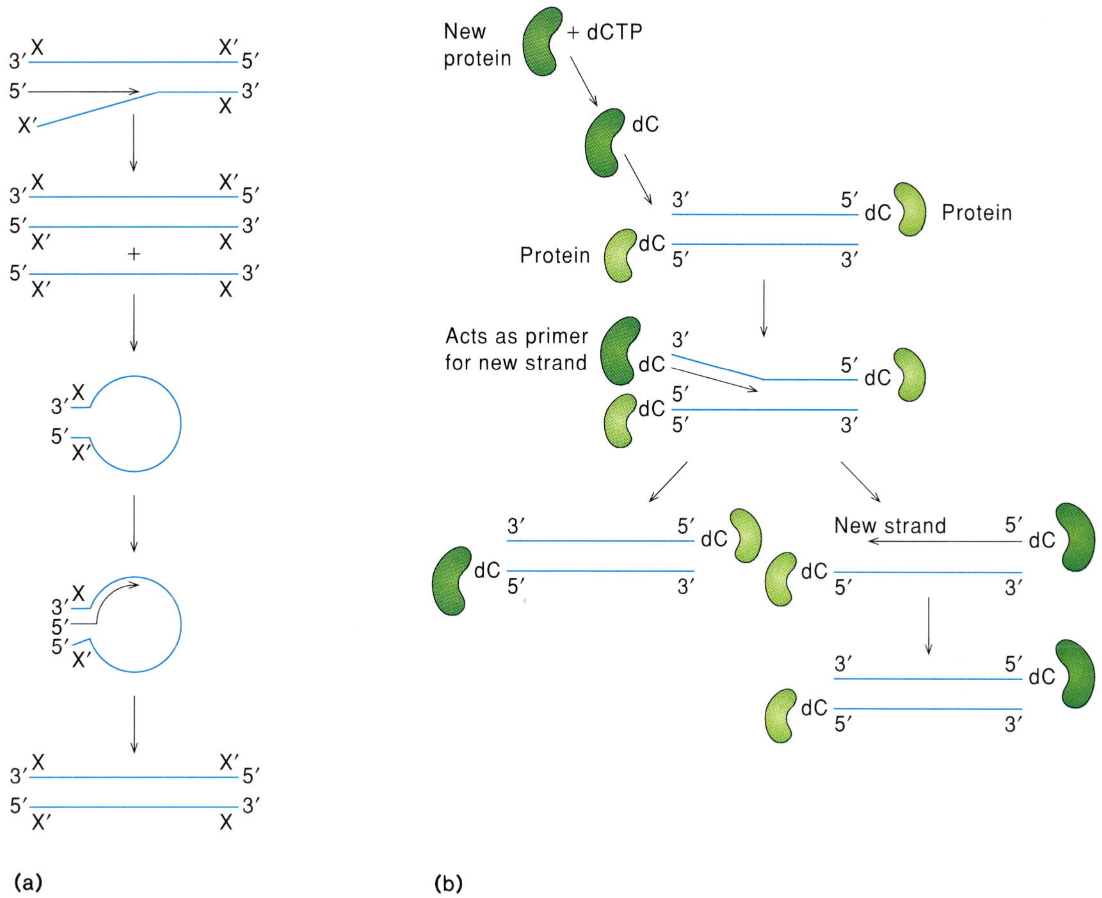

function. The second role for topoisomerase in SV40 DNA replication is to mediate the separation of the newly synthesized daughter duplexes at the completion of DNA replication. In general, the links between the parental strands are not completely removed prior to termination of SV40 DNA synthesis, so the immediate products of replication consist of two circular DNA molecules that are multiply intertwined. The final act of replication is the segregation of these intertwined daughter molecules into two separate unlinked molecules. This reaction is catalyzed by topoisomerase II. Genetic experiments in yeast indicate that the topoisomerase probably plays these same two roles during the replication and segregation of cellular chromosomes.

Adenovirus Uses a Protein Primer

Adenovirus DNA replication occurs by a process that is significantly different from chromosomal DNA replication. Replication initiates by a novel protein priming mechanism, and all daughter strands are elongated by a continuous mode of synthesis such as occurs at the leading strand of a chromosomal replication fork. The biochemical dissection of a soluble *in vitro* system capable of faithfully replicating adenovirus DNA has led to the identification of a number of the proteins involved. Adenovirus DNA replication requires the participation both of virus-encoded replication proteins and of host-cell-encoded factors.

The adenovirus genome encodes three proteins that play central roles in viral DNA replication: the 80-kDa terminal protein precursor (pTP), the 140-kDa adenovirus DNA polymerase (Ad pol), and a single-stranded DNA binding protein (DBP).

The genomes of the human adenoviruses are double-stranded linear DNA molecules containing approximately 35,000 bp. The 5′ terminus of each strand of the viral genome is covalently attached to a virus-encoded protein (TP) with a molecular weight of about 45,000. In addition, the nucleotide sequences at the extreme ends of the genome are identical, but in a reverse orientation (when adenovirus DNA is denatured, the single-stranded DNA can form a panhandle structure; see fig. 26.27). Both of these structural features play important roles in the initiation of viral DNA replication.

Synthesis of adenovirus DNA begins about 8 h after cellular infection by the virus and reaches a maximum about 10 h later. Initiation of DNA synthesis begins asynchronously at one end. DNA synthesis invariably requires both a primer and a template. In the case of adenovirus, the primer for DNA synthesis is not another nucleic acid chain but rather a serine β-hydroxyl group of the viral-encoded pTP protein. The initiation process begins with the formation of a phosphodiester linkage between the serine hydroxyl group of the pTP and dCTP. Formation of the pTP-dCMP complex requires factors present in uninfected cells as well as the 140-kDa Ad pol. The pTP protein active during initiation has a size of 80 kDa. During maturation this protein is cleaved by a virus-encoded protease to a protein of 55 kDa and this latter protein is the one found attached to the 5′ ends of the mature viral DNA.

In vivo studies indicate that adenovirus DNA replication takes place in two stages (see fig. 26.27). In the first stage, DNA synthesis is initiated at either terminus but does not occur simultaneously at the two DNA ends of the duplex viral genome by the protein priming mechanism. The initiation process results in formation of a replication fork that moves from one end of the genome to the other. At each replication fork only one of the two parental DNA strands serves as a template for DNA synthesis. Thus the products of the first stage of replication are a daughter duplex and a displaced single strand. In the second stage of DNA replication, the strand complementary to the displaced single strand is synthesized. The initial step in this process probably involves the circularization of the single-stranded template by annealing of its self-complementary termini. The resulting duplex panhandle has the same structure as the terminus of the duplex adenovirus genome and is presumably recognized by the same initiation machinery that operates in the first stage of replication. Following a second initiation event, complementary strand synthesis proceeds from one end of the template to the other, generating a second daughter duplex. In both stages of adenovirus DNA replication there is only one priming event per nascent daughter strand, so all viral strands are synthesized in a continuous fashion from their 5′ termini to their 3′ termini.

Two other animal viruses, polio virus and encephalomyocarditis virus, have proteins attached to their terminal nucleotide that are probably involved in a protein-priming mechanism of replication initiation.

Herpes Virus Encodes Many of the Proteins Required for the Replication Process

In contrast to SV40, herpes simplex virus (HSV), with its much larger genome, encodes many, if not all, of the proteins that are involved in its DNA replication. The complete set of viral genes necessary has been identified. Several of these purified proteins have functions expected of replication proteins, including a DNA polymerase, a helicase, a primase, a single-stranded DNA-binding protein, and an origin recognition protein.

Bovine Papilloma Virus Replicates Once per Cell Division

Adenovirus, SV40, and HSV are all examples of viruses that normally multiply by productive cytocidal infection. In all of these cases, viral DNA replication begins soon after infection and continues at a high rate until the death of the host cell. In contrast, bovine papilloma virus (BPV) represents an example of a virus that is capable of multiplying as a stable extrachromosomal element. Under normal circumstances the host is not killed. As in the case of SV40, the BPV genome is relatively small and encodes only a small number of proteins involved in DNA replication; viral DNA synthesis strongly depends on host-cell replication proteins. Genetic analysis has provided evidence for a negative control system that apparently ensures that each viral genome is replicated once and only once during each cell cycle. This result suggests that bovine papilloma virus represents an excellent model system for investigating the mechanisms involved in regulating DNA replication.

Retroviruses Are RNA Viruses that Replicate through a DNA Intermediate

RNA viruses that cause tumors (oncogenic RNA viruses) contain an unusual DNA polymerase called reverse transcriptase. This enzyme carries out template-directed polymerization and shares many features with the cellular DNA polymerization reaction. The main difference is that RNA is the preferred template for reverse transcriptase, which catalyzes the polymerization of deoxynucleoside triphosphates only (fig. 26.28). If a primer is added, for example poly(dT), which will base-pair with the poly(A) at the 3′ end of mRNA, then an RNA-DNA hybrid is the initial product of the reaction. A second round of DNA synthesis on the hybrid produces a double-stranded DNA. In this manner, viral RNA sequences can be transcribed into double-stranded DNA sequences called cDNA. The duplex DNA copy (cDNA) of the viral RNA is subsequently integrated into the host chromosome, where it functions in the synthesis of further viral RNA, often causing malignant transformation of the cells. In the laboratory, reverse transcriptase is used extensively for conversion of mRNA species into DNA sequences (cDNA) for cloning and sequencing studies (see chapter 27).

Interestingly, reverse transcriptase has now been demonstrated in bacteria; however, its role there is unknown. G. Fink has made an interesting proposal for one possible role of the reverse transcriptase encoded by the Ty transposable elements in *S. cerevisiae,* namely, that the enzyme could reverse-transcribe mRNA to generate cDNA *in vivo.* Then by homologous recombination, known to occur in yeast, the cDNA would integrate into the genome, replacing any introns at the site of recombination. This proposal could explain the rare presence of introns in the *S. cerevisiae* genome.

Hepatitis B Virus Is a DNA Virus that Replicates through an RNA Intermediate

Hepatitis B is an enveloped virus with a small DNA genome. The genome is circular, but both of its DNA strands are linear. A gap in one strand is bridged by an incomplete complementary strand. The longer strand has a protein bound at its 5′ end. A DNA polymerase present in the mature virus particle can elongate the 3′ ends of the incomplete strands. The unique feature of this virus is that replication involves an RNA intermediate.

Figure 26.28

Synthesis of viral DNA from the viral RNA genome, as shown in a model of the generation of double-stranded DNA carrying two copies of a long terminal repeat (LTR = 3′ ■ 5′). The sequence X is a marker for the plus strand: the complementary sequence X′ occurs on the minus strand. (*a-c*) Synthesis of minus-strand DNA from the genomic RNA template using a tRNA primer (represented by ⌡ in step (*a*). This initial DNA synthesis of minus strand reaches a pause site at the end of the RNA template as shown in (*c*). (*d*) Degradation of the RNA portion of the resulting DNA-RNA hybrid by RNase H partially exposes the newly synthesized DNA. (*e*) Bridge formation between the newly synthesized segment of minus strand and repeated sequences at the 3′ end of genomic RNA permits continued synthesis of the minus-strand DNA. This synthesis of minus-strand DNA continues until it reaches the end of the RNA template. (*f*) At the same time, DNA synthesis of the plus-strand DNA begins in the region of the LTR on the minus-strand DNA using the plus-strand RNA as a primer. (*g*) This synthesis pauses after 300 bases, that is, when it reaches the end of the template. The tRNA primer is degraded by RNase H (*h*). Continued synthesis of plus-strand DNA requires bridge formation between sequences repeated in a completed minus-strand DNA and in the 300-nucleotide fragment. (*i*) Completion of synthesis and tidying up results in a duplex with copies of the LTR at both ends. The DNA duplex is now ready for the next series of reactions, which results in the integration of the viral DNA sequences into the host genome.

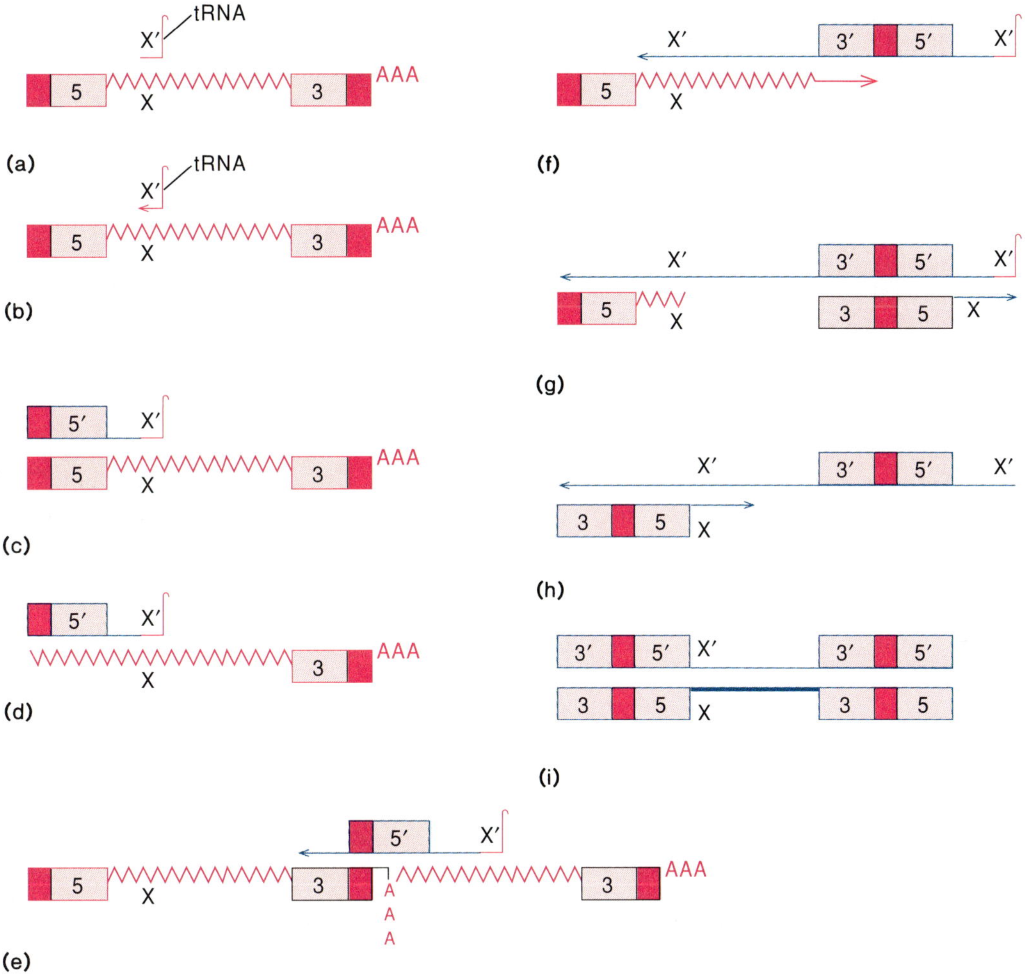

Cellular Systems for DNA Repair

We have seen that chromosomes are usually formed by a single DNA molecule regardless of their size. Such a large molecule makes a very easy target to damage and, in fact, limited damage occurs quite frequently to these large intricate polymers. A single lesion in the DNA molecule could prevent the whole chromosome from replicating and could thus cause the death of the cell. Clearly, it is highly advantageous for a living system to repair the damage. The proofreading functions of the polymerizing enzymes have already been described as a means of correcting some errors (mainly mismatches) made during synthesis. For correcting damage to existing DNA, more elaborate systems exist. DNA repair systems have received increasing attention as it has become clear that most types of repairable damage to DNA are both mutagenic and carcinogenic.

DNA damage is caused by a variety of physical, chemical, and biological agents. Physical agents include UV and ionizing radiation. Damage can be in the form of a missing, incorrect, or modified base or an alteration in the structural integrity of the DNA strands by breaks, cross-links, or dimerization of bases, usually pyrimidines.

Some chemicals that modify the bases in nucleic acids are very specific for one base. Hydroxylamine (NH_2OH) forms a specific adduct with cytosine that can then base-pair with adenine. Nitrous acid, however, is a general reagent used to deaminate bases, converting adenine to hypoxanthine, guanine to xanthine, and cytosine to uracil. Other compounds that cause DNA damage alkylate the bases or form other covalent adducts. Nitrosamine derivatives usually must be converted to more reactive species by biological oxidation before becoming active as base-alkylating reagents. The formation of alkyl adducts may alter the pairing characteristics of the bases. *N*-methyl-*N*-nitrosoguanidine is a particularly potent nitrosamine that has found widespread use in the laboratory for generating mutations in bacteria. It acts with greater efficiency at the growing point during DNA synthesis.

Aromatic polycyclic hydrocarbons, such as 2-acetamidofluorene, benzo(a)pyrene, and the mycotoxin known as aflatoxin B, form adducts with nucleic acids. Most of these compounds are inactive by themselves but can be converted to highly reactive derivatives within organisms. The reactive derivatives are nonselective and can modify most nucleic acid bases. The bulky groups that are introduced by the adducts apparently act by preventing base pairing rather than by causing errors of base pairing, such as those seen with the simpler alkylating agents.

Adjacent pyrimidine bases in a DNA strand form dimers with high efficiency after absorption of ultraviolet light (fig. 26.29). By contrast, purines are quite resistant to damage by ultraviolet. Pyrimidine dimers formed within an otherwise intact DNA duplex have provided a useful substrate to assay for DNA repair. These dimers can be repaired directly by enzymatic photoreactivation (fig. 26.30). The photoreactivation enzyme binds to the DNA containing the pyrimidine dimer and uses visible light to cleave the dimer without breaking any phosphodiester bonds.

Pyrimidine dimers and other forms of DNA damage can also be removed by a general excision repair mechanism. In this form of repair the lesion in the DNA is removed and repaired. The well-defined genetic system of *E. coli* has promoted an understanding of the enzymatic steps in excision repair, particularly in response to UV-induced damage. The genes affecting DNA excision repair fall into three categories: those exclusively involved in repair processes, those involved in both repair and recombination processes, and those also involved in other aspects of DNA replication (table 26.5).

The removal of ultraviolet-induced pyrimidine dimers in bacterial cells is catalyzed by two mechanisms (fig. 26.31). Special glycosylases that recognize abnormal or incorrectly paired bases can cleave N-glycosidic bonds to generate an apurinic or apyrimidinic (AP) site (fig. 26.31). Alternatively, a double incision is made in the strand that carries the lesion by the enzymes of the *uvrA, uvrB,* and *uvrC* system (fig. 26.31*b*). In the case of pyrimidine dimers, one incision is made seven nucleotides 5′ to the pyrimidine dimer, and then a second cut is made three or four nucleotides 3′ to the same dimer. The *uvrD* gene product (shown to be helicase II), together with DNA polymerase I and possibly a single-strand DNA-binding protein, releases the 12–13-nucleotide oligomer generated by the incision of the uvrABC complex. After the release of the oligomer, DNA polymerase I and ligase are required to resynthesize the excised region and ligate the nicks.

Figure 26.29

Structure of a thymine dimer formed in DNA by exposure to short-wavelength ultraviolet light.

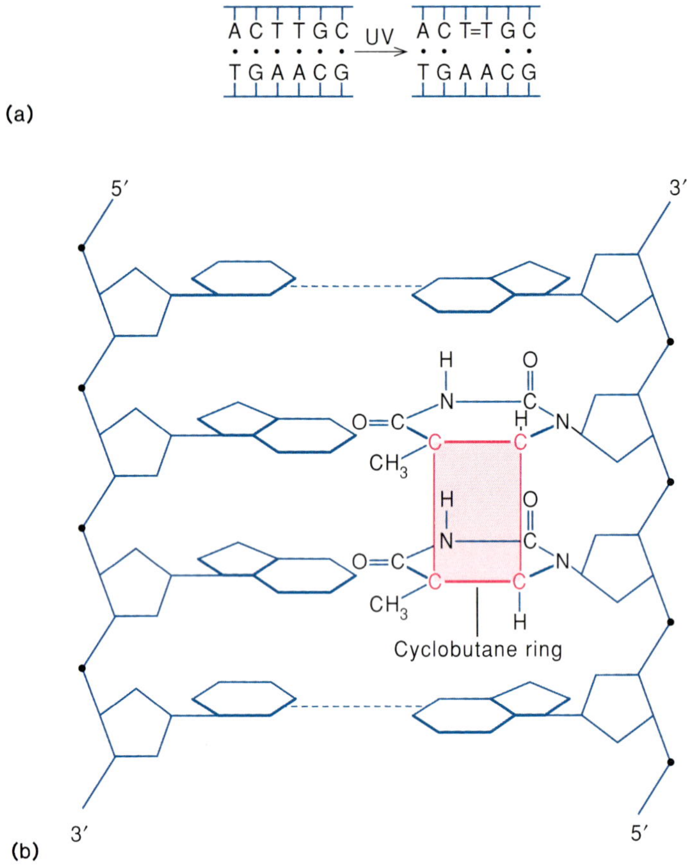

Figure 26.30

Thymine dimers may be monomerized from DNA by enzymatic photoreactivation. In this case no nucleotides are removed in the repair reaction.

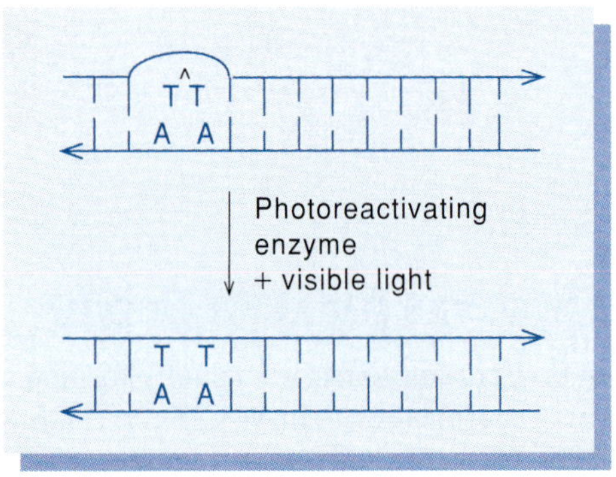

Table 26.5
Some *E. coli* Genes that Affect Responses to DNA Damage

Gene	Map Location	Function
uvr Genes		
uvrA	91	Gene products work together to make initial incision at or near the site of DNA damage.
uvrB	17	
uvrC	42	
uvrD		Helicase
rec Genes		
recA	58	Structural gene for recA protein. *RecA* carries a highly specific protease that is activated by DNA damage.
recB	60	Encodes two subunits of exonuclease V. Believed to make the initial nick required to initiate DNA recombination.
recC	60	
Other Genes		
lexA	90	Controlling gene for *recA* and other SOS functions
polA	85	Structural gene for DNA polymerase I
polC	4	Structural gene for DNA polymerase III (same as *dnaE*)
lig	51	Structural gene for DNA ligase

Figure 26.31

Pyrimidine dimers and other forms of DNA damage can be removed by a general excision repair mechanism. The first reaction in this form of repair involves forming nicks about the damaged region of the DNA. In (*a*) we see the mode of incision of UV-irradiated DNA by the pyrimidine-dimer-specific glycosylase and AP endonuclease activities of *M. luteus* and bacteriophage T4. In (*b*) we see the mode of incision of the uvrABC endonuclease of *E. coli*. (Source: G. Walker, "Inducible DNA repair systems" in *Annual Review of Biochemistry*, 54:425, 1985. Copyright © 1985 Annual Reviews Inc., Palo Alto, Calif.)

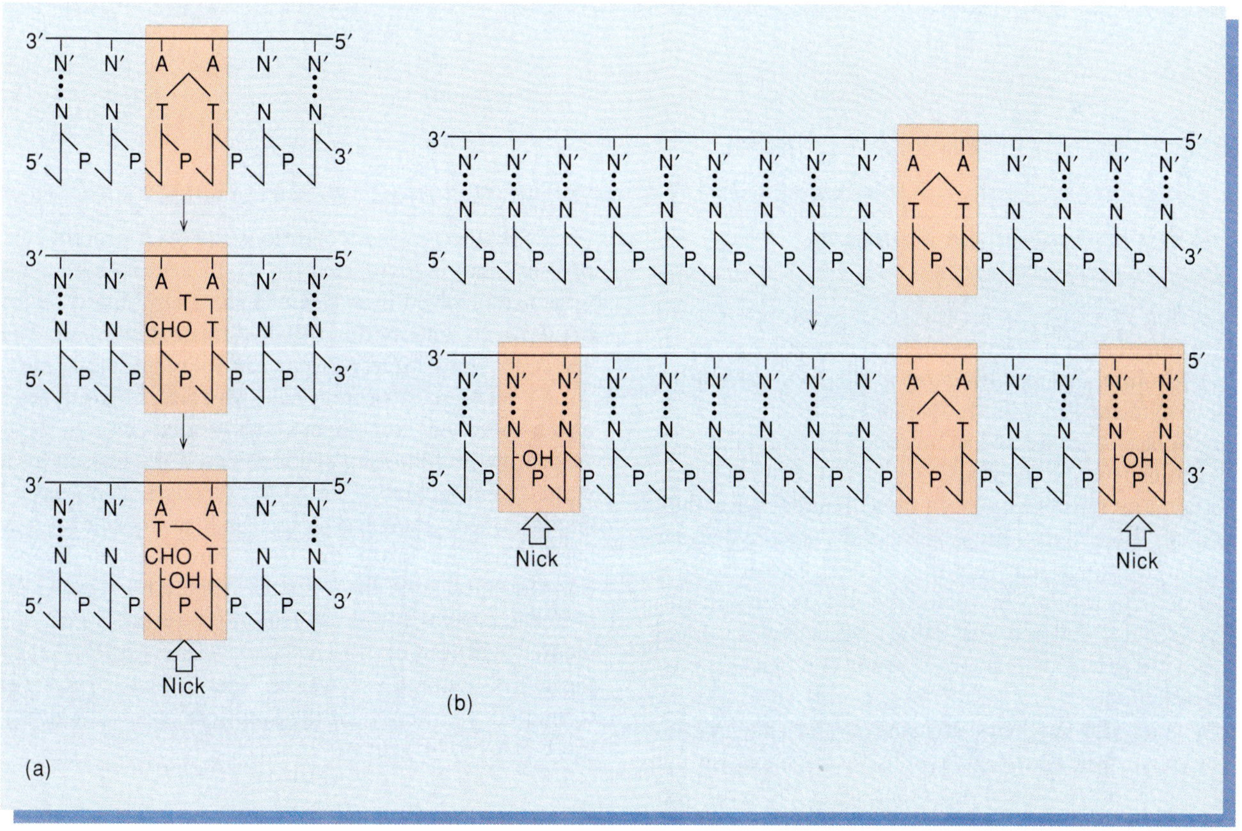

Figure 26.32

Model for the SOS regulatory system. In normally growing cells the SOS functions associated with DNA repair are not expressed. This is because lexA repressor inhibits their transcription. LexA repressor also inhibits its own expression and that of recA. SOS functions are turned on by a series of reactions that starts with DNA damage. DNA damage results in an inducing signal that activates the protease function of recA. This protease cleaves lexA protein, so that all genes that were formerly inhibited by lexA can be expressed. Once the damage is repaired, the level of lexA repressor builds up again and the SOS genes return to their usual repressed state.

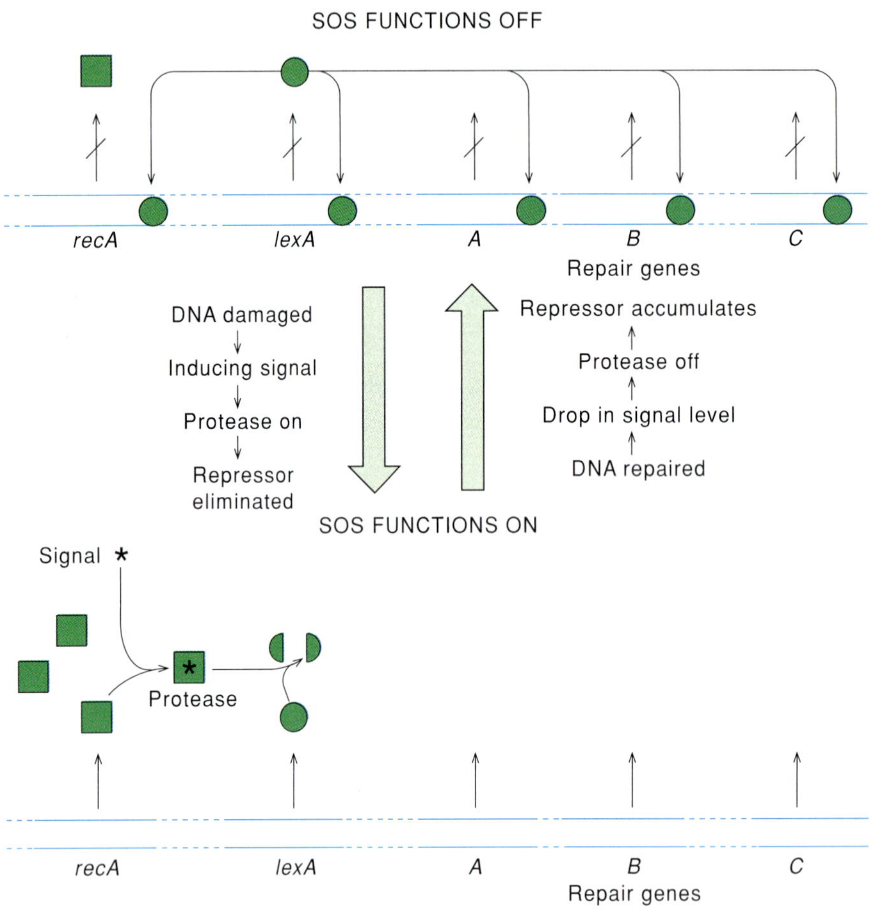

A Special Glycohydrolase Removes Uracil Residues from DNA

Uracil appearing in DNA, as a result of misincorporation of dUTP or deamination of cytosine, can be removed by a uracil-DNA glycohydrolase. The resulting gap is filled with a cytosine residue. Since cytosines deaminate spontaneously at a low rate, this glycohydrolase is considered to be quite important over the course of evolutionary time for preserving the G-C pairs in DNA. This may be a major reason why DNA uses thymine rather than uracil. Both of these pyrimidines have the same hydrogen-bonding potential, but an enzyme that is designed to remove misplaced uracils in the DNA would have less trouble distinguishing between a uracil and a thymine than between a uracil that was correctly paired with an adenine and one that was mispaired with a guanine.

In *E. coli* the synthesis of many of the enzymes involved in the repair process is regulated by the SOS system. At the heart of the SOS system are two genes, *lexA* and *recA* (fig. 26.32). Under normal conditions the lexA protein acts as a repressor binding to approximately 17 genes whose encoded proteins are involved in excision repair, gap repair, double-strand break repair, and methyl-directed mismatch repair. These genes are collectively referred to as *din* (damage-inducible) genes.

The recA protein has two roles. It catalyzes recombination between homologous DNA molecules in *E. coli* (discussed in the following section) and regulates, in an activated state, the induction of the SOS responses. It does the latter by mediating the cleavage of the lexA protein at an—AlaGly—bond. Thus an insult to DNA that leads to DNA damage somehow activates the protease function of recA. When the protease cleaves the lexA protein, it results in a greatly augmented synthesis of proteins associated with DNA repair. Once the DNA damage is removed, the activated recA function is turned off and newly synthesized lexA protein again represses the DNA repair genes.

In humans, the inability to repair DNA damage is associated with rare genetic syndromes. The best known is xeroderma pigmentosum. There are several variations of this condition. People with the disease are unable to repair the DNA damage caused by exposure to ultraviolet light and some chemicals. Individuals with different maladies associated with defective DNA repair mechanisms show different sensitivities to damaging agents. This fact suggests that there are several enzyme systems for DNA repair in humans. Hypersensitivity to DNA-damaging agents may be caused by a defect in a single enzyme in any one of the different pathways used for DNA repair. The multiplicity of defects in repair in different patients has been confirmed by complementation studies carried out with cells grown in culture; there are about six to eight complementation groups, suggesting at least an equal number of loci that encode enzymes involved in repair of DNA damage in human cells.

The Three Types of DNA Recombination

DNA recombination is the process of rearranging DNA segments or chromosomes. In nature, DNA recombination is very widespread; it serves different functions and it occurs in many different ways. There are three basically different types of recombination. The first and most common is called general recombination. This form of recombination takes place between identical or nearly identical segments of chromosomes such as homologously paired eukaryotic chromosomes during meiosis (box 26A), or bacteriophage λ chromosomes during multiplication. The second type of recombination, site-specific recombination, is limited to highly select regions of the genome and to very specific functions. One of the best-understood examples of site-specific recombination is the integration of bacteriophage λ DNA into the *E. coli* host chromosome. The third type of recombination, nonspecific recombination, occurs between nonhomologous regions of chromosomes, requiring little or no site specificity. The main function of this type of recombination might be to bring about a controlled level of mutation.

The details of recombination are fascinating, but they are more appropriately dealt with in a genetics text or a molecular biology text. We will limit ourselves in this section to a few comments about general recombination.

In the 1960s Meselson proposed a three-step model for general recombination: (1) formation of staggered breaks in two parental DNAs; (2) base pairing between single-stranded regions of the two parental types; and (3) repair synthesis (fig. 26.33). Meselson's model explained many experimental results. In fact, it went beyond that. The proposal that base pairing takes place between single-stranded regions of the different parental DNAs provided an explanation for why chromosomes tend to recombine in homologous regions.

The main approach to the enzymology of recombination has been to correlate mutations that affect recombination with actual enzyme activities. The ultimate goal of these studies is to reconstitute a cell-free system with purified enzymes that can carry out recombination. Most of this work with bacteria has taken place in *E. coli*. A considerable amount of work is now being carried out with the yeast *S. cerevisiae*.

Figure 26.33

Meselson model for phage recombination (circa 1964). Phage form staggered breaks. Base pairing leads to an annealed complex with gaps. Gaps are mended by repair synthesis.

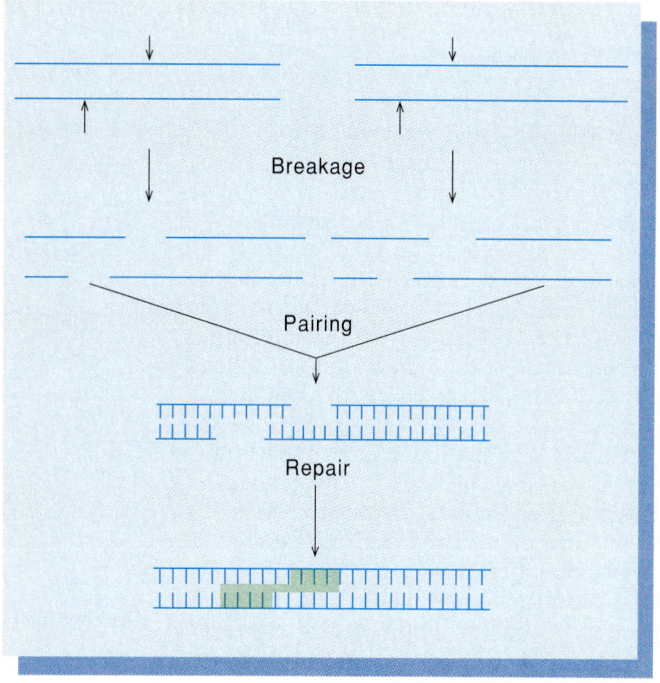

Considering the proposed steps in the recombination process, it seems likely that several activities would be required. These include (1) an endonuclease activity to make the initial incision or incisions in the DNA, (2) an activity to catalyze single-strand invasion of a duplex structure and subsequent unwinding and rewinding of base-paired structures, (3) enzymes for repair, synthesis, and ligation of DNA, (4) an enzyme that removes mismatched bases, and (5) another endonuclease to make scissions that resolve the complex into separate chromosomes.

The search for enzyme activity associated with recombination has been based on analysis of organisms carrying mutations that affect general recombination. Such mutants, known as *rec* mutants, were first discovered more than fifteen years ago, and new ones are still being found. *E. coli rec* mutations, *recA, recB, recC,* and *recD* reduce recombination efficiency to different extents. The *recA* mutants are totally deficient in homologous recombination, while *recB* or *recC* mutants can recombine only about 10^{-4} times as well as wild-type cells. In addition, mutations in two other genes will reverse the defect of *recB* or *recC* mutants; no comparable second-site suppressors of *recA* have been found. This lack suggests that *recA* is absolutely required for homologous recombination but that there may be alternate pathways to recombination that do not require *recB* and *recC*.

26A
BOX

Recombination between Homologous Chromosomes Occurs during Meiosis

Most eukaryotic cells contain two sets of homologous chromosomes, each originating from one of the parents. Before meiosis begins there is a duplication of chromosome mass as in premitosis (see fig. 1.21). Meiosis is divided into two stages called meiosis I and meiosis II (figure 1*a* and *b*). Each cell entering meiosis ultimately gives rise to four cells. Meiosis I involves the separation of homologous chromosomes of the parent diploid cell and the consequent halving in the number of chromosomes per cell. Meiosis II involves an equational separation of identical sister chromatids. During meiosis I homologous chromosomes pair, and frequently there is recombination between them so that they exchange homologous regions (figure 1*c*). Since most genes are represented by two or more alleles, the recombination event gives rise to different combinations of alleles, which are destined for different germ cells.

Recombination is extremely important to the species because it permits the testing of different combinations of alleles, combinations that may prove to be more or less beneficial to the organism than the original ones. Meiotic recombination has also been used by geneticists for gene mapping. Classical mapping of genes is based on the principle that the recombination frequency between alleles of different genes is a function of the distance between them. Thus the physical chromosome length is related to the "genetic length" measured by recombination. The unit of genetic length, defined as the length of chromosome over which there occurs a 1% recombination frequency for a given mating pair, is called a centimorgan (cM).

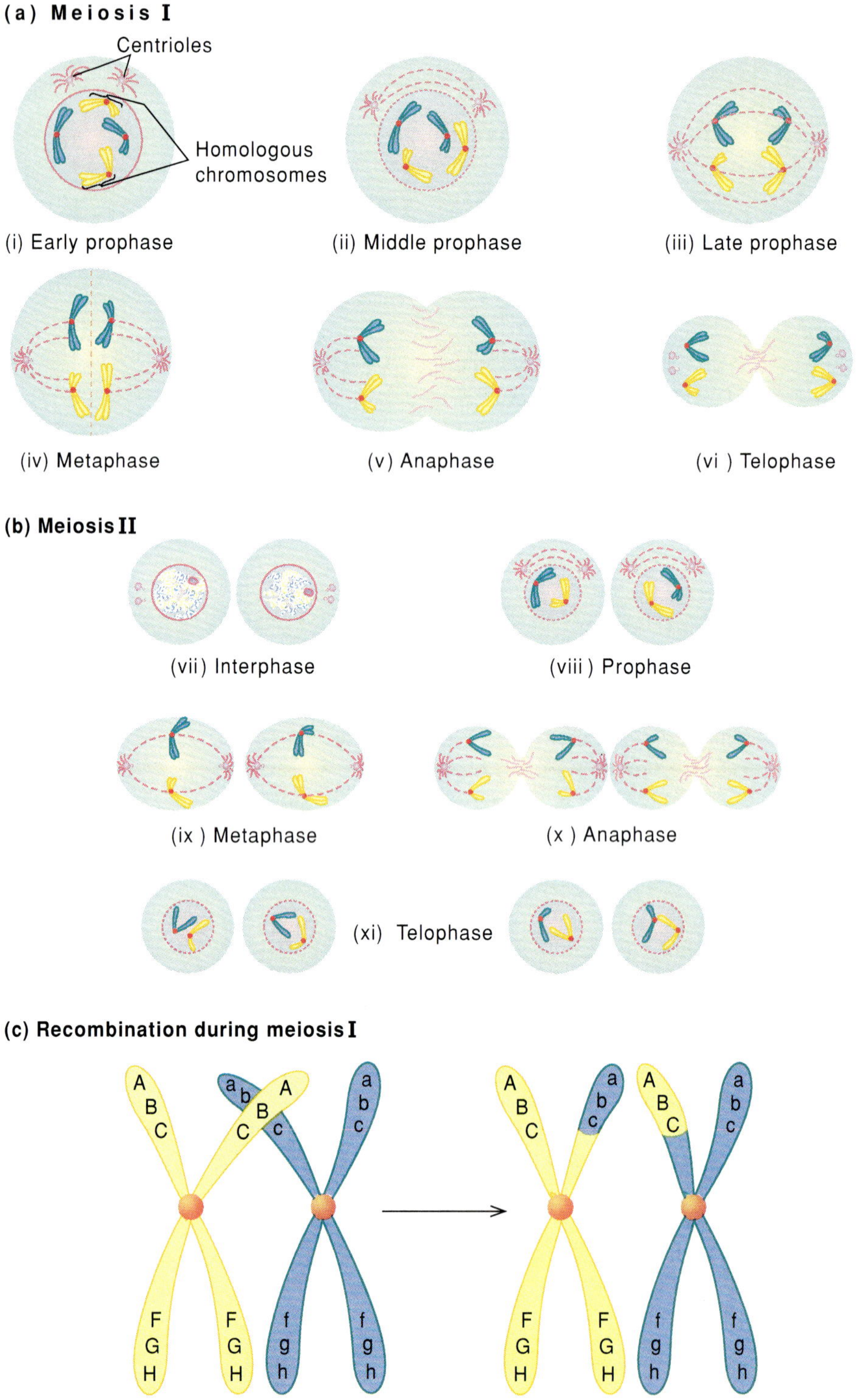

Figure 1

Meiosis. The process of meiosis involves two cell divisions with two segregation cycles, meiosis I and meiosis II. These cycles are pictured for a hypothetical cell containing four chromosomes. (*a*) During the first cycle homologous chromosomes pair and then segregate. (*b*) During the second cycle the sister chromatids from each chromosome segregate. The second cycle is very much like mitosis (see fig. 1.21). Each diploid cell that enters meiosis ultimately yields four haploid cells. These haploid cells are the sex cells for the next round of mating. (*c*) Recombination between homologous chromosomes during meiosis I leads to different combinations of alleles. In this illustration alleles for the same gene are represented by corresponding capital or lower-case letters.

Figure 26.34

Reactions catalyzed by purified recA protein *in vitro*. RecA catalyzes a number of different reactions between DNA strands, all of them involving the unwinding and winding of base-paired structures. (*a*) D-loop formation by interaction between supercoiled circular duplex DNA and single-stranded DNA. (*b*) Partial circular duplex formation by displacement of one strand from a duplex structure. (*c*) Strand exchange between a gapped circular duplex structure and a linear duplex structure. (*d*) Complex formation between two helices, one of which is gapped.

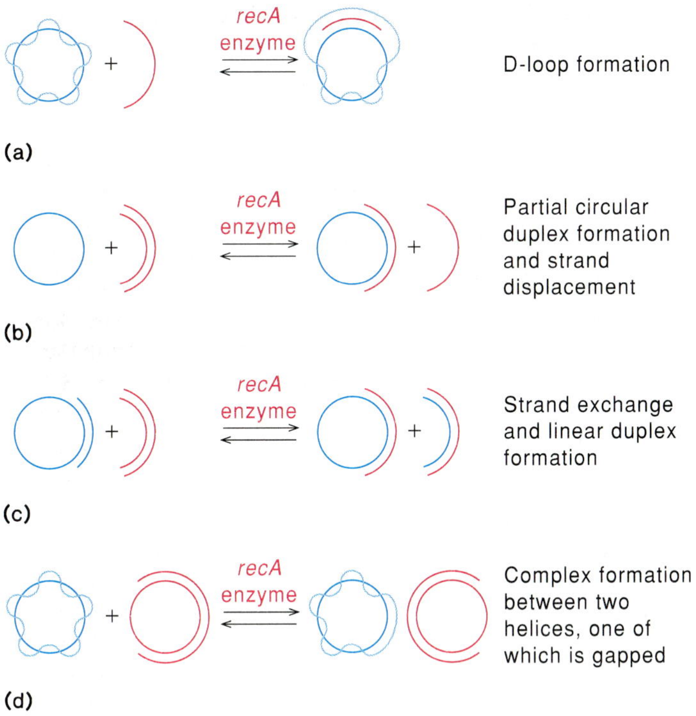

We have discussed the repair role played by the protease function of *recA* (see fig. 26.31). The second class of reactions catalyzed by recA protein indicates a prominent role in the initiation of homologous pairing during genetic recombination. Thus the purified recA protein catalyzes various forms of complex formation exchange between duplex and single-stranded DNAs in reactions that entail the cleavage of ATP (fig. 26.34). D-loop formation can result when the recipient is a circular duplex and the donor is a single strand (see fig. 26.34*a*). Partial circular duplex formation can result from the exchange between a circular single strand and a linear duplex (see fig. 26.34*b*). Strand exchange can occur where a single-stranded fragment is removed in the presence of a double-stranded fragment (see fig. 26.34*c*). Finally, recA protein will also cause a complex to form between two circular helices, provided that at least one input helix is gapped on one strand (see fig. 26.34*d*). Interestingly, while the two helices must contain homologous sequences in order to form a four-stranded structure, the gap may lie in a nonhomologous region. The fact that mutations in *recA* completely eliminate recombination, as well as the finding that the recA-related protein catalyzes strand exchange reactions like those required in all three of the recombination models we have been discussing, makes it highly likely that recA plays a key role in these processes *in vivo*. If this is so, then recA protein is clearly situated at the hub of activities in the recombination complex.

Figure 26.35

DNA unwinding by exoV enzymes. This is best observed *in vitro* in the presence of ATP and Ca^{2+}. The Ca^{2+} inhibits the exonucleolytic activity of the enzyme. The ATP provides the energy that drives the enzyme in a concerted manner into the duplex structure. The duplex unwinds in front of the path of the enzyme and rewinds in back of the enzyme.

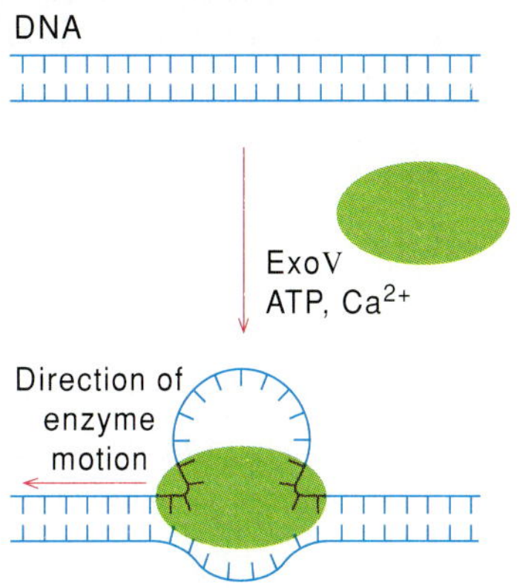

The recA enzyme will not catalyze any recombination event unless at least one free end of a single strand or a DNA duplex is available. Thus it seems unlikely that the recA protein could be involved in making the initial incision(s) in the duplex structure that are required to initiate recombination. There are reasons for believing that the *recB* and *recC* genes are involved in this capacity. First, a mutation in either of these genes reduces recombination by a factor of about 10^{-4}. Second, these two genes together with *recD* encode the subunits of a 330 kDa enzyme exonuclease V (exoV). Surprisingly *recD* mutants, which result in an exoV complex lacking all known enzyme activities, possess an increased rate of recombination *in vivo*. Normally exoV is both a helicase and a nuclease. The helicase activity is best demonstrated *in vitro* under conditions where the nuclease activity is blocked; this can be done by adding Ca^{2+}. When linear duplex DNA is incubated with exoV in the presence of ATP and Ca^{2+}, it begins to unwind at one end. As the unwinding progresses, the single-stranded regions collapse back on each other to re-form a duplex. A double-loop structure is maintained in the vicinity of the migrating enzyme (fig. 26.35).

What is the point of this migrating behavior of exoV? The exoV enzyme may scan the duplex structure to find a site where it would be appropriate to make the strand scission required to initiate recombination. These indications come from detection of an octanucleotide sequence known as *chi*, 5′-G-C-T-G-G-T-G-G-3′, which is associated with regions highly active for recombination. The *chi* sequence does not stimulate recombination in special *recBC* mutants where exoV retains some recombination proficiency but lacks detectable nuclease activity. This observation has led to the suggestion that exoV normally scans the duplex and makes strand incisions in or near regions containing the *chi* sequence.

The recombination process triggered by the initial scissions made by the recBCD complex and mediated by the recA enzyme are not sufficient for recombination between interacting chromosomes. A third enzyme, resolvase, is required for efficient separation of the recombining chromosomes in the recA-chromosome complex. This protein is encoded by the *ruvC* gene in *E. coli,* and is absolutely required for homologous recombination in the bacterium. Its effectiveness has also been demonstrated in a cell-free *in vitro* system.

Incidentally the fact that the recA enzyme carries two enzyme activities, one for homologous recombination and one for regulating the concentration of DNA repair genes including itself (the recA protease, which cleaves the lexA repressor), leads to the suspicion that recA is also more directly involved in DNA repair. Recombination between two homologous chromosomes badly damaged in different regions could lead to one normal chromosome. Although, technically a haploid organism, *E. coli* carries two copies of the bacterial chromosome for a considerable period of time before cell division. Thus the necessary identical pairs of chromosomes are present for homologous recombination.

Summary

This chapter deals with reactions involved in DNA synthesis, degradation, repair, and recombination. The chief points to remember are as follows.

1. DNA replication proceeds by the synthesis of one new strand on each of the parental strands. This mode of replication is called semiconservative and it appears to be universal. DNA synthesis initiates from a primer at a unique point on a prokaryotic template such as the *E. coli* chromosome. From the initiation point, DNA synthesis proceeds bidirectionally on the circular bacterial chromosome. The bidirectional mode of synthesis is not followed by all chromosomes. For some chromosomes, usually small in size, replication is unidirectional.
2. In eukaryotic systems, replication can start at several points (still not well defined) along the chromosome. Replication is usually bidirectional about each initiation site. The termination points of replication are interspersed between initiation sites. In most cases of unidirectional or bidirectional replication, synthesis occurs nearly (but not exactly) simultaneously on both strands of the parent DNA template. Since synthesis can occur only in the 5′→3′ direction on the growing chain, and since the two strands in the parent duplex are oriented in opposite directions, this means that synthesis can occur continuously on only the leading strand. On the other (lagging) strand it must pause for the template to unwind. Synthesis on the lagging strand does not occur continuously but rather in small discontinuous spurts, generating Okazaki fragments.
3. Many proteins are required for DNA synthesis and chromosomal replication. These include polymerases; helicases, which unwind the parental duplex; enzymes that fill in the gaps and join the ends in the case of lagging-strand synthesis; enzymes that synthesize RNA primers at various points along the DNA template; topoisomerases, which permit rotation and supercoiling; and single-strand DNA-binding proteins, which stabilize single-stranded regions that are transiently formed during replication. Most of these proteins have been isolated from whole cells and studied in cell-free systems.
4. In *E. coli,* mutations have been isolated in the genes encoding a number of these enzymes. Many of these mutations are conditional, since the functional enzymes involved are required for DNA synthesis and cell viability. Mutants carrying mutationally altered proteins have been important in confirming their roles predicted from cell-free studies.
5. In eukaryotes, most of the work on enzymes involved in DNA synthesis has been done without mutants. However, mutants are currently being isolated in *S. cerevisiae* that prove as useful as those of *E. coli* to help characterize the mechanism of DNA replication. Considerable progress has been made in studying the *in vitro* replication of animal viruses such as SV40 and adenovirus. The importance ascribed to the enzymes that have been characterized is largely based on a comparison of their properties with similar prokaryotic enzymes whose functions are better understood.
6. Many enzymes that act on DNA are involved in processes other than DNA synthesis. They include DNA repair enzymes, DNA degradation enzymes, and DNA recombination enzymes.

Selected Readings

Bauer, W. R., F. H. C. Crick, and J. H. White, Supercoiled DNA. *Sci. Am.* 243(4):118–133, 1980.

Beese, L. S., and T. A. Steitz, Structural basis for the 3′–5′-exonuclease activity of *E. coli* DNA polymerase I: a two metal ion mechanism. *EMBO J.* 10:25–33, 1991.

Bell, S. P., and B. Stillman, ATP-dependent recognition of eukaryotic origins of DNA replication by a multiprotein complex. *Nature* 357:128–134, 1992.

Blackburn, E. H., Structure and function of telomeres. *Nature* 350: 569–573, 1991.

Bohr, V. A., and K. Wasserman, DNA repair at the level of the gene. *Trends Biochem. Sci.* 13:429–432, 1988.

Bramhill, D., and A. Kornberg, A model for initiation at origins of DNA replication. *Cell* 54:915–918, 1988.

Campbell, J. L., Eukaryotic DNA replication. *Ann. Rev. Biochem.* 55:733, 1986.

Cedar, H., DNA methylation and gene activity. *Cell* 53:3–4, 1988.

Challberg, M. D., and T. J. Kelly, Animal viruses and DNA replication. *Ann. Rev. Biochem.* 58:671–717, 1989.

Cox, M. M., and I. R. Lehman, Enzymes of general recombination. *Ann. Rev. Biochem.* 56:229–262, 1987.

Demple, B., and P. Karran, Death of an enzyme: suicide repair of DNA. *Trends Biochem. Sci.* 8:137–139, 1983.

Dunderdale, H. J., F. E. Benson, C. A. Parsons, G. J. Sharples, R. G. Hoyd, and S. C. West, Formation and resolution of recombination intermediates by *E. coli.* RecA and RunC protein. *Nature* 354: 506–510, 1991.

Fangman, W. F., and B. J. Brewer, Activation of replication origins with yeast chromosomes. *Ann. Rev. Cell. Biol.* 7:375–402, 1991.

Fink, G. R., J. D. Boeke, and D. J. Garfinkel, The mechanisms and consequences of retrotransposition. *Trends Genet.* 2:118–123, 1986.

Holliday, R., A different kind of inheritance. *Sci. Am.* 260(4):60–73, 1989. On DNA methylation.

Itoh, T., and J. Tomizawa, Antisense RNA. *Ann. Rev. Biochem.* 60: 631–652, 1991. Includes an excellent description of the initiation of DNA synthesis for Col E1 plasmid.

Kohlstaedt, L. A., J. Wang, J. M. Friedman, P. A. Rice, and T. A. Steitz, Crystal structure at 3.5 Å resolution of HIV-1 reverse transcriptase complexed with an inhibitor. *Science* 256:1783–1790, 1992.

Kong X-P, R. Onrust, M. O'Donnell, and J. Kuriyan, Three-dimensional structure of the β subunit of *E. coli* DNA polymerase III holoenzyme: a sliding clamp. *Cell* 69:425–437, 1992.

Kornberg, A., and T. A. Baker, *DNA replication,* 2d ed. Freeman, New York, 1991. A magnificent up-to-date, clearly written and thorough treatment by the chairman of the board.

Landy, A., Dynamic, structural and regulatory aspects of site-specific recombination. *Ann. Rev. Biochem.* 58:913–950, 1989.

Linn, D., and W. Maas, Reverse transcriptase–dependent synthesis of a covalently linked branched DNA-RNA compound in *E. coli* B. *Cell* 56:891–904, 1989.

Lohman, T. M., W. Bujalowski, and L. B. Overman, *E. coli* single strand binding protein. *Trends Biochem. Sci.* 13:250–255, 1988.

Maxwell, A., and M. Gellert, Mechanistic aspects of DNA topoisomerases. *Adv. Prot. Chem.* 38:69–107, 1986.

McHenry, C. S., DNA polymerase III holoenzyme of *E. coli. Ann. Rev. Biochem.* 57:519–550, 1988.

Meselson, M., and F. W. Stahl, The replication of DNA in *Escherichia coli. Proc. Natl. Acad. Sci.* 44:671–682, 1958. A classic paper.

Messer, W., and W. Noyer-Weidner, Timing and targeting: the biological function of Dam methylation in *E. coli. Cell* 54:734–737, 1988.

Modrich, P., DNA mismatch correction. *Ann. Rev. Biochem.* 56:435–466, 1987.

Newlin, C. S., Yeast chromosome replication and segregation. *Microbiological Reviews* 52:568–601, 1988.

Ogawa, T., and T. Okazaki, Discontinuous DNA replication. *Ann. Rev. Biochem.* 57:519–550, 1988.

Radman, M. and R. Wagner, The high fidelity of DNA replication. *Sci. Am.* 259(2):40–46, 1988.

Sancar, A., and G. B. Sancar, DNA repair enzymes. *Ann. Rev. Biochem.* 57:29–67, 1988.

Sancar, G. B., and A. Sancar, Structure and function of DNA photolyases. *Trends Biochem. Sci.* 12:259–261, 1987.

Shapiro, J. A., *Mobile Genetic Elements.* New York: Academic Press, 1983.

Smith, G. R., Homologous recombination in *E. coli:* Multiple pathways for multiple reasons. *Cell* 58:807–809, 1989.

Stahl, F. W., Genetic recombination. *Sci. Am.* 256(2):90–101, 1987.

Varmus, H., Reverse transcription. *Sci. Am.* 257(3):56–64, 1987.

Wang, E. H., P. N. Friedman, and C. Prives, The murine p53 protein blocks replication of SV40 DNA *in vitro* by inhibiting the initiation functions of SV40 large T antigen. *Cell* 57:379–392, 1989.

Wang, J. C., DNA topoisomerases. *Ann. Rev. Biochem.* 54:665–697, 1985.

Wang, T. S. F., Eukaryotic DNA polymerases. *Ann. Rev. Biochem.* 60:513–553, 1991.

Problems

1. Explain how Meselson and Stahl were able to demonstrate that *E. coli* replicates its DNA in a semiconservative mode. How would the data appear for the first and second doublings if the mode of replication were dispersive? Does the dispersive mode of DNA synthesis occur in cells? (Explain your answer.)

2. The genetic map below represents the distribution of eight genes (*a–h*) as they occur on a bacterial chromosome. The plot is a graphic representation of the average number of copies for each of the eight genes, for a cell that grows with a doubling time equal to the time required for a complete round of DNA replication.
   (a) Estimate the location of the origin of replication.
   (b) Infer whether replication is bidirectional or unidirectional.

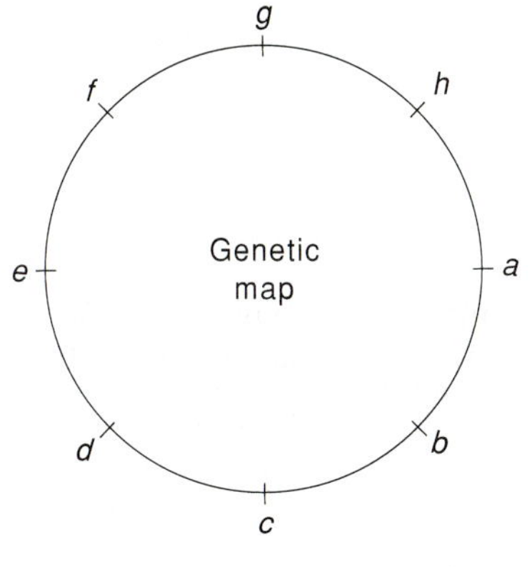

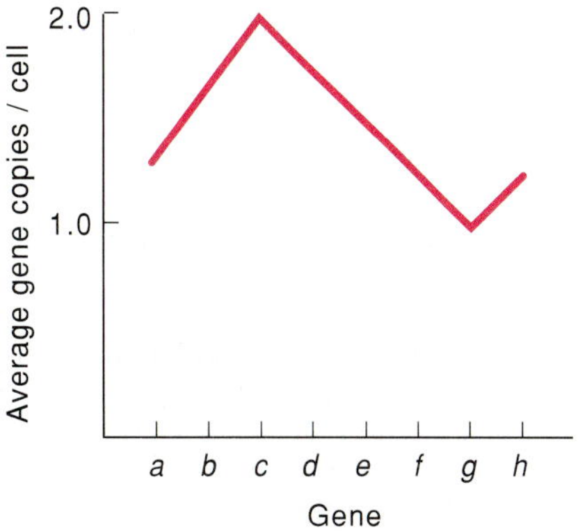

3. Draw the chemical reaction mechanism for the formation of a phosphodiester linkage during DNA synthesis. Discuss the significance of the pyrophosphate product that is formed. What is the significance of the Mg^{2+} requirement?
4. Would you expect the replication of either ϕX174 DNA or the *E. coli* chromosome to be sensitive to inhibition by the antibiotic rifampicin? If inhibition occurs, at what stage during replication does it occur?
5. Draw the chemical reaction mechanism for DNA ligase in *E. coli* (uses NAD as the source of energy and forms a covalent intermediate with an ϵ-amino group of lysine). Why are ligation reactions that require ATP more thermodynamically favorable?
6. *E. coli* has a genomic complexity of about 4×10^6 base pairs (bp) and each replication fork can move at about 10^3 bp per second. How long would it take to replicate the *E. coli* chromosome? With an ample carbon source and ideal growth conditions, cells of *E. coli* can divide in about 20 min. How can this shorter division time occur if the rate of fork migration remains constant at 10^3 bp per second?
7. Humans have about 3×10^9 base pairs of DNA in their genome and the replication forks migrate much slower than in bacteria (about 30 bp per second). How long would it take to replicate the entire genome if it was a single continuous piece of DNA (one chromosome)? How many replication origins would be required to replicate this DNA in an hour?
8. In the graph below, *E. coli* was labeled with radioactive thymidine for a short pulse (10 s) followed by a chase with an excess of nonradioactive thymidine. The DNA was extracted and centrifuged in alkaline sucrose gradients (under high pH conditions the DNA denatures). Explain what these data imply, and interpret these results in light of our current model for DNA replication.

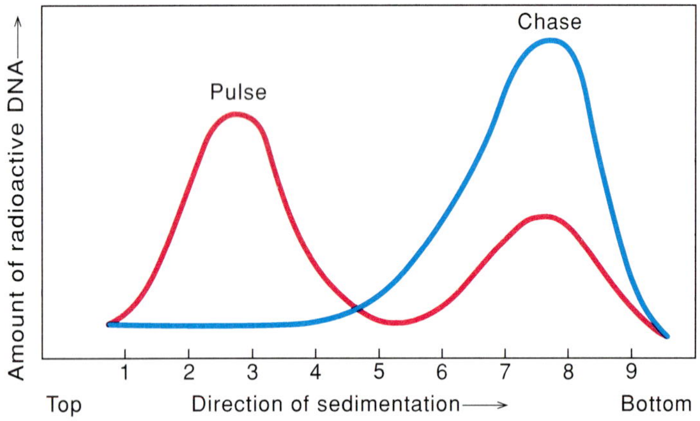

9. Cairns and De Lucia isolated a mutant strain of *E. coli* that had only about 1% of the DNA Pol I activity found in wild-type cells, yet the strain replicated its DNA at a normal rate. Explain how this discovery was important in understanding the role of the different DNA polymerases in replication and repair.
10. All enzymes that make DNA in a template-dependent fashion require a primer. Why does a primer increase the fidelity of DNA synthesis and why is this primer usually RNA?
11. What is the novel priming system for adenovirus DNA replication? Why is this system an advantage to the virus? What other viruses have similar priming systems?
12. Outline in general terms how a retrovirus replicates its RNA. What unique viral protein is used in this replication process? What serves as the primer for retrovirus replication?
13. Eukaryotic DNA is replicated at a slower rate than prokaryotic DNA. One reason may be the requirement for the deposition of histone proteins on DNA (histone synthesis and DNA replication are coupled). Describe a model for the replication of eukaryotic DNA and nucleosome formation.
14. Normal human fibroblasts were grown in culture and then exposed to UV light. A short time later the DNA was extracted and applied to an alkaline sucrose gradient, and the data in graph (*a*) were observed. Another sample of cells was also exposed to UV light but about 12 h were allowed to pass before the DNA was extracted and applied to an alkaline sucrose gradient (graph *b*). Explain these data from what you know about DNA repair. Another sample of fibroblast cells, isolated from a patient with xeroderma pigmentosum, was exposed to UV light and then applied to a gradient after a short time. Would you expect the data to resemble those in graph (*a*) or (*b*)? Why?

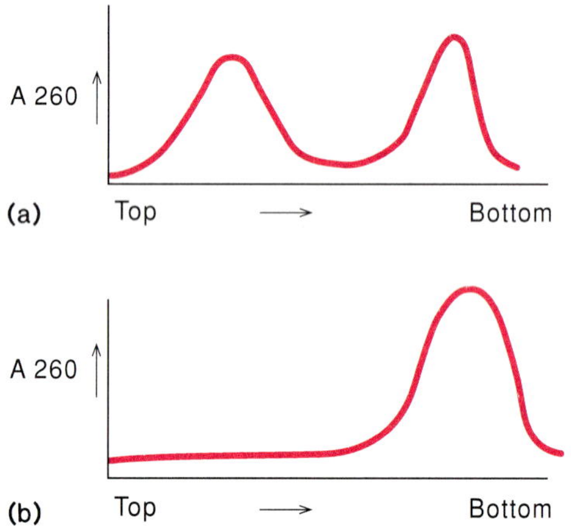

15. Why is the uracil-DNA glycohydrolase very important in DNA repair? Why is thymidine-containing and not uracil-containing DNA the modern product of evolution?
16. What is the role of the recA protein in *E. coli* DNA repair and in recombination? What use are *recA* mutants in biochemical and genetic research?

27

CHAPTER

DNA Manipulation and Its Applications

In recent years, researchers have developed a broad array of techniques to investigate the fine structure of DNA (fig. 27.1). The techniques of DNA manipulation have also proved to be of enormous value in related fields. In particular, they have made it possible for us to map and isolate genes from complex eukaryotes—feats that, for technical reasons, could never have been accomplished by classical genetic methods. Beyond this, they have also enabled us to redesign genes to achieve a variety of practical goals. These newly designed genes may be inserted into the same species from which the original unmodified genes came, or into other species.

In this chapter we will first consider DNA sequencing and the synthesis of DNAs with a predesignated sequence. Then we will explore the different approaches for isolating specific genes or gene segments. Following this, we will examine the methods that are currently available for restructuring existing DNA sequences. Finally, we will look at some of the major advances in our understanding of gene structure and location that have resulted from the new technology. This part of the discussion will focus on two examples: the mapping of the human globin gene family and the mapping of the gene responsible for the genetically inherited disease cystic fibrosis.

Sequencing of DNA

Initial efforts at sequencing nucleic acids were confined to RNA molecules that could be readily isolated in pure form. The first sequence to be determined was that for tyrosine tRNA from yeast. From roughly 100 pounds of yeast, Robert Holley was able to isolate enough of the tyrosine tRNA to carry out a sequence analysis. This historically significant effort required a number of enzymes and chromatographic techniques. We will not, however, elaborate on this accomplishment, because sequencing of RNA is no longer done directly. In fact, the sequencing of RNA has been replaced by the sequencing of cDNA, which results from the transcription of RNA into DNA.

Figure 27.1

Procedures for manipulation of DNA. DNA manipulation is divided into two steps: (1) isolating the selected region(s) of the chromosome that is under study and (2) studying the properties of the isolated DNA. Most of the procedures that are discussed in this chapter are indicated in this figure. First the nuclear (genomic) DNA is isolated. This DNA is degraded into smaller, more manageable fragments by treatment with a restriction enzyme that cleaves at defined sequences. Small, manageable fragments of DNA may also be isolated by reverse-transcribing the messenger RNA to obtain cDNA. Either of these DNAs, genomic or cDNA, is then inserted into a vector for purposes of amplification (either a plasmid or a virus may be used for this purpose). The plasmid mixture containing a mixture of DNAs is then used to transfect cells, which are plated to form clones. Each clone contains many copies of a vector linked to an inserted DNA segment. The mixture of clones is called a cDNA or a genomic DNA library, depending on the source of the inserted DNA. Selective detection procedures are used to find the clones that carry the desired DNA. The plasmids are isolated from these clones, and cells are transfected again with the purified plasmid to obtain a large amount of DNA. DNA isolated by this means can be used in a number of ways: (1) Segments of the desired inserted DNA may be selectively amplified by the polymerase chain reactions (PCR); (2) the DNA may be sequenced; (3) cells may be transfected with the purified plasmid to make protein, to study the effect of the encoded protein, or to change the properties of the cells. Alternatively, the DNA may be changed by directed mutagenesis and the properties of the changes may be investigated. The DNA in conjunction with an appropriate library also provides a useful tool for locating the gene in the chromosome by chromosome walking (and/or jumping) and for isolation of DNA fragments in the adjacent regions of the chromosome.

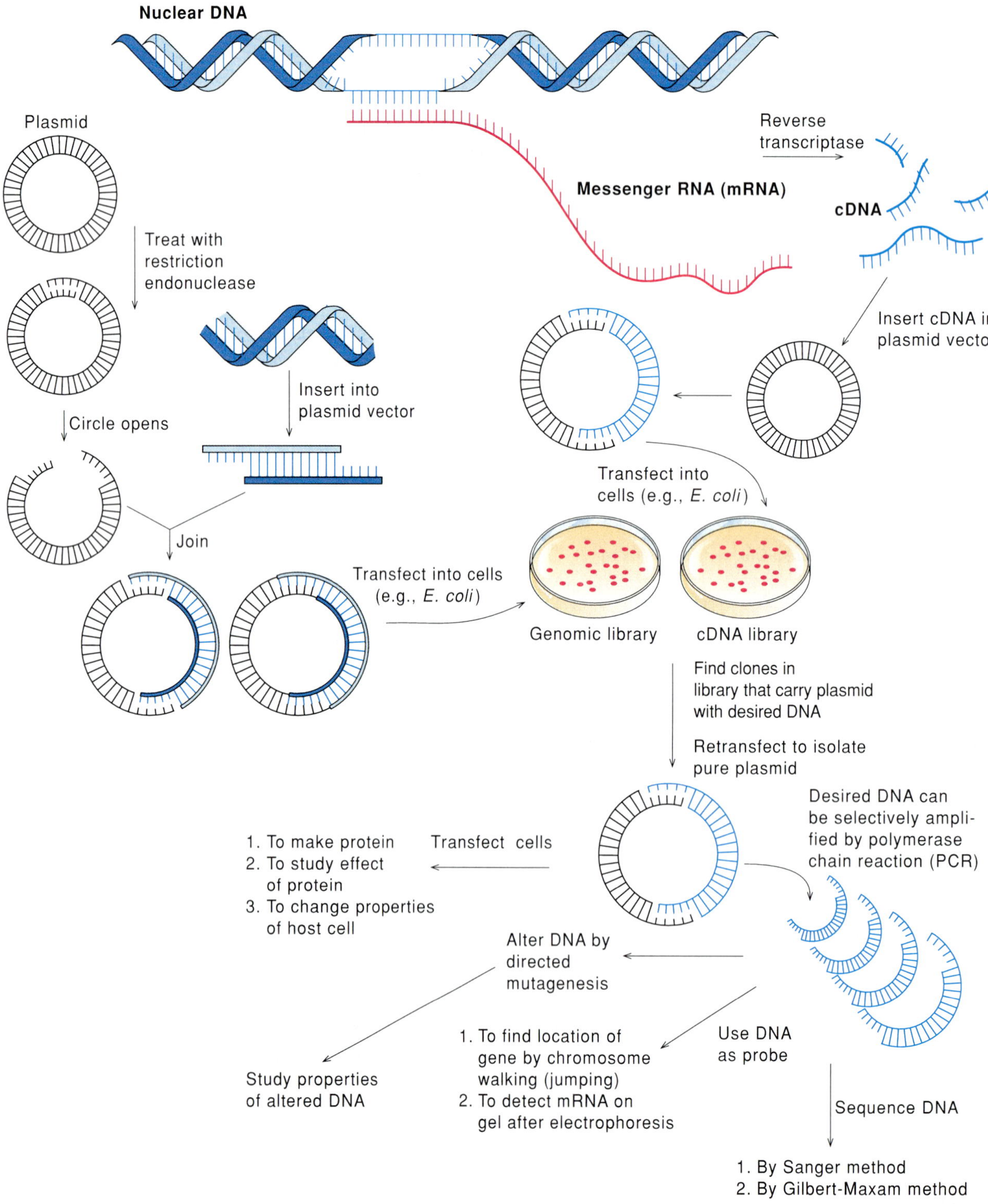

Figure 27.2

Sequencing end-labeled DNA by limited, base-specific chemical cleavage. (*a*) A sequence of three reactions leading to strand cleavage at a guanine. If the entire DNA is subjected to this reaction sequence for a limited time, the result is (*b*), a family of labeled fragments that will be cleaved at different guanine sites.

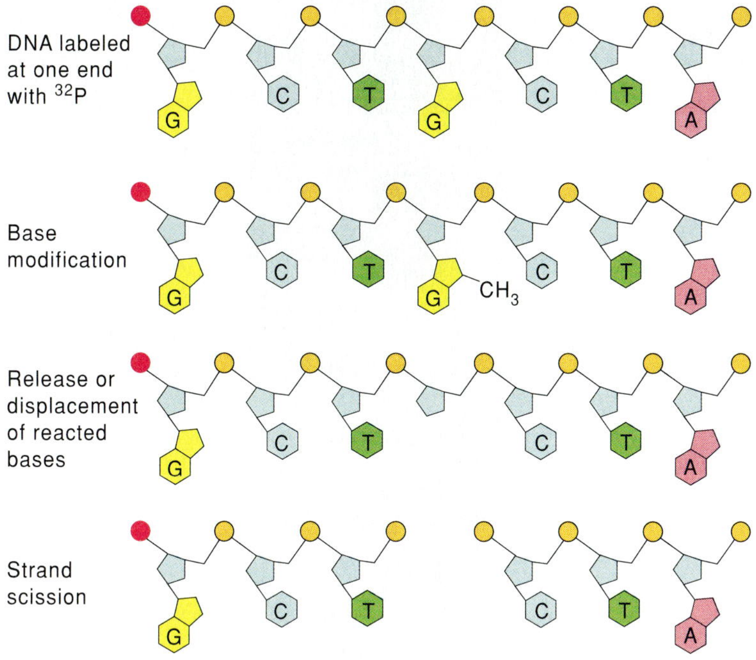

(a)

^{32}pGpCpTpGpCpTpApGpGpTpGpCpCpGpApGpC

Complete molecule

^{32}p

^{32}pGpCpTp

^{32}pGpCpTpGpCpTpAp

^{32}pGpCpTpGpCpTpApGp

^{32}pGpCpTpGpCpTpApGpGpTp

^{32}pGpCpTpGpCpTpApGpGpTpGpCpCp

^{32}pGpCpTpGpCpTpApGpGpTpGpCpCpGpAp

^{32}pGpCpTpGpCpTpApGpGpTpGpCpCpGpApGpC

(b)

Two quite different methods have been developed for sequencing DNA. One method, involving cleavage of preexisting DNA, uses a chemical approach. This was developed by Walter Gilbert and Alan Maxam. The second method, involving premature termination of newly synthesized DNA, uses an enzymatic approach and was developed by Fred Sanger, the same person who received the Nobel Prize for sequencing the first protein, insulin. Both Gilbert and Sanger received Nobel Prizes for their work on DNA sequencing. Both methods merit description here, as they are both useful in different ways.

The Maxam-Gilbert Sequencing Procedure Uses Chemical Methods for Cleavage of Preexisting DNAs

The Maxam-Gilbert sequencing procedure is important both for sequencing and because it provides a technique for determining where proteins bind to the DNA (see chapter 30). For initial sequencing of DNA, either single- or double-stranded molecules can be used.

In the first step a specific fragment of DNA is ^{32}P-labeled at its 5′ end with polynucleotide kinase. This enzyme transfers the $^{32}PO_4^-$ group from γ^{32}P-ATP to the 5′ hydroxyl end of the deoxyribonucleotide chain. If double-stranded DNA is used, both 5′ ends will become labeled, so such a molecule cannot be sequenced directly. Instead the complementary strands or the ends of a doubly labeled DNA must first be separated. This can be done by denaturation followed by gel electrophoretic separation of the individual chains, or by treatment of the DNA with an enzyme (usually a restriction enzyme) that cleaves the molecule into two segments that can be separated by electrophoresis.

Next, the DNA is treated with a chemical reagent that specifically reacts with one of the four bases. Reaction is carried out for a limited period of time, so that on the average only a few residues in the polynucleotide chain react with the reagent. The modified base introduces a linkage amenable to backbone cleavage by subsequent chemical treatment (fig. 27.2). To a first approximation, all bases of a given type in a chain are equally susceptible to modification. The net result is a family of products labeled at the 5′ end with ^{32}P and terminating at the point of cleavage (see fig. 27.2).

Four chemical reactions are used that cleave DNA preferentially at guanines (G > A), adenines (A > G), cytosine alone (C), and cytosines and thymines equally (C + T). When the products of the four reactions are resolved according to size by gel electrophoresis, the DNA sequence can be "read" from the pattern of radioactive bands as an autoradiogram (fig. 27.3), which is obtained by placing a sheet of film over the gel for a suitable exposure time.

Figure 27.3

Autoradiogram of a sequencing gel according to Maxam and Gilbert. Only a portion of the autoradiogram is shown. To obtain this autoradiogram four different reaction mixtures were used. Each reaction mixture started with the same labeled DNA, then was subjected to a different set of limited digestion conditions. In column 1, the digestion conditions led to cleavage at guanines. In column 2, the conditions were adjusted so that fragmentation occurred at both guanines and adenines. In column 3, digestion conditions that resulted in cleavage at cytosines were used, and finally, in column 4, digestion conditions that resulted in approximately equal extents of cleavage at the cytosines and thymines were used. The different digestion products were electrophoresed (from top to bottom in the figure). The smaller the chain size, the farther the product migrated in the gel. It is possible to "read" the sequence by tracing the individual bands one step at a time, starting at the bottom of the gel. The sequence of the DNA for this gel pattern is given in the center of the autoradiogram.

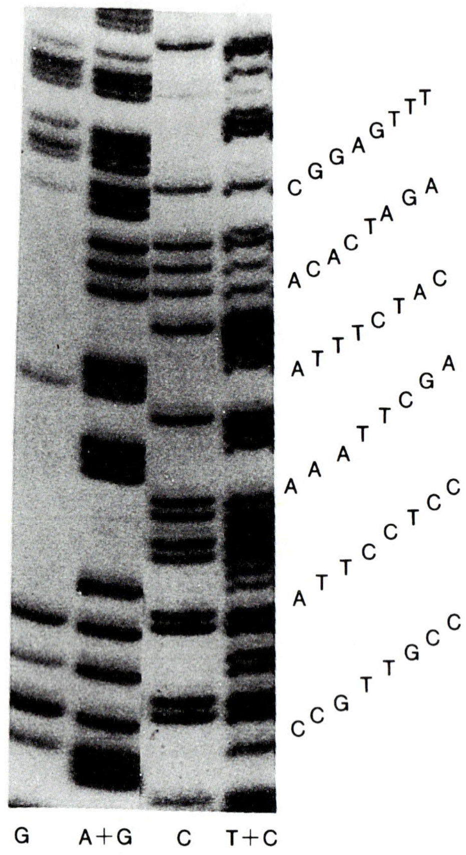

Figure 27.4

Treatment of DNA with dimethylsulfate, resulting in methylation at N-3 in adenine (*a*) and at N-7 in guanine (*b*). Guanine is about five times more reactive than adenine. After reaction for a suitable period of time with the dimethylsulfate, the pH is adjusted to pH 7.0 to permit removal of the modified base and subsequent cleavage of the polynucleotide chain.

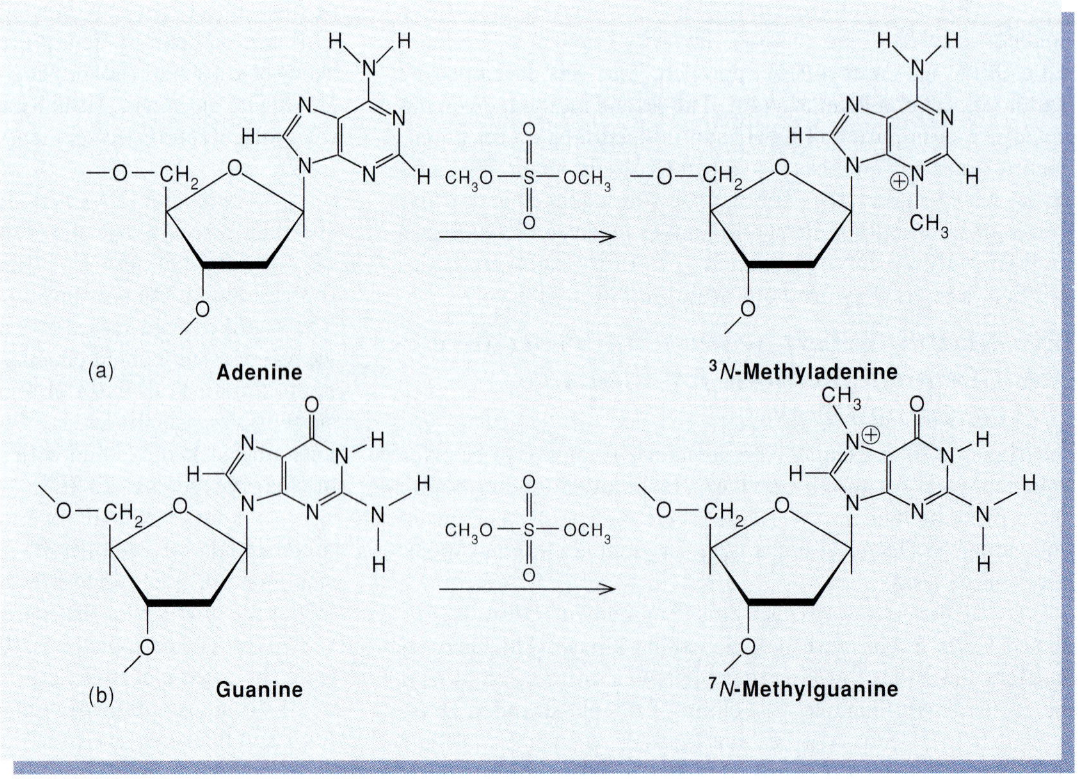

Figure 27.5

Reaction of thymine (*a*) and cytosine (*b*) bases with hydrazine. The products of the reaction are identical. Preferential reaction of cytosine occurs in high salt. In low salt the two bases react at approximately equal rates.

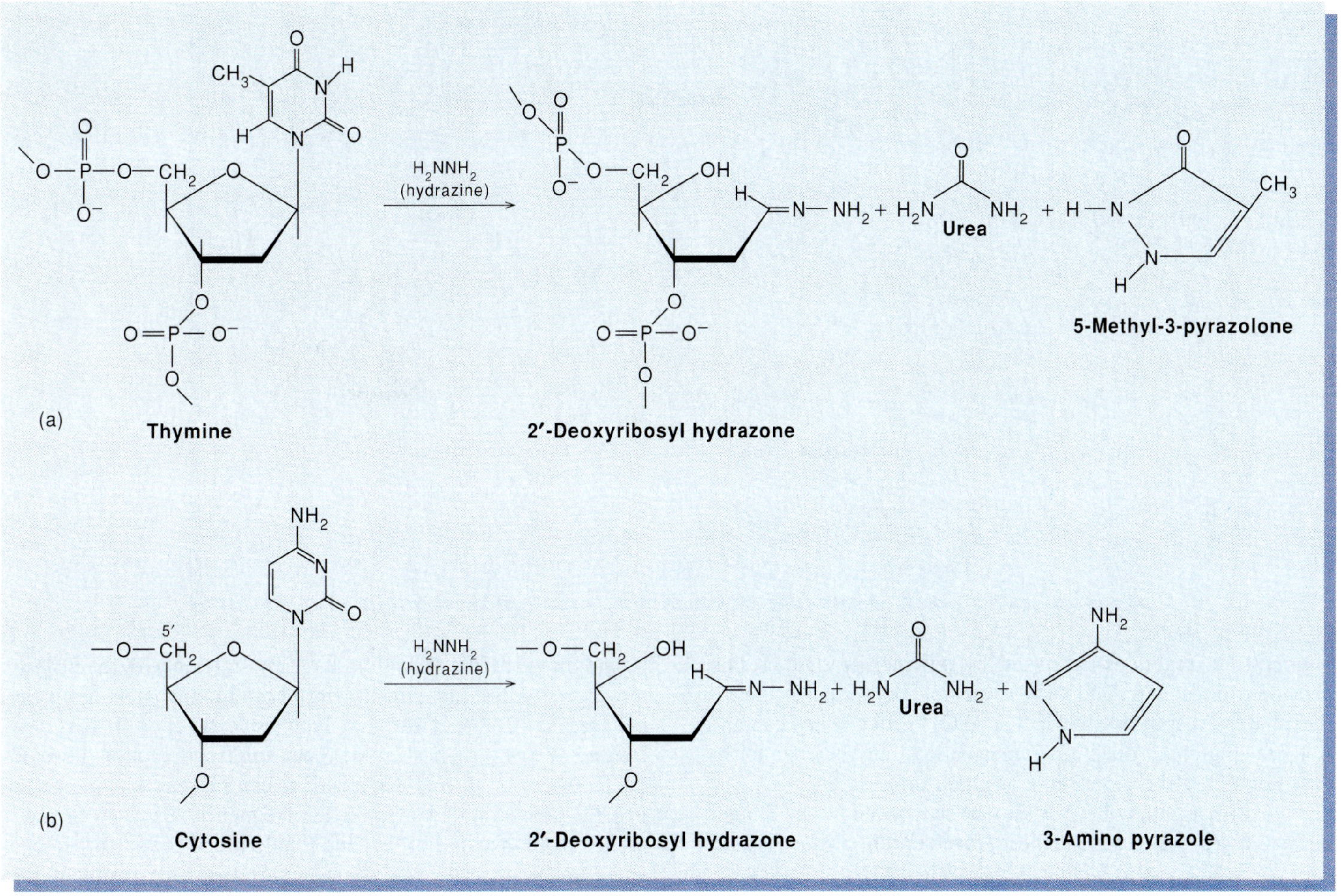

For the purine-specific reaction, an aliquot of the DNA is treated with dimethylsulfate, which methylates the guanines in DNA at the N-7 position and the adenines at the N-3 position (fig. 27.4). The glycosidic bond of a methylated purine is cleaved on heating at neutral pH, leaving the sugar free. Alkali at 90° C will then cleave the sugar from the neighboring phosphate groups. When the resulting end-labeled fragments are resolved on a gel, the autoradiogram contains a pattern of dark and light bands (G > A lane). An adenine-enhanced cleavage can be obtained by treating the methylated DNA with acid, which releases methylated adenine preferentially (A > G lane).

Other aliquots are reacted with hydrazine, which attacks cytosine and thymine bases (fig. 27.5). After a partial reaction in aqueous hydrazine, the phosphate backbone of the DNA is cleaved with 0.5-M piperidine (fig. 27.6). The final gel pattern contains bands of similar intensity owing to the cleavages at cytosines and thymines (C + T lane). However, if 2-M NaCl is included in the hydrazine reaction, the reaction of the thymine residues is suppressed. Then the piperidine breakage produces bands only from cytosines (C lane). Consequently, if the results with and without added hydrazine are compared, C's and T's in the sequence can be distinguished.

Many sequences have been determined by the Maxam-Gilbert method; it is still the tool of choice in cases where the goal is to determine the site of binding of a protein (see chapter 30).

The Sanger Method for Sequencing Uses Newly Synthesized DNAs that Are Randomly Terminated

For pure sequencing the Sanger method is more popular and easier to automate. The method employs chain-terminating dideoxynucleoside triphosphates to produce a continuous series of

Figure 27.6

Cleavage of the hydrazine derivative of the cytosine and the thymine in the presence of piperidine.

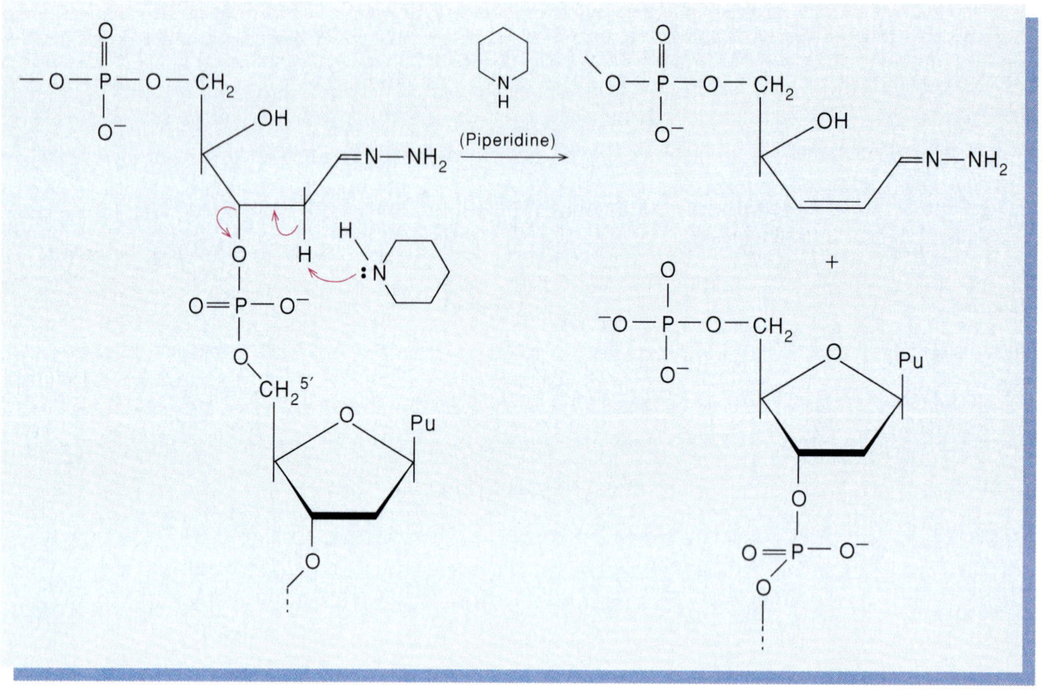

fragments in reactions catalyzed by polymerase. Dideoxynucleoside triphosphates (ddXTPs) resemble deoxynucleoside triphosphates except that they lack a 3′-OH group. They can add to a growing chain during polymerization, but they cannot be added onto and therefore serve as chain terminators.

Either DNA or RNA may be sequenced by the Sanger method. For RNA to be sequenced, reverse transcriptase is used to make a DNA copy of the RNA. When DNA is being sequenced, DNA polymerase is used to make a complementary copy of a primed single-stranded DNA fragment. By choosing an appropriate primer, the region of the nucleic acid that is copied can be predetermined (fig. 27.7).

Synthetic reaction mixtures are then set up, each containing one or more radioactive deoxyribonucleoside triphosphates to label the fragments for detection by autoradiography. Each mixture contains a single, limiting amount of dideoxynucleoside triphosphate to randomly terminate the synthesized fragments at one of the four nucleotides. The products of four separate reaction mixtures, each containing a different dideoxynucleoside triphosphate, are analyzed (see fig. 27.7). Reaction 2, using dideoxyadenosine triphosphate (ddATP), contains all fragments with an A terminus; reaction 2, using ddCTP, contains all C terminations; and so on. Following synthesis, the reaction products are separated from the template by denaturation and fractionated by electrophoresis on polyacrylamide gels. After electrophoresis the positions of the fragments on the gel are detected by autoradiography. The sequence is read directly from the autoradiogram, starting with the fastest-moving (smallest) fragment at the bottom, and moving up the gel (see fig. 27.7). If the first band is in reaction 3, it is a G residue; the next highest band, appearing from reaction 4, would be T; and so on. Up to 800 residues can be read from a single gel. ^{35}S-labeled dXTPs (fig. 27.8) frequently are used in preference to ^{32}P because of the longer half-life of the sulfur isotope and also because it gives sharper patterns, as a result of the shorter pathlength of its β-particle emission.

One of the latest adaptations of the Sanger method uses colored fluorescent derivatives attached to the dideoxynucleotide terminators. The advantage of this approach is that each reaction corresponding to an unknown DNA sequence can be carried out in a single test tube and loaded onto a single column for electrophoresis.

Oligonucleotides of the Desired Sequence Can Be Made in Vitro

Short DNA chains (oligodeoxynucleotides) are essential to make primers in the Sanger sequencing procedure; they are also valuable for other procedures that we will discuss shortly. Chemical methods for making synthetic DNA originated in the laboratory of H. G. Khorana. Khorana's goal was to chemically synthesize an entire gene and then to see whether it would function properly when reinserted into the organism. He wisely chose a small tRNA gene for this study, the gene for the 76-nucleotide tRNA for alanine. His success in the total synthesis of this RNA was a historic achievement and led others to refine the methods and apply them in other areas.

Figure 27.7

The Sanger dideoxynucleoside method of sequencing DNA. (*a*) A suitable template is chosen, and the primer is chosen so that DNA synthesis begins at the point of interest. In addition to the template-primer complex the reaction mixture contains all four radioactive deoxyribonucleoside triphosphates and small amounts of a single dideoxynucleoside triphosphate. The dideoxy compound serves as a chain terminator. (*b*) After synthesis in the presence of DNA polymerase I, the products of the reaction mixture are separated by gel electrophoresis and analyzed by autoradiography. For a given dideoxy compound all fragments terminating with that particular base should give rise to bands on the gel. The interpretation of the gel pattern is given in (*c*). The smallest labeled fragment moves the fastest and appears at the bottom of the gel. (*d*) A typical sequencing film. The sequence begins CAAAAAACGG. (Courtesy of GIBCO-BRL, Life Technologies, Inc., Gaithersburg, Md.)

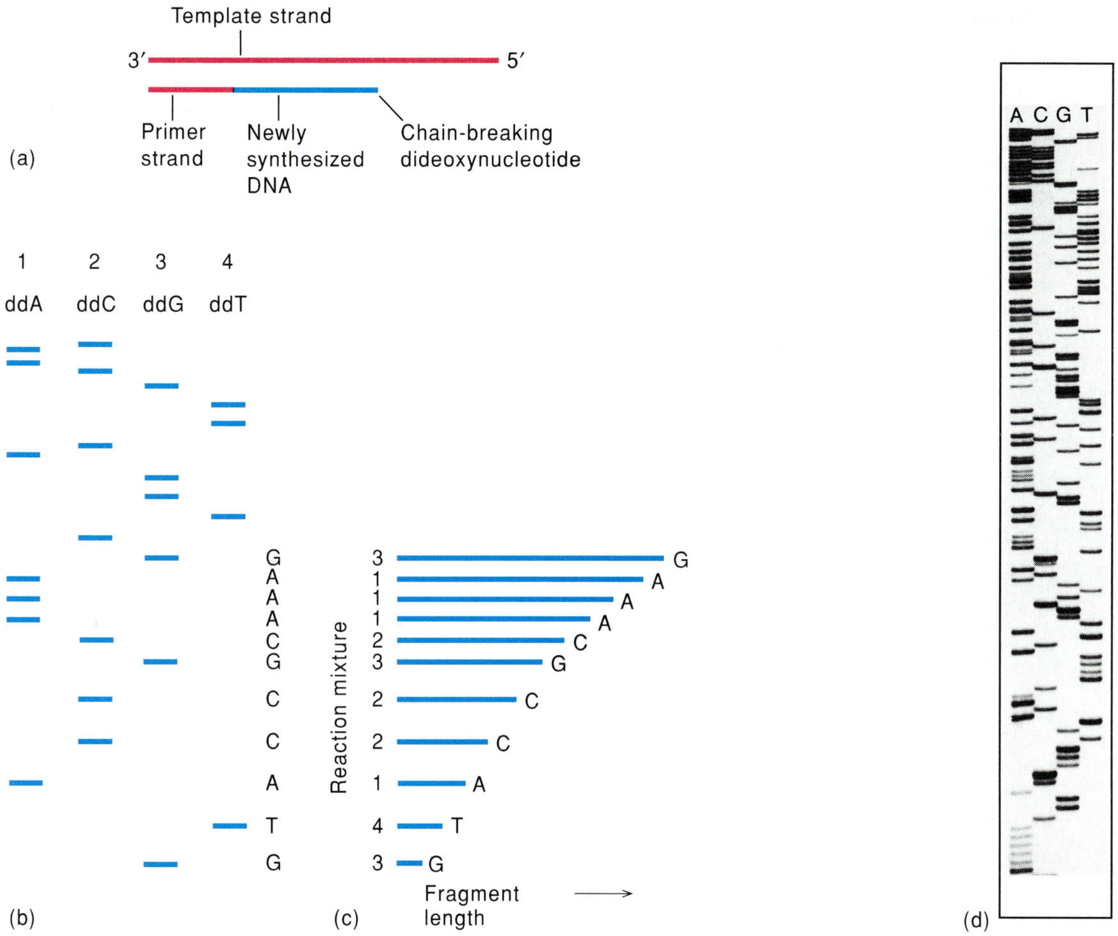

Various chemical methods of synthesis have evolved over the years, but they all have two features in common: (1) 3′ activation is involved in the synthesis instead of 5′ activation as in biosynthesis, and (2) protecting groups are required in intermediate steps to prevent other groups from reacting. Currently, synthesis centers around use of highly reactive phosphite groups, which involve phosphorus in the +3 valence; this is oxidized to +5 phosphorus after incorporation by iodine. The chemistry of the phosphite-triester scheme for synthesizing oligonucleotides is shown in figure 27.9. This method involves sequential treatment of a 5′-O-protected nucleoside with (a) $R'OPCl_2$, (b) a 3′-O-protected nucleoside, and (c) iodine-water. The chain is extended to the desired length by successive repeats of this cycle. After construction of the desired sequence, all phosphoryl protecting groups and the 3′-O-terminal protecting groups are removed (d) to give a synthetic oligonucleotide.

Figure 27.8

Sulfur-labeled substrate is preferred to phosphorus-labeled substrate because of its longer half-life and the shorter pathlength of its ß emission, which gives sharper banded patterns after autoradiography.

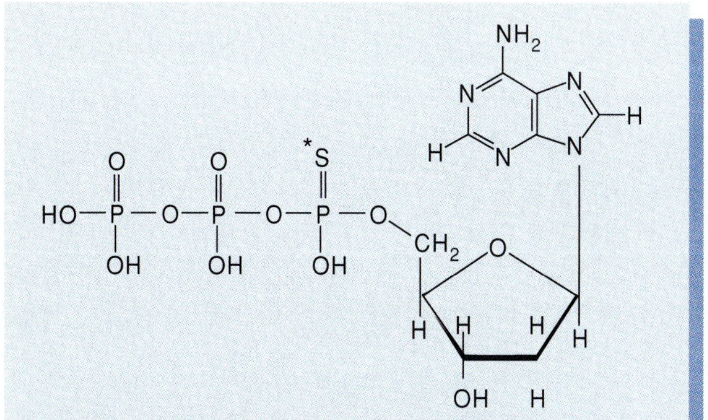

Deoxyadenosine 5′- [α-thio] triphosphate, [^{35}S]-

Figure 27.9

Phosphite-triester method for the synthesis of oligonucleotides. The procedure for making a dinucleotide is shown. This is extended to making oligonucleotides (up to approximately 100 residues) by repeating steps (*a*) through (*c*) in a cyclic fashion.

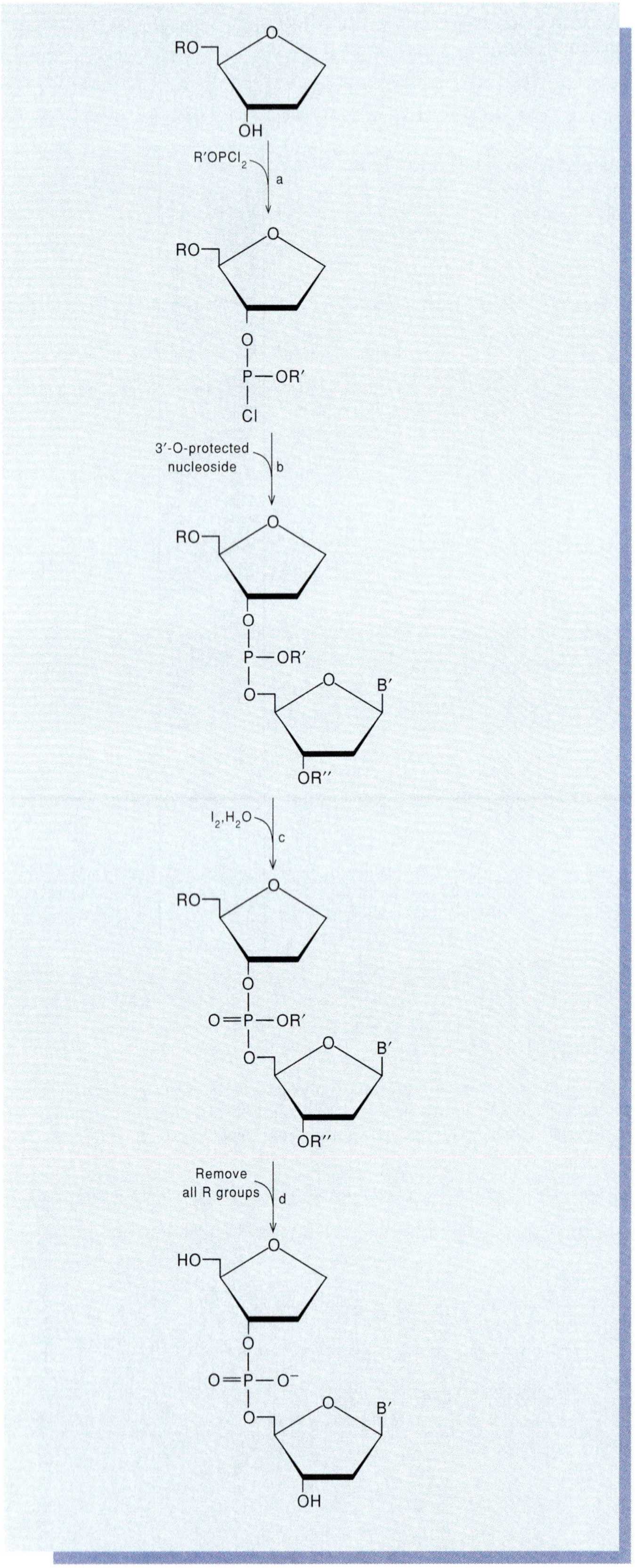

Figure 27.10

Schematic diagram showing how the tRNA gene for alanine tRNA was synthesized. First, small oligonucleotides representing different fragments of the gene were synthesized. Then these were annealed and covalently linked with the help of the enzyme DNA ligase. The process of annealing was repeated until the full gene was assembled.

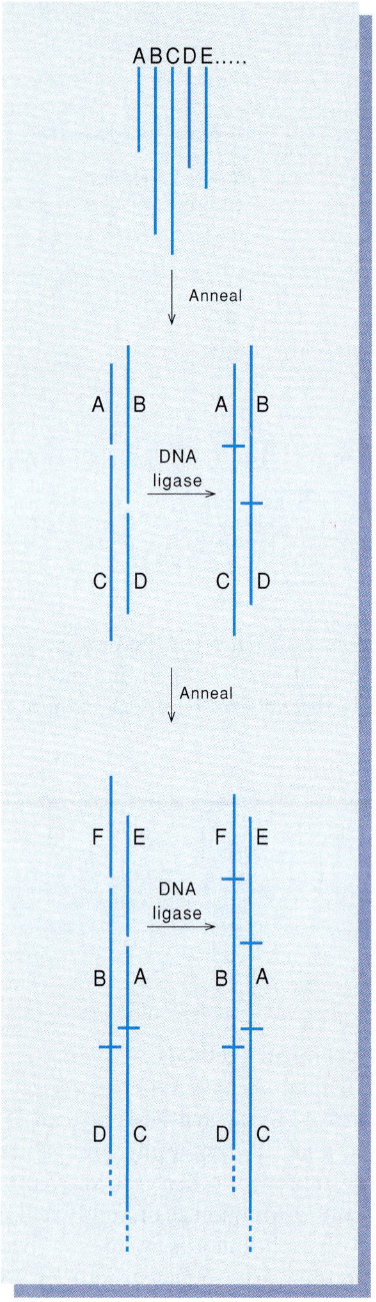

Figure 27.11

Steps in the polymerase chain reaction (PCR). The DNA to be amplified is denatured and annealed with two oligonucleotides that flank the region of interest. These oligonucleotides (or primers) are extended. Extension continues to the ends of the DNA strands. The products are again denatured and annealed to primers for a second round of extension. This process of denaturation, annealing, and primer extension is repeated many times. The primary product of the reaction is duplex DNA, bounded by the sequences of the primers.

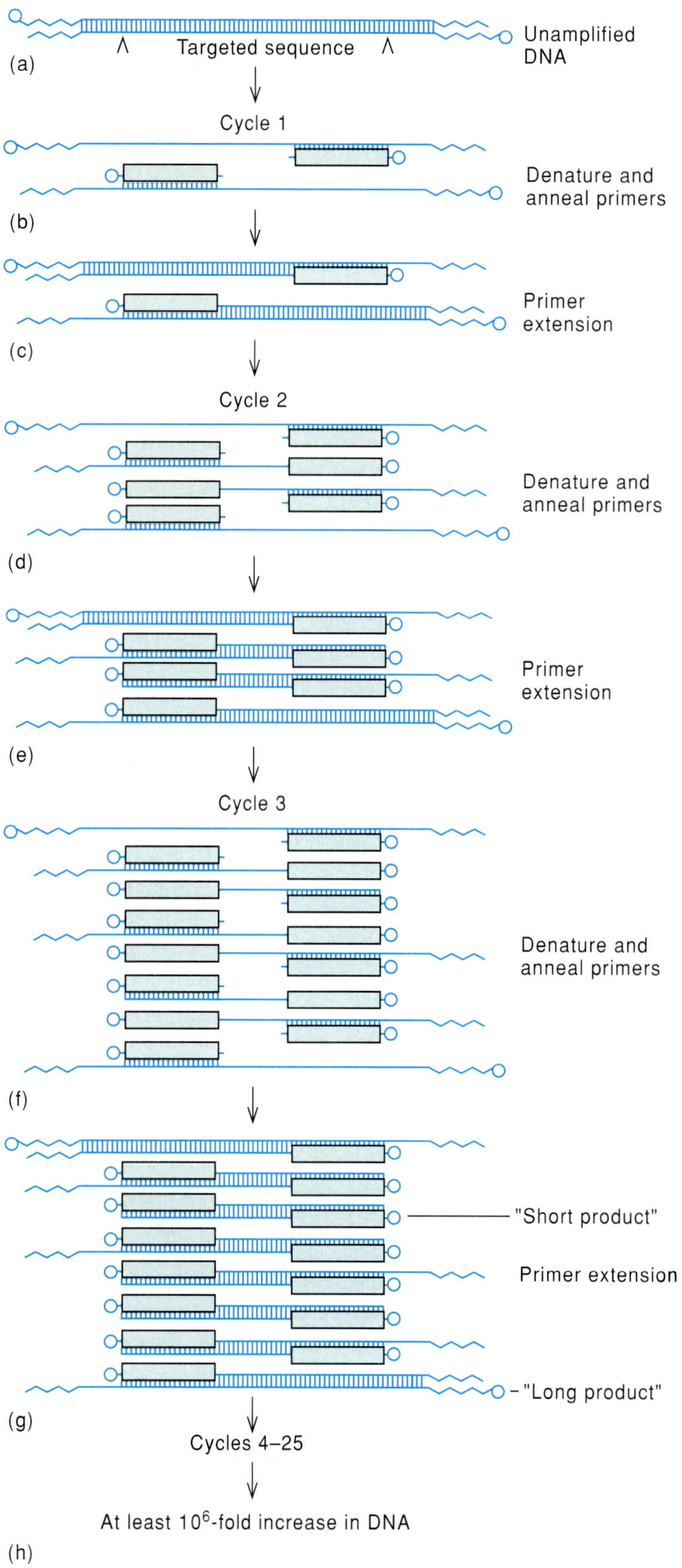

Drawing on the technology developed by Merrifield for the solid-phase synthesis of polypeptides (mentioned in chapter 3), procedures were developed for coupling the 5′-OH group of the initial nucleotide to a resin support, to facilitate the stepwise addition of nucleotides. All mixing and washing steps were thereby greatly simplified and the procedure has become amenable to automation. In fact, it is now possible to program the synthesis of a 100-mer by merely pushing a button for the order of nucleotides.

On occasion, when larger oligonucleotides are desired they can be assembled by synthesizing overlapping, complementary oligonucleotides, which are then annealed and ligated (fig. 27.10).

Sequences of Any Type Can Be Amplified in Vitro *by the Polymerase Chain Reaction*

There are two widely used methods for amplifying defined segments of DNA *in vitro*. One is DNA cloning, which we will describe in a later section. A simpler and more recently discovered procedure is the preferred method where only a limited quantity of the amplified DNA is needed. This method, called the polymerase chain reaction (PCR), entails the use of two oligonucleotide primers, which flank the DNA segment to be amplified (fig. 27.11*a*). The primers must complement opposite strands so that, after annealing, their 3′ ends in effect face each other (fig. 27.11*b*).

There are three steps in the PCR procedure, which are usually repeated many times in a cyclical manner:

1. Denaturation of the original double-stranded DNA sample at high temperature.
2. Annealing of the oligonucleotide primers to the DNA template at low temperature (37° C).
3. Extension of the primers using DNA polymerase.

These steps are illustrated in figure 27.11. Each set of three steps comprises a cycle. The extension products of one primer provide a template for the other primer in a subsequent cycle so that each successive cycle essentially doubles the amount of DNA synthesized in the previous cycle. The result is an exponential accumulation of the specific target fragment to approximately 2^n, where n is the number of cycles. The specific target fragment is also referred to as the "short product" and is defined as the region between the 5′ ends of the extension primers. Each primer is physically incorporated into one strand of the short product.

Other products are also synthesized during the succession of cycles, such as the "long product," of indefinite length, which is derived from the template molecules. However, the amount of long product will increase only arithmetically during each cycle of the amplification process, because the quantity of original template remains constant. At the end of the PCR process the short product is so overwhelmingly abundant in comparison with the long product that for most purposes it does not need purification.

Originally, the PCR technique used the *E. coli* DNA polymerase I enzyme to carry out the primer extension reaction at 37° C. This approach required the addition of fresh enzyme during each cycle, because the thermolabile DNA polymerase I was inactivated by the heat denaturation step used to separate the newly synthesized strands of DNA. The process was slow and expensive, and it resulted in rapid accumulation of denatured enzyme in the sample. To overcome these problems, a thermostable DNA polymerase purified from the thermophilic bacterium *Thermus aquaticus* (abbreviated Taq) was introduced into the reaction. Taq DNA polymerase is unaffected by the denaturation temperature and consequently does not need to be replenished at each cycle. The use of the Taq DNA polymerase has greatly increased the convenience and with it the popularity of the PCR method. The entire process of amplification requires about three hours and requires very few manipulations.

Restriction Enzymes Are Used to Cut DNA into Well-Defined Fragments

In order to sequence DNA, we must first obtain a well-defined piece of DNA. In the case of small DNA viruses, the virus provides a unique polymer of DNA in large amounts. However, even in such cases the DNA is too large to be sequenced directly, and therefore must be broken down into many smaller, well-defined fragments. Systematic breakdown of duplex DNA requires restriction enzymes that cleave DNA at specific recognition sequences. Hundreds of restriction enzymes with different specificities are available, giving a great deal of choice in where to cut the genome. Some representative enzymes and their recognition sites are indicated in table 27.1. Most of these enzymes recognize a sequence of either four or six successive base pairs. The cleavage sites are situated so that a blunt-ended or staggered-ended DNA results from the cleavage reaction. As a rule, the recognition sites are located on an axis of symmetry so that the freshly cut ends have identical structures.

Table 27.1
Recognition Sequences and Cutting Sites of Selected Restriction Enzymes

Enzyme	Recognition Sequence
AluI	A G↓C T
BamHI	G↓G A T C C
BglII	A↓G A T C T
ClaI	A T↓C G A T
EcoRI	G↓A A T T C
HaeIII	G G↓C C
HindII	G T Py↓Pu A C
HindIII	A↓A G C T T
HpaII	C↓C G G
KpnI	G G T A C↓C
MboI	↓G A T C
PstI	C T G C A↓G
PvuI	C G A T↓C G
SalI	G↓T C G A C
SmaI	C C C↓G G G
XmaI	C↓C C G G G

A viral genome cleaved exhaustively with a particular restriction enzyme will usually yield several fragments. Some restriction enzyme cleavage sites for the 5,300-bp SV40 virus genome are shown in figure 27.12. The duplex fragments obtained after cleavage can be separated according to size by gel electrophoresis. Nondenaturing conditions are used so that the duplex strands stay together. The larger a fragment is, the slower it migrates on the gel. After electrophoresis for a time sufficient to separate the fragments, the gel is stained with a fluorescent dye such as ethidium bromide and is viewed under long-wavelength ultraviolet light (long-wavelength UV is used because it does not damage the DNA). Individual fragments may be extracted from the gel for sequencing, PCR amplification, or cloning (described later).

The problem of determining how a set of restriction fragments is normally interconnected may be systematically resolved by determining the sequences of another set of fragments cut with a different restriction enzyme. The overlapping information thereby obtained should permit a determination of the complete sequence of the intact genome. Overlapping information can be obtained also by comparing the digestion products from a DNA sample that has been partially digested with one that has been allowed to go to complete digestion with the same enzyme. These two approaches are used to different extents with different chromosomes.

DNA Cloning

While restriction enzymes permit DNAs to be cut into well-defined fragments, other enzymes permit DNAs from different sources to be linked together to form recombinant DNAs (table 27.2). DNA to be amplified by cloning is linked to a plasmid or a virus that can be extensively replicated. After amplification the DNA of interest is cut from the plasmid or virus with the

Figure 27.12

Cleavage map of the SV40 genome. The zero point of the map is the unique *Eco*RI site. For clarity, the circular genome is shown opened at the R1 site, and the cleavage sites (and resulting fragments) for each restriction enzyme are indicated on a separate line.

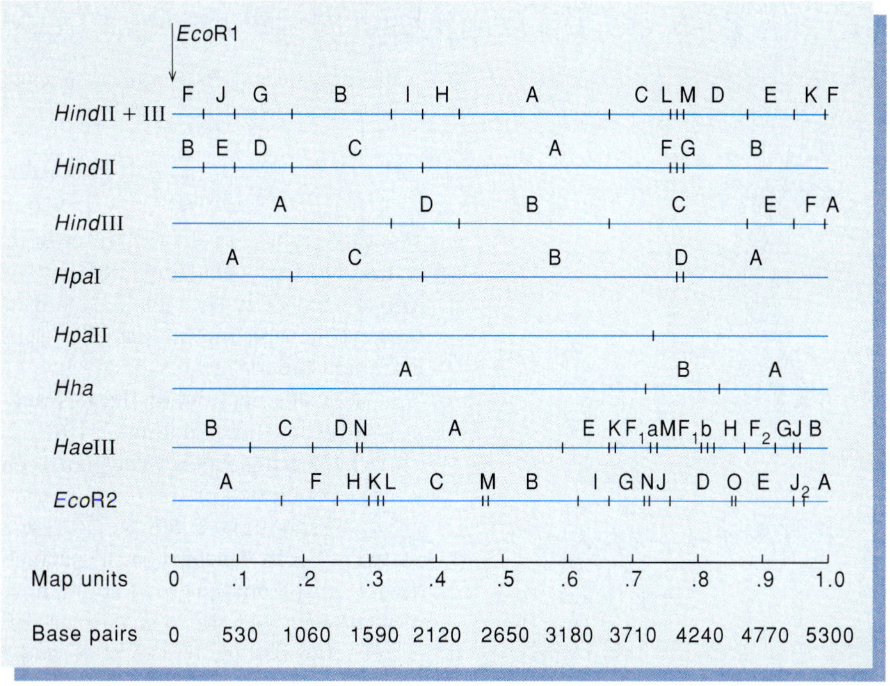

appropriate restriction enzyme and reisolated by electrophoresis. This method of amplification is more work but also opens up more possibilities than the PCR method that we described earlier.

Plasmids Are Used to Clone Small Pieces of DNA

In the basic procedure for DNA cloning, an autonomously replicating plasmid and a DNA insert are cut with a restriction enzyme and then the pieces are annealed and covalently joined by the action of DNA ligase. The resulting recombinant molecules are then transfected into *E. coli,* where they replicate. When plasmid vectors are used, a population of permeabilized cells is bathed in the plasmid DNA containing the inserted DNA. Since only a small number of cells become transfected by this procedure, there needs to be a way to select cells that carry the desired hybrid plasmids. A particularly useful plasmid vector for selecting transfected cells, called pBR322, is itself a hybrid plasmid (fig. 27.13). This plasmid contains two genes, amp^r and tet^r, which confer resistance to penicillin and tetracycline, respectively.

To begin, *Pst*I restriction fragments of foreign DNA may be inserted into the unique *Pst*I restriction site on pBR322 (see fig. 27.13). This is done by digesting pBR322 with *Pst*I, mixing the product of this step with the restriction fragments

Table 27.2
Some Enzymes Used in DNA Recombinant Methodology

Enzyme	Function
Polynucleotide ligase	Joining of two DNA molecules
Polynucleotide kinase	Adding a phosphate to the 5′-OH on a polynucleotide
Terminal transferase	Adding homopolymer tails to the 3′-OH ends of a linear duplex
DNA polymerase I	Repair synthesis
λ exonuclease	Stepwise removal of nucleotides from 5′ ends of a duplex to expose the 3′ ends
Type II restriction endonucleases	Specific cleaving of DNAs to produce staggered ends or blunt ends
Alkaline phosphatase	Removing terminal phosphates from either the 5′ or 3′ end or both
Reverse transcriptase	Synthesizing a DNA copy of an RNA molecule
S1 endonuclease	Removing unpaired single-strand regions

Figure 27.13

Structure of the pBR322 plasmid (*a*) and construction of a hybrid plasmid containing the pBR322 vector and a segment of foreign DNA (*b*). For pBR322 the unique sites for various restriction enzymes are indicated. Also indicated are the locations of the tetracycline (tet^r) and the ampicillin (amp^r) resistance genes and the origin for DNA replication. The hybrid plasmid is constructed by treating the plasmid and the foreign DNA with the *Pst*I restriction enzyme and mixing the two DNAs together in the presence of DNA ligase.

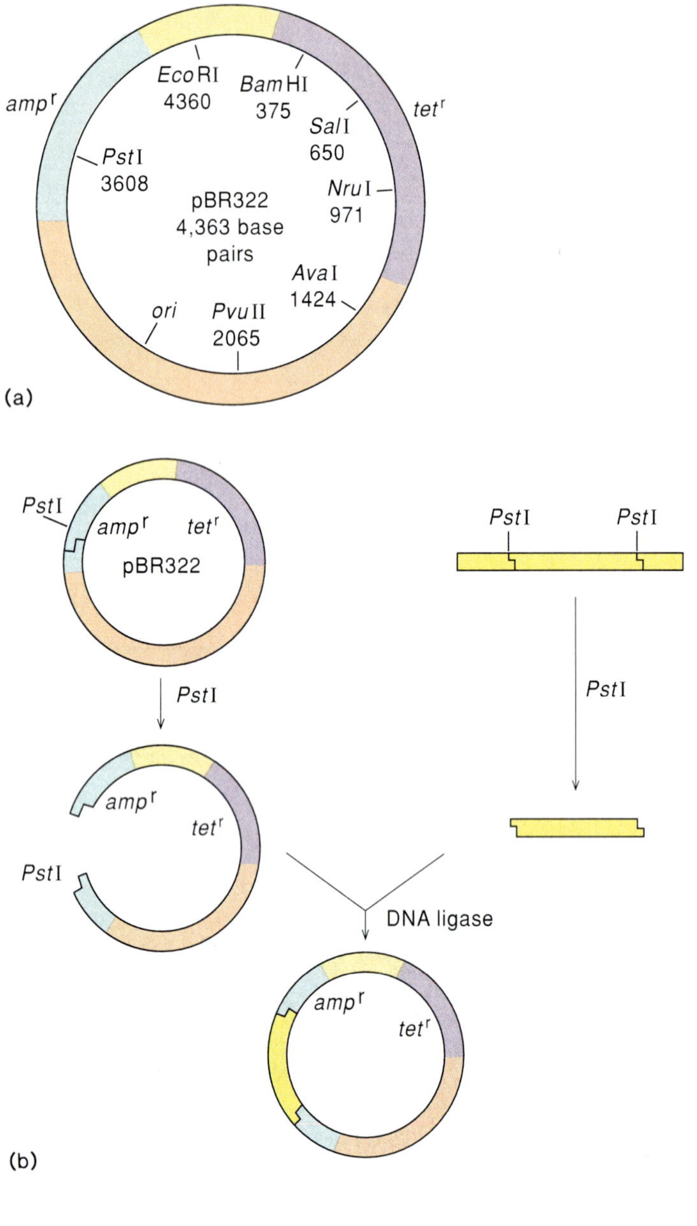

Figure 27.14

Replica plating technique for detecting hybrid plasmid-containing cells.

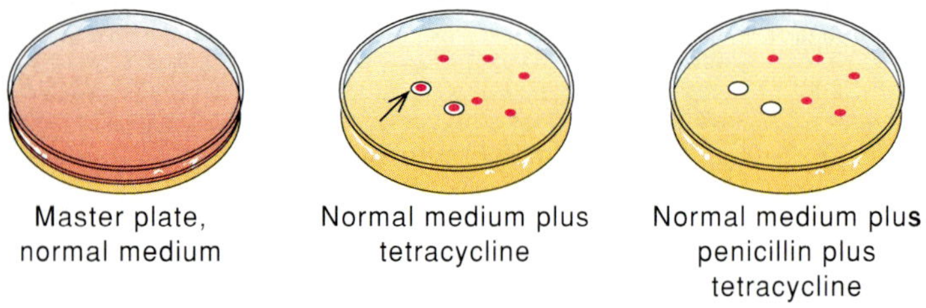

to be cloned at low temperatures, to permit annealing to take place between the two DNAs, and finally ligating the annealed fragments with DNA ligase. The end product will contain some of the original pBR322 and some pBR322 with the inserted foreign DNA. When this mixture is used in transfection, most cells will not be transfected, some will be transfected with pBR322, and some will be transfected with the desired hybrid plasmid. The three types of cells will differ in their drug resistance properties. Normal cells are killed by tetracycline or penicillin. Transfected cells with the DNA inserted in the plasmid are tetracycline-resistant but penicillin-sensitive, since the insert has disrupted the tet^r gene.

Cells containing the desired plasmids can be distinguished from those containing pBR322 by replica plating (fig. 27.14). On semisolid agarose plates containing growth medium, a large population of treated bacteria is spread and allowed to grow overnight (see fig. 27.14). A seemingly homogeneous "lawn" of cells develops on the surface of the gel. Actually, the lawn results from the growth of many microcolonies to the point of confluency. At this point a piece of velvet is lightly pressed against the surface of the plate and this impression is transferred to other agarose plates containing growth medium with tetracycline or penicillin plus tetracycline. Only the transfected cells will produce colonies on the plates containing antibiotics, and because of their small number, each of these will give rise to readily detectable clones. The clones that are present on the tetracycline-containing plates but missing on the penicillin plus tetracycline plates are the ones most likely to contain the desired hybrid plasmids (see fig. 27.14). These clones are usually plucked from the tetracycline plates and retested to eliminate any uncertainty about the original drug testing.

Once their identity is confirmed, the appropriate plasmid-containing cells are grown in liquid culture. After a moderate density of growth is achieved, the plasmid DNA is selectively amplified by overnight growth. Plasmid DNA replication continues for several hours, until each cell contains 1,000 to 2,000 copies of the small circular plasmid DNA. This DNA is readily separable from the host DNA and can be characterized by its rate of migration on gel electrophoresis (fig. 27.15), or other more specific tests, to see whether it contains the inserted DNA sequence. If desired, the inserted sequence may be removed from the plasmid vector by digestion with *Bam*HI, the restriction enzyme used in the initial construction of the hybrid plasmid. The cleaved fragments can be readily separated by gel electrophoresis.

Plasmids such as pBR322 are excellent, easy-to-use vectors, but other types of vectors offer specific advantages. Let us now look at some of these alternatives.

Figure 27.15

Electrophoretogram of restriction enzyme digests of pBR322 and pBR322 with a DNA insert at the *Pst*I site. The insert is assumed to have no internal *Bam*HI restriction sites. In channels A and B the pBR322 is predigested with *Pst*I and *Bam*HI, respectively. The resulting DNA migrates with the same mobility because the plasmid has one site for each of these enzymes and therefore has the same molecular weight. In C and D the hybrid plasmid containing a DNA insert is treated with *Bam*HI and *Pst*I, respectively. In C the hybrid plasmid has been linearized by one cut at the *Bam*HI site in the *amp*[r] gene. It runs more slowly than the pBR322 as it is larger because of the DNA insert. In D the plasmid cuts at two *Pst*I sites located between the pBR322 sequences and the insert sequences. Consequently one segment migrates at the rate of a linearized pBR322 plasmid. The other segment, also linearized, migrates at a rate characteristic of the size of the DNA insert. The electrophoresis is run from left to right; fragments are stained with ethidium bromide and photographed with UV light.

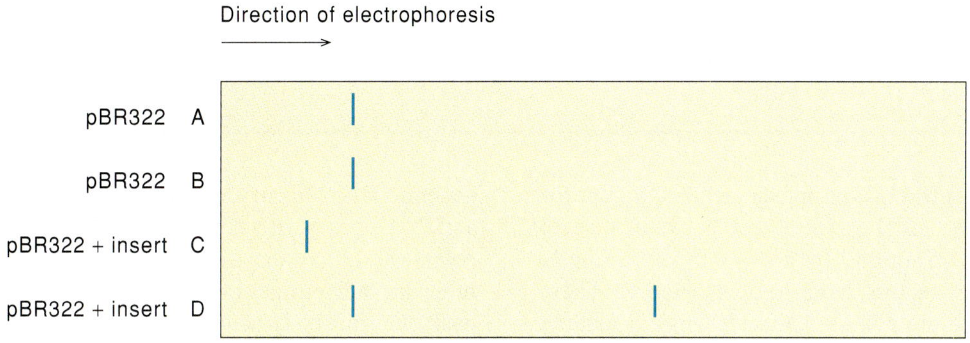

Bacteriophage λ Vectors Are Useful for Cloning Larger Segments of DNA

Bacteriophage λ possesses a number of advantages as a cloning vector. DNA fragments as large as 24 kb can be propagated using this vector. The primary pool of clones can be amplified by limited phage growth as plaques, and the entire collection of phage clones (recognized as clear plaques) can be stored for long periods in a small volume (fig. 27.16).

Since λ phage will not accommodate molecules of DNA that are much longer than the viral genome, this phage cannot be used as a vector for cloning substantial DNA fragments unless a significant portion of the viral DNA is removed beforehand. Fortunately, the central third of the genome contains genes that are not essential for plaque production, and it can therefore be deleted. We now have available a number of bacteriophage λ vectors that accommodate foreign DNA fragments generated by a variety of restriction endonucleases. The recombinant DNA molecules that incorporate some of these vectors can be introduced directly into *E. coli* by transfection. Alternatively, recombinant DNA molecules can be packaged into phage particles and subsequently infected into suitable host cells.

Figure 27.16

The nutrient agar plate contains a continuous lawn of *E. coli* bacteria except for circular clearings that represent phage plaques. Each plaque was originally derived from a single phage particle infecting a single *E. coli* bacterium.

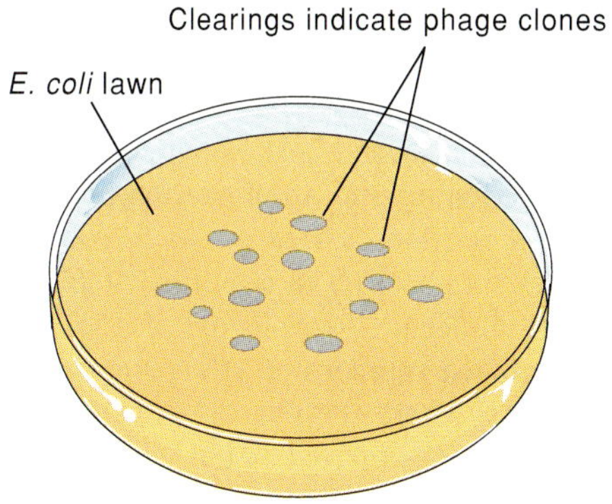

Cosmids Are Used to Clone the Largest Segments of DNA

While plasmids and bacteriophage λ are both useful vectors, the size of the DNA fragments that can be cloned in them is limited. With plasmids, the larger the fragment of foreign DNA inserted, the lower the efficiency of ligation and transfection, and thus the cloning of DNA fragments larger than 15 kb is experimentally difficult. Even in λ vectors, the length of the nonessential region of λ DNA limits fragment size to 24 kb or less. Also, the original λ vectors do not allow propagation of viable bacterial cells that carry the inserted DNA fragment; the insert is propagated as part of a virus that lyses the cell.

For cloning large DNA fragments, the vectors needed are cosmids. The first part of their name, *cos,* comes from the fact that cosmids contain the cohesive ends or *cos* sites of normal λ. These ends are essential for packaging the DNA into λ phage heads. The last part of their name, *mid,* indicates that cosmids carry a plasmid origin of replication like the one found in the pBR322 plasmid.

Cosmids can be used for cloning in the same way as any other plasmid vector. However, because cosmids also contain the *cos* sites, cosmid DNA along with an inserted DNA fragment can be packaged as a λ phage. The result after packaging is a defective but nevertheless infectious phage particle. Once the cosmid and the inserted DNA fragment are introduced by infection into a λ-sensitive cell, the plasmid replicates.

Table 27.3
Theoretical Number of Clones Required to Fully Represent the Entire Genome of Various Organisms

Size of Cloned DNA Fragment (bp)	Genome Size (bp)		
	2×10^6 (e.g., bacteria)	**2×10^7 (e.g., fungi)**	**3×10^9 (e.g., mammals)**
5×10^3	400	4,000	600,000
10×10^3	200	2,000	300,000
20×10^3	100	1,000	150,000
40×10^3	50	500	75,000

Since cosmids lack the entire bacteriophage genome except for the region adjacent to the *cos* sites, these vectors can propagate exogenously derived DNA fragments of up to 40–50 kb in length.

Several cosmid vectors have been developed. These differ in size (and therefore cloning capacity), useful cloning sites, and selectable properties, or markers. For example, the pJB8 vector is a 5.1-kb plasmid containing several possible cloning sites and the amp^r selectable marker. Its cloning capacity is limited to DNA fragments between 30 and 47 kb. A feasible protocol for using this vector is described in box 27A.

Shuttle Vectors Can Be Cloned into Cells of Different Species

Vectors that include replication systems derived from more than one host species are known as shuttle vectors. Such vectors commonly include a replication system able to function in *E. coli* and one that works in a second host, which may be bacterial or eukaryotic. Initial cloning and amplification of the DNA segment to be studied is often carried out in *E. coli* because it is easier to make large quantities in a culture of *E. coli*. The recombinant DNA molecule—consisting of the "bifunctional vector" plus the cloned segment of DNA—is then introduced into the second host, where the purpose is usually to measure the expression of the genes carried by the vector. Shuttle vectors that can replicate in both *E. coli* and yeast are the most common.

Constructing a DNA Library

Cloning can produce a single vector-linked DNA fragment or a whole collection of independently isolated vector-linked DNA fragments derived from a single organism. Such a collection, termed a library, can serve as a good source of well-defined sequences from a given organism, since each clone of a library harbors a purified sequence from that organism. Within the entire library some sequences may be repeated but other sequences may be missing. The ideal library, which can only be approached, represents all of the sequences with the smallest possible number of clones.

As we saw at the outset of this chapter (see fig. 27.1), a library can be prepared in two ways. The genome may be fragmented and ligated to the appropriate vector to produce a genomic DNA library. An alternative approach is to construct a cDNA library, in which the DNA fragments are obtained by reverse transcription from the cellular RNA. Each type of library has advantages and disadvantages, and for a specific purpose one library is usually preferred over the other. In either case, the vast majority of DNA within the library is uncharacterized. Therefore the task of finding the desired genes or sequences within a library is often harder than the task of constructing the library. We will describe ways of selecting the correct clones after we have discussed the two types of libraries in more detail.

A Genomic DNA Library Contains Clones with Different Genomic Fragments

A major problem in constructing a genomic DNA library is to maximize the probability that all segments of the genome will be represented. If the genomic DNA is prepared by cutting with a restriction enzyme, an added problem is the possibility that the enzyme will cleave genes of interest at one or more sites.

To increase the likelihood of isolating desired genes in one piece, different restriction enzymes can be used on different parallel preparations. But even if the genes of interest are not cut by the enzyme(s) chosen, the DNA fragments produced may be inconveniently small to work with. An enzyme that recognizes a sequence of six bases (a 6-cutter) gives an average fragment size of 4,096 bp*, which is a reasonable size for making a plasmid library but much smaller than the size desirable for cloning in λ or a cosmid vector. Therefore, when large randomly generated fragments are desired, the method of choice usually entails making an incomplete digest with a 4-cutter restriction enzyme, which produces overlapping ends that can be readily cloned into the chosen vector as described earlier. The extent of digestion is controlled so that cleavage occurs at only some of the restriction enzyme recognition sites and the average size of the fragments produced is in the desired range. The conditions used thus depend on whether the product is going to be cloned in a plasmid, a λ phage, or a cosmid vector.

Table 27.3 gives the minimum number of clones (that is, the size of the library) required to fully represent the entire genome in a genomic DNA library, as a function of the average

*$(1/4)^6 = 1/4096$

27A
BOX

Construction of a Cosmid

The pJB8 vector is a 5.1-kb plasmid containing a λ cos site, an *amp*ʳ selectable marker, an origin of replication, and several possible cloning sites. To construct it, first one of the staggered-end ligation sites of the plasmid was opened with the appropriate restriction enzyme (step 1). In this case the *Bam*HI site was chosen. The linearized vector was then treated with alkaline phosphatase (step 2) to prevent recircularization of the vector in the subsequent ligation. The remaining procedures were dictated by the goal of this protocol, which was to produce very large cloned segments of DNA. Thus ligation had to yield large DNA fragments that were produced by random cutting (and thus were representative of the entire genome) and that could be inserted into the *Bam*HI site of the cosmid vector. On a random basis, any very large DNA fragment is likely to be cut several times by even a 6-cutter restriction enzyme under conditions of complete digestion. In order to avoid this problem and to obtain maximum randomness of fragments of the appropriate size, the *Mbo*I restriction enzyme was used under conditions of very limited digestion. This enzyme recognizes the four-base sequence of GATC, whereas *Bam*HI recognizes a six-base sequence, GGATCC. The two enzymes produce identical overlapping fragments (see upper right of illustration). Thus the eukaryotic fragments should ligate efficiently with the *Bam*HI restricted cosmid vector (step 3). Because limited digestion conditions were used, only some of the *Mbo*I (and *Bam*HI) sites were cleaved, and these varied from molecule to molecule on a random basis. Thus some molecules in the population were likely to contain an uncleaved enzyme recognition site at any given position. After ligation, the resulting concatamers were packaged *in vitro* into λ particles and introduced into a suitable *E. coli* strain (step 4). Transformants were then selected with ampicillin.

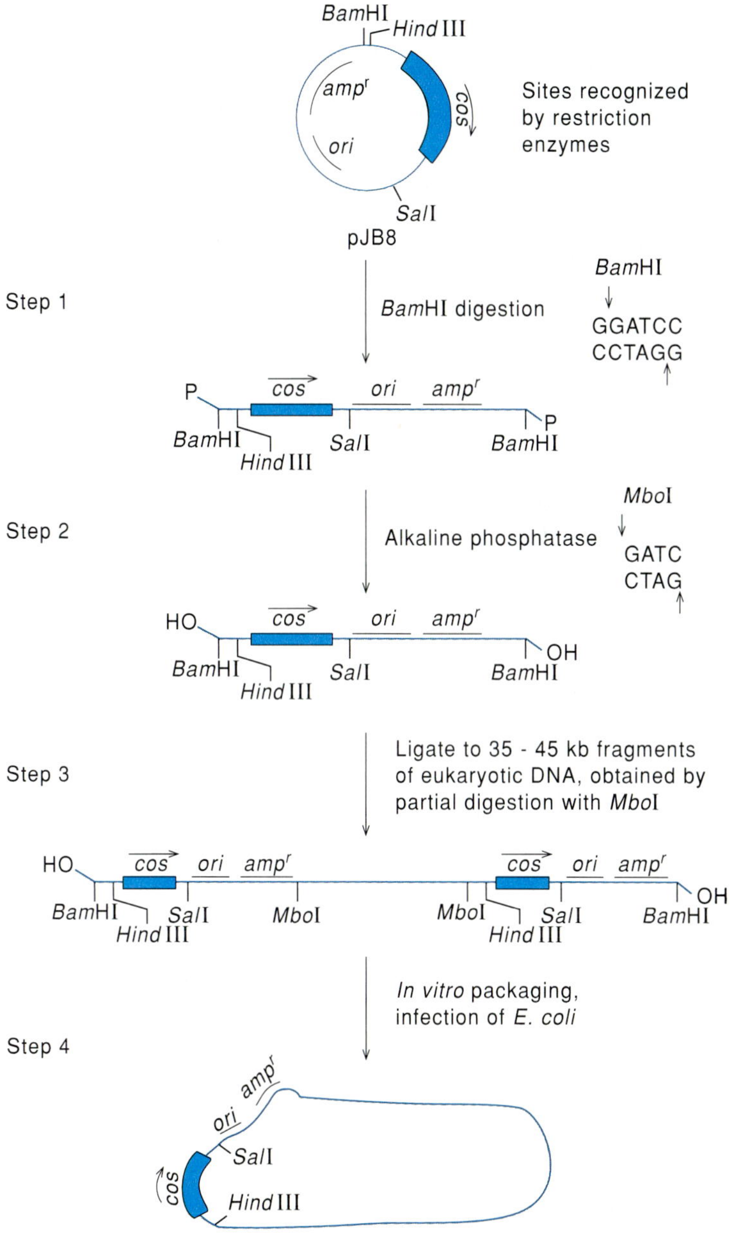

Figure 1

A plasmid containing a λ *cos* site is linearized by treatment with the BamHI restriction enzyme (step 1). The linear structure is treated with alkaline phosphatase to remove the 5′ terminal phosphates (step 2). Eukaryotic restriction fragments are ligated to the linearized plasmid in the presence of a large excess of the latter (step 3). The concatamer shown, which contains two *cos* sites, packages sufficiently into a λ phage that can be introduced by infection into an *E. coli* cell. The transfected structure readily circularizes (step 4). (From Geoffrey Zubay, *Genetics*. Copyright © 1987 Benjamin/Cummings Publishing Company Inc., Menlo Park, Calif. Reprinted by permission of the author.)

size of the cloned fragments and the size of the genome. Since DNA fragments in a population are cloned on a random basis, there is a 50% chance of finding a given single-copy gene in a library of the indicated size. A clone bank should be three to ten times the minimum size to give a high probability that a particular segment will be represented.

A cDNA Library Contains Clones Reflecting the mRNA Sequences

A cDNA library consists of a collection of clones that contain DNA copies of the cellular or organismic RNA. If the RNA is obtained from a differentiated multicellular organism, then the library varies in composition according to the type of cell used

Table 27.4
Distribution of Population of mRNA Molecules of Typical Eukaryotic Cells

Abundance Class	Fraction of the Total mRNA Population in Each Abundance Class	Number of Different mRNA Sequences in Each Abundance Class	Number of Copies of a Unique mRNA Sequence Present per Cell
High	22%	30	3,500
Medium	49%	1,090	230
Low	29%	10,670	14

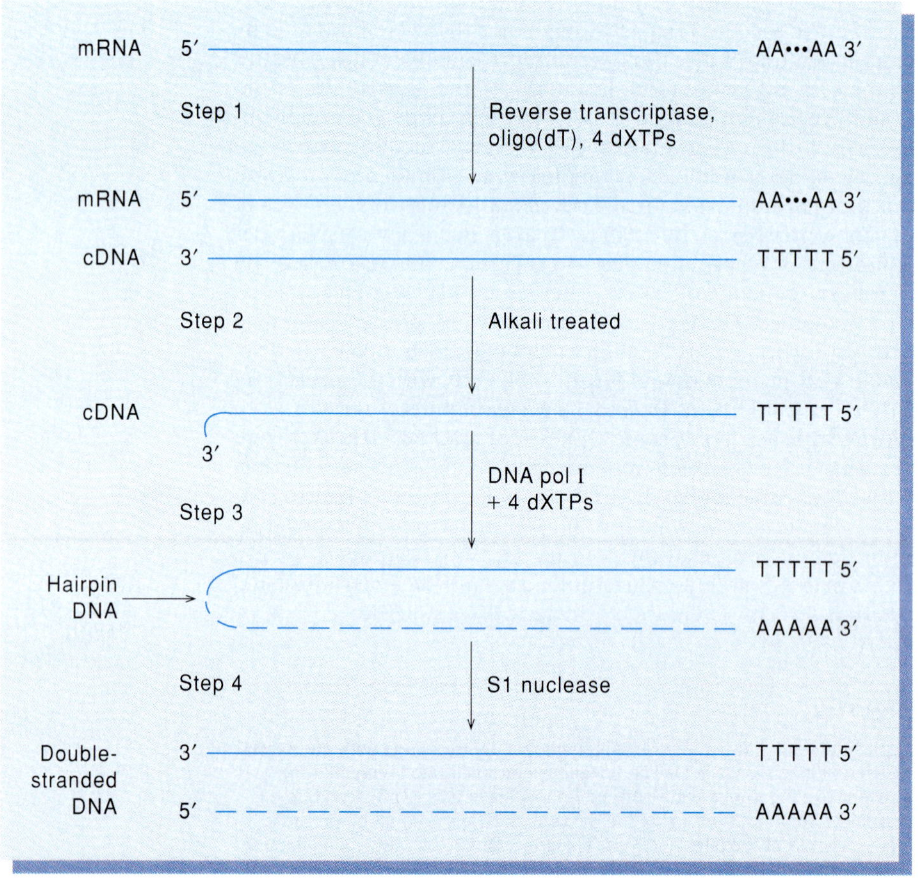

Figure 27.17

Formation of a cDNA duplex from a poly(A) mRNA. Step 1: To the poly(A) mRNA are added reverse transcriptase, oligo(dT), and the four-deoxynucleotide triphosphates. In favorable circumstances this step results in a full-length DNA-RNA hybrid duplex. Step 2: The RNA is removed and degraded by alkali treatment. Step 3: The remaining single-stranded DNA serves as both template and primer for the synthesis of a complementary DNA strand. The newly synthesized DNA is linked to the primer template DNA by a hairpin turn. Step 4: The hairpin is cleaved with S1 nuclease.

as the RNA source and also the physiological state of the cell. This variation is a reflection of the relative abundances of particular messenger RNAs made by different cell types. If a cDNA species corresponding to a particular gene product is desired, it is often possible to select a cell type known to synthesize a large amount of the corresponding mRNA. Thus pituitary cells can be used if cDNA encoding growth hormone is desired, and liver cells can be used if a serum albumin cDNA is the goal. mRNAs present in low amounts will clearly require the screening of a larger library than mRNAs present in medium or high abundance (table 27.4).

Sometimes some prepurification of the crude RNA is possible to increase the chances of finding a particular cDNA in the library. A technique commonly used with RNA from eukaryotic cells is to purify the crude RNA by chromatography over a cellulose column containing oligo(dT) chains. This procedure enriches for mRNA because most eukaryotic mRNAs contain a 3′-poly(A) tail that sticks to the oligo(dT) chains on the column.

The poly(A) tail not only aids in the fractionation step, but it also provides an important feature for the subsequent synthesis of the cDNA (fig. 27.17). The crude poly(A)-mRNA

Figure 27.18

Insertion of cDNA into pBR322 plasmid by the homopolymer tailing method. The cDNA duplex is tailed with terminal transferase and dCTP. pBR322 is opened with the *Pst*I enzyme and tailed with G residues. The two DNAs spontaneously anneal when mixed. The complex is transfected directly allowing for the repair synthesis and ligation to take place *in vivo.*

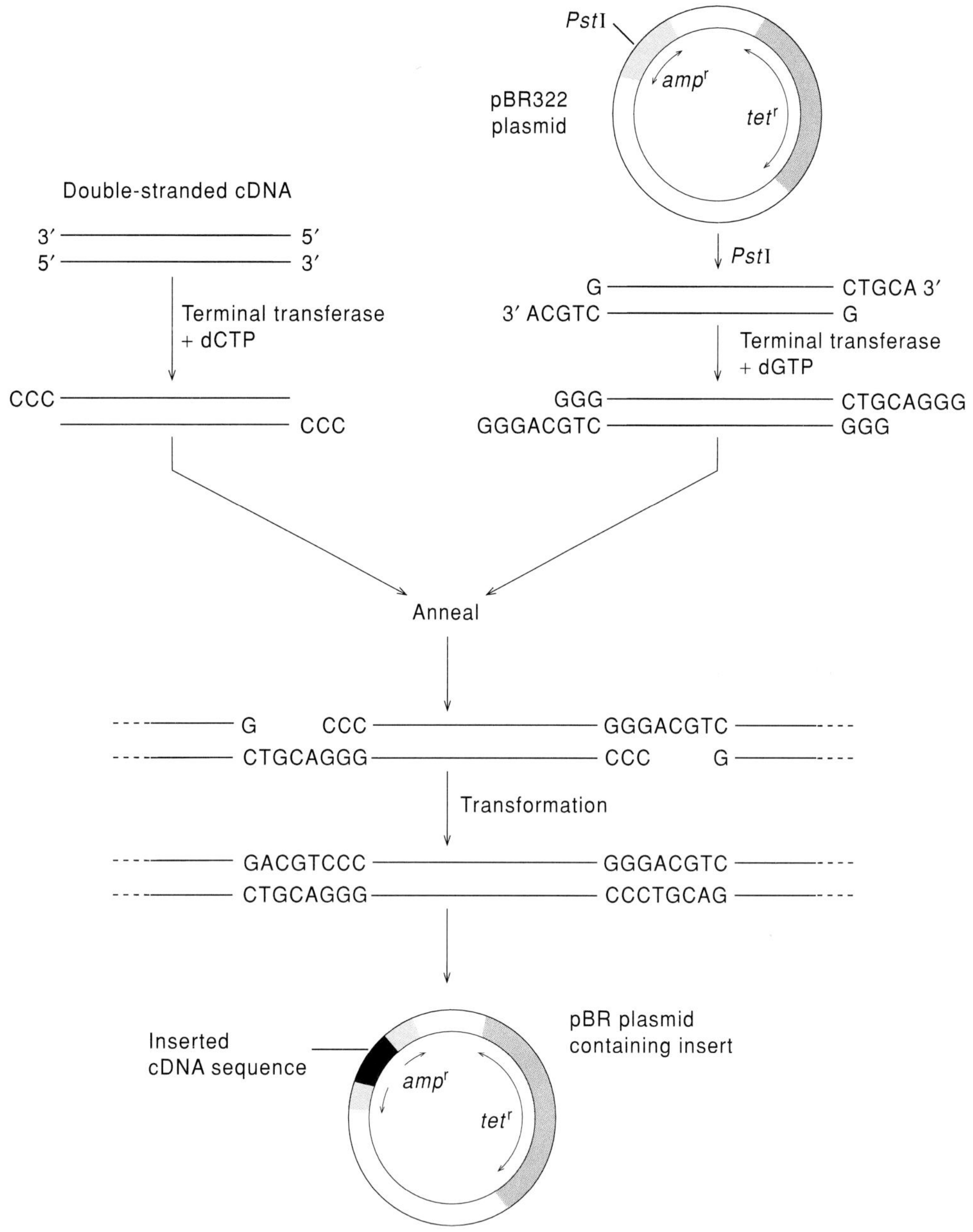

fraction serves as a template for the synthesis of the first strand of the cDNA, using the enzyme reverse transcriptase (step 1, fig. 27.17). When poly(A)-mRNA is the template, a short poly(dT) chain makes an ideal primer as it complexes with the poly(A)-containing portion at the 3′ end of the RNA. Synthesis of the polynucleotide chain then proceeds along the mRNA template in the 5′ direction. The product of this reaction is an RNA-DNA hybrid duplex, which, in favorable cases, extends the full length of the mRNA. The RNA strand is removed from the duplex by alkali treatment (step 2, fig. 27.17) and the re-isolated single strand of enzymatically synthesized DNA is used as a template for the synthesis of a complementary second strand of DNA by DNA polymerase I (step 3, fig. 27.17). This enzyme also has a requirement for primer; commonly the 3′ end of the first strand of DNA seems to be able to loop around and serve as a primer and template for its own synthesis, leading to the formation of a duplex DNA molecule having its strands joined together at one end like a hairpin. The next step (step 4, fig. 27.17) in the preparation of a cDNA duplex involves scission of the hairpin. This can be done by brief treatment with S1 nuclease, which acts as an endonuclease on single-stranded regions of the DNA.

After S1 cleavage, appropriate ends are attached to the newly synthesized duplex to make it suitable for insertion into a plasmid or other vector. There are two ways of doing this. One is called homopolymer tailing. The double-stranded cDNA is tailed with one base, and a suitable plasmid vector such as pBR322 is tailed after opening with the complementary base. A detailed protocol for this is shown in figure 27.18. The double-stranded cDNA is 3′-tailed with C residues. The plasmid is

Figure 27.19

Insertion of cDNA into pBR322 plasmid by the linker method. The strategy here is to open up the plasmid with a restriction enzyme that makes staggered cuts and to attach linkers that contain the same recognition site to the cDNA. After the linkers are attached to the cDNA, the duplex is treated with the same restriction enzyme (*Pst*I) to expose the overhangs. The two DNAs are mixed together and ligated. After transfection, cells containing the hybrid plasmids are recognized by tetracycline resistance and ampicillin sensitivity. Identification of the insert is discussed in the text.

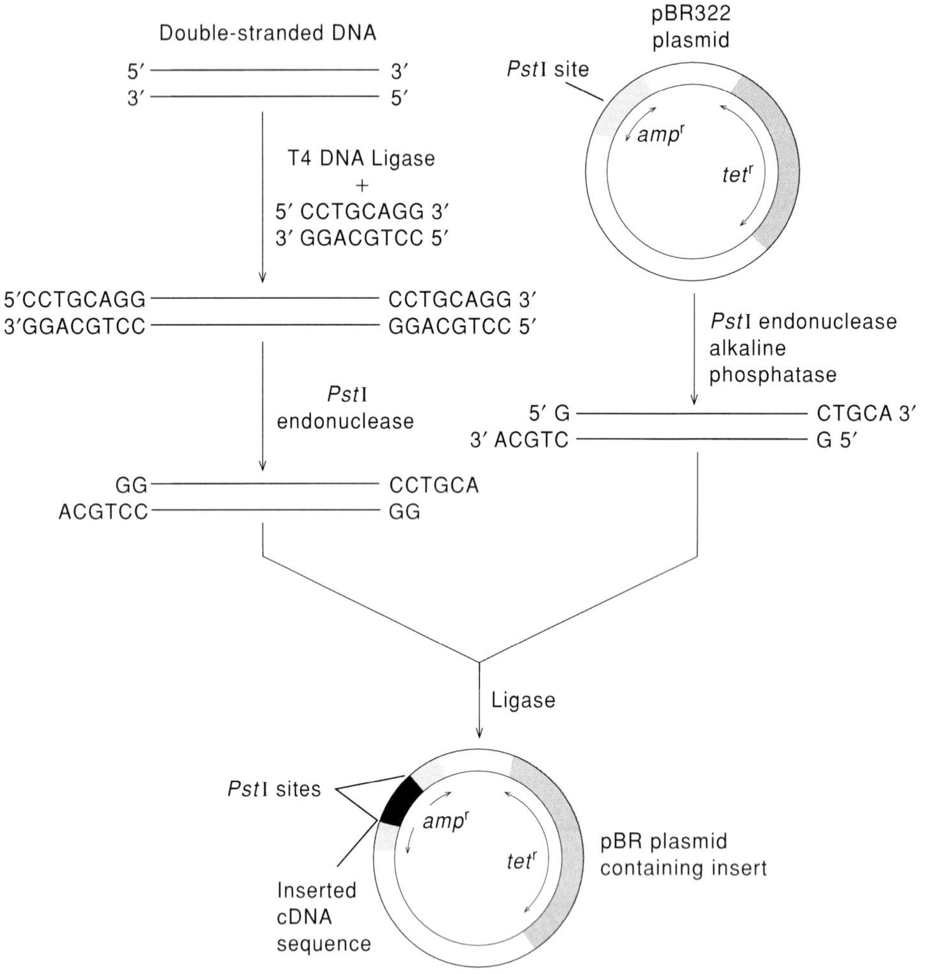

linearized by treatment with *Pst*I restriction endonuclease at a unique site in the *amp*^r gene. This treatment results in a linear duplex with a four-base 3′-OH overhang, TGCA. This is 3′-tailed with G residues. The two-tailed duplexes are mixed together, whereupon they spontaneously anneal. The annealed complex can be directly transfected, leaving the final folding-in to take place in the bacterial cytoplasm.

Another way of inserting the cDNA into a vector is to tail it with linkers. Linkers are synthetic single-stranded oligonucleotide segments (6, 8, 10, or 12 bases in length) that self-associate to form symmetrical blunt-ended double-stranded molecules containing the recognition sequence for a particular restriction enzyme. Figure 27.19 shows an eight-base linker (CCTGCAGG) containing a *Pst*I recognition site. This linker self-associates to produce an eight-base blunt-ended duplex structure that adds to the double-stranded cDNA in the presence of T4 ligase. The resulting product is treated with *Pst*I to produce the characteristic 3′ overhang. The plasmid, linearized with *Pst*I, and the two DNAs are mixed and reacted with ligase to produce plasmid with the insert. Considerable plasmid will reclose without incorporating the insert, but this material can be distinguished from plasmid with the insert because the latter has lost its ampicillin resistance. Alternatively, the linearized plasmid DNA can be treated with alkaline phosphatase to remove the 5′ phosphates from its termini. Phosphate removal prevents the plasmid from recircularizing. When the terminally dephosphorylated duplex is reacted with the cDNA insert in the presence of ligase, a circular structure with two nicks is formed. Upon introduction into a bacterial cell, these nicks are mended.

Although a cDNA library requires more work to prepare than a genomic DNA library, it has major advantages for certain purposes:

1. Since only a small fraction of the genomic DNA of eukaryotes is expressed at any one time in any one cell type, the variety of clones is greatly reduced. If the desired genetic sequence is among those cloned, screening is much easier.

2. cDNA contains only a small number of repetitive sequences, whose presence can complicate screening when cloned DNA is used as a probe. A probe with repetitive sequences is likely to anneal to any structure that has similar repetitive sequences, whereas the goal is usually to use a probe that will anneal to a unique sequence in the genome.

3. Because intervening sequences or introns* present in eukaryotic DNA are not present in the RNA (and consequently in the cDNA), the cloned cDNA frequently can be used in *E. coli* or other microorganisms to obtain the proteins encoded by the eukaryotic gene. A DNA containing introns is not properly expressed in *E. coli.*

4. A single cDNA clone is pure in the sense that it contains only the sequences corresponding to a single mRNA. For this reason the results obtained when using the cDNA probe are often easier to interpret than data from probing experiments that employ cloned genomic DNA segments, which sometimes may contain two or more different genetic sequences.

There Are Numerous Approaches to Picking the Correct Clone from a Library

A library can contain thousands or even tens of thousands of different kinds of clones (see table 27.3), and the screening needed to isolate a clone that has the DNA of interest can take a lot of time. Cloned DNAs do not usually include easily selectable genetic markers, such as a gene conferring resistance to a specific antibiotic.

Most currently used procedures for screening large numbers of colonies for plasmids or phage that contain specific DNA inserts are variants of the colony hybridization method developed by Grunstein and Hogness. This procedure makes use of a specific radioactive probe that contains some sequences complementary to those in the DNA of interest. The colonies to be screened are first grown on agar Petri plates (fig. 27.20). A replica of each plate is made on another agar plate, which is stored for reference. A replica is also made on a nitrocellulose filter. The colonies formed on the filter are lysed and the contents denatured simultaneously by treatment with sodium hydroxide. After heating, the denatured DNA is fixed on the filter at each site where a colony was located. The DNA on the filter is then hybridized with a radioactively labeled nucleic acid probe complementary to the specific DNA sequence to be selected. The presence of hybridized probe at sites occupied by DNA derived from colonies that include the DNA fragment of interest is detected by autoradiography on film. The colony whose DNA hybridizes with the nucleic acid probe can then be picked from the reference plate, which contains a viable bacterial colony at a corresponding location.

A parallel procedure developed by Benton and Davis makes it possible to screen bacteriophage λ recombinant DNA clones by hybridization to single plaques *in situ.* The number of phage plaques that can be placed on a single Petri plate is much larger than the number of individual bacterial colonies

Figure 27.20

Colony hybridization procedure used to identify bacterial clones harboring a plasmid containing a specific DNA. Step 1: Replica-plate the colonies containing plasmids onto nitrocellulose paper. Step 2: Lyse cells with NaOH and fix denatured DNA to paper. Step 3: Hybridize to ^{32}P-labeled DNA carrying desired sequence and autoradiograph the product. Locations of desired DNA should be emphasized in autoradiograph. Clones carrying desired plasmids (circled) may then be isolated from a corresponding agar replica plate carrying untreated colonies.

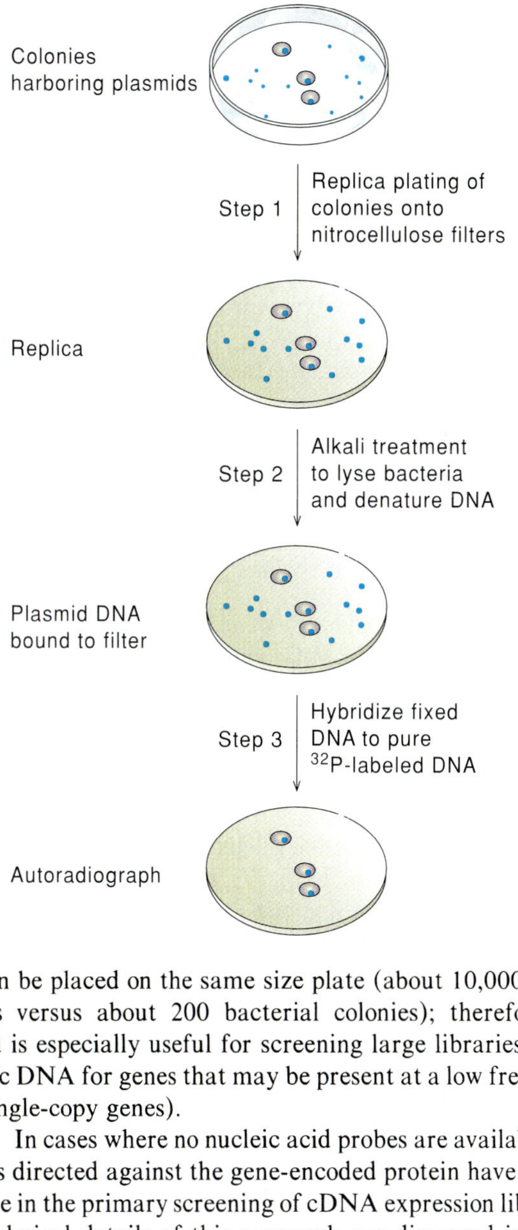

that can be placed on the same size plate (about 10,000 phage plaques versus about 200 bacterial colonies); therefore the method is especially useful for screening large libraries of eukaryotic DNA for genes that may be present at a low frequency (i.e., single-copy genes).

In cases where no nucleic acid probes are available, antibodies directed against the gene-encoded protein have proved valuable in the primary screening of cDNA expression libraries. The technical details of this approach are discussed in references given at the end of this chapter.

Cloning in Systems Other Than *E. coli*

Despite the success and broad applications of *E. coli* cloning systems, gene products cannot always be made in this bacterium. Sometimes they are not synthesized in their entirety, or they are rapidly broken down after synthesis. In addition, for

*Introns are noncoding regions located between coding regions. They are much more common in eukaryotes than prokaryotes (see chapter 28).

the study of certain processes indigenous to other species (e.g., photosynthesis, antibiotic production) it is often necessary to use a host bacterial species that carries out the process naturally, rather than *E. coli.*

Effective gene cloning systems are available for a variety of bacterial hosts, including *Bacillus subtilis, Streptomyces* species, and *Agrobacter tumefaciens.* Cloning systems have also been developed for eukaryotic hosts. In this section we will consider examples of cloning in three eukaryotic systems—the yeast *Saccharomyces cerevisiae,* mammalian cells in tissue culture, and plant cells.

Yeast Is the Most Popular Eukaryotic Cell for Cloning

Simple and generally useful techniques exist for isolating and amplifying virtually any yeast gene in *E. coli.* Moreover, by transformation these genes can be returned to the yeast cell. Such procedures have greatly facilitated the molecular dissection of yeast genes; they have also provided a variety of new approaches to gene mapping.

Yeast gene isolations usually begin by taking fragments of total yeast DNA and cloning them into an *E. coli* vector, either λ bacteriophage or a plasmid. After transfection or transformation, respectively, *E. coli* cells containing specific regions of the yeast DNA are selected. Cells containing the desired fragments are then amplified, purified, and sometimes modified, after which the fragments may be returned to yeast by transfection.

The first isolation of a gene encoding a protein in yeast took advantage of the fact that many yeast genes function in *E. coli.* For example, a plasmid containing an origin for replication in *E. coli* and the yeast *LEU2* gene was selected from a yeast plasmid library by its ability to restore to a leu^- mutant of *E. coli* the ability to grow in the absence of added leucine. The purified clone containing the *LEU2* gene could then be returned by transfection to a yeast cell defective in *LEU2.* Subsequent analysis showed that stable transformants contained the *LEU2* gene, integrated at the homologous site in the yeast genome. Thus the yeast recombination system, like the *E. coli* recombination system, appears to have the capacity to guide a gene introduced by transfection to a site of homology. Selection methods now exist for isolating mammalian cells showing homologous recombination of transfected sequences (see Selected Readings article by Mansour et al.).

The transformation frequency in yeast, which is usually quite low, may be greatly increased by incorporating special yeast DNA sequences into the plasmid used for transfection. The best-known of these special sequences derives from the so-called 2μ plasmid found in most *Saccharomyces* strains. Addition of a fragment of the 2μ plasmid into a plasmid containing a selectable gene, such as *LEU2,* increases the transformation frequency a thousandfold or more. High-frequency transformation using all or part of the 2μ plasmid is usually due to autonomous replication of the plasmid, which is maintained in multiple copies. Other types of high-frequency transformation have been observed when certain yeast sequences, referred to as autonomously replicating sequences (ARS), are an integral part of the transforming plasmid. These probably contain a yeast chromosomal origin for DNA synthesis (see the discussion of ARS sequences in chapter 26).

Integrative transformation with specially constructed plasmids has opened the door to many forms of gene mapping that would be much more difficult to do by conventional mapping techniques. In addition, it has permitted the construction of unusual gene arrangements. For example, if a yeast-integrating plasmid contains a gene to be transposed and a segment of yeast DNA homologous to another location on the yeast gene, then it should be possible to transpose the selectable gene to another position on the genome.

Studies on Cloned Genes in Mammals Start with Tissue Culture Cells

Mammalian cells from various sources can be adapted for growth as single cells in liquid culture or on plates. Such cells can be formally treated like bacteria or yeast; single cells give rise to genetically homogeneous colonies. Cultured cells have proved to be effective recipients of cloned DNAs.

The simplest procedure for transfection of mammalian cells utilizes purified DNA. The DNA is mixed with calcium chloride and sodium phosphate to give a finely divided calcium phosphate–DNA precipitate. Treated cells appear to take up the DNA by endocytosis. Transfection efficiencies, measured as the fraction of treated cells that become transfected, vary from 10^{-4} to several percent by this method.

Cultured mammalian cells permit many types of studies that would be impractical to carry out on whole animals. However, only a few mutants provide readily selectable genetic markers; this limitation creates a problem in selecting for transformed colonies. One of the best systems available for selection involves the thymidine kinase (*tk*) gene of herpes simplex virus. Mutants that are tk^- have been isolated for a number of cell types, including mouse, rat, and human. Such cells cannot grow on the selective medium known as HAT. The tk^- cells can be transformed to tk^+ cells by plasmid DNA that contains the herpes simplex virus *tk* gene. Other genes of interest can be inserted into the same plasmid vector so that selection of tk^+ cells after transformation leads to a high probability of coselection of adjacent genes.

One of the most versatile experimental systems using DNA transfection is the mouse teratocarcinoma cell (TCC) system. With this system it is possible to transfect cells in tissue culture as in other tissue culture systems and then to implant the cells into a growing embryo so that the cells become part of the animal. By this procedure we can study the effects of cells altered by transfection in any organ of the whole mammalian organism.

The teratoma is a unique type of tumor found in many kinds of mammals. It is composed both of neoplastic cells, such as occur in other tumors, and also of many kinds of differentiated cells. A typical teratoma may include nerve cells, muscle cells, blood cells, skin cells, and other differentiated cells all mixed together with neoplastic stem cells. Only the stem cells are neoplastic, producing more stem cells or more differentiated cells, usually both. In many ways the stem cells behave like embryonic cells. They can be dissociated into single cells and cultured *in vitro* like bacterial cells. Cells grown from the culture can be reintroduced into the animal by subcutaneous implantation, in which case they produce a tumor. Most remarkably, they can be introduced into an early embryo (blastocyst) to produce a hybrid chimera where, in favorable cases, all of the tissues possess some cells from the tumor parent (fig. 27.21). Even egg cells have been isolated that are derived from the tumor parent. These egg cells, when fertilized by normal sperm, result in progeny that could truly be said to have a tumor for a mother. This example shows that the neoplastic stem cells are capable of reverting to completely normal behavior when subjected to the embryonic environment.

Taking advantage of this quite remarkable result, Bea Mintz and her co-workers have shown that teratocarcinoma (TCC) stem cells can function as vehicles for the introduction of specific recombinant genes into mice. For this purpose, TCC cells were first grown in single-cell tissue culture, and cells with a thymidine kinase deficiency were selected and treated with DNA containing the human β-globin gene and the thymidine kinase (*tk*) gene. Then tk^+-transformed cells were selected with the help of HAT medium as explained earlier. Further tests revealed that the majority of the transformants also had copies of the human β-globin gene. Thus, although there is no facile selection procedure for the human β-globin gene, it can be successfully cotransfected by linking it to the *tk* gene. Such altered TCC cells can then be introduced into early embryos to engineer new mouse strains carrying the human β-globin gene. This experiment shows that virtually any gene can be transfected into the whole animal by this two-step procedure.*

Figure 27.21

Manipulation of mouse teratoma cells. The cells from a teratocarcinoma may be dispersed and grown in tissue culture. These cells can be injected into an embryo (A), in which case the resulting animal is a chimera in which some cells come from the original parents and others arise from the cells injected into the blastocyst. (B) Alternatively, these cells may be implanted subcutaneously, in which case the animal develops a tumor at the site of implantation.

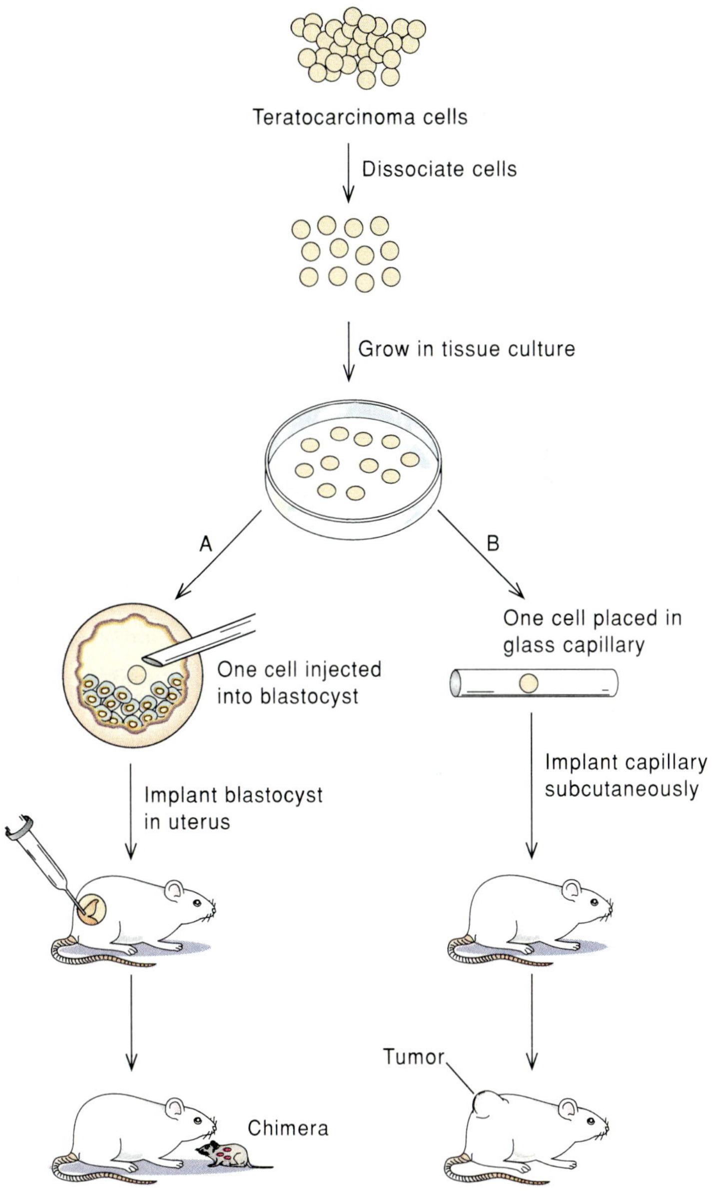

Oncogenes Can Be Selected from a Genomic Library by Subculture Cloning

Sometimes a gene can be identified when it confers an obvious phenotype on a mammalian cell. However, even in such cases indirect selection procedures often must be used because of the limitations in the cloning vehicles currently available for mammalian cells. In a strategy called subculture cloning, a genomic library of mammalian DNA is made in *E. coli* and then clones from the library are tested in subgroups in mammalian cells until one exhibiting the desired properties is identified. The DNA of interest is then propagated in *E. coli*. This selection method is illustrated in figure 27.22 for the isolation of an oncogene from a chicken lymphoma. A library containing 200,000 phage clones was divided into sublibraries, and the DNA of the clones in each sublibrary was introduced by transfection into mouse fibroblast cells growing in tissue culture. Cellular transformation was assayed by the appearance of a clone or clones of rapidly dividing cells (foci) superimposed on a background of slowly dividing, nontransformed cells. The sublibrary that assayed positively was further subdivided and retested. This procedure was repeated until a clone was identified that conferred the rapid growth properties associated with oncogene transformation.

Subculture cloning depends on the sensitivity of the assay used for detecting a small number of positive clones in a heterogeneous population. Alternatives to subculture cloning

*Embryonal stem cells (ES cells) are frequently used in place of TCC cells.

Figure 27.22

Subculture cloning procedure. Subculture cloning is used in successive steps to isolate the clone that contains the gene of interest. In this case the gene of interest was an oncogene originating from a chicken lymphoma and identified by the rapid growth characteristics it confers on transformed cells. A λ library was made with the chicken lymphoma nuclear DNA. The library was subdivided into ten approximately equal lots (sublibraries). From each lot the DNA was isolated and used in a transformation assay. Lot 2 scored positively. This lot was further subdivided and the procedure was repeated again and again until a pure plaque carrying the oncogene was isolated.

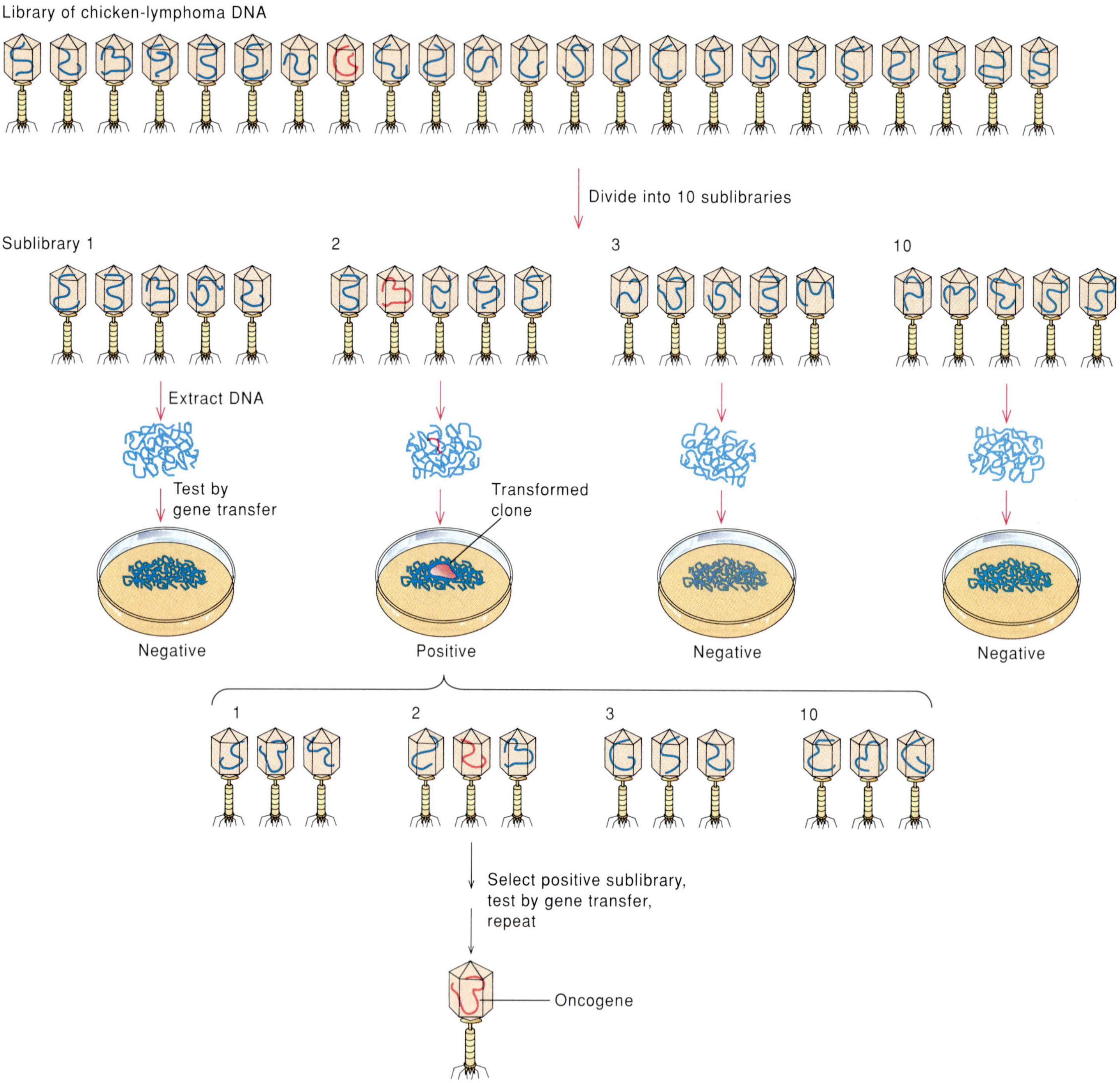

have been devised, but these methods still clone genes in bacteria and then detect them in mammalian cells. However, papilloma viruses, which can function as true plasmid vectors in mammalian cells, may permit the direct cloning and selection of genes in mammalian cells.

Cloning in Plants Has Been Accomplished with a Bacterial Plasmid

Tissue culture systems for plants are hard to manipulate and grow for extended periods of time. The difficulties of cloning DNA in plants are compounded by the fact that there is no plasmid that can be introduced directly into plant cells. Much plant DNA cloning uses a naturally occurring host-vector system involving *Agrobacter tumefaciens,* a pathogenic soil bacterium that causes crown gall tumors in a wide variety of dicotyledonous plants. (It is noteworthy that this is the only known case of naturally occurring nucleic acid transfer between a prokaryote and a eukaryote.) These plant tumors are caused by tumor-inducing (TI) plasmids in some agrobacteria; related bacteria that lack the plasmid are not tumorigenic.

The TI plasmids are large circular DNA duplexes ranging in size from 160 to 240 kb. Bacterial infection of the plant leads to development of a tumor, usually near the junction of root and stem, as a consequence of a plasmid segment called T-DNA. T-DNA is integrated into the nuclear genome of the plant cells, where it alters the metabolism of the tumor cells so that small molecules known as opines are synthesized in large quantities. Different TI plasmids encode genes that lead to the synthesis of different types of opines. While these compounds are of no apparent use to the plant's cells, they provide a growth substance for the agrobacteria that contain the TI plasmid. Other genes carried by the TI plasmid confer upon the host bacteria the ability to metabolize the T-DNA-encoded opine and use it as a substrate to promote bacterial growth.

It is impractical to use the entire TI plasmid as a cloning vector because of its size. However, foreign genes can be transferred into plants by introducing them into the T-DNA segment of TI, reinserting the T-DNA into TI, and then letting the natural biological process of T-DNA transfer to the plants cells do the rest (see box 27B). Some plant cells carrying part or all of the TI plasmid can regenerate whole plants, so that it is possible to measure the effects of transfected genes on any tissues of the whole plant.

Site-Directed Mutagenesis

By combining different procedures of DNA manipulation, we now can make discrete changes in a given gene. This technique, called site-directed mutagenesis, is one of the most important manipulations in modern genetics. The first site-directed mutagenesis studies were carried out by Shortle and Nathans in 1978 with the help of the mutagen sodium bisulfite, which deaminates C residues and thus converts them into U residues. Directed mutagenesis as it is practiced today is based on the chemical synthesis of a deoxyoligonucleotide that contains discrete changes in its sequence from that normally observed in the genome under investigation. These changes may be single base or multibase, and they may involve base changes, base deletions, or base additions.

There are many variations of site-directed mutagenesis. We can start out with a circular single-stranded DNA and anneal it to a synthetic primer DNA carrying the desired changes (fig. 27.23). After extending the primer, we transfect the resulting product. Finally we select clones of cells containing the plasmid with the desired changes.

For most site-directed mutagenesis, however, the polymerase chain reaction (PCR) will probably be the method of choice in the future. The use of PCR for this purpose requires

Figure 27.23

Scheme for oligonucleotide-directed mutagenesis of double-stranded circular plasmid DNA. Supercoiled plasmid circles are nicked in one strand and rendered partially single-stranded by treatment with exonuclease. The gapped circles are hybridized with a homologous oligodeoxynucleotide carrying, by design, some mismatches. *In vitro* DNA synthesis, primed in part by the oligodeoxynucleotide, leads to heteroduplex plasmid circles. (Source: G. Dalbadie-McFarland et al., "Oligonucleotide-directed mutagenesis as a general and powerful method for studies of protein function" in *Proceedings of the National Academy of Sciences USA* 79:6408-6412, 1982. Copyright © 1982 National Academy of Sciences, Washington, D.C.)

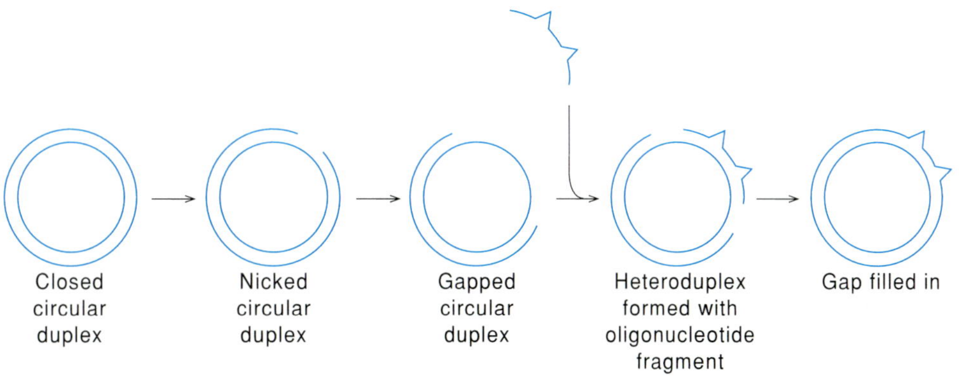

27B

BOX

Construction of a T1 Plasmid Derivative Suitable for Cloning

The following procedure is quite complex, but it introduces some new and very worthwhile principles of cloning. First, a segment of T-DNA that included a preselected restriction enzyme cleavage site was cloned into pBR322 for replication in *E. coli* (step 1). The plant gene to be cloned was covalently linked to a selective genetic marker, the kanamycin resistance (*kan*^{r}) gene (step 2). The segment containing both the *kan*^{r} gene and the plant gene was spliced into a T-DNA segment cloned in the pBR322 plasmid (step 3), which was used for amplification of the plant gene in *E. coli*. However, since pBR322 does not replicate in agrobacteria, the segment containing the T-DNA and the *kan*^{r} gene was inserted into the plasmid pRK290, which replicates in either bacterial host (step 4). The modified T-DNA fragment, now attached to pRK290, was introduced by transformation into an agrobacterium that contained a T1 plasmid (step 5).

Transformed cells retained both the T1 plasmid and several copies of the pRK290 plasmid. The T-DNA segment in pRK290 recombines *in vivo* with the homologous T-DNA region in the T1 plasmid, thus inserting the modified T-DNA segment (step 6). But this event occurs spontaneously at a very low frequency, so it was necessary to devise a way of screening for those cells that contained the T1 plasmids with a modified T-DNA segment. A strategy referred to as plasmid eviction was used to accomplish this. It was reasoned that if pRK290 could be evicted under conditions requiring kanamycin resistance, the most likely survivors should be T1 plasmid recombinants containing the modified T-DNA segments with the *kan*^{r} gene. It is known that when two closely related plasmids cohabit the same cell, incompatibility between them results in the loss of one. R751MG2, a plasmid closely related to pRK290 but carrying a gentamycin resistance gene

Figure 27.24

Illustration of a general method of mutagenesis using PCR. Primers are represented as short lines with arrowheads pointing in the 3′ direction. —^— indicates a mismatched base.

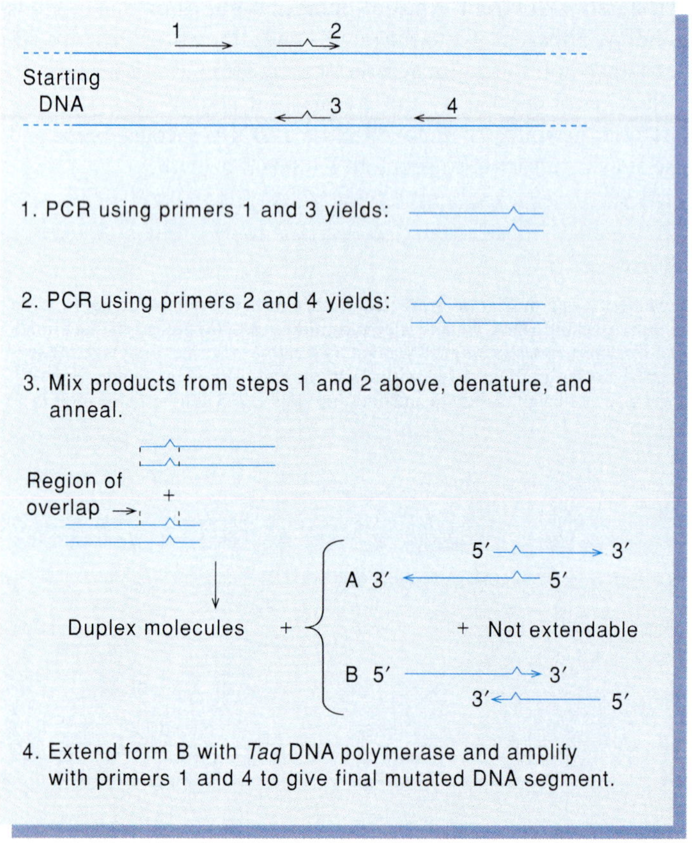

a piece of duplex starting DNA, two outside flanking primers that are perfect complements to segments of opposing strands, and two complementary primers with the desired changes in their sequence as diagrammed in figure 27.24. The steps followed parallel the steps of PCR amplification.

PCR amplification can be coupled with classical cloning methods using cloning vectors. For this purpose the PCR product should contain restriction sites at its ends that are suitable for cloning. Thus PCR amplification and cloning need not be thought of as alternatives for certain purposes but as complementing one another to give a greater variety of approaches.

Recombinant DNA Techniques

The human globin family of proteins is a paradigm for studying differential gene activity during development and the molecular basis of genetic disorders in gene expression. Hemoglobin is a tetramer containing two α-like and two β-like subunits (see chapter 5). These proteins are encoded by a small number of genes that are expressed sequentially during development. The (*gent*^{r}) and lacking a T-DNA segment, was introduced by conjugation with *E. coli* containing pRK290 and the T1 plasmid. To direct the process so that pRK290 was evicted instead of R751pMG2, the conjugation was done in the presence of both gentamycin and kanamycin (step 7). Only cells that carried both of the drug resistance genes could survive; these cells had lost the pRK290 plasmid, but they retained the *kan*^{r} gene in T1 plasmids that had acquired the T-DNA/*kan*^{r} gene segment by recombination. The R751MG2 plasmid was also present. The T1 plasmid in the agrobacterium could then be used for transfer of the desired gene to a plant cell system.

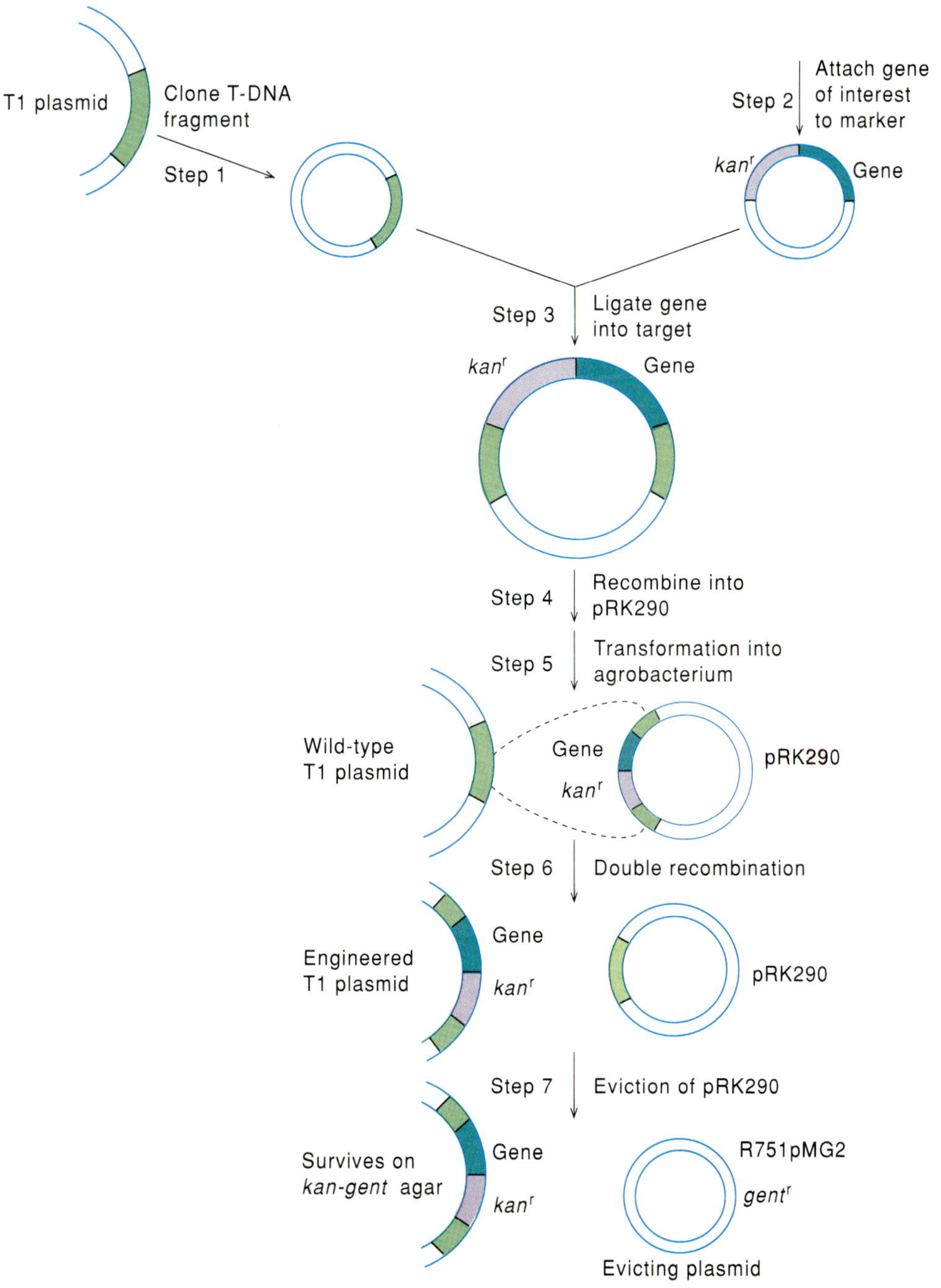

Figure 1

Construction of T1 plasmid for cloning. (From Geoffrey Zubay, *Genetics*. Copyright © 1987 Benjamin/Cummings Publishing Company Inc., Menlo Park, Calif. Reprinted by permission of the author.)

Figure 27.25

Changes in types of hemoglobin observed in early development. A single switch in gene expression is observed for α-like chains. Two switches in gene expression are observed for ß-like chains. The corresponding tetrameric hemoglobin molecules observed at different stages in development are also indicated.

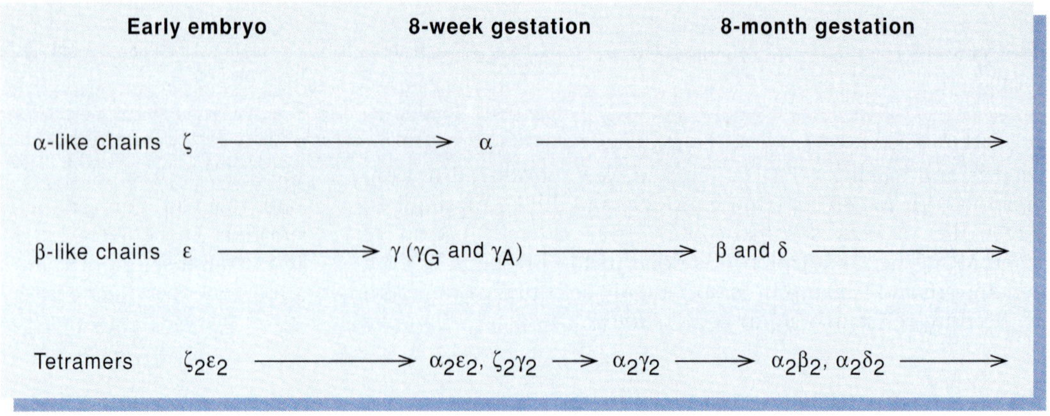

Figure 27.26

The chromosomal localization and genomic organization of the human globin genes. The α- and ß-globin gene complexes are positioned on chromosomes 16 and 11, respectively. For each complex, the arrangement of genes on the chromosome is depicted above, and the general structure of the major gene is shown below, together with the location of the intervening sequences or introns (IVS) and codon numbers. Coding regions are shown by solid boxes and IVS regions by open boxes.

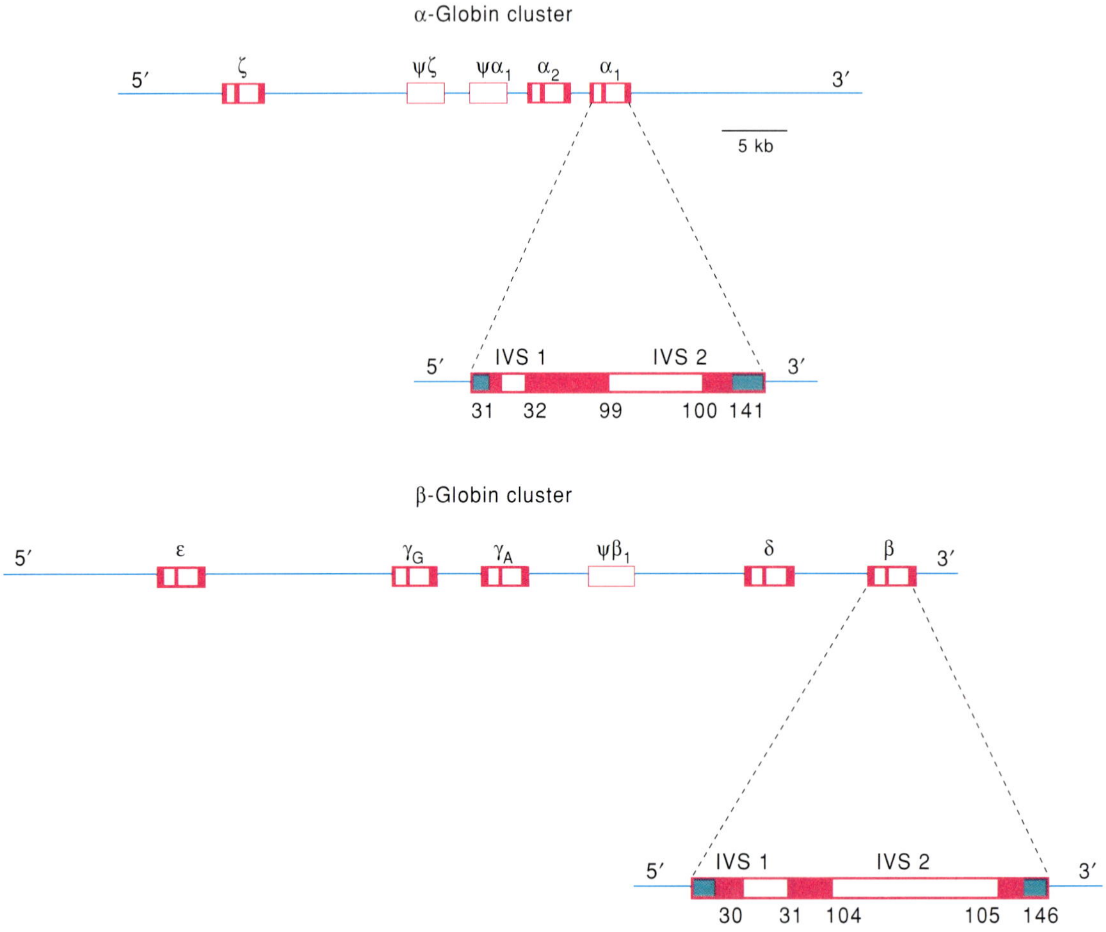

information summarized in figure 27.25 indicates that the α-like and β-like globin gene families have coordinated programs for expression: Two switches in gene expression (embryonic to fetal to adult) are observed for the β-like genes, while a single switch results in activation of adult α-globin production early in fetal life. A combination of classical and recombinant DNA techniques has been used to show that the α-like genes are located in a single cluster on chromosome 16, while the β genes are located in a single cluster on chromosome 11 (fig. 27.26).

cDNA Probes Were Used for Detecting the Genomic Globin Sequences

The first time that the technique of mapping with recombinant DNA probes was applied, it was to the human globin genes. The procedure used was that of DNA hybridization in solution. Radioactive human α- and β-globin cDNA probes reanneal to form stable duplexes with the corresponding human genomic DNA sequences, and do not cross-react with each other or with mouse globin genes. The presence of duplex material in a hybridization reaction containing human globin cDNA and hybrid cell genomic DNA thus indicates the presence of the human globin gene DNA in the hybrid. However, the liquid hybridization procedure has two drawbacks. First, it requires large amounts of DNA. More important, in many cases the donor gene probe does cross-react with the homologous recipient cell gene, so that interpreting the assay is difficult. This disadvantage limits the range of genes for which this mapping procedure is useful.

The Southern blotting procedure avoids both difficulties (fig. 27.27). In this procedure, genomic DNA isolated from somatic cell hybrids is digested with restriction endonucleases to yield specific DNA fragments. These are fractionated according to size by means of agarose gel electrophoresis. The fractions are then denatured, transferred to nitrocellulose sheets, and hybridized with isotopically labeled DNA probes. Noncoding sequences, flanking and intervening, in a structural gene diverge rapidly, so that even genes highly conserved between two species can usually be resolved.

Isolation of the human β-globin gene was greatly simplified because it was possible to obtain the mRNA in essentially pure form from reticulocytes, the cells that carry vast quantities of hemoglobin. From this mRNA a cDNA probe was made and this probe was used to isolate the gene from a genomic library. All members of the human genomic library that anneal to the radioactive cDNA probe were isolated and each of these was cross-hybridized. Restriction-site mapping to each of these DNAs, followed by sequence analysis, resulted in a complete description of the β-globin gene. This analysis showed that the β-globin gene was appreciably larger than the β-globin messenger RNA; in addition to containing regions that are present in the final mRNA, the gene contains two intervening regions not represented in the mRNA sequences (see fig. 27.27). We will have more to say about intervening sequences in the next chapter.

Figure 27.27

The steps involved in assaying by Southern blotting. The DNA to be analyzed is digested with a restriction enzyme (1). The resulting fragments are electrophoresed on an agarose gel (2). The DNA fragments on the gel are transferred to a cellulose nitrate sheet by placing the cellulose nitrate sheet next to the gel and passing solvent through the gel into the sheet. Flow of the solvent is maintained by blotting the far side of the cellulose nitrate sheet with paper towels. The DNA, first denatured with alkali, flows with the solvent but gets stuck in the sheet (3). The sheet is hybridized to radioactively labeled DNA containing the gene sequence of interest (4). The hybridized sheet is autoradiographed to determine the location of the labeled restriction fragment on the gel (5).

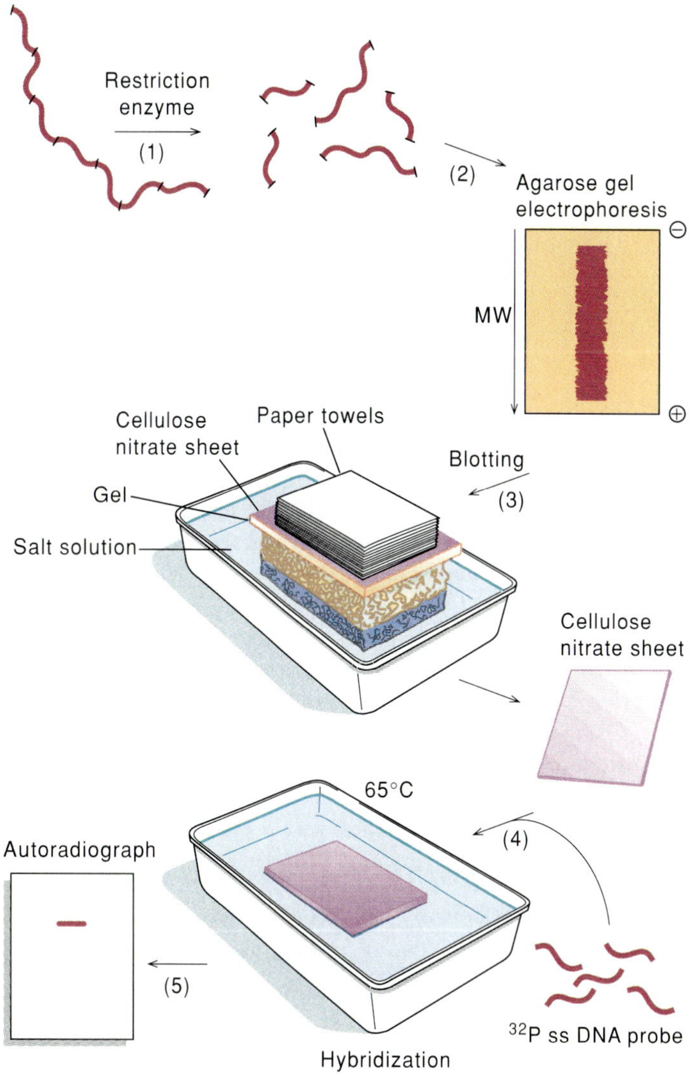

Chromosome Walking Was Used to Identify and Isolate the Regions Around the Adult Globin Gene

Further studies of the β-globin gene resulted in an analysis of the region surrounding the gene. The original cDNA probe could detect only the members of the genomic library that contained sequences homologous to those present in the probe. In order to explore the region flanking the β-globin gene, members of the

Figure 27.28

The linkage map of the human ß-globin gene locus as shown by the structural analysis of overlapping λ genomic clones. Both λHßG1 and λHßG3 clones contained the entire ß-globin gene. Other clones detected by "walking" led to the discovery of other ß-globin-like genes. These included four genes that are expressed and two pseudogenes that are not expressed. The genomic segments of the clones isolated are shown together with the cleavage sites for the enzyme *Eco*R1. The numbers on the top line indicate the size of the fragments in kilobase pairs.

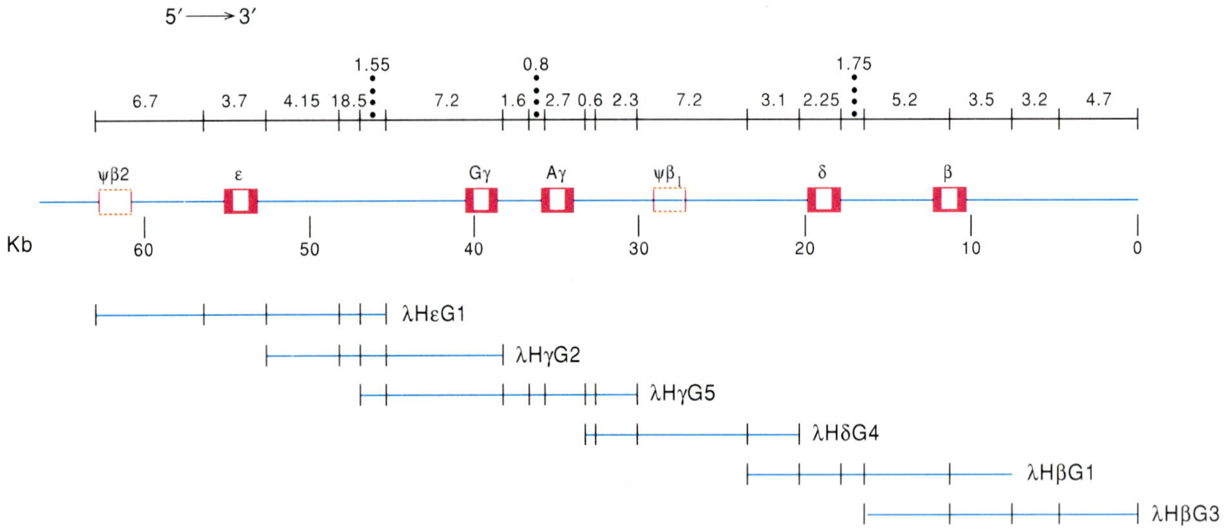

genomic library that hybridized with the original cDNA were themselves converted into radioactive probes, and these were used to locate additional members of the library that contained sequences flanking the β-globin gene. This process led to the detection of sequences bordering the sequences of the β-globin gene. A cyclic repetition of the process resulted in a gradual extension of sequence information in and around the β-globin gene. Using the library in this cyclic manner to extend the map is known as chromosome walking (fig. 27.28). In the case of the β-globin gene the walk was quite rewarding; it was discovered that all other β-globinlike genes were present nearby.

DNA Polymorphism Has Been Used for Prenatal Diagnosis of Sickle-Cell Anemia

A gene represented by two or more different alleles within a population is said to be polymorphic. Allelic differences are caused by sequence differences; in other words, the DNA itself is polymorphic. A DNA polymorphism might be in the coding (protein-specifying) region or in a portion of the gene that regulates the amount of gene product or elsewhere. In fact, DNA polymorphism appears to be a general phenomenon that results in sequence differences between homologous regions of DNA in different individuals. These differences can be detected even when no other differences can be found.

Operationally, then, DNA polymorphism should be recognizable by sequence differences. But rather than directly sequencing a gene to detect polymorphisms, differences called restriction fragment length polymorphisms (RFLPs) can be used to highlight rapidly the regions of sequence differences between individuals. The technique depends on the presence of a polymorphic marker—a particular restriction enzyme recognition site present in some individuals but not in others. That site will cause the length of fragments cut by a given restriction enzyme to be smaller than it is in individuals not possessing the marker.

Kan and Dozy in 1978 were the first to discover allele-specific linked polymorphism in the globin genes. Using the *Hpa*I restriction enzyme, they showed that the normal β-globin gene was contained within a 7.6-kb *Hpa*I fragment, but that the β-globin gene of sickle-cell anemia, Hb^S, was contained within a 13-kb fragment (fig. 27.29). Further analysis showed that the RFLP in this case resulted from a *Hpa*I restriction site, 5 kb to the 3′ side of the β-globin gene, that was present in the normal case and absent in Hb^S. A possible explanation for why this RFLP was associated with the sickle-cell trait is that the population, after a mutation, diverged into the 7.6-kb type and a few 13-kb types before introduction of the sickle-cell gene. After the Hb^S mutation was introduced into the 13-kb type, it became greatly expanded because of the selective advantage of the sickle-cell trait in certain populations.

Knowledge of this polymorphism could be used for pre- or postnatal diagnosis for the sickle-cell gene. Such information has been of practical value in genetic counseling. Incidentally, it is now possible to diagnose sickle-cell disease with greater precision because the point mutation leading to the Hb^S protein itself produces a recognizable RFLP.

Figure 27.29

Inheritance pattern of an RFLP associated with sickle-cell disease. Humans carry two alleles for the same gene, and each offspring inherits one allele from each of its parents in an entirely random fashion. Normal individuals are homozygous for normal *Hb* alleles; individuals with sickle-cell trait are heterozygous, with the one normal *Hb* allele and one Hb^S allele; and individuals with sickle-cell disease are homozygous for the Hb^S allele. At the top (*a*) we see a three-generation pedigree analysis for a family that carries both the normal and the sickle-cell gene for ß-globin. Males are represented by squares and females by circles. An open pink circle or square indicates an individual who is homozygous normal. A half-filled circle or square (pink/purple) indicates a heterozygous individual with sickle-cell trait. A filled circle or square (purple) indicates a homozygous individual with sickle-cell disease. In (*a*) both sets of grandparents produce a heterozygous individual with sickle-cell trait. Since one of the grandparents is homozygous normal and the other is heterozygous, there is a 50% chance that the grandparent mating will give rise to a heterozygous offspring as shown and also a 50% chance that they will have normal offspring (not shown). The two heterozygous parents have an increased chance of having abnormal offspring because in this mating each parent carries one abnormal gene or allele. There is a 25% chance of a homozygous sickle-cell anemic offspring, a 50% chance of an abnormal offspring with sickle-cell trait, and a 25% chance of a normal offspring. Below the pedigree chart is the electrophoretic pattern of a *Hpa*I digest probed with ß-globin cDNA (*b*). At the bottom we see an interpretation of the normal and abnormal DNAs (*c*).

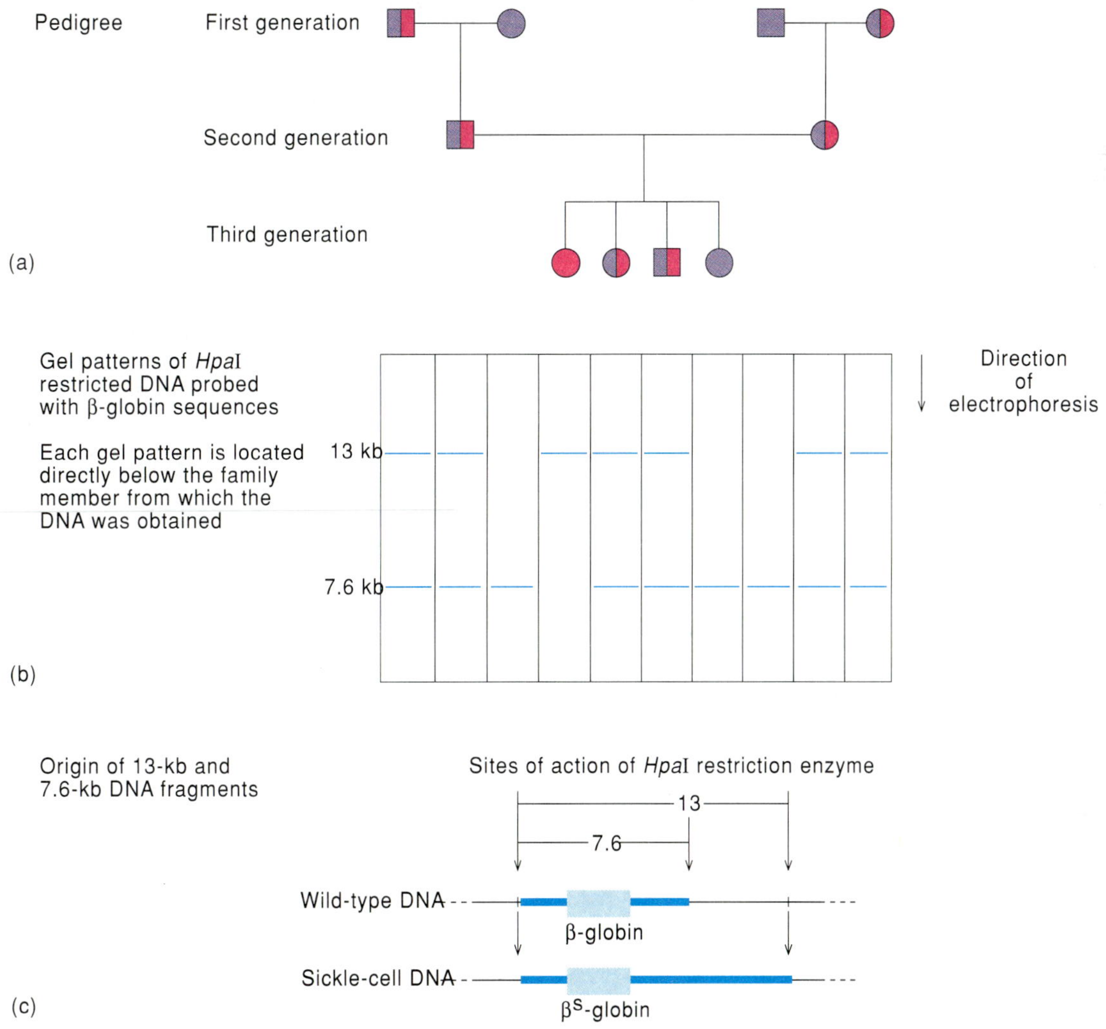

Walking and Jumping Were Both Used to Map the Cystic Fibrosis Gene

By combining the linkage information obtained from RFLP mapping with other DNA manipulation techniques, it has been possible to locate genes causing serious genetic disorders even when these genes are known only from their inheritance patterns. The list of serious disorders that can be linked to single genetic loci is growing. It includes Huntington's disease, Duchenne's muscular dystrophy, polycystic kidney disease, cystic fibrosis, chronic granulomatous disease, peripheral neurofibromatosis, central neurofibromatosis, familial polyposis coli, and multiple endocrine neoplasia. One of the most spectacular achievements in recent times has been identification of the gene causing cystic fibrosis (CF).

Cystic fibrosis is the most common serious genetic disorder in Caucasian populations. The major clinical symptoms include chronic pulmonary disease, pancreatic exocrine insufficiency, and an increase in the concentration of sweat electrolyte. Bearers of this disease, who are readily diagnosed, often die of congestive lung complications before age 30. Pedigree analysis shows that a single gene inherited in autosomal recessive fashion results in the disease syndrome. The frequency of the disease is 1 in 2,000, from which it may be calculated that the carrier frequency is about 1 in 20 (the frequency of the heterozygote for a rare allele is twice the square root of the frequency of the homozygote). By classical genetic analysis (see box 26A), the *CF* locus has been assigned to the long arm of chromosome 7 near the *Met* locus. A map of the region containing the *Met* locus and the *CF* gene is shown in figure 27.30.

In many genetic disorders the analysis must begin before the responsible gene and its protein product are known. Thus it is not possible to locate the gene directly, as in the case of the β-globin gene, and then determine its approximate location in subsequent analysis. For genes such as cystic fibrosis the approximate location is determined by conventional recombination analysis, and then researchers attempt to close in on the gene by the process called reverse genetics.

Detailed genetic mapping by recombination frequency is not practical with human genes that are separated by map distances of less than 1 centimorgan (cM)* because of the small number of test recombinant crosses that are ordinarily available for observation. Unfortunately, 1 cM on the human genome is still equivalent to a physical distance of about 10^6 (1 million) bp. One obstacle to the use of chromosome walking over such long distances is the size limitation of probes. Even cosmids, which provide the largest probes, cannot harbor probes larger than 40 kb, so it would take 25 cosmids end-to-end to span 10^6 base pairs. A walk involving this many probes would take a very long time, if it were possible at all. Another hindrance to such a long walk is that the human genome is sprinkled with segments of repetitive DNA. Those regions interrupt the walk because upon encountering such a region the probe will anneal to many members of the library, which could be situated almost anywhere in the genome. Therefore walking must be replaced by a procedure called chromosome jumping. Probes for jumping carry segments of DNA that are not ordinarily next to one another and that will therefore anneal to two regions that are a considerable distance apart. This distance is determined by how the probes are made.

Suitable jumping probes can be constructed in several ways. In one particularly elegant approach, the genome is first digested with a restriction enzyme that recognizes rare sites in the DNA (fig. 27.31). *Not*I, for example, is a "rare cutting enzyme" that recognizes a sequence of eight bases,

$$GC\downarrow GGCCGC$$

The fragments produced by complete digestion of the human genome with *Not*I are, on average, about 500 kb long. These fragments are circularized in the presence of a small marker DNA, so that thereafter the marker DNA is always located between the ends of the original large fragments. The circular

*Genetic loci 1 cM apart recombine 1% of the time (see box 26A).

Figure 27.30

Map of restriction fragment length polymorphisms (RFLPs) closely linked to the cystic fibrosis (*CF*) gene. The inverted triangle near the right-hand end indicates the location of the ΔF_{508} mutation characteristic of most persons with cystic fibrosis disease. (Source: B-S. Kerem et al., "Identification of the cystic fibrosis gene: Genetic analysis" in *Science* 245:1075, 1989. Copyright © 1989 American Association for the Advancement of Science, Washington, D.C.)

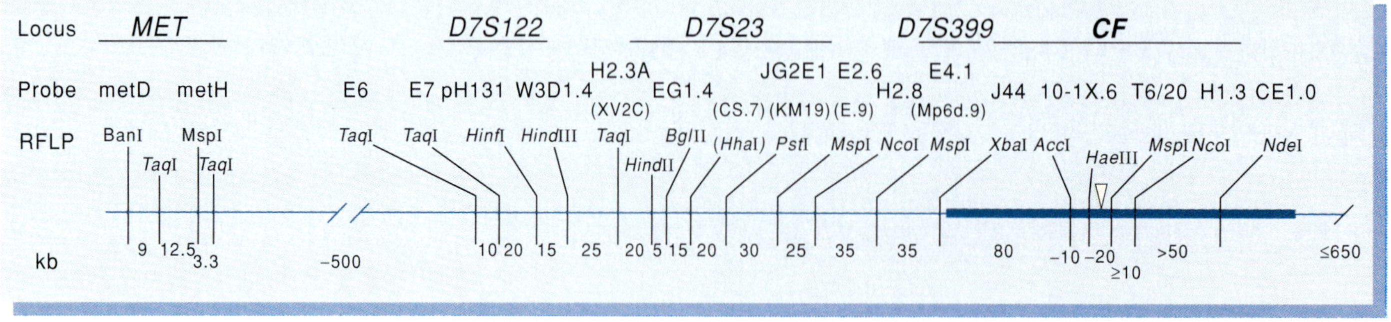

Figure 27.31

Construction of a jumping library (left) and a complementary linking library (right). The jumping library is constructed by exhaustive digestion of total nuclear DNA with a rare cutting *Not*I restriction enzyme. The cut fragments are circularized in the presence of a small gene required for amplification in a special *E. coli* host. The circularized molecules are recut with *Bam*HI restriction enzyme and ligated into a λ vector. The linking library is constructed from the same starting DNA. In this case the DNA is partially digested with *Sau*3A restriction enzyme. The resulting fragments are circularized in the presence of the same small gene required for amplification and the circularized fragments are recut with *Not*I. The linear fragments are ligated into a suitable λ vector and amplified in the special *E. coli* host.

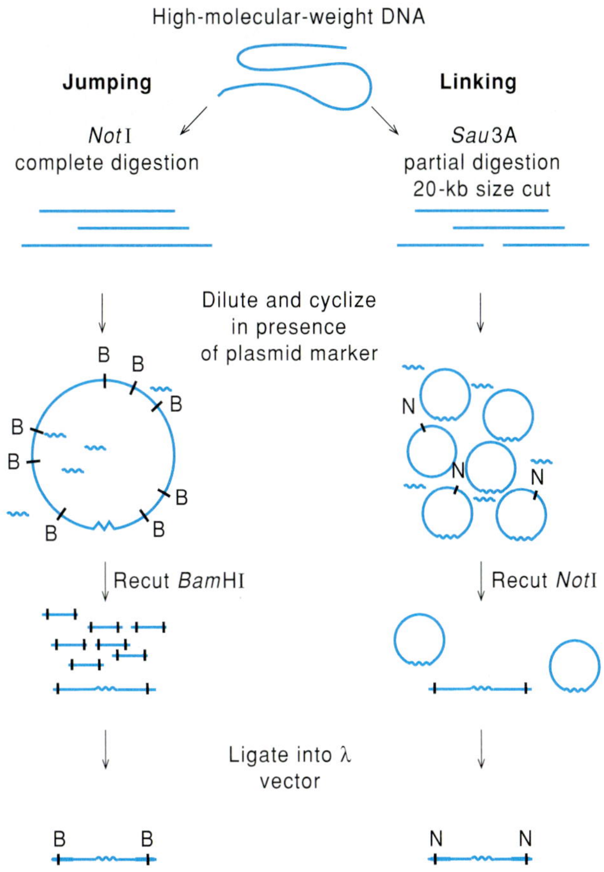

DNA is recut with a restriction enzyme that produces fragments suitable in size for cloning in λ. The marker DNA used contains a gene essential for λ multiplication on a suitable *E. coli* host. In this way the only λ clones that develop are those that carry the marker DNA and hence the ends of the original restriction fragments.

A jumping library is of very limited value on its own because it permits only one jump if used by itself. This is because the members of such a library would not be expected to cross-anneal. Therefore, a jumping library is always used in conjunction with a complementary linking library (fig. 27.31). Starting from the same DNA sample from which the jumping library was made, a partial digest is made with a restriction enzyme that gives considerably smaller restriction fragments, about 20 kb in size. These fragments are circularized in the presence of the same marker DNA that was used to construct the jumping library, and the resulting circular pieces are recut by complete digestion with the *Not*I enzyme. Only the circular pieces containing a *Not*I site are linearized by this treatment, so only these fragments become part of the linking library. The linearization step is followed by ligation of the linear fragments into the λ vector and selection on the same *E. coli* strain that requires the marker DNA for λ replication.

Each member of the linking library carries sequences on both sides of *Not*I restriction sites in the original DNA. In contrast, in the jumping library each member carries sequences from one side of two adjacent *Not*I sites. The jumping library and the complementary linking library are used in strict alternation as illustrated in figure 27.32. Since there are two directions in which jumping can occur initially, the direction is determined by the first annealing. After the first annealing the direction of future jumps is fixed. By using other markers it is possible to determine whether the direction of the first jump was the one of interest. Sometimes several jumps have to be made before the direction can be known.

In the cystic fibrosis investigation, starting from each *Not*I site determined by the jumping-linking library analysis, the sequences surrounding the *Not*I sites were scrutinized with

Figure 27.32

Directional jumping by alternating between a jumping library and a complementary linking library. Because of the way in which these two libraries were constructed (see fig. 27.31) each member of the linking library should contain sequences on both sides of a single *Not*I site, whereas each member of the jumping library should contain sequences on one side of adjacent *Not*I sites. By alternating between the two libraries it should be possible to traverse large regions on the genome relatively quickly and in a unique direction. The process starts with a member of either library that originates from the general region of the cystic fibrosis gene. If the search starts with a member of the jumping library as suggested in the figure, then this member is used to probe the linking library. The member of the linking library found in this way is then used to probe the jumping library, and the process is repeated as many times as necessary.

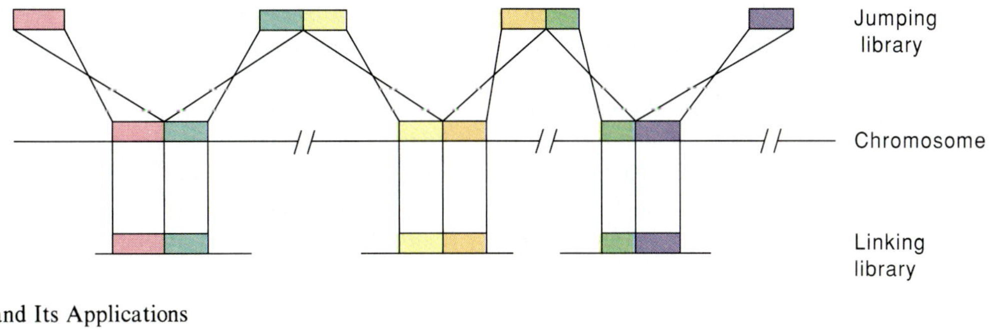

Figure 27.33

Jumping and walking to find the cystic fibrosis gene. Following each jump, the locus defined by the jump was used as a starting point for a chromosome walk. Each DNA segment so found was hybridized to a sweat gland cDNA library until a match was found. It seemed likely that the sweat gland cDNA library would have a good representation of the cystic fibrosis gene transcript because the disease shows sweat gland involvement.

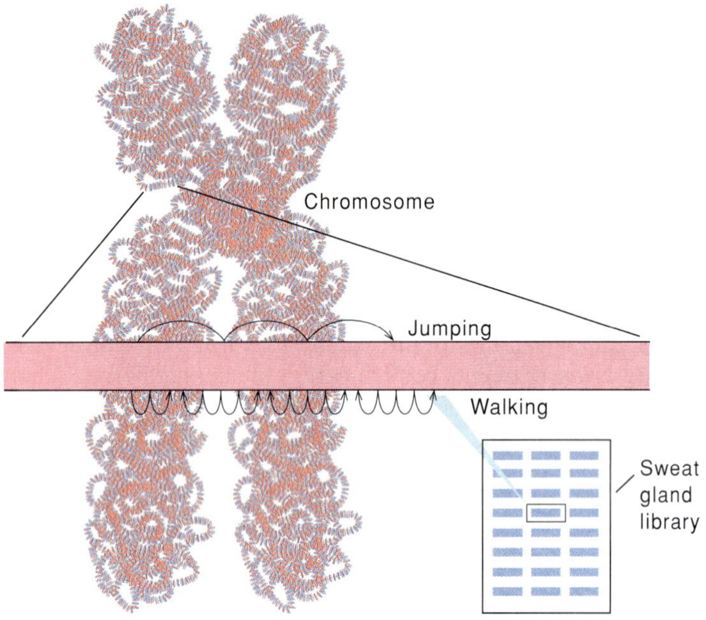

Figure 27.34

Predicted structure of the CFTR protein. Cylinders represent membrane-spanning helical segments. The cytoplasmically oriented NBFs are shown as blue spheres with slots to indicate the points of entry of nucleotides. R represents the large polar domain, which is linked to two halves of the protein molecule. Charged amino acids are shown as small circles with the charge sign. Net charges on the internal and external loops joining the membrane cylinders and on regions of the NBFs are contained in open squares. Potential sites for phosphorylation by protein kinases A or C (PKA or PKC) and N-glycosylation (N-linked CHO) are indicated. (K = Lys; R = Arg; H = His; D = Asp; E = Glu.)

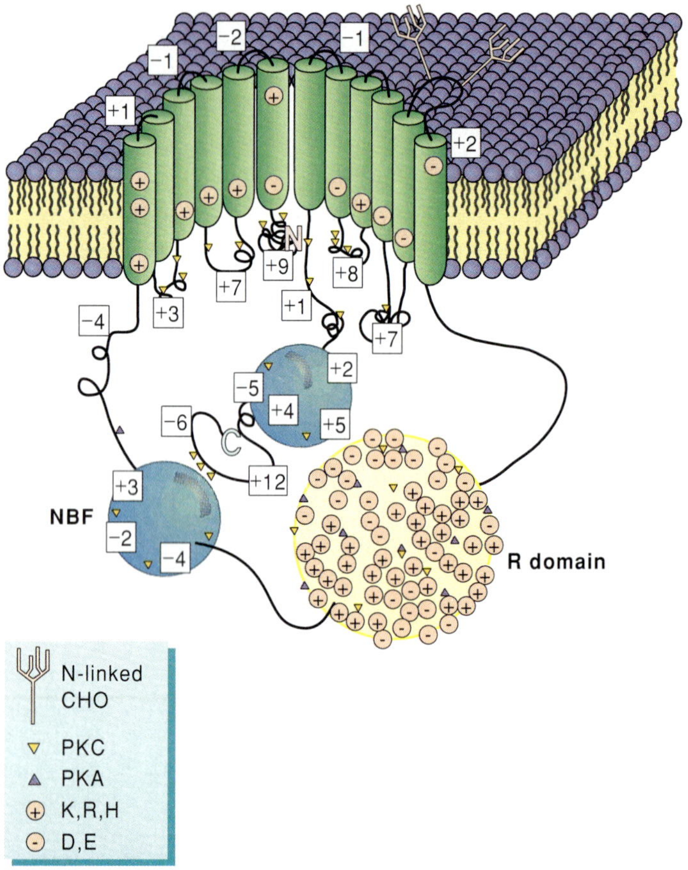

a walking library made from the same starting DNA. The purpose of these local walks was to get as much information as possible about the region, in the hope that by trial and error one of the probes would be the lucky one that overlapped the cystic fibrosis gene. The joint procedure of jumping and walking is illustrated in figure 27.33.

All of this analysis would have been in vain in the absence of some criterion for knowing when the goal of finding the cystic fibrosis gene had been reached. This is where clues from the physiological nature of the condition became useful. In a brilliant strategy a cDNA library was prepared from the messenger RNA fraction of sweat gland tissue. Recall that in cystic fibrosis there is sweat gland malfunction. Therefore it seemed that the mRNA for the cystic fibrosis gene might be well represented in the messenger RNA fraction of the sweat gland cells.

While the walking and jumping process was in progress, each new segment-mapping in the general region of interest was tested against the sweat gland cDNA library. Finally, a member of the walking library was found that annealed with a member of the sweat gland cDNA library. Was this match fortuitous or did it mean the cystic fibrosis gene was in hand? To answer this question the researchers used the cDNA discovered in this way to probe the genomic library, and thereby were able to map a gene that extended over a region of about 250 kb with 23 intervening sequences. In extracts of sweat gland tissue they detected a unique transcript, approximately 6,500 nucleotides in size, that matched the transcript size expected from this gene. The protein predicted from a sequence analysis of this transcript consists of two similar motifs, each with (1) a domain having properties consistent with membrane association and (2) a domain believed to be involved in ATP binding (fig. 27.34). Finally the researchers discovered that most CF patients carry a three-base deletion in this transcript, eliminating a phenylalanine residue from the protein. Thus they reasoned that the protein in cystic fibrosis patients should bear a defect. This defect correlated with the notion that CF patients have a faulty membrane protein that leads to the secretion problems characteristic of the disease. The fact that the abnormal gene is located as close as can be detected by classical genetics to the CF locus added additional support to the notion that the CF gene had been found.

Finding the disease gene has not, of course, meant an end to the problem of cystic fibrosis. However, the characterization of the disease gene will be a tremendous aid in diagnosing carriers and fetuses that are homozygous for the disease gene. It also should be a help in focusing approaches to finding a cure for the disease.

Summary

In this chapter we have described methods for sequencing, synthesizing, cloning, and restructuring DNA, and we have seen how recombinant DNA technology is being used to advance our knowledge of gene structure and function. Our discussion has focused on the following points.

1. To sequence DNA, researchers use chemical methods to cleave specific bases in preexisting DNA, or they carry out DNA synthesis under conditions where the synthesis is interrupted at specific bases.
2. By chemical methods it is possible to synthesize oligodeoxynucleotides up to a length of 100 monomers.
3. A specific segment of DNA can be synthesized *in vitro* by a polymerase chain reaction (PCR), in which short segments of DNA bordering the segment of interest are added to a mixture containing the segment of interest, a special heat-resistant DNA polymerase, and the deoxyribotriphosphate substrates. The DNAs are first denatured, then annealed and then synthesized. This cycle is repeated twenty or more times by raising the temperature to stop synthesis and then lowering the temperature to permit annealing and further synthesis. The outcome is a mixture in which the vast majority of the DNA is newly synthesized DNA bounded by the sequences of the added primers.
4. Another procedure for amplification cuts DNA containing the segment of interest into small pieces with a restriction enzyme. The cut pieces are incorporated into a plasmid or virus vector to be amplified in a suitable host. After growth, the mixture is plated to produce a mixture of bacterial or viral clones. The clones of interest are often identified by hybridizing replica plates of clones with a radioactive probe, then subjecting them to autoradiography.
5. Most cloning has been done in *E. coli*. Yeast is the most used eukaryotic host. Cloning is also possible in a number of plant and animal cells.
6. Mapping with recombinant DNA probes was first applied to the human globin genes. Starting probes were obtained by isolating the globin messenger from reticulocytes and converting it into a cDNA, which was then used to scan a human genomic library for cross-hybridizing members. Once detected and purified, these cross-hybridizing members carrying globin messenger sequences were themselves converted to radioactive probes and used to further scan the genomic library for nearby sequences. Repeating this cycle several times, a process known as a chromosome walk, revealed a region around the adult hemoglobin gene that contained several closely related genes associated with hemoglobin.
7. Frequently alleles of the same gene can be distinguished by restriction site differences in the genes themselves or in nearby locations. Alleles identified in this way are said to show restriction fragment length polymorphism (RFLP). The allele responsible for sickle-cell disease was identified in this way.
8. The cystic fibrosis gene was mapped by chromosome walking and jumping, a newer approach in which the relevant probes contain segments of the genome that are normally located about 500 kb from one another. A cDNA library was made from normal sweat gland tissue, chosen because of the disease's association with abnormal release of sweat salt, which suggested that the sweat gland would contain an abundance of the messenger associated with the gene. By hybridizing the genomic DNA probes with the cDNA sweat gland library, a segment of genome was identified as a candidate for the cystic fibrosis gene. This gene was characterized in detail and found to encode a complex transmembrane protein that carries a specific amino acid change in most persons with cystic fibrosis. This correlation provided overwhelming support for the conclusion that the gene responsible for cystic fibrosis had been mapped and characterized.

Selected Readings

Caruthers, M. H., A. D. Barone, S. L. Beaucage, D. R. Dodds, E. F. Fisher, L. J. McBride, M. Matteucci, Z. Stabinsky, and J. Y. Tang, Chemical synthesis of deoxyoligonucleotides. *Methods Enzymol.* 154: 287–313, 1987.

Cohen, S. N., A. Change, H. Boyer, and R. Helling, Construction of biologically functional bacterial plasmids *in vitro. Proc. Natl. Acad. Sci.* 70:3240–3244, 1973.

Gusella, J. F., DNA polymorphism and human disease. *Ann. Rev. Biochem.* 55:831–854, 1986.

Hunkapiller, T., R. J. Kaiser, B. F. Koop, and L. Hood, Large-scale and automated DNA sequence determination. *Science* 254:59–67, 1991. State of the art on the mammoth project to sequence the human genome.

Jackson, D. A., R. H. Symons, and P. Berg, Biochemical method for inserting new genetic information into DNA of Simian Virus 40: Circular SV40 DNA molecules containing lambda phage genes and the galactose operon of *E. coli. Proc. Natl. Acad. Sci.* 69: 2904–2909, 1972.

Kerem, B., J. M. Rommens, J. A. Buchanan, D. Markiewicz, T. K. Cox, A. Chakravarti, M. Buchwald, and L. C. Tsui, Identification of the cystic fibrosis gene: Genetic analysis. *Science* 245:1073–1079, 1989.

Mansour, S. L., K. R. Thomas and M. R. Capecchi, Disruption of the proto-oncogene *int-2* in mouse embryo-derived stem cells: a general strategy for targeting mutations to non-selectable genes. *Nature* 336:348–352, 1988.

Mullis, K. B., The unusual origin of the polymerase chain reaction. *Sci. Am.* 262:56–65, 1990.

Riordan, J. R., J. M. Rommens, B. S. Kerem, N. Alon, R. Rozmahel, Z. Grezelczak, J. Zielenski, S. Lok, N. Plasvsic, J.-L. Chou, M. T. Drumm, M. C. Iannuzzi, F. S. Collins, and L-C. Tsui, Identification of the cystic fibrosis gene: Cloning and characterization of complementary DNA. *Science* 245:1066–1073, 1989.

Sambrook, J., E. F. Fritsch, and T. Maniatis, *Molecular Cloning: A Laboratory Manual,* 2d ed. A three-volume collection that is thorough and up-to-date.

Sanger, F., Sequences, sequences, and sequences. *Ann. Rev. Biochem.* 57:1–28, 1988. A scientific memoir.

Problems

1. What are the advantages and disadvantages of the two methods of sequencing DNA (Gilbert's chemical method and Sanger's chain termination method)?
2. What are the limitations on sequencing RNA by the indirect method of sequencing the cDNA obtained from the reverse transcription of RNA?
3. What are the major advantages of the polymerase chain reaction (PCR) method for amplifying defined segments of DNA as opposed to the use of conventional cloning methods? How might the PCR method be used to test for infection with the AIDS virus and how would this be an improvement over the antibody test currently used? (The current ELISA test is an indirect test for the presence of antibodies against the HIV proteins.)
4. You have just isolated a novel recombinant clone and purified the desired insert (a 10-kb linear duplex DNA) from the vector. Now you wish to map the recognition sequences for restriction endonucleases A and B. You cleave the DNA with these enzymes and fractionate the digestion products according to size by gel electrophoresis. You observe the following:
   (a) Digestion with A alone gives two fragments, of lengths 3 and 7 kb.
   (b) Digestion with B alone gives three fragments, of lengths 0.5, 1, and 8.5 kb.
   (c) Digestion with A plus B generates four fragments, of lengths 0.5, 1, 2, and 6.5 kb.
   Draw a restriction map for the insert, showing the relative positions of the cleavage sites with respect to one another.
5. Describe the procedure you would use to clone a DNA fragment into the *Bam*HI site of pBR322.
6. A small circular duplex viral DNA with about 5,000 bp contains a 72-bp direct repeat in tandem. Each 72-bp segment contains a cleavage site for the *Eco*RI restriction enzyme that is not present in the rest of the virus. How could you produce a modified viral chromosome with only one copy of the 72-bp segment?
7. How large a genomic library should you construct in order to detect and isolate a 15-kb gene out of a genome containing 3×10^9 bp?
8. Why isn't a primer required to promote second-strand synthesis using the single-stranded DNA produced by reverse transcriptase from an RNA template?
9. If you were interested in isolating a cDNA for human serum albumin, why would you use a cDNA library established from mRNA isolated from liver? If you wanted to isolate the gene for albumin, why could you use a genomic library established from any human tissue?
10. The analysis of DNA regulatory sequences can be undertaken via the technique of DNA-mediated gene transfer, using a gene that provides a functional assay ("reporter gene") for the effect of the putative regulatory sequence on gene expression. How could this be done?
11. Explain how the thymidine kinase gene (*tk*) can be used as a selectable genetic marker for isolating transformed mammalian cells containing DNA of interest.
12. Site-directed mutagenesis is one of the most powerful tools available to the biochemist. What are some of the applications of this technique? How can the PCR method be used to do site-directed mutagenesis, and what is the advantage of this method?
13. Briefly outline how you could map the globin gene family using recombinant DNA techniques.
14. What is the procedure called chromosome jumping? How was this procedure used to map the cystic fibrosis gene?

28

CHAPTER

RNA Synthesis and Processing

In the early 1950s George Gamov suggested that DNA might serve as a template for protein synthesis. However, biochemical investigations showed that in eukaryotes the processes of RNA and protein synthesis are compartmentalized in different parts of the cell—the nucleus and the cytoplasm, respectively. For this reason, it seemed unlikely that DNA could serve directly as a template for protein synthesis.

Base composition studies by Erwin Chargaff, which had provided information that was vital in deducing DNA structure and in suggesting a mechanism for DNA replication, were less helpful in regard to the structure and synthesis of RNA. Total cellular RNA did not reflect the base composition of the DNA. However, Volkin and Astrachan showed that when a DNA bacterial virus, T2 bacteriophage, infects an *E. coli* cell, the newly synthesized RNA (which is exclusively transcribed from the phage DNA) reflects the base composition of the DNA. From experiments such as this, and from an understanding of the DNA duplex structure, there emerged the hypothesis that DNA, by unwinding, serves as a template for the synthesis of RNA.

In the early 1950s Crick proposed what he called the "central dogma" of molecular biology, which states that genetic information flows from DNA to RNA and thence to protein:

$$\text{DNA} \rightarrow \text{RNA} \rightarrow \text{protein}$$

This dogma provided researchers with the theoretical framework for experiments designed to demonstrate the existence of a transcribing enzyme that copies the DNA sequence into an RNA molecule.

The first enzyme discovered that could catalyze polynucleotide synthesis was polynucleotide phosphorylase. This enzyme, isolated by Severo Ochoa and Marianne Grunberg-Manago in 1955, could make long chains of 5′ → 3′ linked polyribonucleotides starting from nucleoside diphosphates. However, no template dependence could be found for this synthesis, and the sequence was only controllable in a crude way by adjusting the relative concentrations of different nucleotides in the starting materials.

Sam Weiss was the first to obtain evidence for a true transcribing activity in cell-free extracts of rat liver (1959). His experimental design was influenced by the theoretical framework provided by Crick and by Kornberg's discovery that *in vitro* DNA synthesis required nucleoside triphosphates for substrates. With crude liver extracts, Weiss was able to demonstrate a capacity for RNA synthesis that was severely inhibited by DNase. This was the beginning of systematic investigations of the biochemistry of RNA metabolism. A continuous expansion of research effort in related areas over the past 30 years has provided us with a wealth of understanding about the transcription process and other aspects of RNA metabolism.

We considered various aspects of nucleotide metabolism in chapter 20. In this chapter we will focus on RNA synthesis and the reactions that follow synthesis, whereby the initially transcribed RNA is modified into a mature functional form. These reactions are summarized in figure 28.1.

Evidence that RNA Sequences Are Derived from DNA Sequences

The base composition studies of Volkin and Astrachan had suggested a template-product relationship between DNA and RNA, and Sol Spiegelman and Benjamin Hall pursued this result a step further. Using ^{3}H-labeled T2 DNA and ^{32}P-labeled RNA that was synthesized after T2 infection of *E. coli* cells, they were able to demonstrate that the newly synthesized RNA forms a specific complex with the viral DNA (fig. 28.2). To show that this RNA contains sequences homologous to those in DNA, advantage was taken of the annealing technique developed by Doty and Marmur. They reasoned that it should be possible to make a DNA-RNA duplex in the same way as a DNA-DNA duplex, by annealing. Since such a complex should contain one strand of DNA and one strand of RNA, it is referred to as a hybrid duplex, and the process of making the duplex is referred to as hybridization. Following hybridization, the product was subjected to CsCl density-gradient centrifugation. (See chapter 26 for a general explanation of this technique and for its application to the study of DNA replication.) After equilibrium was reached, they analyzed various fractions from the centrifuge tube for ^{32}P and ^{3}H, using isotope-counting methods that make it possible to distinguish between ^{32}P and ^{3}H in the same sample. Thus they could determine the locations of DNA and RNA by measuring ^{32}P and ^{3}H in the various fractions. They found that the ^{3}H gave a sharp peak near the middle of the tube, while the ^{32}P gave a broad peak with a maximum near the bottom. RNA is denser than DNA, which accounts for its lower average position in the centrifuge tube. The excitement in this experiment centered around a sharp peak of ^{32}P coincident with the ^{3}H label (see fig. 28.2). This peak indicated the presence of a complex that involved both DNA and RNA. Proof that the complex was specific was obtained by carrying out the annealing experiment with other DNAs. Only when T2 DNA was used did the ^{32}P peak appear. Furthermore, the peak appeared only if the T2

Figure 28.1

Different types of enzyme-catalyzed reactions involving RNA. RNA is best known as the intermediary *messenger RNA* that carries the information from DNA for the synthesis of protein and for other polynucleotides that are directly involved in protein synthesis, *ribosomal RNA* and *transfer RNA*. Certain RNAs possess catalytic activity leading to the evolutionary argument that RNA as template and catalyst preceded protein. The first reaction to be discovered whereby polynucleotides are synthesized was catalyzed by the bacterial enzyme *polynucleotide phosphorylase*. Polynucleotide phosphorylase uses nucleotide diphosphates as substrates and has no template requirement. All other *RNA polymerases* use nucleotide triphosphates as substrates. Double helix DNA is the template that is used in the synthesis of most cellular RNAs. The DNA carries specific sequences of bases called *promoters* that are recognized as RNA polymerase binding sites and initiation points for RNA synthesis. Other sites on the DNA are recognized as termination points for RNA synthesis. A gene always has a promoter and a termination point and the information in between that is transcribed by the RNA polymerase. Most transcripts carry the information for making protein. The enzyme *reverse transcriptase*, which is very widely distributed, is capable of using RNA as a template for the synthesis of DNA. This is a minor reaction in most cells but it may play a highly significant evolutionary role and it is indispensable to the life cycle of certain viruses as we saw in the previous chapter. Another RNA polymerizing enzyme known as *RNA replicase* is used by certain RNA viruses to replicate viral RNAs.

Many enzymes are involved in RNA processing. These enzymes have been named according to the types of reactions they catalyze. *Endonucleases* and *exonucleases* hydrolyze phosphodiester linkages in preformed RNAs. *Splicing enzymes* cleave internal segments and knit the remaining polymers back together. *Base modifying enzymes* alter or replace the original bases present in the polymer at select points. *End addition enzymes* add additional nucleotides at either the 3′ or 5′ ends of the ribopolymer. Recently discovered *editing enzymes* remove certain nucleotides and add additional nucleotides.

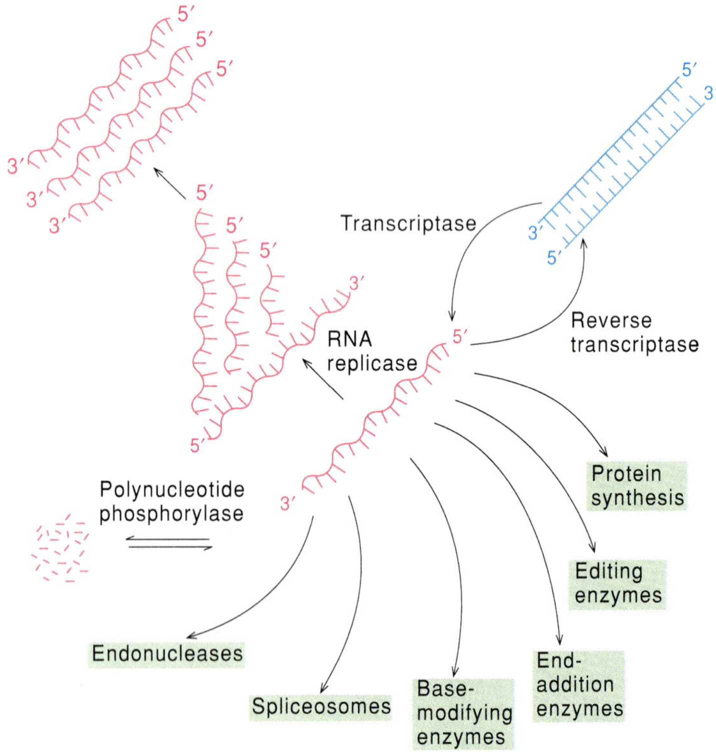

Figure 28.2

Demonstration that phage RNA has sequences complementary to phage DNA. Spiegelman and Hall used ^{3}H-labeled T2 DNA and ^{32}P-labeled RNA, the latter made after T2 infection of *E. coli* cells. The two nucleic acids were mixed together, annealed, and then centrifuged to equilibrium in a CsCl density gradient. In (*a*) the DNA was first heat-denatured and then mixed with RNA and annealed at 65° C prior to centrifugation. In (*b*) the DNA denaturation step was left out. Most of the RNA goes to the bottom of the centrifuge tube because of its high density. The DNA bands about one-third of the way from the top. In (*a*) some RNA also bands at approximately the same location as the DNA, but in (*b*) this is not the case. The comigration of a fraction of the RNA with the DNA is believed to be due to the formation of a specific DNA-RNA hybrid duplex. The RNA is much smaller than the DNA, so the RNA in the hybrid duplex migrates at the density of the DNA. In (*b*) no hybrid duplex forms because the DNA was not denatured before carrying out the annealing process.

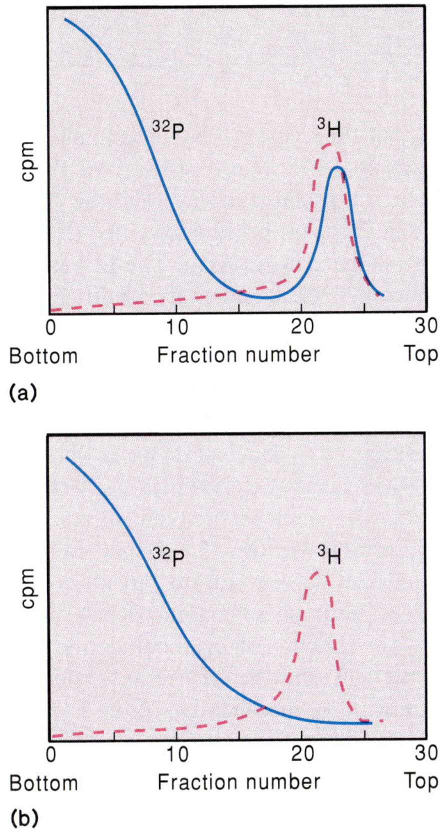

DNA was denatured before the annealing step. This result indicated that the two DNA strands must be separated to make the DNA-RNA complex, as would be expected in an annealing operation.

The hybridization technique developed in Spiegelman's laboratory was to take its place, alongside the DNA-DNA annealing technique developed by Doty and Marmur, among the techniques most often used in molecular biology. It has since been used to characterize RNAs and their location of synthesis on the genome. Historically, that first hybridization experiment gave credence to the proposal that RNA was a direct transcript of a DNA chain.

The Three Major Classes of RNA

Although most RNAs are synthesized as a result of transcription from a DNA template, different strategies are used in their synthesis and in their posttranscriptional modification. The strategy used in a given case is strongly related to the function of the RNA. For this reason, it is important that we discuss the different classes of RNA found in the cell before we consider their mode of synthesis.

There are three major types of cellular RNA involved in protein synthesis: messenger RNA (mRNA), ribosomal RNA (rRNA) and transfer RNA (tRNA). In addition, there are other RNAs, some of which function as primers for DNA synthesis, as component parts of ribonucleases, and in the posttranscriptional modification of eukaryotic mRNA precursors. The properties of the RNAs found in *E. coli* are summarized in table 28.1. You will find it helpful to refer to this table in the course of the following discussion.

Messenger RNA Carries the Information for Polypeptide Synthesis

Messenger RNA is transcribed from DNA by RNA polymerase, and its sequence contains the information for the sequence of amino acids in the protein product. Specific mRNAs are synthesized by the cell in response to the conditions under which it finds itself, and the cell can thereby control the kinds and amounts of proteins it produces. Prokaryotic cells have unstable mRNAs that turn over rapidly, with an average half-life of 1 to 3 min, whereas eukaryotic cells, which do not have to be able to respond as rapidly to changing conditions, often have more stable mRNAs.

The mRNA fraction is heterogeneous in size, ranging from 500 to 6,000 nucleotides in *E. coli*. This range reflects not only the heterogeneity in size of the proteins of the cell, but also the fact that some prokaryotic mRNAs contain the information to encode more than one protein. Such mRNAs are referred to as polycistronic mRNAs, each cistron containing the information for the synthesis of a single polypeptide chain.

In prokaryotes, the messenger is engaged in protein synthesis even before the entire mRNA molecule has been synthesized. This rapid utilization of nascent transcript minimizes the opportunity for processing of the transcript. In contrast, in eukaryotes, where the processes of transcription and translation are sharply divided, the nascent transcript undergoes an elaborate regimen of processing, usually at both ends and frequently internally, before it is transported to the cytoplasm for use as mRNA.

Transfer RNA Carries Amino Acids to the Template for Protein Synthesis

The second major type of RNA is transfer RNA. Its function is to carry amino acids to the mRNA template, where the amino acids will be linked in a specific order. The tRNAs contain both a site for the attachment of an amino acid and a site, the anticodon, that recognizes the corresponding three-base codon on

Table 28.1
Types of RNA in *E. coli*

Type	Function	Number of Different Kinds	Number of Nucleotides	Percent of Synthesis	Percent of Total RNA in Cell	Stability
mRNA	Messenger	Thousands	500–6,000	40–50	3	Unstable ($t_{1/2}$ = 1 to 3 min)
rRNA	Structure and function of ribosomes	3 { 23S, 16S, 5S	2,800 1,540 120	50	90	Stable
tRNA	Adapter	50–60	75–90	3	7	Stable
RNA primers	DNA replication	?	<50	<1	<1	Unstable
RNA component of RNase P	Catalytic	1	377	<1	<1	?

the mRNA (see chapter 29). There are about 50 different kinds of tRNAs in a bacterial cell. Each amino acid is enzymatically attached to the 3′ end of one or more tRNAs by a specific aminoacyl-tRNA synthase that recognizes both the amino acid and the tRNA. This two-step process is discussed in chapter 29.

A great deal is known about the structure of tRNA because the entire nucleotide sequences of a large number of tRNAs have been determined, and the three-dimensional structures of some of them have been obtained by x-ray crystallography. The complete three-dimensional structure of phenylalanine tRNA of yeast, $tRNA^{Phe}$, is shown in figure 28.3. This molecule contains 76 nucleotides in a single chain with four loops and a stem.

Most of the bases are involved in a complex secondary and tertiary structure. The four loops in the molecule are referred to as the TψC loop, the anticodon loop, the D loop, and the variable loop. The variable loop is so named because it contains a different number of bases in different tRNAs. The remainder of the molecule (with the exception of the D loop in some instances) contains the same number of residues in all tRNAs, even though there is considerable sequence variability. The $tRNA^{Phe}$ molecule contains 20 base pairs that are hydrogen-bonded in Watson-Crick fashion. It contains an additional 40 or so hydrogen bonds, most of which are not of the Watson-Crick type. These additional hydrogen bonds and the accompanying base stacking stabilize the tRNA in the complex folded structure shown in figure 28.3. Some of the unusual hydrogen-bonded interactions involved in stabilizing the complex tertiary structure are shown in figure 28.4. These structures involve two to four residues from different regions of the RNA. Not only are the hydrogen-bonding patterns unusual, but in some cases phosphate and the C-2′-OH group participate in the hydrogen bonding. Tertiary structures of most tRNAs are very similar except for the variable loop, even though tRNAs show considerable sequence variability.

Earlier (chapter 25) we stressed the importance of base stacking in stabilizing duplex nucleic acid structures. Despite the irregularity of the hydrogen bonding found in tRNA, base stacking, as assessed by a close scrutiny of the three-dimensional structure, is a prominent feature of the tRNA structure. Only four of the 76 bases in the molecule (D^{16}, D^{17}, U^{47}, and G^{20}) do not participate in stacking. The D^{16} and D^{17} dihydrouracil residues cannot stack under any conditions, because these residues no longer have a planar configuration. It is believed that stacking and hydrogen bonding contribute about equally to the conformational stability of the tRNA.

Another significant feature of the tRNA structure is the high percentage of ribonucleotide bases that differ from the usual four. We will see that tRNA is initially synthesized from the usual four bases, which are then enzymatically modified in many cases by a large variety of enzymes. Some of the more than 30 such unusual bases so formed are shown in figure 28.5. Of the 76 nucleotides in phenylalanine tRNA, twelve are of the modified type. The function of the modification is not known in general. However, in the case of the anticodon region of the tRNAs it is known that modification of the 5′ base can change the ability of the tRNA to read various synonym codons (see chapter 29). Modification of bases in other types of RNA has also been observed, but not to the extent that it occurs in tRNA.

The detailed structural investigations of tRNA greatly broadened our expectations in considering the less-well-understood structures formed by other nuclear and cytoplasmic RNAs. This single example illustrates the fact that nucleic acids with a properly adjusted primary sequence can adopt complex secondary and tertiary structures. *A propos* of this point, Francis Crick once said that transfer RNA is "an RNA trying to look like a protein." The same could be said of the ribosomal RNAs.

Ribosomal RNA Is an Integral Part of the Ribosome

Ribosomal RNA is a structural and functional component of ribosomes, the large ribonucleoprotein bodies that house the machinery for most of the reactions associated with protein synthesis. Ribosomes in prokaryotes are referred to as 70S ribo-

Figure 28.3

The tertiary structure of yeast phenylalanine tRNA. (*a*) The full tertiary structure. Purines are shown as rectangular slabs, pyrimidines as square slabs, and hydrogen bonds as lines between slabs. (Source: G. J. Quigley and A. Rich, "Structural domains of transfer RNA molecules" in *Science*, 194:796, 1976. Copyright © 1976 American Association for the Advancement of Science, Washington, D.C.) (*b*) Nucleotide sequence. Residues that appear in most of the yeast tRNAs and residues that appear to be constantly a purine or a pyrimidine are indicated. Residues involved in tertiary base pairing are shown connected by solid lines. Several of the nucleotides are methylated. These are indicated by a small m.

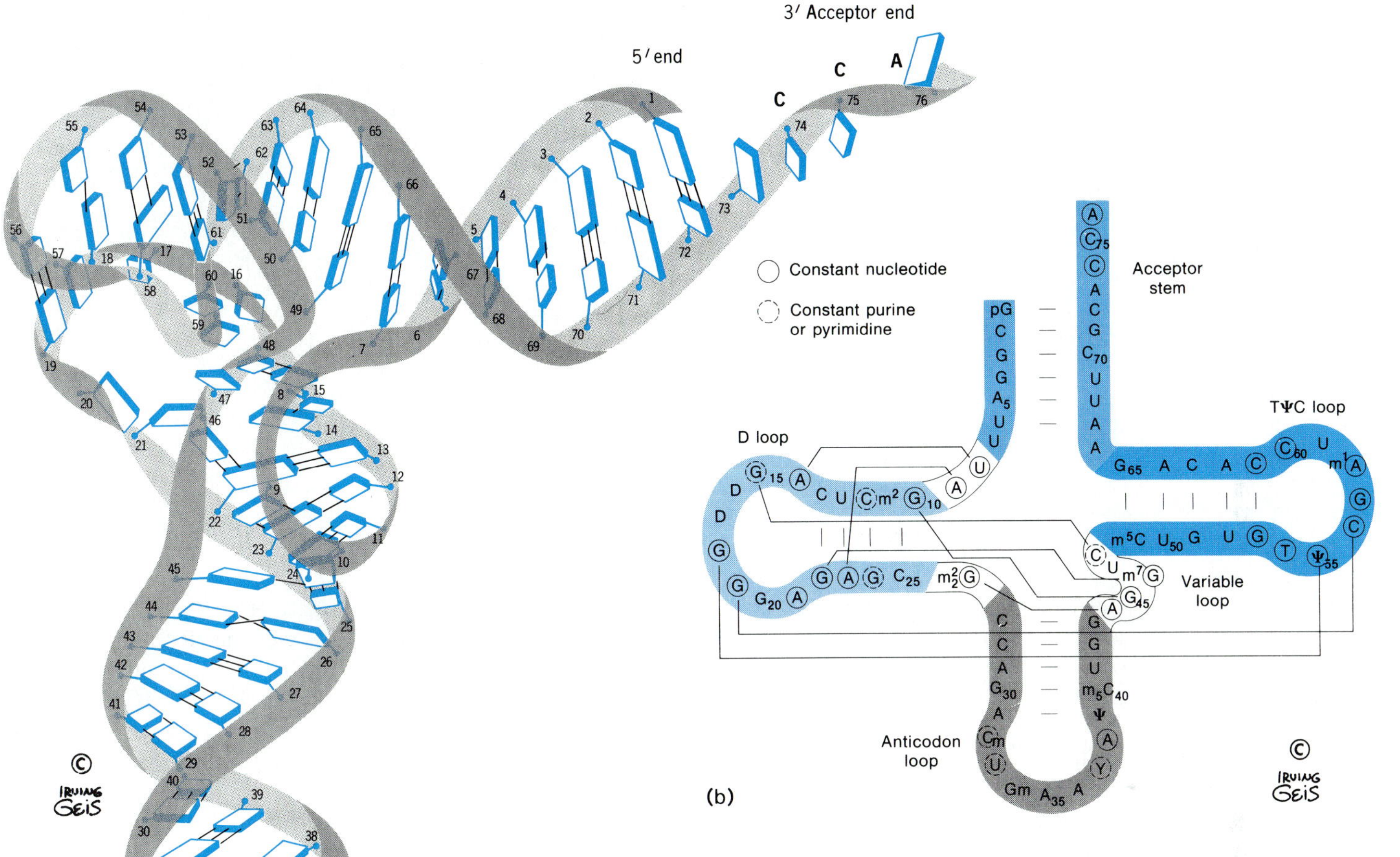

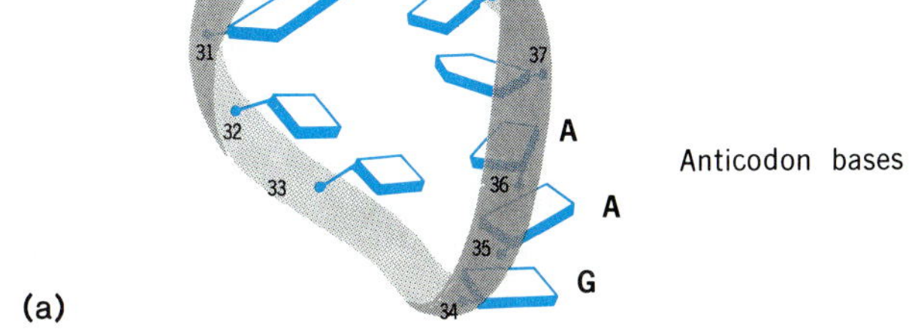

somes, a measure of their rate of sedimentation in a centrifuge and hence their size (S refers to Svedberg units, which are defined in chapter 5). A 70S ribosome consists of two subunits, the 50S subunit and the 30S subunit, each of which is made up of RNA and protein. The 50S subunit contains 23S and 5S rRNAs. The 30S subunit contains a single 16S rRNA (fig. 28.6). Eukaryotic ribosomes are similar in structure, although they are somewhat larger (80S) and contain mostly larger RNAs (25–28S, 18S, 5S, and an additional 5.5 to 5.8S; this additional rRNA corresponds in sequence to the first 150 nucleotides of the prokaryotic large rRNA subunit). Chloroplasts and mitochondria have ribosomes and rRNAs that are distinctly different from those present in the cytoplasm and strongly resemble those of prokaryotes. These RNAs are encoded in the organelle DNA and are transcribed by organelle-specific RNA polymerases.

Ribosomal RNA Secondary Structure Is Highly Conserved

The primary sequences of many rRNAs and their genes have now been determined, and the results have done much to shape our understanding of rRNA structure and function. Sequence analysis has revealed that conservation of secondary structure is the major paradigm of ribosomal RNA evolution. In general, rRNA has evolved by the introduction of complementary base changes that conserve regions of secondary structure. Much evidence indicates that regions that can vary in primary sequence, without the necessity of compensating changes in a complementary region turn out to be single-stranded parts of the rRNA. There are few regions in rRNA where primary sequence is conserved, and these are generally located in single-stranded regions of the molecule. The secondary structures of rRNAs that have been deduced by this method of phylogenetic sequence comparison are shown in figure 28.7.

Figure 28.4

Some of the tertiary hydrogen-bonded interactions found in yeast phenylalanine tRNA. Superscripts refer to base number in the tRNA molecule. It can be seen that there are many hydrogen-bonding arrangements in a tRNA structure that are not found in a duplex DNA structure. (Source: G. J. Quigley and A. Rich, "Structural domains of transfer RNA molecules" in *Science,* 194:796, 1976. Copyright © 1976 American Association for the Advancement of Science, Washington, D.C.)

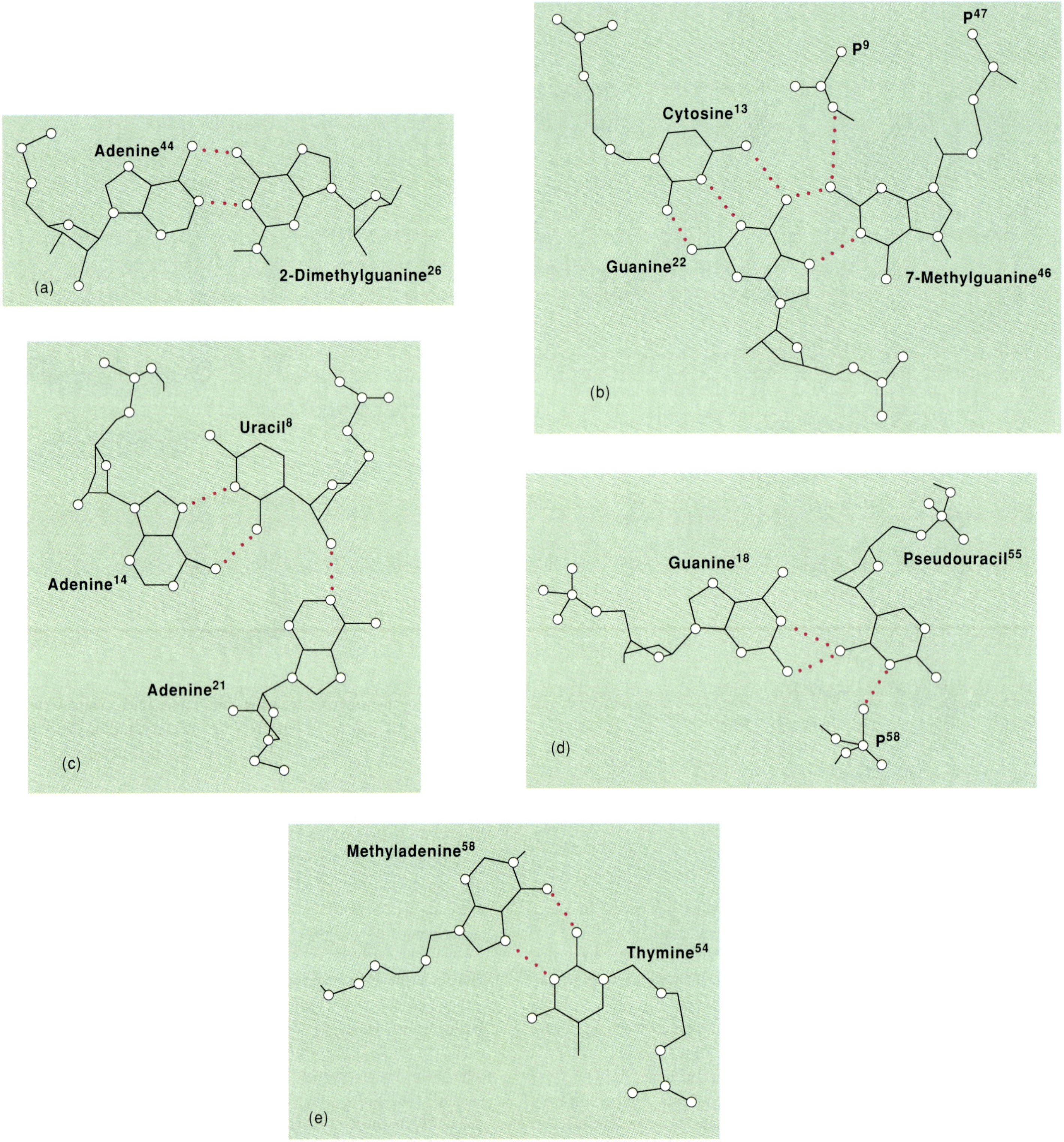

Figure 28.5

The structures of some modified bases found in tRNA. The parent ribonucleosides are shown with the numbering of the atoms in the purine and pyrimidine rings.

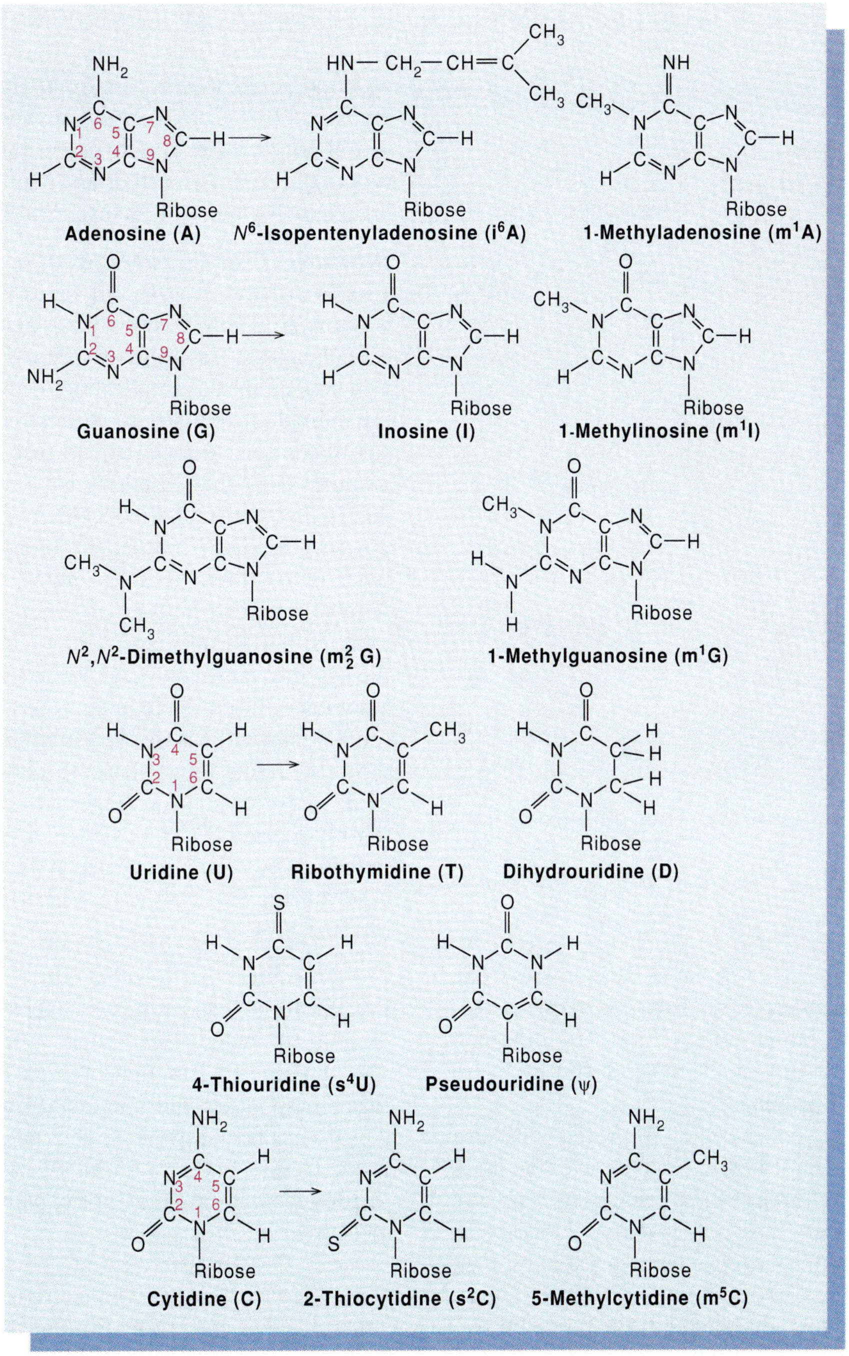

For years, the pattern of rRNA structure observed in common bacteria and in many eukaryotes—the presence of discrete pieces of defined sizes—has been considered an integral feature of ribosome structure. This view is now changing with the discovery that the number and size of the individual pieces of rRNA can vary widely in different organisms. Ribosomal RNAs in some organisms have been found to occur in as many as a dozen separate pieces as the result of transcription from independent genetic units. It now appears that rRNA consists of discrete and functionally essential domains that need not be joined together. Much of the difference in the overall length that distinguishes rRNA from different species appears to arise by insertions or deletions between these individual segments. Thus rRNA seems to be composed of discrete functional domains that can be independently inherited and separated by segments of variable length.

Figure 28.6

Composition of the *E. coli* ribosomes. The 70S ribosome can dissociate into a 50S and a 30S subunit. *In vitro* this can be done by lowering the Mg ion concentration. The individual subunits can be dissociated into their constituent RNAs and proteins by exposure to urea denaturant. Molecular weights are given for the subunits and the proteins, and the numbers of nucleotides are given for the RNAs.

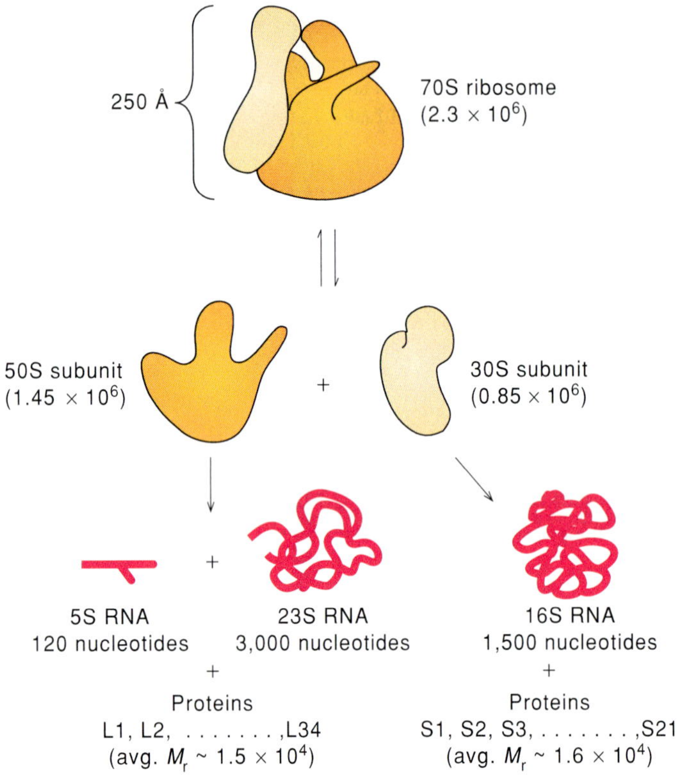

Ribosomes Contain a Discrete Set of Proteins Ribosomes from different species vary from about 30 to 70% protein by mass. This mass is generally distributed among 50 to 100 different proteins, but ribosomes from any single species appear to be composed of a unique set of proteins.

E. coli ribosomes are composed of 56 proteins that are mostly basic in charge and small in size, with an average mass of about 20,000 daltons. Fifty-three of these proteins have unique sequences and occur once per ribosome. Twenty of these proteins belong to the 30S subunit with one copy per subunit (see box 28A). One protein occurs on both the small (S20) and large (L26) subunits. Another protein, designated L7/L12, yields two electrophoretic spots because it is partially acylated at its amino terminal. This protein is present in four copies per 50S as a highly elongated complex that appears to form a stalk on the subunit.

For many years it was generally assumed that the ribosomal proteins are the functionally important components of the ribosomes. However, despite a great deal of effort, the function of individual ribosomal proteins has consistently eluded discovery. The location of some proteins near functional binding sites has been shown by chemical cross-linking of bound ligands. Mutational alterations in others have been shown to cause resistance to antibiotic inhibitors of protein synthesis. Remarkably as many as half of the individual *E. coli* ribosomal proteins, as well as at least some eukaryotic ribosomal proteins, can be completely deleted genetically without inactivating the ribosome. The view is developing that the primary function of ribosomal proteins is to bind and thus assemble the individual domains of ribosomal RNA into their appropriate locations. It would not be surprising to learn that ribosomal proteins play a functional role in the ribosome similar to the role played by the protein component of RNase P, to be discussed later.

Ribosomes Can Assemble Spontaneously The *in vitro* reconstitution of the *E. coli* 30S subunit from purified components was achieved about twenty years ago by Masayasu Nomura and his colleagues. Subsequently, *in vitro* assembly of the 50S subunit was achieved. In addition to the structural components of the individual subunits, assembly required the appropriate ionic conditions and elevated temperature. Nomura and his co-workers demonstrated that assembly is an ordered process in which individual proteins bind to rRNA in a prescribed sequence. Some proteins, primary binding proteins, bind to rRNA in the absence of all others. Other proteins, secondary binding proteins, require the prior binding of one or more primary binding proteins. Still other proteins, tertiary binding proteins, cannot join the assembly until one or more secondary binding proteins have assembled. From such indications Nomura developed an assembly pathway for the 30S subunit (fig. 28.8). He also showed that the assembling ribosome undergoes temperature-dependent conformational changes at specific points along the assembly pathway. The existence of the pathway implies that the individual ribosomal proteins interact with both rRNA and each other during assembly, which in turn implies underlying spatial relationships.

The Fine Structure of the Ribosome is Beginning to Emerge Results from many different experimental approaches are beginning to coalesce to produce a three-dimensional picture of the ribosome that includes the location of its individual structural components and functional sites. The development of this picture has been especially challenging because the ribosome is large, fragile, and structurally complex and has resisted efforts to produce crystals that are capable of giving high-resolution structural information.

The current view of the overall morphology of ribosomes is based largely on electron micrographic studies of the subunits of *E. coli*. From this analysis it appears that both subunits are asymmetric (fig. 28.9). The large subunit is visualized as having several protuberances at one end. A large protuberance in the center is flanked by smaller ones on either side. The small subunit is visualized as having a head, neck, and platform, all at one end of the structure.

The relative location of individual ribosomal proteins within the two subunits has been carefully examined by two different techniques. One method involves determining which ribosomal proteins can be chemically cross-linked to each other, and has yielded an elaborate grid of spatial relationships based

Figure 28.7

Comparison of secondary structures of 16S-like rRNAs from (*a*) eubacteria (*E. coli*), (*b*) archaebacteria (*H. volcanii*), (*c*) eukaryotes (*S. cerevisiae*), and (*d*) a structure showing features common to all sequenced 16S-like rRNAs, including mitochondrial types.

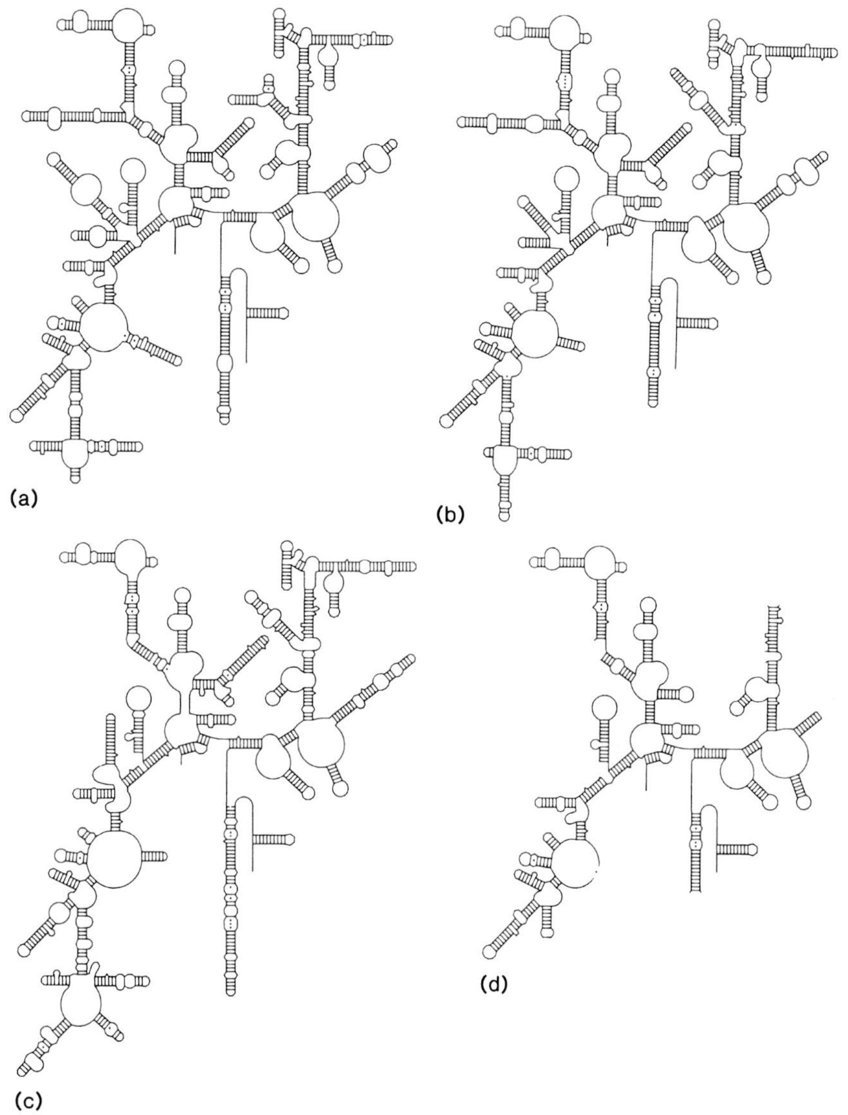

on the frequency of cross-linking. The other method relies on neutron diffraction. This method employs individually deuterated ribosomal proteins and locates their relative centers of mass within the subunit. The two methods of determining the location of ribosomal proteins have yielded a relatively consistent spatial picture.

Recently, information concerning protein locations within the 30S subunit has been combined with the secondary structure model of 16S rRNA and the location of protein binding sites in rRNA to generate an initial, partial three-dimensional picture of how the rRNA and proteins interact in this ribosomal subunit (fig. 28.10).

Other Cellular RNAs Are Involved in DNA Synthesis and RNA Processing

In addition to the three major classes of RNA involved in protein synthesis, chromosomal DNA also carries the information for the synthesis of other, minor cellular RNAs.

DNA synthesis requires a primer—an oligonucleotide that is base-paired with the template strand and provides a 3′ hydroxyl group upon which to start polymerizing a new DNA strand (see chapter 26). The cell uses small RNAs as primers for DNA replication. In some cases in prokaryotic cells these primers are synthesized by RNA polymerase (the main tran-

28A
BOX

Analysis of Ribosomal Proteins

Ribosomal proteins are conveniently analyzed by two-dimensional electrophoresis. Generally, a single system is sufficient to resolve all proteins from a single type of ribosomal subunit, and the mobility of the proteins on the gel is used to identify them. Thus the *E. coli* 30S subunit contains 21 proteins that are designated S (for small) 1 to 21. The 50S subunit proteins are designated L for large.

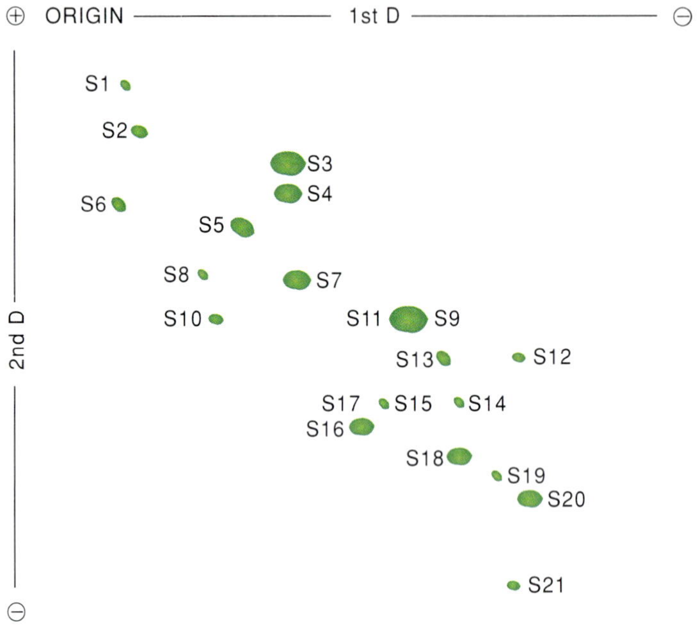

Figure 1

A two-dimensional gel electrophoretogram of *E. coli* 30S proteins. (Courtesy of Bishwajit Nag and Robert R. Traut.)

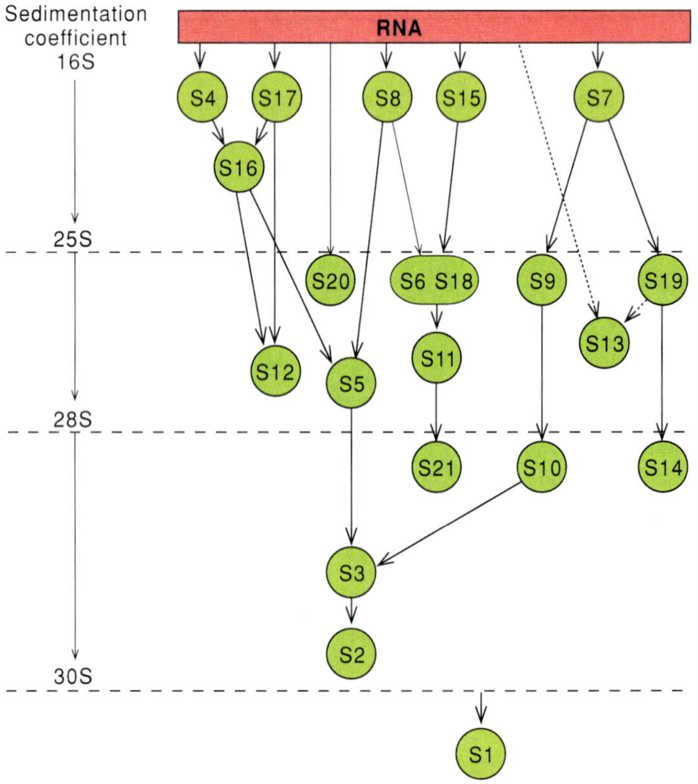

Figure 28.8

Assembly map of the 30S subunit. Initially only five ribosomal proteins associate with the rRNA. The binding of further proteins occurs after these initial proteins have bound, and these proteins make contacts with both the ribosomal RNA and the already bound proteins. The assembly process can be monitored by the changing sedimentation constant of the ribosome. The folding of the rRNA in the completed structure, together with the location of the individual ribosomal protein subunits, is illustrated in figure 28.10.

Figure 28.9

Gross shapes of the *E. coli* ribosomal subunits and the ribosome.

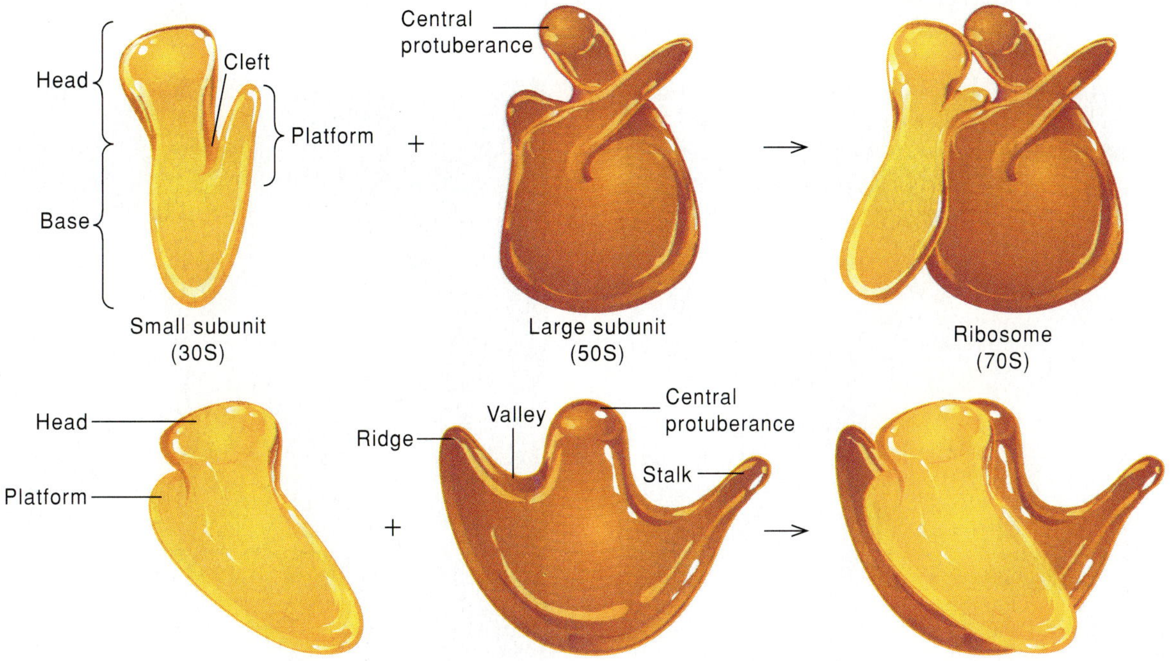

scribing enzyme), but most primers in *E. coli* are made by a special enzyme, called DNA primase, that is encoded by the *dnaG* gene.

Eukaryotic cells contain in their nuclei a unique collection of small RNAs, called small nuclear RNAs (snRNAs), which are complexed with certain proteins to form small nuclear ribonucleoprotein particles (snRNPs). These RNAs have been named U1, U2, U3, . . . , U13 RNA and range in size from 100 to 220 bases. One, U3, is found in the nucleolus, the site of rRNA synthesis. All are very abundant and there may be as many as 1 million copies per nucleus. Their base sequences are highly conserved among organisms, and all seem to contain unusual trimethylguanosine structures at their 5′ ends. Each snRNP contains an snRNA and six to twelve proteins, some of which are common to all snRNPs. The role of all of the snRNPs is not fully known, but they seem to be involved in processing and maturation of RNA precursors including splicing (U1, U2, U4, U5, U6), polyadenylation of pre-mRNA (U11), formation of 3′ ends of histones (U7), and maturation of rRNA (U3).

Transcription in Prokaryotes

There are many differences in synthesis and processing of RNAs in prokaryotes and eukaryotes. We will first deal with the reactions as they occur in *E. coli,* the best studied prokaryote. All DNA-dependent RNA polymerases carry out the following reaction:

$$\text{NTP} + (\text{NMP})_n \xrightarrow[\text{DNA}]{\text{Mg}^{2+}} (\text{NMP})_{n+1} + \text{PP}_i$$

The DNA template strand determines which base will be added to the growing RNA molecule, by base pairing similar to that used to direct the semiconservative replication of DNA. For example, a cytosine in the template strand of DNA means that a complementary guanine will be incorporated into the corresponding position of the RNA. Synthesis proceeds in a 5′ → 3′ direction, with each new nucleotide being added onto the 3′-OH end of the growing RNA chain.

E. coli *RNA Polymerase Transcribes the Three Major Classes of RNA*

Prokaryotic polymerases are used to transcribe all major genes of the prokaryotic genome. They do this in a discriminatory way so that the organism is supplied with just the amounts of appropriate mRNAs, tRNAs, and rRNAs that the organism requires.

The Bacterial RNA Polymerase Contains Five Subunits at the Time of Initiation The purification of substantial amounts of RNA polymerase from *E. coli* has made it possible to study the structure of the enzyme. Richard Burgess, Andrew Travers, John Dunn, and Ekehard Bautz showed that the active enzyme molecule is a pentamer containing four different polypeptide chains with a total molecular weight of about 500,000. The subunits of the enzyme can be separated by electrophoresis on polyacrylamide gels. The four different polypeptide chains, termed β', β, σ^{70}, and α, have molecular weights of 155,000, 151,000, 70,000, and 36,500, respectively. Additional σ-like proteins have

Figure 28.10

Arrangement of components in the *E. coli* 30S ribosomal particle. In (*a*) the relative locations of the ribosomal proteins, numbered 1–21, are shown. (Illustration prepared by Dr. Malcolm Capel from data described in M.S. Capel, M. Kjeldguard, D.M. Engelman, and P.B. Moore, *Journal of Molecular Biology* 200:66–87, 1988.) In (*b*) the conformation of the rRNA is shown. The location of the proteins is indicated by numbers given in the figure. (Illustration prepared by S. Stern, B. Weiser, and H. F. Noller from data described in *Journal of Molecular Biology* 204:447–481, 1988.)

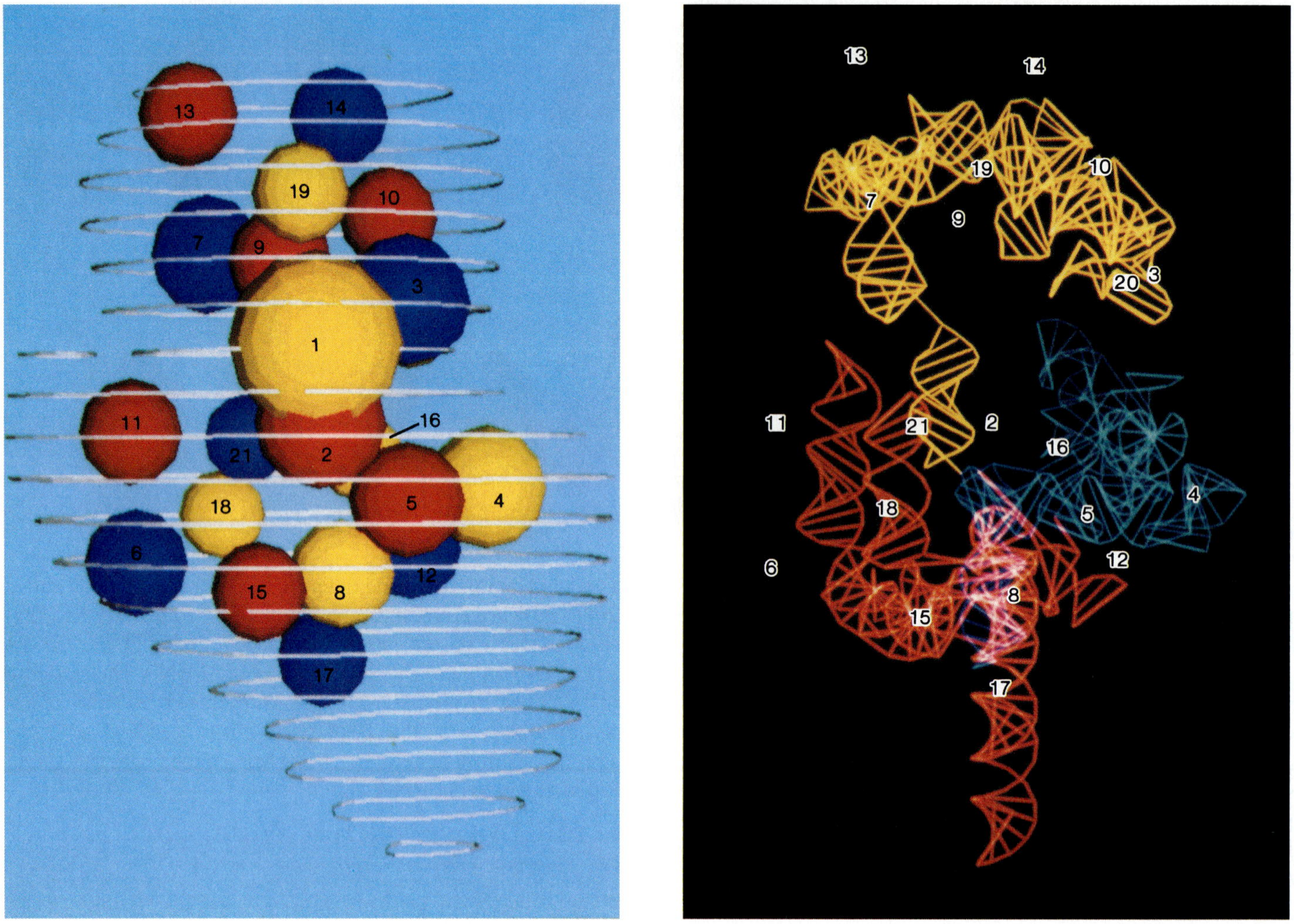

(a) (b)

also been identified, and we will discuss them later. Some of the properties of the polymerase subunits are summarized in table 28.2.

A complex with the subunit structure $\alpha_2\beta\beta'\sigma^{70}$ can carry out the functions necessary for correct and efficient synthesis of RNA and is referred to as holoenzyme. Holoenzyme can be reversibly separated into two components by chromatography on a phosphocellulose column:

$$\underset{\text{Holoenzyme}}{\alpha_2\beta\beta'\sigma^{70}} \rightleftharpoons \underset{\text{Core polymerase}}{\alpha_2\beta\beta'} + \underset{\text{Sigma-70}}{\sigma^{70}}$$

One component, called core polymerase ($\alpha_2\beta\beta'$), retains the capability to synthesize RNA, but it is defective in the ability to bind and initiate transcription at the appropriate initiation sites (promoters) on the DNA. The other component, σ^{70}, has no RNA synthetic activity, but when added back to core polymerase, it re-forms holoenzyme with its ability to bind tightly and selectively at promoters and initiate RNA chains efficiently. Because of its role in binding and initiation, σ^{70} is often referred to as an initiation factor.

Although the precise functions of the various subunits of core are not known, β' is a basic (positively charged) polypeptide thought to be involved in DNA binding. The β subunit is the site of binding of several inhibitors of transcription and is thought to contain most or all of the active sites for phosphodiester bond formation. The α subunit is necessary in order to reconstitute active enzyme from separated subunits and appears to be the site of contact with a number of transcription factors, such as CAP.

Table 28.2
E. coli RNA Polymerase Subunits and Regulatory Factors

Subunit or Protein	Gene Name	Map Position (min)	Polypeptide M_r	No. in Enzyme	Function	Properties
β' (beta')	*rpoC*	90	155,000	1	DNA binding?	Basic
β (beta)	*rpoB*	90	151,000	1	Active site	Acidic
σ^{70} (sigma-70)	*rpoD*	67	70,000	1	Promoter recognition, initiation	Acidic
α (alpha)	*rpoA*	72	36,500	2	Site of activator contact	Acidic
CAP	*crp*	74	23,000	2	Activation	Basic
ρ (rho)	*rho*	85	46,000	6	Termination	Basic
nusA	*nusA*	69	55,000	1	Termination	Acidic

Genetics of the Subunits Since RNA polymerase is an essential cellular enzyme, most mutations in the polymerase genes are lethal. Those that can be studied are the ones that affect the enzyme activity only under certain conditions, e.g., at higher than normal temperatures. As we saw in chapter 26, mutants that can grow normally at one temperature but not at another are called temperature-sensitive mutants. Another class of mutants is composed of those in which the enzyme is resistant to an inhibitor of RNA synthesis, which acts by binding to the enzyme. We will give examples of such inhibitors later in this chapter.

Strains of *E. coli* have been obtained with mutations in the α, β, β', and σ^{70} subunits of RNA polymerase. These mutants have made it possible to map the genes for the corresponding subunits. The circular map of the *E. coli* genome is shown in figure 28.11; on it we show the positions of the genes known to be involved in RNA synthesis, modification, processing, and degradation. It is evident that these genes are scattered throughout the genome. At first it may seem surprising that the genes encoding the subunits of RNA polymerase are not all clustered together. Their dispersion is probably due to the fact that only β and β' are required in equimolar amounts. The genes for β and β' are in the same region at 90 min that contains the genes for two ribosomal proteins. The gene for α is found far from those for β and β', at 73 min on the map, and also is in a region containing genes for several ribosomal proteins. The gene for the σ^{70} subunit is found at 67 min, in a region containing the genes for DNA primase and a ribosomal protein.

Figure 28.11

Genetic map of *E. coli* showing the location of some genes involved in RNA metabolism. The map is divided into 100 minutes. Three types of genes are indicated: (1) genes involved in RNA synthesis, (2) genes involved in regulation, and (3) genes involved in processing and degradation. (Source: B. J. Bachmann, "Linkage map of *E. coli* K12, edition 7" in *Microbiological Reviews*, 47:180, 1983. Copyright © 1983 American Society for Microbiology, Washington, D.C.)

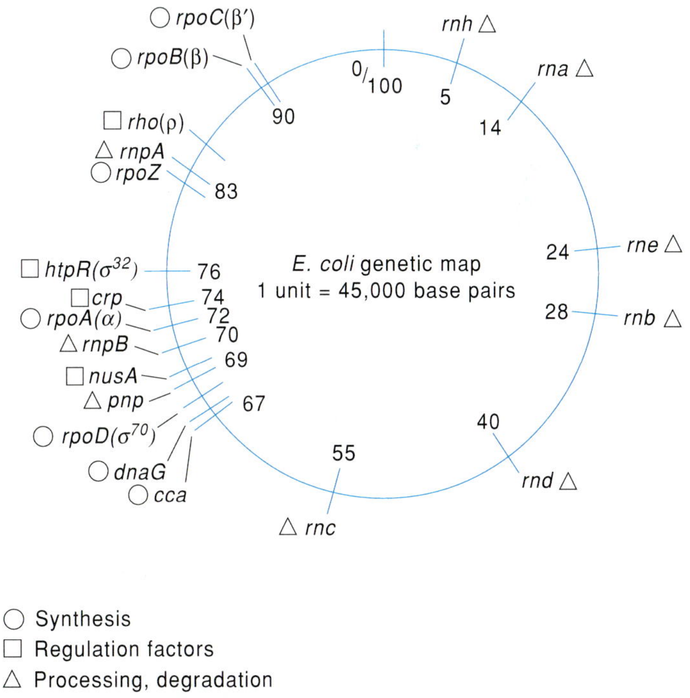

Transcription Involves Initiation, Elongation, and Termination

The overall transcription cycle, involving binding, initiation, elongation, and termination, is shown in figure 28.12. Let us now look at this cycle in some detail.

Transcription Units and Promoter Signals In addition to the sequences that code for proteins, precise sequences along the DNA signal RNA polymerase start and stop sites. The promoter region contains the information that tells the RNA

Figure 28.12

Schematic of the overall transcription cycle.

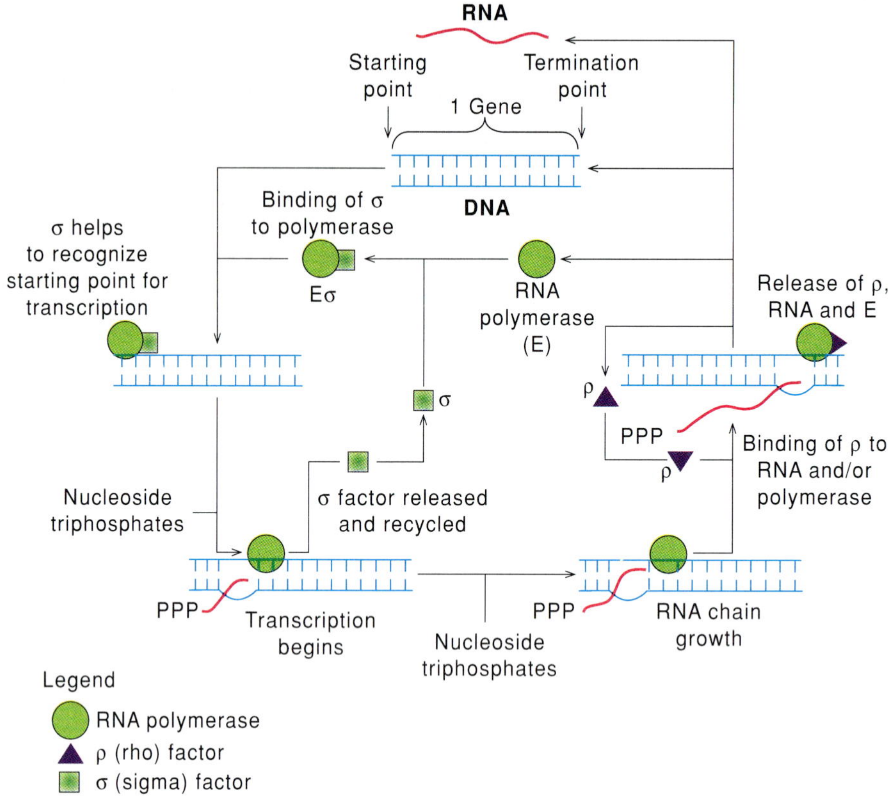

Figure 28.13

Important features of a typical transcription unit. DNA is shown with promoter and terminator regions expanded below. RNA is transcribed starting in the promoter region at +1 and ending after the stem and loop of the terminator. The protein resulting from translation of this RNA is shown above with its N and C termini indicated.

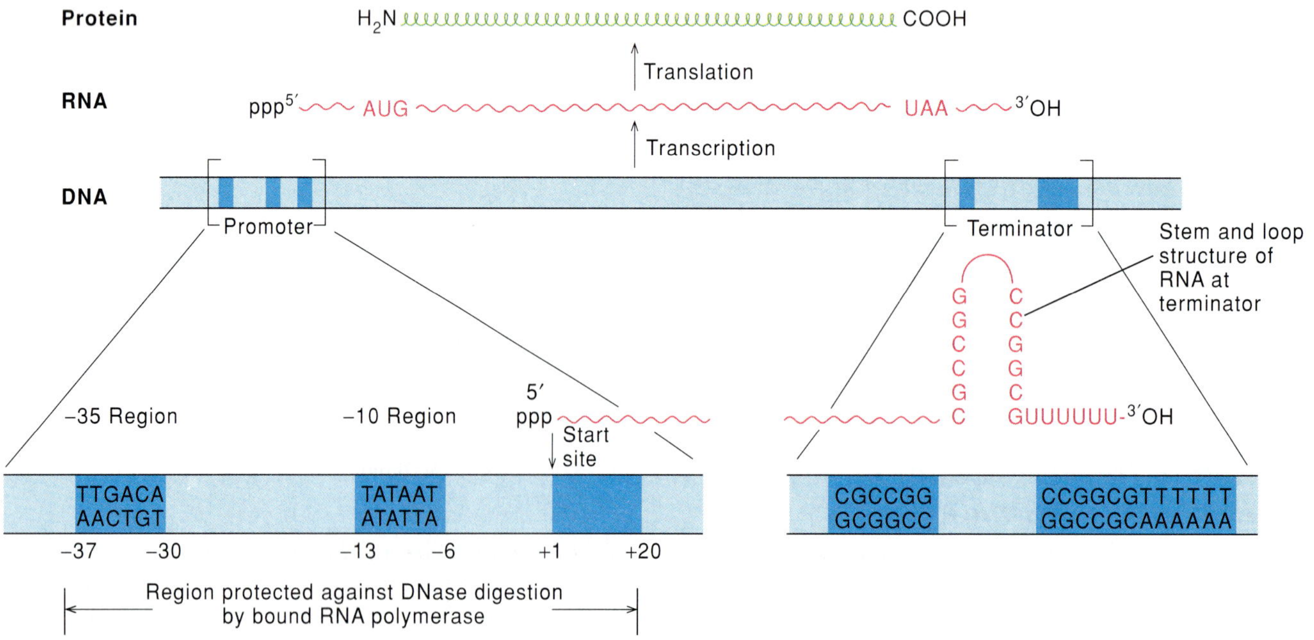

polymerase where to bind, how tightly to bind, and how frequently to initiate an RNA chain. It also often contains sites at which additional regulatory proteins bind; these proteins influence binding and initiation by RNA polymerase. The terminator region contains DNA sequences that cause the RNA polymerase to stop transcribing. Then, either spontaneously or with the aid of a termination factor, the RNA polymerase and RNA are released from the template. The transcribed region, including the start and stop signals, is called a transcription unit or operon. It may include the structural genes for several proteins whose syntheses are controlled coordinately. An example of a transcription unit in *E. coli* is shown in figure 28.13.

A large number of transcription units have been studied, and many promoter and terminator regions have been sequenced. Common features have been identified in the sequence of over 100 different promoters, and in the case of several promoters, further information about DNA bases important for binding has been obtained. Promoter features are summarized in figure 28.13.

RNA polymerase binding at the promoter protects about 60 base pairs of DNA from digestion by DNase. The protected region runs from −40 (40 bp before the RNA chain starting site) to +20 (20 bp after the starting site). Two sites within this 60-bp region, one centered at −35 and one centered at −10, contain specific recognition sequences that are common to most but not all bacterial promoters. Mutations in these two sites affect the ability of RNA polymerase to bind and initiate.

The DNA base sequence of each promoter differs somewhat from the "average" promoter sequence. This is to be expected because promoters differ tremendously in their "strength" or frequency of RNA initiation. Some promoters are very "weak"; for example, the promoter for the lac repressor* produces an mRNA only once in 20 to 40 min. This frequency is estimated as follows: Each *E. coli* contains about twelve molecules of lac repressor. Since each repressor is composed of four identical subunits, this represents about 48 molecules of repressor polypeptide. If each mRNA is translated by about 50 ribosomes before it decays, then only one mRNA is needed per generation of 20 to 40 min. In contrast, the promoters for rRNA genes must be utilized once every second or two in order to produce over 10,000 rRNAs per generation from the seven copies of the rRNA gene. This 2,000-fold difference in promoter strength is encoded in the base sequence of the promoter.

Binding at Promoters In the cell, RNA polymerase transcribes only selected regions of the DNA and makes RNA complementary to only one of the DNA strands in any particular region. This selectivity is possible because holoenzyme is able to recognize and form a stable complex with DNA at specific promoter regions. RNA polymerase is able to form unstable nonspecific complexes at any place on the template, mainly by an ionic interaction with the DNA phosphates, but it either rapidly dissociates and rebinds, or slides along the DNA until it reaches a promoter region. It then forms a moderately stable complex with the promoter DNA, most probably interacting stereospecifically with particular nucleotides in the −35 region of the promoter. These complexes can form at 0° C, where the DNA remains double helical or unmelted, and are referred to as closed promoter complexes (fig. 28.14). The next step involves a conformational change in the polymerase and a temperature-dependent melting of about ten base pairs of DNA from positions −9 to +2. The resulting very stable complex is termed the open promoter complex.

Figure 28.14

Various types of RNA polymerase-DNA binding complexes. A nonspecific complex is formed at any point along the DNA. The closed promoter complex is formed at a polymerase binding site. Following the formation of the closed promoter complex, an open promoter complex is formed at the same site.

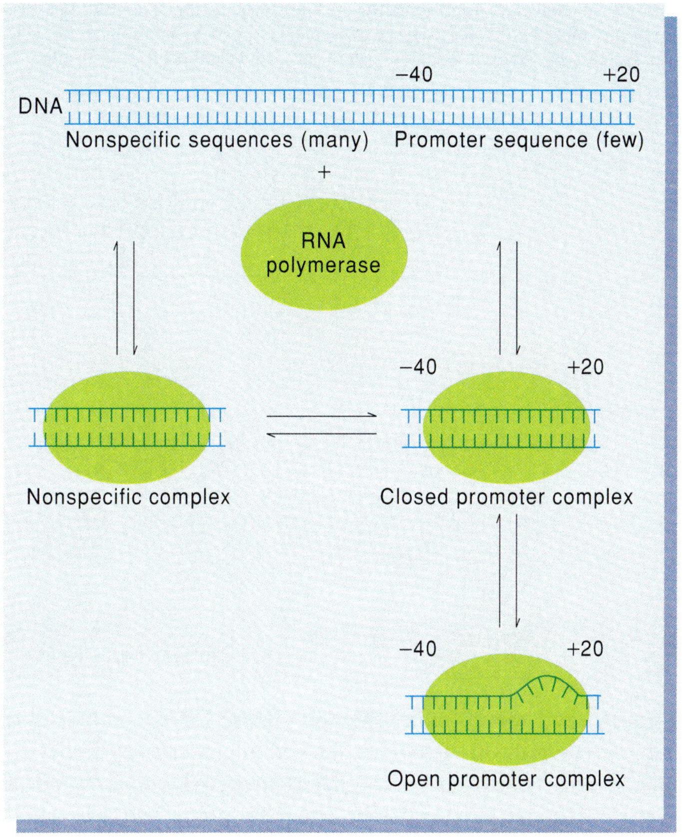

The existence of consensus sequences in both the −35 and −10 regions of the promoter carries with it the implication that these regions are important in forming the open promoter complex with RNA polymerase. One possibility is that both regions are complexed with polymerase in the open promoter complex. Another possibility is that one region serves for initial recognition of the polymerase, but that in the final complex the polymerase binds only to the other region.

Several approaches to establish more precisely the regions of contact (or close approach) between polymerase and DNA were developed in Walter Gilbert's laboratory. One

*The lac repressor will be discussed in chapter 30. It is not important at this point that we know anything about the lac repressor except what is stated here.

Figure 28.15

(*Upper*) Sequence of the T7 A3 promoter, showing the stronger contacts to the RNA polymerase. (▼ = phosphate contacts; ○ and ∧ = purines that the polymerase protects from methylation or whose susceptibility to this methylation is enhanced, respectively; * = methylated purines that interfere with polymerase binding.) The most probable bases for the −10 region and the −35 region are shown above the corresponding regions in the A3 promoter sequence; +1 represents the start of transcription. The minimal region unwound by the polymerase is represented by a separation of the strands. (*Lower*) Planar representation of the cylindrical projection of the DNA molecule (10.5 base pairs per turn), with contacts to the polymerase marked. (● = phosphate contacts ⊛ = methylated purines that interfere with polymerase binding, and other symbols are as in the upper drawing.) Contact regions, strands, front view, back view, and initiation site are indicated. Regions likely to interact with polymerase are shaded with vertical lines. (Source: U. Siebenlist and W. Gilbert, "Contacts between *E. coli* RNA polymerase and an early promoter of phage T7" in *Proceedings. National Academy of Sciences USA* 77:122, 1980. Copyright © 1980 National Academy of Sciences, Washington, D.C.)

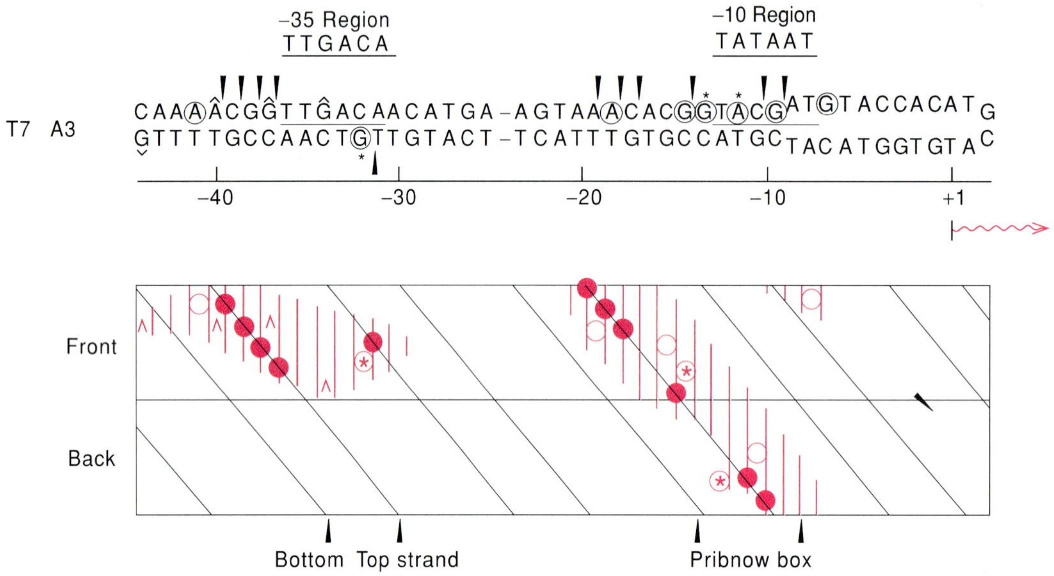

strategy used was to make the polymerase-DNA complex and then determine which regions showed altered susceptibility to specific chemical probes. Another strategy was to carry out a partial reaction with a specific chemical probe and see which sites interfered with polymerase complex formation. For these experiments, two chemical probes have been most useful. The first, dimethylsulfate, methylates double-stranded DNA at the N-7 position of guanine in the major groove and the N-3 position of adenine in the minor groove. The methylated purine will depurinate on heating; alkali then can cause a β-elimination reaction and create a series of breaks in the DNA chain. The second probe, ethylnitrosourea, preferentially ethylates the DNA phosphates. The resulting phosphotriesters serve as cleavage sites when the DNA is exposed to alkali.

The results of these experiments are depicted in figure 28.15. Domains of interaction are focused on the −35 region and the −10 region. This result provides strong support for the conclusion that polymerase binds simultaneously to both of these regions.

From the experiments with dimethylsulfate, evidence emerged that the polymerase produces an unwinding of a segment of the DNA duplex from −9 to +2. Dimethylsulfate, which reacts with the N-1 of adenine and the N-3 of cytosine only in single-stranded DNA, reacted with this region in the presence of polymerase. In the double helix these nitrogens are normally unavailable for reaction.

Another type of experiment was used to establish the thymine contacts with polymerase. For this purpose, thymine was replaced in the DNA by 5-bromouracil. Ultraviolet light normally cleaves DNA at the bromouracils. A nearby bound protein can either protect the DNA from such cleavage or become cross-linked to the DNA at position 5′ of the bromouracil. This approach was particularly valuable in suggesting which of the polymerase subunits are in direct contact with the DNA. Observations for the *lac*UV5 promoter show that the σ^{70} subunit cross-links at position −3 of the sense strand (the sense strand carries the same sequence as the transcribed RNA) and that the β subunit cross-links at position +3 of the sense strand.

The chemical-probe approach used to establish points of close contact between polymerase and DNA also provided useful information on regulatory protein interaction with DNA (see the discussion on CAP and *lac* repressor in chapter 30).

Initiation at Promoters Once the polymerase binds to the promoter and strand separation occurs, initiation usually proceeds rapidly (1 to 2 s). The first, or initiating, nucleoside-5′-triphosphate, which is usually ATP or GTP, binds to the enzyme. The binding is directed by the complementary base in the DNA template strand at position +1, the start site. A second nucleoside-5′-triphosphate (NTP) binds, and initiation occurs upon formation of the first phosphodiester bond by a nucleophilic attack of the 3′ hydroxyl group of the initiating NTP on the α phosphorus atom of the second NTP. Inorganic pyrophosphate derived from the second NTP is a product of the reaction. This process is illustrated in figure 28.16.

The initiation process can be followed during an *in vitro* transcription reaction in several ways. NTP radioactively labeled with ^{32}P in the β or γ phosphate positions can be used in transcription reactions. Since the initiating NTP retains its 5′ triphosphate the beginning of the RNA chain, or the 5′ triphosphate end, will be exclusively labeled. Another way is to add ^{32}P-labeled pyrophosphate ($\overset{**}{pp}_i$) to an otherwise unlabeled reaction and measure initiation by a reaction called pyrophosphate exchange. The labeled pyrophosphate participates in the reverse of the phosphodiester-bond formation reaction, allowing the 3′-terminal nucleotide to be removed and the label to be incorporated into its β and γ positions. Pyrophosphate exchange also can occur during chain elongation:

$$\text{pppApU} + {}^{32}\overset{**}{\text{pp}}_i \underset{\text{Bond formation}}{\overset{\text{Pyrophosphorolysis}}{\rightleftharpoons}} \overset{**}{\text{pp}}\text{pA} + \text{pppU}$$

A third method for measuring initiation depends on the finding that in the absence of the nucleotide needed for the third position on the RNA chain, the dinucleotide formed, pppXpY, dissociates from the active site and the initiation process must start again. This phenomenon is called abortive initiation. For example, if only ATP and [α-^{32}P]UTP are added to a reaction mixture where an RNA chain that begins with the sequences pppApUpCp is being synthesized, the reaction produces the dinucleotide pppA^{32}pU. The production of this dinucleotide can be quantified to measure initiation of this chain. Some promoters *in vitro* undergo considerable abortive initiation, even in the presence of all four nucleoside triphosphates, releasing oligonucleotides of two to eight residues several times before finally succeeding in producing a long RNA chain. Therefore, promoter strength should be equated not strictly with the frequency of initiation at a promoter, but rather with the frequency with which long RNA chains are produced.

Elongation of RNA After initiation has occurred, chain elongation proceeds by the successive binding of the nucleoside triphosphate complementary to the next base in the template strand, bond formation with pyrophosphate release, and translocation of the polymerase one base farther along the template strand. Transcription proceeds in the 5′ → 3′ direction, antiparallel to the 3′ → 5′ strand of the template DNA. Once elongation has produced an RNA chain about ten bases long, the σ^{70} subunit dissociates from the holoenzyme to yield core polymerase, which continues the elongation reaction until a termination signal is reached. The released σ^{70} is available to bind to a free core polymerase and re-form holoenzyme capable of binding at a promoter and initiating a new RNA chain (see fig. 28.12).

In vivo, mRNA chains grow at a rate of about 45 nucleotides per second, nicely matched with the rate of translation of 15 amino acids per second. rRNA is synthesized about twice as fast as mRNA.

Figure 28.16

Details of phosphodiester bond formation. The α, ß, and γ phosphates are indicated on the initiating NTP, which in this case is ATP. The colored ovals represent NTP binding sites on the RNA polymerase. The biochemistry of bond formation in RNA synthesis is very similar to that in DNA synthesis.

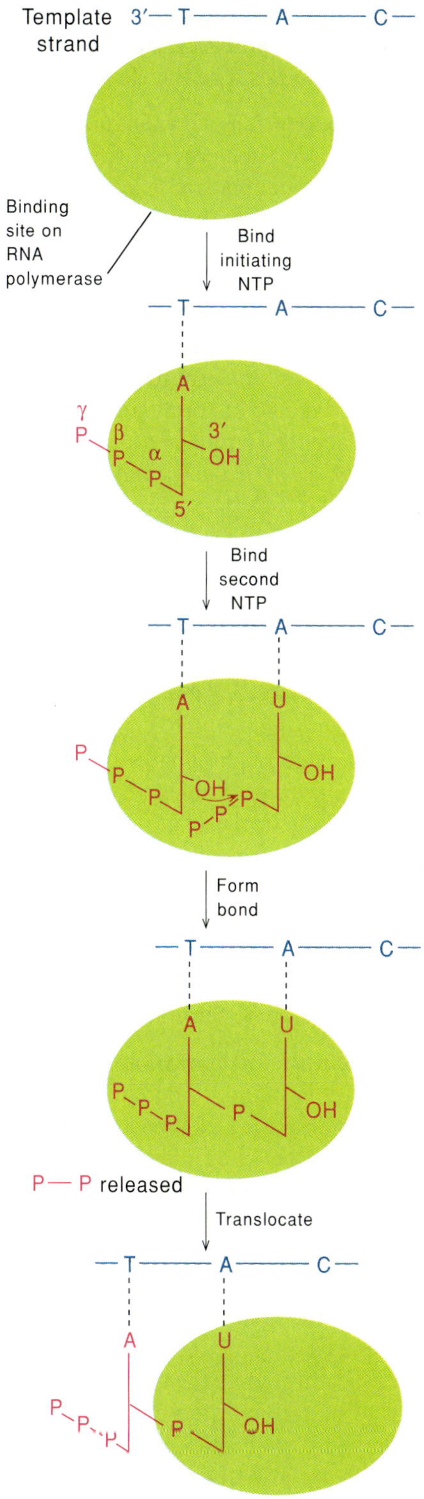

As the polymerase travels along the DNA, it must continually cause an opening or strand separation of the DNA so that a single DNA template strand is available at the active site of the enzyme. For the transcription reaction to be energetically feasible, one base pair must re-form behind the active site for every base pair opened in front of it. It is likely that a short transient RNA-DNA hybrid duplex forms between the newly synthesized RNA and the ten-base-long unpaired region of the DNA and helps hold the RNA to the elongating complex.

Termination of Transcription Termination of transcription involves stopping the elongation process at a region on the DNA template that signals termination and releases the RNA product and RNA polymerase. The majority of terminators that have been studied are similar in that they code for a double-stranded RNA stem-and-loop structure just preceding the 3′ end of the transcript (see fig. 28.13). It is thought that such a structure causes the RNA polymerase to pause or stop elongating.

Two main types of terminators have been distinguished. The first is capable of termination with no accessory factors and contains about six uridine residues following the stem and loop. Apparently, a particularly weak interaction between the 3′ terminal oligo(U) and the oligo(dA) region in the template DNA strand allows the release of the RNA from the transcription complex. The second type of terminator lacks the oligo(U) region and requires termination factor rho for RNA chain release. Rho, discovered by Jeffery Roberts in 1969, is a hexamer of subunits (M_r = 46,000 per subunit). Rho does not bind tightly to DNA or to RNA polymerase. It does bind tightly to RNA, especially cytosine-rich RNA, and in its presence rho hydrolyzes ribonucleoside triphosphates to nucleoside diphosphates. A current theory is that rho binds to sites on RNA and, by hydrolyzing nucleoside triphosphates, either moves along the RNA or winds the RNA around it until it reaches an RNA polymerase that has stopped at a terminator, where rho causes release. Another termination factor, nusA, is an acidic protein with a molecular weight of 55,000 that is able to bind to core polymerase after sigma is released and aid in the release of RNA and RNA polymerase at some terminators.

Other Factors Regulating Transcription As more transcription units are studied in detail, it becomes apparent that the basic processes just described can be modulated in a great number of ways, both by including additional signals in the DNA sequence and by supplying additional regulatory factors. Some of these regulatory mechanisms are discussed in more detail in chapters 30 and 31.

Many transcription units have terminator signals called attenuators preceding the structural gene or between two structural genes. Attenuators sometimes cause termination and sometimes allow read-through to the structural part of the gene. The efficiency of termination at these sites is variable and can be regulated in response to changes in the growth conditions of the cell. Attenuators are also discussed in chapter 30.

Protein factors can inhibit transcription (negative control) or stimulate it (positive control). The binding of a repressor molecule to a specific site (operator) overlapping the promoter region prevents transcription by sterically interfering with RNA polymerase binding. This is an example of negative control. An example of positive control is the stimulation of transcription caused by the binding of the catabolite activator protein (CAP) in the presence of cyclic AMP (cAMP) to sites just adjacent to the promoters of a number of genes coding for enzymes involved in the catabolism of a variety of sugars. In *in vitro* transcription reactions, these promoters bind RNA polymerase holoenzyme only weakly and are greatly activated by the addition of CAP and cAMP.

Although most transcription in *E. coli* appears to utilize σ^{70} as an initiation factor, several additional initiation factors have been discovered that bind to core polymerase to form holo-like enzymes that recognize different promoters. One such factor is the product of the *htpR* gene, which is involved in heat shock regulation. This factor is called sigma-32 (σ^{32}) because its molecular weight is 32,000. It becomes bound to core polymerase in *E. coli* cells that have been subjected to heat shock and directs the polymerase to bind to and initiate at a class of promoters (the heat shock promoters) responsible for high-level expression of a class of a dozen or so heat shock proteins. Another such factor, σ^{54}, is a protein (M_r = 54,000) encoded by the *ntrA* gene that is needed for expression of certain genes implicated in nitrogen metabolism. Recently three additional sigmas have been identified. Sigma-E is thermostable and recognizes the promoter for σ^{32}. Sigma-F (M_r of about 28,000) is involved in transcription of several genes coding for flagellar proteins and enzymes involved in chemotaxis. Sigma S (M_r = 41,500), a product of the *katF* gene, is involved in regulating *katE* and other genes turned on during carbon starvation. The consensus sequences associated with different classes of promoters are given in table 28.3.

Multiple sigmas have also been found in the bacterium *Bacillus subtilis*. Several different σ-like subunits are present in growing cells, and additional ones appear during sporulation to allow expression of sporulation-specific genes. Furthermore, the infection of *B. subtilis* with certain bacteriophages, such as SP01 or SP82, results in virus-coded σ-like factors that bind to core polymerase in place of the major sigma, σ^{43}, and direct RNA polymerase to viral promoters.

Differences between Eukaryotic and Prokaryotic Transcription

Although the basic mechanism by which RNA is synthesized is quite similar in prokaryotes and eukaryotes, there are several important differences. The DNA in eukaryotic cells is complexed to form chromatin (see chapter 25). The chromatin, in turn, is condensed to form chromosomes. Only a small fraction

Table 28.3
Summary of *E. coli* Sigma Factors

Sigma Factor	Gene	Consensus Sequence −35 Region	Consensus Sequence −10 Region	Genes Recognized
σ^{70}	*rpoD*	TTGACA	TATAAT	Most genes
σ^{32}	*rpoH (htpR)*	CCCTTGAA	CCCCAT-TA	Heat-shock regulated
σ^{54}	*rpoN (ntrA)*	$CTGGCACN_5TTGCA$		Nitrogen-regulated
σ^{E}	Not identified	GAACTT	TCTGA	*rpoH, htrA*
σ^{F}	fliA	TAAA	GCCGATAA	Flagellar, chemotaxis
σ^{S}	*rpoS (katF)*	Unknown		*katE,* Starvation regulated

of the chromosome is actively being transcribed at any given time. In transcriptionally active regions, the DNA must be partially exposed and be more accessible to RNA polymerase. The DNA in these exposed regions is found to be sensitive to cleavage by mild treatment with bovine pancreatic DNase I and appears to contain bound RNA polymerase, additional nonhistone proteins, and modified histones. In addition, DNA in active regions in some species is undermethylated when compared with DNA from regions not active in transcription. Although prokaryotic DNA is not complexed with histones, it seems to be coated with small basic proteins; however, most parts of the bacterial DNA are accessible to RNA polymerase binding and transcription.

As we stated earlier, transcription and translation occur simultaneously in prokaryotes. RNA polymerase moves along a gene generating an mRNA chain, and as soon as a bit of mRNA is synthesized, ribosomes bind to it and start translating it. The situation is very different in eukaryotes, where transcription and translation occur in separate compartments of the cell. The nucleus, where the chromosomes are located, is the site of DNA-dependent RNA synthesis. Large RNA precursors to mRNA are synthesized in the nucleus, become complexed with proteins to form ribonucleoprotein particles (RNPs), and then are modified, and processed to form smaller mRNAs that are transported across the nuclear membrane to the cytoplasm, where they are translated into protein. These differences lead to a complex process of modification and processing in the eukaryotic RNAs.

Eukaryotes Have Three Nuclear RNA Polymerases

Unlike prokaryotes, in which all major types of RNA are synthesized by one RNA polymerase holoenzyme, eukaryotic cells have become more specialized in their transcription capabilities. They contain different nuclear DNA-dependent RNA polymerases, each responsible for synthesizing different classes of RNA.

Nuclear extracts can be fractionated by chromatography on DEAE-cellulose to give three peaks of RNA polymerase activity. (The use of column chromatography is explained in chapter 5.) These three peaks correspond to three different RNA polymerases (I, II, and III), which differ in relative amount, cellular location, type of RNA synthesized, subunit structure, response to salt and divalent cation concentration, and sensitivity to the mushroom-derived toxin α-amanitin. The three polymerases and some of their properties are summarized in table 28.4.

RNA polymerase I is located in the nucleolus and synthesizes a large precursor (fig. 28.17) that is later processed to form rRNA. It is resistant to inhibition by α-amanitin at concentrations even greater than 1,000 μg/ml. RNA polymerase II is located in the nucleoplasm and synthesizes large precursor RNAs (sometimes called heterogeneous nuclear RNA or hnRNA) that are processed to form cytoplasmic mRNAs. It is also responsible for the synthesis of most viral RNA in virus-infected cells. It is very sensitive to α-amanitin, being 50% inhibited by about 0.05 μg/ml. RNA polymerase III is also located in the nucleoplasm and synthesizes small RNAs such as 5S RNA and the precursors to tRNAs. This enzyme is somewhat resistant to α-amanitin, requiring about 5 μg/ml to reach 50% inhibition.

All three of these enzymes have been purified extensively and have complex subunit structures. All contain two subunits with molecular weights larger than 120,000 and six to ten smaller subunits. Although each has some unique subunits, some subunits are common to two or three. At present, very little is known about the functions of the individual subunits. The subunit structure for RNA polymerase II is quite similar for enzymes purified from a variety of eukaryotes, including human, calf, mouse, wheat, cauliflower, acanthamoeba, yeast, and slime molds.

The genes for the largest subunits of RNA polymerases I and II from yeast and of RNA polymerase II from *Drosophila* have been cloned and sequenced. Similar amino acid sequences are present in several regions of these three subunits and, surprisingly, are also found in the β' subunit of *E. coli.*

Crude enzyme preparations of RNA polymerases I, II, and III have been shown to be capable of selective transcription of defined DNA templates, initiating at sites known in some

Table 28.4
Comparison of Eukaryotic DNA-Dependent RNA Polymerases

Type	Location	RNAs Synthesized	Sensitivity to α-Amanitin
RNA polymerase I	Nucleolus	Pre-rRNA	Resistant
RNA polymerase II	Nucleoplasm	hnRNA, mRNA	Sensitive
RNA polymerase III	Nucleoplasm	Pre-tRNA, 5S RNA	Sensitive to very high levels
Mitochondrial	Mitochondria	Mitochondrial	Resistant
Chloroplast	Chloroplasts	Chloroplast	Resistant

Figure 28.17

Electron micrograph of regions of the nucleolus of the newt *Triturus* that are believed to be active in transcribing ribosomal RNA. Clarifying diagram below. (Magnification 18,000 ×.) (From O. L. Miller, Jr., and B. R. Beatty, "Visualization of nuclear genes," *Science* 164:955-957, 1969; cover photo of *Science* 23 May 1969. Copyright 1969 by AAAS.)

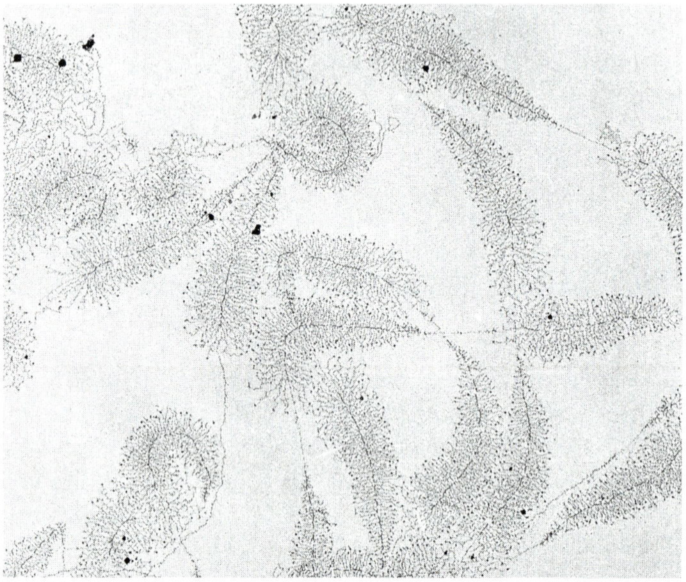

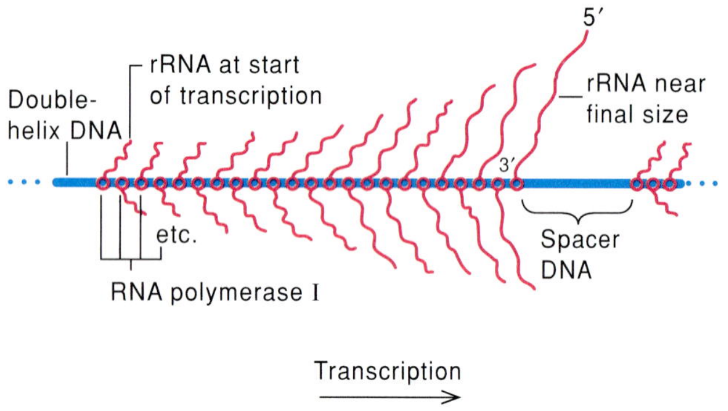

cases to be utilized *in vivo.* Fractionation of these extracts has revealed additional proteins that are required for *in vitro* transcription. These additional proteins are called transcription factors (TFs). With all three polymerases, two or more transcription factors are required, and some or most of these factors must bind to the promoter before the RNA polymerase can bind (fig. 28.18).

Because many eukaryotic genes have been cloned during the last few years, it has become possible to compare the DNA sequences preceding the genes that may act as promoterlike signals for RNA polymerase II. One feature that stands out is a common sequence, TATAAA, often called a TATA box, usually found 25 to 30 bases before the transcription start site in many but not all genes. Certain proteins selectively bind to the TATA box and these proteins appear to be essential for the formation of the initiation complex.

Phil Sharp and Leonard Guarente have shown that four transcription factors are required in addition to polymerase II for initiation from the major late promoter of adenovirus. *In vitro* studies indicate that these factors assemble in an orderly fashion (see fig. 28.18*b*). Thus complexes were generated by sequential binding of TFIID, TFIIA, TFIIB, RNA polymerase II, and TFIIE. TFIIA derived from yeast or mammalian cells forms a complex with yeast TFIID and the TATA element. TFIIB binds to this complex and probably acts as a bridge to the polymerase and the initiation site. Finally TFIIE binds. It is believed that this complex may function for a large number of eukaryotic promoters that contain TATA boxes. Progress in our understanding in this area promises to be accelerated by the observation that many transcriptional factors are interchangeable between yeast and mammals. For investigatory purposes yeast has distinct advantages to mammals, as it is more manipulable both genetically and biochemically.

Several additional promoter regions have been identified that are needed for elevated levels of transcription. These additional regions tend to be gene-specific. Both yeast and vertebrate (fig. 28.19) genes often have regions from about −200 to −40 bases upstream from the transcription start site where upstream activator elements (UASs) are located. These elements bind additional transcription activation factors, which

Figure 28.18

Formation of the initiation complex for transcription for the three major classes of eukaryotic RNA polymerase, Pol I, Pol II, and Pol III. Each polymerase consists of many subunits (not shown). In addition to the firmly bound subunits there are a number of protein factors called transcription factors (TFs) which only associate with the polymerases at the initiation site for transcription. For all three polymerases some of these transcription factors must bind to the promoter before the polymerase can bind. In each case we see an orderly progression of binding of factors and polymerase. In the case of Pol III, three transcription factors are involved for 5S rRNA genes and only two are involved for tRNA genes.

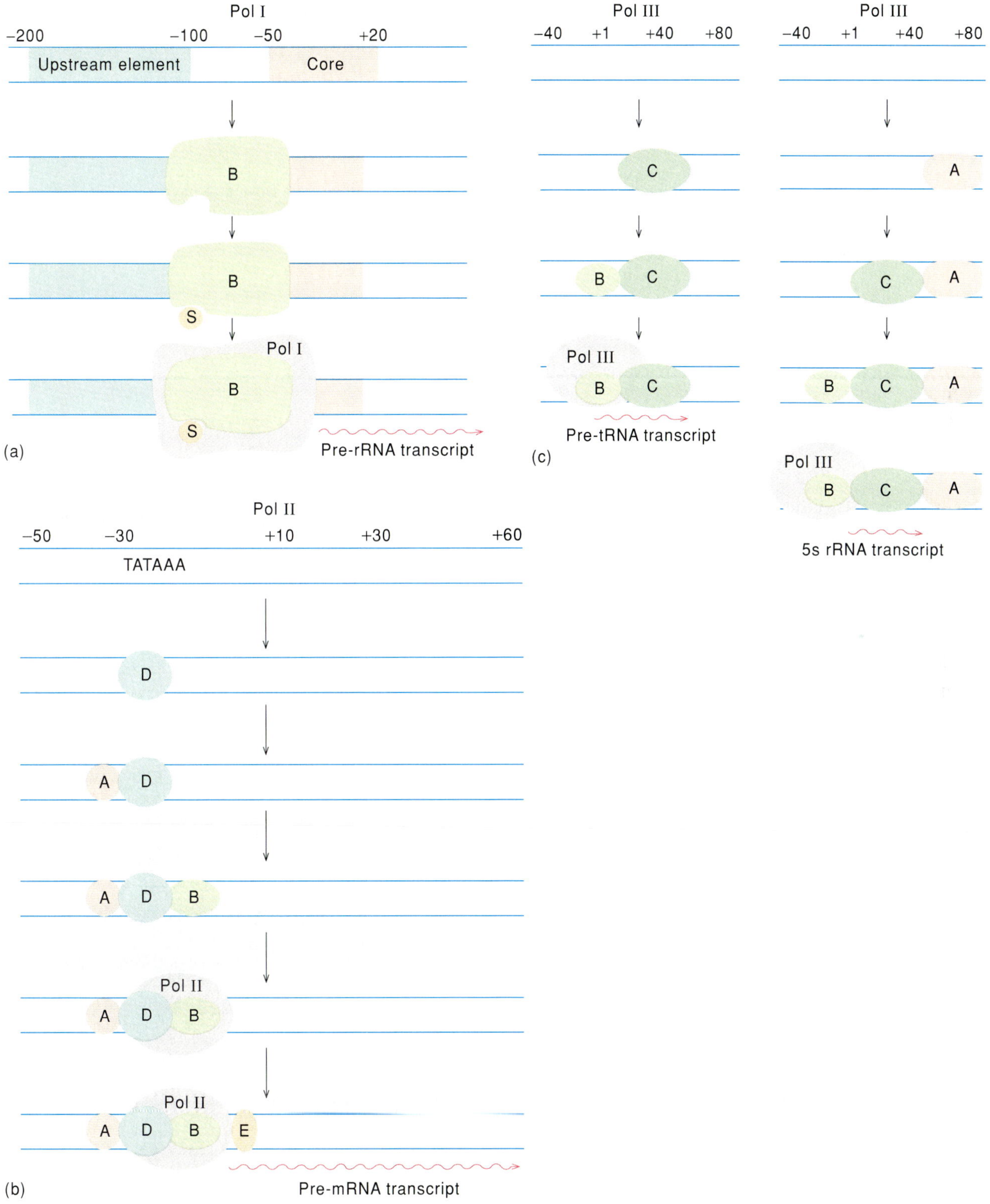

Figure 28.19

Cis elements involved in transcription in yeast (*a*) and in vertebrates (*b*). Upstream activator sequences in yeast are similar to upstream enhancers in vertebrates. Yeast has no parallel to downstream enhancers found in vertebrates.

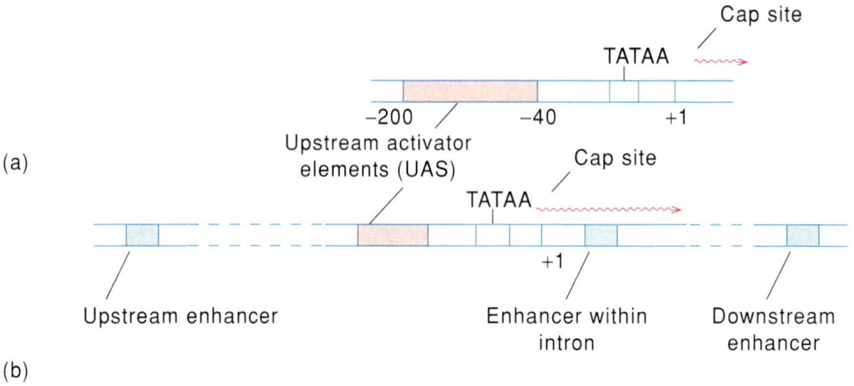

28B BOX

Use of the Footprinting Technique to Determine the Binding Site of a DNA-Binding Protein

In order to confirm the site of action of TFIIIA, the approximate location of its binding site on a 5S gene of *Xenopus borealis* was determined by the so-called footprinting technique. DNA fragments containing the 5S gene were 5′ end labeled with ^{32}P, mixed with the protein, and digested with DNase I. The resulting DNA fragments were electrophoresed on a polyacrylamide gel and autoradiographed. The region of the DNA that was protected from DNase attack by TFIIIA binding appears as a series of blank spots ("footprint") on the autoradiogram. The footprint shows that the protected region is situated between the 45th and the 90th base pairs.

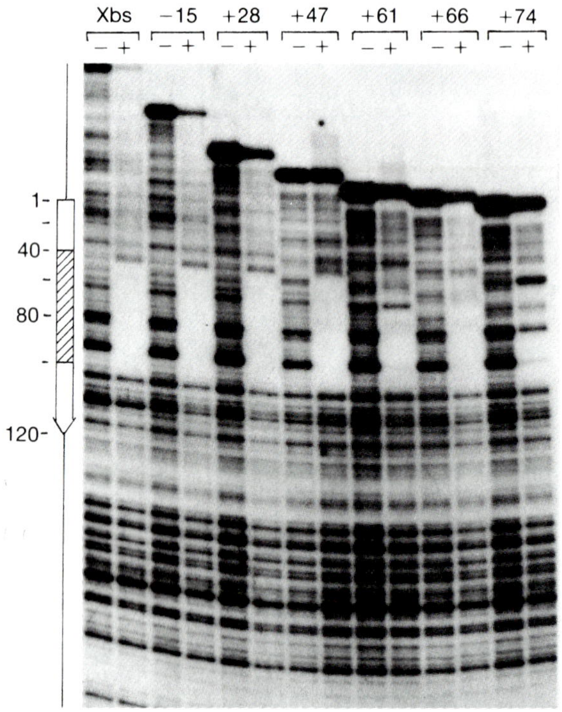

Figure 1

DNase I protection (footprinting) experiment on 5S DNA of *Xenopus*. The diagram on the left indicates the region on the gel that corresponds to the 5S RNA gene. Arrow points in the direction of transcription. Cross-hatched area indicates the region that binds transcription factor protein. Column labeled Xbs refers to an intact gene containing 160 bp of the 5′ flanking sequence, the 5S RNA gene (120 bp), and the 3′ flanking sequence (138 bp). The various deleted 5S DNAs are preceded by 74 bp of the plasmid pBR322 sequence. Numbers in other columns refer to portions of the 5S gene that have been deleted. All samples were subjected to partial digestion with DNase I, then were electrophoresed and autoradiographed. In + columns, transcription factor protein was added before DNase I treatment. (Courtesy Donald D. Brown of the Carnegie Institute of Washington.)

Table 28.5
RNA-Synthesizing Enzymes

	Template	Primer	Molecular Weight of Subunit(s)	Gene Name	Substrate	Inhibition by Rifampicin
Template-dependent						
Enzymes from bacteria						
Holoenzyme (*E. coli*)	DNA	—	155,000 151,000 70,000 36,500	*rpoC* *rpoB* *rpoD* *rpoA*	4 NTPs	Yes
DNA primase	DNA	—	65,000	*dnaG*	4 NTP, 4 dNTP	No
Enzymes from phage or phage-infected bacteria						
T7 RNA polymerase	T7 DNA	—	99,000	T7 *gene1*	4 NTPs	No
N4 RNA polymerase	N4 DNA	?	350,000	Viral	4 NTPs	No
Qβ replicase	Qβ RNA	—	65,000 55,000 43,000 35,000	*rpsA* Viral *tuf* *tsf*	4 NTPs	No
Template-independent						
CCA enzyme	—	3′ end tRNA	45,000	*cca*	CTP ATP	No
Poly(A) polymerase (eukaryotic)	—	3′ end mRNA			ATP	No
Polynucleotide phosphorylase	—	3′ end RNA	86,000 48,000	*pnp*	4 NDPs	No

interact with those that bind at the TATA box (fig. 28.19). In addition, vertebrate DNA contains enhancer elements, which activate specific genes at some distance. They can be found upstream, downstream, and even in the middle of genes. Enhancer elements may influence transcription as far as 10 kb from the gene. As in the case of the UASs, the enhancer elements are binding sites for additional transcription activators. We will have considerably more to say about these additional transcription activator elements in chapter 31, where we focus on the regulation of gene expression.

The promoter region for RNA polymerase I also involves the cooperative binding of additional transcription factors. In this case the transcription factors, of which there are two, bind upstream of the transcription start site (see fig. 28.18*a*).

In contrast to the initiation complex for RNA polymerases I and II, the region necessary for selective transcription of 5S RNA by RNA polymerase III is located in a region 40 to 80 bases downstream of the transcription start site (see fig. 28.18*c*). One of the additional proteins needed for selective transcription of the 5S gene (TFIIIA) binds to this site and in conjunction with two other protein factors (TFIIIB and TFIIIC), directs RNA polymerase III to bind and initiate transcription. The binding site for factor TFIIIA was determined by the DNA "footprinting" technique (box 28B), using a similar approach to that used to determine the location of binding of the *E. coli* polymerase in the T7 promoter (see fig. 28.15).

The Pol III polymerase is also used to transcribe tRNA genes. In this case TFIIIA is not used, but the other two transcription factors are still used (see fig. 28.18*c*).

Chloroplasts Have Their Own RNA Polymerases

In addition to nuclear RNA polymerases, certain organelles encode their own RNA polymerases. The RNA polymerase found in chloroplasts is very similar in structure to the bacterial RNA polymerase, a fact supporting the notion that these organelles are bacterial in origin. By contrast, RNA polymerases found in mitochondria often contain only one protein subunit.

Other RNA Synthesis

Although most RNA in cells is synthesized by cellular DNA-dependent RNA polymerases, there are several other enzymes that are capable of forming phosphodiester bonds and of synthesizing additional RNA in normal cells or virus-infected cells. Some properties of these enzymes are summarized in table 28.5.

DNA Primase Makes Primer RNA for DNA Synthesis

RNA primers used for the initiation of DNA synthesis during replication are made by DNA primase, the protein product of the *dnaG* gene of *E. coli* (also see chapter 25). It usually binds to DNA in association with another protein, the product of the *dnaB* gene, although it can initiate primer synthesis by itself at certain hairpin structures in single-stranded DNA. The primase can use either NTPs or dNTPs as substrates *in vitro* and synthesizes primers 10 to 50 nucleotides long that are complementary to the DNA template. Synthesis is in the $5' \rightarrow 3'$ direction. As we mentioned earlier, RNA polymerase is also capable of synthesizing primers. The use of primase or RNA polymerase to make primer is controlled by the DNA. One or the other is used exclusively in specific instances.

Many Viruses Make Their Own RNA Polymerases

There are three strategies used by different DNA viruses to accomplish transcription of viral DNA. The first type utilizes the host RNA polymerase, in some cases modifying it or synthesizing new promoter-specificity factors to direct it to read the viral promoters. Examples of such viruses are bacteriophage ϕX174, λ, and T4 (of *E. coli*), and SP01 and SP82 (of *B. subtilis*), as well as animal viruses SV40 and adenovirus.

The second type of virus utilizes the host RNA polymerase to transcribe some early expressed viral genes. One of these "early" genes codes for a new RNA polymerase that transcribes exclusively the remaining "late" viral genes. *E. coli* bacteriophages T7, T3, and SP6 are the best-known examples of this type. T7 RNA polymerase is a single polypeptide with a molecular weight of 99,000. It is not inhibited by the antibiotics rifampicin and streptolydigin, which inhibit host RNA polymerase. It recognizes specifically the T7 late promoters, all of which contain a nearly identical sequence of 18 to 22 nucleotides just before the 5′ triphosphate terminal GTP start site. T7 RNA polymerase is also capable of termination at specific points on the template. Thus it seems to be able to carry out the basic polymerization reaction and specific initiation and termination with a much simpler subunit structure than the host holoenzyme. However, with simplicity it loses versatility. Unlike the host polymerase, it is not able to recognize a wide variety of related but nonidentical promoter sequences, nor is it able to be regulated by positive and negative control factors that act at sites of initiation and termination.

A third type of virus, exemplified by bacteriophage N4, carries a viral RNA polymerase within its virion. This enzyme enters the cell along with the viral DNA and transcribes some early viral genes. Some of these genes code for specificity factors that direct the host RNA polymerase to transcribe late genes. Vaccinia virus is another example of a virus that contains a virion-encapsulated RNA polymerase. It shares some of the properties of the N4 enzyme.

RNA-Dependent RNA Polymerases of RNA Viruses The RNA genomes of single-stranded RNA bacterial viruses (such as Qβ, MS2, R17, and f2) are themselves mRNAs. Bacteriophage Qβ codes for a polypeptide ($M_r \approx 55{,}000$) that combines with three host proteins to form an RNA-dependent RNA polymerase (replicase). The three host proteins are ribosomal protein S1 (*rpsA* gene) and two elongation factors for protein synthesis, EF-Tu (*tuf* gene) and EF-Ts (*tsf* gene) (see table 28.5). The Qβ replicase can use the Qβ RNA plus strand as a template to make a complementary RNA transcript (minus strand); following this the minus strand is used as a template to make more viral RNA plus strands (see fig. 28.1). Like the DNA-dependent RNA polymerases, the replicase utilizes ribonucleoside-5′-triphosphates and transcribes in a $5' \rightarrow 3'$ direction. The phage RNA must first act as an mRNA to direct the synthesis of a component of the replicase, since uninfected cells do not have an RNA-dependent RNA polymerase or replicase.

RNA tumor viruses (retroviruses) that infect animal cells exhibit a different replication strategy. They carry in their virion an enzyme that can use the RNA viral genome as a template to synthesize a DNA copy. Since this process is the reverse of transcription, the enzyme is called reverse transcriptase. The result is a DNA-RNA hybrid. A second DNA strand is then synthesized, displacing the RNA strand. Once a double-stranded DNA copy is made, it is integrated into the host genome, and additional virus RNA is synthesized by the host RNA polymerase II in a normal DNA-dependent fashion.

Reverse transcriptase activity has been found in virus-free cultures of yeast and fruit flies, a fact suggesting that this enzyme may be involved in normal cellular metabolism. There is evidence that certain segments of the genomic DNA can jump from one location to another with the combined assistance of RNA polymerase and reverse transcriptase. This capability could provide a major mechanism for evolutionary change. Finally, reverse transcriptase has been found in bacteria and is associated with the synthesis of a molecule containing both RNA and DNA sequences.

3′-End Addition Enzymes Add Nucleotides to tRNA

Three enzymes are known that add ribonucleotides posttranscriptionally to the 3′ hydroxyl end of RNA. None of them are DNA-dependent. One adds the CCA sequence that all tRNAs have at their 3′ ends. The 3′ terminal adenine is the base to which the amino acid is covalently attached by the aminoacyl-tRNA synthase (see chapter 29). The 3′-CCA is relatively unstable and is continually being added when needed by this enzyme, called the CCA enzyme, or tRNA nucleotidyltransferase.

In eukaryotes, 100 to 200 adenosine residues are added to the 3′ end of most mRNAs by a poly(A) polymerase. This addition occurs in the nucleus before the mRNA is processed and transported to the cytoplasm, as we will describe later. The third enzyme capable of adding nucleotides posttranscriptionally to the end of RNA is polynucleotide phosphorylase.

Polynucleotide Phosphorylase Makes Polynucleotides from Nucleotide Diphosphates

Polynucleotide phosphorylase was the first enzyme found that could synthesize long polynucleotide chains *in vitro*. Two forms of the enzyme have been isolated, a form with three identical subunits ($M_r = 86{,}000$) and a form with these subunits and two additional subunits ($M_r = 48{,}000$). Unlike all the other enzymes we have discussed, polynucleotide phosphorylase utilizes ribonucleoside-5′-diphosphates instead of triphosphates as substrates for RNA synthesis. It catalyzes the reaction

$$\text{NDP} + (\text{NMP})_n \underset{}{\overset{\text{Mg}^{2+}}{\rightleftharpoons}} (\text{NMP})_{n+1} + \text{P}_i$$

Polynucleotide phosphorylase does not require a template and randomly incorporates bases into RNA, depending on the relative concentration of the four NDPs in the reaction medium. The enzyme takes advantage of a primer if one is available to provide a 3′ hydroxyl end and synthesizes RNA in the 5′ → 3′ direction. The reaction is readily reversible, and we do not know whether this enzyme plays a role primarily in degradation or in synthesis of RNA in the cell.

RNA Ligase of Bacteriophage T4 Links RNAs Together

Perhaps the most unusual enzyme capable of synthesizing RNA is the bacteriophage T4 RNA ligase. It can link together, in an ATP-dependent reaction, a 3′-OH terminus on an "acceptor" and a 5′-PO_4 terminus on a "donor," as shown below:

$$\text{ATP} + \cdots + \underset{\text{Acceptor}}{\text{XpY-3}'\text{-OH}} + \underset{\text{Donor}}{\text{PO}_4\text{-5}'\text{Zp}\cdots} \rightarrow$$
$$\underset{\text{Ligated product}}{\cdots\text{XpYpZp} + \cdots} + \text{ADP} + \text{PP}_i$$

A template strand is not required to align the ends of the reactants. The donor and acceptor end can be on the same RNA molecule, in which case a circular RNA product is produced. The enzyme has been used to end-label the 3′-OH end of RNA molecules by the addition of the short donor [5′-^{32}P]Cp.

This enzyme can be used to synthesize defined sequences of RNA. The ligase also can accept deoxyribose polymers as donors. We do not know whether this enzyme carries out the ligation reaction *in vivo* after T4 infection of *E. coli*.

Posttranscriptional Alterations of Transcripts

Most major types of RNA synthesized by cellular DNA-dependent RNA polymerases undergo changes before they can carry out their functions (fig. 28.20). The two exceptions are prokaryotic messengers and eukaryotic 5S ribosomal RNA. Two types of changes are sometimes distinguished. Modification involves additions to or alterations of existing bases or sugars. In some cases it involves addition of one or more nucleotides. Processing, strictly speaking, involves phosphodiester bond cleavage and loss of certain nucleotides. These changes are summarized in table 28.6. The types of alterations that pre-rRNA and pre-tRNA transcripts undergo are very similar in prokaryotes and eukaryotes. We will focus on the processing of these two classes of transcripts in *E. coli* because they are better understood. The complex processing of eukaryotic messengers will also be discussed. We will see that splicing of eukaryotic messengers is a unique process, although occasional splicing also occurs for other types of transcripts.

Processing and Modification of tRNA Requires Several Enzymes

Transfer RNAs are processed from larger precursors in both prokaryotic and eukaryotic cells. This processing involves two types of enzymes that can cleave phosphodiester bonds in RNA. Endoribonucleases cleave at internal sites in the RNA, resulting in two smaller RNAs. Exonucleases sequentially remove single nucleotides from one end of the RNA. Some of the processing ribonucleases of *E. coli* are shown in table 28.7.

As an example, the processing of the *E. coli* tyrosine $tRNA_1$ is diagrammed in figure 28.21. The initial transcript has, in addition to the 85 nucleotides of the final product, 41 nucleotide residues at the 5′ end and 225 residues at the 3′ end. The initial transcript probably folds to form the typical cloverleaf tRNA structure. Processing begins when a specific endonuclease, called RNase F, cleaves the precursor at a site three nucleotides beyond what will be the 3′ end of the mature tRNA. Another endonuclease, RNase P, then cleaves the remaining RNA to produce the mature 5′ end. At the 3′ end, exonuclease RNase D sequentially removes the additional nucleotides and stops, leaving the 3′ terminal CCA sequence. (Some tRNAs encode the CCA terminal sequence; some do not.) Individual bases on the tRNA molecule are then modified by a variety of enzymes, including methylases, deaminases, thiolases, pseudouridylating enzymes, and transglycosylases. Some of the modified bases are presented in figure 28.5. In addition, some eukaryotic tRNAs are processed to remove internal RNA sequences, as we will see later.

Processing of Ribosomal RNA Precursor Leads to Three RNAs

Both eukaryotic and prokaryotic cells synthesize large precursors to rRNA that are processed to produce the mature rRNAs. The most detailed studies of the numerous enzymatic steps involved in this process have been carried out in *E. coli*. This processing scheme is summarized in figure 28.22.

The initial transcript is over 5,500 nucleotides long and includes, reading from the 5′ end of the RNA: the 16S rRNA, a spacer region with one or two tRNAs, the 23S rRNA, the 5S rRNA, and in some cases one or two more tRNAs. Extra bases are found preceding and following each of these RNAs. Primary processing events include the endonucleolytic action by RNase III to produce pre-16S and pre-23S rRNAs and then

Figure 28.20

Comparative processing of major transcripts in prokaryotes and eukaryotes.

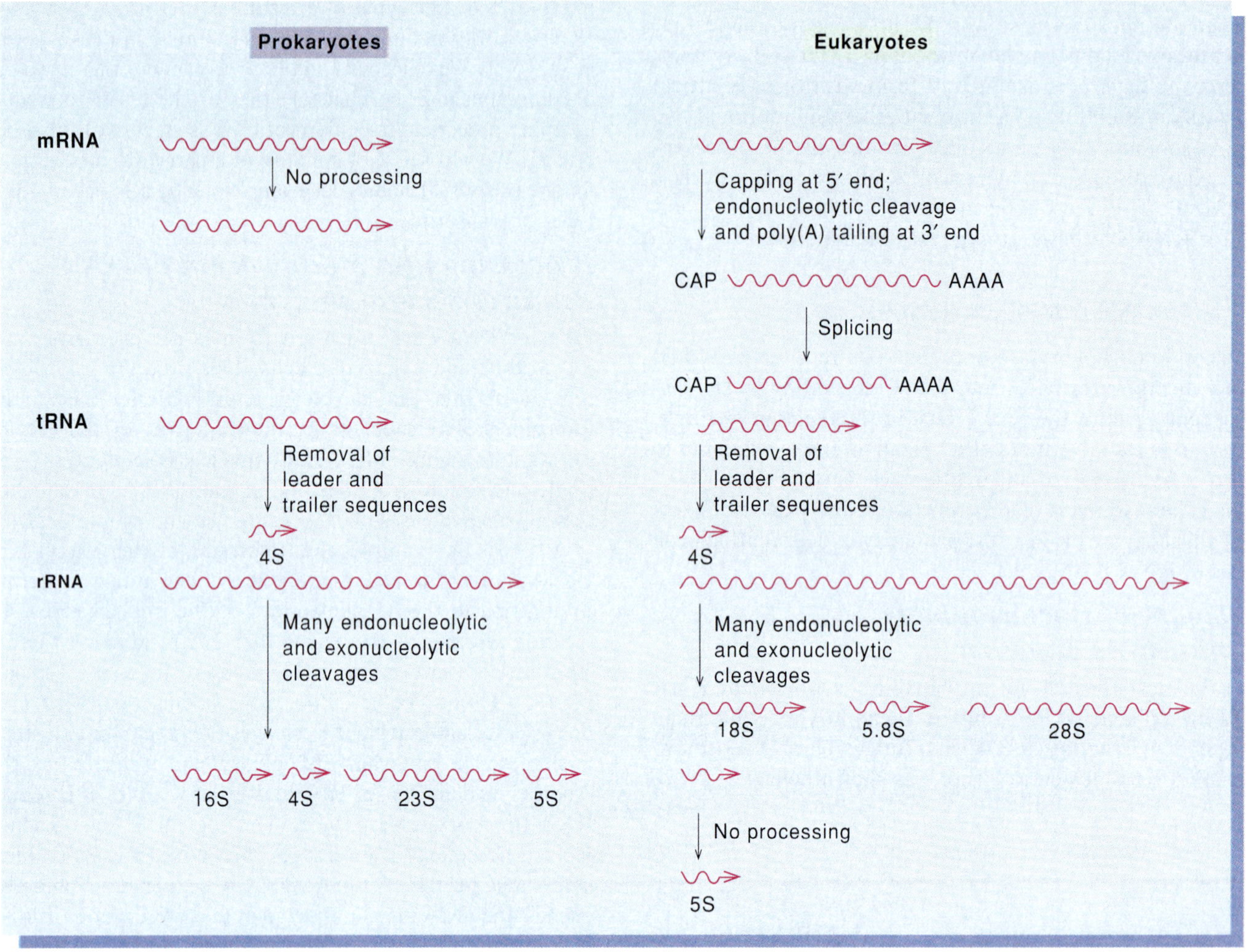

Table 28.6
Summary of RNA Modification and Processing

RNA	Precursor	Modification	Processing	Products
mRNA				
Prokaryotic	None	None? Polyadenylation?	In some cases specific cleavage by endoribonucleases	mRNAs
Eukaryotic	hnRNA	Capping, methylation, polyadenylation	In many cases splicing out introns	mRNAs
rRNA	Pre-rRNA	Methylation	Specific cleavage	16S, 23S, 5S, spacer tRNA (18S, 28S, and 5.8S in eukaryotes)
tRNA	Pre-tRNA	Many modified bases	Specific cleavage by endonucleases, trimming by exonucleases, CCA addition, removal of intervening sequences in eukaryotes	Mature tRNAs

Table 28.7
Representative Enzymes Involved in Processing and Degradation

Enzyme	For *E. coli* Genes: Gene Name	Map Position	Type	Product	Specificity
Processing					
RNase III	*rnc*	55′	endo	3′-OH, 5′-PO_4	Specific, long, double-stranded RNA
RNase D	*rnd*	40′	3′ → 5′ exo	5′-NMPs	Nonspecific, but stops at CCA
RNase E	*rne*	24′	endo		Specific
RNase F	*rnf*	?	endo		Specifically cuts 3′ to tRNA-like structures
RNase P	*rnpA* / *rnpB*	83′ / 70′	endo	3′-OH, 5′-PO_4	Specifically cuts 5′ to tRNA-like structures
RNase M16			endo(s)		Specific, cuts pre-16S to 16S RNA
RNase M23			endo(s)		Specific, cuts pre-23S to 23S RNA
RNase M5			endo		Specific, cuts pre-5S to 5S RNA
Degradation					
RNase I	*rna*	14′	endo	3′-PO_4 oligos	Nonspecific
RNase II	*rnb*	28′	3′ → 5′ exo	5′-NMP	Nonspecific
Polynucleoside phosphorylase	*pnp*	69′	3′ → 5′ exo	5′-NDP	Nonspecific
RNase H	*rnh*	5′	endo		Nonspecific, digests RNA out of RNA-DNA duplex
Bovine pancreatic RNase A			endo	Py-3′-PO_4	Specific, cuts 3′ to pyrimidines
Aspergillus RNase T1			endo	G-3′-PO_4	Specific, cuts 3′ to guanine
Aspergillus S1 nuclease			endo	5′-NMP	Nonspecific, cuts single-stranded RNA or DNA
Bovine spleen phosphodiesterase			5′ → 3′ exo	3′-NMP	Nonspecific
Snake venom phosphodiesterase			3′ → 5′ exo	5′-NMP	Nonspecific

the action of specific ribonucleases to produce the tRNAs and pre-5S rRNA. Secondary processing by endonucleases M16, M23, and M5 results in mature 16S, 23S, and 5S RNAs, respectively. These three endonuclease activities, as well as that of RNase F, have not been completely purified and characterized, and each may represent more than one enzyme. Extra bases on the 3′ end of the tRNAs are removed by exonuclease RNase D. This processing scheme has been deduced by observing the accumulation of intermediates in mutant strains defective in one or more of the nucleases and by cleaving the intermediates *in vitro* with purified or partially purified nucleases.

Processing of Messenger RNA in Eukaryotes Involves Changes at Both Ends and Often Removal of Noncoding Segments in the Middle of Genes

In prokaryotes, many mRNAs function in translation with no prior processing. There is evidence, however, that some are processed by specific endonucleolytic cleavage, often cutting polycistronic messengers into smaller units. In eukaryotes, a much more complex process occurs to produce functional mature

Figure 28.21

Processing and modification of *E. coli* tyrosine tRNA. (T = ribothymidine; ψ = pseudouridine; i⁶A = isopentyladenosine; mG = methylguanosine; s⁴U = thiouridine.) See figure 28.5 for the structures of these modified bases.

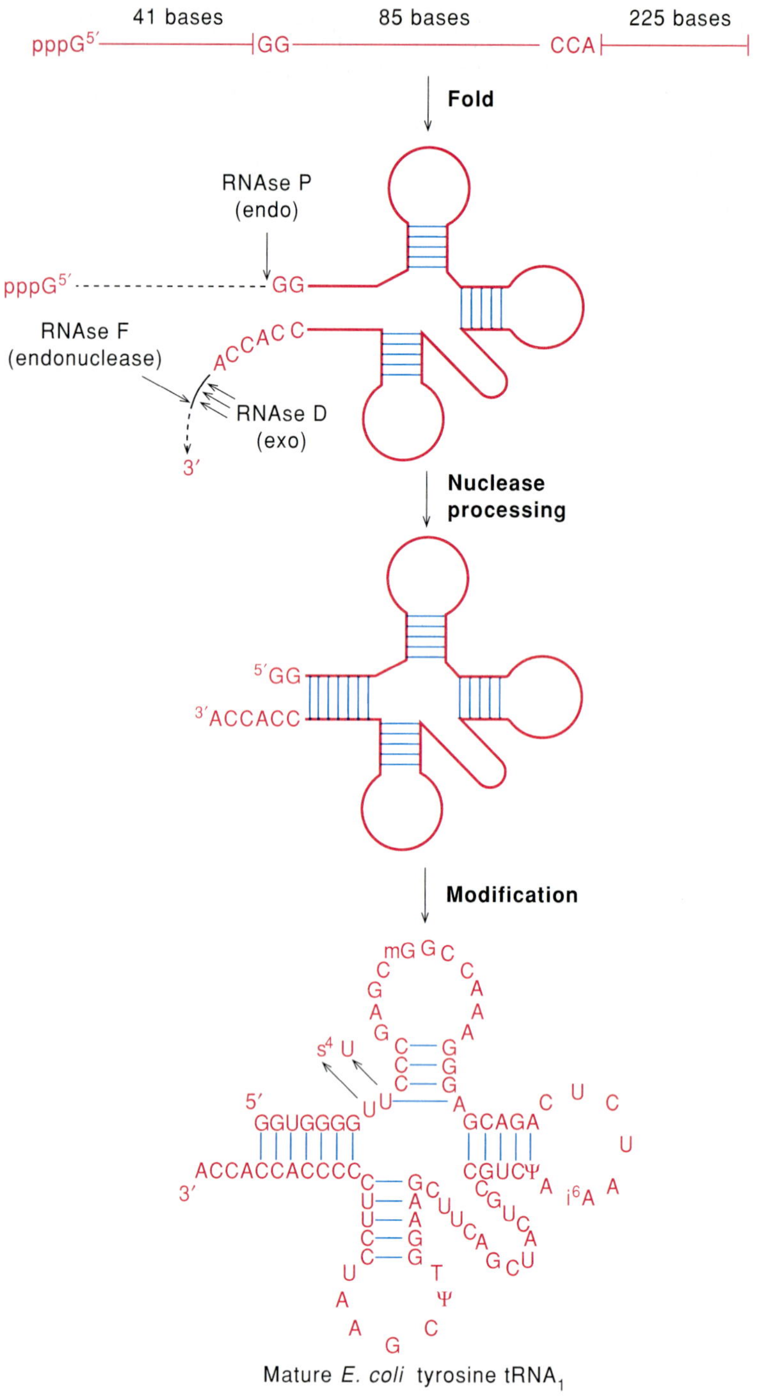

mRNA. Changes occur in sequence. First there is a modification at the 5′ end of the message called capping. Then the message is cleaved by an endonuclease and a poly(A) tail is added to the 3′ end. In many cases, especially in higher eukaryotes, noncoding regions called introns are removed from the middle of genes before they can be used in translation.

Capping Occurs before the Nascent Transcript is Completed
In 1975, Aaron Shatkin and co-workers found that most viral and cellular mRNAs contain an unusual methylated nucleotide at the 5′ terminus. This entire methylated terminal oligonucleotide is called a cap structure (fig. 28.23). The cap structure is formed in the nucleus by a series of enzymatic reactions (fig. 28.24). First, a triphosphatase converts a 5′-triphosphate to a

Figure 28.22

Processing of *E. coli* ribosomal RNA. The ribosomal RNA is transcribed as one long RNA molecule, which contains the sequences for the three ribosomal RNAs and one or two tRNA molecules. There are many processing sites and many different enzymes involved in the processing, as indicated by the vertical arrows and the symbols associated with these arrows. The various nucleases are described in table 28.7. (Source: D. Apirion and P. Gegenheimer, "Processing of bacterial RNA" in *FEBS Letters,* 125:1, 1981. Copyright © 1981 Elsevier Science Publishers B.V., Amsterdam, The Netherlands.)

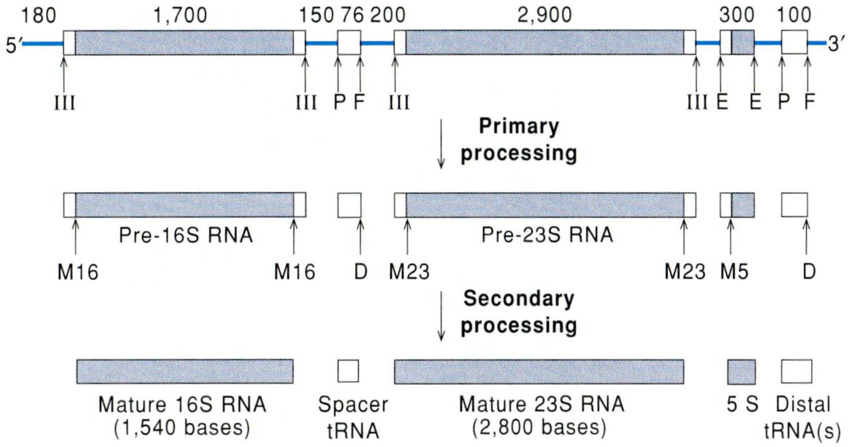

Figure 28.23

Structure of the 5′ methylated cap of eukaryotic mRNA. A 7-methylguanosine (in color) is attached through a triphosphate linkage formed between its 5′-OH and the 5′-OH of the terminal residue in the initial transcript. Note that the 2′-OH groups on the last two bases of the initial transcript have also been modified by methylation (in color). N_1, N_2, and N_3 can be any purine or pyrimidine bases.

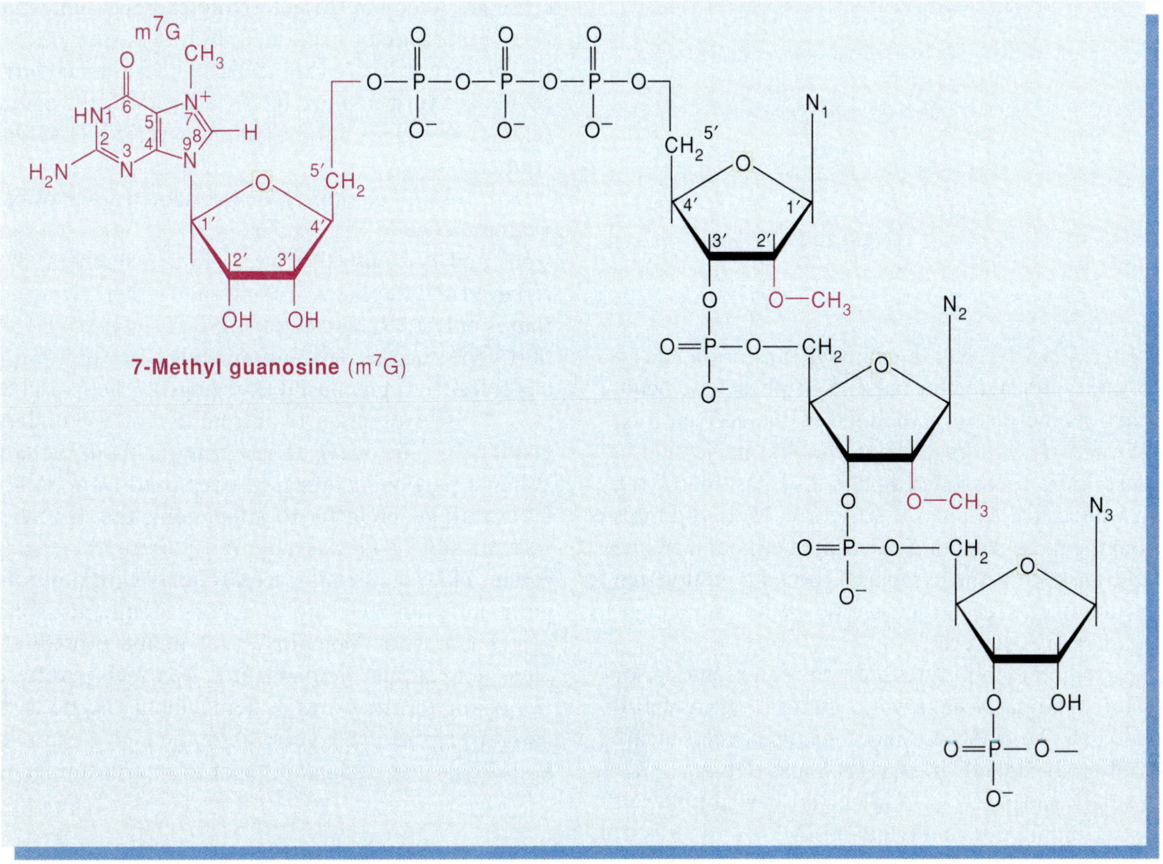

Figure 28.24

Proposed reaction sequence for cap formation in HeLa cells. The enzyme catalyzing each reaction is shown on the left. (Source: S. Venkatesan and B. Moss, "Donor and acceptor specificities of HeLa cell mRNA guanylyltransferase" in *Journal of Biological Chemistry* 255:2835, 1980. Copyright © 1980 American Society for Biochemistry and Molecular Biology Inc., Bethesda, Md.)

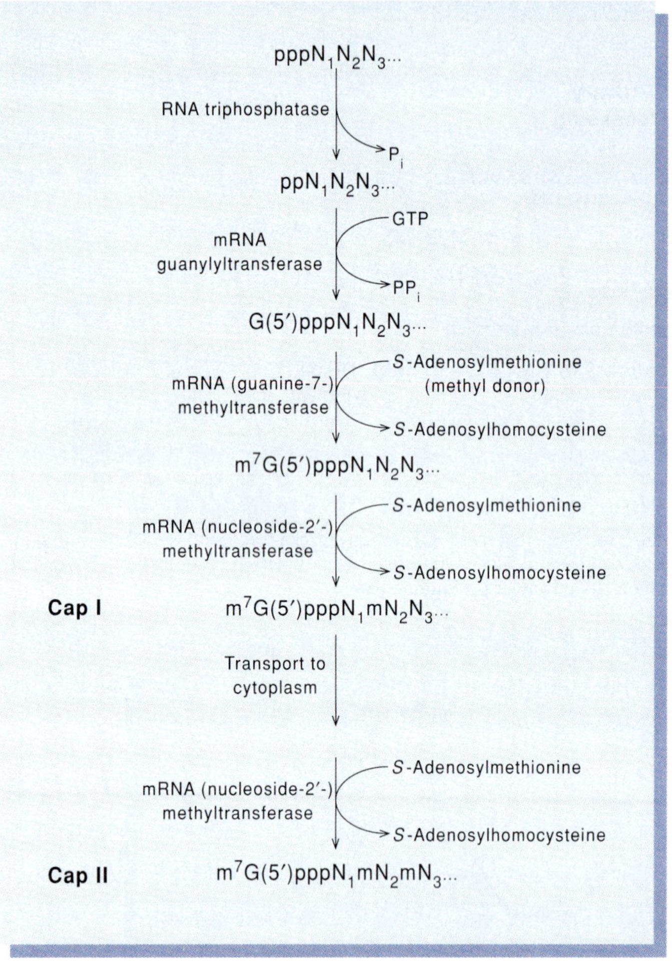

5′-diphosphate terminus. Second, a guanylyltransferase adds a GMP to the 5′ end to form an unusual 5′-5′ triphosphate bond, and this terminal guanosine is methylated in the N-7 position by a guanine-7-methyltransferase. Then the first nucleotide in the initial transcript is methylated in the 2′-0 position of the ribose to form what is called a cap I structure. Methyl groups are derived from the methyl donor *S*-adenosylmethionine. Some mRNAs, after transport to the cytoplasm, become methylated in the 2′-0 position of the second nucleotide of the initial transcript to form a cap II structure.

The cap structure facilitates binding of ribosomes prior to initiation of translation of eukaryotic mRNAs; it may also function to stabilize the mRNA, since uncapped messengers have considerably reduced half-lives. rRNA and tRNA are not capped. Most small nuclear RNAs (snRNAs) have a different cap structure, containing a trimethylguanosine.

Polyadenylation of the 3′ End Most eukaryotic mRNAs found associated with ribosomes contain 50 to 150 adenine nucleotides on their 3′ ends. This poly(A) is not coded by the DNA template but is added to the mRNA before it leaves the nucleus. In several cases studied, the process occurs in at least two steps. First, the RNA is cleaved about twelve nucleotides past an AAUAAA sequence, and then 200 to 250 adenylate residues are added, one residue at a time, by a poly(A) polymerase. After the mature mRNA is transported to the cytoplasm, the poly(A) is shortened somewhat as the mRNA ages. Poly(A) addition does not appear to be essential for transport or translation of all mRNAs, since some eukaryotic mRNAs, including histone mRNAs, do not contain poly(A).

Removal of Intervening Sequences It had long been known that some of the mRNA precursors in the nucleus (hnRNAs) are much larger than the mRNAs found in the cytoplasm associated with ribosomes. Therefore, it came as no surprise to learn that processing occurs to remove parts of the precursor RNAs. What was surprising was the finding made simultaneously by Tom Broker's laboratory and Phil Sharp's laboratory in 1977 that the parts that are removed are not at the ends of the molecules but are interspersed with the coding regions. The phenomenon was first observed for adenovirus transcripts.

One of the early demonstrations of sequence removal resulted from finding that mouse β-globin precursor mRNA did not form a perfect hybrid with DNA complementary to mature mRNA. When the hybrid was observed with an electron microscope, a loop in the heteroduplex appeared, which suggested that the precursor contained internal sequences not present in the mature mRNA (fig. 28.25). These noncoding intervening sequences (introns) are interspersed with the coding sequences (exons) in a large number of eukaryotic RNAs that have been studied.

As an example, the primary transcript of the chicken ovalbumin gene (fig. 28.26) is 7,700 bases long and has 7 introns and 8 exons. The intervening sequences, or introns, are removed and the exons are spliced together, giving a final product that is only 1,872 nucleotides long. It includes 1,158 nucleotides that code for the 386 amino acids of ovalbumin and 714 nucleotides from untranslated regions at the 5′ and 3′ ends.

The function of introns is not yet understood. Some genes (e.g., histones) do not contain them, yet function well. When a particular intron was removed from SV40 virus DNA before using the latter to infect cells, the mRNA that formed was unstable and was not transported from nucleus to cytoplasm. However, similar experiments with other mRNAs have not resulted in abnormal function. Frequently, splice points are correlated with "domains" that define protein structural regions, and similar domains are often seen in different proteins. For example, the exons of hemoglobin encode three structural domains of different types, while the heavy-chain immunoglobulin exons encode domains that are quite similar in structure.

Figure 28.25

Electron micrograph showing mouse ß-globin precursor mRNA (nascent transcript) hybridized with DNA (cDNA) complementary to mature mRNA (upper photo). A control experiment (lower photo) shows that mature mRNA forms a perfect hybrid with DNA complementary to mature mRNA (cDNA) as expected. The intron region in the nascent transcript is indicated by a loop in the upper figure. Schematics indicating the RNA and cDNA are shown on the right. A second small intron is present in the precursor mRNA near the 5′ end and is the reason that the 5′ end of the RNA is not hybridized to the cDNA in the upper photo. (From A. Kinniburgh, J. Mertz, and J. Ross, "The precursor of mouse ß-globin contains two intervening sequences," *Cell* 14:681, 1978. © Cell Press.)

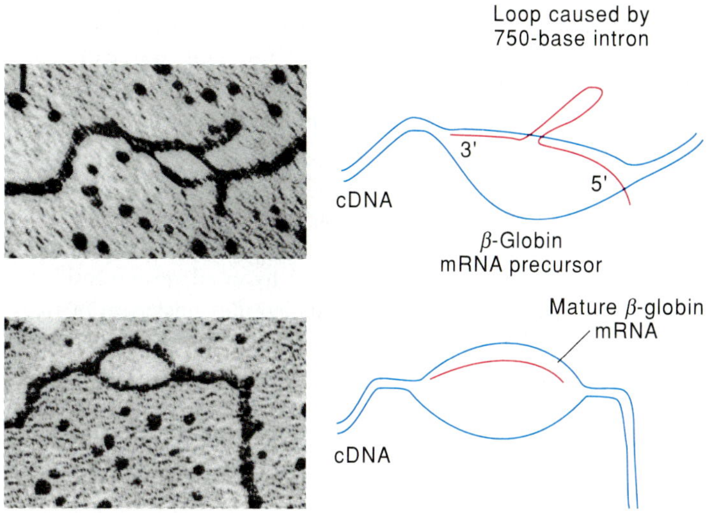

Figure 28.26

Maturation of ovalbumin mRNA. First, the entire ovalbumin gene is transcribed into a precursor RNA, the primary transcript. The transcript is capped at the 5′ end and the poly(A) tail is added at the 3′ end. Then the transcripts of the introns are excised and the adjacent exon transcripts are ligated in a series of splicing steps; an intermediate, from which five of the seven intron transcripts have been eliminated, is illustrated. These steps are accomplished in the cell nucleus. After splicing, mature messenger is transferred to cytoplasm. *L* indicates the 5′ leader region, which is part of the mature RNA that is not translated. (From "Split genes" by Pierre Chambon. Copyright © 1981 Scientific American Inc. All rights reserved.)

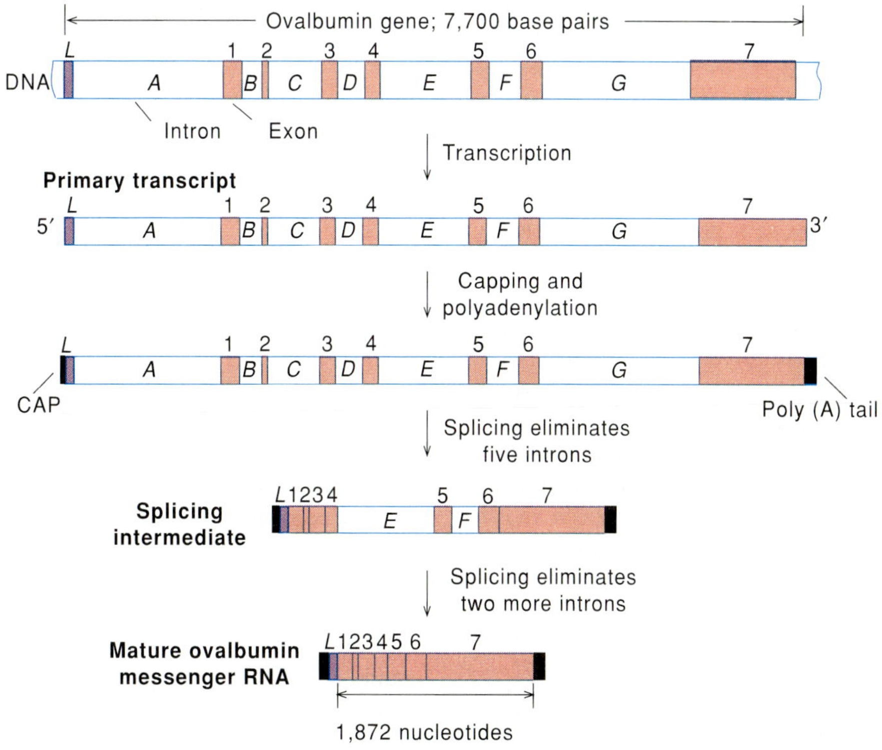

By attaching the promoter for the bacteriophage SP6 RNA polymerase to the β-globin gene, it is possible to transcribe the gene *in vitro* with SP6 RNA polymerase to produce abundant amounts of β-globin pre-mRNA. This source of precursor mRNA has been used to determine the steps and factors involved in splicing. The process (fig. 28.27) involves cleavage of the pre-mRNA at the 5′ splice site to generate the proximal exon and an RNA species containing the intron in a "lariat" configuration connected to the distal exon (fig. 28.27, step 1). The lariat is formed via a 2′-5′ phosphodiester bond, which joins the 5′ terminal guanosine of the intron to an adenosine residue within the intron at a spot 18–40 nucleotides upstream of the 3′ splice site. These two RNA species are probably held together in a noncovalent complex until the next step (fig. 28.27, step 2) in the reaction, which is cleavage at the 3′ splice site to generate the free intron RNA and ligation of the two exons via a 3′-5′ phosphodiester bond. U1 and U2 snRNAs as well as several additional protein factors appear to be necessary for the reactions to occur *in vitro*. These various factors are sometimes seen as a large 40S–60S complex, termed a spliceosome.

Abnormal mRNA splicing appears to be one cause of the human disease β^+-thalassemia. The inefficient splicing of β-globin mRNA precursors in affected individuals seems to be due to mutations in the intron and leads to very low levels of mature mRNA and thus to a β-globin deficiency.

Removal of intervening sequences in eukaryotes is not restricted to mRNA processing. It also occurs in the processing of rRNA and some tRNAs. A temperature-sensitive mutant of yeast has been isolated that accumulates certain tRNA precursors at the nonpermissive temperature. One of these is a tyrosine tRNA precursor that has a 14-base intervening sequence that can be removed by cell extracts *in vitro*. The reaction occurs in two steps and is shown in figure 28.28.

First, a 14-base sequence is removed by an ATP-independent endonucleolytic cleavage to produce two half tRNA molecules. These cleavages produce 3′-PO_4 and 5′-OH termini, unlike the 3′-OH and 5′-PO_4 termini produced by most RNA processing enzymes, such as RNase III or RNase P. The two termini are then ligated in a second reaction that requires ATP.

Some yeast tRNAs have intervening sequences that are up to 60 bases long, but they seem to be processed by the same endonuclease, which appears to recognize a structural feature in the precursor rather than a particular sequence. The function of the yeast intervening sequences is not known, nor is it known whether their splicing is at all similar to the splicing that occurs in mRNA.

RNA Editing Involves the Processing of the Coding Sequences of a Nascent Transcript

In addition to the processing that occurs at the 5′ and 3′ ends of pre-mRNAs and the removal of noncoding regions (introns) from coding regions, there are an increasing number of examples where the coding sequences within the nascent transcripts are changed. In cases where this occurs, the changes, known as RNA editing, are crucial since they often convert a nontranslatable transcript into one that can function as a message.

Figure 28.27

Splicing scheme for pre-mRNA. In step 1 the 2′-OH on an adenosine attacks a phosphate that is 5′-linked to a guanine residue. This leads to a lariat configuration connected to the distal exon. The lariat is formed via a 2′-5′ phosphodiester bond, which joins the 5′ terminal guanosine of the intron to an adenosine residue within the intron, 18–40 nucleotides upstream of the 3′ splice site. In the next step there is a cleavage at the 3′ splice site to generate the free intron RNA and a ligation of the two exons via a 3′-5′ phosphodiester bond. Bases usually found in the region of the splice sites are indicated. The spliceosome is presumed to hold the reacting components in the proper juxtaposition to facilitate the two-step reaction.

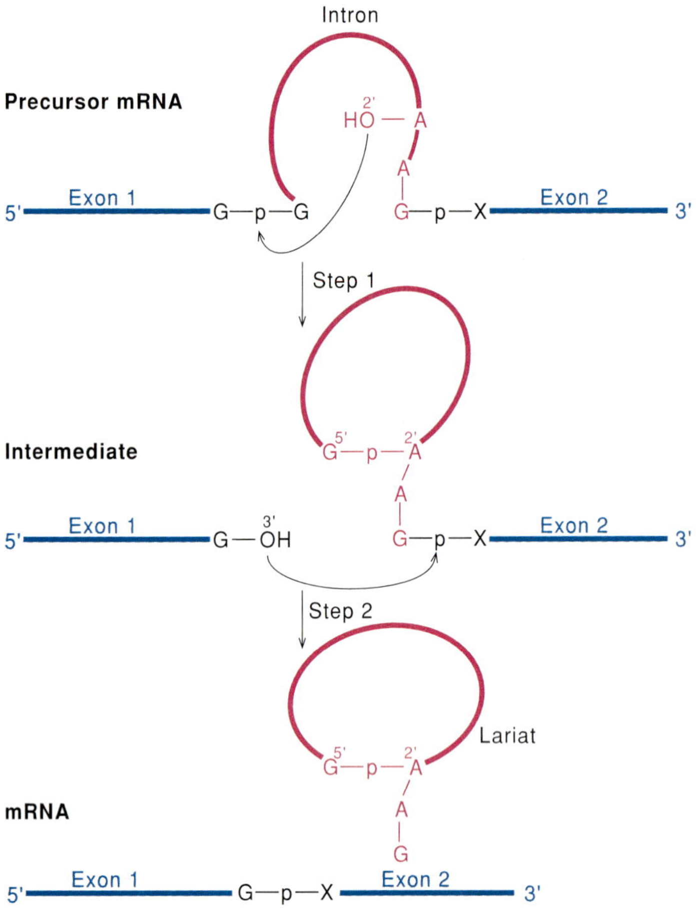

A case in point is the mitochondrial message for cytochrome B oxidase in the flagellated protozoan *Leishmania tarentolae*. In this case several uracil residues are added at different points in the nascent transcript (fig. 28.29). These editing changes involve complex formation between the nascent transcript and a so-called guide RNA. The changes in sequence are dictated by the complementary sequence in the guide RNA in the region where the editing occurs. A string of U's at the 3′ end of the guide RNA is believed to serve as the source for additional U's that are added to the transcript. In addition to the guide RNA, a special enzyme system is required, which is capable of inserting and removing bases at various points in the nascent transcript. Why make a message that requires such complex changes before it can function in translation? One possibility is that this is part of a regulatory mechanism that controls the amount of cytochrome B oxidase.

Figure 28.28

Processing of yeast tyrosine tRNA to remove intervening sequence. First, a 14-base sequence is removed by an ATP-independent endonucleolytic cleavage. The two newly produced termini are then ligated in a second reaction, which requires ATP.

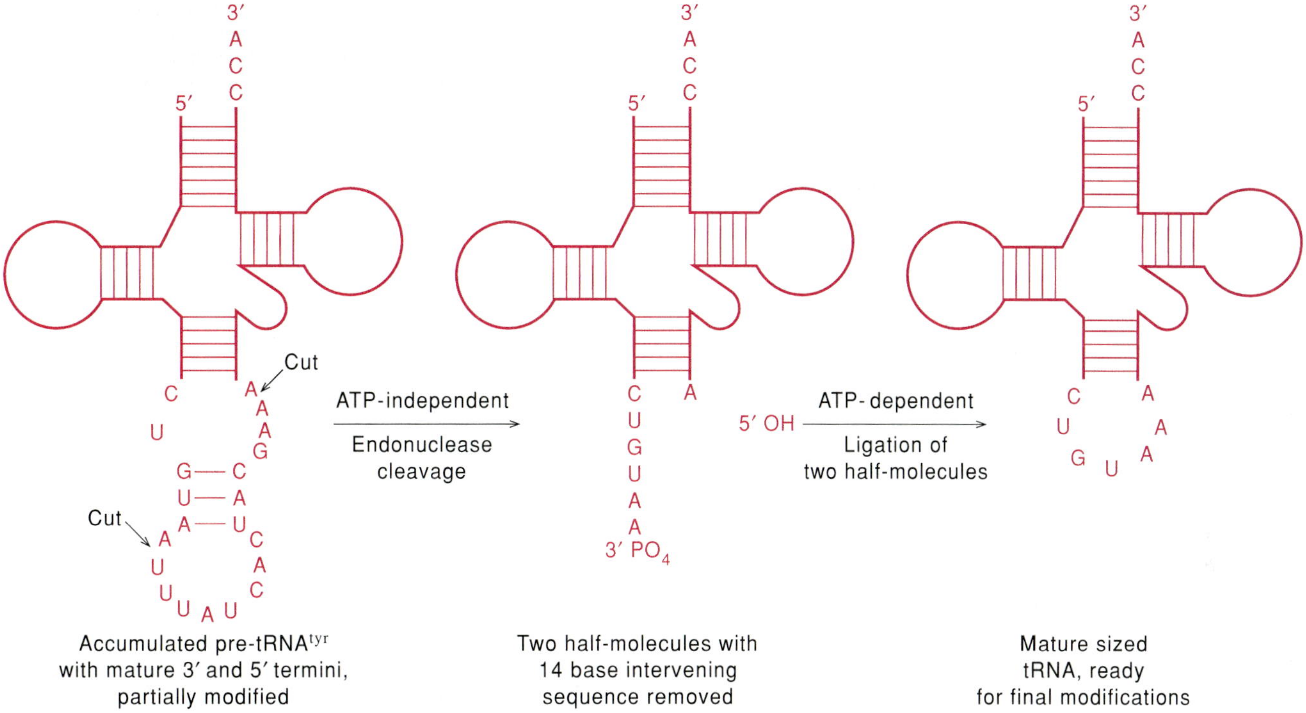

Figure 28.29

Editing of the cytochrome B message in the flagellated protozoan *Leishmania tarentolae*. Editing, which involves the insertion of eleven U residues, creates a translatable RNA where no translation was possible before. Structure shown for mRNA is after editing.

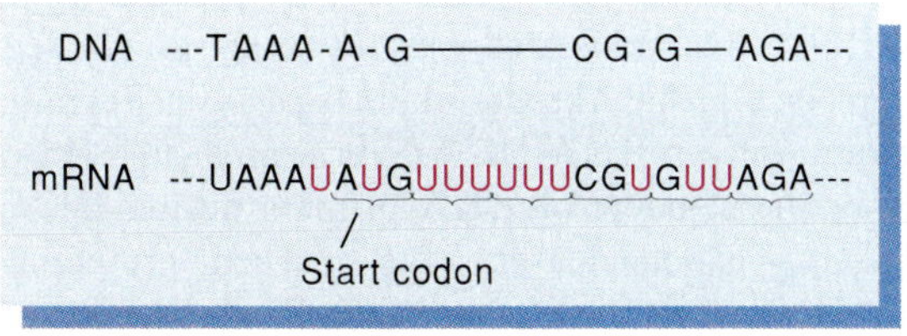

Some RNAs Are Self-Splicing

In 1982, the splicing of rRNA from the protozoan *Tetrahymena* was shown by Tom Cech to involve a startling mechanism. When the pre-rRNA was incubated under the appropriate conditions of salt, Mg^{2+}, and guanosine nucleotide, splicing of the RNA occurred without the help of a protein enzyme! This ability of RNA to act as a catalyst for its own splicing has given rise to the term "ribozyme" and has broadened our definition of enzymes.

The mechanism of the self-splicing reaction is shown in figure 28.30. The first step is a transesterification reaction in which the 3′ hydroxyl group of the guanosine cofactor attacks the phosphodiester bond at the 5′ splice site. The second step is another transesterification reaction in which the 3′ hydroxyl group of the upstream exon attacks the phosphodiester bond at the 3′ splice site and displaces the 3′ hydroxyl group of the intron. The final reaction products are the joined exons and the excised intron.

Since its initial discovery, self-splicing has been found to occur for RNAs from a wide variety of organisms. However, the fraction of unprocessed RNAs that can be shown to undergo self-splicing *in vitro* is quite small in all cases. Certain precursor RNAs that exhibit self-splicing produce intron lariats, just like those seen in the commonly observed splicing reactions that take place in most precursor RNAs with introns. In this case the guanosine nucleotide is not required. This fact suggests that originally RNAs may have been self-splicing, and that the spliceosome components have evolved to improve catalytic efficiency and perhaps to regulate the process.

Self-splicing of RNA shows that in certain cases RNA has enzymelike activity. However, it does not by itself demonstrate true enzyme behavior. Recall that an enzyme is a substance that acts as a catalyst—it accelerates a reaction without itself being consumed in the reaction. More recently Zaug and Cech have shown that a fragment of the self-splicing ribosomal RNA intervening sequence of *Tetrahymena thermophila* can act as an enzyme *in vitro*. This RNA fragment has been shown to catalyze the breakage and rejoining of oligonucleotide substrates in a sequence-dependent manner, with a $K_m = 42\ \mu M$

Figure 28.30

Self-splicing of pre-rRNA from the protozoan *Tetrahymena*. The first step is a transesterification reaction in which the 3′ hydroxyl group of a guanosine attacks the phosphodiester bond at the 5′ splice site. The second step involves another transesterification reaction in which the 3′ hydroxyl group of the upstream exon attacks the phosphodiester bond at the 3′ splice site and displaces the 3′ hydroxyl group of the intron.

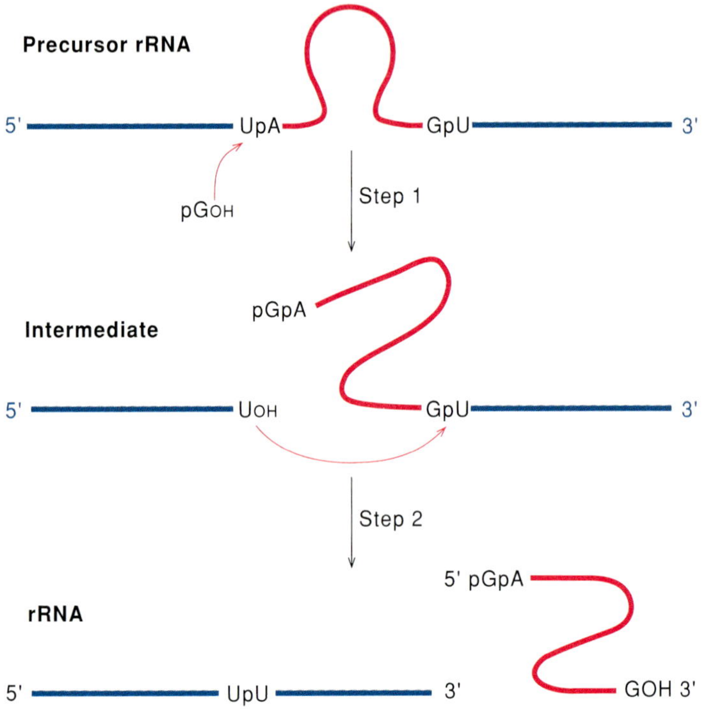

Figure 28.31

Model for the enzymatic mechanism of the 19-base oligonucleotide intervening sequence isolated by Cech from a *Tetrahymena* precursor rRNA. (*a*) The oligonucleotide enzyme (ribozyme) is shown with the oligopyrimidine binding site (RRRRRR), a sequence of six purines near its 5′ end, and a guanine with a 3′ hydroxyl group at its 3′ end. (*b*) The enzyme binds its substrate, a pentacytidylic acid. (*c*) Nucleophilic attack by the terminal 3′-OH of the guanine residue leads to formation of the covalent intermediate and displacement of the tetracytidylic acid. (*d*) A second pentacytidylic acid binds at the enzyme site and, by transesterification, a hexanucleotide is formed.

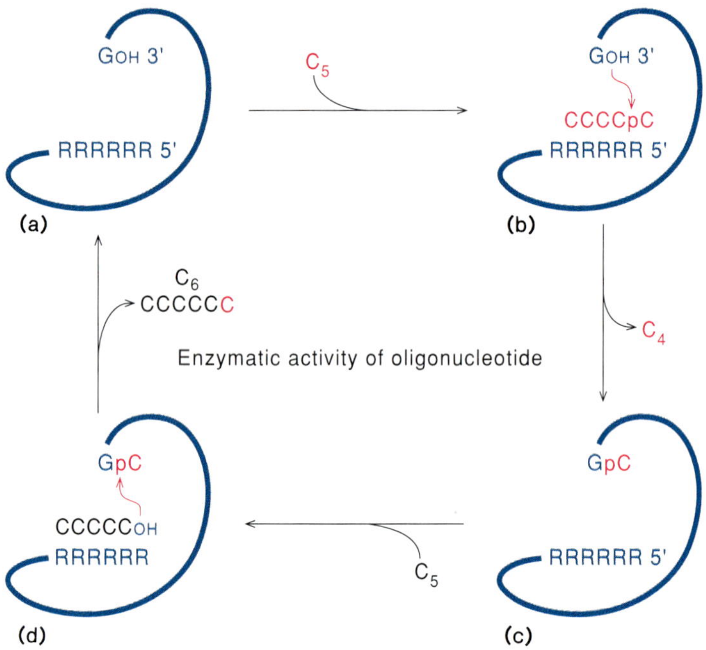

and a $k_{cat} = 2\ \text{min}^{-1}$. With pentacytidylic acid as the substrate, successive cleavage and rejoining reactions lead to the synthesis of higher polymers of polycytidylic acid (fig. 28.31).

The discovery of Cech and his colleagues that RNA can function as an enzyme raises many possibilities both for biochemical reactions as they are currently understood and for reactions that may have played a major role in prebiotic evolution (see later discussion).

Degradation of RNA by Ribonucleases

Although a large number of nonspecific exonucleases and endonucleases have been identified in many organisms, the role of most of them is not understood. Some are extracellular or secreted enzymes that presumably function in breaking down RNA to recycle the purine and pyrimidine bases. Intracellular ribonucleases also may be involved in recycling, as well as in some aspects of processing, as described earlier. Some of the better-known enzymes are listed in table 28.7.

E. coli RNase I is an endonuclease that is located in the periplasmic space between the cell membrane and the cell wall. RNase II is an intracellular 3′ exonuclease that rapidly degrades RNA fragments. Mutants defective in these enzymes were very important in the development of *in vitro* translation systems, since extracts from cells defective in these enzymes allowed mRNA to be translated without being rapidly degraded.

RNase H has the unusual property of degrading the RNA strand from a RNA-DNA hybrid molecule and may be involved in the removal of RNA primers during DNA replication. Some endonucleases have marked preferences for cleavage after certain bases. Pancreatic RNase A cleaves at the 3′ side of pyrimidines, while RNase T1 cleaves only after guanosine residues. Both leave 3′ phosphate termini. Because of their specificity, these enzymes have proven very useful in the analysis of RNA sequences. S1 nuclease digests single-stranded DNA or RNA.

Finally, there are exonucleases that digest in the 3′ → 5′ direction and others that digest in the 5′ → 3′ direction. All known intracellular exonucleases in *E. coli* degrade from 3′ → 5′ and produce 5′-NMPs or 5′ NDPs, which can be readily recycled for use in RNA synthesis.

It is likely that degradation of mRNA proceeds by numerous endonucleolytic cleavages and then 3′ → 5′ exonucleolytic digestion of the resulting fragments. Differences in decay rates of bacterial mRNAs result from differences in mRNA sequence and structure. RNase II and PNPase, the major bacterial exonucleases involved in mRNA turnover, rapidly degrade single-stranded RNA from the 3′ end, but are impeded by 3′ stem-and-loop structures.

Some Ribonucleases Are RNAs

RNase P, which we mentioned earlier, was the first nucleic-acid-containing ribonuclease to be discovered. The enzyme is a ribonucleoprotein composed of both a 20-kDa protein and a 377-nucleotide RNA molecule. The RNA component is catalytically active, whereas the protein component helps to maintain the three-dimensional structure of the RNA.

Certain circular and low-molecular-weight single-stranded plant pathogenic RNAs called viroids can self-cleave at specific sites in the presence of Mg^{2+}. This property is considered important in their replication *in vivo* by a rolling-circle mechanism. All RNAs produced by the rolling-circle mechanism are polymers of indefinite length. The self-cleavage reaction results in linear monomeric molecules that can subsequently be circularized. The self-cleavage site occurs at a specific locus in a secondary structure that has been described as a hammerhead structure (fig. 28.32). Short RNA molecules that anneal to form a hammerhead structure with a consensus sequence undergo self-cleavage as shown in figure 28.32*b* and *c*.

Figure 28.32

Hammerhead structures that undergo self-cleavage. Cleavage occurs at site indicated by arrow. (*a*) The structure formed by a sequence found in a viroid or a virusoid. (*b*) A short circular RNA molecule with the "consensus" sequence that also undergoes cleavage. (*c*) A hybrid molecule, formed by two RNAs, that undergoes a comparable cleavage reaction. In the presence of excess "substrate" the 19-mer functions as an enzyme, since the same 19-mer can cleave many substrates.

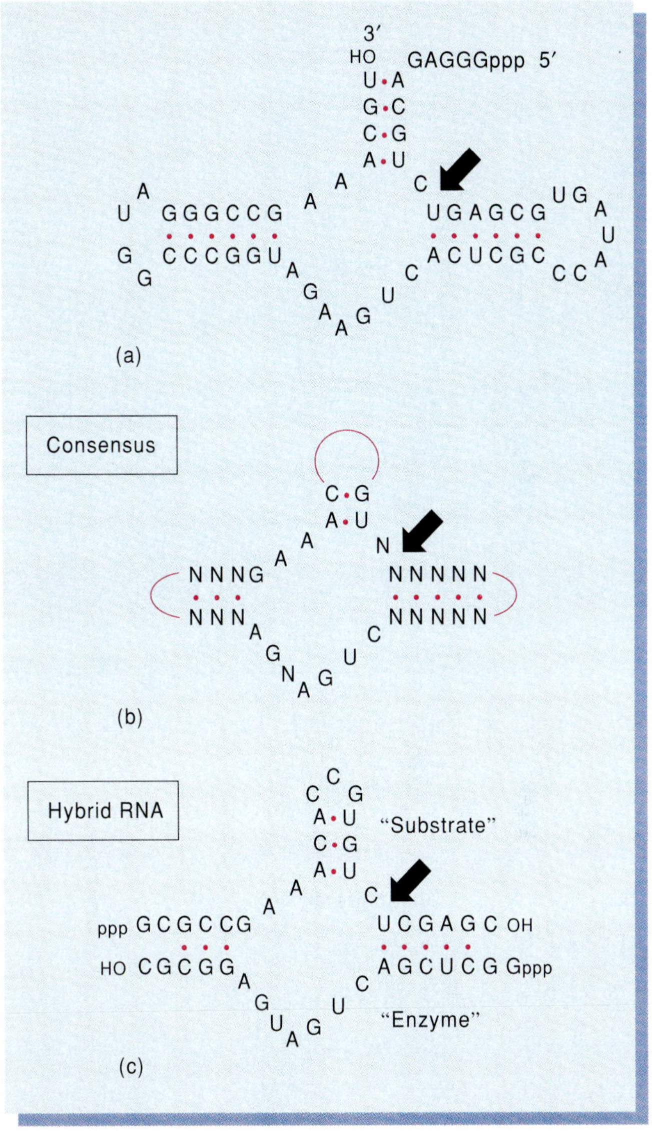

Catalytic RNA May Have Great Evolutionary Significance

For many years prior to the discovery of catalysis by RNA, evolutionists concerned with the origin of life struggled with the question of which came first, nucleic acid or protein. Since protein makes an excellent catalyst but a poor template for transferring information, it seemed likely that nucleic acid and protein must have coevolved. The trouble with this idea is that protein synthesis requires a great number of enzymes. If protein is the only substance that can function as an enzyme, then we clearly have a dilemma. However, if RNA can function as an enzyme, we have a way out. The first systems for protein synthesis could have employed RNA enzymes.

In this same evolutionary context, recall that coenzymes frequently can function as catalysts in the absence of the enzymes with which they are normally associated, although their catalytic activity is usually greatly diminished. In the preprotein period of evolution, it seems likely that coenzymes in conjunction with RNA molecules may have played a far greater role than they do today.

Inhibitors of RNA Metabolism

A large variety of inhibitors of RNA synthesis have been identified. These inhibitors have proved useful in elucidating transcription mechanisms, and some have allowed the isolation of mutant strains with enzymes that are resistant to their inhibition. The inhibitors (fig. 28.33) fall into several classes.

Some Inhibitors Act by Binding to DNA

The best-known example of an inhibitor that binds to DNA is actinomycin D, an antibiotic produced by *Streptomyces antibioticus*. The inhibition of RNA synthesis is caused by the insertion (intercalation) of its phenoxazone ring between two G-C base pairs, with the side chains projecting into the minor

Figure 28.33

Inhibitors of RNA synthesis.

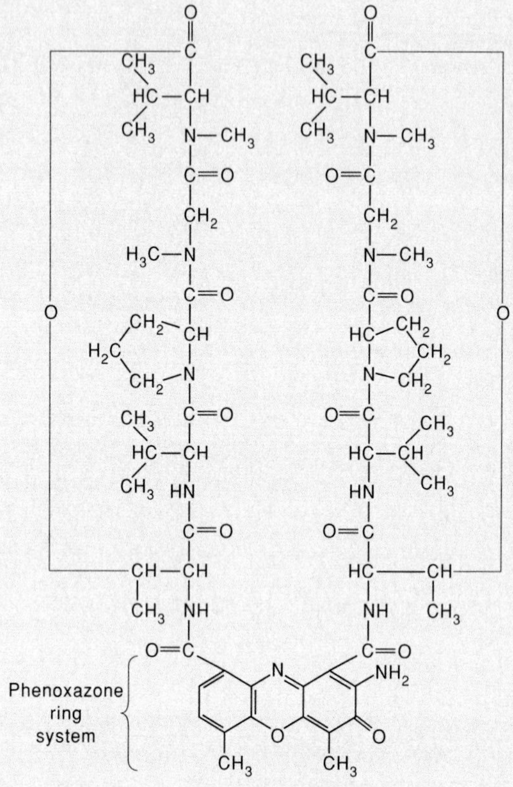

Actinomycin D

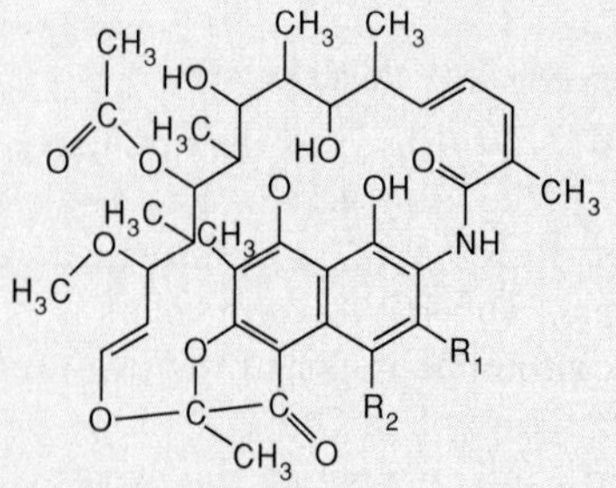

Rifamycin B(R_1 = H; R_2 = O–CH_2–COOH)

Rifampicin (R_1 = CH=N⁺ (piperazine ring) N–CH_3; R_2 = OH)

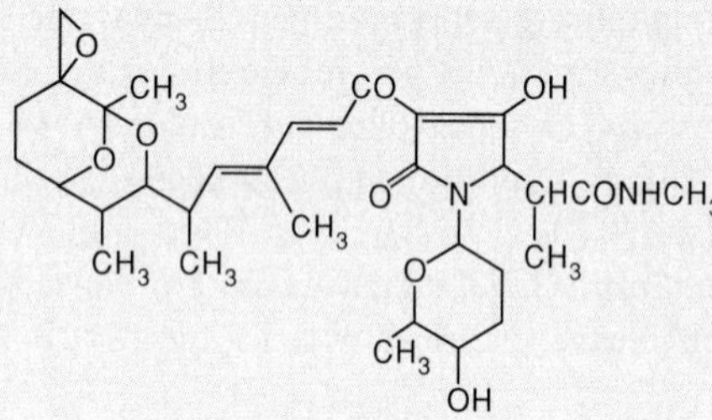

Streptolydigin

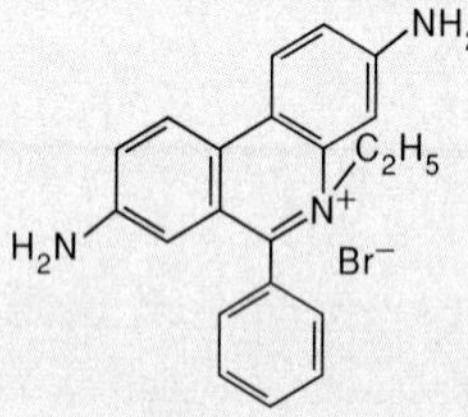

Ethidium bromide

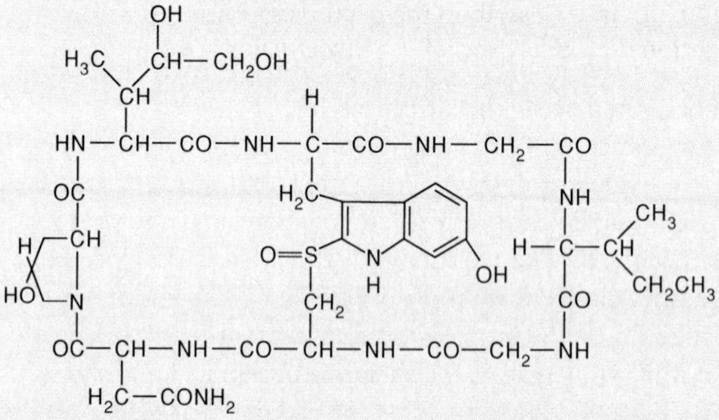

α-Amanitin

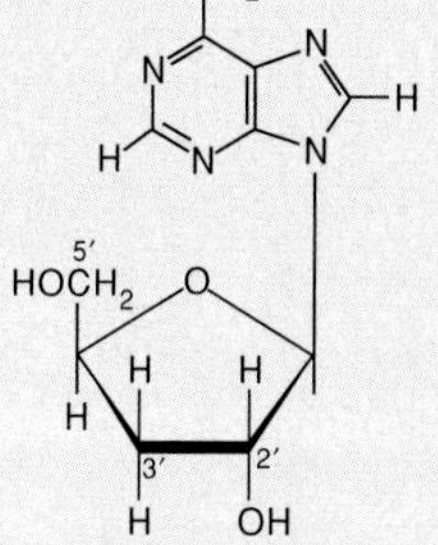

Cordycepin

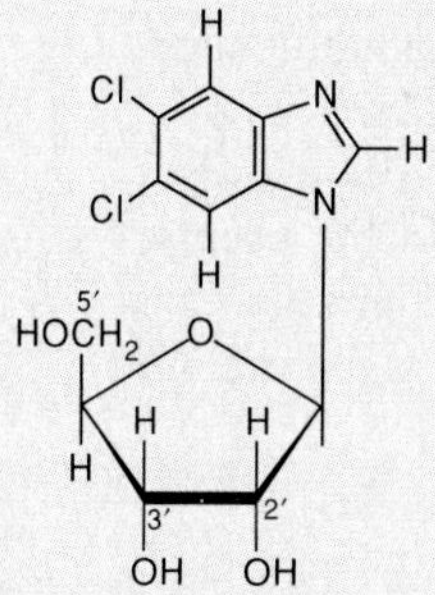

DRB

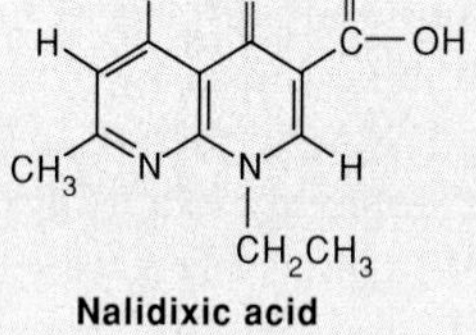

Nalidixic acid

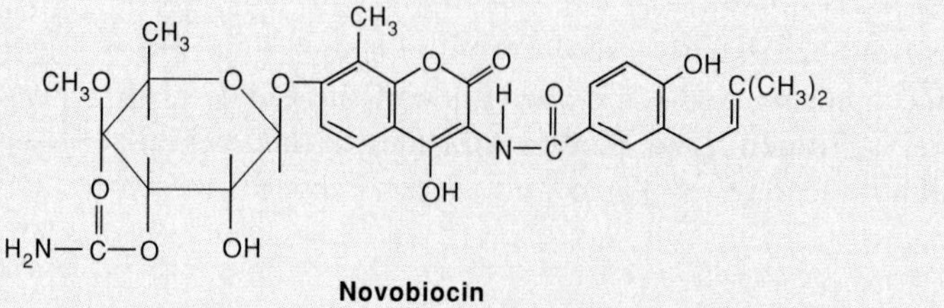

Novobiocin

groove of the double helix, hydrogen-bonded to guanine residues. RNA polymerase binding to DNA that contains actinomycin D is only slightly impaired, but RNA chain elongation in both eukaryotes and prokaryotes is blocked. Ethidium bromide also intercalates into DNA and at low concentrations preferentially binds to negatively supercoiled DNA (see chapter 25). It has been used to selectively inhibit transcription in mitochondria, which contain supercoiled DNA.

Some Inhibitors of Transcription Bind to DNA Gyrase

DNA gyrase is an enzyme in *E. coli* that is needed to maintain the proper degree of supercoiling in DNA. Transcription of some genes seems to be affected by the supercoiled state of the template, and transcription of these genes can be selectively inhibited or stimulated by compounds that interfere with the action of DNA gyrase. Examples of compounds that inhibit DNA gyrase are nalidixic acid, which binds to one of the two gyrase subunits, and coumermycin and novobiocin, which bind to the other (see chapters 25 and 26).

Some Inhibitors Bind to RNA Polymerase

Rifampicin is a synthetic derivative of a naturally occurring antibiotic, rifamycin, that inhibits bacterial DNA-dependent RNA polymerase but not T7 RNA polymerase or eukaryotic RNA polymerases. It binds tightly to the β subunit. While it does not prevent promoter binding or formation of the first phosphodiester bond, it effectively prevents synthesis of longer RNA chains. It does not inhibit elongation when added after initiation has occurred. Another antibiotic, streptolydigin, also binds to the β subunit, but it is able to prevent all bond formation, whether involved in initiation or elongation.

The most useful inhibitor of eukaryotic transcription has been α-amanitin, a major toxic substance in the poisonous mushroom *Amanita phalloides*. The toxin preferentially binds to and inhibits RNA polymerase II. At high concentrations it also can inhibit RNA polymerase III, but not RNA polymerase I or bacterial, mitochondrial, or chloroplast RNA polymerase.

Cordycepin in its 5′-triphosphorylated form is a substrate analog and can be incorporated into growing RNA chains by most RNA polymerases. Cordycepin is a 3′-deoxyadenosine. It causes chain termination, since it does not contain the 3′ hydroxyl group necessary for the formation of the next phosphodiester bond.

Summary

In this chaper we have described the synthesis, transcription, and posttranscriptional modification of the three major classes of RNA. The main points we have covered are as follows.

1. RNA is synthesized in the 5′ → 3′ direction by the formation of 3′, 5′-phosphodiester linkages between the four ribonucleoside triphosphate substrates, analogous to the process of DNA synthesis. The sequence of bases in RNA transcripts catalyzed by DNA-dependent RNA polymerases is specified by the complementary DNA template strand.
2. Some newly synthesized RNA transcripts are the functional species, while others must be modified and/or processed into the mature functional species. Modifying enzymes add nucleotides to the 5′ or 3′ ends or alter bases within the RNA, such as by methylation of specific residues. Specific processing enzymes cleave RNA internally, splice together noncontiguous regions of a transcript, or remove nucleotides from the 5′ or 3′ ends.
3. The major classes of RNA in both prokaryotes and eukaryotes are messenger RNA, ribosomal RNA, and transfer RNA. These distinct classes of RNA play specific functional or structural roles in the translation of genetic information into proteins. Other classes of RNAs transcribed by DNA-dependent RNA polymerases include primers for DNA synthesis, small RNAs that may be directly involved in processing of larger RNA precursors, and still others that may play as yet unidentified regulatory or structural roles in cells.
4. DNA-dependent synthesis of RNA in *E. coli* is catalyzed by one enzyme, consisting of five polypeptide subunits. The complete holoenzyme is composed of four polypeptides (the core enzyme) and an additional polypeptide that confers specificity for initiation at promoter sequences in the DNA template.
5. The steps involved in transcription—binding of polymerase, initiation, elongation, and termination—have been studied in great detail. Elucidation of these steps, as well as the identification and function of the proteins and DNA sequences involved at each step, has been greatly facilitated by the use of specific inhibitors of RNA synthesis, mutants, and a variety of *in vitro* techniques.
6. In eukaryotes, transcription is catalyzed by at least four distinct polymerases, each specializing in the synthesis of a different class of RNA. RNA polymerases I, II, and III are responsible for the synthesis of rRNA, mRNA, and small RNA (such as tRNA) transcripts, respectively. Specific RNA polymerases are responsible for transcription in mitochondria and chloroplasts. Although the basic mechanism of RNA synthesis by these enzymes is similar to that of prokaryotic RNA polymerase, many details of the steps, proteins, and DNA regulatory sequences involved in eukaryotic transcription are less well understood than in prokaryotes.
7. Various other RNA polymerases have been identified and studied. These include DNA primase, specific viral-induced or viral-encoded polymerases, and polymerases that are not dependent on DNA templates for RNA synthesis. Other enzymes involved in RNA metabolism include processing nucleases, degradation nucleases, ligases, and modifying enzymes. Many of these enzymes have been at least partially purified and characterized, and some have proved useful for the *in vitro* manipulation and analysis of RNAs.

8. One of the more interesting processing reactions of nascent transcripts involves the removal of intervening sequences. This type of processing is referred to as splicing. Most splicing reactions appear to require host proteins. However, it is clear that certain splicing reactions require only RNA, a fact demonstrating that the RNA is capable of functioning like an enzyme in the making and breaking of phosphodiester linkages in polyribonucleotides.

Selected Readings

Altman, S., M. Baer, C. Guerrier-Takada, and A. Vioque, Enzymatic cleavage of RNA by RNA. *Trends Biochem. Sci.* 11:515–518, 1986.

Bass, B. L., Splicing: the new edition. *Nature* 352:283–284, 1991.

Bear, D. G., and D. W. Peabody, The *E. coli* rho protein: An ATPase that terminates transcription. *Trends Biochem. Sci.* 13:343–348, 1988.

Bjork, G. R., J. U. Ericson, C. E. D. Gustafsson, T. G. Hdagervall, Y. H. Josson, and P. M. Wikstrom, Transfer RNA modification. *Ann. Rev. Biochem.* 56:263–287, 1987.

Buratowski, S., S. Hahn, L. Guarente, and P. A. Sharp, Five intermediate complexes in transcription initiation by RNA polymerase II. *Cell* 550–561, 1988.

Burtis, K. C., and B. X. Baker, *Drosophila* double sex gene controls somatic sexual differentiation by producing alternatively spliced mRNAs encoding related sex-specific polypeptides. *Cell* 56: 997–1010, 1989.

Cattaneo, R., RNA editing: in chloroplast and brain. *Trends Biochem. Sci.* 17:4–6, 1992. A recent review dealing with RNA editing.

Cech, T., RNA editing: world's smallest introns? *Cell* 64:667–669, 1991.

Cech, T. R., RNA as an enzyme. *Sci. Am.* 225(5):64–75, 1986.

Cech, T. R., and B. L. Bass, Biological catalysis by RNA. *Ann. Rev. Biochem.* 55:599–630, 1986.

Chambon, P., Split genes. *Sci. Am.* 244:60–66, 1981.

Chowrira, B. M., A. Berzal-Herranz, and J. M. Burke, Novel guanosine requirement for catalysis by the hairpin ribozyme. *Nature* 354: 320–323, 1991. In some cases the guanine amino group serves a catalytic role in self-splicing.

Crick, F., Central dogma of molecular biology. *Nature* 227:561–563, 1970.

Dahlberg, A. E., The functional role of ribosomal RNA in protein synthesis. *Cell* 57:525–529, 1989.

Darnell, J. E., Jr., RNA. *Sci. Am.* 253(4):68–78, 1985.

Deutscher, M. P., The metabolic role of RNases. *Trends Biochem. Sci.* 13:136–139, 1988.

Dorit, R. L., L. Schoenbach, and W. Gilbert, How big is the universe of exons? *Science* 250:1377–1382, 1990.

Forster, A. C., A. C. Jeffries, C. C. Sheldon, and R. H. Symons, Structural and ionic requirements for self-cleavage of virusoid RNAs and trans self-cleavage of viroid RNA. *Cold Spring Harbor Symposia on Quantitative Biology* 52:249–259, 1987.

Futcher, B., Supercoiling and transcription, or vice versa? *Trends Genet.* 4:271–272, 1988.

Geiduschek, E. P., and G. P. Tocchini-Valentini, Transcription by RNA polymerase III. *Ann. Rev. Biochem.* 57:873–914, 1988.

Haas, E. S., D. P. Morse, J. W. Brown, F. J. Schmidt, and N. R. Pace, Long-range structure in ribonuclease P RNA. *Science* 254:853–856, 1991.

Hall, B. D., and S. Spiegelman, Sequence complementarity of T2-DNA and T2 specific RNA. *Proc. Natl. Acad. Sci.* 47:137–146, 1964. The first use of RNA-DNA hybridization.

Helmann, J. D., and M. J. Chamberlin, Structure and function of bacterial sigma factors. *Ann. Rev. Biochem.* 57:839–872, 1988.

Hou, Y.-M., and P. Schimmel, A simple structural feature is a major determinant of the identity of a transfer RNA. *Nature* 333:144–145, 1988.

Khoury, G., and P. Gruss, Enhancer elements. *Cell* 33:313–314, 1983.

Leff, S. D., M. G. Rosenfeld, and R. M. Evans, Complex transcriptional units: Diversity in gene expression by alternative RNA processing. *Ann. Rev. Biochem.* 55:1091–1117, 1986.

Lührmann, R., B. Kastner, and M. Bach, Structure of spliceosomal snRNPs and their role in pre-mRNA splicing. *Biochemica et Biophysica Acta* 1087:265–292, 1990.

Nikoliu, D. B., S-H. Hu, J. Lin, A. Gasch, A. Hoffman, M. Horikoshi, N-H. Chua, R. G. Roeder, and S. K. Burly, Crystal structure of TF IID TATA-box binding protein. *Nature* 360:40–45, 1992.

Padgett, R. A., P. J. Grabowski, M. M. Komarska, S. Seller, and P. A. Sharp, Splicing of messenger RNA precursors. *Ann. Rev. Biochem.* 55:1119–1150, 1988.

Patzelt, E., K. L. Perry, and N. Agabian, Mapping of branch sites in trans-spliced pre-mRNAs of *Trypanosoma brucei. Molecular and Cellular Biology* 9:4291–4297, 1989. Cases where the splicing event involves transcripts from different chromosomes.

Pestka, S., J. A. Langer, K. C. Zoon, and C. E. Samuel, Interferons and their actions. *Ann. Rev. Biochem.* 56:757–777, 1987.

Peterson, M. G., J. Inostroza, M. E. Maxon, F. Osvaldo, A. Admon, D. Reinberg, and R. Tjian, Structure and functional properties of human general transcription factor IIt. *Nature* 354:369–373, 1991.

Proudfoot, N. J., How RNA polymerase II terminates transcription in higher eukaryotes. *Trends Biochem. Sci.* 114:105–110, 1989.

Steitz, J. A., "Snurps." *Sci. Am.* 258(6):56–63, 1988.

Stuart, K., RNA editing in mitochondrial mRNA of trypanosomatids. *Trends Biochem. Sci.* 16:68–72, 1991. RNA editing is processing that involves the removal, addition, and modification of nucleotides in the coding regions of nascent transcripts.

Problems

1. One strand of DNA is completely transcribed into RNA by RNA polymerase. The base composition of the DNA template strand is: G = 20%, C = 25%, A = 15%, T = 40%. What would you expect the base composition of the newly synthesized RNA to be?

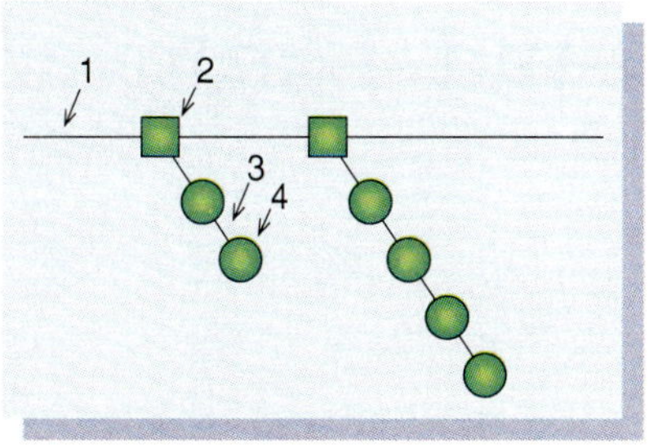

2. The illustration is a schematic drawing representing a portion of Miller's electron microscope picture of transcription and translation in progress in *E. coli.*
   (a) Identify 1, 2, 3, and 4.
   (b) Indicate the 3′ and 5′ ends of the sense (template) strand of DNA.
   (c) Indicate the 3′ and 5′ ends of the mRNA.
   (d) Draw four peptides in the process of being synthesized, indicating relative lengths.
   (e) Indicate the N- and C-terminal ends of the longest peptide.
   (f) Indicate with an arrow the direction in which RNA polymerase is moving.
   (g) What parts of this diagram would be different in eukaryotes?
3. Although 40 to 50% of the RNA being synthesized in *E. coli* at any given time is mRNA, only about 3% of the total RNA in the cell is mRNA. Explain.
4. In *E. coli,* the mRNA fraction is heterogeneous in size, ranging from 500 to 6,000 nucleotides. The largest mRNAs have many more nucleotides than needed to make the largest proteins. Why are some of these mRNAs so large in bacteria?
5. When yeast phenylalanine tRNA is digested with a small amount of RNase (partial digest), such as T2 RNase, almost all of the cleavage sites are found in the anticodon loop. Explain this observation.
6. Speculate on the advantage of having three rRNAs (16S, 23S, and 5S) as part of the same RNA precursor such as found in *E. coli.*
7. In *E. coli* the precise spacing between the −35 and −10 conserved promoter elements has been found to be a critical determinant of promoter strength. What does this suggest about the interaction between RNA polymerase and these elements? What sorts of evidence could you obtain about this interaction by doing "footprint" experiments? (Also explain how you would do these experiments.)
8. What is the maximum theoretical rate of initiation at an *E. coli* promoter, assuming that the diameter of its RNA polymerase is about 200 Å and the rate of RNA chain growth is 45 nucleotides per second?
9. Cordycepin-5′-triphosphate, 3′-deoxy ATP, is a ribonucleoside triphosphate analog that can be bound as a substrate by RNA polymerase. What effect does this analog have on transcription and how could this information be used to determine the direction of transcription?
10. The following hypothetical RNA is a precursor made in a eukaryotic nucleus and its synthesis is inhibited by low levels of α-amanitin:

pppAUUAUGCCGAUAAGGUAAGUA-(N_{100})-
AUCUCCCUGCAGGGCGUAACCAAUAAACGACGACGACGUCACC

Indicate the final processed RNA found in the cytoplasm and point out important features.

11. In early research with intact eukaryotic mRNA, the RNA appeared to have two 3′ ends and no 5′ terminus. Explain these observations.
12. What is unique about the promoter for RNA polymerase III? Diagram how this promoter works.
13. T7 and SP6 RNA polymerases are used to make RNA transcripts *in vitro* as important tools in modern molecular biology. The addition of inorganic pyrophosphatase to these reactions can increase the yield of the RNA product. Explain the increased yield.
14. Draw the chemical reaction mechanism for the formation of a "lariat" in the processing of introns from eukaryotic mRNA.
15. A particular eukaryotic DNA virus was found to code for two mRNA transcripts, one shorter than the other, from the same region on the DNA. Analysis of the translation products revealed that the two polypeptides shared the same amino acid sequence at their amino-terminal ends, but were different at their carboxyl-terminal ends. The longer polypeptide was coded by the shorter mRNA! Suggest an explanation.
16. You are a graduate student working on an *in vitro* processing system for intron removal with a precursor rRNA isolated from *Tetrahymena.* The purified RNA will process by itself without any proteins being added to the reaction. How could you prove that this RNA was autocatalytic and that the reaction was not just due to some protein contamination in your incubation?
17. Draw a chemical reaction mechanism for the first step in the self-splicing reaction of group I introns. (Group I intron splicing requires a guanosine co-factor.)

29

CHAPTER

Protein Synthesis, Targeting, and Turnover

Proteins are informational macromolecules, the ultimate heirs of the genetic information encoded in the sequence of nucleotide bases within the chromosomes. Each protein is composed of one or more polypeptide chains and each peptide chain is a linear polymer of amino acids. The order of the amino acids (see table 3.1) commonly found in the polypeptide chain is determined by the order of nucleotides in the corresponding messenger RNA template. In this chapter we will examine four aspects of protein metabolism (fig. 29.1): (1) the process whereby amino acids are ordered and polymerized into polypeptide chains; (2) posttranslational alterations in polypeptides, which occur after they are assembled on the ribosome; (3) the targeting process whereby proteins move from their site of synthesis to their sites of function; and (4) the proteolytic reactions that result in the return of proteins to their starting material, amino acids.

Figure 29.1

Overview of reactions in protein synthesis. (aa_1, aa_2, aa_3 = amino acids 1, 2, 3.) Protein synthesis requires transfer RNAs for each amino acid, ribosomes, messenger RNA, and a number of dissociable protein factors in addition to ATP, GTP, and divalent cations. First the transfer RNAs become charged with amino acids, then the initiation complex is formed. Peptide synthesis does not start until the second aminoacyl tRNA becomes bound to the ribosome. Elongation reactions involve peptide bond formation, dissociation of the discharged tRNA, and translocation. The elongation process is repeated many times until the termination codon is reached, an event that signals the end of polypeptide synthesis. Termination is marked by the dissociation of the messenger RNA from the ribosome and the dissociation of the two ribosomal subunits. The polypeptide chain sometimes folds into its final form without further modifications. Frequently the folded polypeptide chain is modified by removal of part of the polypeptide chain or by addition of various groups to specific amino acid side chains. The completed polypeptide chain will migrate to different locations according to its structure. It may remain in the cytosol or it may be transported into one of the cellular organelles or across the plasma membrane into extracellular space. Proteins have different lifetimes. Eventually a protein will be degraded, usually down to its component amino acids.

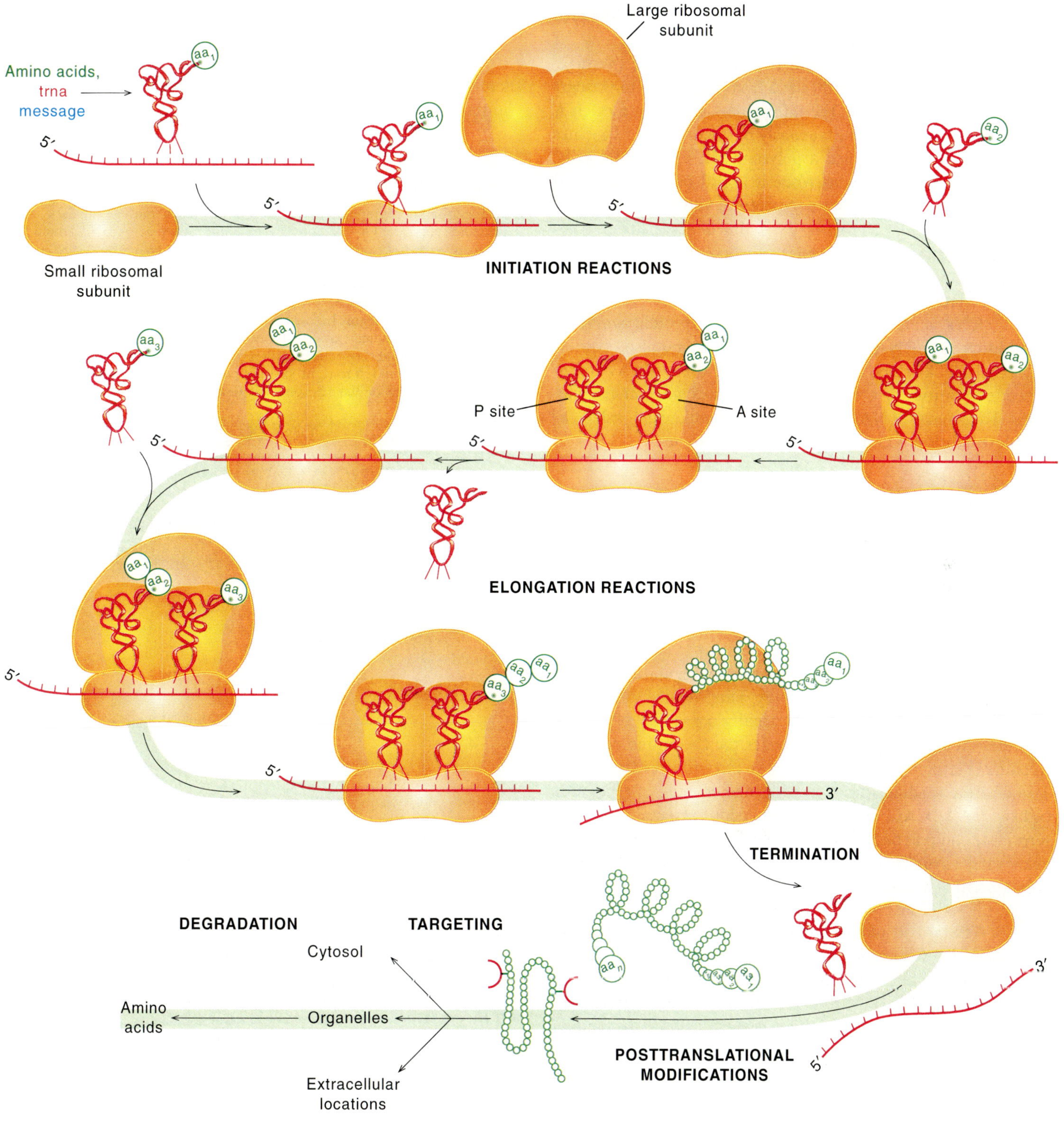

The Cellular Machinery of Protein Synthesis

Amino acids are assembled into polypeptides on ribosomes. Prior to their interaction with messenger RNA (mRNA), amino acids are covalently attached to transfer RNA (tRNA) to form aminoacyl-tRNAs. The aminoacyl-tRNAs attach to specific sites on the mRNA. mRNA contains the instructions for translation in the form of the genetic code that specifies the amino acid sequence of the polypeptide to be synthesized. Each ribosome binds to and moves along the messenger RNA while producing a single polypeptide. The direction or polarity of translation is from the 5′ to the 3′ terminus on the message, while the polypeptide is synthesized from the amino to the carboxyl terminus. As a rule, the information in a single mRNA molecule is translated simultaneously by a number of ribosomes that form a structure called a polysome. In prokaryotic cells, translation of mRNA begins while it is still being transcribed so that assemblies of nascent polypeptides and mRNAs can be seen on chromosomal DNA (fig. 29.2). In eukaryotic cells, transcription and translation are separate and polysomes occur either free in the cytoplasm or bound to membranes in the endoplasmic reticulum (fig. 29.3).

RNA molecules play crucial roles in protein synthesis. In addition to mRNA and tRNA, the ribosome also contains ribosomal RNA (rRNA). The tRNAs and the mRNAs are independent entities or are found in complexes with other molecules. Ribosomal RNAs, on the other hand, exist solely as complexes with proteins in the form of ribosomes.

Messenger RNA Is the Template for Protein Synthesis

The mRNA molecule carries the genetic message in the form of a sequence of nucleotides that determines the order of amino acids in the polypeptide chain. Each amino acid is represented in the mRNA by a sequence of three nucleotides called codons. Codons are arranged in a contiguous reading frame, which is flanked on either side by bases that are not translated. These untranslated regions frequently have roles in regulating the processing and expression of the message. The 5′ end of the reading frame begins with a start codon, usually consisting of the nucleotides AUG, a sequence that codes for the amino acid methionine. Methionine is always used to initiate translation. The 3′ end of the reading frame contains one or more of three stop codons, UAA, UAG, or UGA. Stop codons serve as signals to terminate the polypeptide chain. The 3′ end of the message in eukaryotes usually contains a posttranscriptionally added poly(A) tail. This poly(A) tail has no known role in translation except that it probably increases the lifetime of the message.

The 5′ end of the mRNA plays a special role in the selection of the start codon. This selection process occurs in fundamentally different ways in prokaryotes and eukaryotes. In prokaryotes (fig. 29.4) the mRNA contains a specific ribosome-binding site upstream of the initiating start codon. This binding site allows the ribosome to identify the correct initiating AUG. As we mentioned in chapter 28, prokaryotic mRNAs are frequently polycistronic, which means that they encode more than one polypeptide chain. Internal ribosome binding sites allow for independent initiation at internal reading frames. Eukaryotic mRNAs, on the other hand (see fig. 29.4), are usually monocistronic, encoding only one polypeptide chain. In keeping with this architecture, eukaryotic ribosomes interact with a ribosome entry site at the 5′ terminus of mRNA and move along it by a scanning mechanism to find the start codon. Identification of the ribosome entry site may be partly facilitated by the cap structure (see fig. 28.23) at the 5′ end of the mRNA. Initiation in eukaryotes usually occurs at the first AUG sequence downstream from the ribosome entry site. In multicellular eukaryotes the initiating AUG sequence is selected by a preferred initiation context of adjacent bases.

Figure 29.2

(*a*) Electron micrograph of *E. coli* polysomes. Ribosomes are the dark structures connected by the faintly visible mRNA strand. A DNA strand connecting the polysomes from which the mRNA is being transcribed is visible as a horizontal line. (From O. L. Miller, B. A. Hankalo, and C. A. Thomas, "Visualization of bacterial genes in action," *Science* 169:392, 1970, © 1970 by the AAAS.) (*b*) Line drawing for clarification. In bacteria, translation usually begins before transcription is completed. This results in a DNA-mRNA-ribosome complex as shown. As the mRNA grows, the number of ribosomes associated with it increases. Here the mRNA appears to be growing from left to right. It is not possible to see the growing polypeptide chains on the ribosome because of their small size.

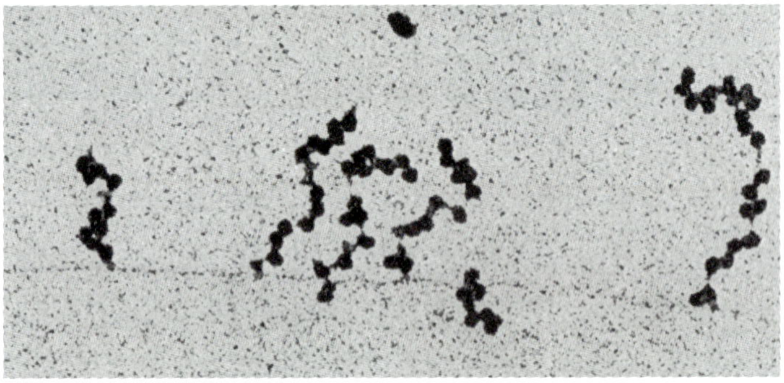

(a)

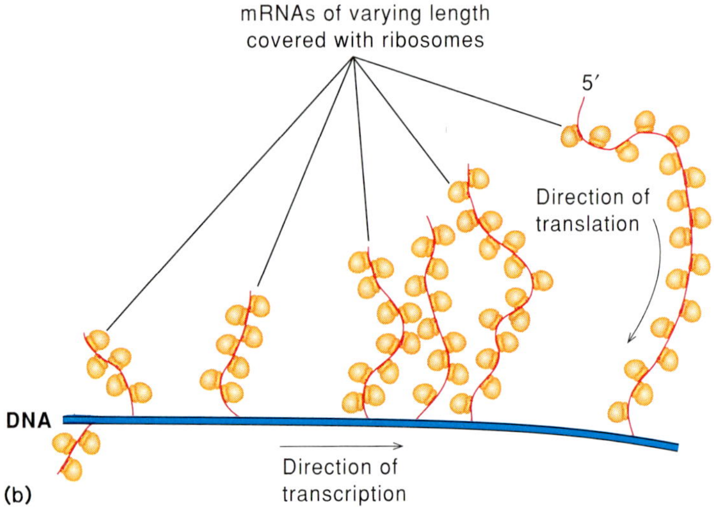

(b)

Transfer RNAs Order Activated Amino Acids on the mRNA Template

Transfer RNAs contain two crucial functional sites—a site for attachment of the amino acid, and a site that interacts with the mRNA during translation on the ribosome. At least one type of tRNA corresponds to each of the twenty types of amino acids that are incorporated into proteins. For accurate translation to occur, tRNAs must be distinguishable from each other by the molecules that recognize these specific types while still being recognizable by the molecules that interact with all tRNAs. A major challenge has been to understand how the similarities and differences in the structures of different tRNAs are related to the functions that they play in protein synthesis. Here we will summarize the structural features that are found in all tRNAs, allowing them to perform their common functions.

tRNAs that correspond to a single amino acid type are known as cognate tRNAs. In writing, the different cognate tRNAs are designated by a superscript. For example, $tRNA^{Phe}$ and $tRNA^{Ser}$ designate the tRNAs that correspond to the amino acids phenylalanine and serine, respectively. Some amino acids have several different cognate tRNAs, which are called isoacceptor tRNAs. Isoacceptor tRNAs are distinguished from each other by subscripts. A single cell typically contains a total of 50 or more different tRNAs.

$tRNA^{Phe}$ was the first tRNA to be crystallized, and its three-dimensional structure is one of the best understood (see fig. 28.3). The other tRNAs are built on a structural theme similar to that of $tRNA^{Phe}$, so that a single figure can describe the common features of their primary, secondary, and tertiary structures (fig. 29.5). All tRNAs are composed of a single polynucleotide chain of from 70 to 95 residues. This chain folds back on itself to form a cloverleaf structure composed of four double-stranded stems and four single-stranded loops. The overall folding is such that the 5′ and 3′ ends of the molecule are brought together to create a stem with seven regular Watson-Crick base pairs. This stem is known as the acceptor stem because during protein synthesis the amino acid is transiently attached to the ribose at this terminus. The 3′ end of the molecule

Figure 29.3

(*a*) Electron micrograph of mammalian rough endoplasmic reticulum. Continuous sheets of membrane create a compartment distinct from the surrounding cytosol. (From S. L. Wolfe, *Biology of the Cell*, 2d ed., Wadsworth, Belmont, Calif., 1981.) (*b*) Clarifying line drawing shows expanded region of endoplasmic reticulum with ribosomes attached to the surface of the membrane facing the cytosol. The term rough endoplasmic reticulum arose because at low magnification the attached ribosomes give the endoplasmic reticulum a rough appearance.

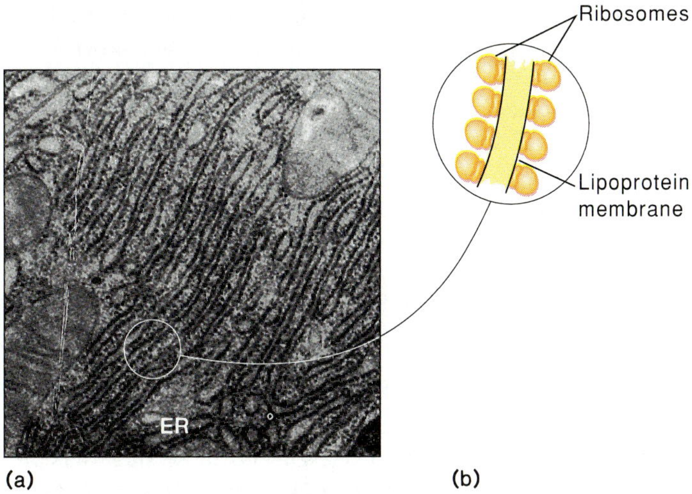

Figure 29.4

Simplified diagram of mRNA structure. (*a*) Typical eukaryotic mRNA. An AUG start codon is located near the 5′ end of the mRNA. The single reading frame ends with one of the three trinucleotide sequences that represents a stop codon. Frequently, but not always, the 5′ end of the mRNA is capped and the 3′ end contains a poly(A) tail. The cap structure is described in figure 28.23. (*b*) Typical bacterial mRNA. Some bacterial mRNAs contain more than one reading frame; some contain only one. Each reading frame contains a start and a stop codon. Recognition of the start codons is facilitated by the presence of a ribosome recognition sequence (Shine-Dalgarno). The space between any two reading frames in the same transcript varies from one transcript to the next.

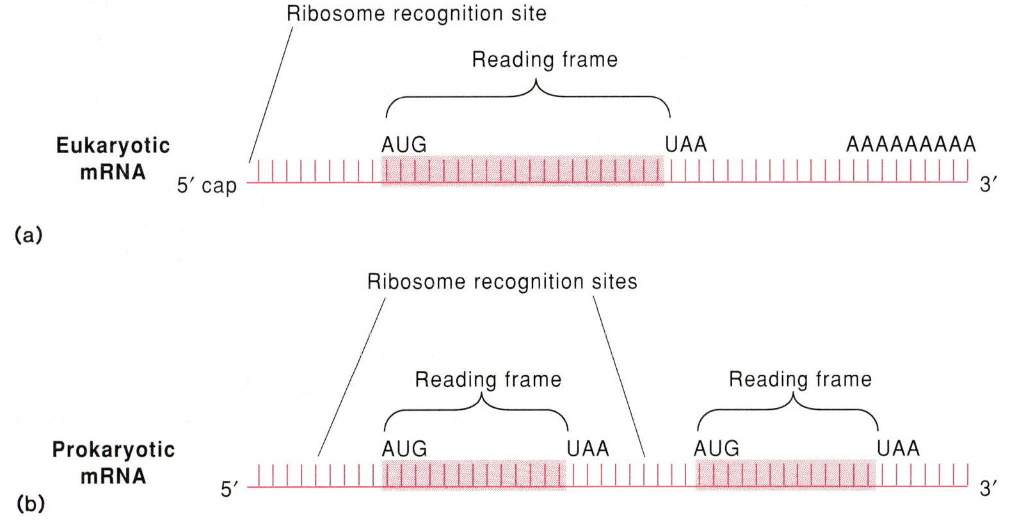

Figure 29.5

The general structure of a tRNA molecule. (*a*) Representation in the form of a cloverleaf which is the clearest way of observing the secondary structure. (*b*) A more realistic drawing of the three-dimensional folded structure. Color coding shows how the various loops in the cloverleaf structure correspond to the parts of the folded structure. For a more detailed drawing showing the types of hydrogen bonds and the structures of the individual bases, see figure 28.3.

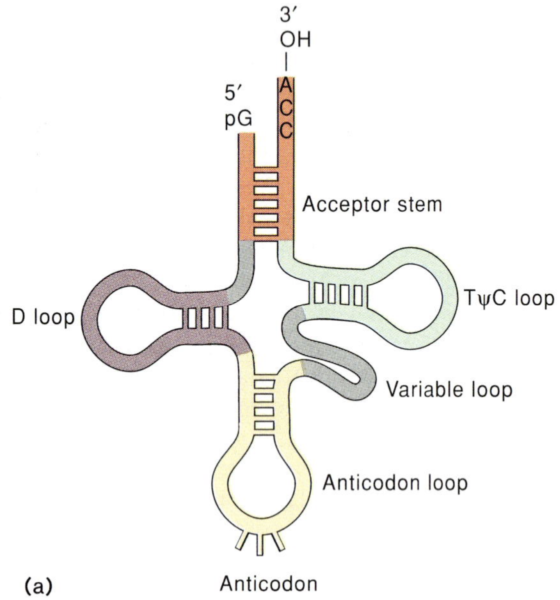

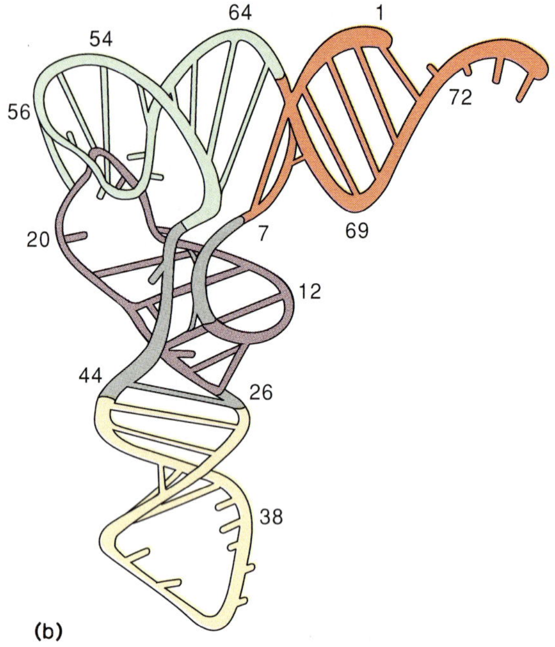

is always terminated with the sequence CCA, which identifies the tRNA to the components that interact with all of these molecules during protein synthesis.

The unpaired loops of tRNA are named according to their unique structural features. Loop I varies in size from 7 to 11 unpaired bases and frequently contains the unusual base dihydrouracil; it is designated the D loop. Loop II contains the three bases known as the anticodon and is therefore designated the anticodon loop. This portion of tRNA plays a key role in translation by pairing with the complementary codon in mRNA and thus serves to align amino acids in their appropriate sequence. Loop III, the variable loop, may contain as few as 3 or as many as 21 bases, making it the major site of size variation in tRNA. Loop IV contains the unusual ribothymidine and pseudouridine bases as part of an invariant sequence; for this reason it is known as the TψC loop.

The secondary cloverleaf structures of all tRNAs in solution fold into a three-dimensional L-shaped structure like that of $tRNA^{Phe}$ (fig. 29.5*b*). This structure is composed of two helical arms joined at right angles. The ribose moiety to which the amino acid is joined is at the end of one arm, identifying it as the acceptor arm. The anticodon is at the end of the other arm, identifying it as the anticodon arm. In the tRNA structures thus far solved, the 3′-terminal ribosome and anticodon are separated by a constant diagonal distance of about 75 Å. This L shape is the hallmark of a tRNA molecule.

Ribosomes Are the Site of Protein Synthesis

We considered the structure of the ribosome in some detail in the previous chapter without referring to those sites that are functionally important in protein synthesis. Here we can localize some of the functional sites on the two ribosomal subunits (fig. 29.6). The mRNA binds to the smaller ribosomal subunit. The peptidyl transferase is an integral part of the 50S subunit,* and the elongation factor EF-G binds to the 50S subunit. The nascent polypeptide chain exits through a channel in the 50S subunit. There are two functional sites on the 50S subunits, called the P site and the A site, where adjacent tRNAs are bound to the messenger. These sites cannot be seen in figure 29.6.

Before we go into a detailed discussion of the reactions that occur on the ribosome, it would be useful to consider the crucial interaction between the anticodon on the tRNAs and the codons on the mRNAs.

The Genetic Code

The concept of a genetic code grew out of the realization that both nucleic acids and proteins are linear polymers made of a limited number of building blocks, plus the knowledge that the structure of proteins is genetically determined. The genetic code is simply the sequence relationship between nucleotides in genes (or mRNA) and the amino acids in the proteins they encode. The coding ratio is the number of nucleotides in an mRNA required to represent an amino acid. Since there are four different

*Recently it has been shown that the peptidyl transferase is part of the rRNA of the 50S subunit.

Figure 29.6

Functional sites on the prokaryotic ribosome. This figure shows a ribosome in the process of elongating a polypeptide chain. The mRNA and the EF-Tu-aminoacyl-tRNA complex are more closely associated with the 30S subunit. The peptidyl transferase and the EF-G elongation factor are associated with the 50S subunit, as is the nascent polypeptide chain.

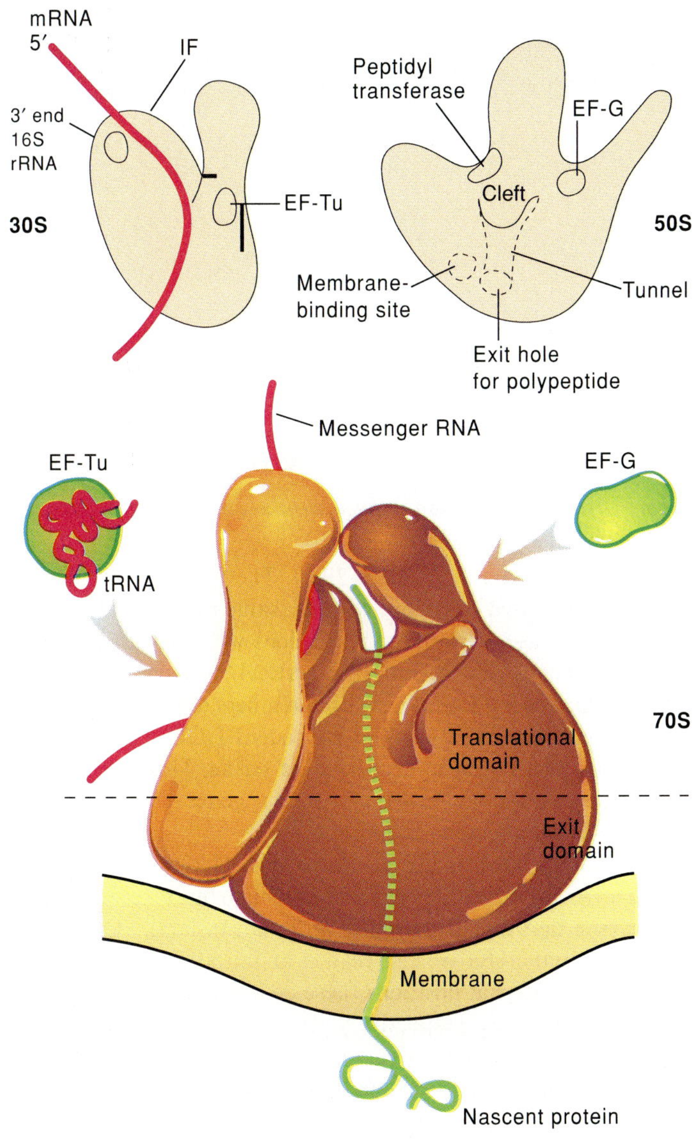

nucleotides in RNA and 20 different amino acids in proteins, the minimum acceptable coding ratio is 3: A singlet code, in which one nucleotide represents one amino acid, could code for only four different amino acids because there are only four different nucleotides in DNA or RNA. A doublet code, in which two nucleotides code for a single amino acid, could code for 16 amino acids since there are 16 possible doublet sequences in RNA. A triplet code, made from four nucleotides taken three at a time, generates a total of 64 possible triplet sequences. This would be more than enough to represent the 20 amino acids found in proteins.

In fact, a triplet code creates a problem of a different sort. What is the use of the extra trinucleotide sequences? Allowing for the three triplets that represent stop signals, we are still left with 61 possible triplets to code for just 20 amino acids. A nondegenerate code, in which a unique triplet codes for each amino acid, would imply that 41 triplet sequences are never found in translation reading frames. On the other hand, a degenerate code, in which all triplets are used, would require that each amino acid be represented, on the average, by more than one triplet sequence. We will return to this point later. For now, we may simply note that in most cases, point mutations in which one base pair in the DNA is replaced by another result either in no amino acid change or in a change to another amino acid. This is the result that we would expect if most of the triplet sequences do in fact represent amino acids.

Another question relates to the location of codons in the translation reading frame. Codons could be arranged in a close-packed side-by-side arrangement. In this event each nucleotide within the translation reading frame would represent a code letter within one, and only one, codon. Alternatively, codons could be separated by one or more "spacer nucleotides." In this event some nucleotides within the reading frame would represent code letters within codons, while others would serve as spacer nucleotides. Finally, we can imagine an overlapping codon arrangement, in which all nucleotides represent code letters and some nucleotides represent code letters in more than one codon. However, biochemical experiments that we will describe later in this chapter favor a nonoverlapping triplet code without spacers. Further genetic experiments in support of this type of code (not elaborated on here) indicate that when a single base pair is added or deleted, a change occurs in the reading frame of the message downstream of the change. Such an alteration in the DNA is known as a frameshift mutation. A frameshift mutation will produce a new termination point because the frameshift changes the reading of the original stop codon and produces a new stop codon.

The Code Was Deciphered with the Help of Synthetic Messengers

In 1961, Marshall Nirenberg and Heinrich Mattaei were attempting to synthesize proteins in cell-free extracts of *E. coli.* These extracts, containing the essential cellular components, were supplemented with nucleotides, salts, and radioactive amino acids, which were used so that the expected small amounts of protein could be detected. For mRNA they chose to use viral RNA from tobacco mosaic virus (TMV), because their goal was to make viral protein. As a control for these first "incorporation" experiments, they needed an mRNA that would not be expected to code for a protein. For this purpose their choice was an artificial polynucleotide produced by Marianne Grunberg-Manago and Severo Ochoa with a newly discovered enzyme, polynucleotide phosphorylase. Initially, Grunberg-Manago and Ochoa had thought that this enzyme might be responsible for cellular RNA synthesis, but they later came to realize that its more likely role was to degrade RNA by a phosphorolysis reaction:

$$RNA_n + nP_i \rightarrow (\text{nucleoside diphosphates})_n$$

As a result of that realization, polynucleotide phosphorylase was relegated to the synthesis of unusual polynucleotides as laboratory curiosities. One such polynucleotide, poly(U), contained only uridine residues. When Nirenberg and Mattaei added this control RNA to their cell-free system, substantial incorporation was evident. They quickly showed that this synthetic messenger specifies the synthesis of a polypeptide containing only phenylalanine residues (polyphenylalanine).

A wave of excitement was set in motion by this discovery. It was rapidly demonstrated that poly(A) promotes polylysine synthesis and poly(C) promotes polyproline synthesis. From these observations it seemed clear that the code words UUU, AAA, and CCC correspond to the amino acids phenylalanine, lysine, and proline, respectively.

Polynucleotide phosphorylase, running in the reverse direction, produces an "RNA" of random sequence whose composition reflects the mixture of nucleoside diphosphates in the reaction mixture. Mixed polynucleotides containing two bases were used in the incorporating system and shown to incorporate a pattern of amino acids that was consistent with a triplet code, but the observed incorporation could not define the code sequence.

The chemical synthesis of short oligonucleotides of known sequence provided a way out of the dilemma. First, Philip Leder and Nirenberg showed that nucleotide triplets cause the specific binding of aminoacyl-tRNA to ribosomes. They cleverly observed that this binding could be readily measured by simply passing a solution containing radioactive aminoacyl-tRNA, ribosomes, and the trinucleotide through a nitrocellulose filter. Ribosomes, along with bound ligands, were adsorbed to the filter and the coding specificity of some triplets was quickly established. Thus UUC, UCU, and CUU were shown to specify phenylalanine, serine, and leucine, respectively. However, many triplets did not give clear binding signals and there was concern that the binding reaction might not display the correct specificity.

The code was ultimately defined by the translation of polynucleotides of repeating sequences. These long-repeat-sequence polynucleotides were produced by chemically synthesizing short, defined-sequence oligomers and then amplifying them enzymatically. When repeating dinucleotides were translated (fig. 29.7), they yielded repeating dipeptides, and when repeating trinucleotides (not shown) were translated, they yielded up to three individual peptides, each containing only a single amino acid. Eventually the sequence of bases from natural messages was correlated with the sequence of bases in the proteins they encoded.

The Code Is Highly Degenerate

A triplet code, one made from four nucleotides taken three at a time, generates a total of 64 different triplet sequences or codons. Three of these codons, as we will see, are utilized to terminate translation and are not generally used to specify amino acids. The remaining 61 codons and the 20 amino acids can be neatly summarized by grouping codons with the same first and second bases into a grid (table 29.1). The four horizontal sections are composed of codons with the same first base. The four vertical sections are composed of codons with the same second base. The boxes representing the vertical/horizontal intersections contain codon families whose members differ only in their 3′-terminal base. For example, the codons UCU, UCC, UCA, and UCG comprise a family encoding serine. Thus the genetic code is degenerate, i.e., one amino acid is generally specified by multiple or synonymous codons.

Figure 29.7

Demonstration that poly(UC) codes for a Ser-Leu repeating peptide. When poly(UC) containing a strict alternating sequence of U and C is used as an mRNA, only serine and leucine are incorporated. Analysis of the polypeptide product indicates that it contains an alternating sequence of serine and leucine.

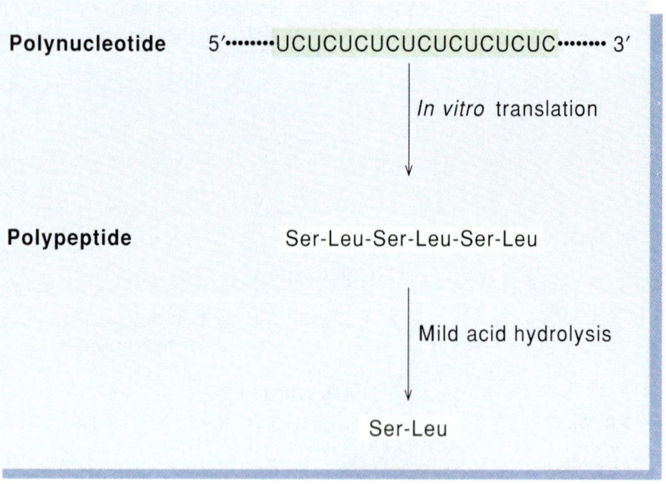

As we have indicated, the codon AUG is the only one that is generally used to specify methionine, but it serves a dual function in that it is also used to initiate translation. Occasionally, GUG and UUG are also read as an initiating codon in bacteria, but in internal positions these codons are always read as valine and leucine, respectively. In eukaryotes, initiation at codons other than AUG is much less frequent than in prokaryotes. Weak initiation occasionally occurs at GUG, CUG, and ACG codons in eukaryotic systems. The UGA triplet also serves a dual function; it is usually recognized as a stop, but on occasion it serves as a codon for selenocysteine (box 29A).

Organisms differ in the frequency with which they utilize synonymous codons. Some synonymous codons are used frequently and some are used infrequently. For *E. coli* and yeast this frequency of use, known as codon usage, correlates with the abundance in the organism of the tRNAs that recognize particular synonymous codons. Also, for the same two species, proteins that occur in the greatest abundance employ high-usage codons more frequently and low-usage codons less frequently than the average proteins. It is not clear that this same trend is followed in multicellular eukaryotes such as *Drosophila* and humans.

Table 29.1
The Genetic Code

First position (5′ end)	Second position: U		Second position: C		Second position: A		Second position: G		Third position (3′ end)
	U		C		A		G		
U	UUU	Phe	UCU	Ser	UAU	Tyr	UGU	Cys	U
	UUC		UCC		UAC		UGC		C
	UUA	Leu	UCA		UAA	STOP	UGA	STOP	A
	UUG		UCG		UAG		UGG	Trp	G
C	CUU	Leu	CCU	Pro	CAU	His	CGU	Arg	U
	CUC		CCC		CAC		CGC		C
	CUA		CCA		CAA	Gln	CGA		A
	CUG		CCG		CAG		CGG		G
A	AUU	Ile	ACU	Thr	AAU	Asn	AGU	Ser	U
	AUC		ACC		AAC		AGC		C
	AUA		ACA		AAA	Lys	AGA	Arg	A
	AUG	Met	ACG		AAG		AGG		G
G	GUU	Val	GCU	Ala	GAU	Asp	GGU	Gly	U
	GUC		GCC		GAC		GGC		C
	GUA		GCA		GAA	Glu	GGA		A
	GUG		GCG		GAG		GGG		G

From Robert F. Weaver & Philip W. Hedrick, *Basic Genetics*. Copyright © 1991 Wm. C. Brown Communications, Inc., Dubuque, Iowa. All Rights Reserved. Reprinted by permission.

Figure 29.8

Examples of standard (*a*) and wobble (*b* and *c*) base pairs formed between the first base in the anticodon and the third base in the codon.

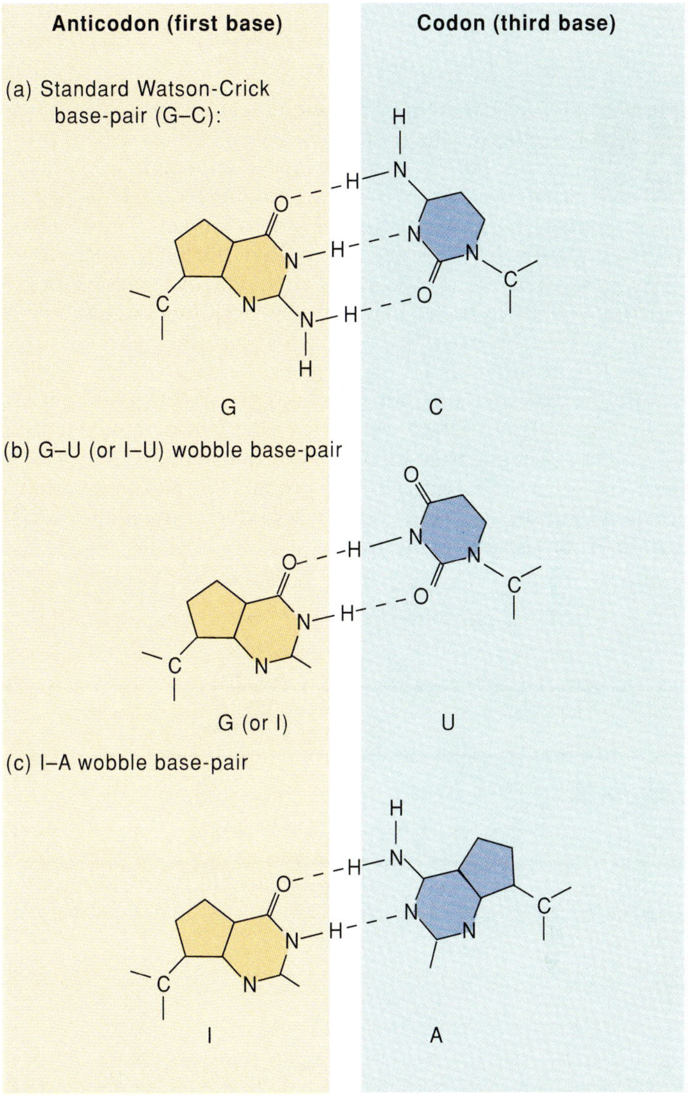

Wobble Introduces Ambiguity into Codon–Anticodon Interactions

In order for 61 triplets to act as codons, tRNAs must interact specifically with each triplet. Strict Watson-Crick base-pairing between codon and anticodon would require 61 different anticodons and, correspondingly, 61 different tRNAs. As the characterization of tRNAs progressed it became clear that in many cases individual tRNAs could recognize more than one codon. In all cases the different codons recognized by the same tRNA were found to contain identical nucleotides in the first two positions and a different nucleotide in the third position (in the 3′ position of the codon). This relationship is the reason that the genetic code can be so neatly arranged as shown in table 29.1, by codon families that differ in their third base.

The 3′-terminal redundancy of the genetic code and its mechanistic basis were first appreciated by Francis Crick in 1966. He proposed that codons and anticodons interact in an antiparallel manner on the ribosome in such a way as to require strict Watson-Crick pairing (that is, A-U and G-C) in the first two positions of the codon but to allow other pairings in its 3′-terminal position. Nonstandard base-pairing between the 3′-terminal position of the codon and the 5′-terminal position of the anticodon alters the geometry between the paired bases; Crick's proposal, labeled the wobble hypothesis, is now viewed as correctly describing the codon–anticodon interactions that underlie the translation of the genetic code.

By inspecting the geometry that would result from different wobble pairings (fig. 29.8) and recognizing that inosine, the deaminated form of adenine, frequently occurs in tRNAs in the 5′ position of the anticodon, Crick grasped the relationship between codon and anticodon. According to this relationship, table 29.2, when C or A occurs in the 5′ position of an

29A

BOX

Site-Specific Variation in Translation Elongation

The unusual amino acid selenocysteine (a derivative of cysteine in which the sulfur atom is replaced by a selenium atom) is an essential component in a small number of proteins. These proteins occur in prokaryotes and eukaryotes ranging from *E. coli* to humans. In all cases, selenocysteine is incorporated into protein during translation in response to the codon UGA. This codon usually serves as a termination codon but occasionally, in some required but unknown context of bases, is used to specify selenocysteine instead.

In *E. coli*, the products of four genes (*selA, B, C,* and *D*) are required for the incorporation of selenocysteine. The product of the *selC* gene is tRNA^{Ser} (a suppressor tRNA whose anticodon is UCA). The first step in the incorporation of selenocysteine is catalyzed by seryl-tRNA synthase. The products of *selA* and *D* function in the subsequent conversion of Ser-tRNA^{Ser} to selenocysteyl-tRNA^{Ser}. The probable pathway is

$$\text{tRNA}^{Ser} \rightarrow \text{seryl-tRNA}^{Ser} \rightarrow \text{phosphoseryl-tRNA}^{Ser} \rightarrow \text{selenocysteyl-tRNA}^{Ser}$$

Incorporation of selenocysteyl-tRNA^{Ser} into protein in response to the UGA codon requires SELB (the protein product of the *selB* gene in *E. coli*). SELB is homologous in sequence to EF-Tu and probably replaces it in translation by specifically recognizing selenocysteyl-tRNA and UGA in the appropriate sequence context. Selenocysteyl-tRNA^{Ser}, in combination with SELB, must be capable of competing with termination factors for the translation of the termination codon when it occurs in the "right" context of bases. This process is known as *site-specific variation* in translation elongation.

Other proteins of unknown function that are homologous to EF-Tu are known to exist. Conceivably, these proteins might participate in the incorporation of other rare amino acids into unique positions in proteins.

Table 29.2
The Wobble Rules of Codon–Anticodon Pairing

5′ Base of Anticodon	3′ Base of Codon
C	G
A	U
U	A or G
G	C or U
I	U, C, or A

anticodon, it can pair only with G or U, respectively, in the 3′ position of a codon. tRNAs containing either G or U in the 5′ or wobble position of the anticodon can each pair with two different codons, while an inosine (I) in this position produces a tRNA that can pair with three codons differing in the 3′ base. Subsequent sequence analysis of many tRNAs has proved this hypothesis as a feature of the translation of the "universal" genetic code (but see the following discussion).

A careful comparison of the wobble rules with the genetic code indicates that the minimum number of tRNAs required to translate all 61 codons is 31. With the addition of tRNA_i^{Met} the total comes to 32. Most cells contain many more than this minimum number of tRNA types.

The Code Is Not Quite Universal

It was originally believed that exactly the same genetic code is utilized by all genetic systems. Initial experiments supported this conclusion, since it was found that some mammalian mRNAs could be faithfully translated in cell-free bacterial systems. We now know, however, that significant variations in the meaning of specific code words occur in many genetic systems. The exact scope and nature of these variations are still being discovered, but the current view is that the variations in genetic meaning all represent divergences from the standard or "universal" genetic code described earlier. In other words, there is no indication that variations in the genetic code resulted from independent origins of the genetic code. Much of the variation in genetic meaning derives from alterations in the pairing of codon and anticodon on the ribosome.

Many of the known variations in the genetic code are found in genes of mitochondria and chloroplasts. It is easy to see why these genetic systems might be more plastic, since they frequently encode only ten to twenty proteins. The remainder of the organellar proteins are derived by importing nuclear gene products.

The tRNAs used to translate mitochondrial mRNAs are entirely derived from mitochondrial chromosomes. The first clue that something was unusual about the mitochondrial genetic code was that only 24 types of tRNA could be found. Recall that according to Crick's rules of wobble, 32 tRNAs minimally are required for the translation of all 61 codons. One possible

Table 29.3
The Genetic Code of Yeast Mitochondria

First position (5′ end)	Second position: U			C			A			G			Third position (3′ end)
U	UUU	Phe	AAG	UCU	Ser	AGU	UAU	Tyr	AUG	UGU	Cys	ACG	U
	UUC			UCC			UAC			UGC			C
	UUA	Leu	AAU*	UCA			UAA	STOP		UGA	Trp	ACU*	A
	UUG			UCG			UAG			UGG			G
C	CUU	Thr	GAU	CCU	Pro	GGU	CAU	His	GUG	CGU	Arg	GCA[b]	U
	CUC			CCC			CAC			CGC			C
	CUA			CCA			CAA	Gln	GUU*	CGA			A
	CUG			CCG			CAG			CGG			G
A	AUU	Ile	UAG	ACU	Thr	UGU	AAU	Asn	UUG	AGU	Ser	UCG	U
	AUC			ACC			AAC			AGC			C
	AUA	Met	UAC[a]	ACA			AAA	Lys	UUU*	AGA	Arg	UCU*	A
	AUG			ACG			AAG			AGG			G
G	GUU	Val	CAU	GCU	Ala	CGU	GAU	Asp	CUG	GGU	Gly	CCU	U
	GUC			GCC			GAC			GGC			C
	GUA			GCA			GAA	Glu	CUU*	GGA			A
	GUG			GCG			GAG			GGG			G

Source: S. G. Bonitz et al., "Codon recognition rules in yeast mitochondria," in *Proceedings National Academy of Sciences USA* 77:3167, 1980.

The codons (5′ → 3′) are at the left and the anticodons (3′ → 5′) are at the right in each box. (* designates U in the 5′ position of the anticodon that carries the —$CH_2NH_2CH_2COOH$ grouping on the 5′ position of the pyrimidine.)

[a]Two tRNAs for methionine have been found. One is used in initiation and one is used for internal methionines.

[b]Although an Arg tRNA has been found in yeast mitochondria, the extent to which the CGN codons are used is not clear.

solution to this conundrum was that mitochondrial genes do not utilize all 61 codons. Another possibility was that the wobble rules might be different for mitochondria. In fact, the latter is the case. Crick's original wobble rules stated that at least two different tRNAs are needed to translate four codon families. In all of these cases (with the exception of the codon family specifying Arg; see the footnote to table 29.3), single tRNAs have been found responsible for specifying all four code words and these tRNAs all contain a U in the "wobble" position of their anticodons. It appears that the mitochondrial ribosome allows these tRNAs to pair with all four members of the codon family. The six mitochondrial tRNAs that pair with the normal two codons contain an altered U in the wobble position and this modification causes them to conform to the normal "wobble" rules.

Another peculiarity of the mitochondrial code emerged from a study of yeast codon usage. By comparing tables 29.3 and 29.1, you will see that the mitochondrial code has several differences in code word meaning. The codons beginning with CU represent Thr instead of Leu, the AUA codon represents Met instead of Ile, and the UGA codon represents Trp rather than a stop signal.

The Rules Regarding Codon–Anticodon Pairing Are Species-Specific

The genetic code differs very little between species. By contrast, there are considerable differences between species in the anticodon translation system of tRNA, as evidenced by the mitochondrial tRNA system. In all systems the bases in the anticodon-codon complex run antiparallel, as in standard double-helix pairing, and in all cases only Watson-Crick-like base pairing occurs between the first two bases in the codon and the opposing bases in the anticodon segment of the tRNA. However, for the 3′ base in the codon, the rules for pairing vary with the species and with the base in question. These rules, summarized in table 29.4, are as follows.

When the 5′ base in the anticodon is a G it can pair with either a U or a C, and this is true in all organisms. When the 5′ base in the anticodon is an A it can pair only with a U in the codon. However, it is rare that an A is found in this position of the anticodon. An A in this position in eukaryotes is usually deaminated to an inosine (I) base, which has an expanded capacity for pairing. Base A can pair only with U, but I can pair with U, C, or A. In eubacteria, deamination is limited to the conversion of the ACG sequence to an ICG anticodon. When C

Table 29.4
Anticodon–Codon Pairing

Anticodon First Base	Codon Third Base	Examples
U	U, C, A, G	Mitochondrial code in family boxes
*U	A, G	Mitochondrial code in two-codon sets
†U	A	Eukaryotes
‡U	U, A, G	Eubacteria in family boxes
C	G	All codes
*C	A	Bacteria, isoleucine codon AUA
G	U, C	All codes
A	U	Rare
I	U, C, A	Eukaryotes, ICG in eubacteria

*, †, ‡ = Various modifications of U (Yokoyama et al., 1985)
*C = modified C

Source: From Thomas Jukes, et al., *Cold Spring Harbor Symposium on Quantitative Biology* 32:775, 1987. Copyright © 1987 Cold Spring Harbor Laboratory Press, Cold Spring Harbor, N.Y.

is the 5′ base in the anticodon it pairs with G only. The only known exception to this rule is found in eubacteria, where the C is covalently modified in the tRNA that recognizes the AUA codon. Thus in this one instance the modified C can pair with a 3′ A in the codon. A U base in the 5′ position of the anticodon shows the greatest variability. An unmodified U can pair with any of the four bases in the 3′ position of the codon. This situation is reflected in the U-family box in mitochondria (table 29.4).

U can also be modified in various ways. In mitochondria one type of modification permits a U to pair with either an A or a G in the two-codon sets. In eukaryotes, another type of U modification limits U to pairing with an A in the codon, and in eubacteria, a third type of U modification permits U pairing with U, A or G but not C in the family boxes.

The Steps in Translation

Protein synthesis is one of the most complex biochemical processes known, involving more than 100 different proteins and more than 30 kinds of RNA molecules. The process begins by the attachment of amino acids to specific tRNA molecules. Subsequent steps take place on the ribosome; amino acids are transported to the ribosome on their tRNA carriers and they do not leave the ribosome until they have become an integral part of a completed polypeptide chain.

Synthases Attach Amino Acids to tRNAs

A unique class of enzymes, called aminoacyl-tRNA synthases, attach amino acids to their cognate tRNAs. This attachment serves two functions: (1) The linkage between the amino acid and the tRNA activates the amino acid, making the subsequent formation of a peptide bond energetically favorable. (2) The tRNA directs the amino acid to a designated location on a messenger RNA so that the amino acid will be incorporated at the appropriate location in the polypeptide chain. Thus aminoacyl-tRNA synthases provide both energy and specificity for protein synthesis.

All synthases act by a common reaction mechanism. This reaction proceeds in two separate steps (fig. 29.9). In the first step the synthase recognizes its corresponding amino acid and its second substrate, ATP, and forms a mixed anhydride bond between the carboxyl group of the amino acid and the phosphate of AMP with the release of PP_i:

$$\text{Amino acid} + \text{ATP} \rightarrow \text{aminoacyl-AMP} + PP_i \qquad \textbf{(1)}$$

The equilibrium constant for this reaction is about 1.0 so that the energy derived from the cleavage of the phosphate anhydride of ATP is conserved in the mixed anhydride. Aminoacyl-AMP remains tightly bound to the enzyme and, as we will soon show, this fact has allowed workers to crystallize this important complex and analyze its structure.

The second reaction catalyzed by the aminoacyl-tRNA synthases results in the attachment of the amino acid through an ester linkage to the 3′-terminal ribose of tRNA:

$$\text{Aminoacyl-AMP} + \text{tRNA} \rightarrow \text{aminoacyl-tRNA} + \text{AMP} \qquad \textbf{(2)}$$

Synthases differ with respect to their site of attachment to tRNA. Some synthases specifically form the 2′-ester, some specifically form the 3′-ester, and still others produce a mixture of the two. The specificity of the synthases was determined by analyzing their ability to act on tRNA derivatives lacking one or the other terminal hydroxyl group. Once esterified to the terminal ribose, the aminoacyl group can migrate between the vicinal 2′ and 3′ hydroxyl groups. Thus, in cells, aminoacyl-tRNAs are mixtures of 2′ and 3′ esters. Only the 3′ derivative is a substrate for the subsequent transpeptidation reaction catalyzed by the ribosome.

The sum of the two reactions catalyzed by aminoacyl-tRNA synthases is:

$$\text{Amino acid} + \text{ATP} + \text{tRNA} \rightarrow \text{aminoacyl-tRNA} + \text{AMP} + PP_i \qquad \textbf{(3)}$$

This overall reaction is reversible but is driven to completion by the subsequent hydrolysis of PP_i to two equivalents of P_i through the action of ubiquitous pyrophosphatases. Thus the formation of aminoacyl-tRNA consumes two equivalents of ATP. The energy that ultimately drives the formation of the peptide linkage during protein synthesis is derived from the ester linkage that joins amino acids to tRNA.

Figure 29.9

Formation of aminoacyl-tRNA. This is a two-step process involving a single enzyme that links a specific amino acid to a specific tRNA molecule. In the first step (1) the amino acid is activated by the formation of an aminoacyl-AMP complex. This complex then reacts with a tRNA molecule to form an aminoacyl-tRNA complex (2).

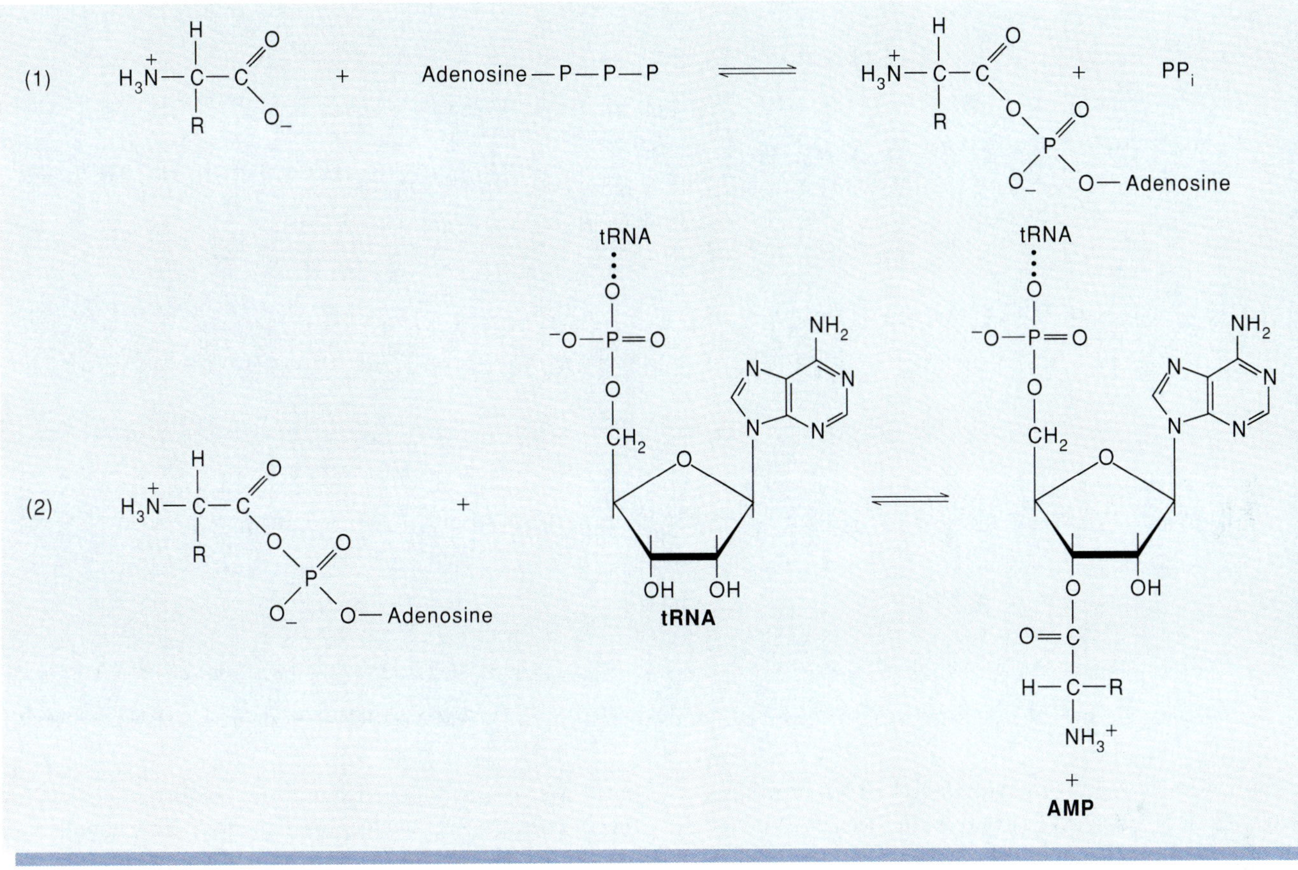

Each Synthase Recognizes a Specific Amino Acid and Specific Regions on Its Cognate tRNA

Even though synthases differ widely in size and subunit complexity, they all appear to be built on a common structural theme. Our understanding of the chemistry of the synthase reaction and the types of the active sites involved in these reactions was advanced substantially by the crystallization and structural solution of tyrosyl-tRNA synthase complexed with the reaction intermediate tyrosyl-adenylate (fig. 29.10). The reaction intermediate is bound in a deep cleft in the enzyme and interacts with it through eleven hydrogen bonds. Six of these bonds are with the AMP moiety and five are with the tyrosyl moiety of the intermediate. The amino acid selectivity of tyrosyl-tRNA synthase is thus determined primarily by the formation of specific hydrogen bonds with the amino acid.

Most cells contain only one synthase for each of the twenty amino acids specified by the genetic code. Each enzyme must be capable not only of recognizing its unique amino acid but also of recognizing the one or more corresponding cognate tRNAs and of discriminating between them and structurally similar noncognate tRNAs. Solving the puzzle of how synthases recognize tRNAs has been one of the major challenges in understanding the nature of the translation of the genetic code itself. The identity relationship between the tRNA and its synthase has, in fact, been called the "second genetic code."

Surprisingly, synthases fall into two categories with respect to the importance of the anticodon as a specificity element that they recognize. Some synthases recognize the anticodon and some do not. This distinction has been demonstrated in several ways. First, some tRNAs can be genetically altered in their anticodon without changing the specificity of their recognition by the cognate synthase. Some suppressor tRNAs (to be discussed later) are formed in this way. In a similar manner, some synthases are capable of recognizing their cognate tRNA even though the anticodon is chemically altered or removed. Other synthases apparently require an unaltered anticodon in order to recognize their cognate tRNAs. Four examples are shown in figure 29.11 (also see box 29B).

Figure 29.10

Recognition of an adenylylated amino acid by the proper synthetase. Shown is adenyl tyrosine bound to the tyrosyl synthetase. A network of H bonds not only stabilizes the reaction but also serves to discriminate between different amino acid residues. MC designates main chain (backbone) carbonyl or amino groups participating in hydrogen bonding.

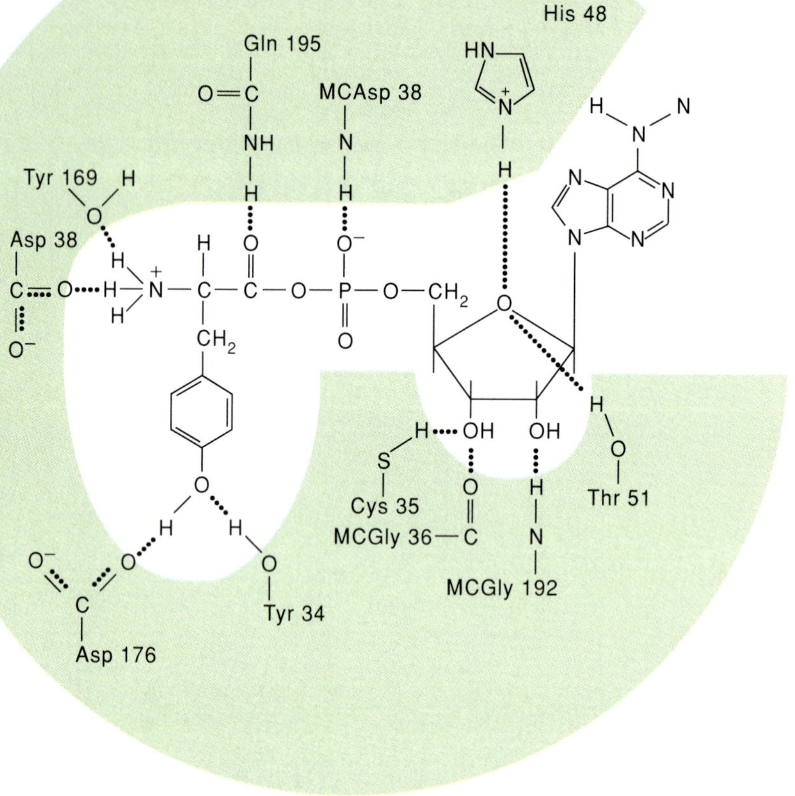

The crystal structure of glutaminyl-tRNA synthase, complexed with tRNA and ATP, has recently been determined by Tom Steitz and his colleagues (fig. 29.12). This accomplishment provides the first structure of a tRNA-protein complex and thus offers important insight into the general nature of the recognition of tRNAs by proteins.

Glutaminyl-tRNA synthase is known to require the integrity of the anticodon loop of tRNA for recognition, and is thus presumed to interact with both ends of the tRNA. In keeping with this interpretation, the synthase is asymmetric and longer (~100 Å) than the tRNA. Moreover, the crystal structure demonstrates the occurrence of significant contact between the protein and bases in the anticodon loop. The additional contacts between the synthase and tRNA appear to occur along the inside of the L-shaped structure. The recognition of the tRNA appears to arise from direct hydrogen bonding between amino acid residues of the protein and bases of the tRNA over a wide region of the tRNA structure. At this stage we do not know which of these protein-nucleic acid contacts are crucial for recognition.

The acceptor stem of the bound tRNA substrate plunges deep into the active site pocket created by the dinucleotide fold, also the site of ATP binding. Surprisingly, binding in the active site induces a significant conformational change in the 3′-terminal CCA acceptor sequence. This conformational change appears to cause the melting of the A-U pair at the end of the acceptor stem in a manner that allows these bases to interact with amino acid side chains of the protein. At this point in our understanding, it seems likely that the interactions that allow synthases to identify their cognate tRNAs are both diverse and complex.

Aminoacyl-tRNA Synthases Can Correct Acylation Errors

Many aminoacyl-tRNA synthases appear to contain a second catalytic site that serves to correct errors by hydrolyzing incorrectly matched amino acids and tRNAs. These proofreading hydrolysis reactions are best understood in the case of isoleucyl-tRNA synthase. The amino acids isoleucine and valine, which differ by only a single methylene group, are among the most difficult to discriminate. Isoleucyl-tRNA synthase is capable of distinguishing these two amino acids at its aminoacylation site but occasionally and mistakenly forms valyl-$tRNA^{Ile}$. When isoleucyl-tRNA synthase encounters valyl-$tRNA^{Ile}$ the hydrolytic active site specifically degrades it:

$$\text{Valyl-tRNA}^{\text{Ile}} + H_2O \rightarrow \text{valine} + \text{tRNA}^{\text{Ile}}$$

Figure 29.11

Identity elements in four tRNAs. Each circle represents one nucleotide. Filled circles indicate nucleotides that serve as recognition elements to the appropriate aminoacyl-tRNA synthase. It is possible that there are other identity elements in these structures still to be discovered.

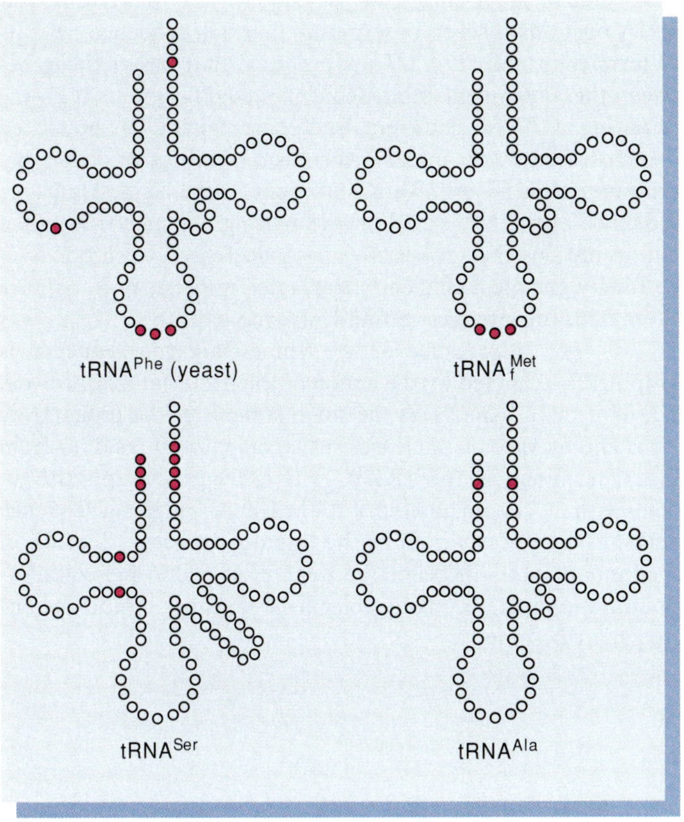

Figure 29.12

Solvent-accessible surface representation of the GlnRS enzyme complexed with tRNA and ATP. The region of contact between tRNA and protein extends across one side of the entire enzyme surface and includes interactions from all four protein domains. The acceptor end of the tRNA and the ATP are seen in the bottom of the deep cleft. Protein is inserted between the 5′ and 3′ ends of the tRNA and disrupts the expected base pair between U1 and A72. (From M. G. Rould, J. J. Persona, D. Söll, and T. Steitz, "Structure of *E. coli* glutamyl-tRNA synthetics complexed with $tRNA^{Gln}$ and ATP at 2.8-Å resolution, implications for tRNA discrimination," *Science* 246:1135–1142, 1989, © 1989 by the AAAS.)

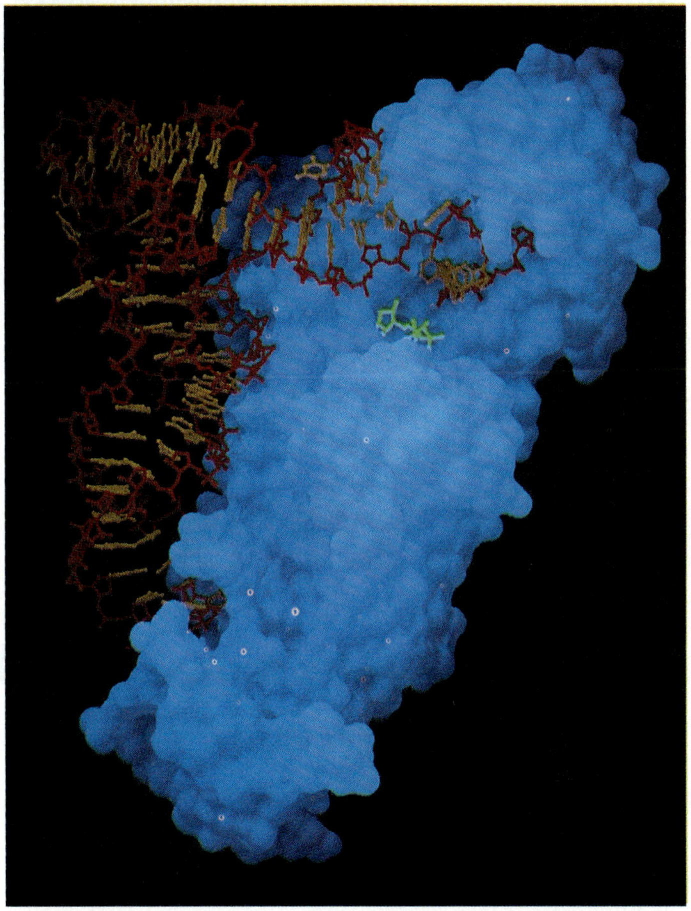

In this way the erroneous incorporation of valine in the place of isoleucine is avoided. The sequential action of aminoacylation and proofreading sites contributes to an overall error frequency of translation of less than 1 in 10,000.

A Unique tRNA Initiates Protein Synthesis

The translation of every protein begins with the incorporation of the amino acid methionine. A unique initiator tRNA, which we will designate $tRNA_i^{Met}$, is responsible for the incorporation of this initiating methionine in all protein-synthesizing systems and it also plays an important role in selecting the appropriate translation start site in mRNA. There are generally only two tRNAs in cells that specify methionine. We will designate the one that is responsible for the incorporation of internal methionines simply as $tRNA^{Met}$. A single methionyl-tRNA synthase is responsible for activating both methionine isoacceptor tRNAs.

The functional discrimination between the two isoacceptor tRNAs is accomplished by protein initiation and elongation factors. Initiation factors recognize $tRNA_i^{Met}$ and elongation factors recognize $tRNA^{Met}$. Clearly, $tRNA_f^{Met}$ and $tRNA^{Met}$ must possess structural features that allow them to be distinguished from all other tRNAs by the single methionyl-tRNA synthase, but they must be sufficiently different that they can be discriminated by the factors.

In prokaryotic systems, a specific formylating enzyme exists that can recognize $tRNA_f^{Met}$ and formylate its amino terminus utilizing N^{10}-formyltetrahydrofolate as the formyl donor. This reaction serves to ensure that the initiator does not participate in the elongation reactions. This recognition step was lost during evolution and is not possessed by eukaryotic systems.

Translation Begins with the Binding of mRNA to the Ribosome

One of the major demands of protein synthesis is to select the appropriate initiating codon, generally AUG, for translation. This is accomplished at the level of the ribosome by the binding of the small ribosomal subunit to mRNA. The recognition of the appropriate start codon is accomplished in different ways in prokaryotes and eukaryotes. This fundamental difference in initiating mechanism is responsible for the fact, noted earlier, that each eukaryotic mRNA generally encodes only one protein, while single prokaryotic mRNAs frequently encode more than one.

In eukaryotes, the AUG sequence closest to the 5′ end of the mRNA usually serves as the start codon for the single protein encoded by each mRNA. The small ribosomal subunit binds at the 5′ end of the mRNA and moves along the strand until it encounters an AUG sequence that is recognized by base pairing with the anticodon of Met-$tRNA_i^{Met}$. In higher eukaryotes, but not in yeast, the recognition of this initiating AUG is facilitated by an initiation context of flanking bases. The preferred initiation context is GCCGCCpurCCAUGG. How this sequence is recognized is unknown, but when the initiation context of the first AUG departs from this sequence the 40S subunit can bypass it and initiate at the next AUG downstream.

In prokaryotic systems, the initiating AUG codon may occur at any point within the mRNA and there may be more than one start site within the same mRNA. How do prokaryotic ribosomes select the appropriate initiating codon from the much more abundant internal AUG sequences? The answer was suggested when Shine and Dalgarno, in the early 1970s, noticed that bacterial mRNAs generally contain a complementary purine-rich region (which has become known as the Shine-Dalgarno sequence) centered approximately ten bases toward the 5′ side of the initiating AUG sequence and that *E. coli* 16S RNA contains a seven-base pyrimidine-rich sequence near its 3′ terminus (table 29.5). They proposed that base pairing between these complementary sequences could serve to align the initiating AUG for decoding. Such base pairing has now been demonstrated as the major mechanism of initiation codon discrimination in *E. coli.* Thus mutations in the Shine-Dalgarno sequence of mRNA that improve pairing enhance translation and mutations that decrease pairing decrease translation. Accordingly, the most efficiently translated mRNAs in *E. coli* are those with the strongest Shine-Dalgarno pairing.

The importance of the Shine-Dalgarno sequence is graphically depicted by the action of the bacterial toxin colecin E3. This toxin inactivates the small subunit of the prokaryotic ribosome by specific nuclease cleavage about 50 residues from the 3′ terminus of 16S rRNA. The cleavage disrupts the sequence that is complementary to the Shine-Dalgarno sequence and thus specifically inhibits the initiation process. Because of the fundamental differences between prokaryotic and eukaryotic initiation just described, colecin E3 does not inhibit the eukaryotic ribosome.

29B BOX

Discrimination of $tRNA^{Ala}$ by Its Synthase

The recognition of $tRNA^{Ala}$ by its synthase has been extensively studied by Paul Schimmel and his colleagues. These investigators have systematically altered the structure of the substrate RNA and created ever-smaller versions of it in order to determine the critical features of its structure that allow the enzyme to recognize it. The features identified in this way have turned out to be surprisingly simple. A single G3:U70 pair in the acceptor stem appears to be a major determinate in the discrimination of this tRNA by its synthase. This relationship was demonstrated in several ways. First, many alterations, except in this single pair, had little or no effect on the recognition of the tRNA by the synthase. Second, substituting this single pair into $tRNA^{Cys}$, which differs in sequence from $tRNA^{Ala}$ at forty other locations, allowed its recognition by alanyl-tRNA synthase. Finally, the synthase was able to acylate a microhelix (a stem and loop containing only seven base pairs patterned after the sequence of the acceptor stem of $tRNA^{Ala}$), provided that it contained the crucial G:U pair. The structure of this synthase-tRNA complex has not been determined, but the crucial recognition, in contrast to that of $tRNA^{Gln}$, appears simple and limited to only a small region of the acceptor stem.

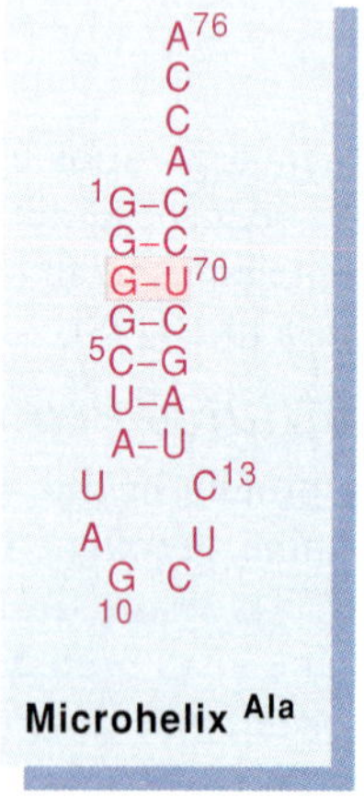

Dissociable Protein Factors Play Key Roles at the Different Stages in Protein Synthesis on the Ribosome

At each stage in protein synthesis on the ribosome—initiation, elongation and termination—a different set of protein factors is engaged by the ribosome. Why do such protein factors, which are crucial to the translation, exist separate from the ribosome? Why must they cycle on and off the ribosome, and why are their functions not performed by firmly bound ribosomal proteins? The answers to these questions appear to lie in the economy of structure that is afforded by the cycling strategy. Since factors cannot interact simultaneously with the ribosome, it makes sense for the ribosome to interact with them sequentially at a single site.

Table 29.5
The Sequence of the 3′ End of *E. coli* 16S RNA and Some Shine-Dalgarno Sequences at the 5′ End of Bacterial mRNAs

	The pyrimidine-rich complement to the Shine-Dalgarno sequence
16S rRNA	3′ · · · HOAUUCCUCCACUA · · · 5′
lacZ mRNA	5′ · · · ACACAGGAAACAGCUAUG · · · 3′
trpA mRNA	5′ · · · ACGAGGGGAAAUCUGAUG · · · 3′
RNA polymerase β mRNA	5′ · · · GAGCUGAGGAACCCUAUG · · · 3′
r-Protein L10 mRNA	5′ · · · CCAGGAGCAAAGCUAAUG · · · 3′
	The purine-rich Shine-Dalgarno sequence; The initiation codon

Protein Factors Aid Initiation

Even though specific differences distinguish the initiation process in eukaryotes and prokaryotes, three things must be accomplished to initiate protein synthesis in all systems: (1) the small ribosomal subunit must bind the initiator tRNA, (2) the appropriate initiating codon on mRNA must be located, and (3) the large ribosomal subunit must associate with the complex of the small subunit, the initiating tRNA and mRNA. Nonribosomal proteins, known as initiation factors (IFs), participate in each of these three processes. IFs interact transiently with a ribosome during initiation and thus differ from ribosomal proteins, which remain continuously associated with the same ribosome.

E. coli has three initiation factors (fig. 29.13) bound to a small pool of 30S ribosomal subunits. One of these factors, IF-3, serves to hold the 30S and 50S subunits apart after termination of a previous round of protein synthesis. The other two factors, IF-1 and IF-2, promote the binding of both fMet-$tRNA_i^{Met}$ and mRNA to the 30S subunit. As we noted before,

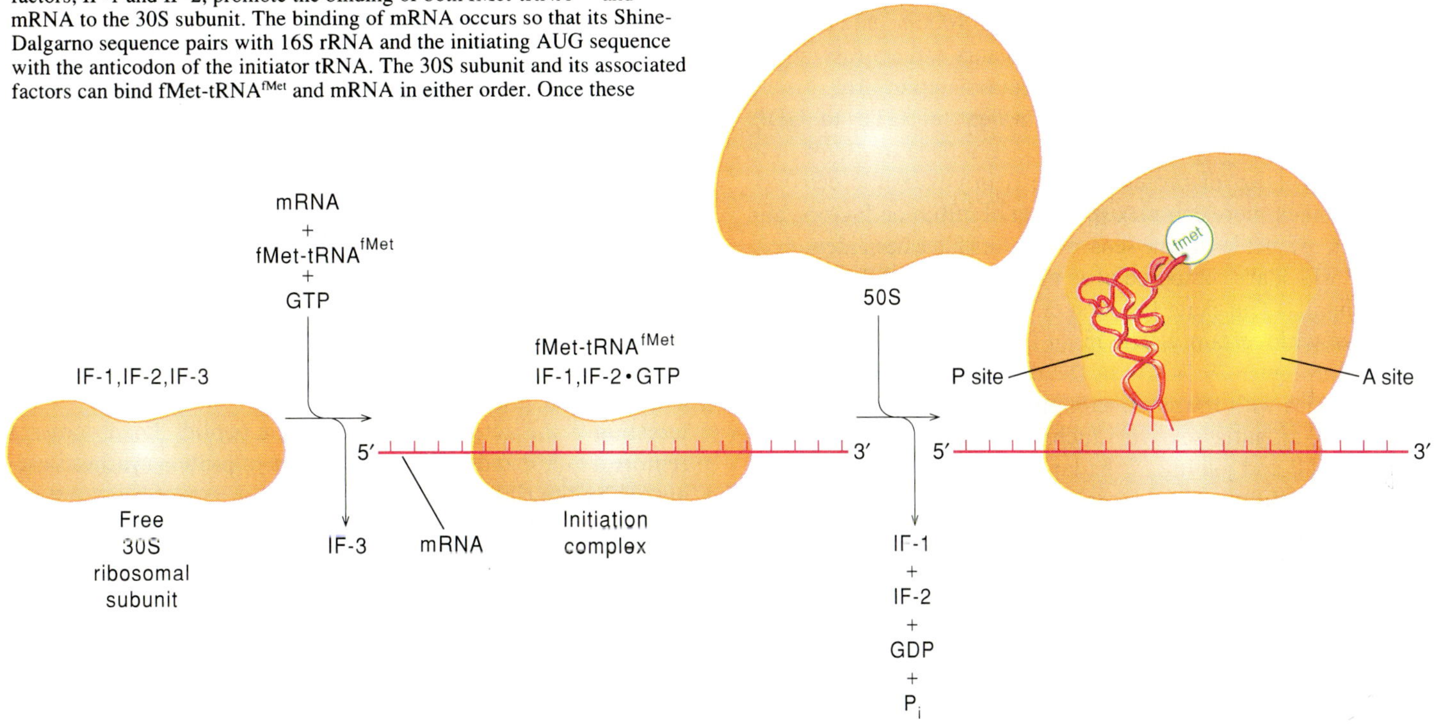

Figure 29.13

Formation of the initiation complex for protein synthesis in prokaryotes. *E. coli* has three initiation factors bound to a pool of 30S ribosomal subunits. One of these factors, IF-3, holds the 30S and 50S subunits apart after termination of a previous round of protein synthesis. The other two factors, IF-1 and IF-2, promote the binding of both fMet-tRNA^{fMet} and mRNA to the 30S subunit. The binding of mRNA occurs so that its Shine-Dalgarno sequence pairs with 16S rRNA and the initiating AUG sequence with the anticodon of the initiator tRNA. The 30S subunit and its associated factors can bind fMet-tRNA^{fMet} and mRNA in either order. Once these ligands are bound, IF-3 dissociates from the 30S subunit, permitting the 50S subunit to join the complex. This releases the remaining initiation factors and hydrolyzes the GTP, which is bound to IF-2.

Figure 29.14

Formation of the initiation complex for protein synthesis in eukaryotes. The reaction begins with the small subunit held apart from the large subunit by an antiassociation factor and ends with the hydrolysis of GTP and joining of the large subunit as in prokaryotes. The intervening reactions are different. A much more complex spectrum of initiation factors (eIFs) is involved, and the exact function of only a few of these factors is known with certainty. The Met-tRNA^{fMet} first binds to the small subunit as a ternary complex with the eIF-2 and GTP. This ternary complex then binds to the 5′ end of mRNA with the aid of several factors, one of which, eIF-4F, contains a subunit that specifically binds to the terminal cap structure of the mRNA. Binding to mRNA is followed by a scanning reaction that moves the small subunit along the mRNA, usually to the first AUG, in a reaction driven by the hydrolysis of ATP to ADP and P_i.

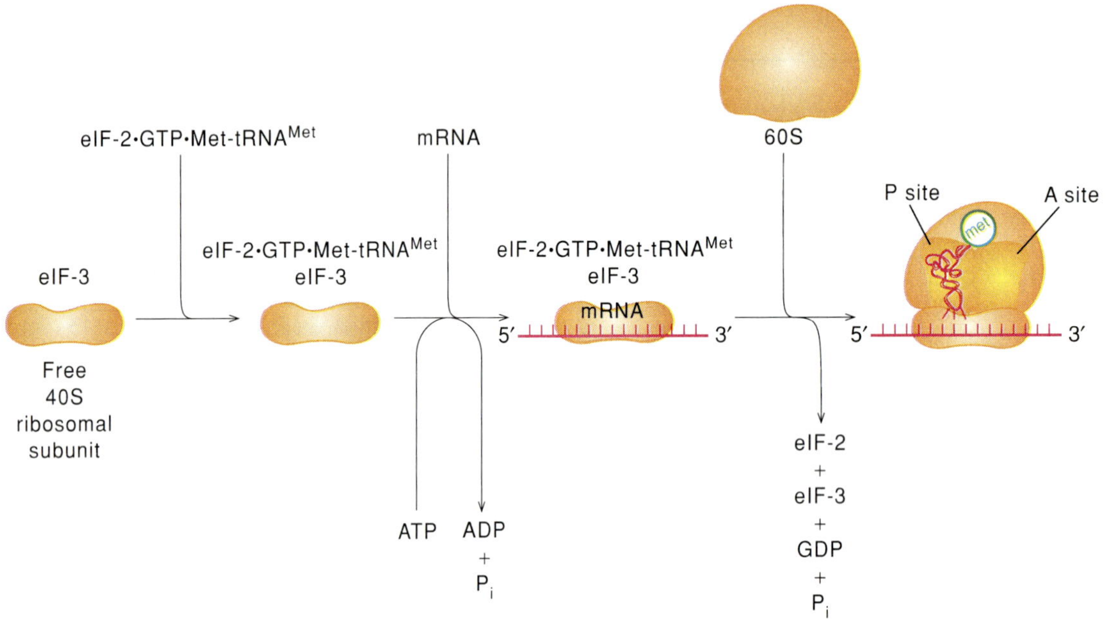

the binding of mRNA occurs so that its Shine-Dalgarno sequence pairs with 16S RNA and the initiating AUG sequence with the anticodon of the initiator tRNA, although this latter interaction has not been rigorously proven. The 30S subunit and its associated factors can apparently bind fMet-tRNA$_i^{Met}$ and mRNA in either order. Once these ligands are bound, IF-3 dissociates from the 30S, permitting the 50S to join the complex. This releases the remaining initiation factors and hydrolyzes the GTP that is bound to IF-2. The initiation step in prokaryotes requires the hydrolysis of one equivalent of GTP to GDP and P_i.

Initiation of protein synthesis in eukaryotic cells requires a much more complex spectrum of initiation factors, abbreviated eIF. At least nine separate factors have been identified, some of which are composed of as many as eleven different peptide subunits. The exact function of only a few of these factors is known with certainty. The main features of the process are outlined in figure 29.14. As in the prokaryotic system, the reaction begins with the small subunit held apart from the large subunit by an antiassociation factor and ends with the hydrolysis of GTP and joining of the large subunit. The intervening reactions are different in prokaryotes and eukaryotes. The Met-tRNA$_i^{Met}$ first binds to the small subunit as a ternary complex with eIF-2 and GTP. The resulting preinitiation complex then binds to the 5′ end of mRNA with the aid of several factors, one of which, eIF-4F, contains a subunit that specifically binds to the terminal cap structure of mRNA. Binding to mRNA is followed by a scanning reaction that moves the small subunit along the mRNA, usually to the first AUG, in a reaction driven by the hydrolysis of ATP to ADP and P_i. The mRNA binding and scanning reactions, which have no counterpart in the prokaryotic system, position the ribosome on the initiating AUG sequence in the appropriate flanking nucleotide sequence.

Three Elongation Reactions Are Repeated with the Incorporation of Each Amino Acid

At the conclusion of the initiation process, the ribosome is poised to translate the reading frame associated with the initiation codon. The translation of the contiguous codons in mRNA is accomplished by the sequential repetition of three reactions with each amino acid. These three reactions of elongation are similar in both prokaryotic and eukaryotic systems; two of them require nonribosomal proteins known as elongation factors (EF). Interestingly, the actual formation of the peptide bond does not require a factor and is the only reaction of protein synthesis catalyzed by the ribosome itself.

The elongation reactions begin with the binding of the aminoacyl-tRNA specified by the codon immediately adjacent to the initiating codon. The binding of this aminoacyl-tRNA is

Figure 29.15

Addition of the second aminoacyl-tRNA to the ribosome complex and the accompanying EF-Tu, EF-Ts cycle in *E. coli*. The purpose of the cycle is to regenerate another protein aminoacyl-tRNA complex suitable for transferring further aminoacyl-tRNAs to the A site on the ribosome.

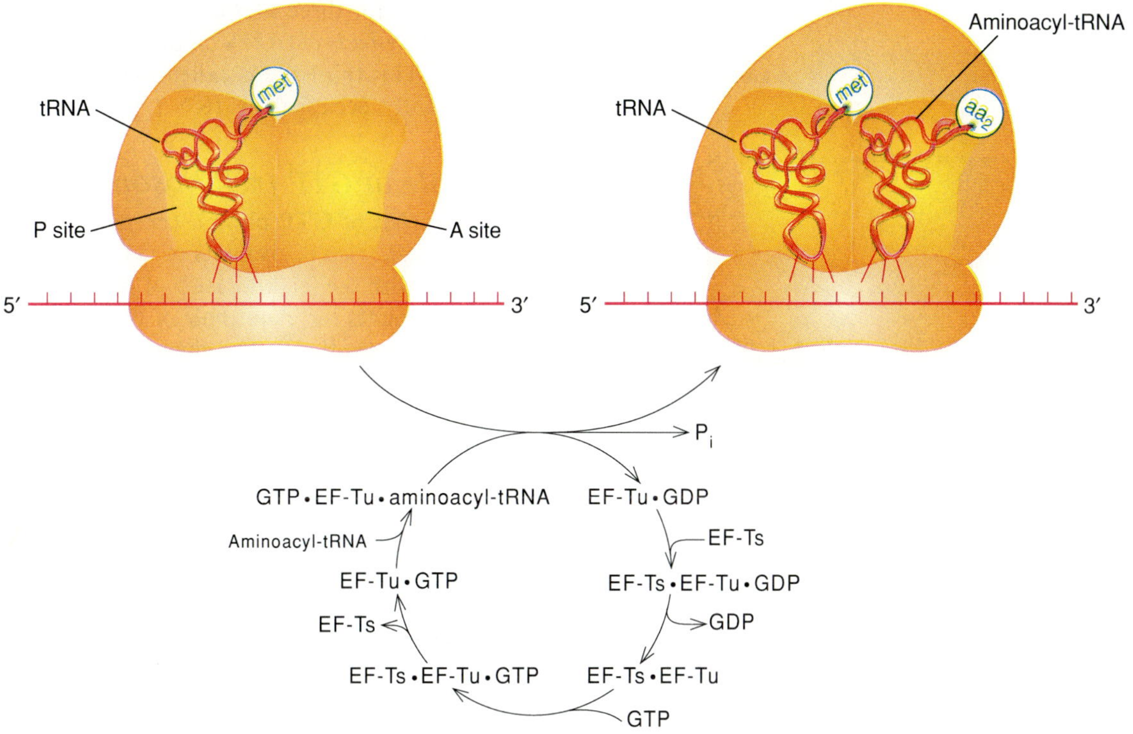

catalyzed by an aminoacyl-tRNA binding factor, designated EF-Tu in bacteria and EF-1 in eukaryotic systems. The factor interacts with the ribosome as a ternary complex bound to both aminoacyl-tRNA and GTP. The binding of this complex to the ribosome is coupled to GTP hydrolysis. A productive complex forms if, and only if, the anticodon of the tRNA in the complex is complementary to the codon bound to the A site on the ribosome. Following the binding of aminoacyl-tRNA, EF-Tu is released from the ribosome as a complex with GDP. A second elongation factor, EF-Ts, catalyzes the regeneration of the EF-Tu-GTP complex so that it can again bind aminoacyl-tRNA. The series of reactions involved in this regeneration is depicted in figure 29.15. EF-1 is a multisubunit protein that combines the functional properties of EF-Tu and EF-Ts.

Peptide bond formation occurs immediately following the dissociation of the binding factor from the ribosome. This reaction is known as transpeptidation and the enzymatic center that catalyzes it is known as peptidyl transferase, although it promotes the conversion of an ester to a peptide bond. Transpeptidation is generally assumed to proceed by nucleophilic attack by the amino group of the incoming aminoacyl-tRNA on the carbonyl of the ester of peptidyl-tRNA with the formation of a tetrahedral intermediate (fig. 29.16).*

*H. F. Noller, V. Hossarth, and L. Zimniak, Ribosomes stripped of all their protein can still catalyze transpeptidation. *Science* 256:14–16, 1992.

The final step in elongation is known as translocation (fig. 29.17). This reaction, like aminoacyl-tRNA binding, is catalyzed by a factor (the translocation factor, known as EF-G in prokaryotic systems and EF-2 in eukaryotic systems) that cycles on and off the ribosome and hydrolyzes GTP in the process. The overall purpose of translocation is to physically move the ribosome along the mRNA to expose the next codon for translation. The mechanical basis of this movement reaction has long eluded understanding. Recent evidence suggests that translocation occurs in two discrete steps that may involve the relative movement of the two ribosomal subunits.

The structural differences between bacterial and eukaryotic elongation factors are highlighted by the selective action that diphtheria has on eukaryotic systems (box 29C).

The Ribosome Binds tRNA at Three Sites

The original model of ribosome function, proposed over 25 years ago by James D. Watson, pictured the ribosome as having two sites of interaction with tRNA. One site was called the P site, because it was visualized as holding the growing peptide chain. The second site was called the A site, because it was visualized as holding the aminoacyl-tRNA prior to peptide bond formation. This two-site model is the minimal stochastic model of ribosome function compatible with the stepwise addition of a single

Figure 29.16

Formation of the first peptide linkage. The formylmethionine group is transferred from its tRNA at the P site to the amino group of the second aminoacyl-tRNA at the A site on the ribosome. This involves nucleophilic attack by the amino group of the second amino acid on the carboxyl carbon of the methionine. The resulting bond formation attaches both amino acids to the tRNA at the A site.

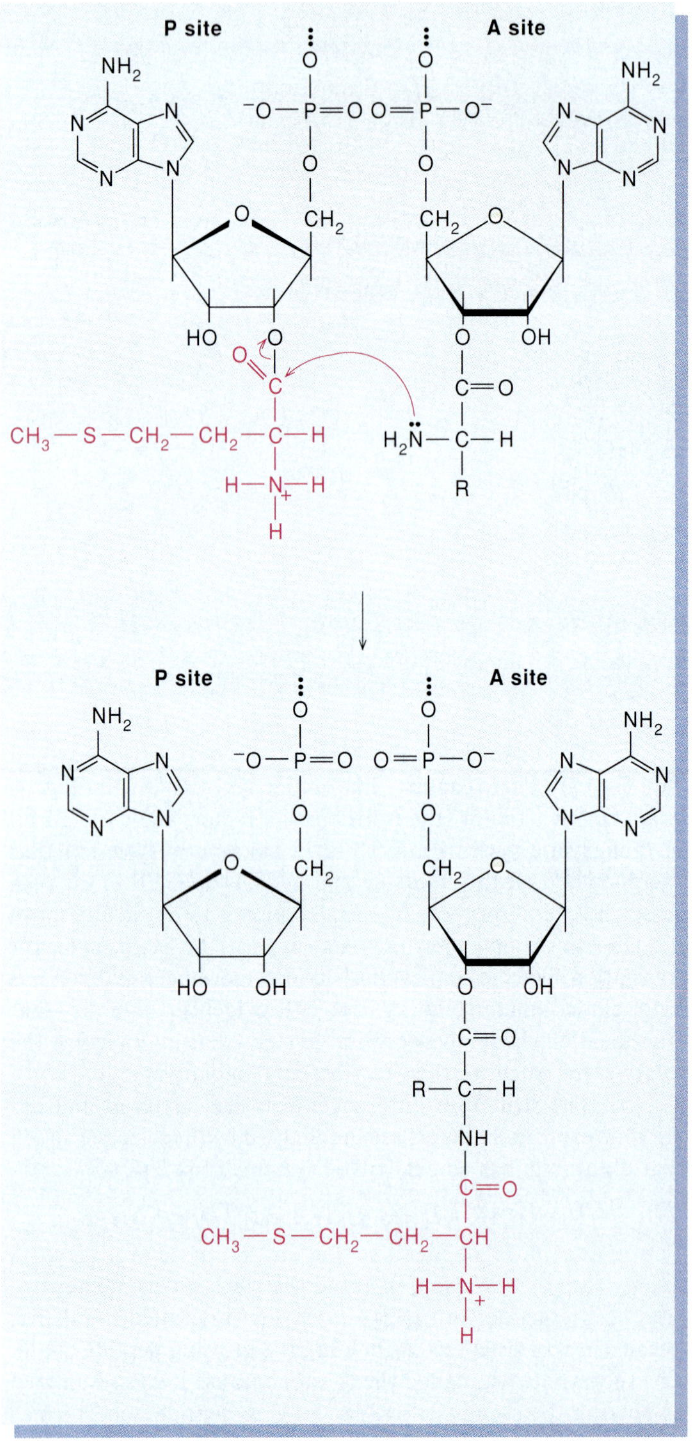

amino acid from a tRNA to a growing chain that is also attached to a tRNA. The model has been vigorously debated over the ensuing years but has resisted attack, largely because of the difficulty of precisely measuring the nature and location of tRNA binding to the ribosome. Recently the two-site model has given way to compelling evidence obtained by Danesh Moazed and Harry Noller for three sites of tRNA binding to the ribosome, based on the effects of tRNA binding on the chemical reactivity of rRNA. This third site is called the E site and is believed to be transiently occupied by tRNA prior to its exit from the ribosome.

The Role of GTP in Ribosomal Reactions

The hydrolysis of GTP plays a conspicuous role in the translation process. Two equivalents of GTP are hydrolyzed during elongation with the incorporation of each amino acid. This hydrolysis accounts for about half of the total energy consumed during protein synthesis. The chemical and functional purposes of GTP hydrolysis are best understood in the case of *E. coli* EF-Tu and EF-G. The sites of GTP binding and hydrolysis are contained entirely on these factors. The interaction of the factor-GTP complex with the ribosome is believed to activate the hydrolytic site. Activation can also be achieved in other ways; hydrolysis by EF-Tu is stimulated by the antibiotic kirromycin and hydrolysis by EF-G is promoted by various organic solvents.

GTP hydrolysis leads to inversion of configuration of the released P_i and is believed to occur by direct attack of water on the γ phosphorus of GTP through an in-line displacement mechanism. There are no covalent intermediates in this reaction. Rather, the change of bound ligand to GDP that results from hydrolysis and release of P_i is believed to change the conformation of the factor. The factor when bound to GTP is thought to be in the "on" configuration, able to bind to the ribosome and through binding to cause either aminoacyl-tRNA binding or translocation. The binding of GDP to the factor puts it in the "off" configuration and causes it to dissociate from the ribosome. Hydrolysis itself is not required for these changes to occur. This point is demonstrated by the fact that the nonhydrolyzable analog of GTP, GMPPCP, containing a methylene bridge instead of an oxygen between the β and γ phosphorus, can substitute for it in the reactions catalyzed by the elongation factors (see fig. 29.18). The use of this analog slows these reactions by delaying the dissociation of the factors from the ribosome.

The translation factors that interact with GTP are members of the G protein superfamily, which includes the signal-transducing G proteins that link membrane receptors with their intracellular targets (see chapter 24) and the Ras proteins that function as growth regulators (see chapter 34). All G proteins bind and hydrolyze GTP and it is believed that they all act by the same general mechanism we have just described. Thus when bound to GTP these proteins are in their "active" configuration and when bound to GDP as a result of hydrolysis they are converted to their "inactive" configuration.

Figure 29.17

The translocation reaction in *E. coli*. The translocation reaction occurs immediately after peptide synthesis. It involves displacement of the discharged tRNA from the P site and concerted movement of the peptidyl-tRNA and mRNA so that the peptidyl-tRNA is bound to the P site and the same three nucleotides in the mRNA. The A site is vacated and ready for the addition of another aminoacyl-tRNA. Translocation in eukaryotes is similar except that the EF-2 factor is involved instead of the EF-G factor.

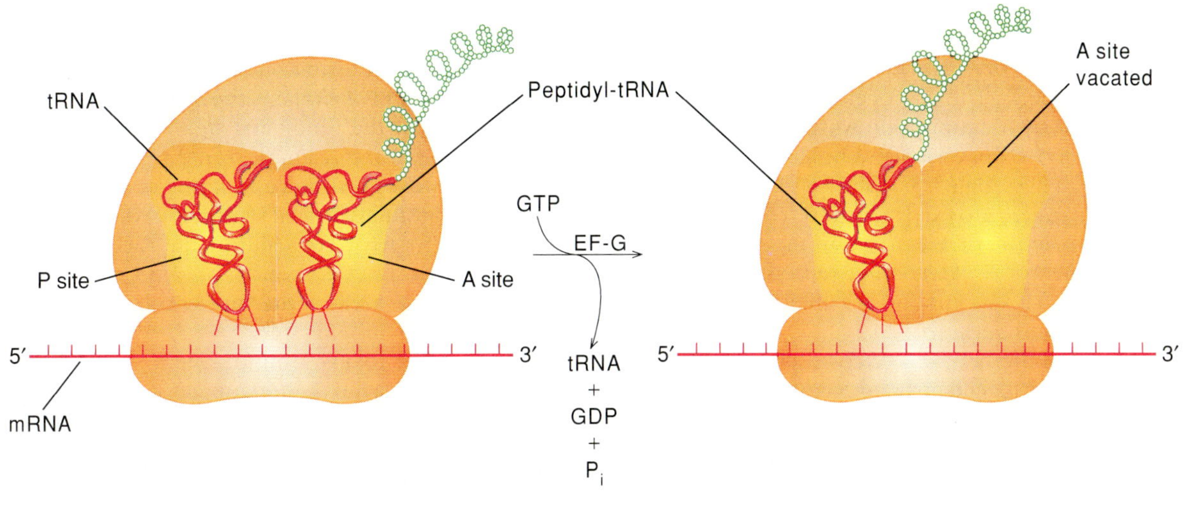

Figure 29.18

The structure of guanylyl methylene diphosphonate (GMPPCP).

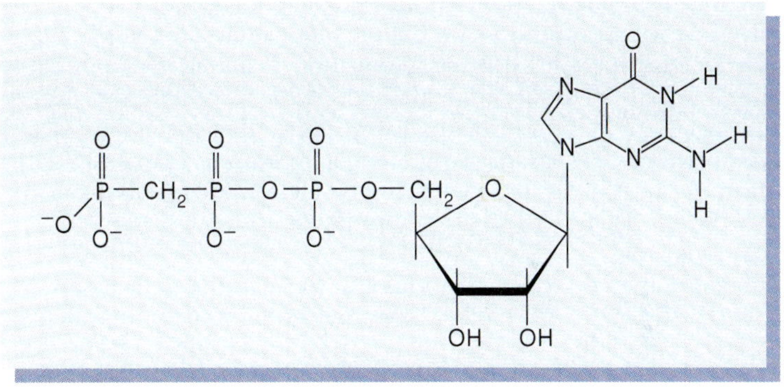

Termination of Translation Requires Release Factors and Termination Codons

The last step in translation involves the cleavage of the ester bond that joins the now complete peptide chain to the tRNA corresponding to its C-terminal amino acid (fig. 29.19). This process of termination, in addition to the termination codon, requires release factors (RFs). The freeing of the ribosome from mRNA during this step requires the participation of a protein called ribosome releasing factor (RRF).

Cells usually do not contain tRNAs that can recognize the three termination codons. In *E. coli,* when these codons arrive in the A site on the ribosome they are recognized by one of three release factors. RF-1 recognizes UAA and UAG, while RF-2 recognizes UAA and UGA. The third release factor, RF-3, does not itself recognize termination codons but stimulates the activity of the other two factors.

The consequence of release factor recognition of a termination codon in the A site is to alter the peptidyl transferase center on the large ribosomal subunit so that it can accept water as the attacking nucleophile rather than requiring the normal substrate, aminoacyl-tRNA (fig. 29.20). In other words, the termination reaction serves to convert the peptidyl transferase into an esterase. This feature of the termination reaction is clearly seen in the simple *in vitro* reaction that occurs when *E. coli* ribosomes are combined with fMet-$tRNA_i^{Met}$, RF-1, and two separate nucleotide triplets, AUG and UAA. Formylmethionine is produced by hydrolysis and this reaction is specifically inhibited by antibiotics, such as sparsomycin, that inhibit peptidyl transferase.

29C

BOX

Mechanisms of Damage Produced by Certain Toxins

A single exotoxin, diphtheria toxin, is responsible for the pathogenesis of *Corynebacterium diphtheriae* and the disease diphtheria. The pathogenic consequences can be prevented by immunization with toxoid, an inactivated form of the purified toxin. Curiously, the structural gene for the toxin is carried by a bacterial virus, called β phage, that must infect the bacterium to induce toxin production. The widespread immunization against diphtheria employed in the United States has caused β phage, but not *C. diphtheriae,* to largely disappear. A catalytically identical but structurally very different exotoxin is produced by *Pseudomonas aeruginosa.*

In the cytoplasm of the cell, the catalytic portion of diphtheria toxin acts as a very specific protein-modifying enzyme. It catalyzes the ADP-ribosylation and consequent inactivation of EF-2 by the following reaction:

$$\text{EF-2} + \text{NAD}^+ \rightarrow \text{ADP-ribosyl-EF-2} + \text{nicotinamide} + \text{H}^+$$

This reaction is reversible when conducted *in vitro,* but under the conditions of pH and nicotinamide concentration that exist in the cell, it is irreversible. Thus diphtheria toxin kills cells by irreversibly destroying the ability of EF-2 to participate in the translocation step of protein synthesis elongation. A number of other protein toxins have subsequently been found to ADP-ribosylate and inactivate cellular proteins involved in other essential cellular pathways. For example, cholera and pertussis toxins ADP-ribosylate and inactivate proteins important to cAMP metabolism.

The enzymatic specificity of diphtheria toxin deserves special comment. The toxin will ADP-ribosylate EF-2 in all eukaryotic cells *in vitro* whether or not they are sensitive to the toxin *in vivo,* but it will not modify any other protein, including the bacterial counterpart of EF-2. This narrow enzymatic specificity has called attention to an unusual posttranslational derivative of histidine, diphthamide, that occurs in EF-2 at the site of ADP-ribosylation (see illustration). While the unique occurrence of diphthamide in EF-2 explains the specificity of the toxin, it raises questions about the functional significance of this modification in translocation. Interestingly, some mutants of eukaryotic cells selected for toxin resistance lack one of several enzymes necessary for the posttranslational synthesis of diphthamide in EF-2 that is necessary for toxin recognition, but these cells seem perfectly competent in protein synthesis. Thus the *raison d'être* of diphthamide, as well as the biological origin of the toxin that modifies it, remains a mystery.

Some other toxins inhibit protein synthesis by inactivating the ribosome through alterations of rRNA. A fungal toxin, α-sarcin, inactivates the ribosome by specific nuclease cleavage of a single phosphodiester bond in a purine-rich region found in the 23–28S RNA of all ribosomes. A second group of toxins, known as the ribosome-inactivating proteins, is abundant in plants of many species. The best-known example of this type of toxin is ricin, the toxic agent in the castor bean. These proteins all act by the curious mechanism of removing a single adenine residue, by N-glycolytic cleavage, from the site in rRNA adjacent to the one that is cleaved by α-sarcin. Both of these modifications of rRNA alter the ribosome's ability to interact with factors, a fact suggesting that this region of rRNA is part of the factor interaction site on the large subunit. One of the ribosome-inactivating proteins, trichosanthin, was originally isolated from a plant tuber as the active principle of an Oriental folk medicine used to induce abortions. More recently, trichosanthin has been found to selectively inhibit viral protein synthesis in HIV-infected T-cells, and it is now undergoing clinical trials as a therapeutic agent for the treatment for AIDS. The biological basis of this selective inhibition is unknown.

Ribosomes Can Change Reading Frame and Jump during Translation

Normally, ribosomes march in lock step down the mRNA, one codon at a time, beginning at the initiating AUG and ending at a termination codon. In this way translation is rigidly maintained in a single reading frame. In the last few years, a number of specific exceptions to this seemingly rigid rule have been discovered in both prokaryotic and eukaryotic systems. In these cases, the correct expression of specific messages requires the ribosome to violate the rule of lock-step translation. One way that this occurs is known as translational frameshifting. Frameshifting generally occurs only one base at a time, but this change or slipping can be in either the $+1$ or -1 direction. Another violation of the rule of lock-step translation involves the failure of the ribosome to translate a substantial portion of mRNA, by skipping as many as 50 bases downstream in mRNA. This process is known as translational jumping. From an informational point of view, translational jumping is analogous to splicing in mRNA except that the information is not removed but is simply ignored by the ribosome. Both translational frameshifting and

Figure 1

Inactivation of the EF-2 factor by diphtheria toxin through the ADP-ribosylation of a modified histidine side chain.

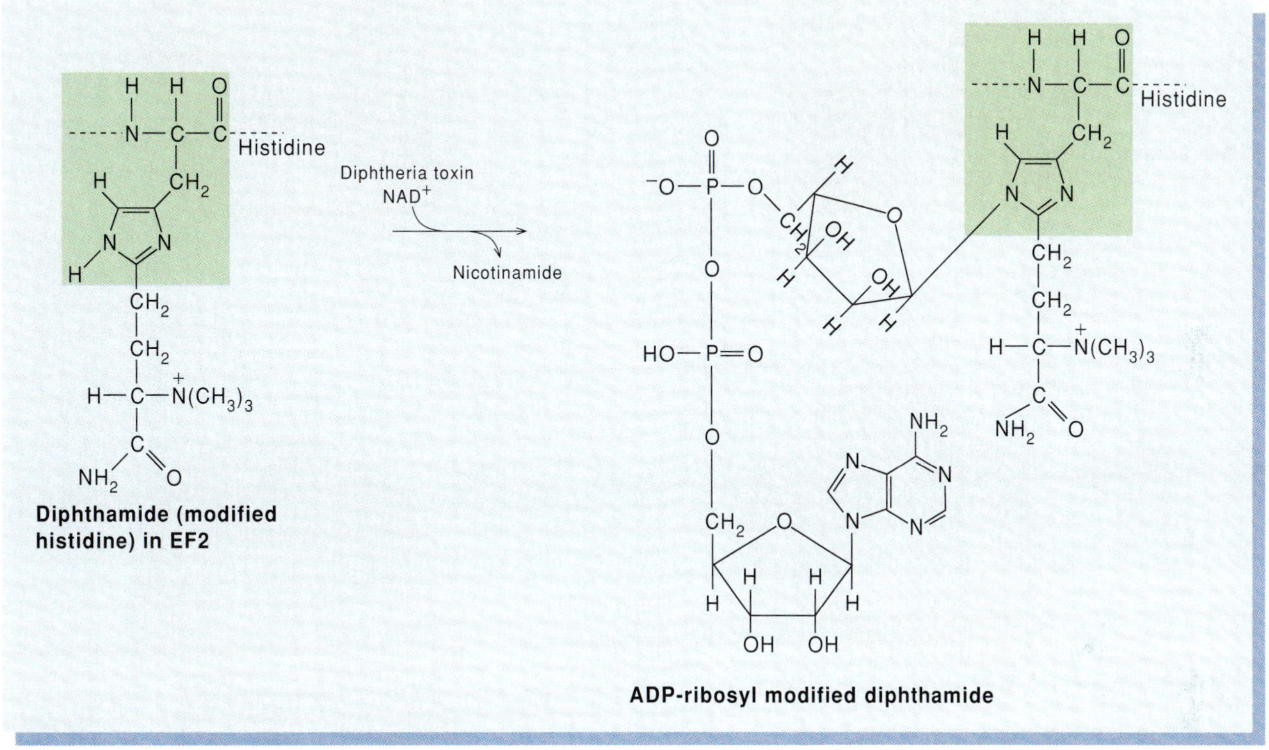

translational jumping occur at specific sites in mRNA and in some cases are known to require the presence of specific sequences and specific structures in mRNA. While the nature of these sequences and structures have been identified in several cases, the mechanisms by which they are read have not been defined.

A notable instance of frameshifting occurs in the gene for the release factor RF-2, which recognizes the UGA stop codon in *E. coli.* This gene does not make RF-2 unless a frameshift takes place at an in-frame UGA sequence. When there is an abundance of the release factor, the frameshift is unlikely to occur because release occurs when the ribosome reaches this point on the messenger. However, when the release factor is in short supply, the ribosome is likely to pause at this site and then undergo a frameshift to make functional release factor. In this example, frameshifting results in a negative feedback mechanism for regulating the amount of release factor that is synthesized.

Translational fidelity of bacterial ribosomes is influenced in a more general way by the binding of antibiotics to the ribosome (box 29D).

Figure 29.19

The release reaction in *E. coli*. The release reaction occurs when the codon adjacent to the anticodon-codon complex is one of the stop codons, e.g., UAA. The stop is recognized by release factor proteins that cause the peptidyl transferase to transfer the nascent polypeptide to water, forming a free polypeptide. Following the release of the polypeptide, the final tRNA and the mRNA dissociate from the ribosome, and the ribosome dissociates into its constituent subunits.

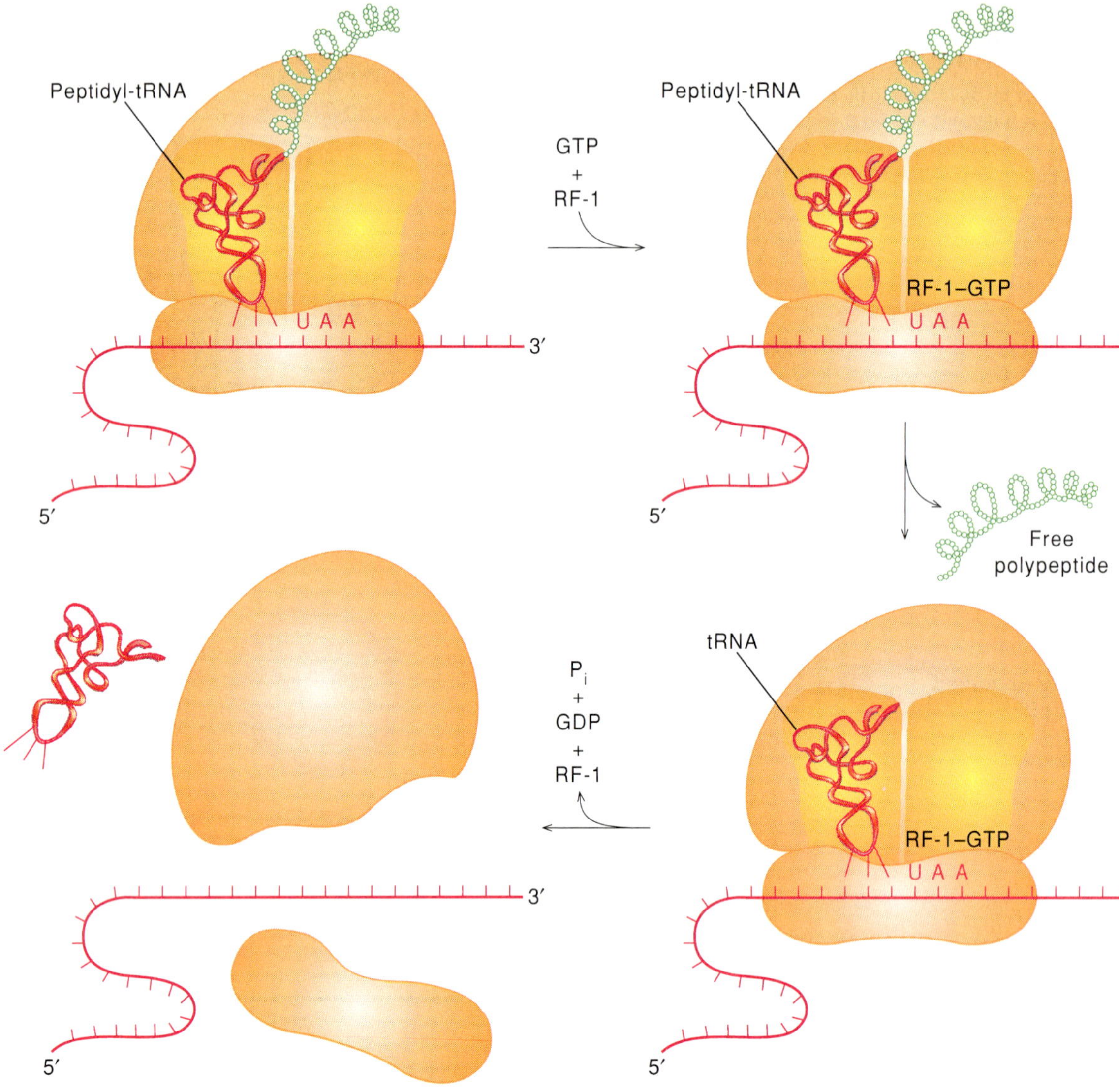

Targeting and Posttranslational Modification of Proteins

Despite the fact that only twenty amino acids (plus selenocysteine and formylmethionine in prokaryotic systems) are known to be directly specified by the genetic code, chemical analysis of mature proteins has revealed hundreds of different amino acids, all of them structural variants on the original twenty. This structural diversity, which greatly expands the chemical lexicon of proteins, results from posttranslational modification of the primary products of translation. Our knowledge of the nature and significance of enzymatic reactions that bring about these important alterations is still very incomplete.

In addition to the modification of amino acid side chains there are many cases where parts of the originally synthesized polypeptide chain are removed during the process of maturation. The types of modification and processing that polypeptide chains undergo is strongly related to the site of protein synthesis and to the mechanisms that are involved in targeting the polypeptide chain to its final destination. In fact, processing frequently begins during polypeptide synthesis and continues for some time thereafter. A major division involving the types of processing reactions that polypeptide chains undergo is related to the site of protein synthesis and the final destination of the protein. Recall (see fig. 5.1) that proteins synthesized on free polysomes either remain in the cytosol or are targeted to the

Figure 29.20

An *in vitro* assay for release factors.

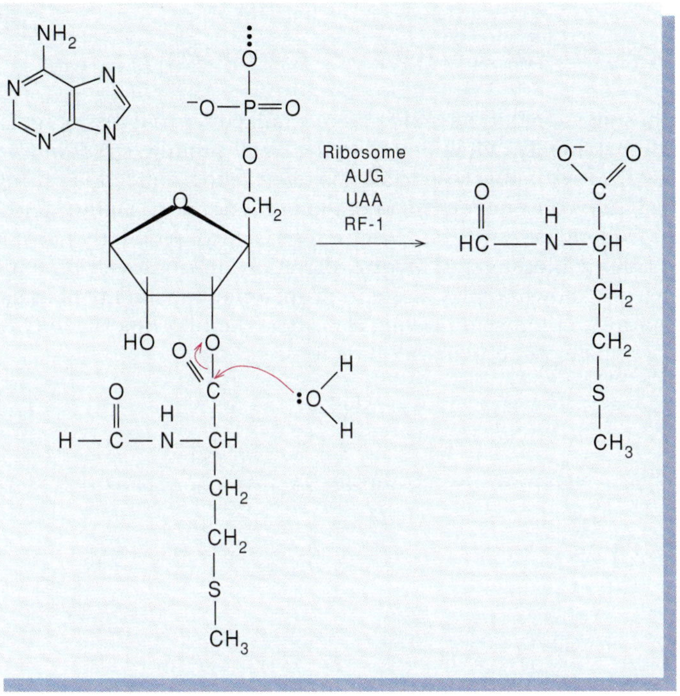

Figure 29.21

Biosynthesis of insulin. Insulin is synthesized by membrane-bound polysomes in the ß cells of the pancreas. The primary translation product is preproinsulin, which contains a 24-residue signal peptide preceding the 81-residue proinsulin molecule. The signal peptide is removed by signal peptidase, cutting between Ala (−1) and Phe (+1), as the nascent chain is transported into the lumen of the endoplasmic reticulum. Proinsulin folds and two disulfide bonds cross-link the ends of the molecule as shown. Before secretion, a trypsin-like enzyme cleaves after a pair of basic residues 31, 32 and 59, 60; then a carboxypeptidase B-like enzyme removes these basic residues to generate the mature form of insulin.

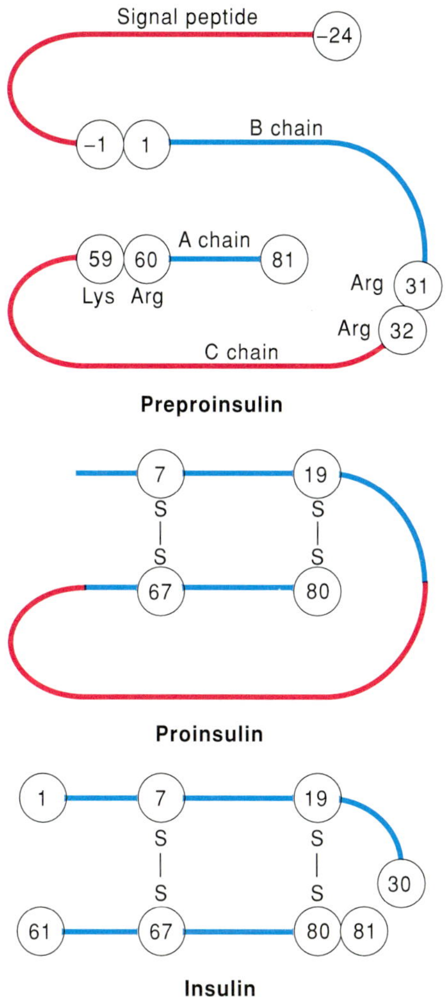

mitochondria, the chloroplasts, or the nucleus. Those that will remain in the cytosol are often ready to function as soon as they are released from the ribosome. Those that must be transported to another site usually undergo substantial modification during the transport process. Many proteins, however, are synthesized on membrane-bound polysomes rather than on free polysomes. In bacteria such proteins are synthesized on polysomes associated with the inner plasma membrane; in eukaryotes, they are synthesized on polysomes associated with the endoplasmic reticulum. Proteins that are synthesized on ribosomes bound to the endoplasmic reticulum either remain in the ER or are targeted to the Golgi, the secretory granules, the plasma membrane, or the lysosomes. In general, proteins that are synthesized on membrane-bound polysomes undergo more extensive modification before they reach their final destination and become fully functional (see chapter 21).

Proteins Are Targeted to Their Destination by Signal Sequences

Despite the apparent complexity of the problem, protein transport in all systems is accomplished by a single, rather simple underlying mechanism; each polypeptide destined for transport contains an amino acid sequence known as a signal or leader sequence that identifies the polypeptide to the appropriate transporting system. This generality was first recognized in the middle 1970s by Gunter Blobel, who articulated the underlying signal hypothesis (see fig. 21.9). Frequently, the signal sequence is cleaved from the parent polypeptide during the transport process. A protein containing a signal sequence that is cleaved upon transport is known as a preprotein. A protein containing a peptide sequence that must be removed for the protein to be active is known as a proprotein, and a protein that contains both of these sequences is known as a preproprotein.

Many protein hormones are synthesized in a preproprotein form. For example, insulin mRNA is translated into a single polypeptide chain, preproinsulin, that contains 84 residues of proinsulin plus a 23-residue signal peptide (fig. 29.21). Within the islets of Langerhans of the pancreas, the N-terminal sequence is removed cotranslationally in targeting proinsulin to the Golgi apparatus, and the two disulfides joining the ends of the molecule are formed. Following this, the

29D
BOX

Antibiotics Inhibit by Binding to Specific Sites on the Ribosome

Many antibiotics (generally, small organic compounds with therapeutic utility) prevent bacterial growth by inhibiting translation. This is not surprising, since translation is both a complex and metabolically essential process. Also, the bacterial ribosome is structurally distinct from the ribosome in the eukaryotic cytoplasm and thus specific bacterial inhibitors can be found.

A great many antibiotic inhibitors of ribosome function belong to the class known as aminoglycoside antibiotics. Of these, streptomycin is the best known and the best investigated. Streptomycin binding produces a variety of functional alterations in the ribosome. One of the first to be recognized was the loss of translational fidelity. When bound to the small subunit, streptomycin distorts its structure so as to allow altered codon–anticodon pairing and the consequent incorporation of incorrect amino acids. Indeed, mutations to streptomycin resistance frequently prevent antibiotic binding and involve alterations in ribosomal proteins that are known to play a role in maintaining the fidelity of translation. Streptomycin binding also alters the ribosome's ability to participate properly in the initiation reactions.

Figure 1

The structures of some antibiotic inhibitors of protein synthesis. All of the inhibitors shown function by binding to specific sites on the ribosome.

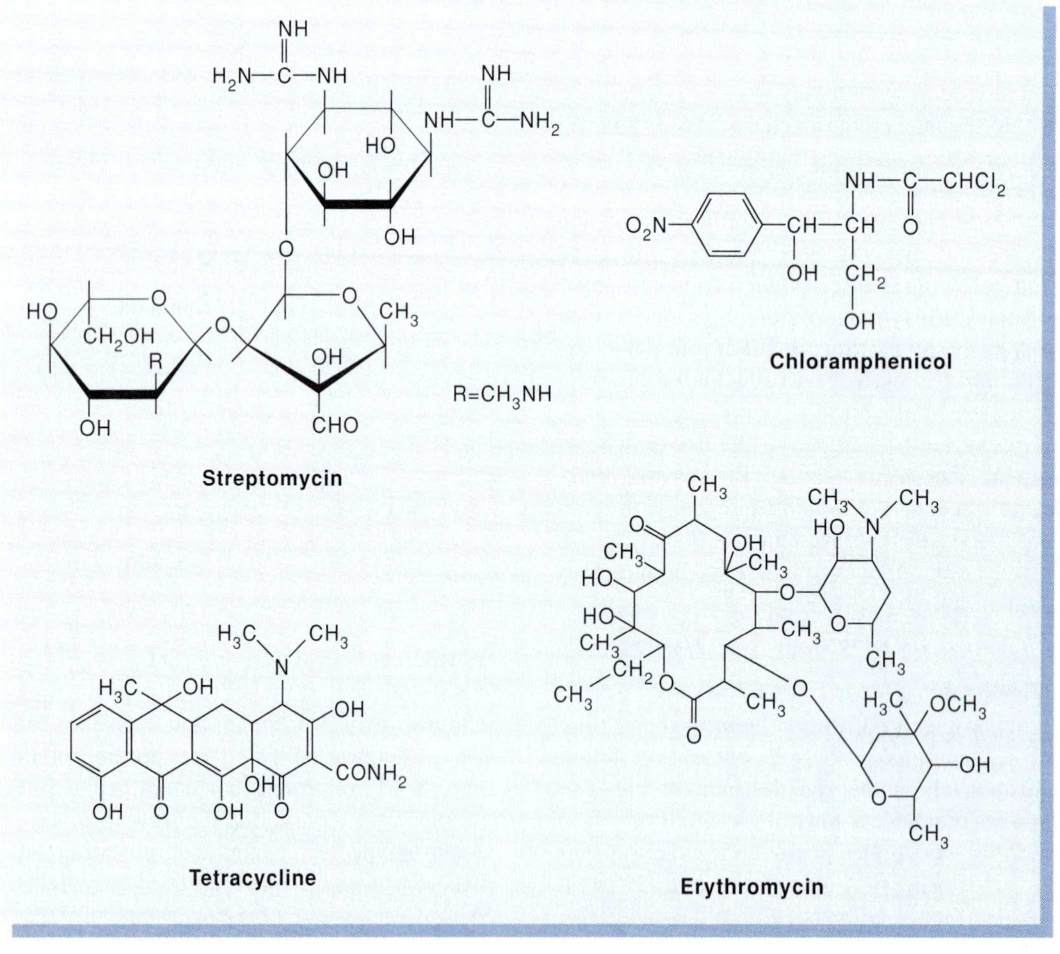

Figure 29.22

Two successive translocations are required to target proteins such as cytochrome c_1 to the intermembrane space. The precursors of cytochrome c_1 have two uptake-targeting sequences at its N terminus. The first targets the polypeptide to the matrix. In the matrix the first target sequence is cleaved by a specific protease. The second target sequence is thereby exposed and targets the polypeptide to intermembrane space, where the second target sequence is removed by another protease. The molecule folds and adds heme to become fully functional. (Source: F. U. Hartl et al., in *Cell* 51:1021-1027, 1987. Copyright © 1987 Cell Press, Cambridge, Mass.; and E. C. Hurt and A. P. G. M. van Loon, "How proteins find mitochondria and intramitochondrial compartments" in *Trends in Biochemical Sciences* 11:204-207, 1986. Copyright © 1986 Elsevier Trends Journals, Cambridge, England.)

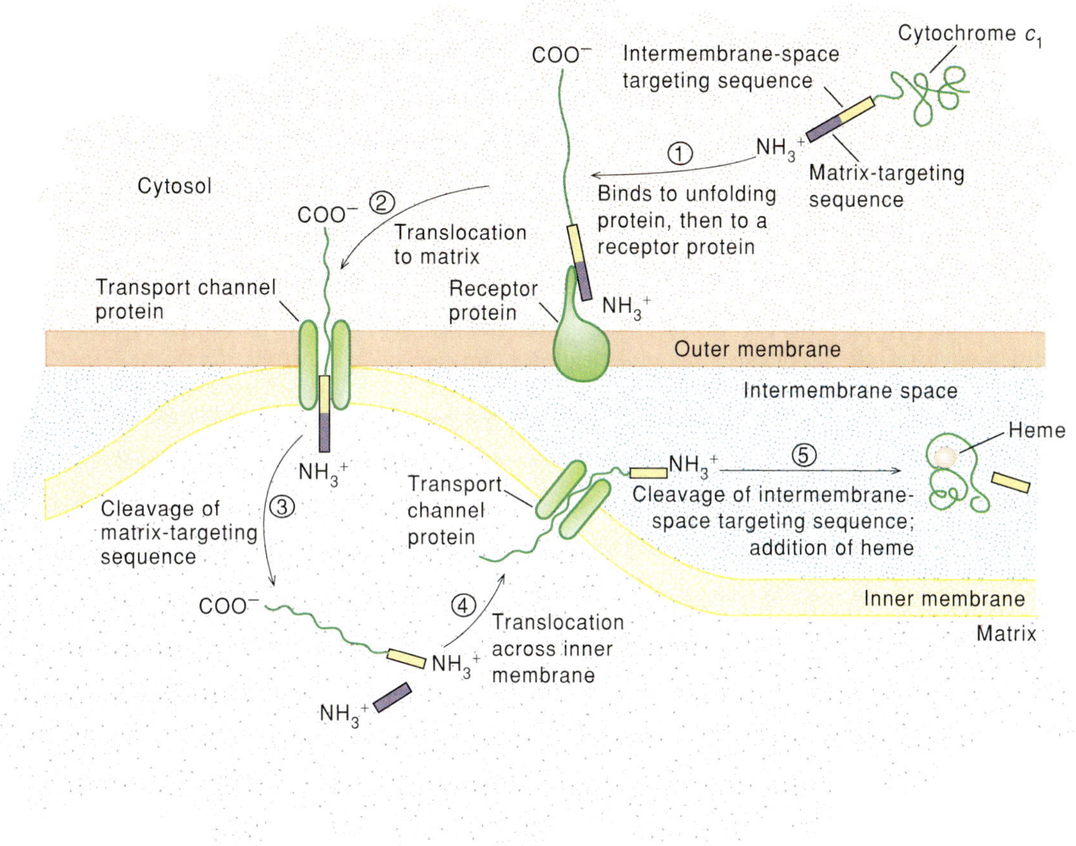

C-peptide region of proinsulin is removed to yield the circulating form of insulin with 51 residues in two disulfide-linked peptides.

Signal sequences usually occur at the amino terminus of polypeptides to be transported. These sequences vary in length from 10 to 40 amino acid residues and are generally characterized by the occurrence of a block of hydrophobic residues. The presence of hydrophobic residues reflects the fact that protein transport is essentially a problem of the movement of proteins across lipid membranes. The hydrophobic residues both target the protein to the appropriate compartment and initiate penetration into the membrane.

The sequence rules that identify signal sequences to individual cellular compartments are not completely clear. The nature of the recognition, however, is phylogenetically conserved because signal sequences generally target proteins to the appropriate compartment in different species. Also, the specificity of the signal is independent of the protein to be transported because cytoplasmic proteins, for example, can be targeted to membrane locations by the addition of appropriate signal sequences.

Some Mitochondrial Proteins Are Transported after Translation

Some proteins, especially those destined for the eukaryotic mitochondria and chloroplasts, are transported after their synthesis on free polysomes is complete. Such transport is known as posttranslational transport. In the case of posttranslational transport it is believed that the polypeptide to be transported must be unfolded from its native folded configuration by a system of polypeptide chain binding proteins (PCBs) before it can pass through the membrane. Posttranslational transport into the mitochondrion requires both ATP and a proton gradient. Presumably the energy from one or both of these sources is used to unfold the protein or separate it from the PCB system in order to allow it to pass through the membrane.

The transport of proteins from the cytoplasm to the mitochondrion is further complicated by the fact that mitochondria themselves are compartmentalized. They have both an outer membrane and an inner membrane (see chapter 15). Some proteins are targeted to either of these membranes, some are targeted to the fluid layer enclosed by the inner membrane, called the matrix, and others such as cytochrome c_1 are targeted to the fluid layer bounded by the inner membrane and the outer membrane, known as the intermembrane space (fig. 29.22). Cytochrome c_1 has two signal sequences located in series at the amino-terminal end of the preprotein. The first is recognized by a specific receptor protein attached to the outer membrane of the mitochondria. This receptor protein guides the polypeptide chain to a transport channel, across which it is transported into the matrix. Once in the matrix, the first signal sequence is removed by a specific protease. A second signal sequence, which is exposed by this cleavage reaction, results in the transport of the polypeptide chain by a similar mechanism across a transport channel into the intermembrane space. In intermembrane space the second signal sequence is removed and the protein folds into its mature configuration, which includes associating with a heme group.

As in the case of mitochondrial proteins, some of the proteins of chloroplasts are synthesized directly in the organelle, while others are synthesized in the cytosol and must be transported. The mechanisms for such transport are very similar to those observed for transported mitochondrial proteins.

All nuclear proteins are synthesized on free polysomes in the cytosol. In contrast to the transport processes that involve specific amino-terminal sequences that are cleaved, more complex structural features are recognized in nuclear proteins and they are somehow selectively transported in an intact state into the nucleus.

Eukaryotic Proteins Targeted for Secretion Are Synthesized in the Endoplasmic Reticulum

The endoplasmic reticulum (ER) is the largest membrane-bounded organelle in a typical eukaryotic cell (see fig. 21.10). The ER is a continuous network of tubules and cisternae extending throughout the cytoplasm, with a total surface area many times that of the plasma membrane. Most of the ER is studded with ribosomes to form the rough endoplasmic reticulum (RER). The ribosomes of the RER are the site of synthesis of membrane and secretory proteins and are the starting point for the protein secretory pathway. The membrane and lumen of the ER contain a characteristic set of proteins that function to process secretory and membrane proteins. After only a short time in the RER, secretory proteins are transported, by a process of vesicle budding and fusion, to the Golgi apparatus and from there to the cell surface, secretory granules, or lysosomes. As you will recall, the Golgi apparatus is a flattened stack of membranes that is the primary site of protein targeting as well as a principal site of carbohydrate addition to form glycoproteins (see chapter 21).

Let us review the way in which proteins that are destined for assembly into membranes, for secretion, or for targeting to other areas of the cell enter the lumen of the ER (see fig. 21.9). The initial targeting of nascent polypeptides to the ER membrane results from the cotranslational recognition of a signal sequence by a ribonucleoprotein complex called the signal recognition particle or SRP. The SRP is a complex that consists of six different proteins and a single 300-residue RNA molecule designated 7S RNA. Exactly how this complex functions is not clear, but it appears to be a structure especially adapted to recognize and bind to the signal sequence when the total nascent chain has reached a length of about 90 amino acid residues. Upon binding, and by an unknown mechanism, the SRP arrests the translation of the nascent chain while it searches for a specific receptor known as the SRP receptor or "docking protein." The SRP receptor is an integral membrane protein that protrudes on the inner face of the ER membrane. When docking is complete, the SRP dissociates from the ribosome, which is now bound by the signal sequence of its nascent polypeptide to the SRP receptor. After dissociation of the SRP, the translational block is relieved and translocation of the nascent polypeptide across the membrane begins. It has recently been discovered that GTP is essential for the docking maneuver and that its hydrolysis is probably essential for the subsequent release of SRP. Furthermore, sequencing of one of the protein subunits of SRP, SRP54, has revealed that it is homologous to the GTP binding domain of the G proteins. Thus signal sequence recognition by SRP may operate in a manner analogous to the GTP-dependent functioning of the G protein family.

Numerous nonmembranous proteins are retained by and function in the ER. Retention of these proteins is accomplished by a surprisingly simple mechanism. Soluble ER proteins all share the common C-terminal sequence, Lys-Asp-Glu-Leu (or KDEL in single-letter language). There appears to be a protein receptor bound to the ER membrane that recognizes and binds this sequence. The passive nature of this retention mechanism is illustrated by the fact that if the four C-terminal residues are removed from an ER protein, it is no longer retained by the organelle.

Proteins that Pass through the Golgi Become Glycosylated

Glycosylation is a conspicuous modification of proteins as they pass through the Golgi apparatus (see chapter 21). Many of the cell-surface and secretory proteins produced here are glycoproteins. Because of the diversity of the glycosylation reactions that occur in the Golgi and the fact that this organelle is the principal site of protein transport, it seems likely that the carbohydrate moieties attached to proteins are responsible for targeting them to their destinations.

One clear-cut example supporting this conclusion is the role of mannose-6-phosphate residues in targeting proteins to the lysosome. Digestive enzymes destined for the lysosome are identified by the presence of mannose-6-phosphate by binding to a specific mannose-6-phosphate receptor protein. The critical step in targeting glycoproteins from the Golgi to the lysosome is their recognition by the enzymes that catalyze the two-step phosphorylation of terminal mannose residues in their N-linked oligosaccharide side chains. Failure to phosphorylate the mannose residues results in the secretion of lysosomal enzymes. This malfunction, you may remember, is associated with the syndrome known as I-cell disease, a condition that leads to the crowding of lysosomes with damaged proteins that it cannot degrade (see chapters 6 and 21).

Processing of Collagen Does Not End Until after Secretion

Recall that collagen is an extracellular matrix protein that serves as a major constituent of many connective tissues (see figs. 4.10 to 4.13). Collagen fibrils have a distinctive banded pattern with a periodicity of 680 Å. Individual fibrils are composed of three polypeptide chains wound around one another in a right-handed helix with a total length of 3,000 Å. Each of the polypeptide chains in the triple helix has a repetitious tripeptide sequence,

Figure 29.23

Major events in the posttranslational processing of collagen.

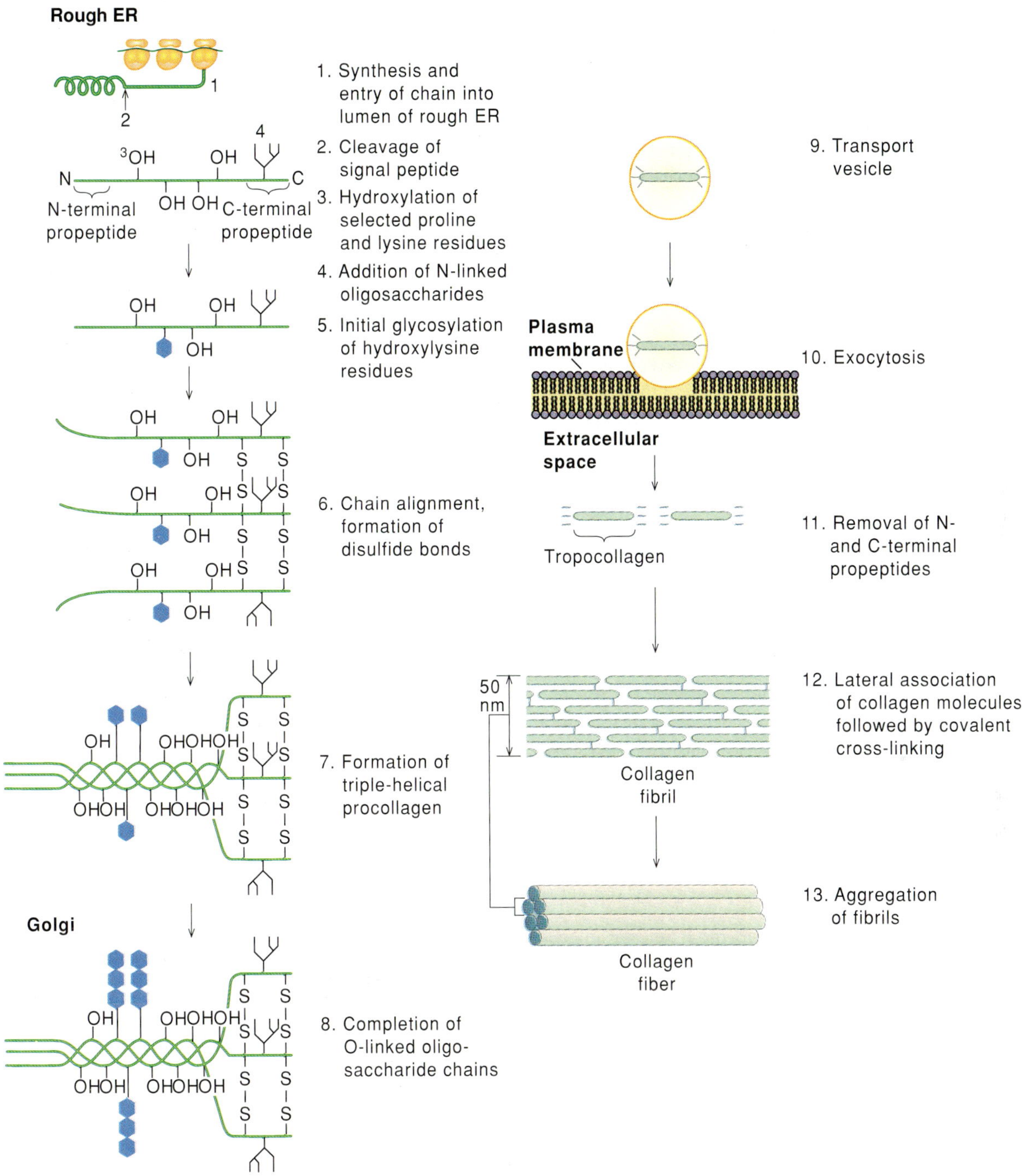

Gly-X-Y, where X is frequently a proline and Y is frequently a hydroxyproline. The latter amino acid is not one of the twenty that are specified genetically, so it must be formed posttranslationally by a modification of some of the prolines.

Since collagen is a secreted protein, we know that it must follow the route of synthesis that starts on a ribosome bound to the endoplasmic reticulum (fig. 29.23). This is where the fun begins, as we find that the nascent collagen polypeptide has extensive N and C termini (150 and 250 amino acids, respectively, for type I collagen) that are not found in the mature collagen. The function of these extensions appears to be to facilitate the initial interaction of chains in triplets and to stabilize the triple helices once they have been formed. The first modification reaction to take place in the endoplasmic reticulum is hydroxylation of specific proline residues and some lysine residues by two different enzyme systems. Glycosylation begins soon thereafter with O-glycosylation of certain hydroxylysine residues and N-glycosylation of certain asparagine residues. Next, the modified polypeptide chains, in clusters of three, form interchain disulfides near the C termini. This brings the chains in close proximity, thereby facilitating the winding reaction that leads to triple helix formation. The winding reaction proceeds in the C to N direction. The triplex helix procollagen molecule is packaged into a secretory vesicle. Following exocytosis, the ends of the procollagen molecule are removed by extracellular proteases. Collagen fibrils form by the spontaneous association of the mature collagen molecules.

Bacterial Protein Transport Frequently Occurs during Translation

The problem of protein transport in bacterial systems is relatively simple. Polypeptides synthesized in the cytoplasm may function there, may be inserted into the plasma membrane, or may be secreted by being passed through this membrane.

Most noncytoplasmic bacterial proteins are targeted into or across the plasma membrane while they are being synthesized on the ribosome. This process is known as cotranslational transport. Cotranslational secretion of proteins in *E. coli* involves a group of proteins encoded by the *sec* (for secretion) genes. Some of these proteins are membrane proteins that function by recognizing and binding the signal sequence of nascent polypeptide chains as they emerge from the ribosome. As a consequence of cotranslational transport and the action of the sec proteins, most bacterial ribosomes engaged in the synthesis of secreted proteins are tethered to the cytoplasmic face of the plasma membrane by their nascent peptide chains. A leader peptidase that is an integral membrane protein cleaves the leader sequence from the secreted polypeptide.

Protein Turnover

Cellular proteins are continuously being formed and degraded. At first glance, continuous degradation appears to be wasteful. However, protein degradation is of major biological importance in regulating protein levels, in protecting against the accumulation of abnormal proteins, in controlling growth and development, and in allowing adaptation to changing environmental conditions.

Although many proteolytic enzymes are known to exist, we are just beginning to understand how they operate to govern the levels and types of proteins within cells—to distinguish those proteins that must be degraded from those that must be preserved.

The Lifetimes of Proteins Differ

The level of a protein within a cell is determined by the balance between its rates of synthesis and degradation. As a consequence, changes in protein levels can be brought about by changes either in synthetic or degradative rates. Moreover, a rapid rate of degradation ensures that the concentration of a protein rises or falls rapidly when its synthetic rate is changed.

The degradation of cellular proteins, like the decay of radioisotopes, follows first-order kinetics. Therefore, the rate of proteolytic degradation is conveniently defined by a protein's half-life, the period of time required for it to be 50% degraded. This characteristic of protein degradation should be contrasted with the finite lifetime of an erythrocyte, which is the result of an aging process. For the purpose of degradation, each molecule is drawn at random from the protein population and therefore is selected for proteolysis independent of its age.

The half-lives of eukaryotic proteins are different and distinct. Representative examples are listed in table 29.6. It is readily apparent that degradative rates of individual proteins within a single cell can vary over a wide range. Generally, at least in the mammalian liver, enzymes that occupy important metabolic control points are degraded most rapidly, while especially long-lived proteins are rarely the sites of metabolic control. In addition, it has been established that degradative rates of specific proteins can vary with changes in physiological conditions. Thus the protein degradative mechanism is in some way tuned to metabolic control.

Structural Features Can Determine Protein Half-Lives

How are individual proteins selected for hydrolysis by the proteolytic machinery? Two sequence characteristics have been identified as correlating with rates of protein degradation.

Several years ago it was observed through sequence analysis that rapidly degraded liver proteins, those with half-lives less than 2 h, nearly all contain regions of their sequence that are rich in the amino acids proline, glutamate, serine, and threonine. These regions, involving from 10 to 60 amino acid residues, have been designated PEST sequences on the basis of the single-letter designations of their constituent amino acids. Presumably, the PEST sequences create structural domains that are recognized by proteolytic enzymes.

A second structural feature that has been correlated with protein degradative rates is the N-terminal residue of the mature form of cytoplasmic proteins. This relationship, summarized in table 29.7, has become known as the N-terminal rule. Here the presumption is that the amino-terminal residue is at least partly responsible for recognition by the degradative machinery. Proteins that are degraded according to the N-terminal rule are believed to be recognized by the ubiquitin-ATP-dependent pathway, which we will describe shortly.

Abnormal Proteins Are Selectively Degraded

A very important function of protein degradation is to protect the organism against the consequences of intracellular accumulation of abnormal proteins. Bacterial and eukaryotic cells do not accumulate structurally altered proteins because such proteins are rapidly degraded. Thus it appears that essentially all cells possess a degradative system that is capable of recognizing "abnormal" proteins.

Table 29.6
Half-Lives of Some Proteins in Mammalian Cells

Enzyme	Half-Life (h)
Rapidly degraded	
1. c-myc, c-fos, p53 oncogenes	0.5
2. Ornithine decarboxylase	0.5
3. δ-Aminolevulinate synthase	1.1
4. RNA polymerase I	1.3
5. Tyrosine aminotransferase	2.0
6. Tryptophan oxygenase	2.0
7. β-Hydroxylβ-methylglutaryl coenzyme A reductase	2.0
8. Deoxythymidine kinase	2.6
9. Phosphoenolpyruvate carboxykinase	5.0
Slowly degraded	
1. Arginase	96
2. Aldolase	118
3. Cytochrome b_5	122
4. Glyceraldehyde-3-phosphate dehydrogenase	130
5. Cytochrome *b*	130
6. Lactic dehydrogenase (isoenzyme 5)	144
7. Cytochrome *c*	150

The features of this degradative system are best seen in *E. coli,* where altered intracellular proteins can be readily manipulated. For example, it has been known for many years that incomplete chains of β-galactosidase are rapidly degraded even though the completed protein is entirely stable. In a similar way, many protein alterations that result from miscoding induced by streptomycin, or by the incorporation of amino acid analogs, also fail to accumulate because the protein products that contain them are rapidly degraded. This system presumably limits expression of heterologous proteins in bacteria that are degraded because they cannot form their native structure.

Another manifestation of the defense against abnormal proteins is seen in the heat shock response. This is a defense reaction, involving programmed changes in gene transcription and translation, that is exhibited by essentially all cells under stressful conditions. The cells reduce their overall rates of gene transcription and translation and for a brief time produce a small repertoire of proteins called heat shock proteins (hsp's). As noted in chapter 4, some hsp's play a role in the folding and transport of proteins during normal cellular function. In the heat shock response it is believed that these proteins protect against the presence of damaged proteins by binding to them and promoting either their refolding or their proteolytic degradation.

Table 29.7
Correlation between Half-Lives of Cytosolic Proteins and Amino Acid Residue at the N Terminal

Amino-Terminal Residue	Half-Life
Stabilizing	
Methionine	
Glycine	
Alanine	
Serine	>20 hours
Threonine	
Valine	
Destabilizing	
Isoleucine	~30 minutes
Glutamate	
Tyrosine	~10 minutes
Glutamine	
Proline	~7 minutes
Highly destabilizing	
Leucine	
Phenylalanine	
Aspartate	~3 minutes
Lysine	
Arginine	~2 minutes

Source: A. Bachmjuir et al., "In vivo half-life of a protein is a function of its amino-terminal residue," in *Science* 234:179, 1986. Copyright © 1986 American Association for the Advancement of Science, Washington, D.C.

Proteolytic Hydrolysis Occurs in Mammalian Lysosomes

The classic studies of Christian DeDuve in the 1960s established that mammalian cells contain a degradative organelle, the lysosome, that is produced in the Golgi and contains a large number of proteases and other hydrolytic enzymes. A similar vacuole containing hydrolytic enzymes is also present in yeast and higher plants. Together, the hydrolytic enzymes in these organelles are capable of completely degrading many macromolecules and delivering their monomeric units to the cytoplasm for further metabolism. The metabolic importance of the lysosome is demonstrated by the existence of various lysosomal storage diseases, in which one or more hydrolytic enzymes is missing. An accumulation of undegraded molecules results from these genetic diseases, frequently with devastating consequences.

A primary function of the lysosome is to digest protein-containing particles derived from the extracellular space. One mechanism of delivery is the process of endocytosis. Endocytosis is the invagination of a group of occupied receptors on the plasma membrane. Most mammalian cells can also engulf large extracellular particles by the less specific processes of pinocytosis and phagocytosis. The endocytic vesicles formed by these processes fuse with lysosomes to form secondary lysosomes where hydrolysis occurs. Lysosomes also function to degrade intracellular proteins, especially under conditions of nutritional deprivation. Under these circumstances, lysosomes can engulf cytoplasmic contents to form autophagic vacuoles and thus recycle cellular proteins.

Lysosomal proteases are called cathepsins, a name derived from the Greek term meaning "to digest." The interior of the lysosome is acidic and the cathepsins, like all lysosomal hydrolases, possess acidic pH optima and exhibit little enzymatic activity at neutral pH. This characteristic protects the cell from autolytic breakdown that might result from leakage of lysosomal contents into the neutral cytoplasm.

Ubiquitin Tags Proteins for Proteolysis

A number of different proteolytic systems are thought to be responsible for the degradation of soluble proteins in the cytoplasm of eukaryotic cells. One of the best understood is that which involves ATP and the protein ubiquitin. Ubiquitin is a small protein of only 76 residues. It occurs universally in eukaryotic cells and is highly conserved in sequence; only three residues distinguish the ubiquitin in yeast and humans. The covalent attachment of ubiquitin to proteins is thought to "tag" them for subsequent hydrolysis by cellular proteases.

Three enzymes participate in the ATP-ubiquitin system that prepares proteins for proteolysis (fig. 29.24). In the first step of this series of reactions, the C-terminal glycine residue of ubiquitin is activated by forming a thiol ester with a specific activating enzyme designated E1. This reaction is driven by the hydrolysis of ATP to AMP and PP_i. Ubiquitin is then transferred to a sulfhydryl group of a second protein (E2) that is a substrate for a group of ubiquitin-targeting proteins designated E3. The E3 proteins are responsible for identifying proteins for degradation (some on the basis of the identity of their N-terminal residue and some because they are "abnormal") and catalyzing the covalent transfer of ubiquitin from E2 to these proteins. This attachment is via isopeptide bonds that join the carboxyl group of the previously activated C-terminal glycine residue of ubiquitin with ϵ-amino groups of lysine side chains of the targeted proteins. Proteins tagged in this way are then recognized and degraded by specific proteases that release ubiquitin so that it can recycle.

ATP plays a conspicuous and rather surprising role in the degradation of proteins. Clearly, the hydrolysis of peptide bonds is a reaction that in and of itself does not require the input of energy. Nonetheless, ATP is required for the action of many proteolytic enzymes, independent of its role in ubiquitination that we have just seen.

One particularly well-understood ATP-dependent protease is the La protease that is the product of the *lon* gene of *E. coli.* This protease, like many ATP-dependent proteases, is a large protein and its ability to hydrolyze proteins is tightly coupled to its ability to hydrolyze ATP. Approximately two ATPs are hydrolyzed to ADP and P_i for each peptide bond that is hydrolyzed. It appears that the hydrolysis of ATP is required to activate the proteolytic active site of La. Other proteases seem to require ATP as an allosteric effecter to activate their hydrolytic sites, but this ATP is not hydrolyzed.

Thus ATP serves at least three independent roles in intracellular proteolysis: (1) it functions in the tagging of proteins through covalent attachment of ubiquitin, (2) it activates proteases such as La through hydrolysis, and (3) it serves as a positive allosteric effecter of other proteases without being hydrolyzed. The presumed function of this energetic requirement is to ensure the fidelity of protein degradation. It is, after all, nearly as important to cellular function to degrade proteins correctly as it is to synthesize them correctly.

Figure 29.24

The ubiquitin marking system targets certain proteins for degradation. At least three enzymes, E1, E2, and E3, are involved in addition to ubiquitin-specific proteases.

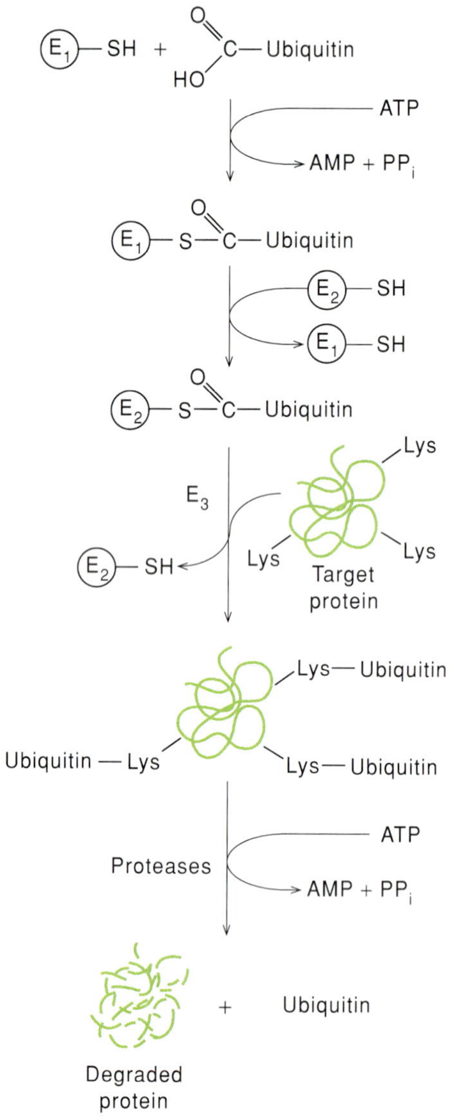

Summary

We have focused in this chapter on the complex mechanisms of protein synthesis. The following points are central to this subject.

1. Three types of RNA carry out protein synthesis: ribosomal RNA, transfer RNA, and messenger RNA. Ribosomal RNA is invariably complexed with many proteins to form ribosomes, on which amino acids are assembled into polypeptides. The amino acids are brought to the ribosomes attached to transfer RNAs. The messenger RNA contains the instructions for translation in the form of the genetic code. Messenger RNAs form transient complexes with ribosomes. Individual aminoacyl-tRNAs bind to specific sites on the messenger RNAs. The interacting site on the messenger is the codon; the interacting site on the tRNA is the anticodon.
2. The part of the messenger that is translated is the reading frame. Eukaryotic messages carry only one reading frame, whereas prokaryotic messengers may carry more than one. In prokaryotes the initiation codons are recognized by a ribosome binding site upstream of the start codon.
3. Most transfer RNAs have common parts and uncommon parts. The common parts facilitate binding of the aminoacyl-tRNAs to common sites on the ribosome. The uncommon sites permit specific reactions with charging enzymes that covalently attach the correct amino acids to the correct tRNA. Another uncommon site on the tRNAs is the anticodon, which leads to specific complex formation with the complementary codon site on the messenger.
4. Attachment of the amino acid to the tRNA is catalyzed by a specific aminoacyl synthase, which recognizes all the cognate tRNAs for a specific amino acid.
5. A unique methionyl-tRNA binds to the initiation codon on all messages.
6. The genetic code is the sequence relationship between nucleotides in the messenger RNA and amino acids in the proteins they encode. Triplet codons are arranged on the messenger in a nonoverlapping manner without spacers.
7. The code was deciphered with the help of synthetic messengers with a defined sequence, by analyzing the types of polypeptide chains that were made when these messengers were used in an *in vitro* protein-synthesizing system.
8. The genetic code is highly degenerate, with most amino acids represented by more than one codon. In many cases the 3′ base in the codon may be altered without changing the amino acid that is encoded.
9. The codon–anticodon interaction is limited to Watson-Crick pairing for the first two bases in the codon but is considerably more flexible in the third position.
10. Translation begins with the binding of the ribosome to mRNA. A number of protein factors transiently associate with the ribosome during different phases of translation: initiation factors, elongation factors, and termination factors.
11. Initiation factors contribute to the ribosome complex with the messenger RNA and the initiator methionyl-tRNA. Elongation factors assist the binding of all the other tRNAs and the translocation reaction that must occur after each peptide bond is made. Termination factors recognize a stop signal and lead to the termination of polypeptide synthesis and the release of the polypeptide chain and the messenger from the ribosome.
12. A large number of antibiotics have been characterized that inhibit protein synthesis. These antibiotics are usually made by a particular microorganism and they inhibit protein synthesis in a broad family of other organisms, mostly bacterial.
13. Specific enzymes catalyze folding after polypeptide synthesis.
14. Proteins are targeted to their destination by signal sequences built into the polypeptide chain. These signals are usually located at the N-terminal end of the protein and are generally cleaved during protein maturation.
15. Posttranslational modifications include many covalent alterations: polypeptide processing, attachment of carbohydrate or lipid groups to specific side chains, and addition of many other low-molecular-weight ligands to side chains.
16. Intracellular protein degradation is not random. Different proteins have quite different half-lives, which are related to specific structural features. Imperfectly folded proteins and polypeptide fragments are frequently degraded most rapidly. In eukaryotes, lysosomes play a major role in protein degradation.

Selected Readings

Bachmjuir, A., and A. Varshavsky, The degradation signal is a short-lived protein. *Cell* 56:1019–1032, 1989.

Beasley, E. M., and G. Schatz, Import of proteins into mitochondria. *Chemtracts* 2:305–317, 1991.

Bjork, G. R., J. U. Ericson, C. E. D. Gustafsson, T. G. Hagervall, Y. H. Jonsson, and P. M. Wilkstrom, Transfer RNA modification. *Ann. Rev. Biochem.* 56:263–288, 1987.

Böck, A., K. Forchhammer, J. Heider, and C. Baron, Selenoprotein synthesis: an expansion of the genetic code. *Trends Biochem. Sci.* 16:463–467, 1991.

Bond, J. S., and P. E. Butler, Intracellular proteases. *Ann. Rev. Biochem.* 56:333, 1987. An overview of the types of proteolytic enzymes that are found in cells and how they may function in biologically important cleavages of proteins.

Brunori, M., M. C. Silvestrini, and M. Pocchiari, The scrapie agent and the prion hypothesis. *Trends Biochem. Sci.* 13:309–313, 1988.

Burgess, T. L., and R. B. Kelly, Constitutive and regulated secretion of proteins. *Ann. Rev. Cell. Biol.* 3:243–294, 1987.

Crick, F. H. C., Codon-anticodon pairing; the wobble hypothesis. *J. Mol. Biol.* 19:548–555, 1966.

Ellis, R. J., and S. M. van der Vies, Molecular chaperones. *Ann. Rev. Biochem.* 60:321–348, 1991.

Englesberg-Kulka, H., and R. Schoulakerk-Schwarz, A flexible genetic code, or why does selenocysteine have no unique codon? *Trends Biochem. Sci.* 13(11):419–421, 1988.

Ferguson, M. A. J., and A. F. Williams, Lipids as membrane tethers for proteins. *Ann. Rev. Biochem.* 57:285, 1988.

Fessler, J. H., and L. I. Fessler, Biosynthesis of procollagen. *Ann. Rev. Biochem.* 47:129–162, 1978.
Finley, D., and A. Varshavsky, The ubiquitin system functions and mechanisms. *Trends Biochem. Sci.* 10:343–347, 1985.
Fox, T. D., Natural variation in the genetic code. *Ann. Rev. Gen.* 21:67, 1987. A review of the exceptions to the "universal" genetic code.
Gold, L., Posttranslational regulatory mechanisms in *E. coli. Ann. Rev. Biochem.* 57:199, 1988. A summary of the mechanisms of initiation and regulation of translation in prokaryotic systems.
Hardesty, B., and G. Kramer, *Structure, Function and Genetics of Ribosomes.* New York: Springer-Verlag, 1986. The third book in the Springer Series in Molecular Biology, describing current research on protein synthesis.
Hershey, J. W. B., Protein phosphorylation controls translation rates. *J. Biol. Chem.* 264:20823, 1989. Describes how protein kinases are thought to regulate translation in eukaryotic systems.
Hershko, A., Ubiquitin-mediated protein degradation. *J. Biol. Chem.* 263:15237–15240, 1988.
Hill, W. E., A. Dahlberg, R. A. Garrett, and J. R. Warner, The ribosome: structure function and evolution. American Society of Microbiology, Washington, D.C., 1990. The fourth book in a series describing current research on protein synthesis.
Hoagland, M. B., M. L. Stephenson, J. F. Scott, L. I. Hecht, and P. Zamecnik, A soluble ribonucleic acid intermediate in protein synthesis. *J. Biol. Chem.* 231:241–257, 1958. Describes the pioneering tracer studies that chart the course of amino acid into polypeptide chain.
Kartles, J. R., and A. I. Hubbard, Plasma membrane protein sorting in epithelial cells; do secretory pathways hold the key? *Trends Biochem. Sci.* 13:181–184, 1988.
Kleinkauf, H., and H. Dohren, Nonribosomal polypeptide formation on multifunctional proteins. *Trends Biochem. Sci.* 8:281–283, 1983.
Kozak, M., The scanning model for translation: An update. *Mol. Cell. Biol.* 8:2737, 1989. The current view of the way in which the eukaryotic ribosome selects the initiating codon, by the originator of the scanning model.
Moore, P. B., The ribosome returns. *Nature* 33:223–227, 1988.
Nirenberg, M. W., and J. H. Mattaei, The dependence of cell-free protein synthesis in *E. coli* upon naturally occurring or synthetic polyribonucleotides. *Proc. Natl. Acad. Sci.* 47:1588–1602, 1961. The landmark paper reporting the finding that poly(U) stimulates the synthesis of polyphenylalanine.
Noller, H. F., V. Hoffarth, and L. Zimniak, Resistance of peptidyl transferase to protein extraction procedures. *Science* 256:1416–1418, 1992. Demonstration that peptidyl transferase is an RNA.
Normanly, J., and J. Abelson, tRNA identity. *Ann. Rev. Biochem.* 58:1029, 1989. A summary of the chemical features of tRNA.
Pain, V. M., Initiation of protein synthesis in mammalian cells. *Biochem. J.* 235:625, 1986. A comprehensive review of the complex process of translational initiation in mammalian systems, emphasizing the mechanism of the process and its regulation.
Parker, J., Errors and alternatives in reading the universal genetic code. *Microbiol. Rev.* 53:273, 1989. A summary of the current knowledge of alternative mechanisms of translation.
Pelham, H. R. B., Control of protein exit from the endoplasmic reticulum. *Ann. Rev. Cell. Biol.* 5:1, 1989. Describes the current picture of how proteins are sorted and transported through the endoplasmic reticulum.
Pfeffer, S. R., and J. E. Rothman, Biosynthetic protein transport and sorting by the endoplasmic reticulum and Golgi. *Ann. Rev. Biochem.* 56:829, 1987. An excellent overview of the major features of protein targeting in eukaryotic cells.
Proud, C. G., Guanine nucleotides, protein phosphorylation and control of translation. *Trends Biochem. Sci.* 11:73–77, 1986.
Rechsteiner, M., S. Rogers, and K. Rote, Protein structure and intracellular stability. *Trends Biochem. Sci.* 12:390–394, 1987.
Rich, A., and S. H. Kim, The three-dimensional structure of transfer RNA. *Sci. Am.* 238(1):52–62, 1978.
Rothman, J. E., Polypeptide chain binding proteins: Catalysts of protein folding and related processes in cells. *Cell* 59:591, 1989. A description of the proteins that are thought to be involved promoting the formation of three-dimensional structure in proteins.
Rould, M. A., J. J. Perona, and T. A. Steitz, Structural basis of anticodon loop recognition by glutaminyl-tRNA synthetase. *Nature* 352: 213–218, 1991.
Schimmel, P., Parameters for the molecular recognition of transfer RNAs. *Biochemistry* 28:2747, 1989. A summary of the relationship between the structure of tRNAs and their recognition by synthases.
Schimmel, P. R., Aminoacyl tRNA synthetases: general scheme of structure-function relationships in the polypeptides and recognition of transfer RNAs. *Ann. Rev. Biochem.* 56:125–158, 1987.
Siegel, V., and P. Walter, Each of the activities of signal recognition particle (SRP) is contained within a distinct domain. *Cell* 52:39–49, 1988.
Thompson, R. C., EFTu provides an internal kinetic standard for translational accuracy. *Trends Biochem. Sci.* 13:91–93, 1988.
Tobias, J. W., T. E. Shrader, G. Rocap, and A. Varshavsky, The N-end rule in bacteria. *Science* 154:1374–1377, 1991.
Von Figura, K., and A. Hasilik, Lysosomal enzymes and their receptors. *Ann. Rev. Biochem.* 55:167–193, 1986.
Webb, R., and L. A. Sherman, Chaperones classified. *Nature* 359: 458–486, 1992.

Problems

1. Compare the translation initiation signals in prokaryotic and eukaryotic systems, and describe those features of each type of mRNA that determine the frequency with which a particular message is translated. What consequences do these differences have for gene organization in the two systems?
2. Transfer RNA molecules are rather large, considering the fact that the anticodon is a trinucleotide. Why is this the case?
3. Draw the chemical reaction mechanism for the formation of an aminoacyl-tRNA.
4. The relationship between tRNAs and their synthases is sometimes called the "second genetic code." Explain.
5. Explain this statement: "The universal genetic code is not quite universal."
6. How much energy is required to synthesize a single peptide bond in protein synthesis? How does this compare with the free energy of formation of the peptide linkage, which is about 5 kcal/mole?
7. Bromouracil is a base replacement mutagen, while acridine causes single-base insertions or deletions. Mutations caused by acridine frequently can be compensated for by secondary mutations several nucleotides distant from the first, while compensation for those caused by bromouracil require changes within the same codon. Explain.
8. Assume that you have a copolymer with a random sequence containing equimolar amounts of A and U. What amino acids would be incorporated and in what ratio, when this copolymer is used as an mRNA?

9. Assuming that translation begins at the first codon, deduce the amino acid sequence of the polypeptide encoded by the following mRNA template:

AUGGUCGAAAUUCGGGACACCCAUUUGAAGAAACAGAUAGCUUUCUAGUAA

10. The effect of single-point mutations on the amino acid sequence of a protein can provide precise identification of the codon used to specify a particular residue. Assuming a single base change for each step, deduce the wild-type codon in each of the following cases.
    (a) Gln ⟶ Arg ⟶ Trp
    (b) Glu ⟶ Lys ⟶ Ile
    (c) Leu ⟶ Ser; Leu ⟶ Val; Leu ⟶ Met
    (d) Thr ⟶ Ile; Thr ⟶ Pro; Thr ⟶ Lys
11. Give examples of several proteins that are not functional without posttranslational modification or processing.
12. Why are PEST sequences more likely to be found in regulatory proteins than in structural proteins?
13. Scientists have tried to isolate the peptidyl transferase from ribosomes for many years without success. It is now thought that this activity is part of the ribosome (large subunit). Discuss this point, in view of what you know about other catalytic RNP complexes (RNA-protein complexes).
14. Explain why toxins such as ricin and α-sarcin are such potent toxins, i.e., act at such low concentrations (one molecule per cell). What is unique about this group of toxic enzymes?
15. The signal recognition particle (SRP) has sometimes been called the "third ribosomal subunit." Do you think this is a valid comment? Explain.
16. Explain why normal hemoglobin is very stable, while the same protein containing the valine analog aminochlorobutyrate is very unstable.

30

CHAPTER

Regulation of Gene Expression in Prokaryotes

In chapter 12 we made the point that a unique aspect of biochemistry is the functional relationship between different reactions. Similarly, in this chapter and the next, we will see that in the regulation of gene expression, the most characteristic aspect of the processes involved is that RNAs and proteins are made and broken down in a way that best suits the needs of the organism. Thus, for example, the special characteristics of prokaryotes dictate the forms of regulation that are most used. Posttranscriptional regulation of messages is rare because the processes of transcription and translation take place in the same cell compartment. Moreover, short messenger lifetimes usually eliminate the need for elaborate mechanisms of translational control.

An overview of the points where gene expression are regulated in prokaryotes is presented in figure 30.1. In this chapter our emphasis will be on major modes of transcriptional regulation in *E. coli* and the bacteriophage λ. We will deal first, and at greater length, with the regulation of gene expression for bacterial genes in *E. coli,* and then with the regulation of gene expression in bacterial viruses, especially λ.

Regulation of Gene Expression in *E. coli*

E. coli contains a single chromosome with about 4.5×10^6 base pairs; this is sufficient to encode about 3,000 genes. Under conditions of active growth, only about 5% of the genome is actively transcribed at any given time; the remainder is either silent or transcribed at a very low rate. When growth conditions change, some active genes are turned off and some inactive genes are turned on. The cell always retains its totipotency (its potential to express any of its genes), so that within a short time—seconds to minutes in most cases—and given appropriate circumstances, any gene can be turned fully on or off.

The level of transcription for any particular gene usually results from a complex series of control elements organized into a hierarchy that coordinates the metabolic activities of the cell. For example, when the rRNA genes are highly active, so are the genes for ribosomal proteins, and the latter are regulated in such a way that stoichiometric amounts of most of the ribosomal proteins are produced. When glucose is abundant, most genes involved in utilizing more complex carbon sources are turned off in a process called catabolite repression. If the glucose supply is depleted and lactose is present, then the genes involved in lactose catabolism are expressed.

Figure 30.1

Gene expression in prokaryotes is controlled at several levels. At the DNA level the number of genes can be controlled by selective DNA amplification or rearrangement. Gross gene amplification is particularly important in the case of viruses. Rearrangement is a rare mechanism of control, which is observed in cases where there is a need to switch between closely related alternative genes for the same function. Regulation of transcription is the most important mode of regulating gene expression in both bacteria and bacteriophages. This form of regulation usually involves the initiation site for transcription, RNA polymerase and regulatory proteins. Translation control mechanisms usually operate at the initiation point for translation. Degradation mechanisms for controlling the amounts of particular RNAs and proteins are also an important type of control of gene expression.

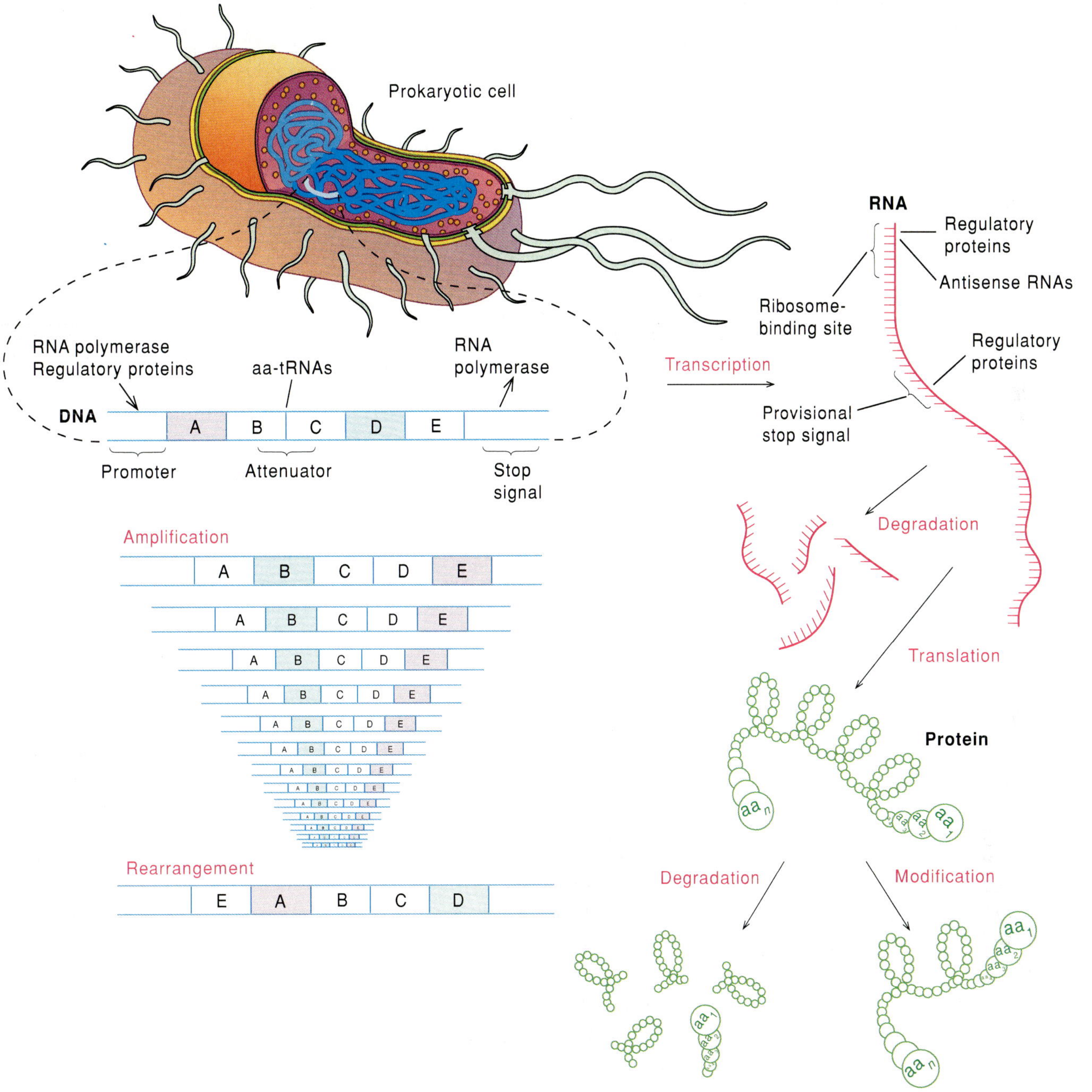

In *E. coli,* the producton of most RNAs and protein is regulated at the transcription level. Rapid response to changing conditions is ensured partly by a short mRNA lifetime—on the order of 1 to 3 min for most mRNAs. Some mRNAs have appreciably longer lifetimes (10 min or longer) and thus the potential for much higher levels of protein synthesis per mRNA molecule. These atypical mRNAs, at least in some instances, also may be subject to translational control.

The Initiation Point for Transcription Is a Major Site for Regulating Gene Expression

The most common way of regulating transcription in bacteria is by controlling the rate of initiation. We described the basic mechanics of transcription in chapter 28. Recall that prior to initiation of transcription, the RNA polymerase holoenzyme attaches to a 35- to 40-nucleotide segment of the DNA called the promoter. The affinity between DNA and RNA polymerase is controlled by a sequence of bases in the DNA. Two main areas of contact are recognized, one centered in the −35 region of the promoter, with a favored sequence of TTGACA, and one centered in the −10 region, with the favored sequence TATAAT. These two hexanucleotide regions are referred to as polymerase binding site 1 (PBS1) and polymerase binding site 2 (PBS2), respectively.

The favored sequences, or consensus sequences, as they are commonly called, were identified by comparing the sequences found in about 400 different promoters. Each base in the consensus sequence represents the base most commonly found in that position. Naturally occurring promoters have never been found with the consensus sequences at both sites. However, artificially constructed promoters bearing the consensus sequence are highly active in transcription. Furthermore, the activity of naturally occurring promoters is frequently proportional to the concordance between their promoter sequences and the consensus sequence. Future research may show that other regions are also important in promoter-RNA polymerase interactions.

From the time it first makes contact with the promoter to the time that it achieves the proper orientation for initiation, the promoter-polymerase complex may go through several metastable states. The final conformation of this complex, adopted immediately before initiation, is referred to as the rapid-start complex or open promoter complex. In this state, the polymerase is in contact with PBS1 and PBS2, and about eleven base pairs—from the −9 to the +1 positions at the origin of transcription—are unpaired (fig. 30.2). Once the rapid-start complex has been formed, initiation of RNA synthesis is rapid, taking only a fraction of a second in the presence of ribonucleotide triphosphates.

The rate of initiation of transcription can be regulated in several ways, all of which influence the rate of formation of the rapid-start complex. The primary sequence of nucleotides in the promoter region is the first factor that has an influence. The closer this sequence is to the consensus sequence, the greater is the affinity of the polymerase for the promoter. Second, for some promoters, negative supercoiling of the DNA serves as an appreciable stimulus to transcription, probably because negative supercoiling facilitates unwinding and unpairing of the double helix (see chapter 25), such as is observed in the −9 to +1 region of the promoter in the rapid-start complex.*

The rate of initiation of transcription also can be altered by changes in the RNA polymerase structure (see chapter 28). These changes include subunit replacement, subunit covalent modification, and small-molecule-induced allosteric transition. When the temperature rises from 30 to 42° C, the usual σ subunit (σ^{70}) is partially replaced by an alternative σ factor (σ^{32}), a shift that changes the types of promoters recognized by the polymerase. In bacteriophage T4 infection, the subunits of the polymerase become ribose-adenylated, lowering the affinity of polymerase for bacterial promoters and raising the affinity for phage promoters. Binding of guanosine tetraphosphate (ppGpp) to RNA polymerase changes the structure of the polymerase so that it has a greatly lowered affinity for rRNA, tRNA, and ribosomal protein promoters and at the same time a somewhat greater affinity for some other promoters.

Finally, the rate of initiation of RNA synthesis can be controlled by auxiliary regulatory proteins that affect the rate of formation of the rapid-start complex in either a positive or a negative way; such regulatory proteins are known as activators or repressors, respectively. Each promoter has a natural affinity for RNA polymerase that is determined by its promoter sequence. Activators increase this affinity, thereby increasing the

*Negative supercoiling does not always stimulate transcription. Some genes are unaffected by the extent of supercoiling. The gene for DNA gyrase is actually much more active when the DNA is relaxed, with no supercoiling. Since DNA gyrase is the enzyme that catalyzes negative supercoiling, this finding indicates that relaxed DNA serves as a signal for the selective synthesis of DNA gyrase.

Figure 30.2

Schematic diagram of DNA conformation in the rapid-start complex. Two regions, PBS1 and PBS2, most important in polymerase binding, are lettered with most favored sequences. Transcription starts at the +1 base pair.

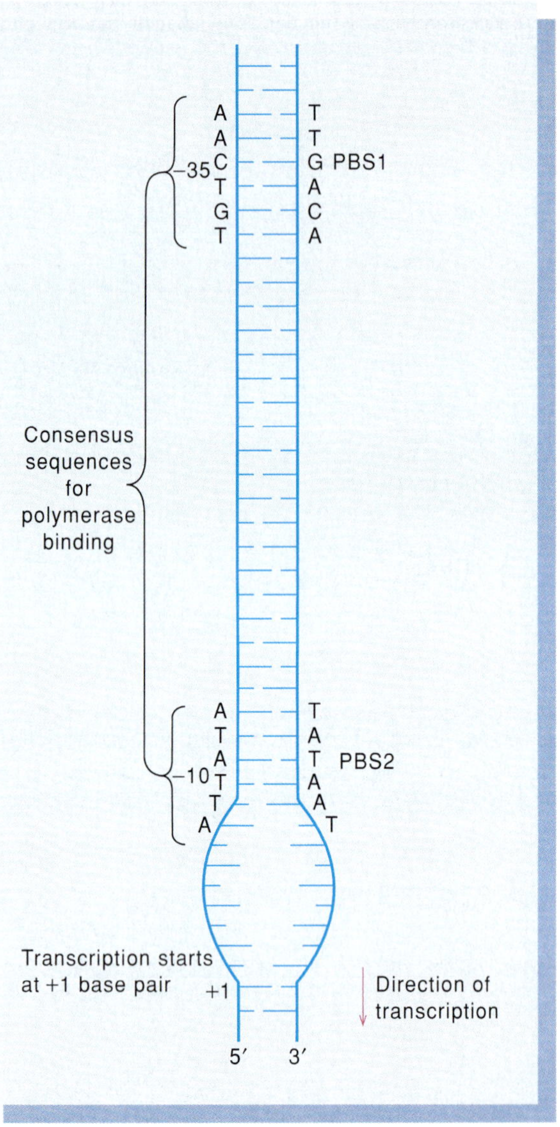

rate of transcription. Repressors have the opposite effect. For example, the lac repressor inhibits productive polymerase binding to the *lac* operon promoter. The CAP protein is an activator that stimulates polymerase binding to the same promoter. Strictly speaking, CAP should be referred to as an apoactivator rather than an activator, because it promotes transcription only when it is complexed to the small molecule coactivator 3′5′-cyclic AMP (cAMP). Let us now look at the ways in which CAP and the lac repressor function.

Regulation of the Three-Gene Cluster Containing the Gene for β-Galactosidase Occurs at the Transcription Level

The stretch of chromosome in *E. coli* that is known as the *lac* operon contains three genes associated with the metabolism of the dissacharide lactose. The story of research into the *lac* operon is one that we will present here in some detail, because it can help give you a sense of the close interplay between genetics and biochemistry and of how important various techniques have been to advances in our understanding of gene expression. Improvement in technical skills and advances in our understanding of the *lac* operon have gone hand in hand. Progress has depended not only on the development of various genetic techniques, but also on elucidation of the basic mechanisms of DNA, RNA, and protein synthesis. More recently, it has depended on cell-free synthesis techniques, use of restriction enzymes for isolation and cloning of small discrete segments of DNA, methods for determining nucleotide sequences, and x-ray crystallography.

The *lac* operon has a molecular weight of about 4×10^6 and thus constitutes about 0.1% of the *E. coli* chromosome. Its DNA is separated into two functional portions: the controlling elements and the structural genes, which code for the three proteins specified by the *lac* operon—β-galactosidase, lactose permease, and thiogalactoside transacetylase (fig. 30.3).

β-Galactosidase (*z* gene) hydrolyzes β-galactosides, in particular lactose, to produce the monosaccharides glucose and galactose. The permease protein (*y* gene) is associated with the β-galactoside active transport system. High concentrations of permease lead to the concentration of galactosides by as much as 100-fold over their concentration in the external medium. The *a* gene codes for thiogalactoside transacetylase. Unlike defects in the other structural genes of the operon, defects in the *a* gene do not affect the ability of the bacteria to grow on lactose. Transacetylase is known to catalyze the transfer of an acetyl group from acetyl-CoA to a thiogalactoside to form an acetylthiogalactoside. Surprisingly, the physiological role of this enzyme is not clear despite all the work that has been done on the *lac* operon.

The elements controlling the operon consist of a promoter locus *p*, an operator locus *o*, and a repressor gene *i*. The operator is the site on the chromosome where the lac repressor binds. The promoter contains a site for RNA polymerase binding and an adjacent site for binding the apoactivator protein CAP. The repressor gene is located near the operon and encodes the lac repressor protein.

Figure 30.3

Different genetic elements of the *lac* operon. The operon contains a control region, the promoter-operator region, and three structural genes, *z*, *y*, and *a*. The *i* gene, a repressor, is also shown. It is not part of the operon, but it is located at an adjacent site on the genome.

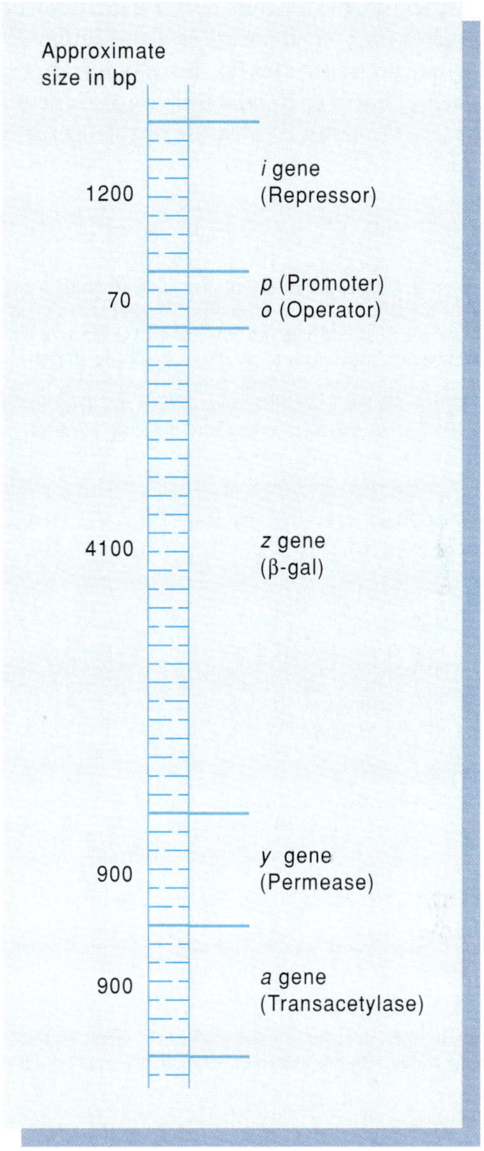

β-Galactosidase Synthesis Is Augmented by a Small Molecule Inducer Wild-type *E. coli* cells grown in the absence of a galactoside contain an average of 0.5 to 5.0 molecules of β-galactosidase per cell, whereas bacteria grown in the presence of an excess of lactose or certain lactose analogs contain 1,000 to 10,000 molecules per cell. Radioactive amino acid has been used as a tracer to show that the increase in enzyme activity observed on induction results from *de novo* protein synthesis. When excess β-galactoside inducer is added, enzyme activity increases at a rate proportional to the increase in total protein within the culture (fig. 30.4). Enzyme formation reaches its maximum rate within 3 min after inducer is added. Removal of inducer leads to cessation of enzyme synthesis in about the same amount of time.

Many compounds have been tested for their capacity to induce β-galactosidase. All inducers contain an intact, unsubstituted galactosidic residue (fig. 30.5). Many compounds that are not themselves substrates for β-galactosidase, such as thiogalactosides, are good inducers (called gratuitous inducers because they are not substrates for the enzyme). No correlation exists between affinity for β-galactosidase and the capacity to induce. Lactose, the natural substrate of the operon, is not the inducer *in vivo*. Rather, allolactose, which is formed as an intermediate in lactose metabolism in the presence of the very limited amount of β-galactosidase that occurs in the uninducer cells, is believed to be the natural inducer (fig. 30.6). The three proteins of the *lac* operon are coordinately induced, i.e., they are induced to the same extent by the same inducer. These results suggest that the receptor molecule for inducer is distinct from the structural components of the operon and that the inducer acts at one site.

Figure 30.4

Effect of inducer on ß-galactosidase synthesis. Differential plot expressing accumulation of ß-galactosidase as a function of increase in mass of cells in a growing culture of *E. coli*. Since the abscissa and ordinate are expressed in the same units (micrograms of protein), the slope of the straight line gives galactosidase as the fraction (*P*) of total protein synthesized in the presence of inducer. (Source: Melvin Cohn, *Bacteriological Reviews* 21:140, 1957. Copyright © 1957 American Society for Microbiology, Washington, D.C.)

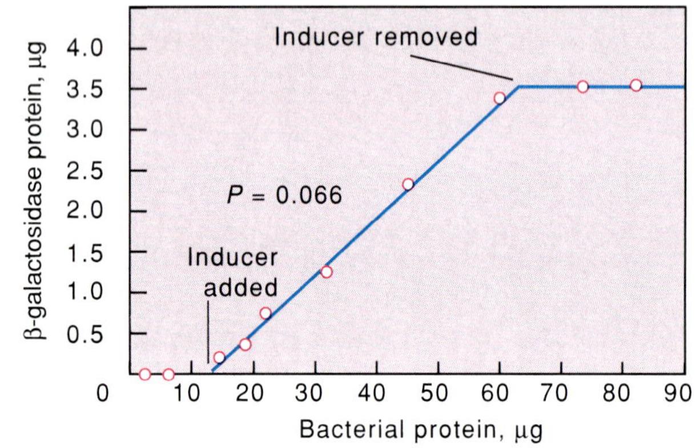

A Gene Was Discovered That Leads to Repression of Synthesis in the Absence of Inducer Two distinct types of mutations have been observed in the lactose system. One class of mutations includes structural gene mutations: (1) β-galactosidase mutations ($z^+ \rightarrow z^-$), expressed as the loss of the capacity to synthesize active β-galactosidase, (2) permease mutations ($y^+ \rightarrow y^-$), expressed as the loss of the capacity to concentrate lactose, and (3) transacetylase mutations ($a^+ \rightarrow a^-$), expressed as the loss of the capacity to form thiogalactoside transacetylase. The other class of mutations involves controlling elements of the operon, such as *i* gene mutations ($i^+ \rightarrow i^-$), expressed as the capacity to synthesize large amounts of β-galactosidase even in the absence of inducer. This type of i^- mutation is called a constitutive mutation. Structural gene mutations usually affect only the enzyme in whose gene the alteration occurs.* In contrast, constitutive mutations in the *i* gene invariably affect the amounts of β-galactosidase, permease, and transacetylase synthesized, but not their structures (at this point you may wish to refer to the review of genetic mutation and concepts in box 30A).

*This is strictly true only when synthetic inducer is used. As we explained earlier, a small amount of β-galactosidase is required to convert lactose to allolactose, the natural inducer.

Figure 30.5

Inducers of the *lac* operon. (*a*) All inducers have the ß-galactoside structure shown, in which R can be a variety of substituents. (*b*) Allolactose is the natural inducer when cells are grown on lactose. (*c*) Isopropyl-ß-D-thiogalactoside (IPTG) is a synthetic inducer useful in the laboratory; the ß oxygen is replaced by a ß sulfur atom. This change prevents hydrolysis by ß-galactosidase. For this reason IPTG is called a gratuitous inducer.

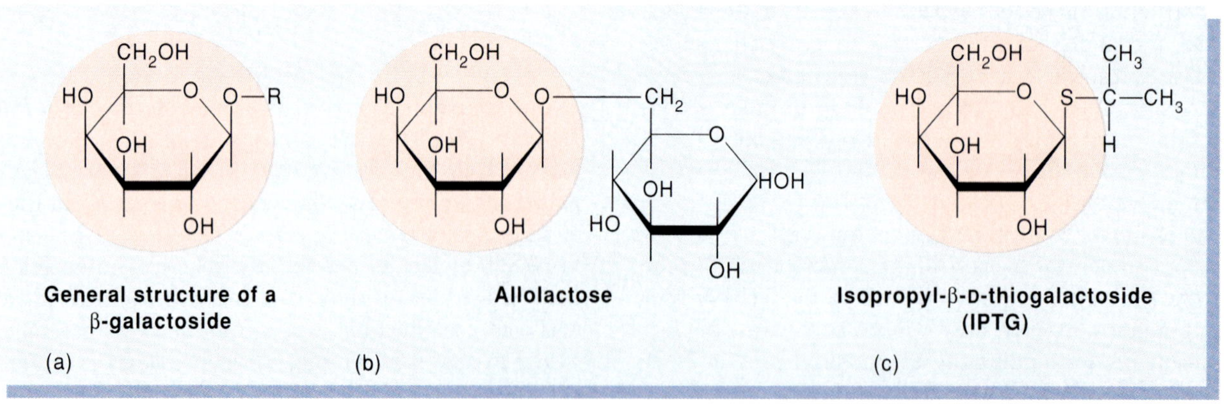

Some of the most informative early studies were performed with partial diploids (merodiploids) that contain the relevant genes on both the cellular chromosome and an F-factor plasmid. Merodiploids of the type $z^+y^-a^-/Fz^-y^+a^+$ or $z^-y^+a^+/Fz^+y^-a^-$ are wild type, that is, they behave normally. They metabolize lactose and form normal amounts of both β-galactosidase and transacetylase. Furthermore, it does not matter which *lac* operon is on the *E. coli* chromosome and which is on the F-factor chromosome. This efficient complementation between structural gene mutations indicates that they belong to independent genes or cistrons. The most significant feature of i^- mutations is that they simultaneously affect all three gene-product proteins, each independently determined by its different structural genes, *z, y,* or *a.*

The study of merodiploids of the types i^+z^-/Fi^-z^+ or i^-a^+/Fi^+a^- demonstrated that the i^+ inducible allele is dominant to the i^- constitutive allele, and that it is active on the same chromosome (*cis*), or on a different chromosome (*trans*) with respect to both *a* and *z* (table 30.1). The fact that it is effective in the *trans* position as well as the *cis* position shows that *i* gene mutations belong to an independent gene, governing the expression of the *z, y,* and *a* genes through production of a diffusible cytoplasmic component. The dominance of the inducible to the constitutive allele suggests that the former corresponds to the active form of the *i* gene.

Further understanding of *i* gene function has come from study of a mutant gene designated i^s. This mutant has lost its capacity to synthesize all structural gene products of the *lac* operon. In merodiploids of the constitution i^s/i^+, the i^s is dominant; that is, the merodiploids cannot synthesize either β-galactosidase or transacetylase even when inducer is present (see table 30.1). The most reasonable explanation for the i^s mutant is that it is an allele of *i* in which the structure of the repressor is changed so that it can no longer be antagonized by the inducer.

A Locus Adjacent to the Operon Is Found to be Required for Repressor Action In the *lac* system, rare dominant constitutive mutants (o^c) were isolated by selecting for constitutivity in cells diploid for the *lac* region, including the *i* gene, thus virtually eliminating the much more frequently occurring recessive (i^-) constitutive mutants. (Since two copies of *i* gene are present in such cells, both *i* genes would have to mutate simultaneously to i^- to give constitutive behavior.) If the probability of an $i^+ \rightarrow i^-$ mutation is 10^{-6}, the probability of two such simultaneous events in the same cell is 10^{-12}. By recombination studies, the o^c mutations were mapped in the *lac* region between the *i* and *z* loci, generating the gene order shown in figure 30.3. Genetic and biochemical evidence has shown that the *o* locus is

Figure 30.6

Conversion of lactose to allolactose, the natural inducer of the *lac* operon. Ultimately lactose is broken down to its constituent monosaccharides, galactose and glucose.

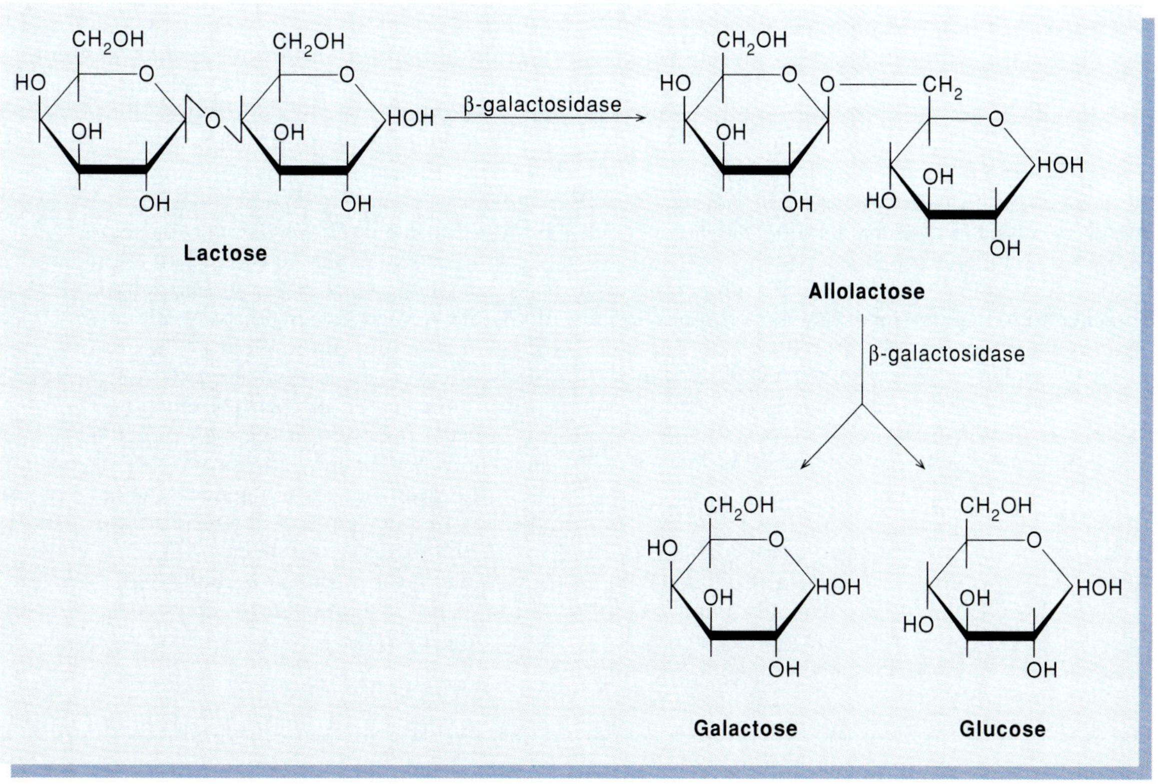

Table 30.1
Synthesis of β-Galactosidase and Galactoside Transacetylase by Haploids and Partial Diploids of *E. coli* Regulator Mutants

Strain No.	Genotype	β-Galactosidase		Galactoside-transacetylase	
		Noninduced	Induced	Noninduced	Induced
1	$i^+z^+y^+$	$<0 \cdot 1$	100	<1	100
2	$i_6^-z^+y^+$	100	100	90	90
3	$i_3^-z^+y^+$	140	130	130	120
4	$i^+z_1^-y^+/Fi_3^-z^+y^+$	<1	240	1	270
5	$i_3^-z_1^-y^+/Fi^+z^+y_U^-$	<1	280	<1	120
6	$i_3z_1^-y^+/Fi^-z^+y^+$	195	190	200	180
7	$\Delta_{izy}/Fi^-z^+y^+$	130	150	150	170
8	$i^sz^+y^+$	$<0 \cdot 1$	<1	<1	<1
9	$i^sz^+y^+/Fi^+z^+y^+$	$<0 \cdot 1$	2	<1	3

Bacteria are grown in glycerol as carbon source and induced, when stated, by isopropyl-β-D-thiogalactoside (IPTG), 10^{-4} M. Values are given as a percentage of those observed with induced wild type. Δ_{izy} refers to a deletion of the whole *lac* region. It should be noted that organisms carrying the wild allele of one of the structural genes (*z* or *y*) on the F factor form more of the corresponding enzyme than the haploid. This is presumably due to the fact that one to two copies of the F*lac* plasmid are present per cell. In i^+/i^- heterozygotes, values observed with uninduced cells are sometimes higher than in the haploid control. This is due to the presence of a significant fraction of i^-/i^- homozygous recombinants in the population. Subscripts in column labeled genotype refer to different mutants. (Data obtained from J. Monod.)

30A
BOX

Genetic Concepts and Genetic Notation

Much of the early work on the *lac* operon was purely genetic. It is essential that certain aspects of genetics be understood. The information in this box should be adequate for those with no prior exposure to genetics except for what they have already encountered in this text.

Genes are specified by one or more small letters in italic. Thus *z* indicates the gene for β-galactosidase and *lac* indicates the operon. Frequently a superscript is appended to the genetic symbol. The two most common superscripts are +, indicating a normal (wild-type) gene, and −, indicating a nonfunctioning (mutant) gene. Different representations of the same gene are referred to as alleles. Thus z^+ and z^- are both alleles of the *z* gene.

Cells that carry a single copy of each gene are referred to as haploids. Cells that carry two copies of each gene are referred to as diploids. Bacteria are haploid cells, as they carry a single chromosome with a unique representation for each gene. Bacterial cells that are partial diploids (merodiploids) may occur naturally or they may be selected for by genetic techniques.

A favored method for constructing merodiploids is to infect the bacterial cell with a virus or a plasmid DNA that carries the extra genes of interest. The F plasmid is commonly used for this purpose. In strict usage, the genetic representation for a cell carrying the *lac* operon on the chromosome and the F plasmid would be $z^+y^+a^+//Fz^+y^+a^+$, where the diagonal lines represent the host chromosome to the left and the plasmid chromosome to the right. As a rule, however, only one diagonal line is used to separate the genetic symbols. Also, for convenience, if all the alleles for a given gene are wild type they may not be shown. Thus $z^+y^+a^+/Fz^-y^+a^+$ and z^+/Fz^- may be taken as representations of the same genetic state in cases where it is known that the *lac* operon is present on both the host and the plasmid chromosome.

Two genetic elements located on the same chromosome are said to be in the *cis* orientation. Two genetic elements located on different chromosomes in the same cell are said to be in the *trans* orientation. In the merodiploid $z^-y^-a^+/Fz^+y^+a^+$, the two mutant genes are in the *cis* orientation. In the merodiploid $z^-y^+a^+/Fz^+y^-a^+$, they are in the *trans* orientation.

A major reason for using merodiploids is to study the interaction between different alleles of the same gene. This often tells us a great deal about how a gene or the gene product functions. The two simplest types of interactions are dominant and recessive. A cell that is z^+/Fz^- behaves like a z^+ cell as far as the metabolism of β-galactosidase is concerned. Therefore the z^+ allele is dominant to the z^- allele, or conversely, the z^- allele is recessive to the z^+ allele.

adjacent to but distinct from the z gene. Thus constitutive o^c mutations affect the quantity of β-galactosidase synthesized but not its structure.

In merodiploids of the type o^c/o^+, β-galactosidase and transacetylase are constitutively synthesized, a result showing that the o^c mutation is dominant to o^+. In merodiploids of the type o^cz^+/o^+z^-, the o^c mutation is dominant; but in o^cz^-/o^+z^+, it is recessive for galactosidase expression. Thus the o^c mutation is dominant only in the *cis* position, i.e., when it is adjacent to a wild-type z^+ gene. From this evidence, François Jacob and Jacques Monod inferred that the $o^+ \rightarrow o^c$ mutations correspond to a modification of the specific repressor-accepting structure of the operator. This modification identifies the operator locus, i.e., the genetic segment responsible for the structure of the operator, but not necessarily the operator itself.

Genetic Studies on the Repressor Gene and the Operator Locus Lead to a Model for Repressor Action The behavior of the various mutations we have just discussed led Jacob and Monod to propose a model for the regulation of protein synthesis. The genetic elements of this model consist of a structural gene or genes, a regulator gene, and an operator locus (fig. 30.7).

1. The structural gene produces a messenger RNA molecule that serves as a template for protein synthesis.
2. The regulator gene (not itself part of the operon) produces a repressor that can interact with the operator locus.
3. The operator is always adjacent to the structural genes it controls.
4. The operator and its associated structural genes are referred to as the operon.

The repressor molecule combines with the operator locus to prevent the structural gene(s) from synthesizing messenger RNA. In induction, the inducer (or antirepressor, a more descriptive term) combines with the repressor to prevent its interaction with the operator; this region of the genome is then free to combine with RNA polymerase. The Jacob-Monod operon hypothesis has provided a tremendous stimulus for investigations directed toward understanding not only the *lac* system, but other genetic regulatory systems as well.

Biochemical Studies Verify the Genetic Predictions Regarding Repressor Action Genetic studies were most important in arriving at a hypothesis to explain how the repressor worked, but biochemical studies were necessary to demonstrate that the hypothesis was correct. To do the biochemical experiments, it was first necessary to isolate repressor.

According to the operon hypothesis, antirepressor is supposed to combine with repressor. In order to isolate repressor, Walter Gilbert and Benno Muller-Hill used ^{14}C-labeled isopropyl-β-D-thiogalactoside (IPTG), one of the strongest known antirepressors (see fig. 30.5), to monitor repressor purification from a crude cell extract. Unlike the natural inducer allolactose, IPTG not only binds strongly to repressor but is completely stable in *E. coli* cells or crude extracts. This characteristic makes it most useful for experimental purposes. A crude cell-free extract was fractionated by standard protein purification procedures, and the fraction containing the repressor that binds IPTG was found. (See box 30B for details of how binding is measured.)

Figure 30.7

Schematic model illustrating the operon hypothesis. This diagram is modified from the original proposed by Jacob and Monod, who thought *i* gene repressor was an RNA rather than a protein. (*a*) The *i* gene encodes a repressor that binds tightly to the operator *o* locus, thereby preventing transcription of the mRNA from the *z*, *y*, and *a* structural genes. (*b*) When inducer is present, it combines with repressor, changing its structure so it can no longer bind to the operator locus. Inducer also can remove repressor already complexed with the *o* locus.

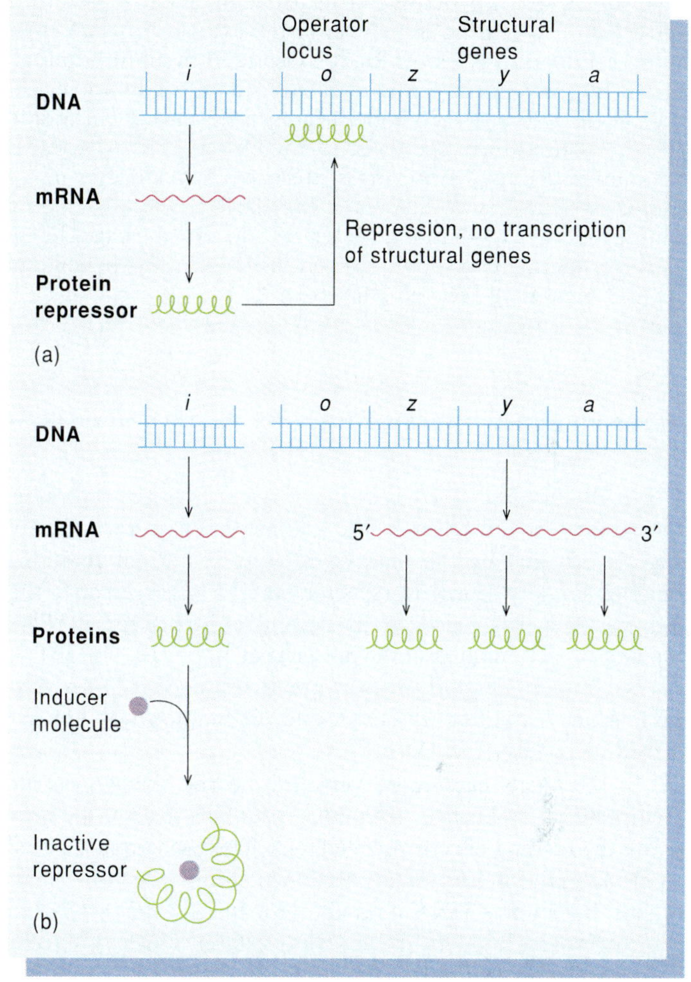

With purified repressor available, the Jacob-Monod hypothesis about binding of repressor to operator could be directly tested. The binding of purified repressor to DNA was studied in two ways. First, ^{35}S-labeled repressor was mixed with λ *lac* DNA (i.e., bacteriophage λ DNA containing the *lac* operon) and centrifuged through a sucrose gradient for a limited time. In the absence of binding, the repressor, detected by its radioactivity, sediments much more slowly than the DNA. In the absence of inducer, some repressor comigrates with the

30B

BOX

Measuring the Binding of Inducer to Repressor

The theory for equilibrium binding is discussed in chapter 10. Here we describe a popular technique for measuring binding of a small molecule to a protein.

Binding of inducer to repressor in any particular repressor-containing extract can be measured by equilibrium dialysis. In this procedure, the extract is placed inside a semipermeable dialysis bag that is impermeable to protein but permeable to IPTG. The extract is allowed to equilibrate with the external solution by gentle agitation in the presence of a buffer containing radioactive inducer. At the end of the dialysis, a certain amount of free inducer (I_f) should be detectable outside the bag. The inducer inside the bag should exceed that outside by the amount that is bound to the protein (I_b).

The strength of binding of repressor to IPTG can also be determined by equilibrium dialysis. When the ratio I_b:I_f is plotted versus I_b, the slope should equal the negative of the formation constant for binding, K_f, where K_f is defined by the equation

$$K_f = \frac{(\text{repressor} - \text{IPTG})}{(\text{repressor})(\text{IPTG})}$$

The formation constant for this complex, using wild-type repressor, is approximately 10^6 M^{-1}, or 10^6 moles^{-1} liters; the K_f for allolactose binding to repressor is about 100 times higher. Variants of wild-type repressor purified to the same degree have also been examined. The mutant repressor i^t, which was isolated from strains that are easier to induce, has about double this affinity. The repressor encoded by i^s, isolated from noninducible strains, shows no affinity for IPTG, as predicted (see main text).

DNA. Because the DNA is so much larger than the repressor, the position of the DNA in the sedimentation pattern is only slightly affected by the binding of repressor. Proof that this binding actually represents specific binding of repressor to the *lac* operator was provided by several experiments: (1) The binding was eliminated in the presence of 1.2×10^{-4}-M IPTG; (2) the binding of repressor was greatly reduced if DNA containing an o^c mutant was used; and (3) binding was not observed for wild-type λ DNA.

A faster and more versatile method, known as the membrane-filter binding technique, was also used to demonstrate the binding of repressor to DNA. This technique is based on the fact that most proteins bind to nitrocellulose filter membranes, but duplex DNA does not. However, DNA complexed to protein does bind to the membrane filter. With this technique, actively labeled DNA is used in conjunction with unlabeled purified repressor. A solution containing the labeled DNA, with or without added repressor, is passed through the membrane filter with the help of mild suction. Only when repressor is present is the label retained by the filter. Other variables were tested, and the same conclusions were reached as when the centrifugation technique was used.

The operator binding site for repressor was further characterized by sequence analysis. This analysis revealed a region of about 36 bases, extending from about −8 to +28, that shows extensive symmetry (fig. 30.8). Twenty-eight of the 36 bases in this region are arranged so that if the double helix is rotated 180° the same sequence is observed. This type of symmetry in DNA is referred to as dyad symmetry. The lac repressor is composed of an even number of identical subunits (four) and most proteins so composed also contain axes of dyad symmetry. Long before the sequences of any operator binding site had been determined or the structure of repressors was known in any detail, Zubay and Chambers had predicted that dyad symmetry would play a major role in the interaction between regulatory protein and DNA. The repressor binding site in this case consists of two symmetrically disposed binding sites on the DNA surface that simultaneously bind to complementary sites on the protein. As we will see, this symmetrical aspect of repressor-operator interaction appears to be common to many DNA–regulatory protein interactions.

Several pieces of evidence add support to the notion that the 36-base region includes the repressor binding site: (1) All eight o^c mutations shown involve base replacements near the center of this region. (2) Repressor binding substantially decreases the reactivity toward dimethylsulfate of several purine bases in this region (see fig. 30.8). Dimethylsulfate reacts with the N-7 position on guanine and the N-3 position on adenine in the double helix (see chapter 27 for a discussion of these chemical reactions). Finally, if all the thymines in the DNA are replaced by bromouracil, a number of the bromouracil bases become cross-linked in the repressor in the presence of ultraviolet light. Taking into account the normal twist of the double helix, all the groups shown by chemical methods to be in the vicinity of the repressor are situated on one side of the double helix. This finding strongly supports the notion that repressor binds on one side of the double helix, over the 36-base region covered by the symmetry axis.

The RNA polymerase and repressor binding sites (promoter and operator) overlap (fig. 30.9), so that the binding of one protein would preclude the binding of the other. Although there are other DNA sequences where the repressor might bind

Figure 30.8

The operator locus (the presumptive repressor binding site). Bases are numbered +1 for the first base transcribed and −1 for the base before that. Regions showing dyad symmetry are underlined and overlined. Arrows indicate point mutations leading to the constitutive phenotype (o^c). Circled bases are those groups that are strongly protected against reaction with dimethylsulfoxide when lac repressor is bound. Shaded circles indicate those groups that become cross-linked to repressor in the presence of ultraviolet light when thymine in the DNA is replaced by 5-bromouracil.

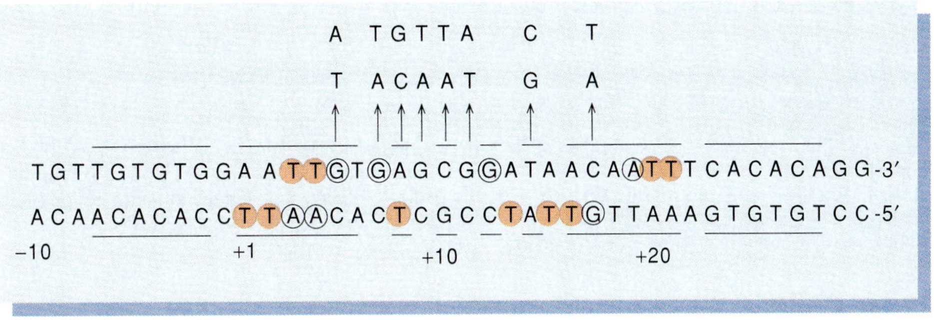

to prevent transcription, most repressors bind so that they overlap the promoter region, thereby preventing the initial formation of the RNA polymerase-DNA complex.

An Activator Protein Is Discovered that Augments Operon Expression In the absence of active repressor or when the repressor is combined with inducer, RNA polymerase can bind to the *lac* operon and transcribe *lac* mRNA. The extent to which it does this depends on other factors in the growth medium, particularly on the presence of sugars other than lactose. For example, when *E. coli* cells are grown in the presence of glucose and lactose as the sole carbon sources, the cells will first utilize glucose and afterwards utilize the lactose. This phenomenon is referred to as diauxic growth. Careful analysis of cells growing under such conditions has shown that during the glucose phase of growth, the *lac* operon is poorly expressed and induction of the operon must occur before the lactose can be utilized. Repression of *lac* operon expression by glucose or its closely related derivatives, such as glucose-6-phosphate or fructose, is known by the general term catabolite repression (or glucose repression). Enzymes of the glycolytic pathway that directly utilize glucose are always present in the cell. The enzymes of the *lac* operon are not needed as long as this more directly utilizable carbon source is available. Many genes in *E. coli* associated with catabolism are subject to this type of repression for the same reason.

A turning point in our understanding of catabolite repression was provided by Sutherland's finding that the level of intracellular cAMP in *E. coli* is drastically lowered in the presence of glucose, from about 10^{-4} to 10^{-7} M. Subsequently I. Pastan and R. Perlman showed that large quantities of cAMP in the growth medium could partially reverse the glucose catabolite repression effect on the *lac* operon. The fact that cAMP has a stimulatory effect on the expression of genes subject to catabolite repression was demonstrated directly in a cell-free system in Zubay's laboratory. The system contained λ *lac* DNA, a cytoplasmic extract of *E. coli,* and the substrates and other low-molecular-weight components necessary for transcription of the DNA and translation of the resulting mRNA. In this system 10^{-3}-M cAMP stimulated β-galactosidase synthesis as much as 30-fold.

Subsequently, Beckwith and his colleagues isolated mutants of *E. coli* that were permanently catabolite-repressed. Such mutants proved to be incapable of expressing a number of genes that required cAMP for activation. The mutants fell into two categories, those that would be phenotypically corrected by growing in the presence of 5×10^{-3}-M cAMP and those that could not. The first class of mutants was defective in the synthesis of cAMP, and the second was defective in the protein(s) with which cAMP interacts to bring about stimulation of β-galactosidase synthesis (as well as that of other catabolite-repressed proteins). Further work showed that these two types of mutants originate in the genes now referred to as *cya* and *crp,* respectively. When the cell-free system was prepared from extracts of crp^- cells, it produced a low level of β-galactosidase that was not stimulated by adding cAMP. The defect could be corrected by adding soluble protein from crp^+ cells. By making use of this *in vitro* complementation assay, Zubay was able to fractionate the crp^+ cells and isolate a single protein that was responsible for the activity.

The protein encoded by the *crp* gene, the apoactivator CAP, is a dimer composed of identical subunits with a molecular weight of 22,000. CAP binds to DNA, and this binding is greatly stimulated in the presence of cAMP. The structural alterations produced in CAP by cAMP binding apparently alter its conformation so that it can form a strong complex with DNA. CAP binds preferentially to the *lac* promoter region, although the selectivity for preferential binding to the specific site is substantially less than is the case for lac repressor.

A series of genetic deletions demonstrate that the site necessary for CAP stimulation of the *lac* operon is in the −50 to −80 base-pair region. Here there exists a 14-pair segment between −53 and −68 that shows dyad symmetry for 12 of the 14 base pairs (fig. 30.10).

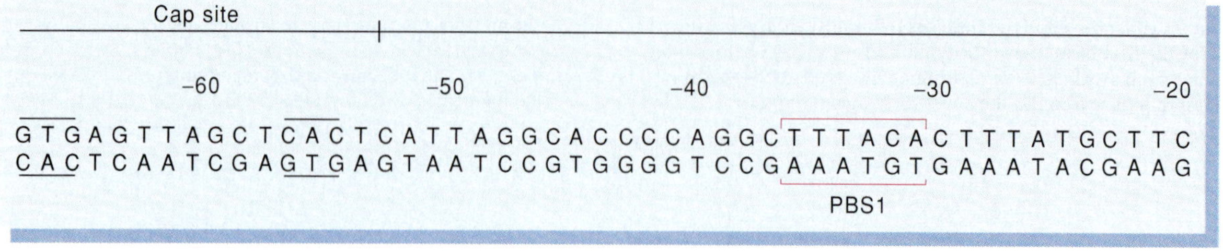

Figure 30.9

DNA sequence in the *lac* promoter-operator region. Bases are numbered +1 for the first base transcribed and −1 for the base before that. CAP and repressor-binding sites are indicated. Regions within these two sites showing dyad symmetry are underlined and overlined. Two regions, PBS1 and PBS2, where polymerase binds most strongly in a number of promoters, are bracketed in color. Note that PBS1 has five of six bases in the idealized sequence (TT-ACA), whereas PBS2 has four out of six (TAT-T). Arrows indicate replacements in the mutant *UV5* promoter that lead to high-level expression in the absence of cAMP-CAP.

As in the case with lac repressor, chemical-probe experiments have been carried out to characterize the region binding to CAP. In these experiments, a small DNA segment containing the CAP-binding site was used. Dimethylsulfate (DMS), which reacts with the adenine N-3 and the guanine N-7 positions, and ethylnitrosourea (EtNu), which reacts with the phosphates, were used. Experiments with the DMS probe showed that CAP binding protects the two outer guanines on each side of the presumptive binding site from methylation, and also that if the DNA is first premethylated, binding of CAP is strongly inhibited. The pattern of protection demonstrates symmetry and interaction in two adjacent, large grooves of the DNA. Experiments with EtNu confirmed the importance of the symmetry and also showed that, like lac repressor, CAP interacts with only one side of the DNA helix, over a region covered by the symmetry axis.

In the absence of bound repressor, the full sequence of reactions in the initiation of transcription is summarized by the following set of equations.

$$\text{cAMP} + \text{CAP} \rightleftharpoons \text{cAMP-CAP}$$

$$\text{cAMP-CAP} + \text{DNA} \rightleftharpoons \text{cAMP-CAP-DNA}$$

$$\text{cAMP-CAP-DNA} + \text{polymerase} \rightleftharpoons \text{cAMP-CAP-DNA-polymerase}$$

First, the coactivator cAMP combines with the apoactivator CAP, which then binds in the −60 region of the promoter. This complex stimulates the binding of RNA polymerase to an adjacent site on the promoter. CAP stimulates binding of RNA polymerase by an affinity between CAP and polymerase. The transcription rate of the *lac* operon in the presence of cAMP-CAP is enhanced 20- to 50-fold, as a direct result of the increased affinity of RNA polymerase holoenzyme for the promoter in the presence of CAP and cAMP.

The overall strategy in creating a promoter sequence responsive to cAMP-CAP activation could be summarized as follows. The nucleotide sequence in the RNA polymerase binding site is adjusted so that polymerase by itself produces a low level of transcription. An adjacent site for binding CAP is created so that when CAP is bound, the additional affinity contributed by favorable contacts between the CAP and the polymerase converts this sequence into a high-level promoter. A mutant (*UV5*) containing two base replacements in the *lac* promoter produces high-level transcription in the absence of cAMP-CAP and no further increase when the latter complex is present. The base replacements in the *UV5* mutant in the PBS2 site (see fig. 30.8) bring the promoter sequence closer to the consensus sequence and therefore should lead to a stronger polymerase binding site in the absence of the cAMP-CAP complex, as is in fact observed.

Figure 30.10

CAP-binding region in the *lac* promoter. Regions showing dyad symmetry that are believed to interact strongly with CAP are overlined and underlined. Circled bases are those that are strongly protected from reaction with dimethylsulfate when CAP is bound. Point mutations L8 and L29, which produce a promoter that is not stimulated in transcription by cAMP, are indicated. Colored dots indicate those phosphate positions which, if ethylated by ethylnitrosourea, block CAP binding. If the area of interest exists as a normal DNA double helix when binding CAP, then most of the groups implicated in CAP binding would appear on one side of the DNA.

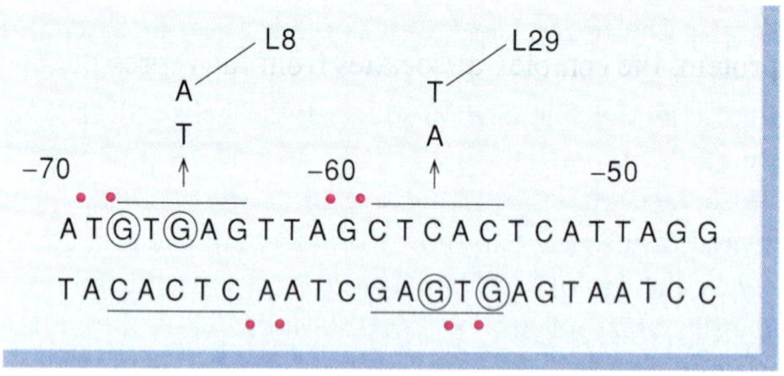

Many Genes Important in Catabolism Are Stimulated by CAP The cAMP-CAP activator is essential for regulating a wide variety of genes involved in catabolism. This complex usually functions in series with a gene-specific regulator so that any particular gene subject to cAMP-CAP activation will be fully expressed only if both control switches are in the "on" position. The galactose (*gal*) operon, which encodes enzymes for galactose catabolism, has a promoter regulated by cAMP-CAP as well as by a specific repressor that is antagonized by galactose.

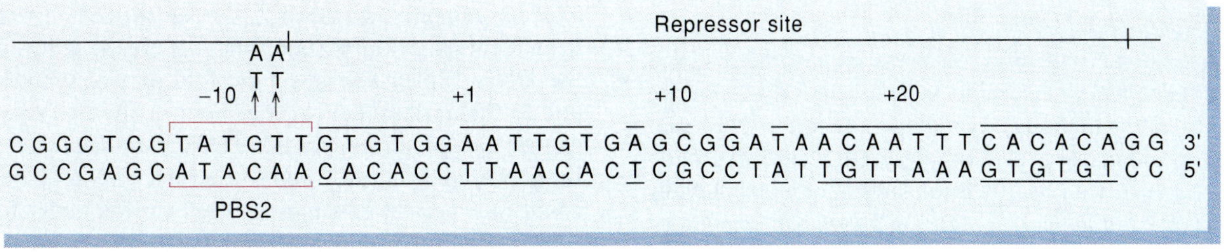

The *gal* operon appears to have a second start signal close by (five bases before the first start signal) that is not subject to cAMP-CAP control; the function of this second promoter is not yet understood. It may be related to the fact that in *E. coli* galactose acts both as a carbon source (catabolite) and as a substrate for cell-wall synthesis (anabolite). The enzymes of the *gal* operon convert galactose to an activated complex, uridine diphosphogalactose (UDPG), that is an intermediate in both the catabolic and the anabolic pathways. Thus the *gal* operon should be responsive to two types of controls that reflect different needs.

The arabinose (*ara*) operon also uses the pleiotropic cAMP-CAP activator and a specific regulator, but here the similarity with *lac* and *gal* ends. The specific regulatory protein that is encoded by the *araC* gene regulates its own synthesis by functioning as a repressor. The sugar L-arabinose serves as both antirepressor and coactivator. When it complexes with the araC protein, the complex dissociates from the repressor binding site and becomes converted into an activator complex that binds at a different site, adjacent to the polymerase binding site for the operon. In this case the binding of cAMP-CAP stimulates the binding of the L-arabinose-araC complex to the adjacent site, which in turn stimulates the nearby binding of RNA polymerase.

The studies on *lac* and *ara* have led to the impression that activators stimulate expression by an energetically favorable interaction between the protein surface of the activator and the protein surface of another activator or the polymerase bound to an adjacent site on the DNA.

It is clear that the repressor proteins discussed thus far interfere with the transcription process in one of two ways: (1) through binding at a site that overlaps the polymerase binding site at the promoter (as in the *lac* operon) or (2) through binding at a site that interferes indirectly with the polymerase binding at the promoter by interfering with CAP binding (as in the *ara* operon). By contrast, the activator complex cAMP-CAP acts by binding at a site adjacent to the polymerase binding site, thereby augmenting polymerase–promoter interaction. In some cases it may be that the most important difference between activators and repressors is not how they bind but where they bind to the DNA.

The Synthesis of the CAP Protein Is Negatively Regulated at the Transcription Level The synthesis of most regulatory genes is itself regulated. We have seen that the synthesis of *araC* is negatively regulated by the araC protein, acting as a repressor for the synthesis of its own messenger. The same is true for the trp repressor that we will discuss a little later. In the case of CAP, synthesis is also negatively regulated, but in this case the mechanism is unusual in that it involves RNA–RNA interaction. A second promoter in the region of the *crp* gene is strongly activated by cAMP-CAP. Transcription from this promoter is initiated twelve nucleotides downstream and on the opposite strand from the initiation site of the *crp* gene. The initial nucleotides of the divergent RNA are complementary to ten of the first eleven nucleotides of the *crp* mRNA (fig. 30.11). The next eleven nucleotides of *crp* mRNA and the 5′ end of *crp* mRNA could produce a structure similar to a ρ-independent terminator (see chapter 28), a change leading to inhibition of *crp* transcription.

Figure 30.11

Model for the inhibition of *crp* transcription by divergent RNA. The proposed RNA-RNA hybrid between the 5′ end of *crp* mRNA and the initial 14 nucleotides of the divergent RNA is shown. The complete structure of the proposed ρ-independent terminator is shown below. The structure consists of the RNA-RNA duplex followed by an A•U-rich RNA•DNA hybrid of 11 bp. Okamoto and Freundlich proposed that the RNA•RNA hybrid causes the polymerase to pause, and as suggested for ρ-independent terminators, the instability of the A•U-rich RNA•DNA hybrid leads to the release of the transcript and termination of transcription. (Source: K. Okamoto and M. Freundlich, "Mechanism for the autogenous control of the *crp* operon: transcriptional inhibition by a divergent RNA transcript," in *Proceedings. National Academy of Sciences USA* 83:5000–5004, 1986. Copyright © 1986 National Academy of Sciences, Washington, D.C.)

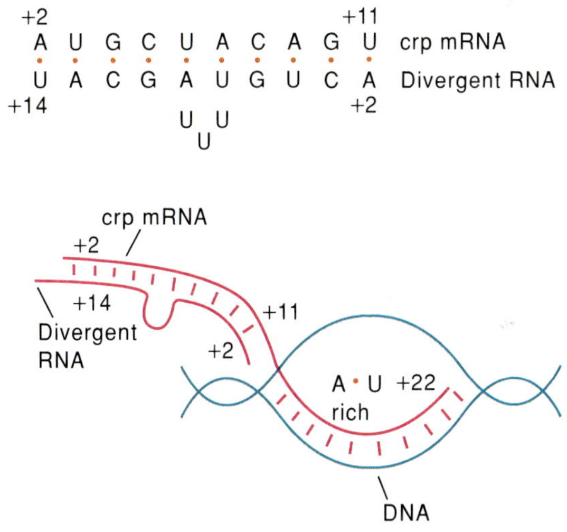

Gene Regulatory Proteins Bind Strongly to Specific Sites on the DNA

Like CAP and the lac repressor, gene regulatory proteins bind strongly and selectively to specific sites on the DNA. Before precise structures for the mutual binding sites on the DNA and the regulatory proteins were known, two predictions were made regarding the general nature of these sites. These predictions were based on certain general features of DNA and regulatory protein structure.

The first feature has to do with symmetry. Many regulatory proteins are composed of an even number of identical subunits. Structures of this composition usually are highly symmetrical; in particular, they usually have at least one twofold axis of symmetry, so that rotation on the axis by 180° results in an identical conformation. If we ignore the sequence of bases, the DNA duplex also has a twofold axis of symmetry that is due to the antiparallel arrangement of the two polynucleotide chains. In a local region of the DNA, the sequence of bases in the DNA could be adjusted so that over a limited region there could be a perfect or almost perfect twofold axis of symmetry. Such a situation would generate two identical sites on the DNA. If one of these sites were used for binding to a regulatory protein, then, given the appropriate dimensions and symmetry, the other site might also be used for binding to the same protein. A hypothetical two-site model of this type is illustrated for the interaction between CAP and DNA in figure 30.12.

The two-site model received its first direct support from sequence data on the lac repressor and CAP DNA binding sites. Both of these sequences showed a significant amount of the required symmetry (see figs. 30.8 and 30.10). Final confirmation of the two-site model resulted from x-ray crystallographic studies on regulatory proteins and their complexes with DNA. We now know that, in many cases, two nearly identical sites on the DNA do indeed interact with two nearly identical sites on the regulatory protein on a twofold axis of symmetry. The clear advantage of the two-site interaction is that it approximately doubles the strength of the interaction between the DNA and the protein.

Another and possibly even more general aspect of the complex formed between DNA and regulatory protein has to do with the location at which the interaction takes place. The formation of a specific complex must involve interaction be-

Figure 30.12

Schematic diagram of the cAMP-CAP-DNA complex. CAP is a dimer composed of identical subunits. The binding site on the DNA is arranged so that rotation of either the DNA or the CAP dimer by 180° on the dyad axis should produce an identical structure. Results from the laboratory of T. Steitz suggest that CAP binding induces a pronounced bend (90° or more) in the DNA.

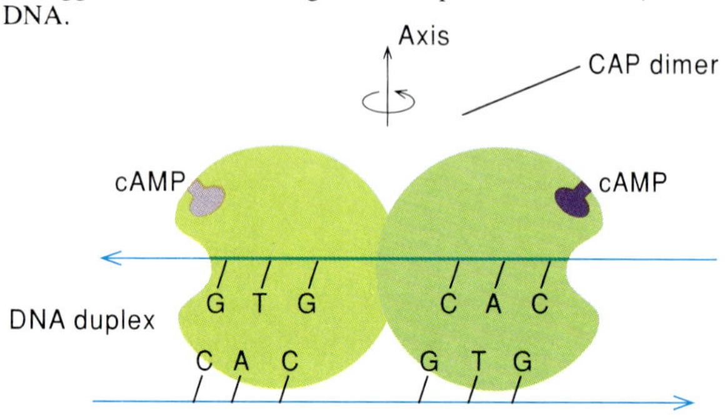

Figure 30.13

The structures of three regulatory proteins. They all possess twofold axes of symmetry, and the protruding helical cylinders that interact with adjacent major grooves on the DNA (red) are spaced about 34 Å apart. N and C labels indicate the N and C termini of the polypeptide chains.

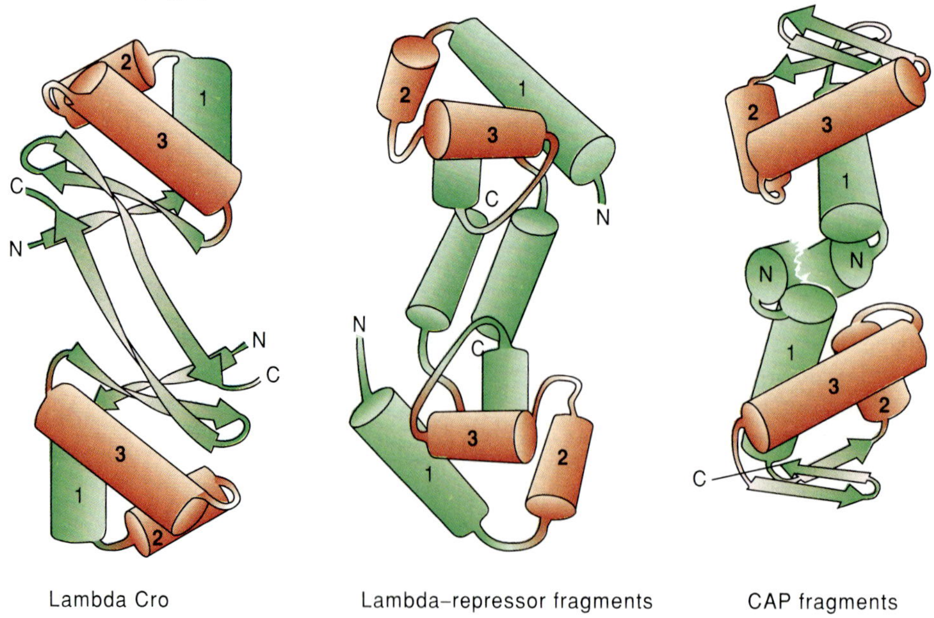

tween specific sites on the protein and specific sites on the DNA. On the DNA it seems likely that the interaction involves sites on base pairs exposed in the major groove. This is because the major groove is much more accessible for interaction with a large molecule than the minor groove. In fact, the major groove is just the right size to accommodate an α helix, whereas the minor groove could not do so. Thus we might expect interaction between DNA and a regulatory protein to involve interaction between a protruding α-helix segment of the regulatory protein and the major groove of the DNA.

Progress beyond these predictions to an understanding of the specific interactions between DNA and protein took more than a quarter of a century of intensive effort by a few highly skilled groups of x-ray crystallographers. We now know the detailed molecular structures of a number of regulatory proteins.

Figure 30.14

Model for location of α_3 and α_2 segments when cro binds to DNA.

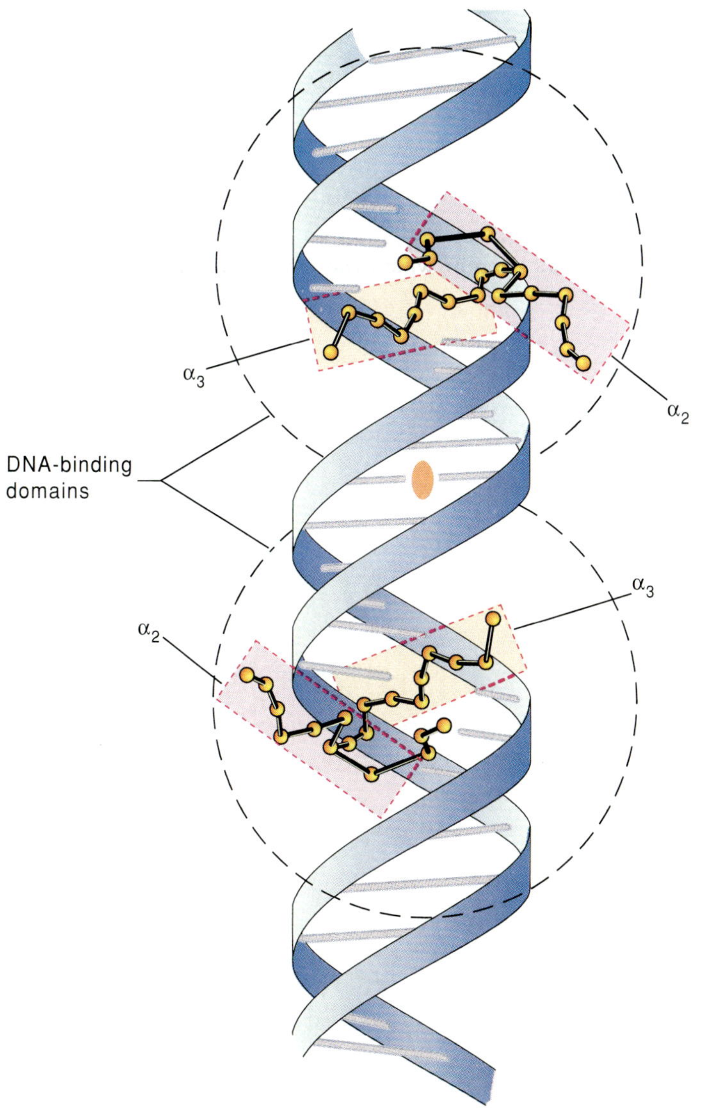

This list includes CAP, the λ repressor, the phage 434 repressor, the 434 cro protein, and the tryptophan repressor. The structures of these regulatory proteins show surprisingly similar features (fig. 30.13). All of the regulatory proteins possess two-fold axes of symmetry and two protruding cylinders of α helix spaced 34 Å apart, which is just the distance necessary for them to interact with adjacent major grooves of the DNA. All regulatory proteins of this type also possess at least one additional segment of α helix; the overall structure is referred to as the helix-turn-helix motif because of the arrangement between the helix segment that interacts with the DNA (the recognition helix) and the additional segment of helix. Although the recognition helices in the protein dimer are spaced at the ideal distance for interaction with adjacent major grooves, their orientation is different for each protein. Only in the case of cro would the recognition helix run parallel to the major groove (fig. 30.14). Despite this lack of perfect orientation of the helix segments in the other regulatory proteins, the amino acid side chains are sufficiently long that interaction between the bases in the major groove and the amino acid side chains should be possible.

Although the helix-turn-helix motif is the most common one thus far found for specific DNA-binding proteins in prokaryotes, it is not the only one. Another radically different motif that has been observed for the met and arc repressors involves a dimeric molecule in which two antiparallel polypeptide chains in the extended β configuration are tucked into the large groove where they interact with the DNA. We will see (see chapter 31) that additional DNA-binding motifs have also been discovered in eukaryotes, which are quite different from both of these.

Final proof of the manner in which the regulatory protein interacts with the DNA was obtained by cocrystal studies on the regulatory protein and the relevant region of the DNA made synthetically. Such studies confirmed the primary interaction with the major groove and permitted a detailed analysis of the specific interactions between the DNA and the protein.

In most of these complexes with protein, the DNA maintains a conformation that is close to the B form. The regulatory proteins interact with the DNA on a two-fold axis of symmetry so that the protein binds on one side of the DNA in two adjacent major grooves. The affinity between the protein and the DNA results partly from nonspecific contacts between amino acid side chains on the protein and the sugar-phosphate backbone of the DNA and partly from specific contacts made between the amino acid side chains and the base pairs of the DNA. In addition to its size, the major groove has one further advantage for interaction. The hydrogen-bonding groups on the base pairs that are exposed in the major groove present a much greater number of possibilities for interaction with complementary hydrogen-bonding groups of the amino acid side chains of the regulatory proteins. (Figure 25.9*c* shows which groups are exposed in the major and minor grooves.) The specific contacts between the amino acid side chains of the λ repressor are illustrated in figure 30.15. In all cases the specific contacts involve hydrogen-bonding groups on the purines in the major grooves

Figure 30.15

The specific interactions between λ repressor and one half of its operator binding site. Identical contacts are made in the other half of the binding site. The numbers associated with the amino acid side chains refer to the distance of the amino acids from the amino-terminal end of the protein. Nucleotides are numbered from the central dyad at the operator, which passes through base pair O. N_6 and N_7 positions on adenines and N_7 and O_6 positions on guanines involved in H bond formation are labeled. Numbering of positions in purine rings is described in figures 1.6, 20.2, and 20.12. (Source: T. Steitz, *Quarterly Reviews of Biophysics* 23:236, 1990. Copyright © 1990 Cambridge University Press, Cambridge, England.)

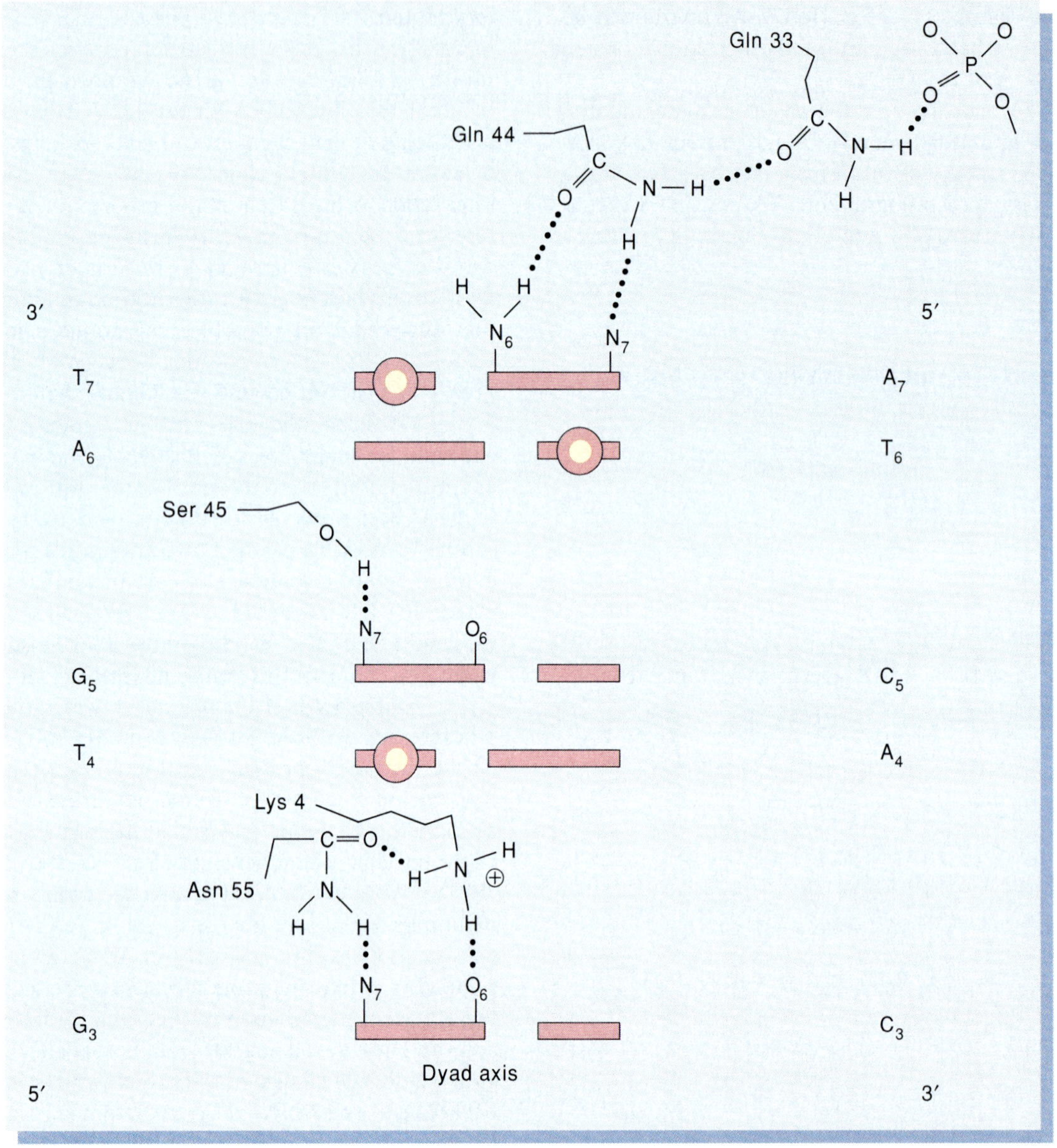

of the DNA with polar amino acid side chains emanating from the interacting α-helix segment. The amino acid side chains make hydrogen bonds directly with the hydrogen-bonding groups on specific purines.

A common feature shared by many but not all regulatory proteins is their sensitivity to small-molecule effectors. For example, the lac repressor adopts two conformations, one that is favored in the absence of small-molecule effectors like allolactose, and one that is favored when binding to allolactose. In the former conformation the repressor has a high affinity for the *lac* operator; in the latter it does not. A similar situation exists for CAP, except that in this case the role of the small-molecule effector is reversed. That is, the binding of cAMP converts the protein from a structure with a low affinity to one with

Figure 30.16

The tryptophan operon, indicating the size of the different genes, the polypeptide chains, the resulting enzyme complexes, and the reactions catalyzed by the enzyme complexes.

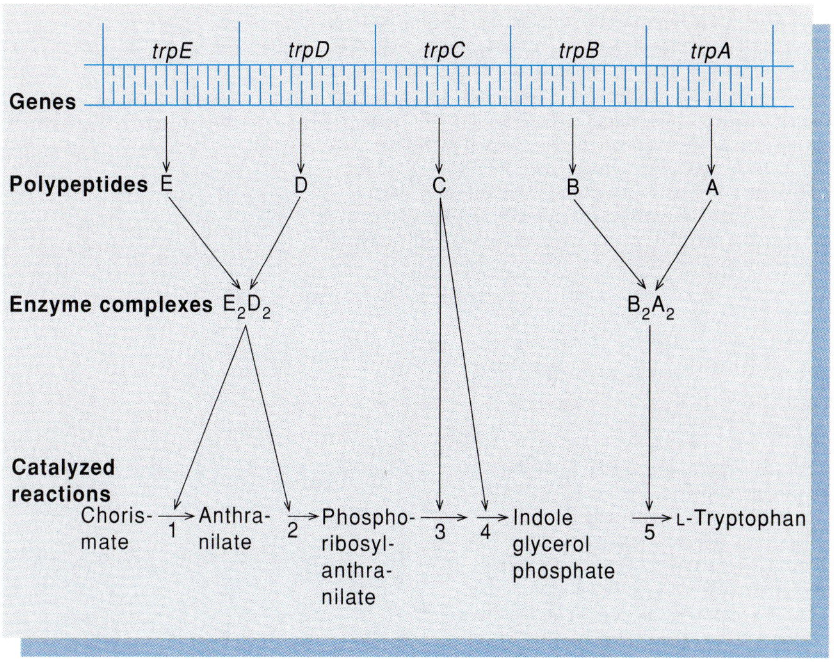

a high affinity for a specific site on the DNA. Although the binding sites for cAMP in CAP are known from structural studies, we still have no idea why the binding of cAMP influences the binding of CAP to the DNA in such a profound way.

Before leaving this section on the binding of regulatory proteins, it is important to point out that sometimes the binding of a regulatory protein can have a major effect on the DNA structure. For example, in the case of the binding of CAP to DNA the alteration in DNA structure is quite large. The first evidence for this came from studies of the electrophoretic mobility of complexes between CAP and segments of DNA containing the CAP-binding site. They migrated in an electric field at an anomalously fast rate as though the DNA was effectively smaller after complex formation than before. It was proposed that this behavior is due to the formation of a bend in the DNA binding to CAP. Seven years later, the crystal structure of this complex has confirmed that a bend is formed of greater than 90 degrees. It seems likely that this major change in the DNA tertiary structure must be important in the action of CAP on gene expression, but it is too early to say precisely what this effect would be.

Enzymes that Catalyze Amino Acid Biosynthesis Are Regulated at the Level of Transcription Initiation

Of all the genes concerned with anabolic processes, the genes uniquely involved in synthesis of the amino acid tryptophan are the best understood. Our understanding is mostly a result of the efforts of Charles Yanofsky and his colleagues, who have used a wide variety of genetic and biochemical techniques to probe the complexities of this system. Expression of the genes involved in tryptophan biosynthesis is regulated at the level of transcription. The way in which transcription is regulated in a biosynthetic operon is quite different from the way in which it is regulated for the catabolic genes. These differences reflect different strategies for assuring that the cell has sufficient amounts of a particular substrate required for protein biosynthesis.

Wild-type *E. coli* can synthesize all twenty types of amino acids from simpler substrates, but for most of them, it does so only when they are not available in adequate amounts from the external growth medium. For example, production of enzymes for tryptophan synthesis is sharply reduced when the external tryptophan supply is high. Lowering the available L-tryptophan selectively stimulates synthesis of the messenger RNA and the five polypeptide chains that form the enzymes associated with the biosynthesis of tryptophan from chorismate (fig. 30.16). The five contiguous structural genes are transcribed as a single polygenic messenger RNA approximately seven kilobases in length. Initiation of transcription is regulated in part by the interaction of the tryptophan aporepressor, the protein product of the *trpR* gene, with its target site on the DNA, the *trp* operator, *trpO* (fig. 30.17). Binding of L-tryptophan to the aporepressor causes a structural alteration essential for strong specific binding to the *trpO* locus. Like lac repressor, the trp repressor binds at a site that overlaps the RNA polymerase binding site, and the region where the repressor binds contains a number of bases arranged with dyad symmetry (fig. 30.18). It seems likely that the bases on the dyad-symmetry axis are

Figure 30.17

Schematic diagram of the repressor control of *trp* operon expression. The *trp* promoter (*P*) and *trp* operator (*O*) regions overlap. The *trp* aporepressor is encoded by a distantly located *trpR* gene. L-tryptophan binding converts the aporepressor to the repressor that binds at the operator locus. This complex prevents the formation of the polymerase-promoter complex and transcription of the operon that begins in the leader region (*trpL*). Only a fraction of the transcripts extend beyond the attenuator locus in the leader region. The regulation of this fraction is discussed in the text.

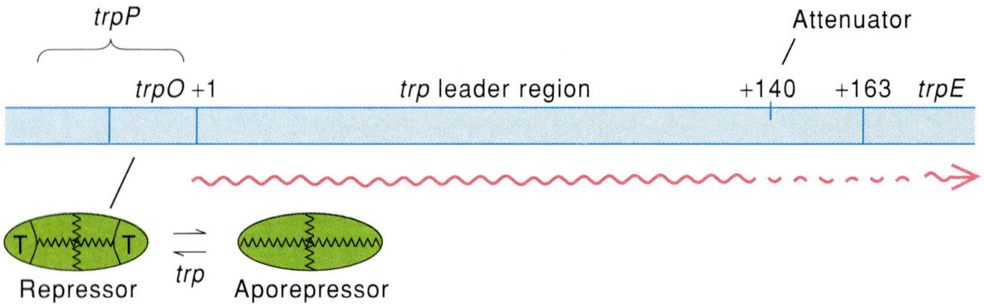

Figure 30.18

The promoter-operator region of the tryptophan operon. Two regions, PBS1 and PBS2, where polymerase binds are bracketed. Regions within the repressor-binding site showing dyadic symmetry are underlined and overlined. Single-base changes that lead to operator constitutive (o^c) mutants are indicated below the duplex.

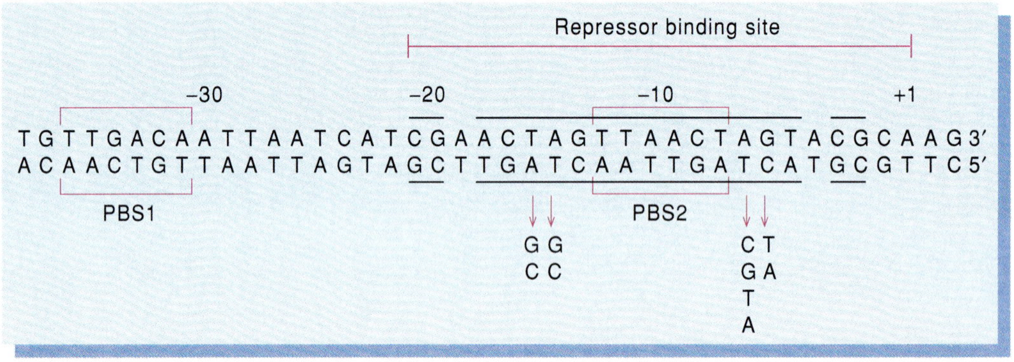

strongly involved in the binding site for repressor and probably reflect similar symmetry in the repressor itself. The repressor is composed of four identical subunits, each with a molecular weight of 12,400.

The most significant difference between the action of the trp and lac repressors relates to the function of the small-molecule effector. In the case of lac, the effector molecule allolactose acts as an antirepressor, causing release of repressor from the operator; in the case of trp, the effector molecule L-tryptophan acts as a corepressor, stimulating the binding of repressor to the operator. It should be obvious that the difference in action of these small-molecule effectors, whose concentrations dictate the level of operon activity, is well suited to the different metabolic needs of the cell satisfied by the two operons.

In Many Cases Transcription of Amino Acid Biosynthetic Operons Is Also Regulated after the Initiation Point by Attenuators

From this point on, the close parallel between the regulation of the *trp* operon and the *lac* operon ends. The *trp* operon has no positive control system like cAMP-CAP, but it regulates transcription in another way. This is a provisional stop signal called an attenuator, located 141 bases downstream from the initiation site for transcription. Only a fraction of the messenger RNAs (0.1 to 0.9, depending on metabolic conditions) that are initiated transcribe through this provisional stop signal to the end of the operon. The existence of the stop signal was first suspected when it was discovered that a genetic deletion of some of the bases between the initiation site for transcription and the first structural gene (*trpE*) raised the level of expression of the operon 8- to 10-fold. This was true even in strains with a defective repressor gene (*trpR*$^-$ strains). How could this be, and what could be the mechanism of action? An important clue was provided by sequence analysis of the 162 bases in the *trp* leader region, that is, the region between the initiation site for transcription and the initiation site for translation of the first structural gene (fig. 30.19). This leader region contains a potential initiation codon (bases 27 to 29), two tandem *trp* codons (bases 54 to 59), and a terminator codon (bases 69 to 71). A so-called leader peptide of 14 amino acids would result from translation of this region.

There are numerous reasons for believing that translation of the leader peptide up to or through the *trp* codons regulates attentuation of mRNA transcription. First of all, selective starvation of cells for tryptophan relieves attenuation and permits most RNA polymerase molecules to read through the leader region. The only other amino acid that relieves attenuation of the *trp* operon when it is lacking is arginine; arginine starvation is about 80% as effective as tryptophan starvation. Notice that an *arg* codon is located adjacent to the two *trp* codons in the leader region. Most telling of all is the finding that a leader mu-

Figure 30.19

The leader region for the tryptophan operon. The region of the leader RNA containing the hypothesized leader polypeptide is shown. The translation start of the trpE protein is also shown.

Leader polypeptide

Leader start site — +27 … +54 … +69

MET-LYS-ALA-ILE-PHE-VAL-LEU-LYS-GLY-TRP-TRP-ARG-THR-SER-

pppAAGUU ~ AUG AAA GCA AUU UUC GUA CUG AAA GGU UGG UGG CGC ACU UCC UGAAACGGGC~

trpE start site — +163 — trpE protein

MET-GLN-THR-GLN-LYS-PRO

+141

GCGUAAAGCAAUCAG ~ CCUAAUGAGCGGGCUUUUUUUUGAACAAAAUUAGAGAAUAACA AUG CAA ACA CAA AAA CCG

tation resulting in the replacement of the AUG start codon by AUA, which should eliminate translation of the leader peptide, also prevents transcription beyond the attenuator.

Other experiments indicated that the fraction of $tRNA^{Trp}$ that is charged with an amino acid is a crucial factor in the attenuation response. This has been examined *in vivo* by comparing the *trp* operon enzyme levels in *trpR*$^-$ strains that are otherwise normal with strains that are defective in some respect in charged $tRNA^{Trp}$ (table 30.2). Such structural defects in $tRNA^{Trp}$ or in the charging enzyme elevates expression, probably by permitting polymerase to transcribe through the attenuator.

These results support the hypothesis that transcription read-through requires partial translation of the leader sequence. However, only if the translation pauses or stops in the region where the *trp* or *arg* codons occur is read-through favored. A careful examination of the secondary-structure possibilities in the attenuator region suggests why this is so. The leader region RNA between bases 50 and 141 has the potential to form a variety of base-paired conformations. Figure 30.20 illustrates the most likely secondary structures that form in terminated *trp* leader RNA. These are based on analysis of regions of the transcript that show resistance to RNase T1 digestion under mild conditions and the base pairing established by studies of defined oligonucleotides. Four regions of base pairing that can form three stem-and-loop structures have been proposed. Region 1, which includes the tandem *trp* codons and the leader peptide translation stop codon (bases 54 to 68), can base-pair with region 2 (bases 76 to 91). Although region 2 (bases 74 to 85) also should be able to base-pair with region 3 (bases 108 to 119), stem-and-loop 2 · 3 has not been observed *in vitro,* presumably because stem-and-loop 3 · 4 and stem-and-loop 1 · 2 form preferentially. Region 3 (bases 114 to 121) can base-pair with region 4 (bases 126 to 134). The existence of this stem-and-loop is inferred from the resistance of the GC-rich region from residue 107 to the 3′ end of the transcript to RNase T1 digestion. The stem-and-loop structure formed between regions 3 and 4, followed by a sequence of U residues, is a common structure in transcription termination (see chapter 28). Hence we would expect that conditions under which this structure is preserved would favor transcription termination. In support of this conjecture, a number of single-base replacement mutations have been isolated that lead to mispairing in the 3 · 4 stem, and all of these lower the level of transcription termination in the leader to some extent.

The absence of translation of the leader would not perturb the 1 · 2 and 3 · 4 structures (fig. 30.21*a*). Consistent with this fact, changing the initiator codon of the leader peptide by a single base prevents read-through, as we discussed earlier. On

Table 30.2
Attenuation of the *trp* Operon Is Affected by *trpT, trpX,* and *trpS* mutations in *trpR*$^-$ Strains

Strain (*trpR* Derivative)	*trp* Enzyme Levels (Normalized)	
	Wild Type (*trpa*$^+$)	*trp*Δ*LD102 (trpa*$^-$*)*
Wild type	1.0	7.1
trpT$_{ts}$	7.1	—
trpX	4.5	6.7
trpS9969	2.7	7.5

All strains were *trpR*$^-$ (derepressed) and grown in the presence of excess tryptophan; *trp*Δ*LD102* deletes the attenuator, *trpT*$_{ts}^-$ is a temperature-sensitive lesion in the $tRNA^{Trp}$ gene. *trpX*$^-$ leads to a defect in the modification of the adenine adjacent to the anticodon sequence of the tRNA. *trpS9969* is a mutant in the tryptophenyl-tRNA synthase gene; the resulting enzyme has one-hundredth the affinity for tryptophan shown by the wild-type enzyme, resulting in much lower levels of charged $tRNA^{Trp}$ than normal.

Figure 30.20

Proposed secondary structures in the leader RNA. Four regions, labeled 1, 2, 3, and 4 at the left, can base-pair to form three stem-and-loop structures (1.2, 2.3, and 3.4). The arrows in the main figure mark the RNAse T1 cleavage sites.

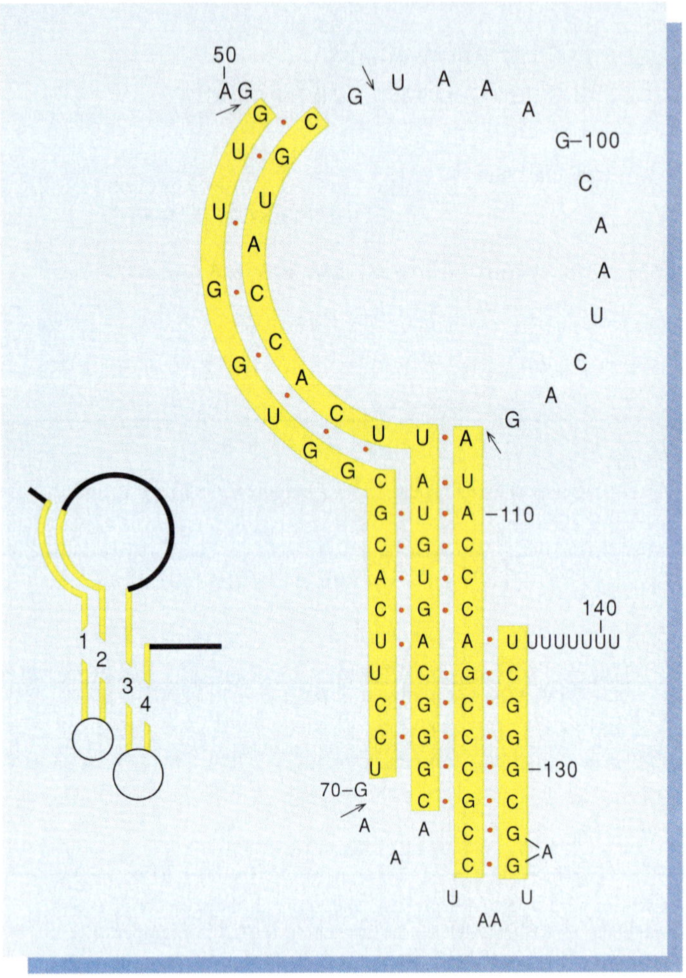

Figure 30.21

Model for attenuation in the *trp* operon, showing ribosome and leader RNA. (*a*) Where no translation occurs, as when the leader AUG codon is replaced by an AUA codon, stem-and-loop 3.4 is intact and termination in the leader is favored. (*b*) Cells are selectively starved for tryptophan so that the ribosome stops prematurely at the tandem *trp* codons. Under these conditions, stem-and-loop 2.3 can form, and this is believed to lead to the disruption of stem-and-loop 3.4, permitting attenuation. (*c*) All amino acids, including excess tryptophan, are present so that stem-and-loop 3.4 is present.

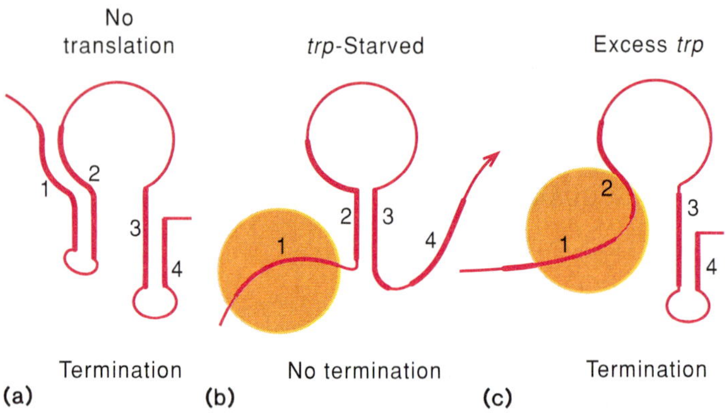

the other hand, selective starvation resulting from either tryptophan or arginine deprivation stimulates transcription read-through. Probably this is because the ribosome stalls in the region of the *trp* or *arg* codons (bases 54 to 62). The resulting rupture of the base-pairing 1 · 2 structure would make region 2 available for pairing with region 3. This would encourage disruption of the stem-and-loop 3 · 4 structure and thus permit transcription read-through (fig. 30.21*b*). In the presence of an adequate supply of all amino acids, translation would proceed beyond this critical region, so that region 2 would not be available for base pairing and the 3 · 4 loop would be maintained, an arrangement favoring transcription termination at the attenuator (fig. 30.21*c*).

This attenuator mechanism of control is amazingly simple, since it requires no proteins other than those normally used for transcription and translation. We would therefore expect the same mechanism to be used repeatedly for other operons involved in amino acid biosynthesis. An indication that the attenuator mechanism might be functioning would be the finding that the biosynthetic enzymes for a particular amino acid were increased when charging of the cognate tRNA was reduced. Indeed, for the several other amino acid biosynthetic pathways in *E. coli* for which tRNA charging is involved in regulation, attenuator mechanisms have been found. Analysis of the leader region preceding the structural genes for the biosynthesis of three such amino acids in *E. coli,* phenylalanine, histidine, and leucine, reveals structures with the features of this type of control mechanism: (1) leader sequences containing several consecutive codons for that amino acid followed by (2) sequences that can be organized into stem-and-loop structures that could signal termination. Even the multivalently controlled* threonine operon (*thr*) and isoleucine and valine operons (*ilv*) of *E. coli* exhibit this mechanism of expression. The leader sequence of the *thr* operon specifies a 21-amino-acid peptide containing 8 threonine codons and 4 isoleucine codons, and that of the *ilv* operon specifies a 32-amino-acid peptide, of which 14 are leucine, isoleucine, or valine. Comparable studies of one of the genes for arginine biosynthesis (*argF*), however, failed to show attenuation. This gene, which is required for arginine biosynthesis, appears to be exclusively regulated by a repressor mechanism. By contrast, the *his* operon has no repressor and appears to be regulated exclusively with an attenuator.

In view of these findings, why are both types of transcription controls used in the *trp* and certain other operons? Two conceivable advantages have been suggested. First, for a system

*The *thr* operon is repressed in the presence of excess threonine and isoleucine. Repression of the *ilv* operon requires leucine, isoleucine, and valine in excess.

with two metabolic switches, the amplification factor between fully on and fully off should be the product of what could be achieved with either switch alone. Second, under some conditions, for example, during very rapid protein synthesis, it is possible that a higher level of amino acid is required to keep the tRNA at an adequately charged level. Finally, we should note that the *trpR* protein has other roles; it is also a repressor for the *trpR* gene and for the *aroH* gene, which encodes one of three isozymes that catalyze the first reaction in aromatic biosynthesis.

Genes for RNA Polymerase and Ribosomes Are Coordinately Regulated

Over 100 genes in *E. coli* participate in the synthesis of the RNAs and proteins that comprise the enzymatic machinery for transcription and translation. The relevant gene products make up between 20 and 40% of the dry cell mass. In rapidly growing cells, about 85% of the RNA is ribosomal, 10% is tRNA, and most of the remainder is mRNA. The various RNAs and proteins are produced according to their need. For ribosomes and tRNAs, the synthesis rate is roughly proportional to the cell growth rate; the relative amounts of the three rRNAs (16S, 23S, and 5S), the 60 or so tRNAs, and the 50 ribosomal proteins is consistent with the stoichiometric needs for making ribosomes. RNA polymerase, which contains four different subunits (see chapter 28), is produced in equimolar amounts of β and β', approximately equal in sum to the number of α subunits. The dissociable σ subunit of RNA polymerase is produced in somewhat less than stoichiometric amounts. The synthesis of RNA polymerase subunits is loosely coupled to the synthesis of ribosomal protein subunits. Demands for large amounts of all these RNAs and proteins and precise regulation of their relative amounts have led to complex gene arrangements and controls at both the transcription and translation levels.

Control of rRNA and tRNA Synthesis by the* rel *Gene Under conditions of rapid growth, *E. coli* cells contain about 10^4 ribosomes. The maximum rate of reinitiation at the ribosomal gene promoter is about one per second. In rapid growth, *E. coli* can duplicate once every 20 min, which would allow for the synthesis of only about 1,200 molecules of rRNA if there were only one gene for ribosomal RNA. In fact, there are seven copies for ribosomal RNA operons in the bacterial chromosome, which makes it possible for rRNA synthesis to maintain the necessary pace under conditions of rapid growth. These operons are dispersed at seven locations around the circular *E. coli* chromosome. Each operon is transcribed into one long transcript that is processed into one each of 16S, 23S, 5S, and usually at least one 4S transfer RNA (fig. 30.22). The order of the genes in the operon, starting from the initiation site for transcription, is 16S, 4S, 23S, and 5S. In some of the operons, additional 4S genes for tRNA are located downstream from the 5S gene. All the known ribosomal RNA operons appear to have two strong promoters located in tandem; this arrangement probably permits a more rapid rate of reinitiation for the genes than would otherwise be possible.

Figure 30.22

A typical rRNA (*rrn*) operon contains two promoters and genes for 16S, 23S, and 5S rRNA and a single 4S tRNA gene. The four fully processed RNAs are derived from a single intact 30S primary transcript.

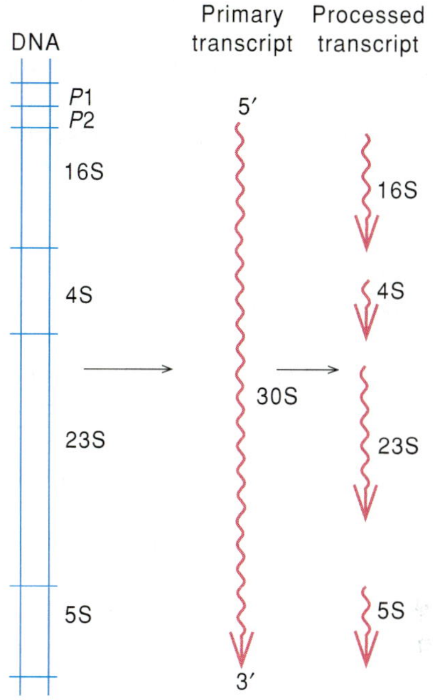

As we stated earlier, rRNA synthesis is usually maintained at a rate that is proportional to the gross rate of protein synthesis. In a normal wild-type cell, when protein synthesis is limited (e.g., by amino acid availability), Michael Cashel and John Gallant showed that there is a rapid rise in ppGpp concentration from about 50 to 500 μM (fig. 30.23). Concomitantly, there is an abrupt cessation of rRNA synthesis. This is part of the phenotype known as the stringent response: First, amino acid deprivation or other factors that slow down protein synthesis provoke an increased rate of accumulation of ppGpp; this in turn leads to an inhibition of rRNA synthesis.

Observations on different mutants indicate that the concentration of ppGpp is regulated by a careful balance between its rate of synthesis (controlled by the *rel* gene) and rate of breakdown (controlled by the *spoT* gene). Correlated observations indicate that the synthesis of rRNA is inversely proportional to the ppGpp concentration. In a *relA* mutant cell, neither the rapid rise in ppGpp concentration nor the cessation of rRNA synthesis is seen when amino acids are removed. The *relA* gene encodes a protein that is required for ppGpp synthesis. If amino acids are reintroduced into the growth medium of a wild-type culture, the ppGpp concentration falls rapidly (half-life about 20 s), and the rate of rRNA synthesis rises rapidly. In a mutant called *spoT,* the normal rise in ppGpp level is observed on amino acid starvation, but the level of ppGpp falls much more slowly on readdition of amino acids to the growth medium. Correlated with this behavior, the rate of rRNA synthesis in a *spoT* mutant also increases very slowly on readdition of amino acids.

The synthesis of ppGpp has been studied both in crude cell-free extracts of *E. coli* and in a partially purified system to determine what factors influence its rate of synthesis. It was found that ppGpp is synthesized on the ribosome from GTP in the presence of the protein encoded by the wild-type *relA* gene. Maximum ppGpp synthesis occurs in the presence of ribosomes associated with mRNA and uncharged tRNA with anticodons specified by the mRNA. If the uncharged tRNA bound to the ribosome acceptor site (A site) is replaced by charged tRNA, the rate of ppGpp synthesis is greatly lowered. If uncharged tRNA anticodons are not complementary to mRNA codons exposed on the ribosome for protein synthesis, ppGpp synthesis does not occur.

Cell-free synthesis studies also strongly support the notion that ppGpp directly inhibits RNA synthesis. Thus in a cell-free system the DNA-directed synthesis of rRNA with *E. coli* RNA polymerase is strongly and selectively inhibited by 100 to 200 μM ppGpp. Such experiments have led to the hypothesis that ppGpp, by binding to RNA polymerase, alters its structure so that it has a lowered affinity for rRNA promoters.

The *in vivo* and *in vitro* studies on ppGpp and rRNA have resulted in a model for how the level of amino acid charging of tRNA controls the rate of rRNA synthesis (fig. 30.24). First, uncharged tRNA that is codon-specific for the exposed codons on the mRNA becomes bound to the ribosome acceptor site, creating a situation unfavorable for protein synthesis but favorable for ppGpp formation. Second, the ppGpp diffuses and binds to RNA polymerase, thereby lowering its affinity for rRNA promoters.

Figure 30.23

Guanosine tetraphosphate (ppGpp) concentration under normal conditions, after amino acid starvation, and after readdition of amino acids (• = wild-type cells; ▲ = *relA* cells; and ■ = *spoT* cells).

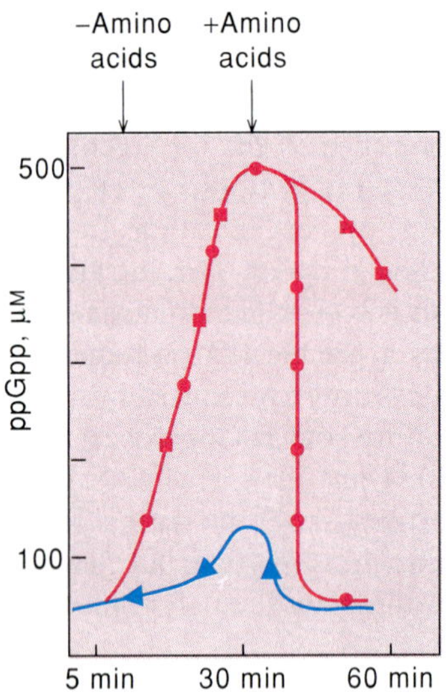

This alteration of the RNA polymerase by ppGpp affects the ability of RNA polymerase to interact with promoters in a differential way. For the promoters of the rRNA operons, the polymerase-promoter interaction is strongly inhibited by ppGpp. For certain other promoters the effect of ppGpp is actually stimulating. As first observed in the cell-free system, ppGpp stimulates expression from the *lac* and *trp* operons. In much more recent studies, Cashel has found that the biosynthetic operons associated with arg, gly, his, leu, lys, phe, ser, thr, and val biosynthetic enzymes have an absolute requirement for ppGpp. The role of ppGpp in these examples seems clear. An absence of amino acid leads to ppGpp buildup which then stimulates the genes required for biosynthesis of the amino acid.

The observations on ppGpp's role in rRNA synthesis show that this small molecule is an important control factor regulating rRNA synthesis, but it does not eliminate the possibility that other factors affect the level of rRNA. *In vivo* and *in vitro* evidence indicates that the inhibitory effect of ppGpp on transcription extends to most tRNA and ribosomal protein genes. Ribosomal protein gene expression also appears to be regulated at the translational level.

Figure 30.24

Schematic diagram of ppGpp synthesis and the hypothesized mechanism for its action. ppGpp is synthesized on the ribosome when there is a peptidyl-tRNA on the P site of the ribosome and an uncharged tRNA on the A site. The ppGpp probably inhibits rRNA synthesis by complexing with the RNA polymerase.

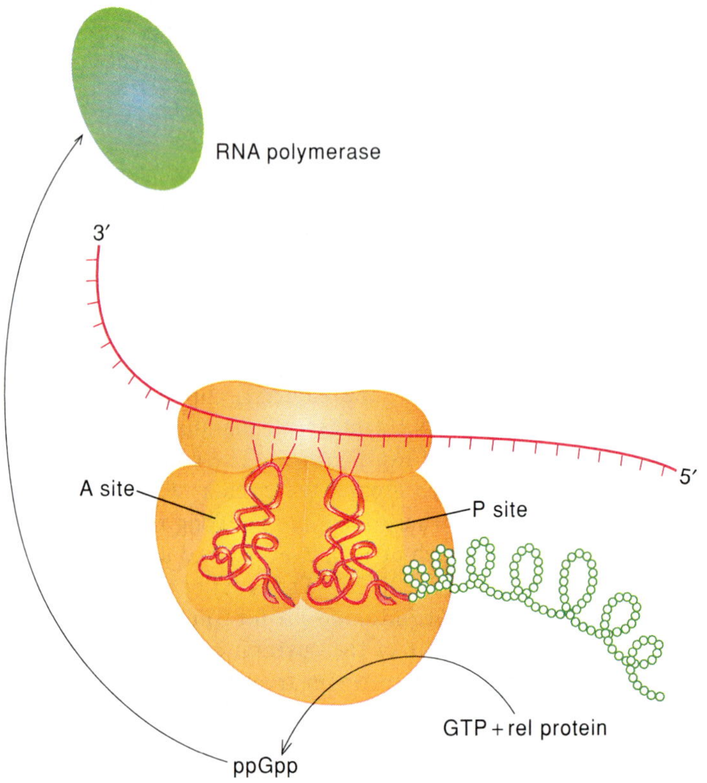

Translational Control of Ribosomal Protein Synthesis In exponentially growing *E. coli* cells, synthesis rates of all ribosomal proteins (except L7, L12, and S6) are identical and coordinately regulated. Nomura and his co-workers have suggested that free ribosomal proteins inhibit the translation of their own mRNA and that as long as the assembly of ribosomes removes ribosomal proteins, the corresponding mRNA escapes this feedback inhibition. This hypothesis has been tested *in vitro* using a protein-synthesizing system with various template DNA molecules carrying ribosomal protein genes; it has been tested *in vivo* by examining the effect of overproduction of certain ribosomal proteins on the synthesis of other ribosomal proteins using various recombinant plasmids. By these means it was found that certain ribosomal proteins selectively inhibit the synthesis of other ribosomal proteins whose genes are part of the same operon as their own; this autogenous or self-imposed inhibition occurs at the level of the translation of the mRNA rather than at the level of transcription of mRNA. Figure 30.25 illustrates how this scheme works for the regulatory protein L1. L1 and L11 are encoded by the P_{L11} operon. L1 can form a complex with either the 5′ end of its own mRNA or with 23S rRNA. It binds more strongly to the 23S rRNA. However, if there is an excess of L1 over the available 23S rRNA, then L1 will bind to the 5′ end of the mRNA, thereby inhibiting the synthesis of both L1 and L11. In this way the amounts of L1 and L11 protein synthesized are kept in register with the amounts of rRNA synthesized. Other operons carrying ribosomal protein genes are believed to be regulated in a similar manner.

A meaningful overall pattern of control appears to be emerging that is consistent with the requirement to balance the relative amounts of rRNA and ribosomal proteins synthesized. Most of the ribosomal protein genes occur in clusters or operons with promoters at one end (e.g., fig. 30.26). In some cases, individual operons are regulated by one of the encoded ribosomal proteins; in other cases, operons appear to be subdivided into "units of regulation," and individual units of regulation are regulated by one of the gene products (e.g., the P_{SPC} operon in fig. 30.26).

Figure 30.25

Schematic diagram explaining autogenous inhibition by L1. L1 can form a complex either with the 5′ end of its own mRNA or with 23S rRNA. If there is an excess of L1 over the available 23S rRNA, then L1 will bind to the 5′ end of the mRNA and inhibit both L1 and L11 synthesis. The inhibition of L1 translation by this binding depends on the fact that the translation of L1 and L11 is somehow coupled.

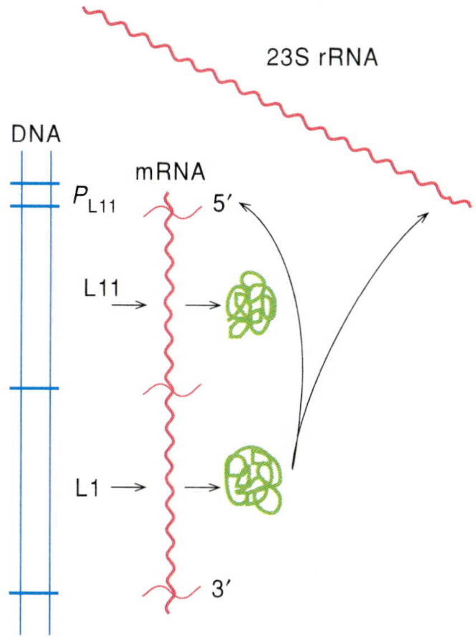

Regulation of RNA Polymerase Synthesis Is Coordinated with Ribosomal Protein Synthesis The genes for the RNA polymerase subunits β and β' (*rpoB* and *rpoC,* respectively) and L7/L12 and L10 (*rplJ* and *rplL,* respectively) belong to a single transcription unit (fig. 30.27). The gene for the α subunit (*rpoA*) is cotranscribed with four ribosomal protein genes coding for the proteins S13, S11, S4, and L17 (see fig. 30.26). This suggests that the regulatory system for RNA polymerase subunit synthesis and that for ribosomal protein synthesis are shared. Even more fascinating is the finding that the gene for the σ subunit of RNA polymerase (*rpoD*) is located on the same operon as the gene for DNA primase (*dnaG*) and the small ribosomal subunit protein S21 (*rpsU*). This operon, known as the σ operon, clearly encodes proteins essential to DNA, RNA, and protein synthesis, which suggests that regulation of the operon may be important to balance all three of these biosynthetic processes.

Despite the use of a common promoter, the synthesis of the RNA polymerase proteins and certain nearby ribosomal proteins is not always coordinately regulated. We will discuss only the β operon, because it is better understood. For brief times, synthesis of L7/L12 and L10 is inhibited *in vivo* under conditions of amino acid deprivation, while the synthesis of β and β' is not. Furthermore, the RNA polymerase inhibitor rifampicin causes a transient stimulation of β and β' synthesis, but not of L10 and L7/L12. This evidence for lack of coordinate regulation implies the existence of a separate regulatory device for

Figure 30.26

A giant cluster containing 25 r-protein genes and the gene for RNA polymerase. This cluster is divided into three operons with promoters P_α, P_{SPC}, and P_{S10}. The corresponding messages are subject to autogenous inhibition at the translation level by S4, S8, and L4 protein, respectively. S8 regulates only genes downstream from L5; these include S14, S8, L6, L18, S5, L30, and L24. Translation is also autogenously regulated by one or both of these proteins. Parentheses around S19 and L22 indicate that the order of genes is uncertain.

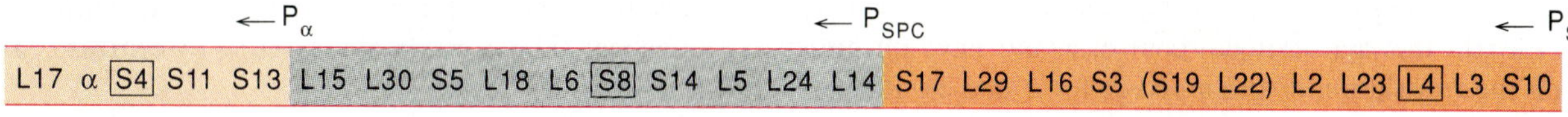

Figure 30.27

DNA and transcripts in the β operon. Only a small number of transcripts read through the attenuator (ATT) located 69 nucleotides beyond the 3′ end of the *rplL* gene. The read-through transcripts are cleaved by RNase III about 200 nucleotides beyond the end of the *rplL* gene. L10 ribosomal protein and RNA polymerase are believed to inhibit translation of the promoter-proximal and promoter-distal messages, respectively. L10 may only inhibit its own synthesis. Normally about five times as much L7/L12 is synthesized at L10. Ribosomes can initiate independently at the initiation site for L7/L12, and it is possible that L7/L12 may regulate this initiation. Bruckner and Matzura have shown that another promoter, two genes upstream from the P_{L10} promoter, may be the true initiation point for this operon *in vivo*.

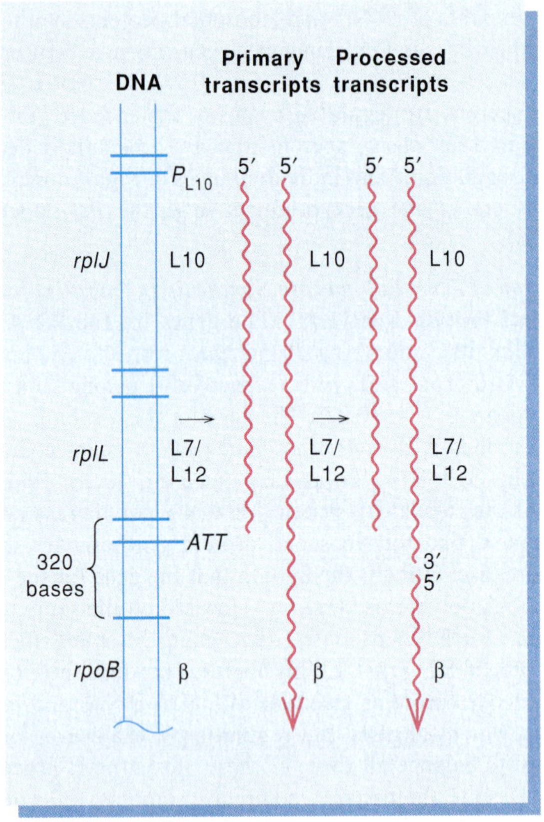

controlling L10 and L7/L12 and for controlling β and β' subunit synthesis. There are 320 untranslated bases between the *rplL* gene and the downstream *rpoB* gene (see fig. 30.27); this is a suspiciously large segment for an intercistronic region, suggesting that it functions as more than just a spacer between cistrons. Under normal growth conditions, only 20% of the transcripts that initiate at the L10 promoter P_{L10} read through the entire operon. The remainder terminate 69 nucleotides beyond the 3′ end of the *rplL* gene. The nonattenuated transcript is normally cleaved by RNase III in the intercistronic region at a point about 200 nucleotides beyond the end of the *rplL* gene. This cleavage divides the mRNA for the operon into two segments, one encoding the ribosomal proteins and the other the two RNA polymerase proteins. The L10 ribosomal protein is an autogenous regulator, inhibiting translation of its own synthesis. Clearly this translational control would have no direct effect on the translation of the message for β and β'. Recently it has been discovered that RNA polymerase holoenzyme inhibits the translation of β and β'. Thus both segments of the operon appear to be regulated at the translational level by different specific autogenous regulatory proteins.

The finding that RNA polymerase inhibits the synthesis of its own subunits immediately suggests an explanation for the differential *in vivo* effect of amino acid deprivation (increased ppGpp) and rifampicin on expression of the ribosomal protein genes and the polymerase protein genes. Whereas ppGpp or rifampicin should inhibit initiation of transcription at the P_{L10} promoter, inhibiting both ribosomal protein and polymerase subunit synthesis, the binding of either rifampicin or ppGpp to RNA polymerase could interfere with its ability to act as an autogenous inhibitor. This interference should stimulate β and β' synthesis and affect all the messages for β and β' already synthesized, as well as any that are in the process of being made.

Overview of Gene Regulation in E. coli

We have seen that in most cases gene expression in *E. coli* is regulated at the level of transcription. However, there are notable exceptions. For example, the case of the ribosomal protein genes, where autogenous regulation plays an important controlling role at the translational level. Controls for the expression of any particular gene are frequently multiple and organized into a hierarchy. In the case of the *lac* operon, for example, transcription is positively regulated by the cAMP-CAP complex that also regulates a large number of catabolic genes in a similar manner; the *lac* operon is also negatively regulated in a highly specific way by the *lac* repressor and lactose.

Bacterial regulatory systems are geared toward a rapid response to changing environmental conditions. Since environmental conditions may change in either direction, the regulatory responses must also be of a highly reversible type. Bacterial systems achieve this flexibility by maintaining low levels of regulatory proteins, which are controlled by the concentration of small-molecule effectors. The concentration of the effector molecules can vary over wide limits.

For the remainder of this chapter we will turn our attention to bacteriophage control systems. The patterned responses of phage systems have led to very different strategies for control of gene expression. By and large, phage systems are regulated by the concentrations of the regulatory proteins themselves; these regulatory proteins are not subject to the influence of small-molecule effectors.

Host-Virus Relationships

Before leaving the subject of bacterial gene regulation, let us review what we know about some of the parasites of *E. coli*. Most bacterial parasites can be classified as plasmids or viruses. Plasmids are DNA molecules that have no extracellular form; they are the ultimate parasites because they are totally dependent on host enzymes for their survival. Therefore it is not surprising that plasmids replicate in moderation, using the normal host RNA polymerase, and interfere little with normal host metabolism. The lifestyle of viruses is, by contrast, far more varied and complex.

Table 30.3
Sizes of Some *E. coli* Phages and Genomes

Phage	Particle Shape	$M_r(\times 10^6)$	DNA Length (bp × 10³)	Structure of Chromosome	Special Properties of Genome
φX174	Icosahedron	1.8	5.4[a]	Single strand, circular	Duplex replicative form
M13	Filament	2.1	6.4[a]	Single strand, circular	Duplex replicative form
T7	Head, short tail	26.5	40.0	Duplex, linear	Terminal redundancy
λ	Head, tail	32.0	48.6	Duplex, linear	Cohesive ends form replicative circles; lysogenic
P1	Head, tail	58.6	88.0	Duplex, linear	Terminal redundancy; circularly permuted
T4	Head, tail	110.0	166.0	Duplex, linear	Terminal redundancy; circularly permuted
R17	Icosahedron	1.0	3.0[a]	Single strand, linear	RNA (not DNA)
Mu		25.00	38.00	Duplex, linear	Replicates in the integrated state
T5	Head, tail	75.0	113.0	Duplex, linear	One strand interrupted

[a]For single-stranded phages, length is for bases (b) rather than base pairs (bp).

Viruses fall into three categories: lytic, nonlytic, and temperate. When lytic viruses infect a cell, they invariably redirect its metabolism to suit their own needs. Upon completion of a life cycle that produces mature virus particles, a virus-encoded enzyme disrupts the cell wall, the virus particles escape, and the cell dies. Lytic viruses include bacteriophage T4 and bacteriophage T7 (table 30.3), which redirect host cell metabolism in different ways. Infection by T4 modifies the host RNA polymerase in a manner that favors T4 phage transcription over host transcription. By contrast, when T7 infects a cell, the first few phage genes to be transcribed use the host RNA polymerase unchanged. Then the remainder of the T7 genes are transcribed by a T7-encoded RNA polymerase. This polymerase is so active that the host polymerase cannot compete effectively with it for substrate. Hence all genes requiring the host polymerase for expression are turned off by default. Like T4, the T7 virus produces a lysozyme that disrupts the cell wall, permitting the viruses to escape.

Nonlytic viruses are relatively rare (see M13 in table 30.3). These viruses are able to replicate in moderation and exit from the cell without disruption of the cell wall or cell membrane.

The most interesting type of behavior is shown by temperate viruses such as λ (fig. 30.28). Upon infecting a sensitive host, λ can adopt either an active lytic state or a dormant lysogenic state. In the active lytic state λ behaves very much like T4. The dormant state is regulated in a unique manner.

Figure 30.28

Electron micrograph of bacteriophage lambda. (Courtesy of Roger Hendrix.)

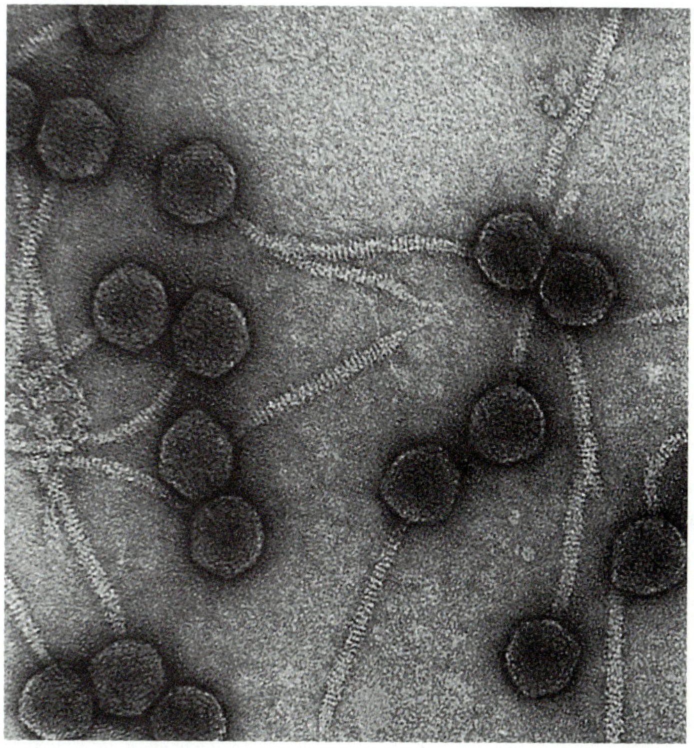

Figure 30.29

A segment of the λ genome containing the P_{RM} and nearby P_L and P_R promoters. Transcription from P_{RM} results in cI synthesis. cI binds near the promoters. An expanded view is shown of the three cI binding sites flanking the P_{RM} and P_R promoters. Transcripts that can originate from the three promoters are shown by the wavy lines. The O_{R1}, O_{R2}, and O_{R3} sites are binding sites for the cI repressor.

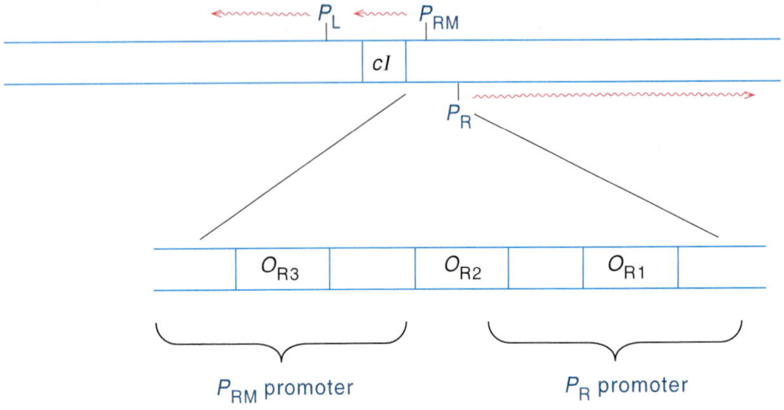

The Dormant State of λ Is Maintained by a Repressor

About half the time, when λ infects a cell it adopts a dormant lysogenic state, in which the virus is linearly integrated into the host genome. The lysogenic state is maintained by a moderate amount of the λ-encoded cI repressor. This protein binds as a dimer at two locations on the virus genome, near the promoter for early leftward transcription, P_L, and near the promoter for early rightward transcription, P_R (fig. 30.29). The cI repressor binding site near P_R overlaps the polymerase binding site at the P_R promoter. This binding site is a complex consisting of three adjacent binding sites for the cI repressor dimer, O_{R1}, O_{R2}, and O_{R3}. At moderate concentrations (about 100 copies of cI repressor per cell) O_{R1} and O_{R2} are occupied. At higher concentrations of cI the O_{R3} site is also occupied. Binding at O_{R3} prevents overproduction of cI, which would be wasteful and possibly make it difficult to induce the active lytic form of λ under the appropriate conditions. Binding of repressor at O_{R1} and O_{R2} has two effects: It inhibits early rightward transcription and it stimulates the promoter for repressor maintenance, P_{RM}. Thus at moderate concentrations, the cI repressor functions simultaneously as a repressor and an activator. It represses rightward transcription because it binds at a site that overlaps the polymerase binding site for this promoter. On the other hand, it activates the P_{RM} promoter because it binds at a site that makes favorable contact with the polymerase at the P_{RM} promoter.

As long as a moderate concentration of the cI repressor is present, λ remains dormant. Exposure of lysogenic cells to ultraviolet light or DNA-damaging drugs such as mitomycin C leads to activation of the virus. This is because DNA damage results in activation of the host recA protease, which cleaves the cI repressor and destroys its capacity to function as a repressor. The activated recA protease cleaves the cI repressor in much the same way that it cleaves the lexA repressor (see fig. 26.32).

Events That Follow Infection of E. coli *by Bacteriophage λ*

The infection process by λ starts by attachment of the mature virus particle to the bacterial membrane and injection of the viral chromosome (fig. 30.30). This is followed by circularization catalyzed by the cellular DNA ligase enzyme. Only two genes are expressed at this time, the regulatory genes *N* and *cro* from the P_L and the P_R promoters, respectively. The N protein combines with the host RNA polymerase, permitting extension of the early right and early left transcripts. This extension leads to expression of other regulatory genes—*cII, cIII,* and *Q*—as well as the genes for DNA replication and recombination. At this point the infectious process can proceed in either of two mutually exclusive directions, the lytic direction or the lysogenic direction. If the lysogenic route is followed, most early genes are turned off and only the regulatory gene *cI* and the integration gene *int* are turned on. If lysis proceeds, all early genes are turned off and the head, tail, and lysis genes are turned on.

In all, λ has six regulatory genes—*cI, cII, cIII, cro, N,* and *Q*. The developmental processes accompanying λ infection correlate with the roles of the proteins encoded by these six regulatory genes (table 30.4) and the factors that govern the level of their expression.

Very Early Expression Results in Two Short Transcripts, One from P_R *and One from* P_L

Six promoters have been identified in λ—P_L, P_R, P_{RE}, P_{int}, P_{RM}, and $P_{R'}$. Only two of of these promoters, P_L and P_R, are recognized by the unmodified host RNA polymerase (fig. 30.31). These promoters lead to the expression of the N and cro regulatory proteins. The transcript originating from the P_L promoter is terminated at the t_{L1} terminator just on the other side of the *N* gene. The transcripts originating from the P_R promoter are

Figure 30.30

Overview of λ development.

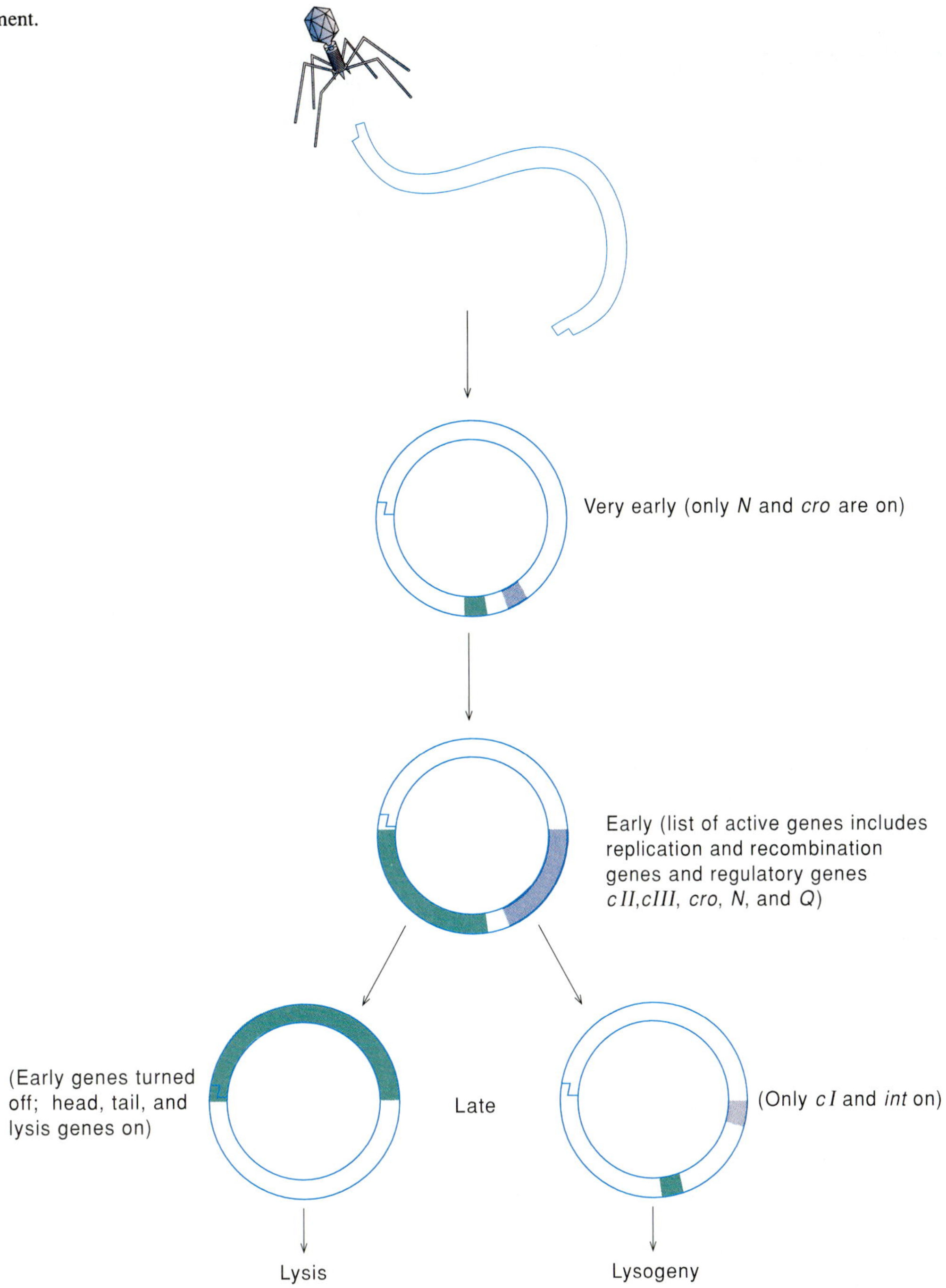

mostly terminated at t_{R1}, just on the other side of *cro*. Some rightward transcripts are not terminated at t_{R1}; these are terminated at t_{R2}.

Early Expression Leads to Extension of the Very Early Transcripts

The N protein resulting from very early expression is an antiterminator. The presence of N protein leads to extension of the transcripts originating from the P_L and the P_R promoters and the resulting expression of a number of early genes. From leftward transcription the gene for regulatory protein cIII and the recombination genes are expressed. From rightward transcription the DNA replication genes and genes for the regulatory proteins cro, cII, and Q are expressed.

The utilization of the N protein is fascinating (fig. 30.32). In the early right and early left transcripts there are sites where the RNA polymerase pauses during the process of elongating the transcripts. At these points, known as *nut* sites (for N utilization), the N protein combines with the RNA polymerase, altering its properties so that it no longer recognizes the early termination sites.

Table 30.4
Regulatory Proteins of λ

Protein	Function
cI	At low concentrations represses P_R and P_L and activates P_{RM}; at high concentrations represses P_R, P_L, and P_{RM}
cII	Activator for P_{int} and P_{RE}
cIII	Stabilizes cII
cro	At low concentrations represses P_{RM}; at high concentrations represses P_{RM} and P_R
N	An antiterminator at t_{L1}, t_{R1}, and t_{R2}
Q	An antiterminator at t_{6S}

Figure 30.32

Early transcripts. The early left transcript is initiated from the P_L promoter. In the absence of N protein this transcript terminates at t_{L1}. In the presence of N protein the polymerase picks up an N protein at the N utilization site, nut_L. This makes it possible for the polymerase to transcribe through t_{L1}. The early right transcript tends to terminate at t_{R1} unless N protein is present. In this case the N protein becomes bound to the polymerase at the nut_R site.

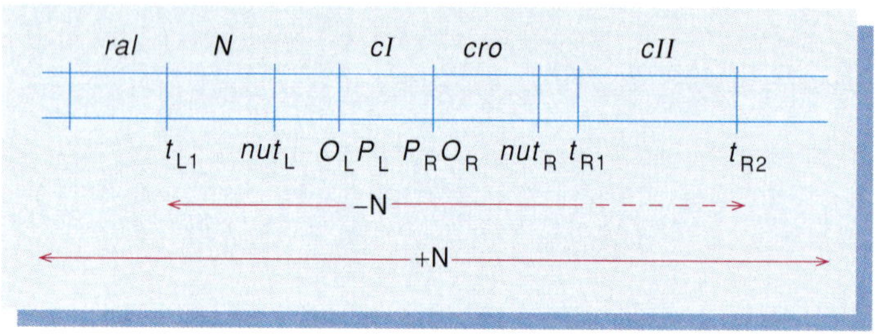

Figure 30.31

Bacteriophage lambda DNA. *Top:* Circular map indicating locations of main control genes and early and late functions. *Bottom:* Expanded region containing control genes and main promoters and terminators. Arrows indicate main transcripts and conditions under which they are active. In the absence of N protein, about half the transcription beginning at P_R reads through t_{R1} to t_{R2} (indicated by dashed portion of arrow). More information on regulatory proteins and promoters is given in table 30.4.

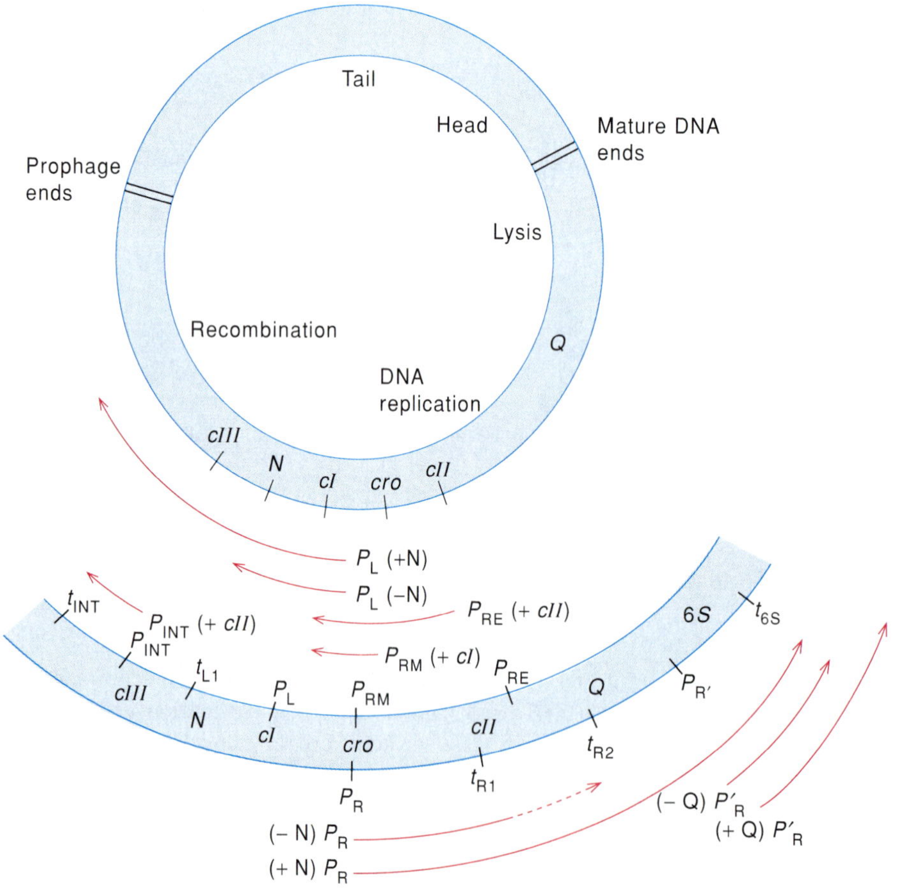

Figure 30.33

A segment of the λ genome containing the P_{RM} and nearby P_L, P_R, and P_{RE} promoters. Transcripts that can originate from the three promoters are shown by the wavy lines. The P_{RE} and P_{RM} require different activators for expression. The transcript from P_{RE} is ten times more effective in cI expression because it has a Shine-Dalgarno sequence for ribosome attachment.

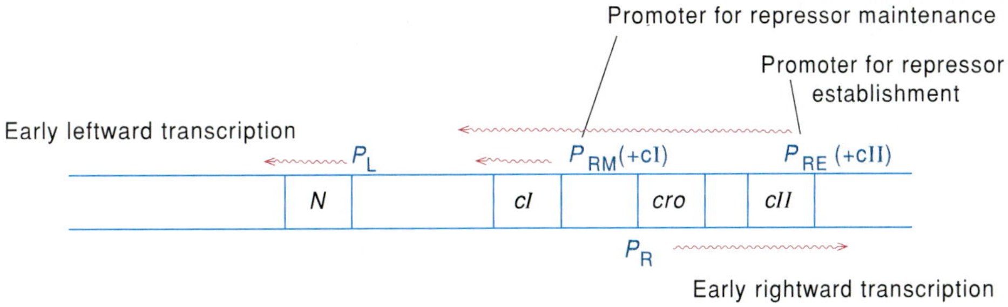

Late Expression Along the Lysogenic Pathway Results from Elevated Expression of the cII Regulatory Protein

Late expression is the phase during which commitment to the lytic or the lysogenic pathway occurs. The choice depends on the level of cII protein that develops shortly after infection. At high levels of cII the lysogenic pathway is followed: at low levels of cII the lytic pathway is followed. Precisely what controls the level of cII expression relative to other events is unclear. To some extent the level of cII expression depends on cIII expression, since the main function of cIII is to stabilize the otherwise quite unstable cII protein. The crucial importance of cII is easier to understand. The cII protein is a positive activator for the P_{RE} and the P_{int} promoters (fig. 30.33). The P_{RE} promoter leads to synthesis of the cI repressor (RE stands for repressor establishment). The cI protein is a negative regulator of the P_L and the P_R promoters. Thus the presence of low to moderate levels of cI shuts off both of the main early promoters. The P_{int} promoter, as the name implies, is responsible for synthesis of the int protein, which is required for λ integration. Integration completes the lysogeny process, and thereafter the cI protein keeps both the P_L and the P_R promoters in the "off" position. Eventually this repression silences the P_{RE} promoter also.

Since the P_{RE} promoter was responsible for the synthesis of cI in the first place, we may well ask what happens to the level of cI protein in the lysogenic cell. Under these conditions the cI gene is expressed exclusively from the less efficient transcript* originating from the P_{RM} promoter. This promoter is inactive unless cI is already present. As we have already indicated, cI at moderate levels activates the P_{RM} promoter. At very high levels, cI inhibits further cI production. In this way a level of about 100 copies of cI protein per cell is maintained. Only a traumatic incident can disturb the dormant lysogenic state. This is exactly what happens when the cell is exposed to DNA-damaging conditions. Under these conditions the pool of cI proteins is completely destroyed and the dormant bacteriophage enters the lytic cycle.

*This transcript is less efficient because the start codon is so close to the 5′ end that there is no Shine-Dalgarno sequence to favor ribosome binding.

Late Expression Along the Lytic Pathway Occurs When the Buildup of cI Repressor Is Slow

The lytic pathway is followed when the buildup of cI repressor does not occur fast enough to shut down early transcription. The extension of the rightward transcript, stimulated by the N protein, leads to expression of the *Q* gene, producing a key regulatory protein for late transcription. Like N, the Q protein is an antiterminator. Its action, which takes place at the t_{6S} terminator, leads to transcription of all of the late genes including the lysis gene and the genes for the head and tail proteins of the mature phage. At late times other factors encourage rightward transcription from the $P_{R'}$ promoter.

Cro Protein Is Essential to the Lytic Cycle

The cro protein and the cI protein bind to the same sites. Despite this fact, the cro protein is required for lysis, whereas the cI repressor is required for lysogeny. These requirements can be shown with mutants. A *cI* mutant invariably undergoes lysis, whereas a *cro* mutant can lysogenize but cannot complete the lytic cycle. This remarkable difference in the behavior of cro and cI is due to the fact that they bind to the same sites but with different relative affinities. Cro binds preferentially to O_{R3} and less strongly to O_{R1} and O_{R2}. The binding of cro to O_{R3} turns off cI expression from the P_{RM} promoter. At high concentrations cro binds to one or more of the other sites, turning off the P_R promoter. The P_L promoter is also turned off at high concentrations of cro. The most important physiological effect of cro is the turning off of the P_{RM} promoter. In the absence of cro, cI expression increases because of the increase in the number of cI gene copies resulting from early replication of viral DNA during the lytic cycle. This buildup will turn off early rightward and early leftward transcription before sufficient transcripts have been made to ensure a start on the lytic pathway. The cro protein prevents the buildup of cI protein that would otherwise result from the increase in gene copies. The binding properties of cI and cro to the tripartite operator shared by P_{RM} and P_R is shown in figure 30.34.

Figure 30.34

Segment of the λ genome showing the three operators O_{R1}, O_{R2}, and O_{R3} around the P_{RM} and the P_R promoters. The cI and cro regulatory proteins bind to these operators with different relative affinities. The net result of these differing affinities is that cI is required for lysogeny and cro is required for the lytic cycle.

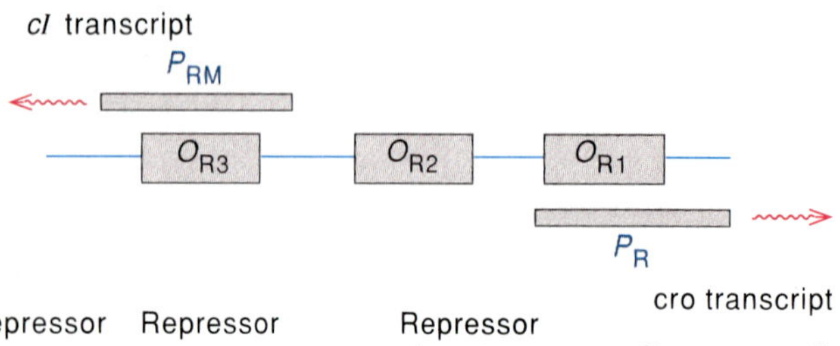

Repressor type	Repressor conc.	Repressor location			Repressor effect
cI	Low	−	+	+	Activates P_{RM}; represses P_R
	High	+	+	+	Represses P_{RM} and P_R
cro	Low	+	−	−	Represses P_{RM}
	Intermediate	+	+	−	Represses P_{RM}, P_R, and P_L
	High	+	+	+	Represses P_{RM}, P_R, and P_L

Summary

In this chapter we have discussed the regulatory systems of the *E. coli* bacterium and the λ bacteriophage. The main points in our presentation are as follows.

1. *E. coli* carries about 4,400 genes. Only a small fraction of the genome is actively transcribed at any given time. But all of the genes are in a state where they can be readily turned on or turned off in a reversible fashion. The level of transcription is regulated by a complex hierarchy of control elements.
2. In the most common form of control, expression is regulated at the initiation site of transcription. There are several ways of doing this, all of them revolving around protein or small-molecule factors that influence the binding of RNA polymerase at the transcription start site.
3. The *lac* operon, a cluster of three genes involved in the catabolism of lactose, exemplifies both positive and negative forms of control that influence the rate of initiation of transcription. Jacob and Monod identified the repressor as a negative control element that is *trans* dominant and the operator as a *cis* dominant site for binding the repressor. Transcription is initiated by RNA polymerase at the promoter, which overlaps the operator site on one side of the three structural genes of the *lac* operon. The tight complex between repressor and operator prevents initiation and it is broken when lactose is present. The lactose is readily converted to allolactose, which binds to the lac repressor. This changes the structure of the repressor so that it dissociates from the DNA.
4. Initiation of transcription proceeds at a greatly increased rate when cyclic AMP is present. This is because cyclic AMP forms a complex with the CAP activator protein, which then binds at a site adjacent to the polymerase binding site. The CAP protein enhances polymerase binding at the adjacent site by cooperative binding.
5. Lactose is the substrate of the enzymes of the *lac* operon. In the absence of lactose, there is no use for enzymes of the *lac* operon.
6. CAP and cAMP activate a large number of genes in *E. coli* that are concerned with catabolism. When glucose is present, the cAMP is greatly lowered and the *lac* operon is expressed at a very low level, even when lactose is present. This is because glucose is a more readily metabolizable carbon source than lactose.
7. The *trp* operon contains a cluster of five structural genes associated with tryptophan biosynthesis. Initiation of transcription of the *trp* operon is regulated by a repressor protein that functions similarly to the lac repressor. The main difference is that the trp repressor action is subject to control by the small-molecule effector tryptophan. When tryptophan binds the repressor, the repressor binds to the *trp* operator. Thus the effect of the small-molecule effector here is opposite to its effect on the *lac* operon. When tryptophan is present, there is no need for the enzymes that synthesize tryptophan.
8. The *trp* operon has a control locus called an attenuator about 150 bases after the transcription initiation site. The attenuator is regulated by the level of charged tryptophan tRNA, so that between 10 and 90% of the elongating RNA polymerases transcribe through this site to the end of the operon. Low levels of *trp* tRNA encourage transcription through the attenuator.
9. Ribosomal RNA and protein synthesis are both controlled at the level of initiation of transcription. This is a result of the direct binding of guanosine tetraphosphate, ppGpp, to the RNA polymerase. This binding decreases the affinity of RNA polymerase for the initiation sites of transcription. Guanosine tetraphosphate is synthesized when the general level of amino-acid-charged tRNA is low.

10. The synthesis of ribosomal proteins is regulated at the level of translation. Certain ribosomal proteins bind to specific sites on the ribosomal RNAs or their own mRNAs. In the absence of the ribosomal RNAs, they bind to their own mRNAs, which inhibits their translation. This form of translational control seems to regulate the rate of synthesis of ribosomal proteins so that it does not exceed the rate of ribosomal RNA synthesis.
11. Viruses borrow heavily on the host enzymatic machinery to obtain energy for synthesis, as well as for replication, transcription, and translation. The virus infective cycle is strongly irreversible. Virus infection is followed by the gradual turning on of viral genes. Viral enzymes are the first viral gene products; in late infection, the virus structural proteins are favored. The irreversible lytic cycle of the virus is directed by a cascade of controls.
12. In λ the host RNA polymerase is used throughout. Regulation is achieved through a series of repressors and activators, as well as two viral proteins that bind directly to the RNA polymerase. The viral proteins that bind to the polymerase modify it so that it can transcribe through provisional stop signals.
13. When λ phage infects an *E. coli* cell, it does not always produce viral progeny. Sometimes it integrates its genome into the host genome and replicates only as the host genome replicates. This so-called lysogenic state can be disrupted by DNA-damaging conditions such as exposure to UV radiation. Under these conditions the dormant viral genome enters the active replication cycle.
14. Bacterial regulatory proteins are controlled by small-molecule effectors; viral regulatory proteins are not. Bacterial genes are regulated in a highly reversible manner; viral genes are usually turned on only once.

Selected Readings

Anderson, J. E., M. Ptashne, and S. C. Harrison, Structure of the repressor-operator complex of bacteriophage 434. *Nature* 306: 846–852, 1987.

Brennan, R. G., and B. W. Matthews, The helix-turn-helix DNA binding motif. *J. Biol. Chem.* 264:1903–1906, 1989. A mini review.

Gilbert, W., and B. Muller-Hill, Isolation of the lac repressor. *Proc. Natl. Acad. Sci.* 56:1891–1898, 1966. First isolation of a repressor protein.

Gold, L., Posttranscriptional regulatory mechanisms in *Escherichia coli. Ann. Rev. Biochem.* 56:199–234, 1988.

Goodrich, J. A., and W. R. McClure, Competing promoters in prokaryotic transcription. *Trends Biochem. Sci.* 16:394–396, 1991. Two or more bacterial promoters are often found in close proximity, and may compete for the binding of RNA polymerase.

Green, P. J., O. Pines, and M. Inouye, The role of antisense RNA in gene regulation. *Ann. Rev. Biochem.* 55:569–597, 1986.

Helman, J. D., and M. J. Chamberlain, Structure and function of bacterial sigma factors. *Ann. Rev. Biochem.* 57:839–872, 1988.

Jacob, F., and J. Monod, Genetic regulatory mechanisms in the synthesis of proteins. *J. Mol. Biol.* 3:318–356, 1961. A classic paper.

Kustu, S., A. K. North, and D. S. Weiss, Prokaryotic transcriptional enhancers and enhancer-binding proteins. *Trends Biochem. Sci.* 16:397–401, 1991. First discovered in eukaryotes, enhancers have now been found to exist for a number of prokaryotic genes.

Nomura, M., R. Gourse, and G. Baughman, Regulation of the synthesis of ribosomes and ribosomal components. *Ann. Rev. Biochem.* 53:75–117, 1984.

Ptashne, M., *A Genetic Switch: Gene Control and Phage* λ. Cambridge, Mass.: Cell Press, and Palo Alto, Calif.: Blackwell Scientific, 1987.

Ptashne, M., A. D. Johnson, and C. O. Pabo, A genetic switch in a bacterial virus. *Sci. Amer.* 247(5):128–140, 1982.

Simons, R. W., and N. Kleckner, Biological regulation by antisense RNA in prokaryotes. *Ann. Rev. Genet.* 22:87–600, 1988.

Steitz, T. A., Structural studies of protein-nucleic acid interaction: the sources of sequence-specific binding. *Quarterly Reviews of Biophysics* 23:205–280, 1990. A very readable and very thorough review of the subject with excellent illustrations, written by one of the pioneers in the field.

Storz, G., L. A. Tartaglia, and B. N. Ames, Transcriptional regulator of oxidative stress-inducible genes: Direct activation by oxidation. *Science* 248:189–194, 1990.

Stwinowski, Z., R. W. Schevitz, R.-G. Zhang, C. L. Lawson, A. Joachimiak, R. Q. Marmorstein, B. F. Luisi, and P. B. Sigler, Crystal structure of trp repressor/operator complex at atomic resolution. *Nature* 335:321–329, 1988.

Weintraub, H., Antisense RNA and DNA. *Sci. Am.* 262(1):40–46, 1990.

Wolberger, C., Y. Dong, M. Ptashne, and S. C. Harrison, Structure of phage 434 Cro/DNA complex. *Nature* 335:789–795, 1988.

Yanofsky, C., Operon-specific control by transcription attenuation. *Trends Genet.* 3:356–360, 1987.

Zubay, G., M. Lederman, and J. DeVries, DNA-directed peptide synthesis III. Repression of β-galactosidase synthesis and inhibition of repressor by inducer in a cell-free system. *Proc. Natl. Acad. Sci.* 58:1669–1675, 1967. Showing that repressor works in a cell-free system.

Zubay, G., D. Schwartz, and J. Beckwith, Mechanism of activation of catabolite-sensitive genes: A positive control system. *Proc. Natl. Acad. Sci.* 66:104–110, 1970. First isolation of an activator protein.

Problems

1. What set of data originally led Jacob and Monod to suggest the existence of a repressor in *lac* operon regulation?
2. In a cell that is z^-, what would be the relative thiogalactoside transacetylase concentration, compared with wild type, under the following conditions?
   (a) After no treatment.
   (b) After addition of lactose.
   (c) After addition of IPTG.

3. Consider a negatively controlled operon with two structural genes (*A* and *B*, for enzymes A and B), an operator gene (*O*), and a regulatory gene (*R*). The first line of data below gives the enzyme levels in the wild-type strain after growth in the absence or presence of the inducer. Complete the table for the other cultures.

Strains	Uninduced		Induced	
	Enz A	Enz B	Enz A	Enz B
Haploid strains:				
(1) $R^+O^+A^+B^+$	1	1	100	100
(2) $R^+O^cA^+B^+$				
(3) $R^-O^+A^+B^+$				
Diploid strains:				
(4) $R^+O^+A^+B^+/R^+O^+A^+B^+$				
(5) $R^+O^cA^+B^+/R^+O^+A^+B^+$				
(6) $R^+O^+A^-B^+/R^+O^+A^+B^+$				
(7) $R^-O^+A^+B^+/R^+O^+A^+B^+$				

4. In a diploid situation would you expect a *crp*$^+$ to be dominant to a *crp*$^-$? Referring to *lac* expression, describe the phenotypes of *crp*$^-$, *crp*$^+$, and *crp*$^+$/*crp*$^-$.
5. Although *E. coli* promoters generally conform to a rather well-defined consensus sequence, no perfect match to this consensus has ever been observed in a naturally occurring promoter. Suggest an explanation.
6. What is the advantage of using IPTG as an inducer of the *lac* operon?
7. The lac repressor has an "on" rate constant for the binding of the *lac* operator (when cloned into λ) of about 5×10^{10} $M^{-1}\ s^{-1}$. This value is much greater than the calculated diffusion-controlled process, which is about 10^8 $M^{-1}\ s^{-1}$ for a molecule the size of the lac repressor. Explain why this repressor binding works better than expected.
8. Explain how histidine biosynthesis is controlled in *E. coli.*
9. A mutation in the *trp* leader region is found to result in a reduction in the level of *trp* operon expression when the mutant is grown in rich medium. However, when the mutant is grown in a medium lacking glycine, a stimulation in the level of trp enzymes is observed. Explain these observations. What would you anticipate would be the effect of growing the mutant in a medium lacking both glycine and tryptophan?
10. In *E. coli* there are no pools of free rRNAs or ribosomal proteins floating around in the cell even when the bacteria are grown at different growth rates. Explain how *E. coli* coordinates the biosynthesis of the ribosome.
11. Explain why rRNA synthesis is slow to resume when amino acids are added back to a *spoT* mutant in *E. coli.*
12. Describe the principal differences between patterns of control of gene expression used by bacterial host and bacteriophage systems.
13. In the infection of the host cell by λ, cI repressor favors lysogeny, while cro repressor favors the lytic cycle, even though both repressors bind to the same sites. Explain the basis of their different effects.
14. How is the synthesis of the CAP protein regulated? What is unusual about this regulation?
15. Gene regulatory proteins in bacteria were predicted (before their precise structure was known) to interact in the major groove of DNA by a two-site model of binding. Describe the data that showed this model to be correct.

31

CHAPTER

Regulation of Gene Expression in Eukaryotes

All eukaryotes carry their DNA in a membrane-bounded organelle, the nucleus. This creates a fundamental difference between the way eukaryotes and prokaryotes carry out transcription and translation. In prokaryotes the processes are strongly coupled and occur concurrently; in eukaryotes transcription is completed before the start of translation, and the nascent transcripts are subject to an elaborate regimen of posttranscriptional controls. There are other differences that affect the number and types of regulatory mechanisms. Eukaryotic cells usually contain many more genes, than prokaryotes distributed on several chromosomes. Multicellular eukaryotes contain cells of different types, which are limited in the types of genes they can express. These differences result in increasingly complex mechanisms for gene regulation in eukaryotes (fig. 31.1).

In this chapter we will emphasize modes of gene regulation that are unique to eukaryotes. First we will consider a unicellular eukaryote; then we will take a look at some of the complex problems associated with multicellular eukaryotes that show extensive cellular differentiation.

Gene Regulation in Yeast, a Unicellular Eukaryote

A cell of the budding yeast *Saccharomyces cerevisiae* occupies more than ten times the volume of an *E. coli* cell. Its genome is about four times the size of that of *E. coli,* and is distributed among seventeen chromosomes. Yeast cells are nearly spherical in shape, divide by budding, and are bounded by a cytoplasmic membrane that is surrounded by a thick polysaccharide wall (fig. 31.2). The cells contain a nucleus, an endoplasmic reticulum, ribosomes, a Golgi apparatus with secretory vesicles, and several types of granular and vesicular inclusions. The amount of endoplasmic reticulum and the number of mitochondria and ribosomes fluctuate with growth conditions suggesting that regulatory devices are at work.

The separation of transcription from translation permits elaborate schemes for the processing of nascent transcripts prior to their translation. Nascent transcripts destined to become mRNA are modified by capping the 5′ end and are tailed by 3′-terminal polyadenylic acid, (see chapter 28). Splicing occurs for some transcripts, mainly ribosomal protein messages and certain tRNAs. In mature messages the 5′ proximal initiator

Figure 31.1

Overview of some of the unique features associated with regulation of gene expression in higher organisms. Eukaryotes may be unicellular, as in the case of yeast, or multicellular, as in the case of vertebrates. Regulation of gene expression in all eukaryotes bears a close resemblance to regulation of gene expression in simple unicellular prokaryotes. Yet there are some striking differences. Thus multicellular eukaryotes are composed of differentiated cells that can express only a limited number of the total genes in their chromatin. And most of the chromatin in any particular cell type is highly condensed (heterochromatic), but a small fraction of the chromatin is expanded or swollen (euchromatic). It is the latter portion of the chromatin that has the potential to be expressed. The regions of the chromatin that are euchromatic are different in different cell types. To convert a region of the genome from the potentially active to the active state requires the binding of regulatory proteins to the promoter. The role of these proteins is to create an environment favorable for the binding of RNA polymerase. Here we come to one of the most striking differences between prokaryotes and eukaryotes. In prokaryotes the regulatory proteins usually bind in the immediate vicinity of the RNA polymerase. In eukaryotes the regulatory proteins may be bound to the DNA at several places, ranging from a point adjacent to the polymerase binding site to points a thousand or more bases away. Although the explanation for how distantly bound regulatory proteins can influence gene expression is not totally clear, the consensus is that chromosome folding is the most likely way of bringing such proteins into the immediate proximity of the RNA polymerase binding site. The function of all of the regulatory proteins binding to a single promoter of the DNA is to promote formation of the transcription initiation complex.

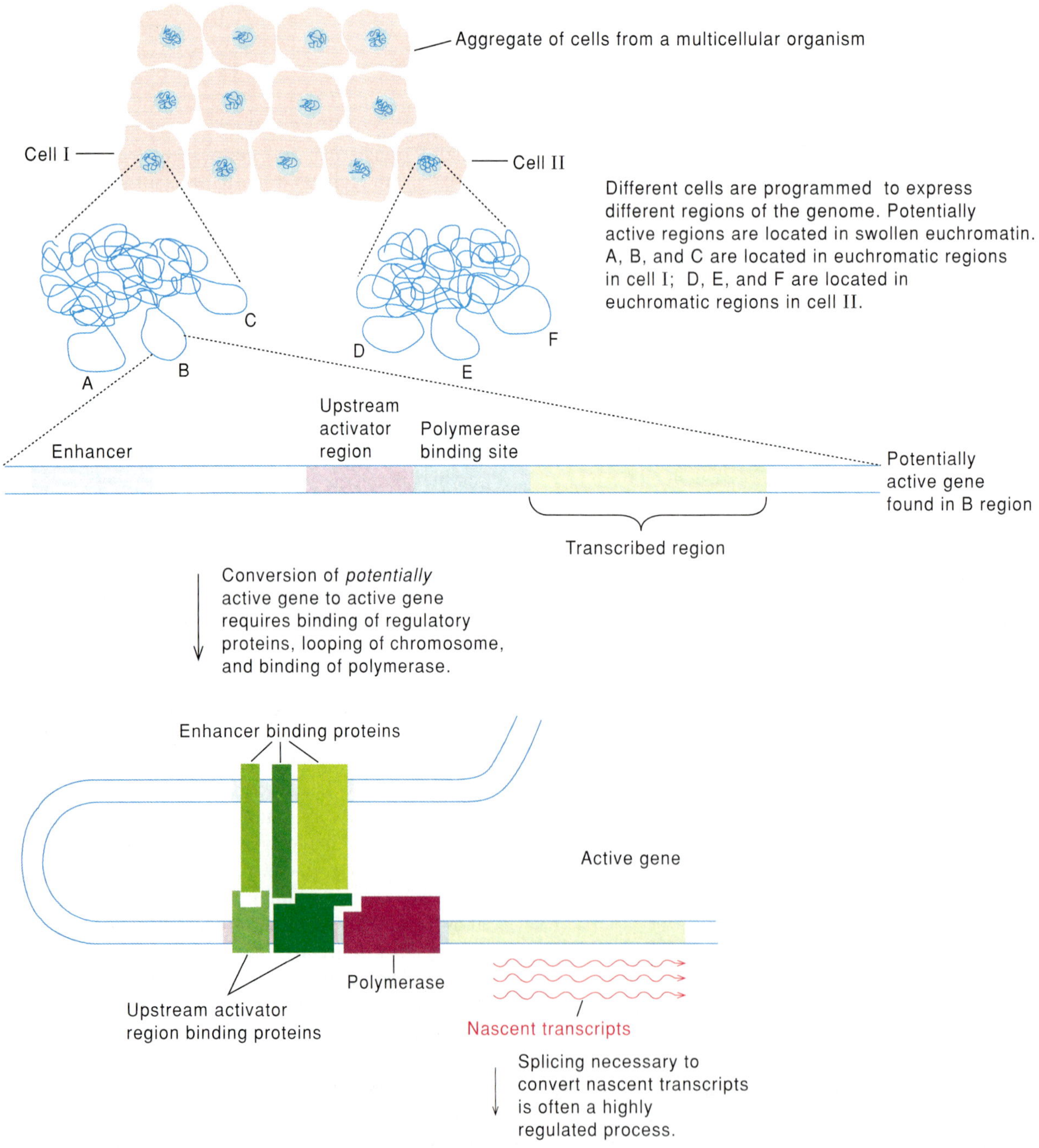

Figure 31.2

Diagram of a haploid yeast cell. A yeast cell contains many of the organelles characteristic of a typical eukaryotic cell. Duplication occurs by a budding process. The bud gradually grows until, just before pinching off, it contains a nuclear equivalent of chromosomes, as well as some mitochondria and other elements present in the cytoplasm. Mechanical strength of the yeast cell is guaranteed by a thick cell wall.

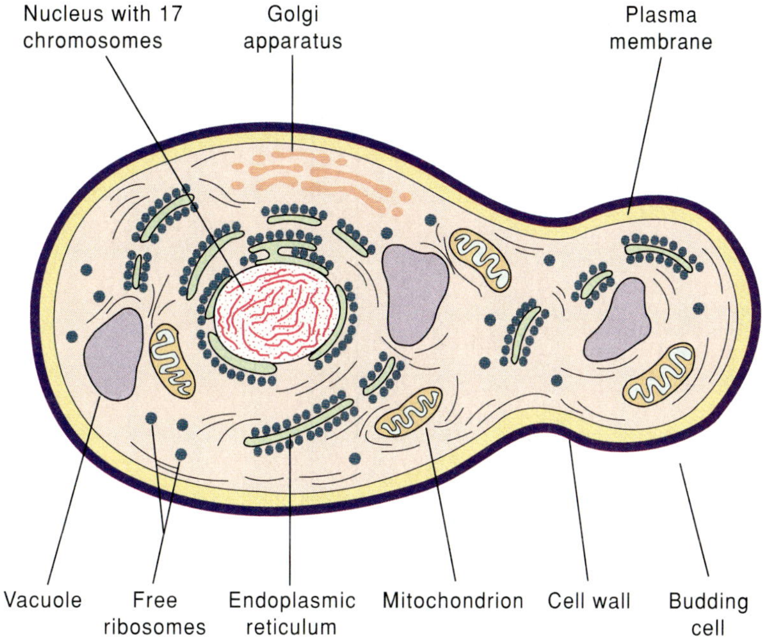

codon (AUG) is usually recognized as a translation start (see chapter 29). This restriction eliminates the value of polycistronic mRNAs and probably accounts for the greater dispersal of genes serving common pathways that are found in yeast compared with bacteria.

A comparison of the gene arrangements and messages for histidine biosynthesis in *E. coli* and *S. cerevisiae* illustrates some of the contrasting modes of gene arrangement and gene expression in these two organisms. In *E. coli,* all ten histidine biosynthetic enzymes are encoded by a single polygenic mRNA, used for the synthesis of the corresponding individual polypeptide chains. In yeast, genes for three of these activities (steps 2, 3, and 10 in the biosynthesis) are clustered at the *HIS4* locus, yielding a single transcript that specifies a single polypeptide chain of molecular weight 9.5×10^4. This polypeptide chain folds into a multifunctional enzyme that carries out three enzymatic functions (fig. 31.3) that in *E. coli* are executed by separate proteins. It is not uncommon to find multifunctional polypeptide chains in eukaryotes for functions that are carried out by two or more polypeptide chains in *E. coli.*

Galactose Metabolism Is Regulated by Specific and General Control Factors in Yeast

In yeast four enzyme activities are required for galactose utilization, each of them encoded by a unique gene. Three of these genes, *GAL7, GAL10,* and *GAL1,* are clustered on chromosome

Figure 31.3

The *HIS4* gene encodes three enzyme activities. It produces a single transcript that results in a single enzyme. The enzyme is multifunctional, carrying out steps 2 (B), 3 (A), and 10 (C) in histidine biosynthesis.

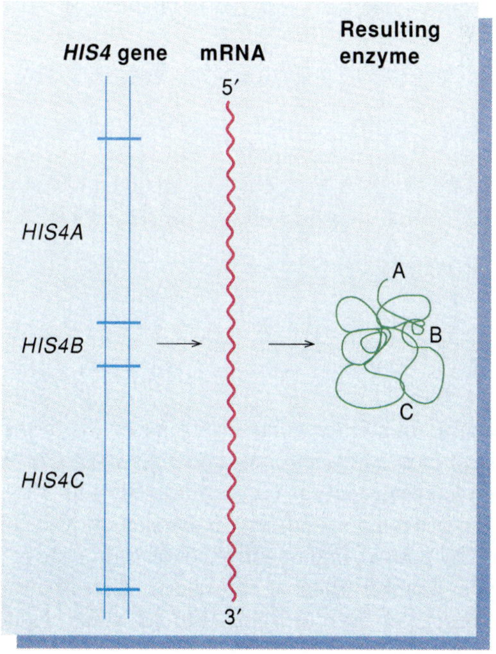

Figure 31.4

Model for the regulation of enzymes of the *GAL* system. (*a*) Three structural genes synthesize distinct mRNAs and enzymes (transferase, epimerase, and kinase). Arrows next to the genes indicate the direction of transcription. Synthesis requires the GAL4 protein. However, it is inactive in the absence of inducer because of complex formation with the GAL80 protein. The GAL80 protein does not prevent the GAL4 protein from binding to specific sites on the DNA, but it does prevent GAL4 protein from activating transcription once bound. (*b*) GAL4 protein becomes active when the GAL80 protein is removed by adding inducer, which binds to the GAL80 protein.

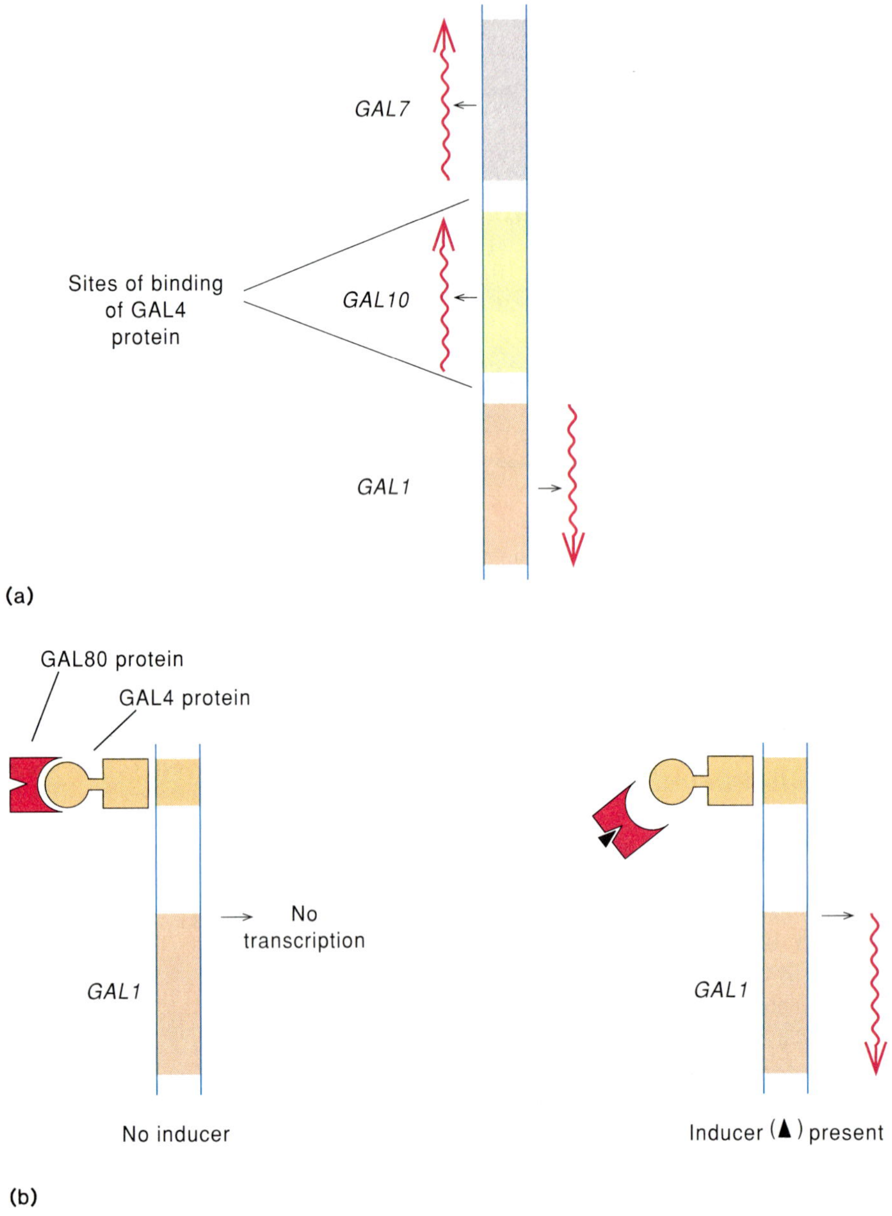

II, whereas the fourth, for galactose transport, is specified by a gene (*GAL2*) located on chromosome XII. (Note: in yeast the normal wild-type genotype is capitalized and the mutant genotype is written in lower case.) Expression of these four genes is regulated by positive and negative controls that are of a highly specific type. Transcription of the *GAL1, GAL7,* and *GAL10* genes is increased several thousandfold when galactose is present.

Specific Controls Affect Expression of the* GAL *Genes Despite their clustering and their coordinate induction by galactose, the *GAL7, GAL10,* and *GAL1* genes are not transcribed into a single mRNA molecule (fig. 31.4). The *GAL7* and *GAL10* genes are transcribed from the same DNA strand. However, the *GAL1* gene, approximately 600 base pairs from *GAL10*, is transcribed from the complementary DNA strand. Discrete transcripts are generated for the synthesis of three distinct proteins that contain the different enzyme activities: transferase (GAL7), epimerase (GAL10), and kinase (GAL1).

The yeast system is ideally suited for genetic studies as the organism can be grown in either the haploid or the diploid state. Genetic analysis of the *GAL* system indicates two *trans*-acting gene products that regulate expression, GAL4 and GAL80. Recall from our observations on prokaryotes that *trans*-acting gene products usually are proteinaceous. Most *gal4* mutants are uninducible for the *GAL* structural genes in the haploid state, but inducible in the diploid state when paired with a *GAL4* wild type. This suggests that the active GAL4 product is probably a positive control protein like CAP in the *lac* operon. Rare *GAL4*[c] mutants result in constitutive expression of the *GAL* genes. This suggests that the structure of the GAL4 protein is modified so that its function is no longer repressible. Most *gal80* mutants also give rise to constitutive expression of the *GAL* genes. This suggests that the active form GAL80 is a repressor of *GAL* gene expression. Rare *GAL80*[s] mutants are uninducible in the haploid state, or in the diploid state where they are paired with wild-type *GAL80*. It is believed that wild-type GAL80 protein binds the inducer galactose and that this binding converts the protein to an inactive form that does not inhibit the action of the GAL4 protein. The GAL80[s] protein appears to have lost the site for galactose binding. The results of the different mutants for *GAL1* and *GAL10* gene expression are summarized in table 31.1.

Table 31.1
Phenotypes of Haploid and Diploid Yeast with Regulatory Gene Mutations

Genotype	GAL10 (epimerase)[a] Noninduced	Induced	GAL1 (kinase) Noninduced	Induced
Wild type	−	+	−	+
gal4	−	−	−	−
gal4/GAL4	−	+	−	+
gal80	+	+	+	+
GAL4[c]	+	+	+	+
GAL80[s]	−	−	−	−
GAL80[s]*/GAL80*	−	−	−	−

[a]In the induced state, galactose is added to the growth medium. It should be noted that GAL10 and GAL1 respond in identical fashion to the different mutations.

Based on the genetic studies with the different regulatory gene mutations, a model has been proposed for the mechanisms of regulation of the *GAL* system (see fig. 31.4*b*). In this model the GAL4 protein binds to a site upstream of the gene and promotes RNA polymerase II-dependent transcription of the target genes, and the GAL80 protein represses transcription by binding to the GAL4 protein. Induction entails dissociation of the GAL4-GAL80 protein complex, permitting the GAL4 protein to function as an activator. Induction results from the binding of galactose to the GAL80 protein, altering its structure so that it has a lowered affinity for the GAL4 protein.

The proposed model requires the presence of *cis*-acting sequences upstream of the target genes. These GAL4 protein-binding sequences are termed upstream activating sequences (UASs). The GAL4 protein binds to four related 17-bp dyad symmetric sequences upstream from *GAL1* and *GAL10*, activating their (divergent) transcription. The specific DNA sequences to which the GAL4 protein binds have been determined by reacting the DNA with dimethylsulfate and then following the methylation pattern by nucleotide sequencing to determine which guanine bases are protected by the regulatory protein. The sequences that showed protection in a wild-type *GAL4* strain were not protected in a *gal4* mutant derivative. Further manipulations of the promoter region showed that only one of the 17-bp dyad sequences is needed for normal regulation.

The model for how GAL4 protein functions depicts a complex protein, with binding sites for DNA and GAL80 protein as well as a site involved in the activation of transcription. Deletion studies were performed to monitor whether these different sites could be allocated to different parts of the 881-amino-acid GAL4 protein. These studies showed that a mutant protein containing only the first 98 N-terminal amino acids binds DNA but cannot bind the GAL80 protein or activate transcription. Additional mutant studies indicate that amino acids 148–196 and 768–881 are required to activate transcription and that amino acids 851–881 are required for GAL80 protein binding.

Further studies of the GAL4 protein showed that different parts of the GAL4 protein and the lexA bacterial repressor protein are functionally interchangeable. To demonstrate this, Mark Ptashne and his co-workers carried out a domain-swap experiment in which the DNA-binding domain of the lexA regulatory protein from *E. coli* was inserted in place of the GAL4 DNA-binding domain. As you might expect, this protein was ineffective at inactivating the wild-type *GAL1* gene for transcription (fig. 31.5). However, if the DNA-binding site for the GAL4 protein was replaced by the DNA-binding site for the lexA protein, then this hybrid protein functioned in transcription of the *GAL1* gene in the normal way.

The DNA-binding domain of GAL4 protein was once thought to have the structure of a simple zinc finger, which we will discuss later (see fig. 31.20). Now, however, it appears that the DNA-binding domain of this protein is more complex—that it involves two zinc ions and six cysteines (fig. 31.6).

Catabolite Gene Products Are Regulated by Repression and Inactivation

A complicating factor in the regulation of the *GAL* genes emerges from two observations regarding yeast that is grown on glucose: (1) The expression of the *GAL* genes is strongly inhibited even in the presence of galactose, and (2) the GAL4 protein does not protect its target DNA sequences from dimethylsulfate. Your first reaction may be to think of the catabolite repression effect that glucose has on *lac* operon expression. However, closer inspection shows that there are significant differences.

Figure 31.5

Ptashne's domain-swap experiment. The GAL4 protein normally binds at a site upstream from the *GAL1* gene. In the absence of GAL80 this activates transcription from the *GAL1* gene. If the DNA-binding domain of GAL4 is replaced by the DNA-binding domain of the lexA protein, there is no transcription from the *GAL1* gene because the hybrid regulatory protein does not bind at the appropriate site. However, if the DNA-binding site is also changed to that of the DNA-binding site for the lexA protein, then the *GAL1* gene becomes active again.

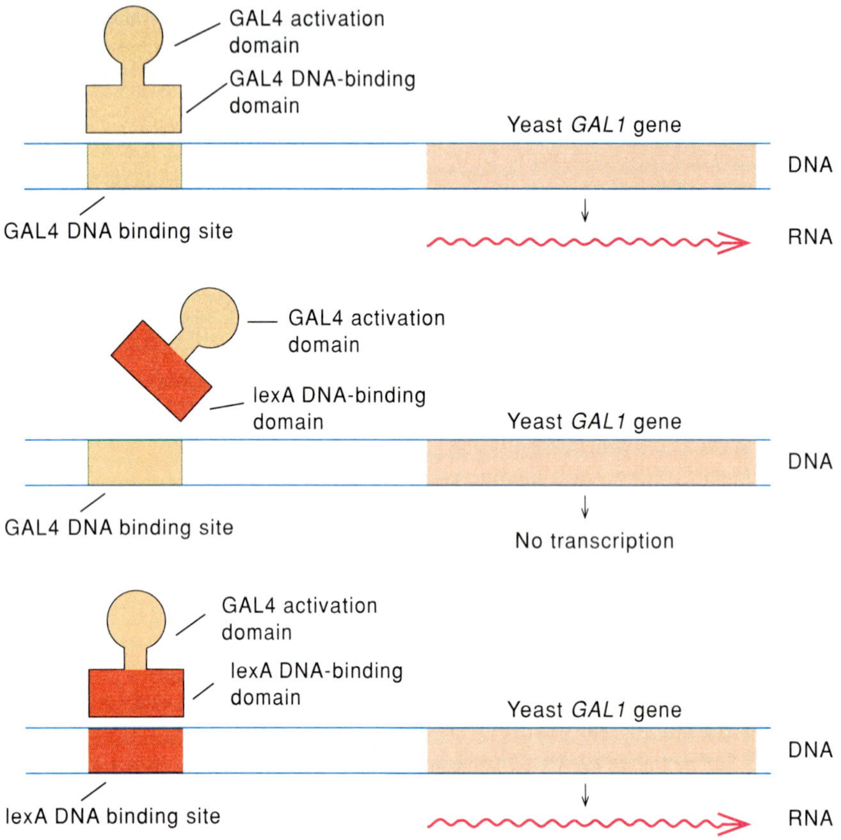

Figure 31.6

Proposed structure for the zinc thiolate cluster of GAL4. (Source: B. L. Vallee and D. S. Auld, "Zinc coordination, function, and structure of zinc enzymes and other proteins," in *Biochemistry* 29:5647, 1990. Copyright © 1990 American Chemical Society, Washington, D.C.)

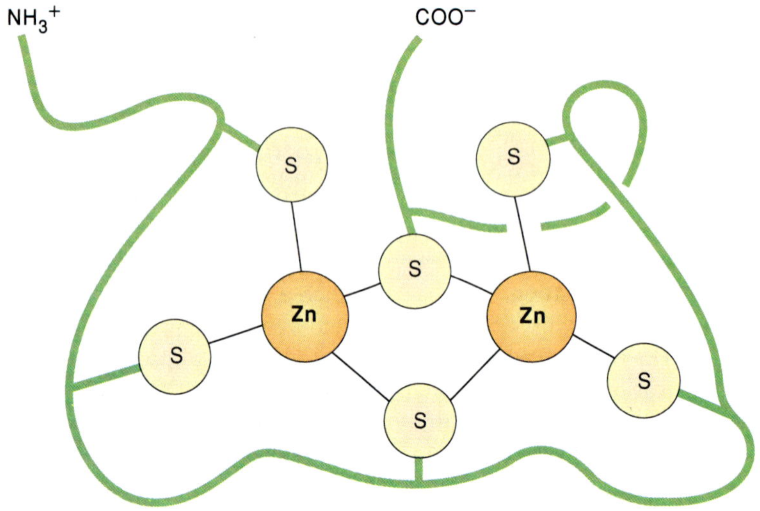

First of all, it is clear that in the yeast system cAMP is not involved, because the effects are the same even in yeast strains that are defective in cAMP synthesis. Second, the presence of glucose not only inhibits the synthesis of the GAL structural proteins, it inhibits the activity of the GAL proteins that are already present. The former effect is called catabolite repression; the latter effect is called catabolite inactivation. The mechanisms underlying these effects of glucose are being intensively investigated.

Positive and Negative Controls of Transcription Are Common in Yeast

In addition to the *GAL* system, the *cis*-acting regulatory regions of many other yeast genes have been investigated. The gene encoding iso-1-cytochrome *c* (*CYC1*) is under positive control; the gene *HAP1* regulates *CYC1* dependence on heme. HAP1 protein also regulates the gene *CYC7*, closely related to *CYC1*, as well as the genes *CYT1* (cytochrome c_1) and *CTT1* (catalase). Another *cis*-acting element, *UAS2*, is regulated by three proteins, HAP2, HAP3, and HAP4, that form a heteromeric duplex to the sequence CCAAT. The transcription of *HIS3* and *HIS4* (involved in histidine biosynthesis) is derepressed by starvation for histidine or other amino acids (general amino acid control). Starvation of yeast for any one of ten amino acids derepressed the synthesis of more than 30 enzymes in nine different amino acid biosynthetic pathways. These coregulated genes have two to four copies of a *cis*-active DNA sequence, 5′-TGACTC-3′, in far upstream regions. Several *trans*-acting genes function in a regulatory hierarchy in the starvation response. The gene *GCN4* functions most directly as a positive regulator of transcription by interaction of the GCN4 protein with the TGACTC sequence in the promoter regions of the genes regulated by this general control protein. The synthesis of GCN4 itself is regulated in a novel way at the translational level. The GCN4 mRNA has four AUG codons at different points in its extensive leader sequence. Starvation conditions, and *trans*-acting factors interact to regulate translation from the fourth translation initiation codon that leads to GCN4 protein synthesis.

Figure 31.7 depicts the *cis*-acting sites of several yeast genes. In all cases these include TATA boxes (T). The open boxes indicate the sites that mediate positive control of transcription. For example, the site that interacts with the GAL4 protein that regulates *GAL10* gene expression is located more than 100 base pairs upstream of its TATA box. The *CYC1* gene has two homologous sites, at about −275 and −225, that mediate activation by heme. UAS control sites can be located at considerable distances from the genes they influence. Furthermore, they can be oriented in either direction relative to these genes. Hence the control sites appear to differ from most controls found in prokaryotes, but are quite similar to enhancer control elements found in multicellular eukaryotes (see chapter 13 and the discussion later in this chapter).

How different are the controlling regions in yeast from bactiera? Yeast genes often have several controlling sites that are variable in their spacing. The multiple UASs for a given gene permit a variety of regulatory transcriptional responses, yet the TATA box still determines the site for transcription initiation. In yeast the canonical TATAAA sequence can be replaced by closely related sequences such as CATTTATT.

Figure 31.7

Location of control elements for a select group of yeast genes. The genes have all been found to contain upstream activation sites (UASs) or upstream sites of repression. Locations of TATA boxes and initiation sites are also indicated. In some cases there is more than one initiation site.

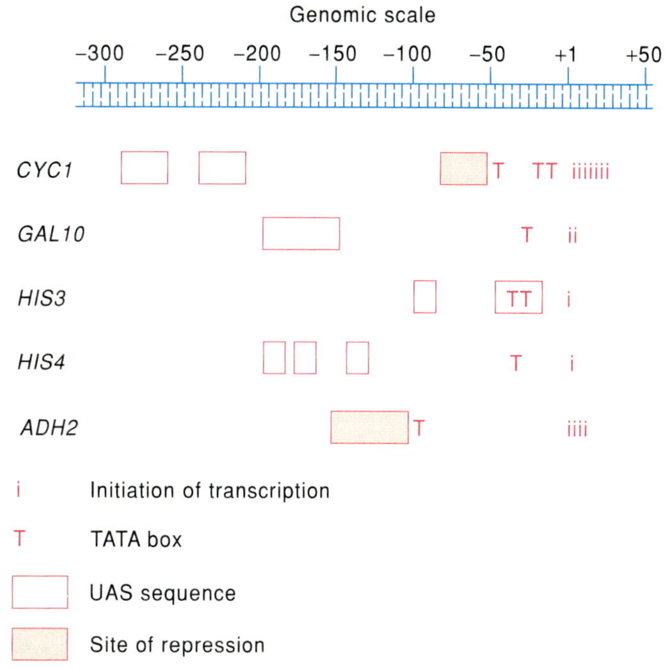

Specific negative regulators that bind directly to DNA are relatively rare in yeast. An example is the protein encoded by one of the genes of the *MATα* locus discussed in the next section.

Mating Type Is Determined by Transposable Elements

Yeast has two haploid cell mating types, *MATα* (or simply *α*) and *MATa* (or *a*) that can fuse to form a single diploid (*a*/*α*) cell. Diploid cells can grow indefinitely as diploid cells or they can undergo sporulation and by meiosis convert back into an equal number of *a* and *α* cells. The haploid mating type of *S. cerevisiae* is determined by the *MAT* locus. Haploid cells with the homothallism (*HO*) gene are able, at almost every cell division, to switch between the *a* and *α* mating types. This efficient switching generates progeny of opposite mating types from the progeny of a single cell, which readily fuse to produce *a*/*α* diploids that do not switch or mate, but can sporulate when starved. Haploid heterothallic (*ho*) strains interchange their mating types rarely, on the order of 10^{-6} events/generation. During interconversion of the mating type, the DNA sequences of *MAT* are replaced by either the *MATα* or *MATa* sequence copied from one of the two silent storage copies *HMLα* or *HMRa* (fig. 31.8). When the sequences are transposed from *HMLα* to *MAT*, the *α* mating type is expressed; transposition from *HMRa* to *MAT* leads to expression of the opposite, or *a*, mating type.

Figure 31.8

Yeast chromosome III, showing the *HML*, *MAT*, and *HMR* loci. *W* is a region common to *MAT* and *HML*, but not *HMR* (~750 bp). *X* is a region found at *MAT*, *HMLα*, and *HMRa* (~700 bp). Y_a is a specific substitution found at *MATa* and *HMRa* (~600 bp). Y_α is a specific substitution found at *HMLα* and *MATα* (750 bp). Z_1 is a region found at *MAT*, *HMLα*, and *HMRa* (~250 bp). Z_2 is a region found at *MAT* and *HMLα* (~70 bp).

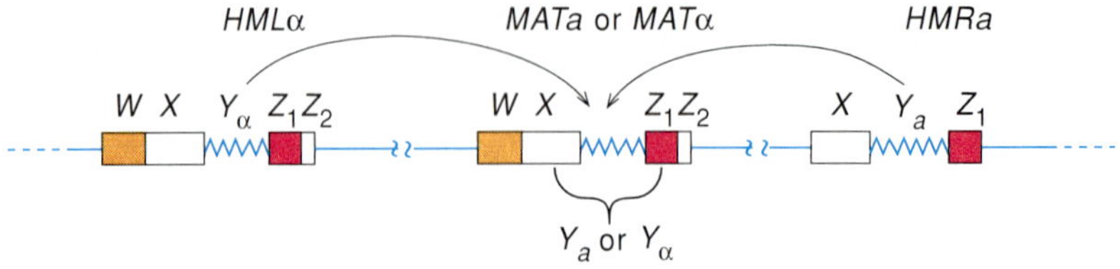

Figure 31.9

Structure of mating loci determinants in yeast. The genetic regions *HMLα* and *HMRa* are normally silent, whereas *MATa* or *MATα* are active. *MATa*, which contains the Y_a segment, expresses transcript *a*1. *MATα*, which contains the Y_α segment, expresses two transcripts, α_1 and α_2. The structures in and around the Y_a and Y_α segments are the same at the storage locus and the expression locus. The inactivity at the storage loci results from far upstream *cis*-acting elements not present at *MAT*. These elements, in conjunction with the four *SIR* gene products, negatively regulate expression of mating type genes at the storage loci.

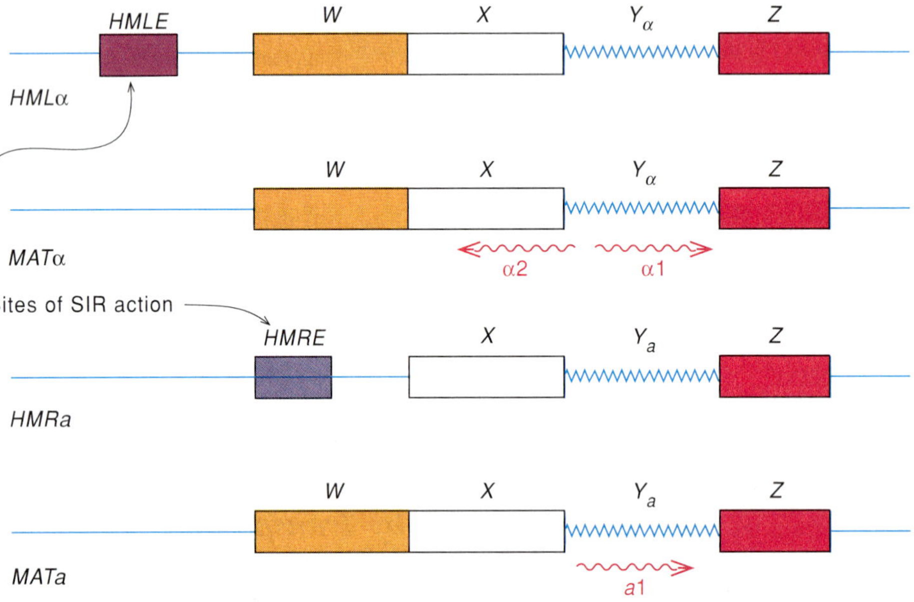

Two immediately interesting questions arise about regulation associated with mating-type switching of haploid yeast strains: (1) What regulates mating-type switching? And (2) why are the *α* and *a* sequences expressed only when they are present at the *MAT* locus? Partial answers are available to both of these questions. In both cases it appears that *trans*-acting control elements are involved.

The *MATa* and the *MATα* sequences that are stored at the *HML* and *HMR* loci are expressed only when they are present at the *MAT* locus. The explanation for this behavior was revealed by the finding of four unlinked genes, *SIR1–SIR4*. In all *sir* mutant strains the *MATa* and *MATα* sequences at *HML* and *HMR* that are normally silent express as they do at *MAT*. The *trans*-acting regulatory factors, encoded by the *SIR* genes, act as repressors at "silencer" sites (*HMLE* and *HMRE*) adjacent to *HMR* and *HML*. Remarkably, these silencer sites are more than a kilobase away from the regions of transcription initiation (fig. 31.9). The silencer sites are not found at or near to the mating type locus (*MAT*), which is why the *a* and *α* coding sequences are expressed when transposed to *MAT*. The *cis*-acting *HMRE* sequence is able to repress in either orientation and over distances 2,600 bp away from the transcription initiation site.

The important question remaining is, How does a single genetic locus determine the haploid yeast-cell mating type? Haploid cells that carry *MATα* behave as *α* cells; cells that carry *MATa* behave as *a* cells; and, finally, diploid cells that carry both *MATα* and *MATa* do not express any haploid specific genes. To explain the key role of *MAT* it was proposed that the *MAT* locus encodes regulatory proteins that control unlinked genes

Figure 31.10

Regulation of *a*-specific genes (*a*sg) and α-specific genes (αsg) in α and *a* haploid cells. In an α cell, α2 inhibits *a*sg and α1 activates αsg. In an *a* haploid cell, the *a*1 transcript made at *MATa* does not appear to exert any regulatory function. (⊕ = positively regulated; ⊖ = negatively regulated.)

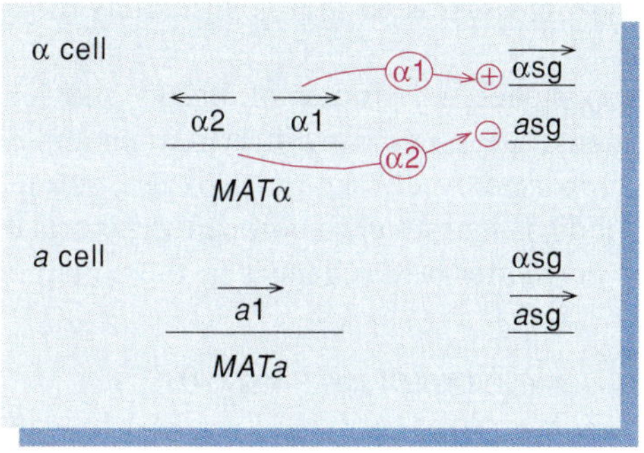

Figure 31.11

Regulation of meiosis in yeast. Haploid cells are not able to initiate meiosis because they express *RME1*, which is a negative regulator of meiosis. Diploid cells are able to initiate meiosis because they make a complex of α2 and *a*1 that inhibits the synthesis of the RME1 product.

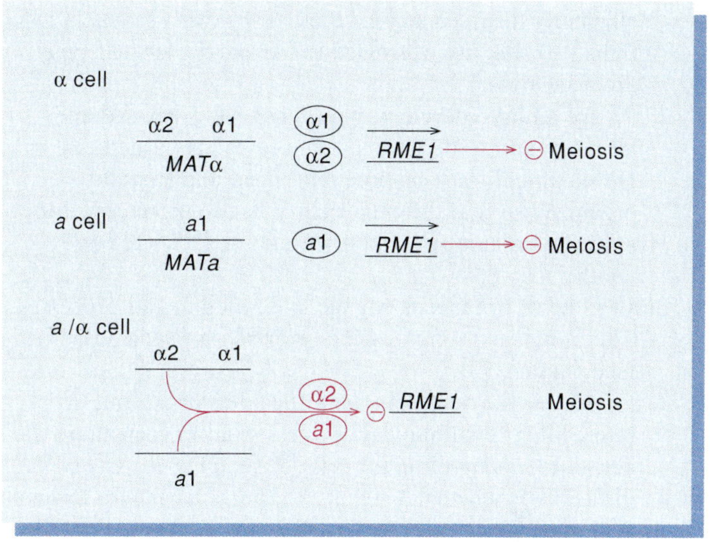

that determine cell type. Subsequently it was shown that the *MATα* locus contains two genes, *MATα1* and *MATα2*, that encode the α1 and α2 regulators, respectively. According to Ira Herskowitz and his co-workers, α1 turns on expression of the α-specific genes and α2 turns off expression of the *a*-specific genes (fig. 31.10). A number of recent observations support this model. For example, the mRNA of *STE6* (an *a*-specific gene) is synthesized in cells that lack the α2 product. It has also been shown that α2 is a sequence-specific DNA-binding protein that recognizes a 32-bp operator sequence upstream of the *a*-specific genes *STE2, STE6, MFA1, MFA2,* and *BAR1*. When this operator is placed between the *CYC1* UAS and its transcription initiation site, α2 brings about repression, or negative control *in vivo.* Even when the operator is placed upstream of the UAS sequence it still represses expression of *CYC1,* but to a lesser degree. Thus the operator need not overlap with essential promoter sequences (as in prokaryotes) to permit repression by α2.

In addition to beginning to understand the regulatory mechanisms that give the haploid cells their specific properties, researchers have made considerable progress in determining the mechanism that regulates meiosis in diploid cells. Diploid *a*/α cells do not mate but can undergo sporulation and meiosis when the diploid cells are starved for nitrogen and have a poor carbon source. It is believed that the reactions leading to meiosis are triggered by a complex of the α2 and *a*1 proteins, which represses the haploid-specific gene *RME1* (regulator of meiosis), allowing the cells to initiate meiosis (see fig. 31.11).

The control of mating type in yeast illustrates one use of mobile genetic elements. Mobile genetic elements are commonly found in both prokaryotes and eukaryotes, but they rarely insert at specific sites as they must in this case. Because of their random insertion behavior, most mobile genetic elements are more apt to be vehicles for evolutionary change than mechanisms for regulating specific genes as in yeast mating-type switching. In higher organisms, mating type (sex) is controlled in part by gene dosage. For example, in humans, males carry one X chromosome whereas females carry two X chromosomes. The mechanism used by yeast and other fungi is more easily influenced by factors in the external environment and may be more suitable for a unicellular organism, which is more exposed to its surroundings.

This concludes our discussion of the regulatory behavior of yeast. There are many ways in which *E. coli* and yeast are quite similar. All of the regulatory processes are highly reversible, so that each cell retains its totipotency, that is, its potential to express any of its genes. This situation does not hold true in more complex eukaryotes. The model developed by Jacob and Monod to explain the regulation of the *lac* operon of *E. coli* at the genetic level (that is, the interaction of a regulatory protein with a specific region of DNA) remains one of the most important conceptual developments in gene expression and development.

General Patterns of Gene Regulation in Multicellular Eukaryotes

In terms of the types of regulatory mechanisms used, yeast is probably typical of the relatively simple, undifferentiated eukaryote. Even in yeast we see the beginning signs of cell differentiation processes and hormonal systems for intercellular communication that typify more complex multicellular eukaryotes. Yeast differentiates into three cell types—the *a* and α haploid types and *a*/α diploids. Furthermore, haploid yeast cells secrete peptides known as pheromones that affect cells of the opposite mating type and change their pattern of gene expression so as to favor cell-to-cell contact and mating. Such similarities between yeast and more complex eukaryotes have made yeast valuable as a model system for studying cellular differentiation and hormonal control, but they do not eliminate the need to study these processes more directly in higher forms.

A number of unique regulatory problems must be confronted in higher eukaryotes:

1. The arrangement of DNA in the nucleus must be efficiently disposed so that those few genes (probably less than 1%) that are expressed in any particular cell type are accessible.
2. As the embryo develops, new genes are expressed and already expressed genes in some cases are turned off. Development must be precisely timed and spatially organized so that different cell types are produced in the appropriate numbers and orientation relative to one another.
3. Even adult organisms engage in developmental processes. The immune system provides examples of such activities (see chapter 33).
4. Finally, in the developing and the adult organism, intercellular communication is essential to coordinate the activities among different cells. A complex battery of diffusible biochemical signals known as hormones trigger responses at the cell membrane in some cases and at the nucleus in others (see chapter 24).

In this chapter we will consider some aspects of these problems, and we will treat additional aspects in chapters 33 and 34.

Nuclear Differentiation Starts in Early Development

In the mature adult state, a multicellular eukaryote contains many cells, each with a potential for the constitutive or inducible expression of only a small, select fraction of the total genome. Since most differentiated cells in the adult contain the normal amount of genomic DNA, the expression of only a fraction of the genome in any particular cell type suggests that the average nucleus may have lost its totipotency.

The first direct evidence that nuclei irreversibly differentiate during development came from Briggs and King in 1952. Working with frog eggs, they showed that the original nucleus from an unfertilized egg could be replaced by the diploid nucleus from a developing animal (fig. 31.12). If the inserted nucleus was taken from the donor very early in its development (at the blastula stage), a mature frog developed. If, however, the nucleus came from a later stage of development, no growth or only limited growth ensued. These nuclear transplantation experiments demonstrated that during development the nucleus assumes a pattern of expression that is difficult or impossible to reverse even when the nucleus is returned to the environment of the egg cytoplasm.

In 1962 Gurdon took this type of experiment one step further by showing that adult frogs could be developed by injecting enucleated eggs with nuclei from the intestinal epithelium of tadpoles. The success frequency was much lower than when nuclei from earlier stages of development were used. Nevertheless, such results showed the influence of the cytoplasm on nuclear expression; they also demonstrated that in certain cases nuclei already tentatively committed to a specific pathway of development can be reprogrammed by placing them in a different environment. The most important lesson to be gained from such experiments is that the nucleus tends to assume a stable pattern of expression that is ultimately dictated by its surroundings.

These pioneering studies of Briggs and King and of Gurdon, demonstrating nuclear differentiation and dedifferentiation, are supported by a broad range of genetic and biochemical findings indicating that chromosomal structural differences often can be correlated with changes in the potential for gene expression.

Chromosome Structure Varies with Gene Activity

In interphase cells, chromosomes exist are composed of chromatin. Chromatin can assume either a compact form known as heterochromatin or a swollen form known as euchromatin. Most nuclei contain both types of chromatin. The heterochromatin is generally concentrated around the nucleolus and the inside of the nuclear envelope. Some chromosomes or parts of chromosomes are always heterochromatic (constitutive heterochromatin); others are heterochromatic only during certain times of the cell cycle or in certain cell types (facultative heterochromatin). The ratio of euchromatin to heterochromatin increases with increasing protein synthetic activity. Many studies

Figure 31.12

Nuclear transplantation technique. The nucleus of an unfertilized egg is mechanically removed with a micropipette (left), and a nucleus from a blastula cell is removed and microinjected into the enucleated egg.

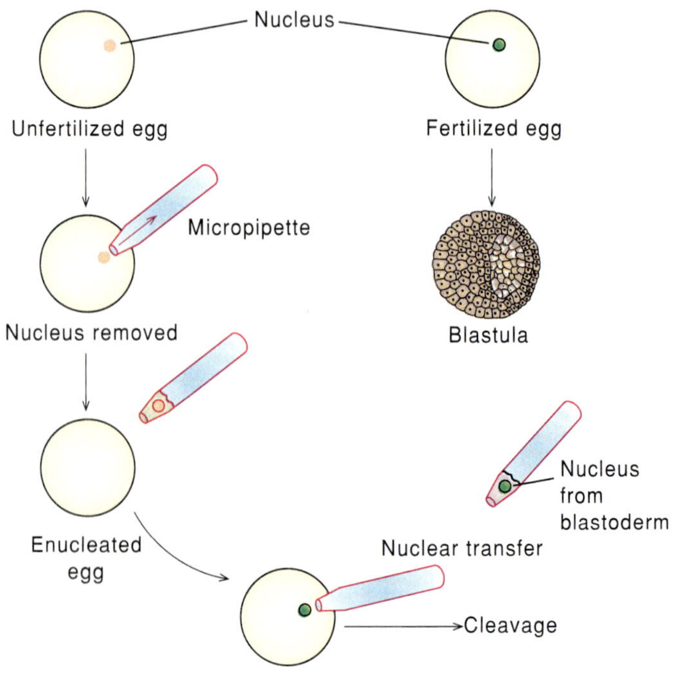

indicate that euchromatin is relatively active in RNA synthesis and heterochromatin is relatively inactive. Although the euchromatic state seems to be necessary for a high level of transcription, for most genes it is not sufficient. This has been demonstrated by autoradiography of cells grown briefly in the presence of ^{3}H-labeled RNA precursors. Those regions of the genome that are transcriptionally active pick up the label; the extent of labeling is proportional to the amount of RNA synthetic activity. Some euchromatic regions appear active by this test, whereas others do not.

Giant Chromosomes Permit Direct Visualization of Active Genes Polytene chromosomes are unusually large chromosomes found in the salivary glands of certain insects. Because of their size, they are convenient vehicles for studying the relationship between chromosome structure and activity. Polytene chromosomes are produced by the repeated replication of interphase chromosomes without separation, a process resulting in a large number of chromosomes that remain laterally aligned. For example, the DNA content in the giant salivary gland cells of *Drosophila melanogaster* may be as much as 1,000 times that of other cells in the fruit fly. Microscopically, the stained chromosomes appear as cross-banded extended bodies (fig. 31.13). There are about 5,000 bands in *Drosophila,* and it is tempting to associate single bands with single genes. However, the average DNA content found in each band is 30,000 base pairs, which is considerably more DNA than necessary to encode the average protein (about 1,000 base pairs).

Transcription in polytene chromosomes usually is associated with local swellings of the bands, called puffs (see fig. 31.13). Sometimes puffing results in a broadening and lengthening, and sometimes the extended DNA projects laterally into loops that combine to form a large ringlike structure. As a rule, a puff originates from a single band, but it can result from swelling of one or more bands. The extent of incorporation of precursors into a puff is approximately proportional to the size of the puff. A detailed investigation of the puffing patterns as a function of the physiological state of the organism shows that different bands become activated, swollen, and transcriptionally active in a sequentially related, tissue-specific fashion.

The correlation of puffs with genetic functions has focused on two types of gene products, secretory proteins and proteins formed in response to heat shock. Certain puffs in *Drosophila* can be artificially induced by the insect hormone ecdysone, which is instrumental in regulating development. Some puffs are induced directly by exposure to the hormone, whereas other puffs are induced indirectly, depending for puffing on the gene products of the directly induced puffs. This dependence is shown by adding drugs that inhibit protein synthesis, with the result that secondary puffs do not form.

In Some Cases Entire Chromosomes Are Inactive In some cases entire chromosomes can be heterochromatized and stay that way through successive cell divisions. Male and female mammals are distinguished by the fact that females carry two X chromosomes, whereas males carry only one. Invariably, one of the X chromosomes in the somatic tissue of females is inactivated and condensed into a heterochromatic state known as a Barr body (fig. 31.14). The consequences of X-chromosome inactivation are readily observed in females who are heterozygous for an X-linked mutation.

For example, the enzyme glucose-6-phosphate dehydrogenase (G-6-PD) is encoded by an X-linked gene. A female that is heterozygous for this gene may carry two alleles that produce electrophoretically distinct forms of the same enzyme (A and B). When isolated cells from a skin biopsy are cloned,

Figure 31.13

A segment of giant chromosome from the midge larva of *Rhyncosciara* at different stages during development. The arrows and connecting lines indicate comparable bands. Changes in the extent of swelling of different regions reflect the activity of those regions in transcription. (Source: M.F. Breuer and C. Pavin, *Chromosoma* 7:275–280, 1946. Copyright © 1946 Springer-Verlag, Heidelberg, Germany.)

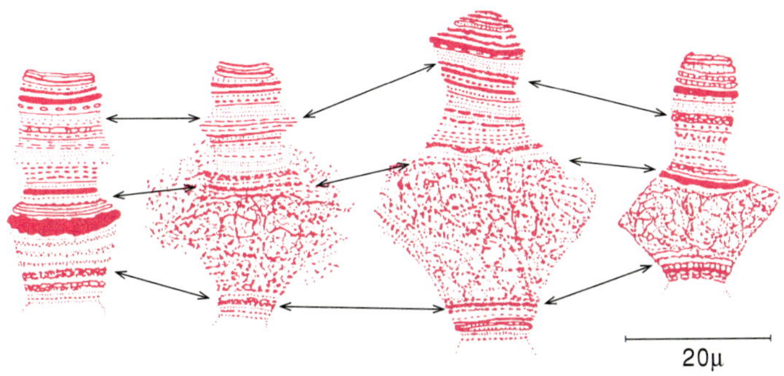

Figure 31.14

Diagram of nuclei obtained from cells in the mucous membrane of the human mouth. (*a*) Nucleus from a female, showing one Barr body (arrow). (*b*) Nucleus from a male, with no Barr body. (*c*) Nucleus from an XXX female, showing two Barr bodies. (*d*) Nucleus from an XXXX female, showing three Barr bodies.

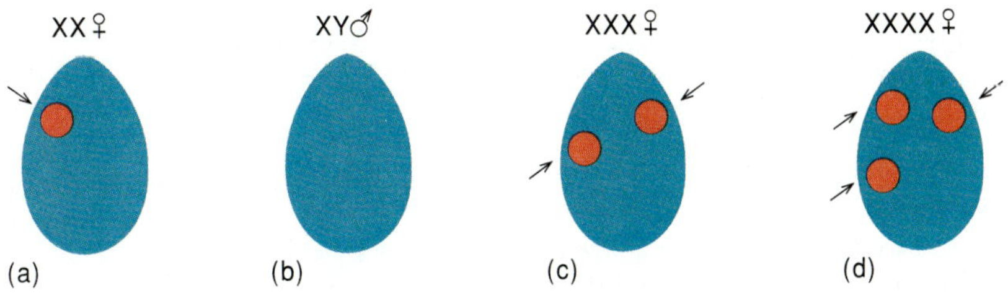

each clone contains either the A or the B form of the G-6-PD enzyme, but never both. If every skin cell of a single organism were to be analyzed for the enzyme, large homogeneous patches expressing one or the other of the G-6-PD alleles would be found. This pattern indicates that the decision as to which X chromosome of the female will be inactivated is made at the multicellular stage in the embryo. Once the decision is made, that X chromosome remains inactive through successive cell divisions. There is an equal chance that the X chromosome from either parent will be inactivated. In genetically abnormal cells that contain three or four X chromosomes, two or three Barr bodies, respectively, can be found (see fig. 31.14).

X-chromosome inactivation provides a simple means of maintaining equal amounts of active X-linked genes in both males and females. This form of so-called dosage compensation is not found in all species. For example, in *D. melanogaster,* neither of the X chromosomes in the female cells is condensed into a Barr body. In this species, dosage compensation, which regulates the activity of specific alleles, operates by a different mechanism so that many alleles in the female X are about 50% as active as the corresponding alleles in the male. However, not all alleles in the *Drosophila* X chromosome are regulated in this way, so that for certain genes twice as much gene product is produced in females as in males.

Chromosomal inactivation takes a different form in the ciliated protozoan *Tetrahymena.* In the vegetative phase each cell contains two nuclei, a micronucleus and a macronucleus. The micronucleus is transcriptionally inactive; it divides mitotically during normal cellular duplication and it undergoes meiosis during the sexual phase of the life cycle. The macronucleus contains much more DNA than the micronucleus, but only about 10% of the sequences present in the micronucleus are represented in the macronucleus. Clearly, the average sequence in the macronucleus is represented many times. The macronucleus divides in an imprecise fashion during normal cell divisions. It is completely lost during meiosis and therefore must be generated anew following each sexual phase.

All nuclear transcripts are made in the macronucleus. From its size and limited sequence complexity, it is clear that the macronucleus must contain many copies of selected regions of the genetic nucleus. How the number of copies made from different genes is regulated is unknown. Selective hybridization studies on rDNA show that the micronucleus contains about 20 copies of the rDNA genes, whereas the macronucleus contains about 200. These observations on *Tetrahymena* demonstrate that even in a unicellular eukaryote the percentage of the nuclear DNA that is transcriptionally active is quite small (about 10%). This intricate compartmentalization and amplification of transcriptionally active regions of the genome in the vegetative nucleus provides a mechanism for the organism to regulate the number of genes according to its needs.

Chromosomal Rearrangement Is Sometimes Required for Gene Expression There are infrequent cases in bacteria and in yeast where DNA rearrangements are essential for gene expression. For example, in yeast we saw that mating type requires DNA rearrangements, and the mechanisms are fairly well understood. We have also seen that in *Tetrahymena* wholesale rearrangements of the genome leading to a vegetative nucleus are essential for all forms of transcription.

In the cells of multicellular eukaryotes it is doctrine that the arrangement of genes in the chromosomes is not disturbed in going from the germ line to various differentiated cells. This simply means that we have not detected rearrangements for the most part. The possibility is still open that certain key rearrangements are involved in differentiation processes, and we must keep our eyes open for such possibilities. In one case, that of immunoglobulin genes, we know that intricate DNA rearrangements in the form of DNA splicing play a key role in determining which antibody genes are expressed. This topic is discussed in chapter 33.

Biochemical Differences between Active and Inactive Chromatin

Much of the preceding discussion on chromatin structure indicates that active chromatin exists in a more swollen state than inactive chromatin. Several biochemical changes accompany the transition from condensed to swollen chromatin. These changes include a redistribution of nucleosomes along the DNA duplex, chemical modification of histones, alteration in the pattern of nonhistone chromosomal protein binding, and chemical modification of the DNA.

DNA in Active Chromatin Is More Susceptible to DNase Degradation The greater accessibility of transcriptionally active chromatin has been elegantly demonstrated by Groudine and Weintraub for the hemoglobin and ovalbumin genes in the chicken. They isolated chromatin from chicken erythrocytes, in which the hemoglobin genes were recently very active, and the oviduct, in which the ovalbumin gene is very active. They then treated these two chromatins with DNase I, using an assay that measured gross DNA degradation as well as that of specific genes. In both cases the rate of DNA degradation from the active gene was much greater than from the average DNA. In erythrocyte chromatin the globin gene DNA was rapidly degraded, whereas in oviduct chromatin the ovalbumin gene DNA was rapidly degraded. Extensive regions surrounding the active genes were also quite susceptible to DNase I hydrolysis.

Histones Are Often Modified in Active Chromatin Histones undergo transient modifications that may have a general effect on transcription. The amino group in the side chain of lysine is subject to acetylation and methylation. Methylation also modifies certain arginine and histidine residues. Histones H3 and H4 are the main targets of these reactions, but H2A and H2B are also affected. Modification peaks during S phase, when the DNA is being duplicated. Certain serine groups in histone H1 are phosphorylated, with the maximum in phosphorylation occurring at a later time during mitosis. These covalent modifications all reduce the net positive charge of the histones, which could lower the affinity between the histones and the DNA.

Figure 31.15

Histone H2A can become linked to ubiquitin to form UH2A. The linkage involves the α-COOH on the C-terminal glycine of ubiquitin and the side-chain amino group on a lysine residue of the histone.

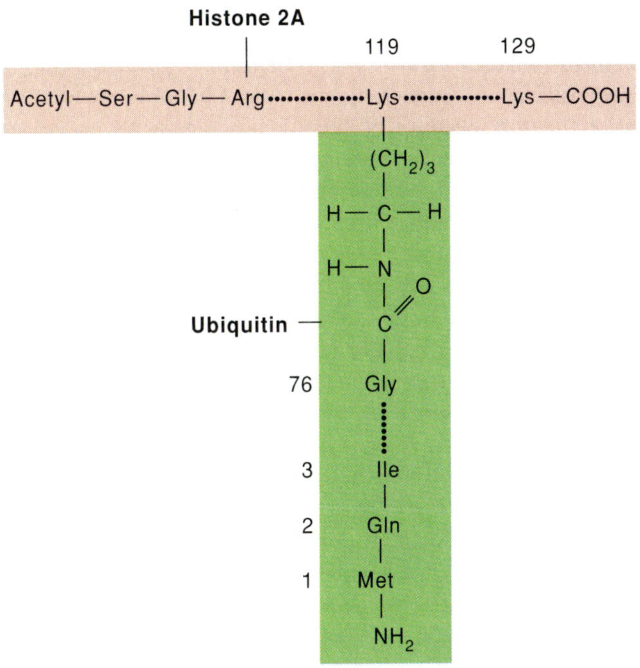

Figure 31.16

Hemimethylated DNA. The unmethylated C residue in the indicated sequence is destined to be methylated shortly after DNA replication.

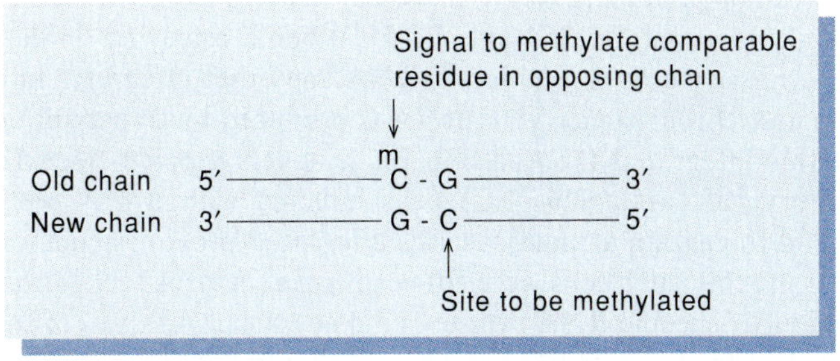

Some histone H2A is modified by combination with a 76-residue protein called ubiquitin. The complex forms between the C-terminus glycine residue of ubiquitin and the side-chain amine group of the lysine at position 119 of H2A (fig. 31.15). Between 5 and 15% of the H2A molecules exist in this type of complex. Although ubiquitin has a greater tendency to be associated with euchromatin, it is released from all chromatin at some time during mitosis. We have seen that ubiquitin is often associated with proteins prior to degradation (see chapter 28); however, there is no evidence that histones linked to ubiquitin are fated for degradation.

Active Chromatin Is Associated with High-Mobility Group (HMG) Proteins In addition to the histones and ubiquitin, chromatin is associated with a group of proteins known as the high-mobility group (HMG) proteins, so named because of their high electrophoretic mobility in low-pH polyacrylamide gels. The four major HMG proteins in calf thymus tissue—HMG1, 2, 14, and 17—have molecular weights of less than 30,000. The HMG proteins can be selectively removed from the chromatin with 0.35-M NaCl, evidence that they are bound less firmly than the histones. The major HMG proteins, HMG1 and HMG2, contain approximately 25% basic and acidic amino acid residues. Amino acid sequence studies show that the charge distribution within the HMG molecules is nonuniform, with major clusters of acidic amino acids. The HMG1 protein contains an extraordinary sequence of 41 continuous aspartic and glutamic acid residues.

The total nuclear concentration of HMGs is low relative to that of the histones (about 3%). Their distribution within the chromatin suggests that they are located in the transcriptionally active regions of the chromatin. Two of the nonhistone proteins, HMG14 and HMG17, are preferentially released from chromatin on partial digestion with DNase I. The notion that HMG proteins are associated with active chromatin regions is supported by reconstitution experiments. Selective removal of HMG from chicken erythrocyte chromatin by 0.35-M NaCl extraction eliminates the selective sensitivity of the globin gene to DNase I. The globin gene in such a preparation regains its DNase I sensitivity on reconstitution with HMG14 and HMG17. This result suggests that there are factors in the active regions of the genome that favor HMG protein binding, and that HMG protein binding enhances the activity of already active genes.

DNA Methylation Is Correlated with Inactive Chromatin Methylation of cytosine in CpG dinucleotides may play a regulatory role in the gene control of eukaryotes. In chapter 25 we noted that 5-methylcytosine (5-MC) is the main modified base in vertebrate DNA. Most of the 5-MC occurs in the dinucleotide CpG. Indeed, in mammals and birds approximately 50 to 70% of all such dinucleotide sequences are modified. In this connection it is noteworthy that CpG sequences occur much less frequently than we would expect on a statistical basis. The mechanism of methylation has been explored by DNA transfection of certain tissue-culture-grown cells. If the DNA used in transfection is methylated, the pattern of methylation is maintained through many cell duplications. Likewise, when unmethylated DNA is used in transfection, a nonmethylated pattern is maintained. These results indicate that under many circumstances methylation is passively maintained by a signal that recognizes hemimethylated DNA. If the C residue in a CpG dinucleotide is methylated immediately after semiconservative DNA replication, only the parental DNA chain is methylated; the passive methylation system signals methylation of the corresponding C residue in the new DNA chain (fig. 31.16).

The extent of methylation of a gene correlates with its ability to transcribe. A most elegant study of this correlation comes from the laboratory of Rudy Jaenisch, in work on the expression of Moloney murine leukemia virus (MuLV). MuLV is a tumor-producing RNA virus that belongs to the retrovirus family (see chapter 25, table 25.4). When a retrovirus infects a permissive cell, it produces a DNA copy that integrates into the host chromosome. Viral RNA is produced by transcription of this integrated DNA, using the host RNA polymerase II. Embryonal carcinoma (EC) cells infected with MuLV were found to contain as many as 100 integrated proviral genomes. Despite this, such cells were not permissive for virus replication, and virus-specific RNA synthesis did not occur. Analysis of the DNA in the infected EC cells showed that the proviral genomes had become highly methylated. Evidently, EC cells are different from many other cell lines in that they rapidly methylate unmethylated (or hypomethylated) DNA. Strong support for the idea that methylation is the cause of this inactivation came from experiments in which the EC-infected cells were pretreated with 5′-azacytidine, which inhibits DNA methylation; EC-infected cells treated with this inhibitor produced infectious viruses.

Given that DNA methylation usually reduces transcription, two important, closely related questions remain unanswered: How is methylation regulated *in vivo?* And how does methylation interfere with transcription? We know that methylation does not interfere with the elongation phase of RNA synthesis; therefore it seems likely that methylation blocks initiation. The binding of polymerase and other regulatory proteins at the initiation locus is sensitive to any modification of these nucleotides that may have taken place. The precise regulation and inhibition mechanisms, however, await further elucidation.

Thus far we have been dealing with aspects of chromosome structure that influence gene expression. You may be wondering how those differences in chromosome structure originate. We cannot give a complete answer, but it seems likely that *cis*-acting DNA sequences and *trans*-acting regulatory proteins are the primary factors that affect gene expression and chromosome structure in the interphase nucleus.

Influence of DNA-Regulatory Protein Interactions on Gene Expression

The primary sequences in the noncoding regions of eukaryotic DNA contain (1) signals for the binding of RNA polymerase, (2) gene regulatory proteins, (3) transcription start sites, and (4) potential coding regions for small peptides. In chapter 29 we saw that several proteins in addition to RNA polymerase II are required for the initiation of transcription at the adenovirus major late promoter; this promoter is believed to share many characteristics with other eukaryotic promoters. In addition to signals for regulatory protein binding in the immediate vicinity of the polymerase binding site, there are other signals, which bind at some distance from the promoter, that have a major influence on gene expression. In yeast, sequences of this type are termed UAS sequences (see our earlier discussion) since their action appears to be confined to regions located upstream from the promoters they influence. *Cis*-active sequences operating over great distances were first discovered in animal viruses. Such sequences are called enhancers because they enhance expression of their associated genes. Enhancer sequences are important and rather unusual transcription signals that play a prominent role in gene expression of multicellular eukaryotes.

Enhancer Sequences Are General Stimulators of Transcription Enhancer signals cannot be understood in conventional terms because of their location relative to the promoter(s) they influence. Enhancer signals, like UAS signals in yeast, have the ability to stimulate transcription at appreciable distances from where they are located in the genome (200–5,000 bp). Enhancer signals are even more versatile than typical UAS signals in that they are effective downstream as well as upstream from the promoters they stimulate. Moreover, some enhancer elements are equally effective in either orientation on the DNA. Thus inversion of an enhancer element frequently does not destroy its effectiveness.

Enhancers have been found in many systems; some enhancers are active in most cells, while others are quite tissue-specific. Enhancers are crucial for regulating the expression of many genes in multicellular eukaryotes. Possibly enhancers account for the difference between euchromatic and heterochromatic regions of the chromatin.

The first evidence for enhancers came from George Khoury's studies on simian virus 40 (SV40), which infects monkey cells. The SV40 genome is a circular duplex with about 5,200 base pairs. In SV40-infected cells, viral RNA synthesis is divided into early and late phases. Early transcription originates from a block of sites illustrated in figure 31.17. There is a TATA box, typical of many eukaryotic gene promoters (see chapter 28) about 27 bp upstream from the transcription start site. Immediately upstream of the TATA box the SV40 promoter contains a series of three repeated GC-rich regions that bind a regulatory protein known as SP1. SP1 protein is a widespread activator found in vertebrates. Further upstream from the transcription start site there is a tandemly repeated 72-bp sequence (-116 to -188 and -189 to -261 from the 5′ end of the messenger). Removal of one of these sequences has no effect on transcription, but removal of both of the 72-bp sequences drastically lowers early transcription.

Surprisingly, the precise position or orientation of the 72-bp segment is not critical for stimulating transcription. Thus the 72-bp segment remains effective after inversion or after translocation further upstream or downstream from the transcription start site. Foreign genes inserted into DNA containing the SV40 enhancer are frequently stimulated in the same way as the SV40 early region, a result demonstrating that this enhancer's effect is gene-specific.

Other animal viruses carry segments with a similar activating function, even though the sequences are quite different. Peter Gruss and his colleagues have shown that the SV40 72-bp repeat may be functionally replaced by a segment comparable in size from either end of the DNA cloned from Moloney murine sarcoma provirus (MSV). Enhancers have also been

Figure 31.17

Region in and around the early transcription start sites of the SV40 genome. The base-pair number on the circular genome is indicated. There are several transcription start sites. One cluster of start sites (early), used initially after infection, is located about 27 bp downstream from the TATA box. The other cluster of start sites (late early) is used at later times. This cluster of start sites is located upstream of the TATA box. One of the products of early transcription is the protein known as large T antigen. This protein binds in and around the core *ori*. T antigen inhibits early transcription. The upstream positive control *cis* elements include the tandem 72-bp enhancer sequences and three tandem 21-bp sites. The latter function in conjunction with host-encoded protein known as SP1.

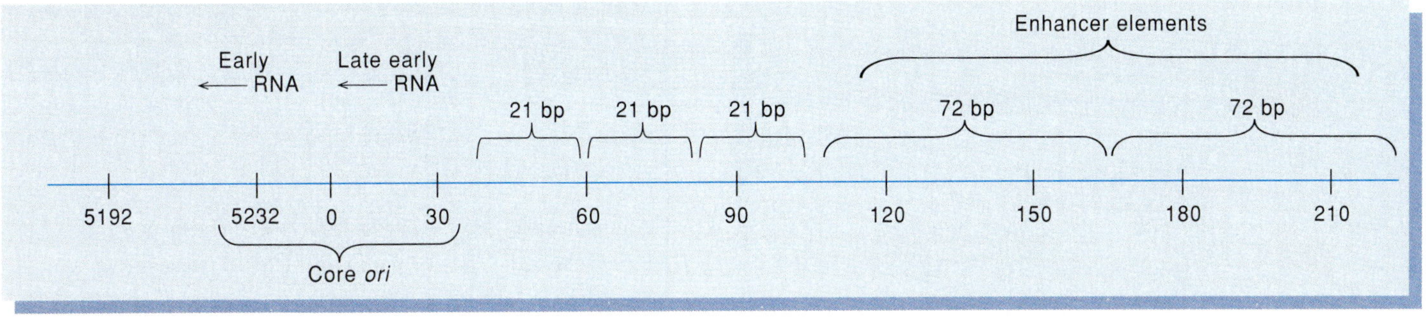

found in the host genome. For example, a tissue-specific enhancer has been found in the intron region of an immunoglobulin gene (see chapter 33); it is considered to be tissue-specific since it stimulates expression of nearby genes only in antibody-forming lymphocyte cells.

Further analysis of the 72-bp SV40 enhancer shows that it is composed of elements 15 to 20 bp in length that bind one or more protein factors specifically. These elements, called enhansons, are important when separated from one another. It is believed that different enhancers are composed of different combinations of enhansons. This system permits a great variety of responses in different cell types and for different genes in the same cell because many diverse combinations of enhansons can be generated from a limited number of enhansons. The general picture of the enhancer that is emerging is that of a complex multicomponent segment of DNA that can bind protein factors in a tissue-specific manner (fig. 31.18).

This type of modular construction of enhancer from enhansons may also be characteristic of promoters, which in addition to the polymerase binding site contain sequences where a number of protein factors that bind immediately next to or near to the polymerase. The basic distinction between enhancers and promoters is operational. An enhancer is able to stimulate transcription at a considerable distance; a promoter and its component modules are only able to stimulate transcription in the immediate vicinity.

Conventional regulatory proteins usually bind to specific sites on the DNA or to other specific DNA-binding proteins. It is not certain how the proteins that bind to enhancers influence transcription over long distances. However, several proposals have been made: (1) The enhancer element may function as a point of attachment to a structural component of the nucleus to stimulate transcription; (2) it may serve as an initial binding site for some factor required for transcription, which must subsequently move along the duplex to the initiation point for transcription; or (3) folding or looping of the chromosome may bring the *cis*-bound enhancer proteins into close contact with the gene it stimulates (fig. 31.19). Currently the third possibility is favored. But even if the third mechanism is the one most frequently used, the other two may be used in place of or in conjunction with the third mechanism for some genes.

Figure 31.18

Enhancers are frequently composed of two or more enhansons. Each enhanson contains two or more proteins binding side-by-side to specific sites on the DNA and with affinities for one another. An enhancer that contains two enhansons and two protein factors can be arranged in 16 ways. An enhancer composed of two enhansons with three possible protein factors can be arranged in 77 ways if all arrangements and combinations of protein factors are permitted. The power of combinations permits a great deal of variety in enhancer construction with a very limited number of protein factors.

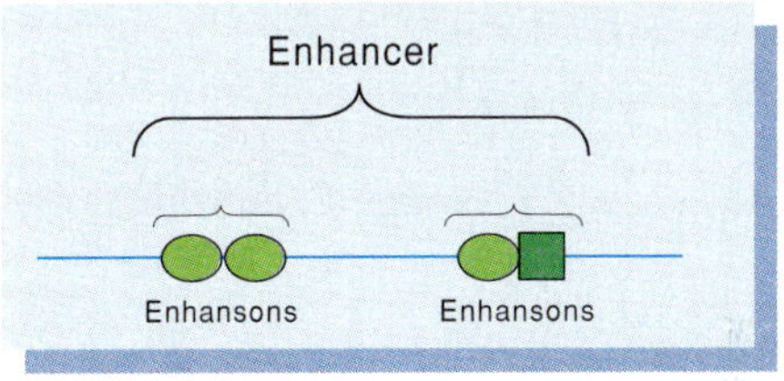

Many DNA Regulatory Proteins Contain Two Domains: One for Binding DNA and One for Binding to Another Protein

Many transcription activators and transcription repressors have two domains, one for binding to DNA and one for binding to a transcription-related protein. The two-domain structure was most effectively demonstrated by Ptashne's domain-swapping experiment (see fig. 31.5). Each domain comes in many varieties.

DNA-Binding Domains of Transcription Factors On the basis of current evidence, the helix-turn-helix is the most commonly used DNA-binding motif found in bacteria. This motif is also used extensively by eukaryotes (table 31.2). For example, it is found in a number of proteins that regulate transcription during

Figure 31.19

Possible mechanisms for enhancer action. (1) The enhancer draws the associated gene to the nuclear matrix, where it is more accessible to the transcription apparatus. (2) The enhancer is the initial binding site for an element that subsequently moves. (3) The enhancer folds or loops, depending on its polarity, to bind with other promoter elements.

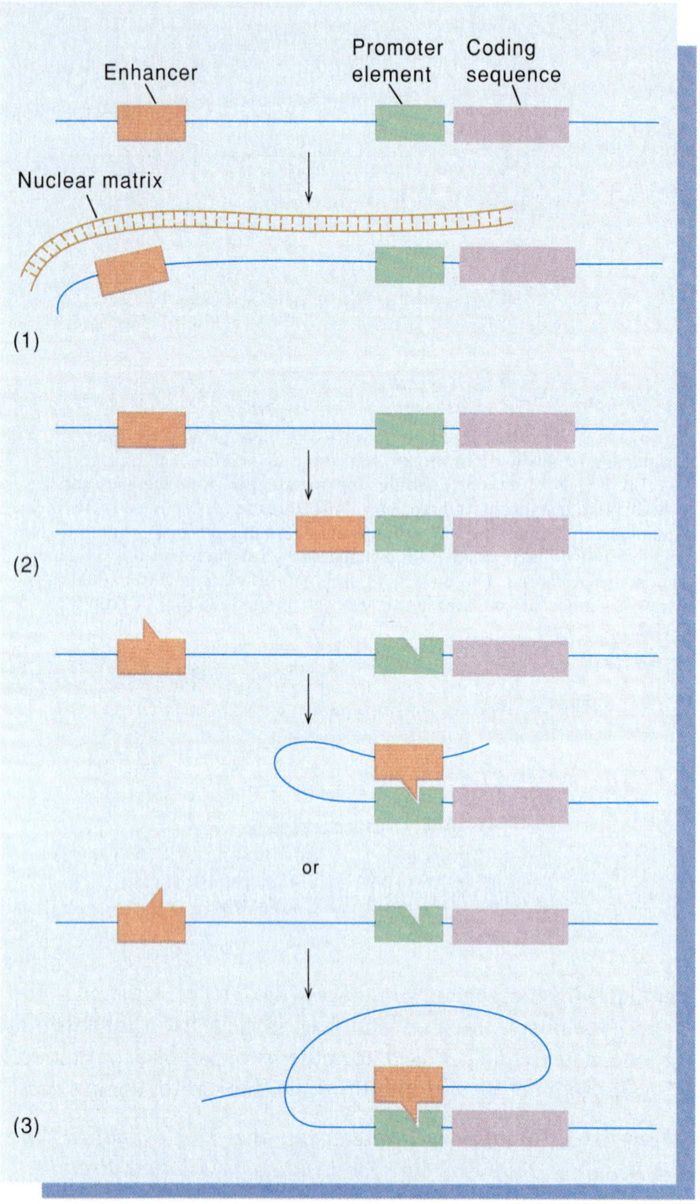

early development in *Drosophila.* These proteins carry a characteristic 180-base segment known as a homeobox (box 31A). Although most helix-turn-helix proteins in bacteria bind as dimers, the homeobox-containing proteins bind as monomers. Since dimers bind more strongly than monomers, other factors being equal, we must ask why monomers are used in some cases. Possibly it is because transcription activators that bind as monomers interact with one or more additional proteins in place of an identical monomer.

Table 31.2
DNA Regulatory Proteins Found in Eukaryotes

Regulatory Protein	Source	Function
Helix-turn-helix		
MATα1	Yeast	Activates α-specific genes
MATα2	Yeast	Inactivates *a*-specific genes
MAT*a*1	Yeast	Combines with α2 to repress haploid-specific genes
Antennapedia	*Drosophila*	Homeotic gene
Ultrabithorax	*Drosophila*	Homeotic gene
Engrailed	*Drosophila*	Segment-polarity gene
Fushi tarazu	*Drosophila*	Pair-rule gene
Zinc finger[a]		
GAL4	Yeast	Galactose-dependent activator
TF IIIA	*Xenopus laevis*	Required for Pol III transcription of 5S RNA
Krüppel	*Drosophila*	Gap genes
Hunchback	*Drosophila*	Gap genes
Glucocorticoid steroid receptor	Vertebrates	Positive and negative acting; in many but not all cell types
Leucine zipper		
GCN4	Yeast	Activates genes for enzymes that make amino acids
C/EBP	Mammals	Activates genes in liver and other cells; limited cell distribution
c-Fos/c-Jun	Mammals	Growth regulation

[a]The category in which we should place a number of DNA regulatory proteins is uncertain because we do not know the precise three-dimensional structures of these proteins. There is a tendency to classify a protein in a specific category on a limited amount of composition and sequence data. For example, the GAL4 protein was once thought to be a zinc finger (see fig. 31.20), but current indications are that this protein involves two zincions and six cysteines in a more complex structure (see fig. 31.6).

In eukaryotes an even more popular DNA-binding motif is found in certain zinc metalloproteins. The role of zinc ion (Zn^{2+}) in this so-called zinc finger motif is to stabilize the tertiary structure of the recognition helix in the DNA-binding protein. This it does by forming a tetrahedral complex with histidine and cysteine residues. The most common complexes are formed with two histidines and two cysteines (the C_2-H_2 motif).

31A
BOX

Localization of eve *and* ftz *Transcripts in the Wild-Type Developing Embryo*

Localization of the *eve* and *ftz* transcripts in *Drosophila* embryos is determined by *in situ* labeling with a ^{32}P-labeled specific DNA probe. All embryos are oriented so that anterior is to the left and dorsal is up. In dark-field photomicrographs of tissue autoradiograms (see illustration), the labeling pattern is seen only near the dorsal and ventral edges of the embryos because the labeling is done on sections cut through the middle of the embryos. Keep in mind that cells are concentrated near the surface in the early embryo. *Eve* transcripts are located in adjacent anterior and posterior regions of every other segment, giving rise to seven bands in all (top). The same is true of *ftz* except that complementary regions are labeled (middle). This point is made most clear by the labeling pattern observed in a double-labeled section, where the embryo is simultaneously labeled with both probes (bottom). (Photomicrographs made by K. Harding and M. Levine.)

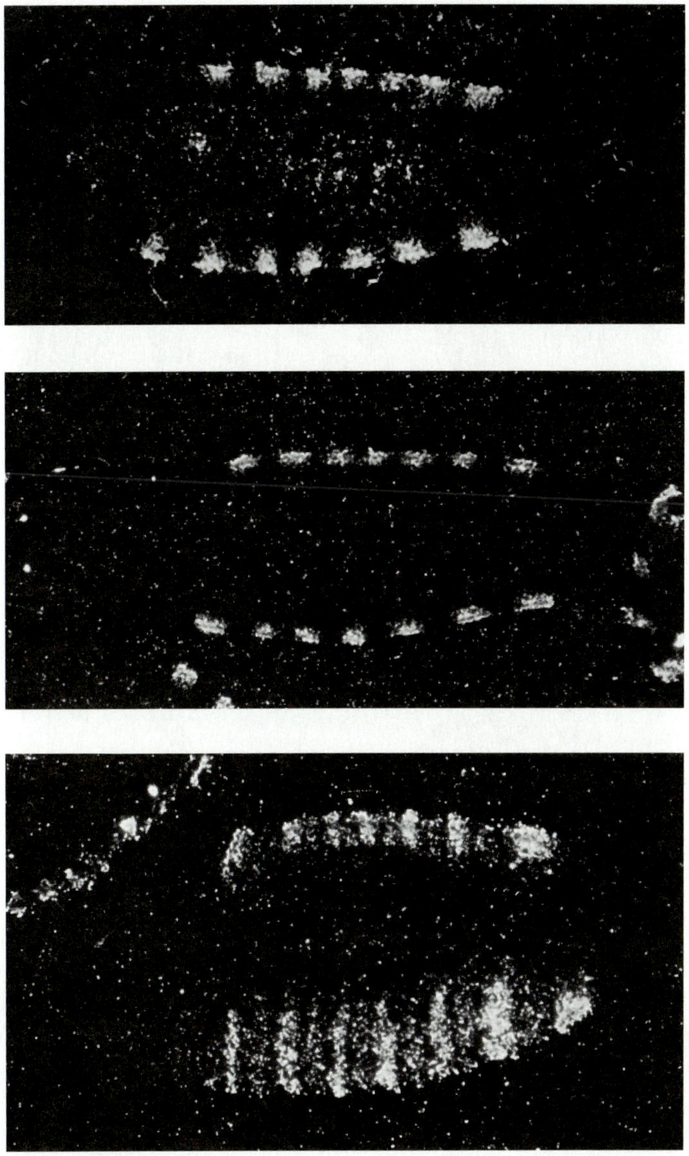

Complexes with four cysteines (the C_2-C_2 motif) are also well known. Zinc ion (Zn^{2+}) is a virtually unique divalent cation for making structural complexes with proteins for several reasons: (1) It has a strong affinity for polarizable nitrogens and sulfurs in amino acid side chains; (2) it is completely stable in the +2 valence state; and (3) in tetradentates, zinc ion (Zn^{2+}) always forms a complex with the precise geometry of a tetrahedron.

Proteins that use the C_2-H_2 and the C_2-C_2 motifs differ in several ways. The C_2-H_2 motif was first identified for the TFIIIA transcription factor that binds to the internal control region of the 5S rRNA gene in *Xenopus laevis* (see chapter 28). In this protein a sequence of 30 amino acids is repeated consecutively nine times. Two-dimensional NMR studies have shown that each 30-residue repeating unit folds to form an independent domain with a single zinc ion tetrahedrally coordinated between an antiparallel β sheet and a short thirteen-amino-acid α helix. Regulatory proteins with the C_2-H_2 motif have the same general tertiary structure (fig. 31.20*a*). In all cases the DNA-binding portion of the structure results from interaction

Figure 31.20

Schematic representation of three types of DNA-binding motifs found for regulatory proteins in eukaryotes. (*a*) The C_2-H_2 zinc finger found Xfin from *Xenopus laevis*. (Adapted from M. S. Lee et al., *Science* 245:645, 1989.) (*b*) The C_2-C_2 zinc finger found in the glucocorticoid receptor. (Adapted from T. Hard et al., *Science* 249:157, 1990.) (*c*) The leucine zipper motif. (Adapted from C. R. Vinson et al., *Science* 246:911, 1989.) The helical regions that make contact with the DNA are colored in red; the rest of the polypeptide chains are colored in green except for the zinc fingers, where cysteine sulfurs are colored in yellow and histidine nitrogens are colored in blue. The knobs on the leucine zipper represent leucine side chains.

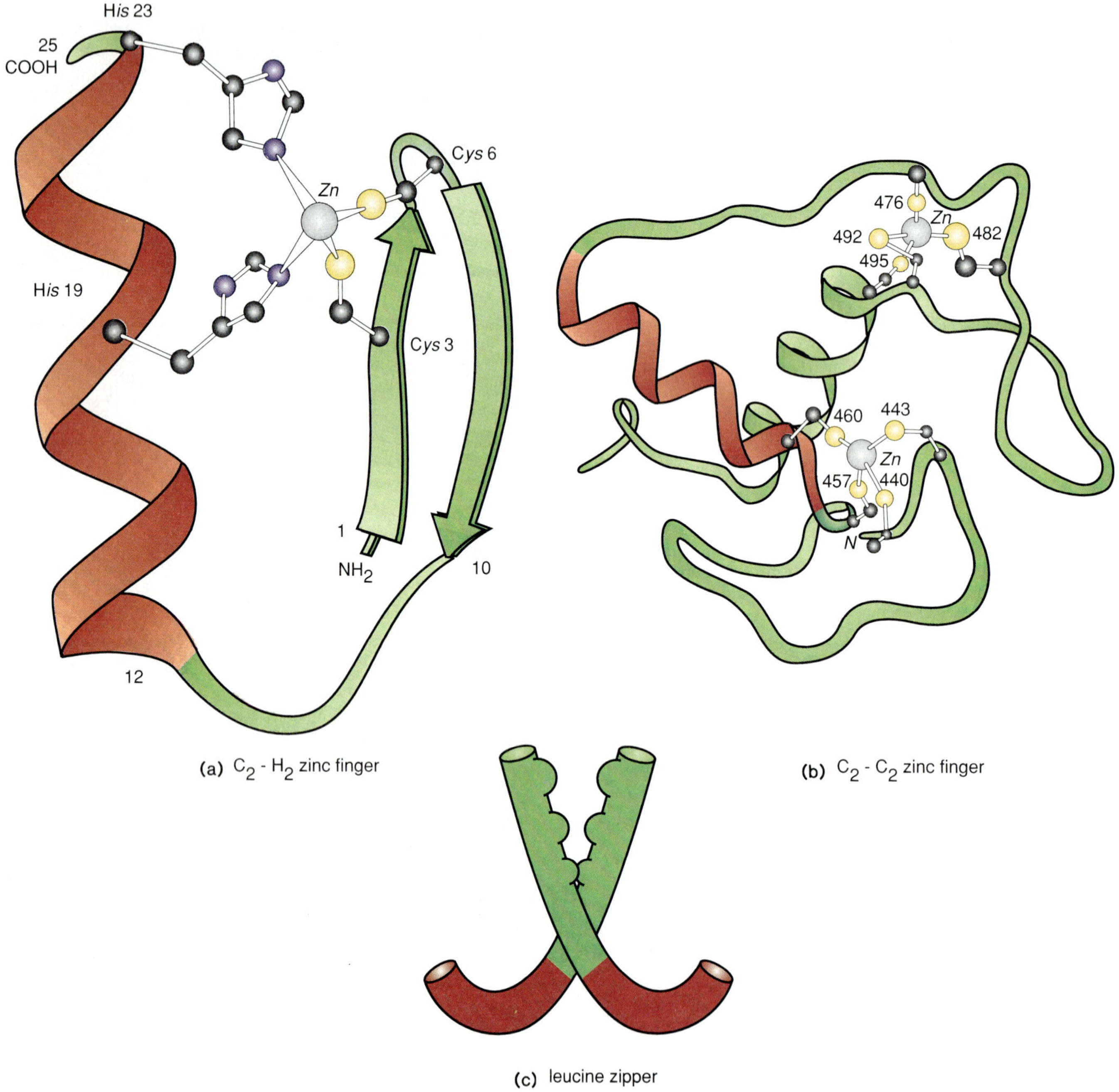

between the DNA base pairs in the large groove of the DNA and amino acid side chains on the exposed side of the recognition helix.

The C_2-C_2 motif is exemplified by the DNA-binding domains of steroid hormone receptors. There are several differences between proteins bearing this motif and those with the C_2-H_2 motif: (1) The antiparallel β sheet is used only in the C_2-H_2 motif; (2) in the C_2-C_2 motif the recognition helix extends beyond the zinc ligands, whereas in the C_2-H_2 motif it is located between the metal ligands; (3) the C_2-H_2 zinc fingers fold to form independent modules with no direct structural interaction between adjacent fingers, whereas in C_2-C_2 receptors two zinc-binding motifs are folded together to form a single structural domain; and (4) proteins containing the C_2-H_2 motif bind to DNA as monomers, whereas proteins containing the C_2-C_2 motif bind as dimers.

A third type of zinc binding motif has been found in a set of yeast activators, including GAL4, with two closely spaced zinc ions sharing six cysteines with their DNA-binding domains (see figure 31.6). It should be recalled that the yeast transcriptional activator GAL4 binds cooperatively to four related 17 base-pair sequences within an upstream activator sequence to activate the transcription of the GAL1 and GAL10 genes (see figure 31.4). The regulatory protein dimerizes through the interaction of two α-helical segments that form a short coiled-coil structure. The dimerization element makes contact with the narrow groove. The main recognition modules lie in the major groove separated by about one-and-one-half turns of the DNA helix and centered over the CCG triplets in the 17 base-pair consensus sequence (5′-CGGAGGACTGTCCTCCG-3′). It is conjectured that there may be another protein involved in the recognition complex.

Another type of DNA-binding domain was first described for the mammalian enhancer binding protein C-EBP. Proteins of this class show a primary sequence similarity consisting of a highly conserved stretch of about 30 amino acids with a substantial net basic charge, immediately followed by a region containing four leucine residues positioned at intervals of seven amino acids. The latter segment, named the leucine zipper by Steve McKnight, is required for dimerization and for DNA binding. Dimerization is favored by hydrophobic interactions between closely apposed leucine side chains (see fig. 31.20*c*). The adjacent helical regions, oriented in parallel, twist slightly around each other as in the classical coiled-coil conformation (see fig. 4.6). The basic regions adjacent to the leucine zipper interact in the large groove of the DNA; both proteins contribute their basic regions to form the DNA-binding domain. Dimerization is not confined to identical monomers. Many of the dimers that contain the leucine zipper motif are heterodimers. Thus the leucine zipper motif may be thought of as a general way of bringing two proteins together, but the effectiveness of this union will be determined by how and where they interact with specific segments on the DNA.

One of the best known heterodimers resulting from leucine zipper interaction is that formed between the Jun and Fos proteins. Both of these proteins are known to have the potential for activating a wide variety of host genes in mammals. The Jun protein can form homodimers or heterodimers, but the Fos protein can only form a heterodimer. Fos forms a heterodimer with Jun. Correlated with these dimer forming properties we find that Fos protein does not bind to the DNA except in the form of a heterodimer. By contrast the Jun protein can bind to DNA in the form of a homodimer or a heterodimer.

Thus far we have focused on the DNA-binding domains of regulatory proteins. The transcription activation domains of regulatory proteins are less well characterized. These domains may interact with the RNA polymerase directly or with other regulatory proteins. What makes an effective transcription activation domain? Different regulatory proteins appear to possess different types of groups for this purpose and some more than one type of grouping. Acidic, glutamine-rich, and proline-rich domains are implicated in the activation domains of a number of regulatory proteins.

In this section we have focused on some of the better known facts about regulatory proteins. This is an explosive field in which new discoveries are being made at a furious clip. It should be anticipated that we will know much more about this subject in the near future.

Posttranscriptional Modes of Regulating Gene Expression

Thus far we have concentrated on regulatory mechanisms for gene expression that occur at the transcriptional level. There are additional modes of regulation that are very important in eukaryotes that occur after transcription has taken place. These are divided into three categories: those that involve (1) RNA processing and those that involve (2) regulation of translation or (3) processing of the polypeptide chain.

Alternative Modes of Splicing Are Probably Very Widespread We discussed the basic mechanisms of RNA splicing in chapter 28. RNA splicing was first discovered in the transcription of adenovirus DNA, where different reading frames are connected to the same 5′ end. Thus we have known about the phenomenon of alternative splicing as long as we have known about splicing itself. Alternative splicing occurs for eukaryotic viruses such as SV40 and polyoma and it is also a common phenomenon in eukaryotic genes that contain multiple exons in their nascent transcripts. It is clearly a regulatory phenomenon in viruses, since we see a shift in the types of splicing as virus infection progresses. For eukaryotic genes it is also a regulatory phenomenon, since different modes of splicing are seen in different types of differentiated cells of the same multicellular organism.

The various patterns observed for alternative splicing are reviewed in fig. 31.21. Splicing usually involves segments of a single transcript, but on occasion we encounter transsplicing,

Figure 31.21

Patterns of alternative RNA splicing. Constitutive exons are shown in full color; alternative sequences (orange) and introns (solid lines) are spliced according to different pathways (dotted lines). Alternative promoters (TATA) and polyadenylation signals (AATAAA) are indicated. (Source: R. E. Breitbart et al., "Alternative splicing: A ubiquitous mechanism for the generation of multiple protein isoforms from single genes," in *Annual Review of Biochemistry* 56:467–495, 1987. Copyright © 1987 Annual Reviews Inc., Palo Alto, Calif.)

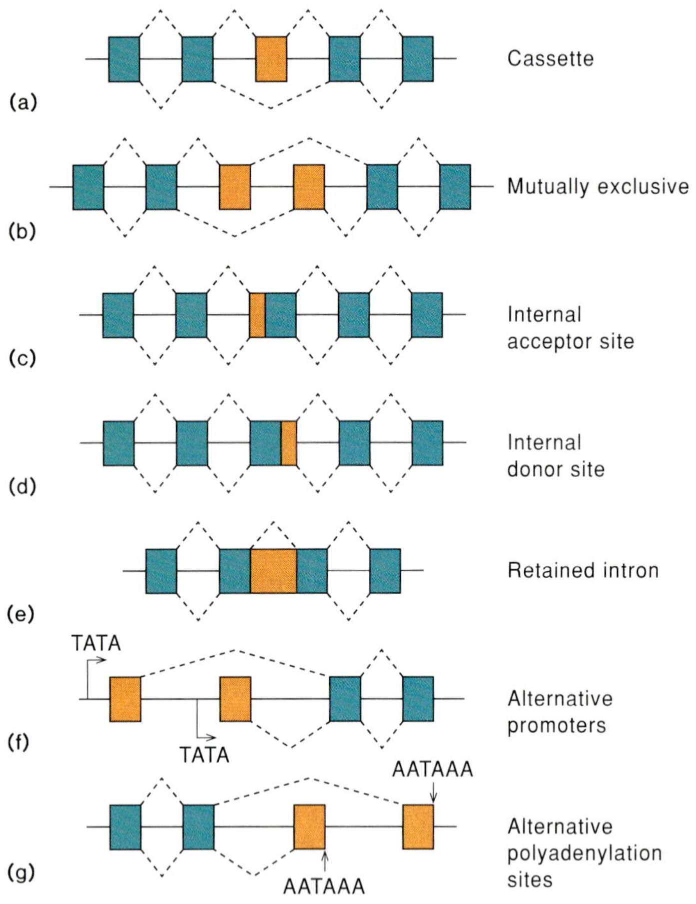

in which two transcripts participate in a common splicing operation. In the most common splicing situation, a promoter is present at one end of the transcript and the combination of exons that is used in the mature mRNA varies according to the pattern of splicing. In some cases one or more exons are excluded from the message by selective splicing. In other cases part of an intron is fused to one of the exons to make the final message. Splicing can also be influenced by the choice of the promoter or the choice of the polyadenylation site. In the latter two cases, parts of the upstream or downstream regions of the gene are excluded from the initial transcript and so the splicing of the transcript is limited to the RNA remaining in the nascent transcript.

The possibilities for alternative splicing are enormous and it is currently believed that this is a major mechanism for expanding the variety of proteins available to the organism in multicellular eukaryotes. One particularly elaborate example of this is seen in the gene for tropomyosin in vertebrates. Recall that tropomyosin is an important component of vertebrate striated muscle (see fig. 5.14). The polypeptide chain composition of the tropomyosin found in striated muscle involves seven splices in the nascent transcript from the gene (fig. 31.22). Variants of tropomyosin resulting from alternative splicing are found in other tissues of the same organism. It seems likely that the type of tropomyosin made in any given tissue is the type most suitable for the needs of that tissue. Mechanistically, it seems likely that some of the components involved in the splicing must also be tissue-specific.

Bingham and his co-workers have uncovered another role for splicing in *Drosophila.* In three different genes they have observed that most of the introns are removed from pre-mRNA molecules, leaving one or two introns intact. In this state the pre-mRNAs are nonfunctional. Only when the final introns are removed can they function as messages. Apparently the final splicing operations are regulated by the protein encoded by the messenger. If that protein is present above a certain concentration, then it prevents any further splicing of the pre-mRNAs. Bingham views this as a type of feedback inhibition that may be fairly common in *Drosophila* and perhaps other species.

Translation and Polypeptide Processing Also Affect Gene Expression After messenger formation there are still other ways in which the amount and types of finally synthesized protein can be affected. The initial polypeptide can be processed in various ways so that different polypeptides or proteins are expressed in specific tissues. We described such a possibility (see fig. 24.6) for the processing of the precursor polypeptide preproopiomelanocortin. This polypeptide is processed in different ways in the anterior and intermediate lobes of the pituitary gland to give rise to unique hormones in the two tissue types.

Because of the long lifetime of eukaryotic messengers it seems likely that translation-level controls play a major role in regulation of eukaryotic gene expression. Despite this belief, few mechanisms have yet been elucidated.

Inactivation of eukaryotic translation factors by covalent modification is one of the few mechanisms known to regulate the gross rate of translation. Specific protein kinases have been identified that phosphorylate and inactivate both eIF-2 and EF-2. The significance of the phosphorylation of EF-2 as a regulatory mechanism of the elongation rate is still not clear, but the phosphorylation of eIF-2 appears to be a general mechanism of controlling translation initiation in many cells.

The regulation of translation through the phosphorylation of eIF-2 is best understood as it operates in the rabbit reticulocyte. Two protein kinases specific for the *a* subunit of eIF-2 have been purified from reticulocytes. One of these kinases, termed the heme-regulated inhibitor repressor (HRI), serves to coordinate the rate of hemoglobin synthesis (more than 90% of the total protein synthesized in the reticulocyte is hemoglobin) with the availability of hemin (the precursor of the heme group in hemoglobin). Hemin binds to and inhibits the activity of HRI and this inhibition enhances the rate of globin synthesis (fig. 31.23).

Figure 31.22

Alternative modes of splicing of the tropomyosin gene transcript in different tissues. (Source: R. E. Breitbart et al., "Alternative splicing: A ubiquitous mechanism for the generation of multiple protein isoforms from single genes," in *Annual Review of Biochemistry* 56:467–495, 1987. Copyright © 1987 Annual Reviews Inc., Palo Alto, Calif.)

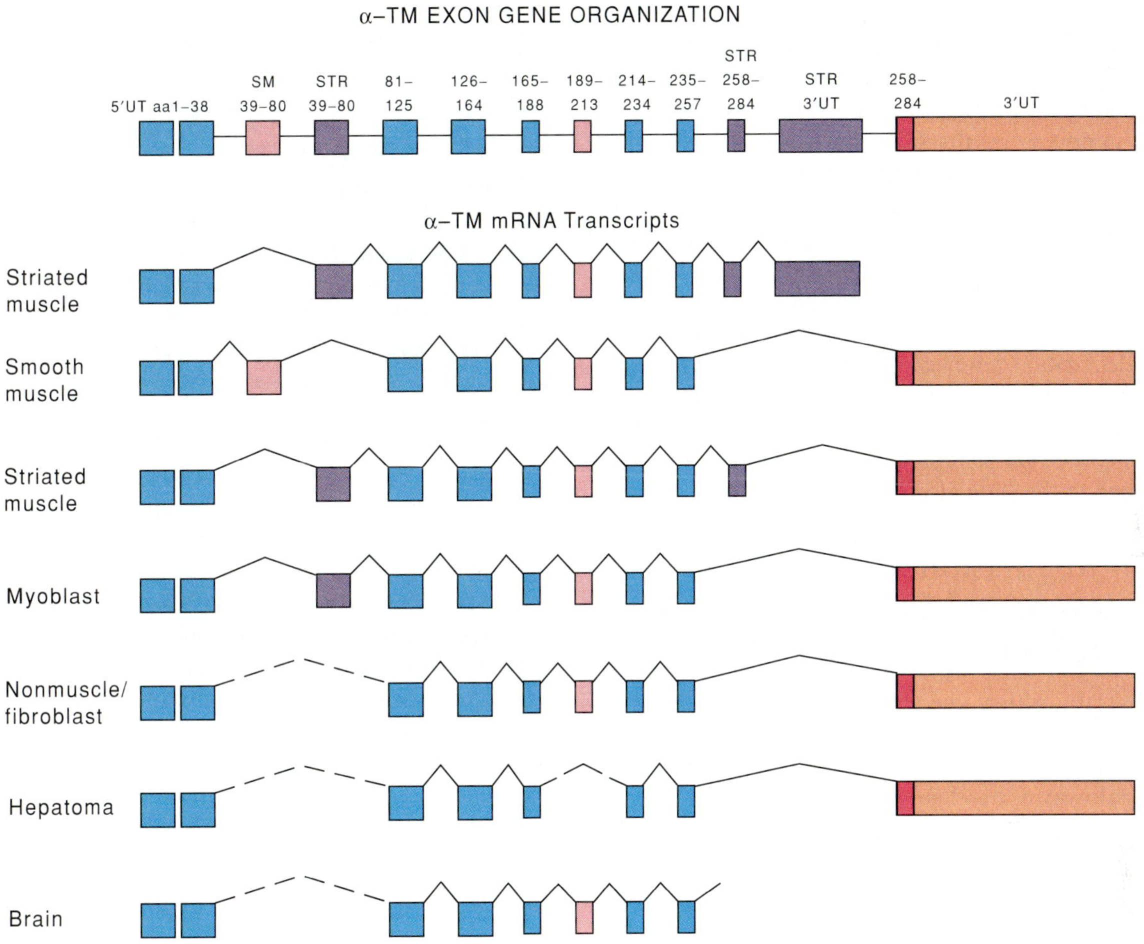

Figure 31.23

Regulation of protein synthesis in the rabbit reticulocyte. The vast majority of the protein synthesized in the rabbit reticulocyte is hemoglobin. The gross rate of protein synthesis in the reticulocyte is controlled indirectly by the concentration of heme. Heme inactivates a kinase that would otherwise inactivate the initiation complex involving eIF-2 and eIF-2B. The kinase phosphorylates the eIF-2 factor, making it impossible for the eIF-2–eIF-2B complex to exchange GDP for GTP.

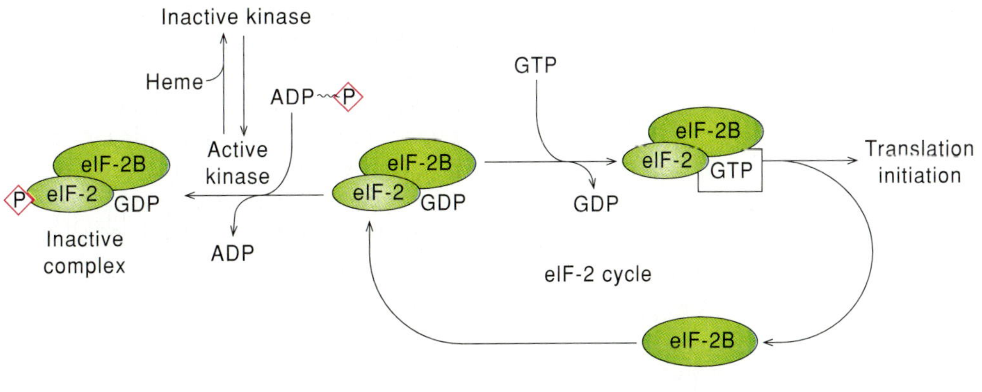

The second eIF-2-specific kinase appears to be present at low levels in most mammalian cells. This kinase is activated by double-stranded RNA and is therefore known as the double-stranded RNA-activated inhibitor (DAI). DAI may play a role in defending cells against invasion by viruses.

HRI and DAI are differnt proteins, but both phosphorylate the same amino acid residue in the *a* subunit of eIF-2 and, in consequence, both kinases inhibit protein synthesis by the same mechanism. Phosphorylated eIF-2 is capable of catalyzing the binding of Met-$tRNA_f^{Met}$ to the ribosome, but it does so in a stoichiometric rather than a catalytic manner. This mechanism of inhibition results from the fact that eIF-2 dissociates from the ribosome as a complex with GDP and, in order to recycle, the bound GDP must exchange for GTP in a manner very similar to the exchange cycle involving the *E. coli* elongation factors EF-Tu and EF-Ts. Phosphorylated eIF-2 binds to, but is unable to dissociate from, the guanine nucleotide-exchange factor (GEF) that catalyzes the exchange of GTP for GDP. Since cells contain fewer copies of GEF than eIF-2, all GEF can be sequestered by partial phosphorylation of eIF-2, and protein synthesis initiation ceases for lack of GEF rather than eIF-2. Thus the rate of protein synthesis initiation is exquisitely sensitive to the state of eIF-2 phosphorylation.

Patterns of Regulation Associated with Developmental Processes in Multicellular Eukaryotes

As cells proliferate within the embryo, they differentiate; changes are triggered by the genes that they express and the proteins that they synthesize. Cells that are initially capable of following any pathway of development become committed to a particular pathway. Most pathways have many branchpoints, so when partially committed cells reach a branchpoint they may differentiate further, down a more specialized pathway. This process of gradual commitment, repeating itself many times, gives rise to a complex pattern of cell lineages.

It seems likely that the pivotal events in the evolution of a differentiated cell reflect changes in gene expression that result from a complex hierarchy of controls. The key to understanding differentiation, therefore, is to identify the regulatory factors that are responsible for the controls involved in differentiation and to explain how they act.

We will focus on regulatory events that occur in development. First we will discuss some of the classical studies of gene expression in the embryogenesis of sea urchins and amphibians, and then we will turn to some of the developmental events in fruit flies, which have the best-characterized developmental system.

During Embryonic Development in the Amphibian, Specific Gene Products Are Required in Large Amounts

During early development, when cell divisions are occurring rapidly, there is an increased demand for certain products associated with the genome (histones) and the translation apparatus (ribosomes). Special regulatory mechanisms ensure that these gene products are produced in the required amounts and that they are not overproduced in the more mature organism.

Ribosomal RNA in Frog Eggs Is Elevated by DNA Amplification

In eukaryotic organisms, ribosomal RNA genes are present in clusters of hundreds to thousands of copies. The nascent transcript, a 45S molecule, undergoes an elaborate series of processing reactions to produce three kinds of ribosomal RNAs—28S (25–28S), 18S (17–18S), and 5.8S (5.5–5.8S). Multiple copies of ribosomal RNA genes are organized into tandem arrays separated by nontranscribed spacers, which vary in length in different species from about 2,000 to 30,000 bp. *In situ* hybridization studies, using radioactive rRNA and autoradiography, demonstrate that the ribosomal RNA genes are localized around the nucleolus, where rRNA processing and ribosome assembly take place. Germ cells and, in particular, immature eggs or oocytes often raise the level of rDNA per nucleus by amplification to satisfy the high demand for rRNA during the very rapid early cleavage stages. This type of amplification has been most thoroughly investigated in the frog *Xenopus laevis* (fig. 31.24), in which amplification increases the number of rRNA genes about 1,000-fold. These extrachromosomal genes are transcribed during oogenesis and subsequently discarded. The amplification mechanism is not understood. An entirely different mechanism elevates the level of 5S rRNA synthesis during oocyte maturation.

5S rRNA Synthesis in Frogs Requires a Regulatory Protein

Recall that eukaryotic ribosomes contain the small rRNA components (5S and 5.8S). While the 5.8S rRNA is contained in the 45S transcript with the other large rRNAs, the 5S exists independently in multiple gene copies that are organized into simple tandem multigene families. In amphibians the 5S rRNA genes are organized into different multigene families that are under developmental control. For instance, in *X. laevis* the normal haploid genome contains about 20,000 copies of the 5S rRNA genes, organized into three different multigenic families. The genes of two major families, comprising about 98% of all the 5S rRNA genes, are expressed only in growing oocytes. The genes of the third family, existing in about 400 copies, are active in most (somatic) cells and growing oocytes. This developmental control enables the oocyte to accumulate 5S rRNA at rates 1,000-fold higher than is possible in somatic cells.

Figure 31.24

Time course of rDNA amplification in the *Xenopus laevis* oocyte. The amount of rDNA per cell is plotted against time. Note that the ordinate is a logarithmic scale. The sharp drop in rDNA at fertilization is thought to be due to dilution of the amplified rDNA copies by cell division, rather than to their destruction. (Source: A. P. Bird, "Gene reiteration and gene amplification" in *Cell Biology*, vol. 3, *Gene Expression: The Production of RNAs*, edited by L. Goldstein and D. M. Prescott. Copyright © 1980 Academic Press, New York.)

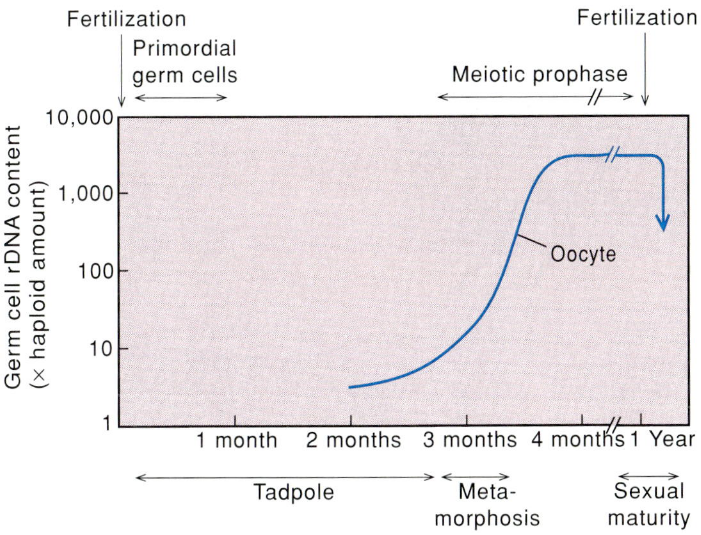

In chapter 28 we noted (see fig. 28.18*c*) that Pol III, which transcribes the 5S genes, requires the binding of three transcription factors to the 5S promoter. These factors are known as TFIIIA, TFIIIB, and TFIIIC. The abundance of TFIIIA appears to be much greater in the oocyte than in most somatic tissue. As a consequence, only in the oocyte is there enough TFIIIA to bind to all of the 5S promoters. Competition binding experiments in which both somatic and oocyte 5S gene DNAs are exposed to limited amounts of the TFIIIA protein show that the regulatory protein binds considerably more firmly to the somatic 5S genes. This fact probably explains why the somatic 5S genes are uniquely active in somatic tissues when the amounts of TFIIIA protein are much lower. Under a variety of conditions, large amounts of 5S rRNA are synthesized in the oocyte, but not in somatic cells; this difference is strongly related to the concentration of TFIIIA regulatory protein in the two situations.

Incidentally, it appears that TFIIIA protein has another function. It interacts with 5S rRNA to form a stable 7S nucleoprotein complex that accumulates in the cytoplasm of young oocytes. In growing oocytes, 5S rRNA is not incorporated immediately into ribosomes; rather, it is synthesized actively for several weeks before 18S and 28S rRNA synthesis begins. In addition to its storage function, it has been proposed that TFIIIA protein also serves a transfer function, facilitating the transport of 5S rRNA to the nucleolus, where ribosome assembly takes place.

Figure 31.25

Molecular maps of histone gene repeat units of a sea urchin and a fruit fly. The histone gene organizational maps are calibrated in base pairs. The mRNA coding regions, including their leader and trailer sequences, are depicted by the colored boxes. The direction of transcription is indicated by the arrows.

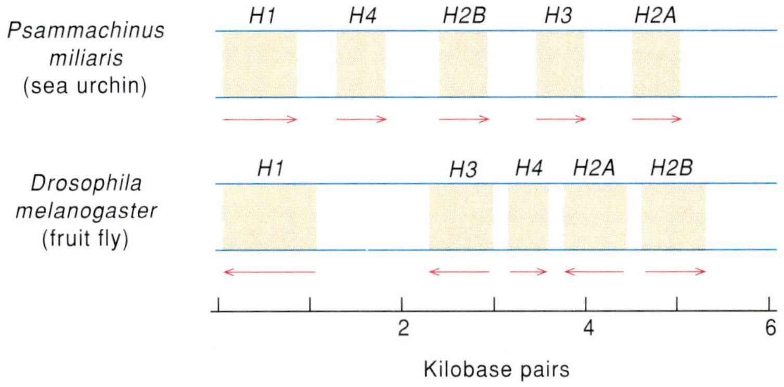

Demands for Histones Are Met by Gene Amplification and Storage

Only a few gene families that encode mRNAs exist consistently in multiple copies. One such family includes the histone genes. Multiple copies of the histone genes are easy to rationalize in terms of the needs of most cells, which must make a large quantity of histones over short periods of time. Histones are usually but not always synthesized during S phase, when the DNA is being replicated in approximately equal-weight proportions to the histone. The five different types of histone genes invariably occur in clusters containing transcribed and nontranscribed regions (fig. 31.25). Each cluster is usually repeated many times in a tandem array.

Most of the detailed structural analysis of histone genes has been carried out in sea urchins. Cells of these organisms have several hundred copies of the genes for histones, whereas humans, mice, and *Drosophila* have only 10–50 copies. It seems likely that histone genes are reiterated in a very high frequency in the sea urchin because very large amounts of histones are needed during early embryogenesis, when cell cleavages occur once every 30 minutes. Indeed, those histone genes that are active during early embryogenesis constitute a distinct class. The set of histone genes that function late in embryogenesis of the sea urchin are present in only 5–12 copies per genome; unlike the clustered, tandemly arranged early histone genes, they are dispersed and irregularly arranged. Specific transcription factors are involved in class-switching between activating early and late histone genes.

In *X. laevis* it takes, on average, 14 min for the number of cells to increase from 50 to 5,000 during early embryogenesis. Since the frog has a larger genome than the sea urchin, its demand for histones should be even greater during early embryogenesis. Nevertheless, the reiteration of histone genes is less

Figure 31.26

Steps in the development of *D. melanogaster*. The fertilized egg contains a single zygotic nucleus (*a*). This divides (*b*) every ten minutes. After eight divisions, when there are 256 nuclei, nuclear migration toward the outer cortex structure begins (*c*). Eventually all the nuclei form a monolayer on the cortex surface. The first nuclei to become enclosed are the pole cells, which will become germ cells in the adult organism (*d*). The fully formed blastoderm contains about 6,000 cells, which form a monolayer around the cortex (*e*). Even before a cell membrane has formed around the nuclei, the embryo has become functionally divided into a segmented structure (*f*). During gastrulation (*g*), there is continued cell duplication, a folding of sheets of cells, and mass migration of segments. Eventually this structure hatches into the first larval stage (*h*). The remaining stages between the larva and the adult fly (*i*) are not illustrated. In (*f*) through (*i*), various segments are labeled. Three segments, Md, Mx, and Lb, fuse to make the head structure. Thoracic segments T1–T3 and abdominal segments A1–A8 retain their segmental appearance in the adult.

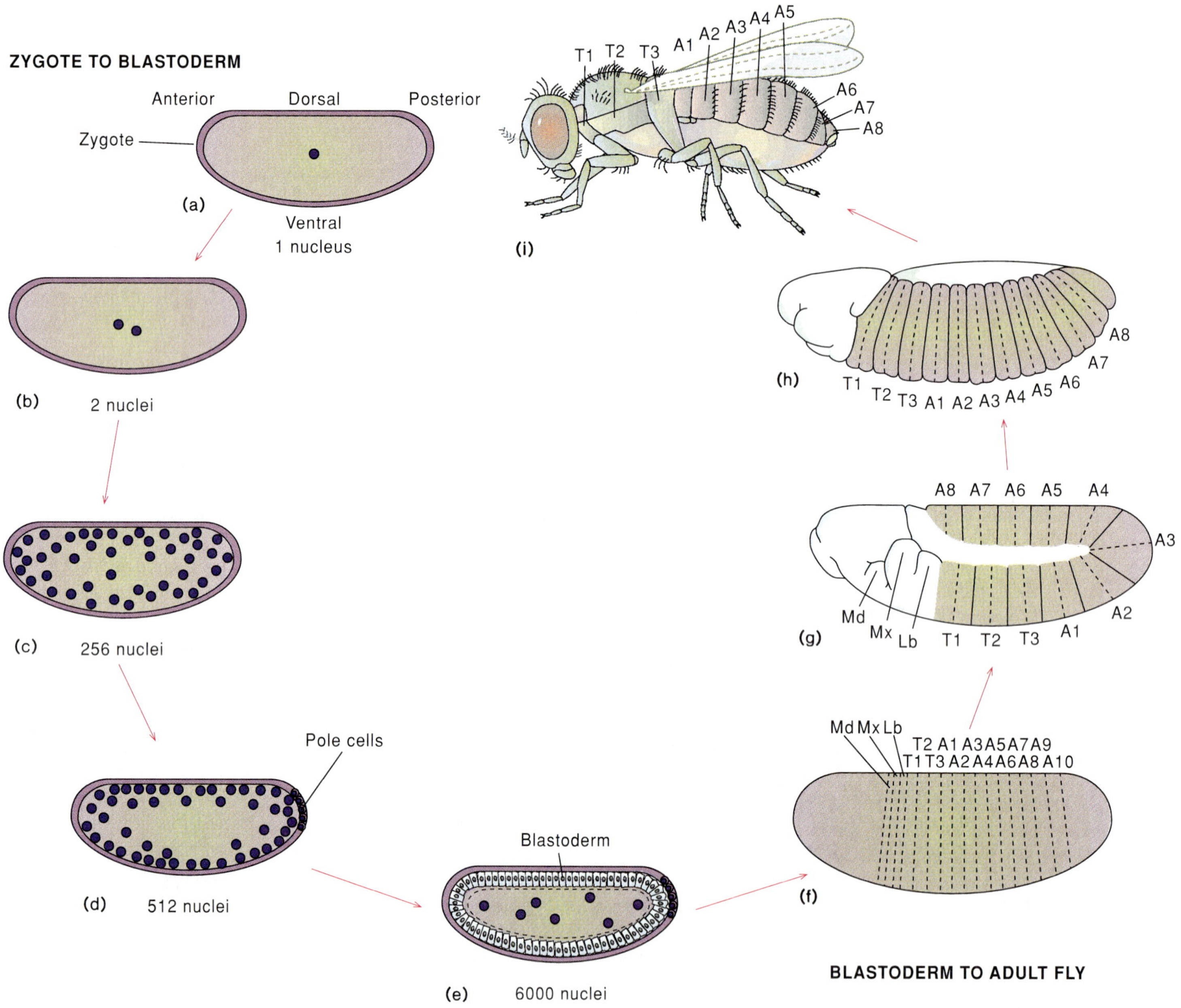

pronounced in the frog than in the sea urchin, because the histone requirements of the frog are met by a large accumulated store of histone proteins and histone mRNA in the egg.

Early Development in Drosophila *Leads to a Segmented Structure That Is Preserved to Adulthood*

Whereas sea urchins and frogs have proved valuable in the study of isolated embryonic events and the structure and composition of the embryo, they have not been very useful for understanding the overall pattern of early embryonic developments. For this purpose it has been necessary to turn to other organisms to identify the genes involved in regulating development. Many fundamental aspects of developmental biology have been studied using the techniques of genetics and molecular biology in *D. melanogaster.*

Some steps in the development of *Drosophila* are shown in figure 31.26. Beginning shortly after fertilization, the new diploid nucleus undergoes a rapid series of divisions with no segregation of nuclei into separate cells. After the eighth division, when there are 256 nuclei (see fig. 31.26*c*), the nuclei begin to migrate to the periphery of the egg cytoplasm. After another nuclear doubling, cell membranes form around a group of cells at the posterior end of the egg; these cells are progenitors of germ cells for the subsequent generation. The remaining nuclei continue to divide until there are about 6,000 nuclei at the periphery (see fig. 31.26*c*). At this point, about two hours after fertilization, membranes are formed, separating the nuclei into a monolayer of cells. The resulting structure, known as the blastoderm, is essentially a cell monolayer enclosing the yolk.

The blastoderm divides into fourteen different segments: Md, Mx, Lb, T1–T3, and A1–A8 (see fig. 31.26*f*). Three segments, Md, Mx, and Lb, become part of the head structure. The remaining segments become subdivided into two compartments, with anterior and posterior parts. During gastrulation there is a continued cell duplication, folding of sheets of cells, and mass migration of segments (see fig. 31.26*g*). The embryo eventually hatches into the first larval stage (see fig. 31.26*h*). Cells within the larva are of two types. About 80% of them are fully functional in the larva. The remaining 20% are embryonic precursors to adult tissues rather than larval tissues. These latter cells form packets within the larva, called imaginal disks, that are arrested in development until pupation (fig. 31.27), when a single hormone ecdysone triggers their differentiation into specific adult structures.

Early Development in Drosophila *Involves a Cascade of Regulatory Events*

There are two distinct phases in *Drosophila* embryogenesis; the first precedes cellularization of the blastoderm and is associated with a cascade of interacting regulators; the second occurs after cellularization and depends on intercellular signals that must be carried, at least in part, by membrane-bound proteins. We will focus on events occurring in the first phase because they are better understood.

Figure 31.27

Imaginal disks in a mature larva and the structures they lead to in the adult fly.

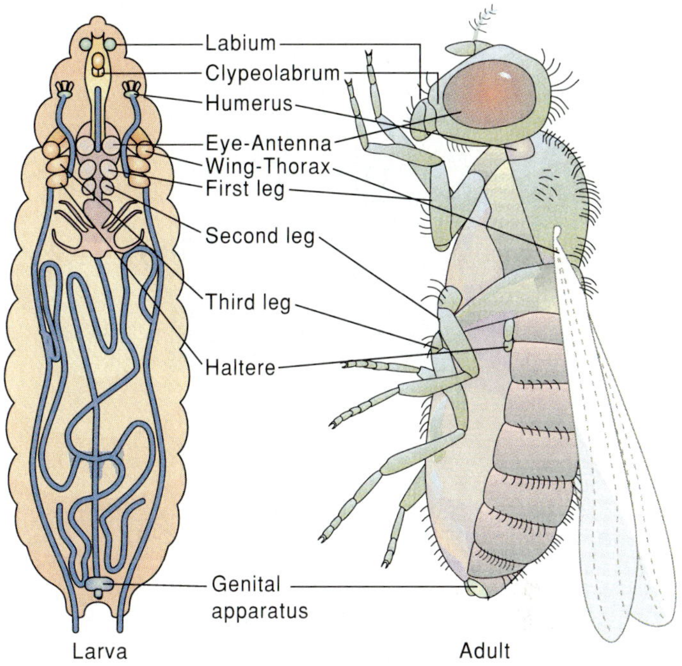

Recall that in the infectious cycle of the λ bacteriophage, each stage is characterized by the expression of specific gene products under the control of one or more regulatory proteins. A new phase in the development of the bacteriophage results from the synthesis of one or more new regulatory proteins. A hierarchy determines the order in which the different regulatory proteins make their appearance. Early development in *Drosophila* is very similar except for its greater complexity. In fact, some of the earliest regulatory proteins in the developing oocyte are supplied by surrounding cells. Another difference in *Drosophila* development is that the large size of the oocyte permits the establishment of gradients within a single cell. The regulatory protein encoded by a messenger RNA injected at the anterior end of the oocyte is likely to exist in highest concentration at the anterior end and at lowest concentration at the posterior end. Similarly, mRNAs injected into the oocyte at the posterior end establish regulatory-protein gradients in the opposite direction. These gradients are crucial for regulating the expression of genes involved in early development. Some gradients are also established in the perpendicular direction, on the dorsal-ventral axis, but we can overlook these for present purposes.

***Three Types of Regulatory Genes Are Involved in Early Segmentation Development in* Drosophila** The first insight into the genetic system directing *Drosophila* development came from the discovery, of bizarre mutations that affect the body plan (fig. 31.28). For example, the mutation *Antennapedia* results in a pair of extra legs sprouting from the head in place of antennae.

Figure 31.28

Abnormal phenotypes resulting from homeotic mutations. *Antennapedia* results in a pair of extra legs sprouting from the head in place of antennae. *Bithorax* results in an extra pair of wings appearing where normally halteres would appear.

Another mutation, *Bithorax,* results in an extra pair of wings appearing where normally there should be much smaller appendages called halteres. These mutations occur in regulatory genes known as homeotic genes.

In addition to homeotic genes there are two other types of genes that influence the segmentation pattern; these are called maternal-effect genes and segmentation genes (table 31.3). Maternal-effect genes are so called because they affect the phenotype only according to the information present in the female parent. Maternal-effect mutations occur in genes responsible for establishing the anterior-posterior axes in the young embryo. Segmentation mutations affect the number and polarity of the body segments. Homeotic mutations are more specific than segmentation mutations; they result in changes in structures that are uniquely associated with individual segments or subsegments.

Table 31.3
Regulatory Genes Involved in Segmentation Development

Maternal-effect genes—establish gradients
Anterior
bicoid (*bcd*)
Posterior
oskar (*osk*)
Dorsal-ventral
dorsal (*dl*)
Segmentation genes
Gap genes—define four broad regions in the egg
hunchback (*hb*)
Krüppel (*Kr*)
Knirps (*kni*)
Pair-rule genes—define seven bands
runt (*runt*)
hairy (*h*)
fushi tarazu (*ftz*)
even skipped (*eve*)
paired (*prd*)
odd paired (*opa*)
Segment-polarity genes—define fourteen bands
engrailed (*en*)
wingless (*wg*)
gooseberry (*gb*)
Homeotic genes—specify structures associated with individual segments
BX-C locus
ANT-C locus

Analysis of the Genes That Control the Early Events of Drosophila *Embryogenesis*

Once we have found a mutation that appears to affect development, we must have a way to detect the presence or absence of the related gene product in the developing embryo. A principal tool for doing this is autoradiography using ^{32}P-labeled specific DNA probes. The embryo is labeled by *in situ* hybridization with the radioactive probe specific for the transcript of a gene (see box 31A). Both the location and the intensity of the labeling give an indication of the level of expression of the gene. In some cases the protein product relating to a specific transcript can also be detected by use of a specific antibody-labeling technique. In what follows we will trace the appearance of some of the major regulatory genes that are involved in early development.

Maternal-effect Gene Products for Oocytes Are Frequently Made in Helper Cells The first regulatory gene products to play a significant role in *Drosophila* development are active before fertilization. The bicoid (*bcd*) gene is a major determinant of the anterior-posterior pattern. Like many other maternally expressed genes, *bcd* is transcribed in the ovary in specialized cells that form a cluster around the future anterior pole of the developing oocyte. The bcd RNA synthesized in these so-called nurse cells passes into the oocyte through cytoplasmic canals and becomes localized at the anterior pole of the oocyte

Figure 31.29

Key events in the expression of developmental regulatory genes in the blastoderm embryo. Events take place in the order shown over a period of a few hours. Measurements are based on the use of probes for detecting the transcripts of various regulatory genes. Description of genes is given in table 31.3. (Source: P. W. Ingham, "The molecular genetics of embryonic pattern formation in *Drosophila,*" in *Nature* 335:25–34, 1988. Copyright © 1988 Macmillan Magazines Ltd., London, England.)

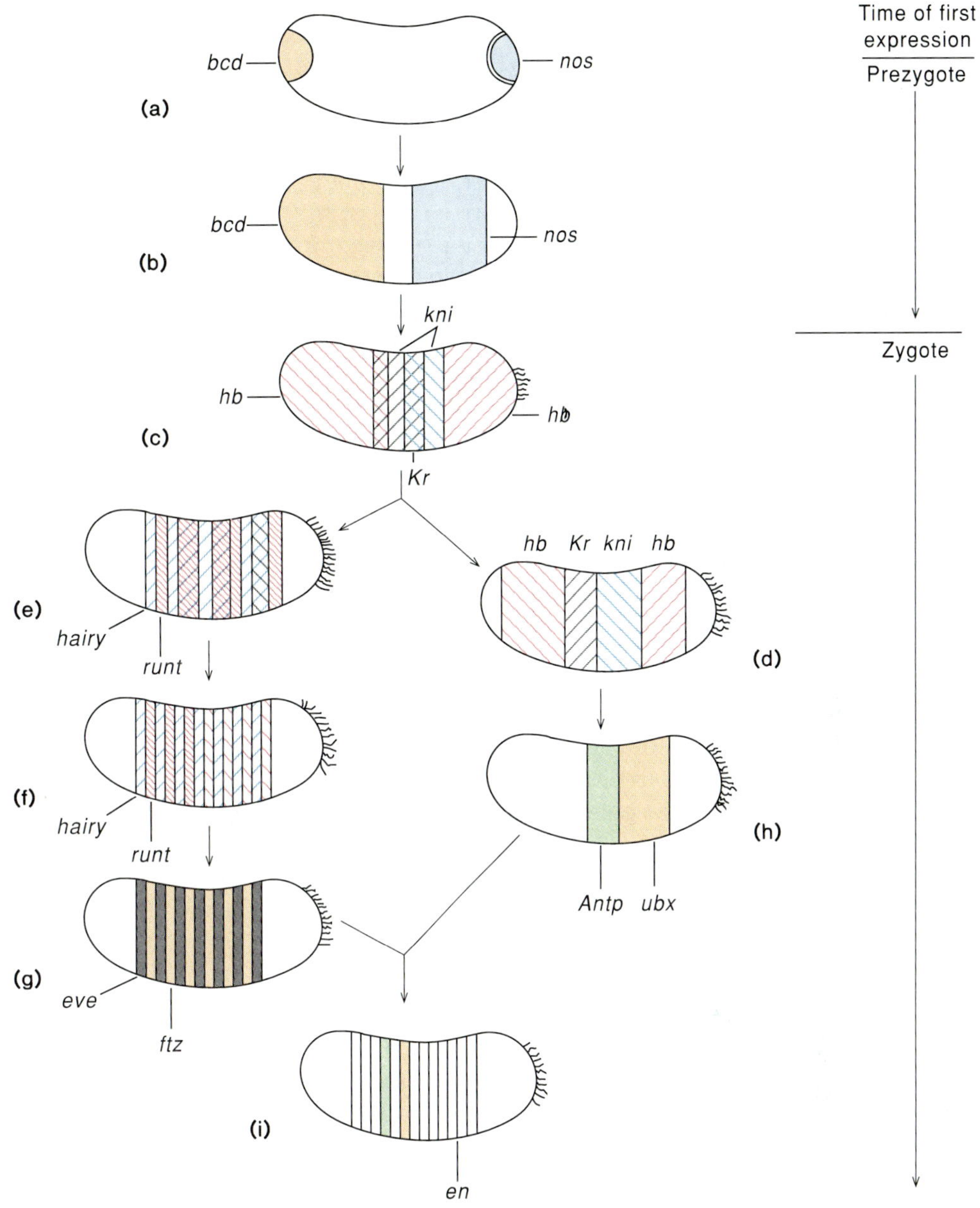

(fig. 31.29). Similarly, the oskar (*osk*) group genes, such as *nos,* are involved in the synthesis and deposition of products that become localized at the posterior pole of the egg (see fig. 31.29*a*). Females lacking a functional copy of the *bcd* gene produce eggs that develop into embryos with no head or thorax. The same effect can be achieved by removal of material from the anterior pole of a normal egg. Embryos produced by female mutants for *nos* develop normal head and thoracic segments, but lack the entire abdomen.

The pattern-forming process continues upon fertilization; as the *bcd* RNA is translated, the protein diffuses from the anterior pole so that it becomes distributed over about half of the length of the egg (see fig. 31.29*b*). Simultaneously, information encoded and localized at the posterior pole during oogenesis by members of the *osk* group begins to move forward.

By the beginning of the precellular blastoderm stage there are gradients for at least two different maternally encoded products along the anterior-posterior axis of the embryo. This quantitative information is now transformed into qualitative differences in the form of region-specific gene expression, by a process requiring interaction between the maternally derived products and the zygotic genome that resulted from fertilization.

Gap Genes Are the First Segmentation Genes to Become Active The first segmentation genes to become active in the zygote are members of the gap class, so-called because their mutants lack major regions of the body, thereby creating a gap in the antero-posterior pattern. The three members of this class are hunchback (*hb*), Krüppel (*Kr*), and knirps (*kni*). These genes are expressed in two division cycles prior to cellularization. Expression of *hb* is restricted to two regions, one extending from the anterior pole to 50% of the egg length, the other from the posterior pole to about 25% of the egg length (see fig. 31.29*c*). Initially the *Kr* gene is expressed in a single broad band in the middle of the embryo, whereas *kni* is expressed in two distinct domains, one anterior and one posterior to the *Kr* band. These transcriptional domains for *hb, Kr,* and *kni* are influenced by the maternally derived information encoded by the *bcd* and *osk* group genes. Thus mutants that lack *bcd* activity do not express *hb,* whereas *Kr* is extended anteriorly in an unusually broad domain. The absence of the posterior maternal determinants results in the extension of the *Kr* domain posteriorly. These observations suggest that the *bcd* and *osk* group genes both act to repress transcription of *Kr* in the anterior and posterior regions, respectively, thereby restricting its region of expression to the central portion of the embryo. In contrast, *bcd* appears to act as a positive regulator of *hb,* and *osk* appears to be a positive regulator of *kni.* In normal embryos the transcriptional domains of *hb, Kr,* and *kni* narrow with time, giving rise to sharp boundaries of expression. This process is driven by negative effects between the gap genes (see fig. 31.29*d*), *hb* and *Kr* mutually repressing one another and *kni* acting as a negative regulator of *Kr.* Thus the establishment of stable domains of gap gene expression is a two-step process; first, a differential response to graded levels of maternal determinants, and second, a mutual repression effect, leading to the generation of stable boundaries between adjacent domains.

The gap gene products regulate the position-specific expression of other genes, those belonging to the pair-rule class of segmentation genes.

Periodic Gene Expression Is Initiated by Pair-Rule Genes A common feature of pair-rule genes is their transient expression in seven bands during the period of cellularization of the blastoderm. Despite this similarity, each pair-rule gene is unique in its pattern of expression. *Runt* and *hairy* are initially expressed throughout the embryo (see fig. 31.29*e*); a restricted pattern of expression of these two genes begins earlier than for other pair-rule genes.

By the beginning of the last interphase before cellularization, transcripts of both *hairy* and *runt* localize to two complementary series of seven bands that encircle the embryo (see fig. 31.29*f*). The generation of these banded patterns is a key event in the pattern-forming process. The sorts of interactions taking place probably involve the local enhancement of expression of one or another gene in response to local levels of one or more gap gene products. Such interactions result in a patchy pattern of *hairy* and *runt,* which is further refined and stabilized into a regular banded pattern by their mutual repression (see fig. 31.29*f*), in a manner analogous to the stabilization of the gap gene expression domains.

The complementary bands of *runt* and *hairy* are instrumental in the refinement of the spatial patterns of expression of other pair-rule genes.

Figure 31.30

A region of the blastoderm showing three of the fourteen parasegments and the initial and ultimate spheres of expression for certain pair-rule and segment-polarity genes (all fourteen parasegments are shown in figure 31.32). The dashed regions for *prd, opa, ftz,* and *eve* show a region of expression that occurs only at early stages, before the narrowing process. The circles represent cells in the early stage in the blastoderm, when the parasegments are only four cells in width.

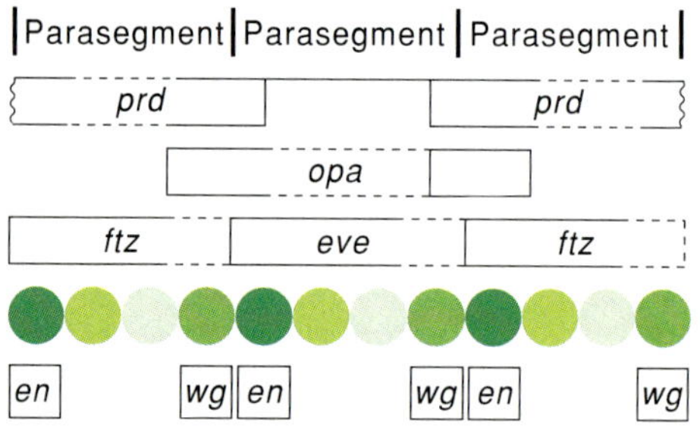

Hairy and Runt Proteins Function as Negative Regulators for the Expression of Other Pair-Rule Genes *Hairy* and *runt* act as negative regulators for the expression of the pair-rule genes *fushi tarazu* (*ftz*) and *even skipped* (*eve*) respectively (see fig. 31.29*g*). Both *ftz* and *eve* encode homeo-domain proteins, a strong indication that they act as transcriptional regulators of genes of the homeotic and segment-polarity class (box 31B). The interactions between *hairy* and *runt* and between *ftz* and *eve* generate complementary sets of seven bands expressing one or the other gene (the locations of *ftz* and *eve* expression are shown in figure 31.29). When they first become visible, each band is four nuclei wide. At this stage, each *ftz* and *eve* band coincides with the boundaries of a so-called parasegment. A parasegment is a unit that is out of register with a morphological segment but coincident with the domains of activity and function of the homeotic genes (fig. 31.30). At the same time as the *ftz* and *eve* patterns are resolved, the pattern of expression of a third homeobox-containing pair-rule gene called *paired* (*prd*), is also changing (see fig. 31.30 for the pattern of *prd* expression). Initially *prd* is expressed in seven broad bands also, but in this case the bands are each about six nuclei wide. This pattern is then transformed into one of fourteen bands by the elimination of expression from the central region of each of the original bands. As with *ftz* and *eve,* the generation of this pattern depends on the activity of *hairy* and *runt.* Thus, in a process resembling the establishment of the gap gene domains, information encoded by two components is transformed into a more complex form by the differential response of the various secondary pair-rule genes (*ftz, eve, prd*) to this information.

***The Boundaries of the Parasegments Are Set by Wingless* (Wg) *and Engrailed* (en)** The combinations of pair-rule gene activities define the states of particular blastoderm cells. These cells subsequently serve as reference points for the elaboration of the pattern within individual parasegments during the period of

31B

BOX

The Homeobox Sequence

One common feature of all homeotic genes, and of some maternal-effect and segmentation genes as well, is that they possess a common 180-base segment in the 3′ exon. The base homology ranges between 60 and 80%, depending on the specific genes being compared, and the amino acid homology is even higher (up to 87%). The high basicity of the amino acid sequence in this region—called the homeo box—and other structural features suggest that this region encodes a DNA-binding protein that regulates gene expression by binding to specific sites on the DNA. Indeed, we see sufficient amino acid homology between the homeo box and the helix-turn-helix motif in many bacterial regulatory proteins to suggest that the gross structures of these very distantly related regulatory proteins are quite similar.

The structure of the homeodomain is built up from three helices connected by short loop regions. The three helices are formed by residues 10 to 21, 28 to 38, and 42–58. Helices 2 and 3 are part of the helix-turn-helix motif in which the structure of the region 30 to 50 is virtually identical to the helix-turn-helix motif found in bacteria.

One of the most exciting findings relating to the homeo box is that very similar sequences have been identified in many other animals, including frogs, mice, and even humans. This finding suggests that there are homeo box proteins in a wide range of organisms, possibly playing similar roles in regulating developmental processes.

Figure 1

The homeo box sequence found in three *Drosophila* regulatory proteins. The sequence of amino acids along the main, continuous line is that found in the *Antp* homeo box. At points where the sequence of ftz proteins is different, the differences are shown above the corresponding Antp proteins, and at points where the sequence of ubx proteins is different, they are shown below the corresponding Antp proteins. Certain proteins isolated from the human and mouse embryos have segments with similar sequences. Note the high basicity of the sequence, which should favor electrostatic binding to nucleic acid.

				5					10					15					20
			Thr																
Arg	Lys	Arg	Gly	Arg	Gln	Thr	Tyr	Thr	Arg	Tyr	Gln	Thr	Leu	Glu	Leu	Glu	Lys	Glu	Phe
Ser			Gly																

				25					30					35					40
					Ile							Asp			Asn			Ser	
His	Phe	Asn	Arg	Tyr	Leu	Thr	Arg	Arg	Arg	Arg	Ile	Glu	Ile	Ala	His	Ala	Leu	Cys	Leu
	Thr		His										Met		Tyr				

				45					50					55					60
Ser															Ser			Asp	Arg
Thr	Glu	Arg	Gln	Ile	Lys	Ile	Trp	Phe	Gln	Asn	Arg	Arg	Met	Lys	Trp	Lys	Lys	Glu	Asn
															Leu				Ile

limited cell proliferation that follows. In particular, the pair-rule genes have been shown to define the domains of expression of the segment polarity genes *en* and *wg*.

At the end of the cellularization process and the onset of gastrulation, transcripts of both the *en* and *wg* genes accumulate in fourteen narrow bands along the anterior-posterior axis of the embryo, patterns that represent the first evidence of its segmental organization. At this time each band of *en* expression is only one cell wide. Each of the *en* bands represents the anterior boundary of a parasegment. The establishment of alternate *en* bands requires the combined activities of different sets of pair-rule genes. In the case of the odd-numbered bands, the *en* expression cells are those that express both *eve* and *prd,* whereas the even-numbered bands require the expression of both *ftz* and another pair-rule gene, *odd-paired* (*opa*).

In contrast to this positive control of *en, wg* expression is repressed by the combination of eve and ftz proteins. At the end of blastoderm the domains of eve and ftz narrow, so that single-cell-wide bands come to separate them, and it is these cells that initiate expression of *wg*. The net result is the generation of two sets of cells that are adjacent to one another and mark the anterior and posterior limits of each of the parasegments (see fig. 31.30). These sets of cells, one expressing *en* and the other expressing *wg*, subsequently serve as reference points for the specification of position in each developing parasegment.

The positive and negative effects established between maternal-effect and pair-rule genes are summarized in figure 31.31. This diagram tells an incomplete story. The most serious gap in our knowledge is how the gradients of maternal-effect and gap genes help to deliver the signal for the establishment of the strict periodic patterns of expression seen for most of the segmentation genes. It seems unlikely that the signals in the form of gradients could ever institute such a pattern without other factors being involved. Possibly some of the signals for expression of pair-rule genes are established in the oocyte before fertilization.

Figure 31.31

Influence of developmental regulatory genes on one another. An arrow with a plus sign at the arrowhead indicates a positive effect on expression, whereas a minus sign at the arrowhead indicates a negative effect on expression. The gap genes are essential for the establishment of the hairy and runt banded patterns, but the precise way in which they do this is unclear.

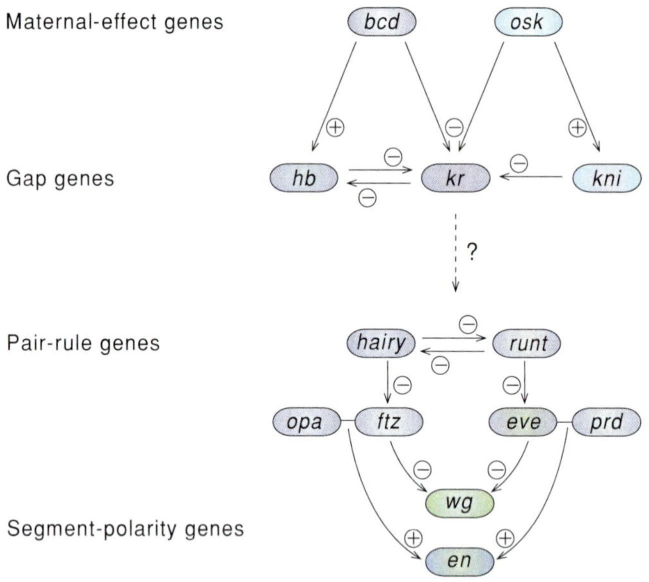

Homeotic Genes Specify Parasegment Character By the time cellularization of the blastoderm is complete, different cells express different combinations of pair-rule genes (fig. 31.32). At this stage, the embryo is subdivided into a series of repeating units. Despite their similarity, each parasegment is programmed to follow a unique pathway of differentiation as evidenced by subsequent events. Most probably the expression of the homeotic genes is influenced by the periodic information generated by the pair-rule and the segment-polarity genes, superimposed on the gradients of information deposited by the maternal genes and the gap genes. We do not know what information turns on specific homeotic genes in specific parasegments, but we do know a most exciting beginning of a story that should unfold in the near future.

Most of the homeotic genes are clustered in two giant loci, the *Antennapedia* (*ANT-C*) and *bithorax* (*BX-C*) complexes. The *ANT-C* locus occupies about 100 kb of the genome and the *BX-C* locus occupies about 300 kb of the genome. Both of these loci are situated on chromosome 3. The *BX-C* locus is subdivided into three sections, *Ubx*, *Abd-A*, and *Abd-B*, according to which body parts are affected by different mutations. The *Ubx* region affects thoracic segments T2 and T3 and the anterior portion of the A1 segment. The *Abd-A* region affects the posterior part of A1 and abdominal segments A2 to A4. Finally the *Abd-B* region affects the abdominal region from A5 to A8. The regions affected by different parts of the *ANT-C* locus include the head and the three thoracic segments (fig. 31.33). It is fascinating that the arrangement of the genes in

Figure 31.32

Pattern of expression of some pair-rule and segment-polarity genes. When the pattern of expression of maternal-effect genes and gap genes is superimposed on this regularly repeating pattern, it gives each region of the blastoderm a unique mixture of developmental regulatory genes.

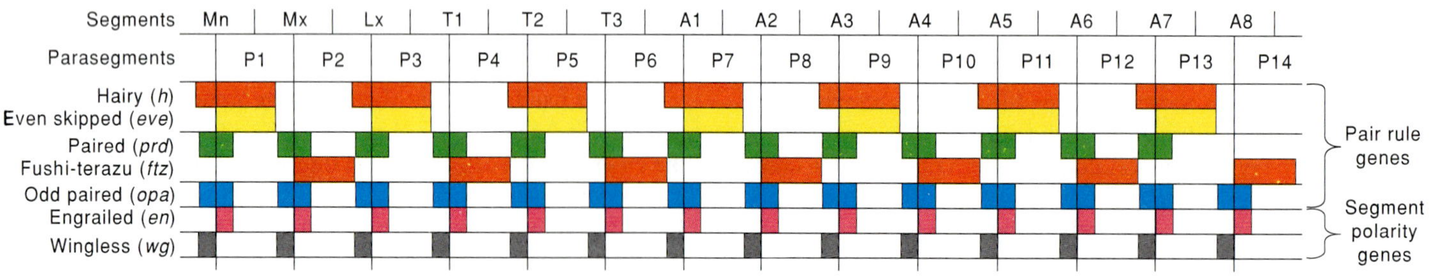

the two homeotic loci is colinear with the body plan; this colinearity may be related to the temporal mode of expression of the different genes in the two loci. Thus transcription of a region the size of *ANT-C* would take about two hours; the temporal expression of promoter proximal and promoter distal regions of the loci could be crucial in assuring the correct specification of body parts during development. Another aspect of gene arrangement and the body plan that is probably related to this factor is that a mutation at a given locus usually results in the conversion of the segment in question to the phenotype of the adjacent segment in the anterior direction. It is as though proceeding from the anterior to the posterior direction results in a stepwise increase in the amount of the locus that gets expressed.

The *ANT-C* and *BX-C* loci both contain very large introns and very few reading frames. These genes first become transcribed just prior to cellularization of the blastoderm, at about the same time as the pair-rule genes. Initially they are expressed at rather uniform levels in broad, overlapping domains. The boundaries of these domains are defined both by the maternal-effect organizing activities and by the gap genes. For example, transcripts from both of the known promoters of the *ANT-C* locus accumulate throughout the *Kr* domain. The longer transcript, P1, depends absolutely on *Kr* expression, whereas the shorter transcript, P2, appears to be defined by a combination of *bcd, osk,* and *hb* expression. Similarly, the initial *Ubx* domain is dependent on activation by the *osk* group activity and repression by *hb* activity. These broad regional differences are modulated by interactions with the early pair-rule pattern to generate a series of unique parasegmental states. Expression of *Scr, Antp,* and *Ubx* is specifically elevated in the cells of parasegments 2, 4, and 6 respectively, where the ftz protein is active. These enhanced levels of transcription are subsequently maintained throughout embryogenesis.

The activity of the homeotic gene domains is dependent in part on interactions between the genes themselves. In general, the most posteriorly expressed genes (such as *AbdA* and *AbdB* of the *BX-C*) repress the expression of the more anteriorly expressed genes (such as *Ubx* and *Antp*). In addition to generating differences between parasegments, homeotic genes are also expressed differentially between germ layers. The differential expression serves to modulate the basic pattern that underlies the organization of each segment.

Cell–Cell Interaction Is Important in the Elaboration of the Developmental Pattern in Parasegments We have seen how the cells at the boundaries of each parasegment are specified, which is reflected by their expressing particular segment-polarity genes. Although it is possible that the intervening blastoderm cells likewise become determined in response to pair-rule gene signals, (genetic studies of specific mutants) suggest that they remain unspecified at this stage. Specification of cells within each parasegment would then occur as the cells divide in response to signals mediated by cell–cell interaction. For example, as cells divide they would assume states dependent on their apposition to *wg*-expressing (or *en*-expressing) cells. Further divisions would allow for the intercalation of additional cell states. Such a process requires the existence of signaling molecules and receptors capable of transducing and receiving information between cells. Although we do not know too much about the specific apparatus used for the transmission of such signals in *Drosophila,* it presumably employs the types of structures that are used in the transmission of hormonal signals (see chapter 24) as well as the type of signals that are involved in B-T cell interaction in antibody formation (see chapter 33).

Figure 31.33

Genetic maps of the antennapedia complex (*ANT-C*) and the bithorax (*BX-C*). The colored dashed lines specify the regions affected by mutations in a given locus. The bithorax complex is shown in greater detail. By lethal complementation assay, Sanchez-Herrero has shown that the *BX-C* is divided into three complementation groups: *Ubx, Abd-A,* and *Abd-B*. Within each of these complementation groups there are nonlethal point mutations (such as *abx, bx, bxd,* and *pbx* in *Ubx*), which cover specific phenotypes in the regions encompassed by the dashed lines. *Ubx* is required for the development of the posterior compartment of the mesothorax (T2p) through the anterior compartment of the first abdominal segment (A1a). The *Abd-A* function is required for morphogenesis of A1p through A4, and *Abd-B* is required for A5 through A8. At least three essential homeotic functions have been assigned to the antennapedia complex: *Dfd, Scr,* and *Antp*. These are shown on the genetic map below the schematic of the fly. The primary domains of *ANT-C* function are indicated. *Antp* function is required for proper segment morphogenesis of the thorax. Analysis of *Scr*$^-$ and *Dfd*$^-$ mutant embryos suggests that these genes are required for the differentiation of the prothorax and posterior head regions.

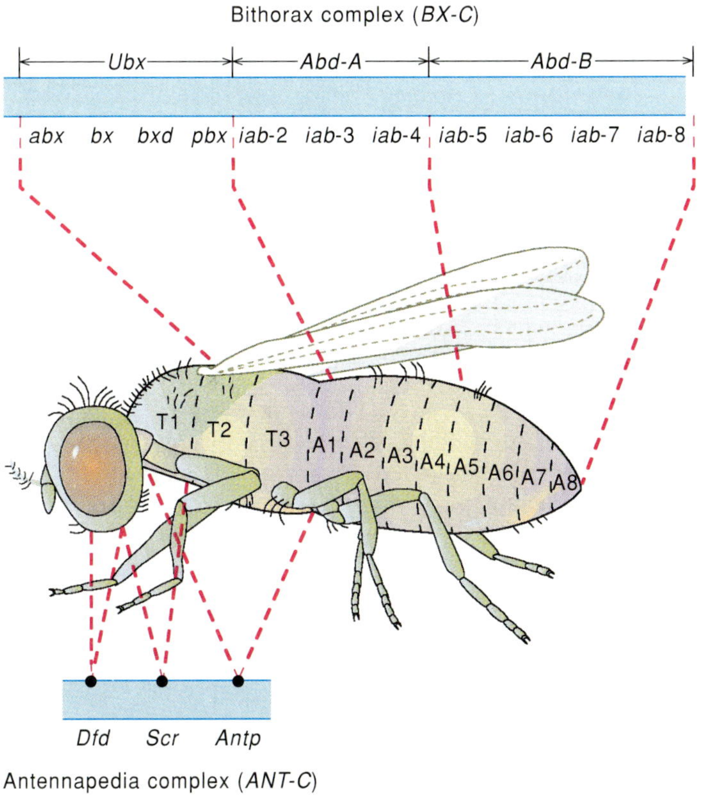

Summary

Some mechanisms of gene expression are found in eukaryotes but are rarely, if ever, seen in prokaryotes. Still, *trans*-acting regulatory proteins that bind to *cis* effector sites on the genome are present in eukaryotic systems as well as in the *E. coli.* The best-understood unicellular eukaryote is the budding yeast *S. cerevisiae.* Gene regulation, particularly of development, can be quite complex in multicellular eukaryotes. Our discussion in this chapter has focused on the following points.

1. In eukaryotes, individual genes encode transcripts for a single polypeptide chain, such as the yeast genes for histidine biosynthesis and galactose catabolism. Although functionally related genes are clustered, each gene has its own promoter.
2. In yeast, genes of the *GAL* system are under the joint control of two *trans*-acting genes that encode regulatory proteins, *GAL4* and *GAL80.* GAL4 protein is an activator that binds at an upstream site and GAL80 is a repressor that inhibits GAL4 action by binding to it. The ability of the GAL80 protein to bind to GAL4 is lost in the presence of galactose, which binds to the GAL80 protein, causing a reversible allosteric change in its structure.
3. Many catabolic genes in yeast, *GAL* genes included, are subject to glucose effects, which selectively degrade enzymes and repress gene expression. The genes and gene products most affected by glucose fulfill comparable catabolic needs for the organism that can be more effectively met by the catabolism of glucose. Glucose-induced repression in yeast may be mediated by the phosphorylation of the relevant regulatory proteins.
4. Yeast has two haploid cells of opposite mating types and one nonmating diploid cell that results from the fusion of haploid cells of opposite mating type. The mating type is determined by the *MAT* locus. Information for the mating type is stored at other, silent loci and is expressed only if it is transposed to the *MAT* locus. Mating-type information is not expressed at the storage loci because of a complex repressor system of proteins interacting in *cis* fashion over a considerable distance at the control centers of the storage loci.
5. The greatest difference between the regulatory systems in yeast and *E. coli* is that yeast regulatory proteins can bind at a long distance from the RNA polymerase binding site and still be effective.
6. Complex multicellular eukaryotes differentiate irreversibly so that different cell types express a different profile of genes. Genes that are expressed are usually associated with swollen chromatin. Proteins found in active regions of the genome show characteristic modifications.
7. Enhancers are elements of the genome that generally stimulate transcription. They resemble yeast UAS sequences in yeast in that they can function over long distances. In fact, they can function over even greater distances than UAS sequences and they are effective in either orientation, either upstream or downstream from the promoter. A possible mechanism for enhancer and UAS function over long distances is that the chromosome folds to bring the proteins bound at the enhancer site in close proximity to other regulatory proteins or RNA polymerase bound at the promoter.
8. A typical enhancer is composed of a cluster (usually two or three) of *cis* sites called enhansons; each enhanson contains binding sites for a unique combination of regulatory proteins. Enhansons must be clustered to be effective.
9. Most regulatory proteins contain a domain for binding to a specific site on the DNA and another domain for binding to some other protein or proteins.
10. Special combinations of regulatory factors give rise to developmental patterns of cell differentiation. Some organisms must amplify genes to keep pace with the high demand for certain gene products, especially in early development. Histones, ribosomal proteins, and ribosomal RNA genes are amplified.
11. The existence of many kinds of regulatory mutants has helped to advance our understanding of early development in the fruit fly, *Drosophila melanogaster.* Regulatory gene products are proteins that activate or repress other genes. Early development in *Drosophila* is a sequence of events in which different regulatory proteins gradually come into play in cascade fashion, controlling a wide range of enzymes and structural proteins and also influencing each other. Up to the blastoderm stage the nuclei in a developing *Drosophila* embryo are not separated by cellular membranes. As a result, the regulatory proteins and other gene products may diffuse freely from their site of synthesis to other nuclei in the embryo. At the late blastoderm stage, the nuclei become cellularized. From this point on, the influence of regulatory proteins made in one cell must be exerted on another cell at the level of the cell membrane.

Selected Readings

Achneider, R. J., and T. Shenk, Impact of virus infection on host cell protein synthesis. *Ann. Rev. Biochem.* 56:317–332, 1987.

Atchison, M. L., Enhancers: mechanisms of action and cell specificity. *Ann. Rev. Cell. Biol.* 4:127–153, 1988.

Binétruy, B., T. Smeal, and M. Karin, Ha-Ras augments c-Jun activity and stimulates phosphorylation of its activation domain. *Nature* 351:122–127, 1991.

Bingham, P. M., T. Chou, I. Mims, and Z. Zachari, On/off regulation of gene expression at the level of splicing. *Trends Genet.* 4:134, 1988.

Brietbart, R. E., A. Andreadis, and B. Nadal-Ginard, Alternative splicing: a ubiquitous mechanism for the generation of multiple protein isoforms from single genes. *Ann. Rev. Biochem.* 56:467–495, 1987.

Brown, D. D., How a simple animal gene works. In *The Harvey Lectures,* Series 76, pp. 27–44. New York: Academic Press, 1982.

Chandler, V. L., B. A. Maler, and K. R. Yamamoto, DNA sequences bound specifically by glucocorticoid receptor *in vitro* render a heterologous promoter responsive *in vivo:* A steroid specific enhancer. *Cell* 33:489–499, 1983.

Chen, H-Z., T. Hoey, and G. Zubay, Purification and properties of the *Drosophila* zen protein. *Mol. Cell. Biochem.* 79:181–189, 1988. First evidence that a homeobox protein binds to DNA as a monomer.

Duncan, I., The bithorax complex. *Ann. Rev. Genet.* 21:285–320, 1987.

Evans, R. M., The steroid and thyroid hormone receptor superfamily. *Science* 240:889–895, 1988.

Forsburg, S. L., and L. Guarente, Communication between mitochondria and the nucleus in regulation of cytochrome genes in the yeast *Saccharomyces cerevisiae. Ann. Rev. Cell. Biol.* 5:153–180, 1989.

Gabrielsen, O. S., and A. Sentenac, RNA polymerase III (C) and its transcription factors. *Trends Biochem. Sci.* 16:412–416, 1991.

Gehring, U., Steroid hormone receptors: Biochemistry, genetics and molecular biology. *Trends Biochem. Sci.* 12:399–402, 1987.

Gehring, W. J., The molecular basis of development. *Sci. Am.* 253(4):152–162, 1985.

Gehring, W., Homeo boxes in the study of development. *Science* 236:1245–1252, 1987.

Giniger, E., and M. Ptashne, Cooperative binding of the yeast transcriptional activator *GAL4. Proc. Natl. Acad. Sci.* 85:382–386, 1988.

Gruenberg, D. A., S. Natesan, C. Alexandre, M. Z. Gilman. Human and *Drosophila* homeo domain proteins that enhance the DNA-binding activity of serum response factor. *Science* 257:1089–1095, 1992.

Guarente, L. P., Regulatory proteins in yeast. *Ann. Rev. Genet.* 21:425–452, 1987.

Guarente, L., UASs and enhancers: Common mechanism of transcriptional activation in yeast and mammals. *Cell* 52:303–305, 1988.

Gurdon, J., Egg cytoplasm and gene control in development. The Croonian Lecture, 1976. *Proc. R. Lond. B.* 198:211–247, 1977.

Hanes, S. D., and R. Brent, A genetic model for interaction of the homeodomain recognition helix with DNA. *Science* 251:426–430, 1991.

Harrison, S. C., A structural taxonomy of DNA-binding domains. *Nature* 353:715–719, 1991.

Herskowitz, I., Life cycle of the budding yeast *Saccharomyces cerevisiae. Microbiological Reviews* 53:536–553, 1988.

Hinnebusch, A. G., Involvement of an initiation factor and protein phosphorylation in translational control of GCN4 mRNA. *Trends Biochem. Sci.* 15:148–152, 1990.

Johnson, P. F., and S. L. McKnight, Eukaryotic transcriptional regulatory protein. *Ann. Rev. Biochem.* 58:799–839, 1989.

Karlsson, S., and A. W. Nienhius, Developmental regulation of human globin genes. *Ann. Rev. Biochem.* 54:1071–1108, 1985.

King, T., and R. Briggs, Serial transplantation of embryonic nuclei. *Cold Spring Harbor Symp. Quant. Biol.* 21:271–290, 1956.

Klevit, R. E., Recognition of DNA by Cys_2His_2 zinc fingers. *Science* 253:1367–1393, 1991.

Lai, E., and J. E. Darnell, Jr., Transcriptional control in hepatocytes: a window on development. *Trends Biochem. Sci.* 16:427–429, 1991.

Lamb, P., and S. L. McKnight, Diversity and specificity in transcriptional regulation: the benefits of heterotypic dimerization. *Trends Biochem. Sci.* 16:417–433, 1991.

Marmorstein, R., M. Carey, M. Ptashne, and S. C. Harrison, DNA recognition by GAL4: structure of a protein-DNA complex. *Nature* 356:408–414, 1992.

Marzluff, W. F., and N. B. Pandey, Multiple regulatory steps control histone messenger-RNA concentrations. *Trends Biochem. Sci.* 12:49–51, 1988.

McClintock, B., Controlling elements and the gene. *Cold Spring Harbor Symp. Quant. Biol.* 21:197–216, 1956.

McKinney, J. D., and N. Heintz, Transcriptional regulation in the eukaryotic cell cycle. *Trends Biochem. Sci.* 16:430–434, 1991.

Melton, D. A., Pattern formation during animal development. *Science* 252:234–241, 1991.

Miner, J. N., and K. R. Yamamoto, Regulatory crosstalk at composite response elements. *Trends Biochem. Sci.* 16:423–426, 1991.

Mitchell, P. J., and R. Tjian, Transcriptional regulation in mammalian cells by sequence-specific DNA binding proteins. *Science* 245:371–378, 1989.

Nevins, J. R., Transcriptional activation by viral regulatory proteins. *Trends Biochem. Sci.* 16:435–439, 1991.

Ptashne, M., How gene activators work. *Sci. Am.* 260(1):41–47, 1989.

Raghow, R., Regulation of messenger RNA turnover in eukaryotes. *Trends Biochem. Sci.* 12:3358–3360, 1987.

Roeder, R. G., The complexities of eukaryotic transcription initiation: regulation of preinitiation complex assembly. *Trends Biochem. Sci.* 16:402–407, 1991.

Schler, A. F., and W. J. Gehring, Direct-homeo domain-DNA interaction in the autoregulation of the *fushi tarazu* gene. *Nature* 356:804–806, 1992.

Schwabe, J. W. R., and D. Rhodes, Beyond zinc fingers: Steroid hormone receptors have a novel structural motif for DNA recognition. *Trends Biochem. Sci.* 16:291–296, 1991.

Singer, S. J., Intercellular communication and cell-cell adhesion. *Science* 255: 1671–1677, 1992. Considers aspects of regulation we didn't have time to cover.

Stark, G., M. Debatisse, E. Giulotto, and G. M. Wahl, Recent progress in understanding mechanisms of mammalian gene amplification. *Cell* 57:901–908, 1989.

Struhl, K., Molecular mechanisms of transcriptional regulation in yeast. *Ann. Rev. Biochem.* 58:1051–1077, 1989.

Struhl, K., Helix-turn-helix, zinc-finger and leucine-zipper motifs for eukaryotic transcriptional regulatory proteins. *Trends Biochem. Sci.* 14:137–140, 1989.

Weinberg, R. A., Finding the anti-oncogene. *Sci. Am.* 259(3):44–51, 1988.

Problems

1. On the basis of what you know about genes in prokaryotic and eukaryotic cells, define a gene. Make sure that your definition is brief and concise. Does your definition have any limitations or problems?
2. Why is attenuation control in eukaryotes unlikely?
3. A mutation in *GAL4*^{c} of yeast leads to constitutive galactose fermentation in haploid cells. Describe the nature of this mutation at the molecular level. Would you expect expression of *GAL* genes to be responsive to glucose repression in this mutant?
4. Explain how starvation of yeast for any one of ten amino acids derepresses the synthesis of more than 30 enzymes in nine different amino acid biosynthetic pathways (general amino acid control).
5. How would you expect a deletion of *HMLE* to affect the expression of mating-type genes in yeast? Compare this effect with the deletion of the $\alpha 2$ gene from *MAT*α. Consider both homothallic and heterothallic backgrounds.

6. Expression of some genes in eukaryotic cells can be induced by treatment with 5-azacytidine. Explain how this happens.
7. Briggs and King were able to grow a differentiating embryo from an egg of *Rana pipiens* whose chromosomes were replaced with a single diploid nucleus from another embryo. What does this tell us about the state of the nucleus in the developing embryo?
8. Hemophilia is an X-chromosome-linked disorder. Bearing in mind the phenomenon of X-chromosome inactivation, suggest an explanation for the observation that females who are heterozygous for the defective gene are essentially asymptomatic.
9. High salt weakens the interaction of histones with DNA but has little affect on the binding of many regulatory proteins. Explain this observation in terms of how these molecules interact with DNA.
10. List some characteristics that distinguish active from inactive chromatin.
11. In the *Xenopus* oocyte a large number of ribosomes are made in a short time to handle the rapid demand for cell growth during cleavage stages. How is this large amount of rRNA made in such a short time?
12. When mammalian cells in culture are treated with the antifolate methotrexate (see chapter 20), cells can be selected that are resistant to high levels of this toxic compound. The enzyme dihydrofolate reductase becomes elevated about 1,000-fold in these resistant cells. What mechanism can account for such a large amount of this enzyme being made and how does this protect the cell from the toxic effects of methotrexate? How could you test for the molecular mechanism of this drug resistance?
13. Given that specific subsets of homeotic genes are required for the development of specific segments of the *Drosophila* embryo, suggest a possible mechanism whereby the necessary spatially restricted pattern of homeotic gene expression might be achieved. Incorporate the observed effect of homeotic mutations upon segment morphology into your model.
14. Explain how a DNA sequence (enhancer sequence) located 5,000 base pairs from a gene transcription start site can stimulate transcription even if its orientation is reversed.
15. Speculate on some possible models to explain tissue-specific alternative splicing such as that seen with variants of tropomyosin.
16. Discuss the types of structural motifs found in eukaryotic transcription factors.

Physiological Biochemistry

7

PART

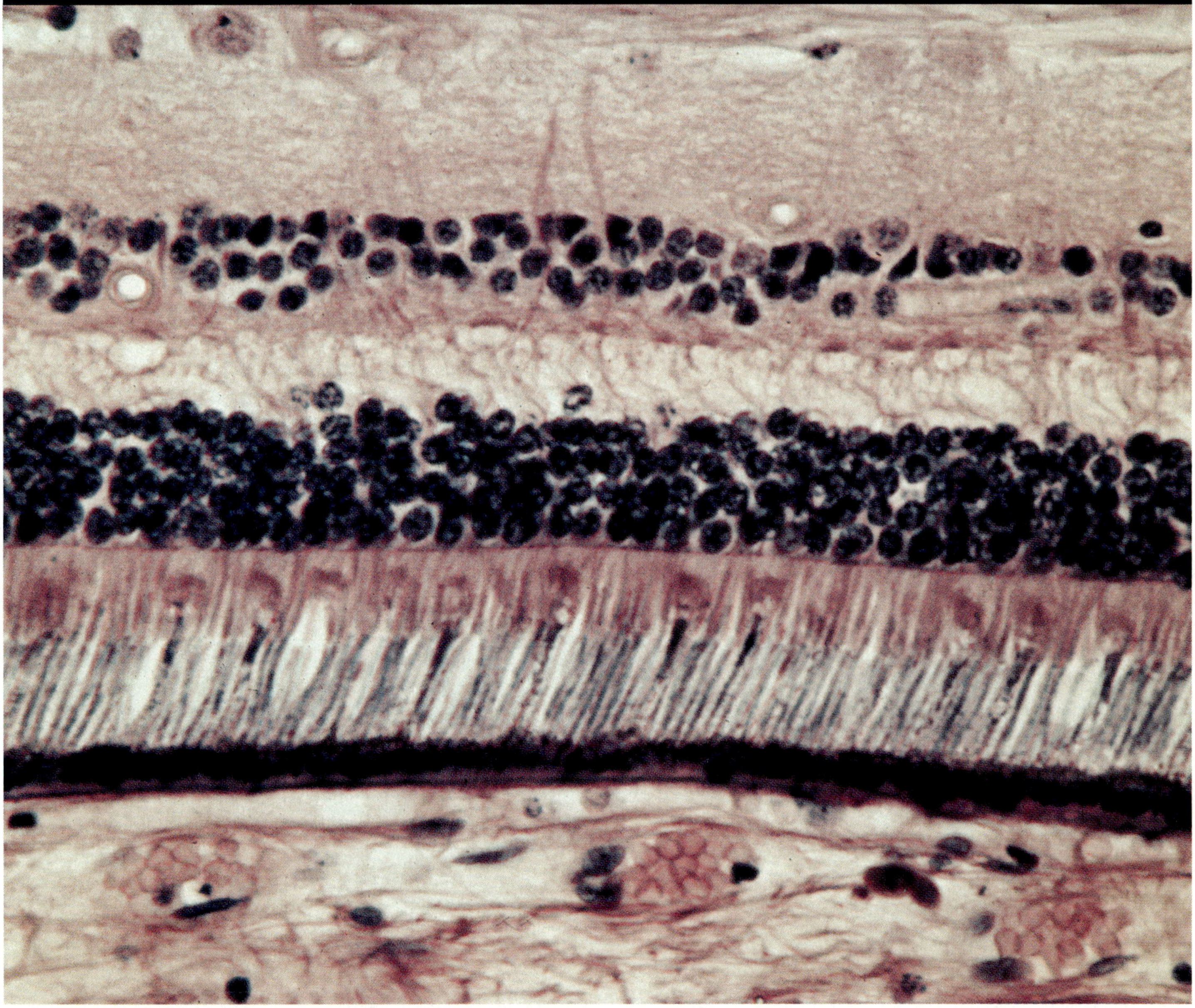

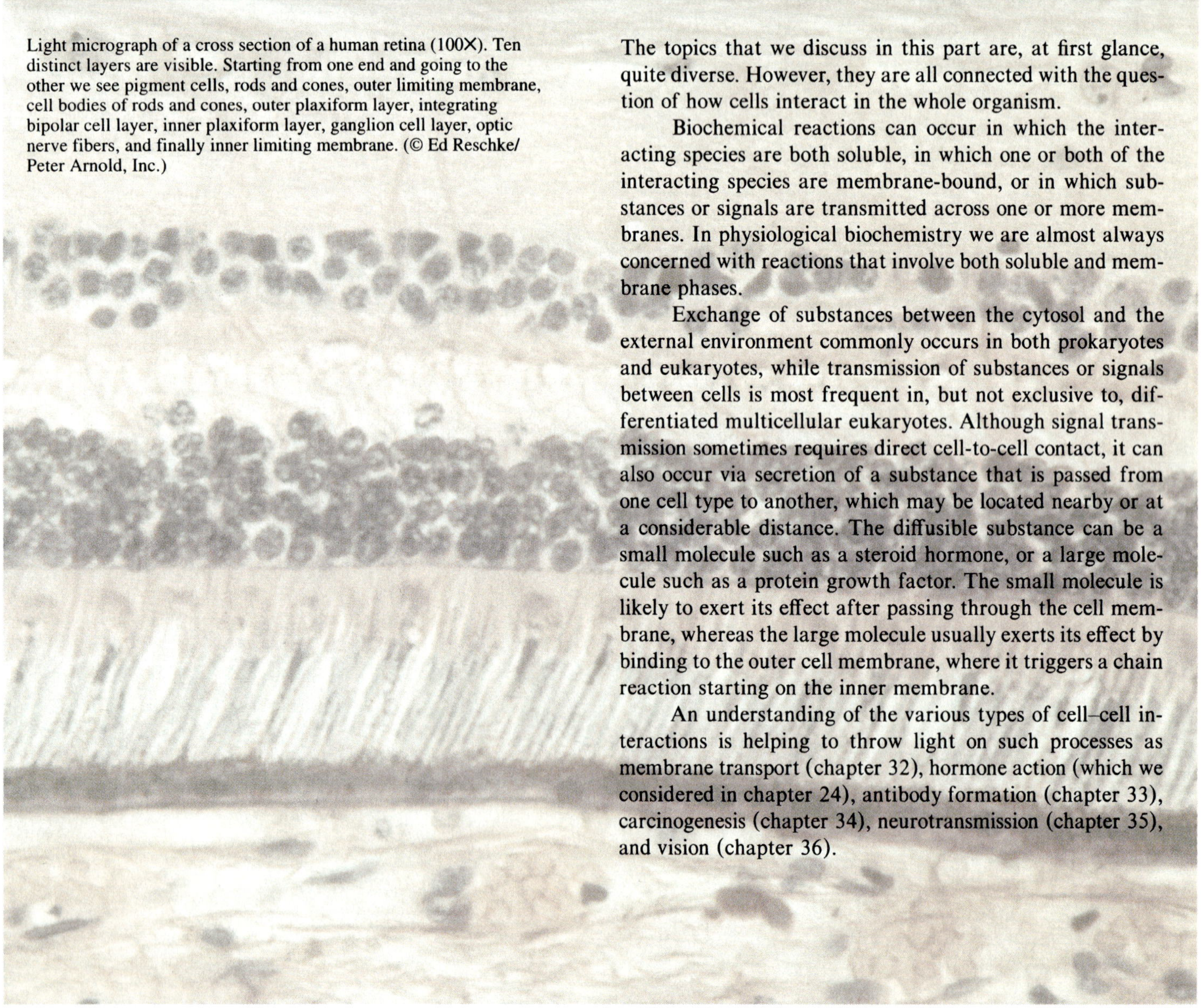

Light micrograph of a cross section of a human retina (100×). Ten distinct layers are visible. Starting from one end and going to the other we see pigment cells, rods and cones, outer limiting membrane, cell bodies of rods and cones, outer plaxiform layer, integrating bipolar cell layer, inner plaxiform layer, ganglion cell layer, optic nerve fibers, and finally inner limiting membrane. (© Ed Reschke/ Peter Arnold, Inc.)

The topics that we discuss in this part are, at first glance, quite diverse. However, they are all connected with the question of how cells interact in the whole organism.

Biochemical reactions can occur in which the interacting species are both soluble, in which one or both of the interacting species are membrane-bound, or in which substances or signals are transmitted across one or more membranes. In physiological biochemistry we are almost always concerned with reactions that involve both soluble and membrane phases.

Exchange of substances between the cytosol and the external environment commonly occurs in both prokaryotes and eukaryotes, while transmission of substances or signals between cells is most frequent in, but not exclusive to, differentiated multicellular eukaryotes. Although signal transmission sometimes requires direct cell-to-cell contact, it can also occur via secretion of a substance that is passed from one cell type to another, which may be located nearby or at a considerable distance. The diffusible substance can be a small molecule such as a steroid hormone, or a large molecule such as a protein growth factor. The small molecule is likely to exert its effect after passing through the cell membrane, whereas the large molecule usually exerts its effect by binding to the outer cell membrane, where it triggers a chain reaction starting on the inner membrane.

An understanding of the various types of cell–cell interactions is helping to throw light on such processes as membrane transport (chapter 32), hormone action (which we considered in chapter 24), antibody formation (chapter 33), carcinogenesis (chapter 34), neurotransmission (chapter 35), and vision (chapter 36).

32

CHAPTER

Mechanisms of Membrane Transport

In chapter 7 we considered the current state of knowledge about the structure of biological membranes. Two of the most important conclusions that we drew were that membranes are both dynamic and asymmetric structures. As we will see, these properties allow transmembrane processes to be carried out, often, but not always, in a unidirectional manner. The most prominent of these processes are the transport of nutrients and inorganic electrolytes into the cell and the export of toxic substances and waste materials out of the cytoplasm. In both instances, the problem is usually the same: How do small, hydrophilic metabolites traverse the hydrophobic lipid bilayer at rates sufficient to maintain the growth needs of the cell? Often this problem is solved by integral membrane proteins, which, by analogy with enzymes, specifically interact with their substrates, but which, unlike enzymes, have as their primary purpose the movement of a molecular species across the membrane rather than the catalysis of a chemical reaction.

The fairly recent development of general techniques for the purification of membrane proteins and of the powerful tools of molecular biology has greatly increased our understanding of the molecular mechanisms of transport proteins, or permeases, as they are often called. In this chapter, we will consider the theory, energetics, and mechanisms of transmembrane transport in cells.

A number of processes rely on transport, directly or indirectly, for their proper functioning. Examples of these seemingly diverse phenomena include nerve impulse propagation, cellular behavior and differentiation, and sensory reception and transduction. Indeed, it should become apparent that nearly all cellular and organismal functions are dependent on biological transport. As in chapter 7, examples from both prokaryotic and eukaryotic worlds will show us that many fundamental principles concerning the biochemistry of transport are probably valid throughout the living kingdom.

Figure 32.1

Different types of transport across membranes. The membrane—whether it is a plasma membrane or another type of membrane, such as those encompassing cellular organelles—imposes a barrier to the free movement of molecules. Five basically different types of processes are used to transport substances across membranes. The first, *simple diffusion,* does not require any special apparatus; it requires only that the substance being transported be soluble in the domain of the membrane bilayer. *Facilitated diffusion* is similar to free diffusion in not requiring energy. It differs from free diffusion in that a special pore or carrier (permease) is used to mediate the transport across the membrane. For transport processes that require energy, such as transport against a concentration gradient, energy is required. Such processes are referred to as active transport. In *primary active transport,* energy is directly expended in the process of transporting the substance across a membrane. This energy is donated in various ways in different active transport processes, for example, by ATP or PEP hydrolysis, by light energy, or by electron flow. In *secondary active transport,* a substance from one active transport process may be used to drive the transport of another substance. Finally, in *group translocation,* the transported substance is chemically modified by the membrane permease during the transport event. This type of transport generally requires energy as well, either directly or indirectly.

Although the focus in this figure is on transport of substances into the membrane-enclosed lumen, transport of substances in the reverse direction is frequently of great importance.

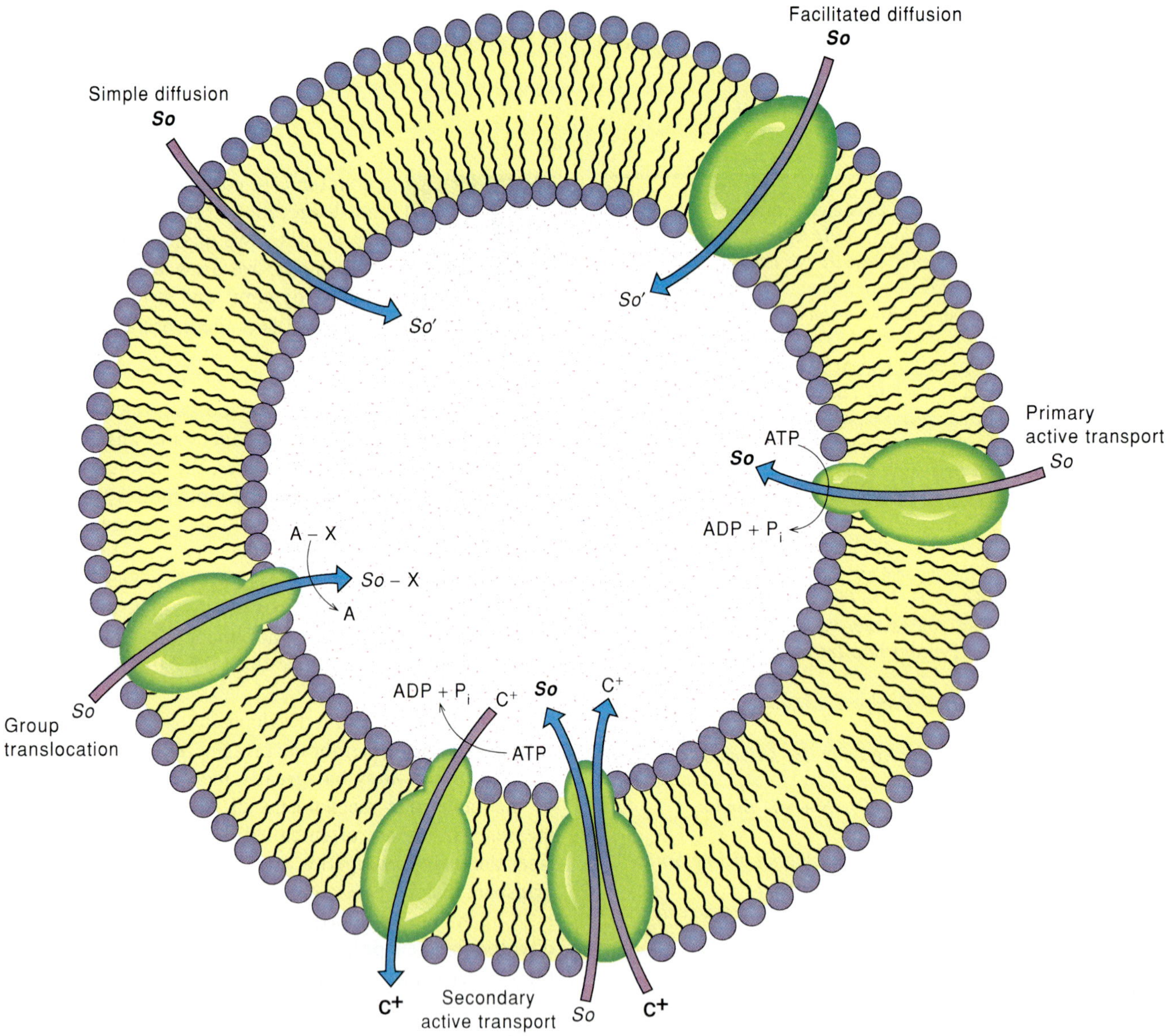

The Theory and Thermodynamics of Biological Transport

Before we consider specific examples of transport across biological membranes, we will discuss the thermodynamics of this process as well as specific ways in which such a transfer can occur. With this foundation, we can assign certain predicted characteristics to each kind of transport mechanism (fig. 32.1). And, on the basis of these predictions, we can design experiments that will give useful information regarding the events responsible for the transmembrane movement of a solute.

Most Solutes Are Transported by Specific Carriers

In principle, a solute could move across a barrier such as a membrane by either of two fundamentally different mechanisms: simple (or passive) diffusion or carrier-mediated transport. In the first mechanism, the molecule in question must be somewhat soluble in the hydrophobic phase of the bilayer, while in the second mechanism, the solute's journey across the membrane is facilitated by another molecule, in most cases a protein. For these reasons, simple diffusion is usually of importance in biological transport systems only for molecules having a largely hydrophobic character or in membrane systems possessing nonspecific aqueous pores, as we will see in a later section. An exception to this rule, however, is water, which traverses many biological membranes by simple diffusion. Most other hydrophilic molecules, which comprise the majority of species transported into and out of cells, are usually recognized by transport systems that include one or more protein carriers.

In order to illustrate the characteristics of simple diffusion, let us consider two dilute solutions of a solute *So*, one twice as concentrated as the other, which are separated by a barrier that is permeable to *So*, as in figure 32.2. As a consequence of random thermal motion, a certain proportion of the molecules originally present on one side of the barrier (one-quarter in figure 32.2) will move to the other side in a specified length of time t. This flux is bidirectional, so that after time t, the ratio of the concentration of *So* on the left side of the barrier to that on the right will have been reduced from 2:1 to 7:5. After many such time intervals, the ratio will approach 1:1, or equilibrium. The number of molecules of *So* that move from left to right in time interval t is therefore directly proportional to the initial concentration of *So* on the left side:

$$v \propto [So] \quad (1)$$

while a similar relationship holds true for movement in the opposite direction:

$$v' \propto [So'] \quad (2)$$

where v and v' refer to the rates of transfer from left to right and right to left, respectively, and $[So]$ and $[So']$ are the initial concentrations of the solute on the left and right sides of the barrier. The net rate of transfer V from the more concentrated to the more dilute solution is therefore the difference between these two component rates:

$$V = v - v' \quad \text{or} \quad V \propto [So] - [So'] \quad (3)$$

Hence, at equilibrium, $V = 0$.

If the barrier in figure 32.2 is, for example, a biological membrane of finite thickness l, we must also take this into account in the expression for overall transfer rate:

$$V \propto \frac{[So] - [So']}{l} \quad (4)$$

Finally, the rate of transfer of *So* will depend on such factors as temperature, viscosity of the solvent, and the solubilities of *So* in the solvent and in the membrane. In a given membrane system, these factors often can be held relatively constant, so that we can express relationship (4) as an equation:

$$V = \frac{D([So] - [So'])}{l} \quad (5)$$

where D is a constant, the diffusion coefficient, which has the dimensions of area per unit of time (e.g., cm^2/s). Equation (5), called Fick's law, shows that the net rate of diffusion is directly proportional to the concentration gradient of *So* across the membrane barrier.

It is apparent from equation (5) that in a system in which *So* traverses the membrane by simple diffusion, a plot of V versus $[So] - [So']$ should give a straight line that passes through the origin with a slope of D/l (fig. 32.3*a*). In practice, it is often possible to measure V as a function of the concentration gradient of *So* across a biological membrane by methods

Figure 32.2

Illustration of simple diffusion in a two-compartment system divided by a semipermeable membrane. After time t, one-fourth of the molecules present originally on one side of the membrane have diffused by random thermal motion to the other side. After many such intervals, equilibrium is approached. The molecules in this illustration are identical, but different colors are used to indicate their initial locations. In simple diffusion, the number of molecules diffusing from one side to the other in any given time interval is directly proportional to their concentration on the first side. (Source: K. D. Neame and T. G. Richards, *Elementary Kinetics of Membrane Carrier Transport*. Copyright © 1972 Blackwell Scientific Publications, Oxford, England.)

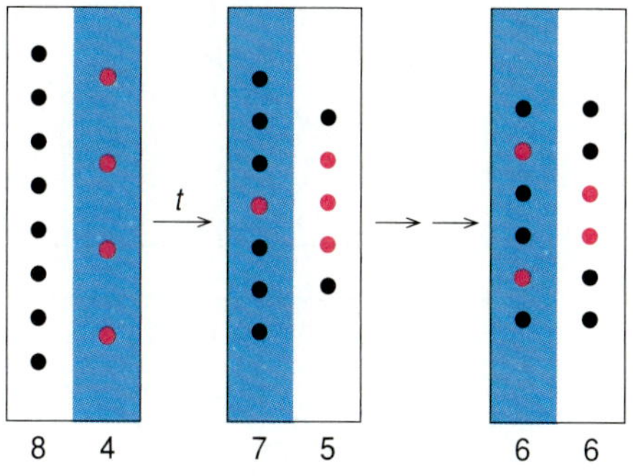

Figure 32.3

Initial rate kinetics of simple diffusion (*a*) and carrier-mediated transport (*b*). In (*b*), [*So*] and [*So′*] refer to the concentration of transported solute molecule on two sides of the membrane. Plots of velocity versus the difference in substrate concentrations on the two sides of the membrane give a straight line with a slope of *D*/*l* in (*a*) and a rectangular hyperbola approaching a maximum velocity (V_{max}) at high [*So*] − [*So′*] in (*b*).

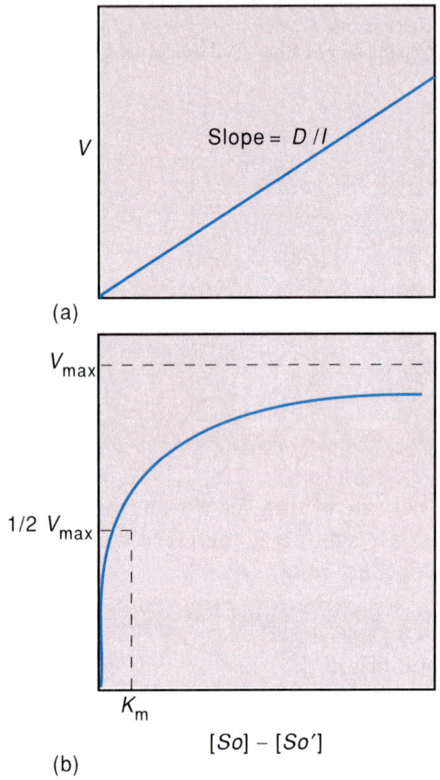

Figure 32.4

Double-reciprocal (Lineweaver-Burk) plots of carrier-mediated transport and simple diffusion. For diffusion, the line passes through the origin, while for carrier-mediated processes the *x* intercept is equal to $-1/K_m$ and the *y* intercept is equal to $1/V_{max}$, as shown in (*a*). In (*b*), the transport kinetics for a cell transporting a given solute by both simple diffusion and a carrier-mediated system are illustrated. The line shown is a combination of the processes plotted in (*a*) and passes through the origin. In the case of two carrier-mediated systems, a similar line would be obtained that would intersect the ordinate above the origin. (Source: K. D. Neame and T. G. Richards, *Elementary Kinetics of Membrane Carrier Transport*. Copyright © 1972 Blackwell Scientific Publications, Oxford, England.)

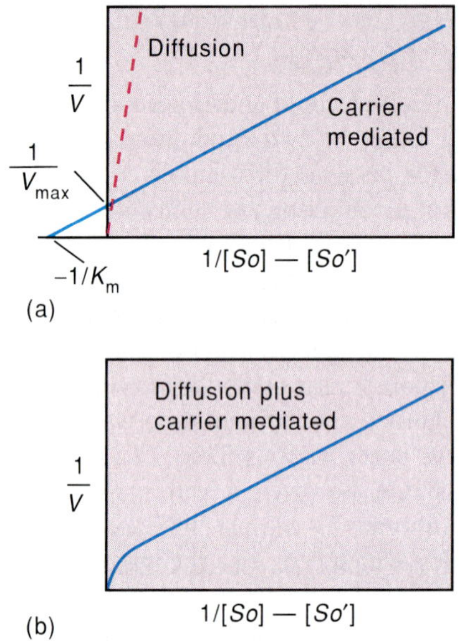

that we will consider later in this chapter. Provided that our data include a wide range of solute concentration gradients, a linear plot of *V* versus [*So* − *So′*] is evidence that *So* is taken up by simple diffusion.

The uptake kinetics shown in figure 32.3*a* for simple diffusion are, in fact, seen only in a minority of cases for transport of molecules across cytoplasmic membranes of cells. More often, the curve approximates that of a rectangular hyperbola, as illustrated in figure 32.3*b*. In this case, at high [*So*] − [*So′*] the transport rate approaches a limiting value that is termed V_{max} by analogy with steady-state enzyme kinetics (chapter 8). This saturation phenomenon implies that uptake of *So* is dependent on a discrete number of carriers in the membrane, all of which become filled at high concentration gradients of *So*. The situation is formally analogous to saturation of an enzyme by its substrate, and the rate equation for this process, carrier-mediated transport, is essentially the same as the Michaelis-Menten equation derived in chapter 8:

$$V = \frac{V_{max}[So]}{K_m + [So]} \quad (6)$$

In the derivation of equation (6), [*So′*] is assumed to be negligible (just as is [*P*] in enzyme kinetics) and is taken to be the initial concentration of *So* inside the cell or vesicle being studied. Experimentally, this situation is usually easy to attain, so that [*So*] equals the concentration initially present on the outside, [*So′*] = 0, and *V* is an initial rate of transport of *So* into the system enclosed by the membrane. As in enzyme kinetics, K_m is the solute concentration at which $V = \frac{1}{2}V_{max}$ (fig. 32.3*b*). Note that the value of K_m, although often assumed to be a measure of the affinity of a carrier for its "substrate," is really a ratio of rate constants and is not necessarily equivalent to the dissociation constant of the solute from its carrier.

From what we have said, it is clear that in many cases, by studying the rate of transport, *V*, as a function of [*So*], we can distinguish simple diffusion from carrier-mediated transport. Often these data are plotted in a double-reciprocal fashion (Lineweaver-Burk plot, chapter 8). Plots of $1/V$ against $1/[So]$ yield straight lines in both instances, but the line for simple diffusion passes through the origin, while that for carrier-mediated transport intersects the ordinate at $1/V_{max}$ and the abscissa at $-1/K_m$ (fig. 32.4*a*). This difference, of course, reflects the fact that carrier systems are saturable (V_{max} is finite), while simple diffusion systems are not ($V_{max} = \infty$; $1/V_{max} = 0$).

Unfortunately for biochemists interested in studying a single biological transport system, double-reciprocal plots for the uptake of a specific solute into a cell, or into vesicles derived from cells, are sometimes nonlinear. Such behavior usually indicates that more than one process is involved in the transport of this substance into the cell. The different processes may involve both simple diffusion and carrier mediation, or both may be carrier-mediated (fig. 32.4*b*). In either case, however, it is first necessary to find experimental conditions under which only the system of interest is functional before we can draw conclusions concerning the transport mechanism. We may accomplish this by working at solute levels recognized only by one carrier, by studying mutants that lack one system, by using specific inhibitors, or even by purification and reconstitution of the carrier of interest (see the following discussion).

Finally, we can in many cases distinguish carrier-mediated transport from simple diffusion by its stereospecificity. For example, in a membrane system that is permeable to glucose only by simple diffusion, both D- and L-glucose will traverse the membrane at equal rates. In contrast, many biological membranes contain carriers specific for glucose that recognize only the D-stereoisomer. In these cells, the initial transport rate for D-glucose is much higher than that for the L isomer and shows saturation kinetics. In fact, the rate of transport of L-glucose is often used in such instances to correct the total rate observed for diffusion in order to obtain the "true" carrier-mediated rate for the D isomer. Stereospecificity in biological transport, as in enzyme catalyzed reactions, is a consequence of the highly selective configuration of the "active site" of the transport protein for its "substrate."

Simple and Facilitated Diffusion Require No Energy

In the preceding section we saw that a solute will tend to diffuse "down" its concentration gradient (from a region of higher to a region of lower concentration) spontaneously, until equilibrium is reached. Such a process requires no outside input of energy into the system and thus occurs with a negative change in free energy ΔG. This change is a consequence of the fact that the equilibrium situation ($[So] = [So']$ in fig. 32.2) is more "random" or disordered than the starting situation (in which $[So] > [So']$) and thus has relatively more entropy S. Because $\Delta G = \Delta H - T\,\Delta S$, an increase in entropy results in a negative free energy change (chapter 2). The magnitude of ΔG can be calculated from the expression

$$\Delta G = 2.3RT \log \frac{[So']}{[So]} \tag{7}$$

for the situation illustrated in figure 32.2. Net transport from left to right will occur spontaneously in this system only if $[So']/[So] < 1$, while at equilibrium, $\Delta G = 0$ ($[So'] = [So]$).

In biological systems, both simple diffusion and facilitated diffusion are transport mechanisms that result in net transport of a solute only in the direction of negative ΔG, i.e., usually from higher to lower concentrations. In contrast to simple diffusion, facilitated diffusion is a carrier-mediated process and hence is saturable (fig. 32.5). Both it and simple diffusion generally result in equilibration of a solute across a membrane, unless the solubility properties of the solute are significantly different on both sides of the membrane or the solute has a net charge. If the solute is charged, both types of diffusion can lead to an equilibrium situation in which $[So] \neq [So']$ if there is an independently maintained electrical potential, $\Delta\psi$, across the membrane (also see chapter 15):

$$\Delta G = 2.3RT \log \frac{[So']}{[So]} + ZF\,\Delta\psi \tag{8}$$

In this equation, Z is the net charge on So, and F is the Faraday constant. Thus at equilibrium ($\Delta G = 0$), unequal concentrations of So can exist on either side of a membrane as long as the term $ZF\,\Delta\psi$ is not equal to zero (fig. 32.6). The relationship between free energy and membrane potential is very similar to that between free energy and electromotive potential discussed in chapter 15.

Figure 32.5

Schematic illustration of diffusion processes in biological transport. In both simple and facilitated diffusion, the concentrations of the solute on both sides of the membrane are the same at equilibrium, and no metabolic energy is expended.

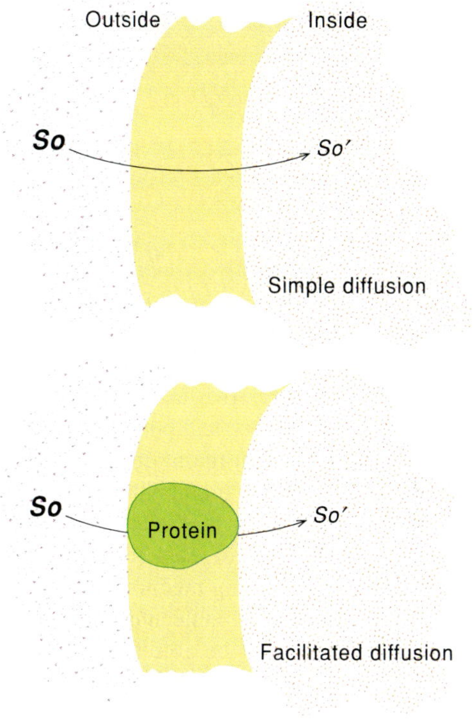

Transport against a Concentration Gradient Requires Energy

Both simple and facilitated diffusion are examples of energy-independent biological transport mechanisms. Although they occur in fundamentally different ways, neither requires the input of energy, nor can either lead to concentration of the solute on

Figure 32.6

At equilibrium, a *charged* solute, So^+, may be unequally distributed across a biological membrane if an independently maintained potential exists across it. This situation can occur even if So^+ is transported by simple or facilitated diffusion. Bold type indicates higher concentration.

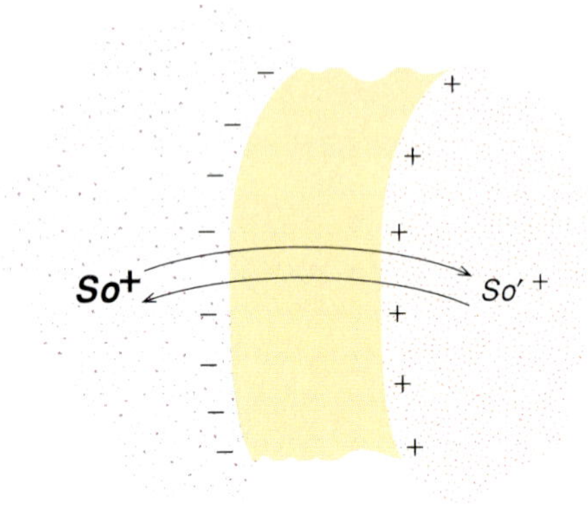

one side of the membrane relative to the other in the absence of a charged solute and an electrical potential across the membrane. Most cells, however, can maintain relatively high intracellular concentrations of some metabolites and ions relative to the extracellular environment. For example, many cells have an intracellular level of K^+ that is at least 30 times the concentration of this ion in the surrounding medium. It therefore follows that if carriers for accumulated molecules, such as K^+, exist in the cell membrane, they must be able to transport their substrates in the direction of a positive ΔG, i.e., up a concentration gradient of the solute. In order to accomplish this thermodynamically unfavorable movement, such processes must be coupled to events that have negative ΔG values, i.e., that can yield energy to "push" the solute "uphill." Concentrative accumulation of a solute inside a cell, organelle, or vesicle with the consequent expenditure of metabolic energy is termed active transport. In a related process, group translocation, energy for the transmembrane accumulation of a solute is provided by chemical modification of the molecule as it passes through the membrane.

Active Transport and Group Translocation Are Used to Concentrate Solutes

The use of metabolic energy to drive active transport implies that the transport system itself must somehow be able to harness the energy in a particular form and use it to do work, that is, to transport a solute against a concentration gradient. The exact molecular mechanisms by which these interconversions take place are not completely known for any system; however, a number of clues are emerging that we will consider later in this chapter. Because systems that carry out active transport are invariably carrier-mediated, it seems likely that in most cases integral membrane permeases themselves directly participate in this energy interconversion process. A major direction of research in membrane-transport biochemistry has therefore been to study the energetics of active transport and, whenever possible, to isolate and characterize the proteins that comprise a particular system.

Two basic types of active transport have been recognized. In primary active transport, energy is provided directly by the hydrolysis of ATP, by electrons flowing down an electron transport chain, or by light. Examples of these include the Na^+-K^+ ATPase of animal cells, which pumps K^+ into the cell and Na^+ out of the cell at the expense of ATP; the respiratory electron transport chain, which pumps H^+ out of the mitochondrion and many bacterial cells (chapter 15); and the light-driven H^+ pump of *Halobacterium,* bacteriorhodopsin. In secondary active transport, ion gradients across the membrane, themselves created by other active transport systems, are used to drive the concentrative uptake of other ions or metabolites. Because an ion gradient is a form of potential energy, a process involving its collapse releases this energy, which can be used to perform work. In mitochondria and bacteria, the potential energy stored in gradients of H^+ is used during respiratory metabolism to synthesize ATP (chapter 15). As we will see, gradients of H^+ and other ions also are an important source of energy for active transport in both prokaryotic and eukaryotic cells (also see chapters 15 and 16).

Related to primary active transport is the process of group translocation. While molecules transported by active systems are translocated across the membrane without chemical changes occurring, in group translocation the transported solute is chemically modified on its journey through the membrane. The best-documented example of group translocation occurs in many bacteria, which convert sugars taken up from the surrounding medium into sugar monophosphates simultaneously with their transport through the membrane. Energy is expended in the process, since the phosphoryl donor is the high-energy phosphate glycolytic intermediate, phosphoenolpyruvate. In this case the permease for the sugar is also an enzyme that catalyzes both the transport and the phosphorylation of its substrate. Figure 32.7 illustrates schematically the differences between primary and secondary active transport and group translocation, while table 32.1 lists representative well-characterized transport systems of both animal and bacterial cells.

Energy-Coupling Mechanisms in Biological Transport

It is the ultimate goal of the biochemist interested in transmembrane transport mechanisms to understand at the molecular level the physicochemical properties of a transport protein that allow it to carry out its function. In this section, we will briefly describe approaches commonly used to gain information about the structures of permease proteins and the relationship of these structures to the transmembrane transport processes catalyzed by them. We will also show how these systems can be used to investigate energy-coupling mechanisms at the molecular level. In doing so, we will place emphasis on some of the best-characterized biological transport systems. With this as a background, we will then venture one step further and attempt to explain how the structures and properties of integral membrane permeases permit them to carry out the myriad of transport processes so vital to the cell.

Figure 32.7

Schematic illustrations of primary active transport (*a*), secondary active transport (*b*), and group translocation (*c*). In (*a*), the energy source can be ATP, as shown, or electron transport or light can be used. In (*b*), active accumulation of *So* is energized by cotransport of an ion that flows down its concentration gradient. In (*c*), *So* is accumulated as a derivative, *So-X*, which is formed by a chemical reaction catalyzed by the carrier during transport though the membrane.

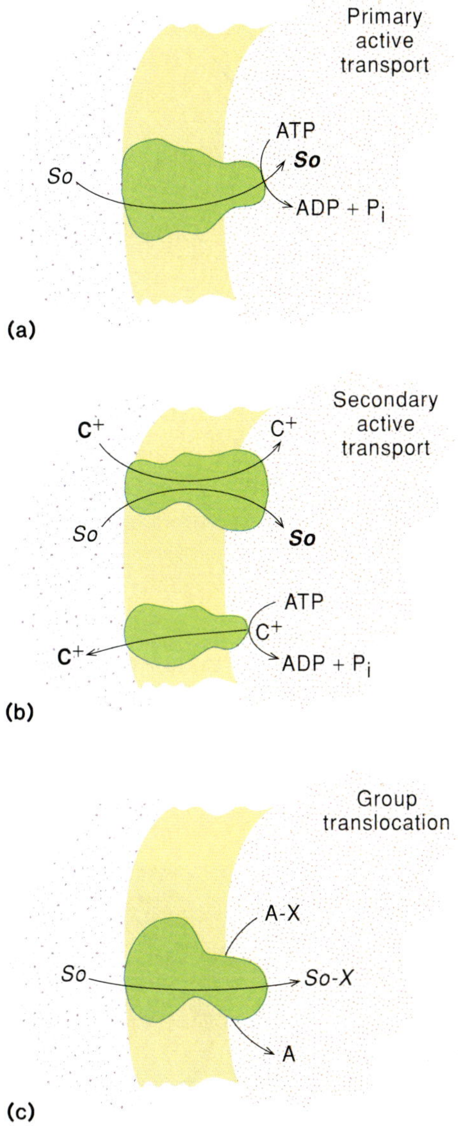

Table 32.1
Representative Well-Characterized Biological Solute Transport Systems

I. Nonstereospecific Channel Proteins	
A.	Muscle acetylcholine receptor
B.	Nerve Na^+ channel
C.	Nerve K^+ channel
D.	Porins in the outer membranes of Gram negative bacteria
E.	Glycerol permease in *E. coli*
F.	Major intrinsic protein of mammalian lens
II. Single-species Facilitators (Uniporters or Antiporters)	
A.	Glucose transporters in red cells, brain, liver, muscle, etc.
B.	Anion antiporters in red cells and various tissues
C.	Na^+/H^+ antiporters in bacteria and animal tissue cells
D.	ADP/ATP antiporters in mitochondria
III. Two-species Facilitators (Symporters)	
A.	Lactose permease in *E. coli* (lactose:H^+)
B.	Melibiose permease in *Salmonella* (melibiose:Na^+)
C.	Xylose and arabinose permeases of *E. coli* (pentose:H^+)
D.	Galactose, glucose, and maltose permeases in yeast (sugar:H^+)
E.	Glucose transporters in certain epithelial cells (glucose:Na^+)
F.	α-Amino acid transporters in tissue cells (amino acid:Na^+)
G.	NaCl/KCl symporters of animal tissues
IV. ATP-driven Active Transporters	
A.	Na^+, K^+, H^+, and Ca^{++} ATPases in eukaryotic cells and K^+ ATPases of bacteria (P-type)
B.	Vacuolar H^+ ATPases of eukaryotes; also found in bacteria (V-type)
C.	H^+ and Na^+ F_1F_0 ATPases in mitochondria and bacteria (F-type)
D.	ATP-driven, binding-protein-dependent permeases of bacteria (ABC-type)
E.	Multidrug resistance glycoproteins of animal cells (ABC-type)
V. Light and Electron Flow Driven H^+, Na^+, or Cl^- Transporters	
A.	Bacteriorhodopsin of *Halobacterium* (H^+ translocating)
B.	Halorhodopsin of *Halobacterium* (Cl^- translocating)
C.	Cytochrome oxidases of mitochondria and bacteria (H^+ or Na^+ translocating)
VI. Group Translocation	
A.	Mannitol, β-glucoside, and *N*-acetylglucosamine Enzyme II of the *E. coli* phosphotransferase system (one-component systems)
B.	Glucose, sucrose, and glucitol Enzyme II–III pairs of the *E. coli* phosphotransferase system (two-component systems)
C.	Mannose Enzyme II-P—II-M—III complex of the *E. coli* phosphotransferase system (three-component systems)

The Molecular Genetic Approach to Transport

Until recently, the amino acid sequences of very few permease proteins were known. With the advent of rapid gene cloning and DNA sequencing techniques (chapter 27), we now know the primary structures of literally dozens of biological solute transport proteins. These amino acid sequences first of all allow us to group these proteins into evolutionarily related families. We now know that the thousands of permeases probably fall into only about a dozen permease classes, and that representatives of these classes cross phylogenetic lines (see table 32.1). Second, the primary structures of these proteins permit the construction of models that allow us to predict the folding of these proteins in the membrane (see chapter 7). We can then test these predictions by a variety of methods. Third, comparison of the primary structures of related permeases allows us to predict which residues may be necessary for catalytic function. In general, those residues and regions of the protein that are most strongly conserved in a group of diverging proteins are believed to be important for tertiary structure and/or function. Finally, by employing the technique of site-directed mutagenesis, we can change residues that are predicted to be important for function to other aminoacyl residues and assess the effects of such changes on catalytic activity. The use of these new molecular genetic approaches, coupled to biochemical analyses, has greatly expanded our knowledge concerning transport protein mechanisms during the past decade.

Isotopes, Analogs, and Specially Prepared Vesicles Are Used to Study Transport

In order to measure the rate of transport of a solute into a cell, we must have some method for easily identifying it and distinguishing molecules that have indeed crossed the membrane. Most commonly, researchers use radioactively labeled solutes for this purpose, and time-dependent uptake of radioactivity into a cell is evidence that the compound is being transported through the cell membrane. With many animal cells grown in culture, it is often sufficient to separate intracellular label from that remaining in the medium by quickly washing the cells (which often adhere to the vessel in which they are grown) with buffer or medium that is free of the solute being tested. With bacteria and small, free-living eukaryotes such as yeast and protozoans, cells can be quickly separated from the medium by centrifugation or filtration.

One of the problems associated with measuring transport rates, especially in whole cells, is the fact that subsequent metabolism of the transport substrate may affect the rate observed. In order to uncouple transport from metabolism, nonmetabolizable analogs of some solutes have been used. Examples of compounds that have been employed to study the transport of glucose, a common carbon and energy source for many cells, are methyl-α-glucoside and 2-deoxyglucose (fig. 32.8). Both compounds cannot be metabolized beyond the hexose-phosphate stage by most cells. Another technique that has been used, especially in bacteria, is the isolation of mutants that are able to transport a given solute, but that lack one or more enzymes for the further metabolism of the compound. By these procedures, it is often possible to determine many characteristics of the transport system itself, without any complications owing to other reactions that take place after the solute enters the cell.

Finally, vesicles derived from the membrane system of interest are often used by biochemists interested in membrane transport phenomena. Cytoplasmic membrane vesicles can often be isolated from eukaryotic cells after mild homogenization or N_2 cavitation (chapter 7), which leave many of the intracellular organelles intact. In bacteria, the use of membrane vesicles was pioneered in the 1960s by H. R. Kaback, who developed a method, similar to that shown in figure 7.23, for preparing membrane vesicles from *E. coli*. Most of these vesicles (fig. 32.9) appear to be "right side out"; that is, the orientations of most integral membrane proteins are the same as those in the intact

Figure 32.8

Methyl-α-glucoside and 2-deoxyglucose are two analogs that are useful in the study of glucose transport in many cells. Usually they are not metabolized beyond the hexose-phosphate stage, which avoids the complications of subsequent reactions that might affect the measurement of transport rates.

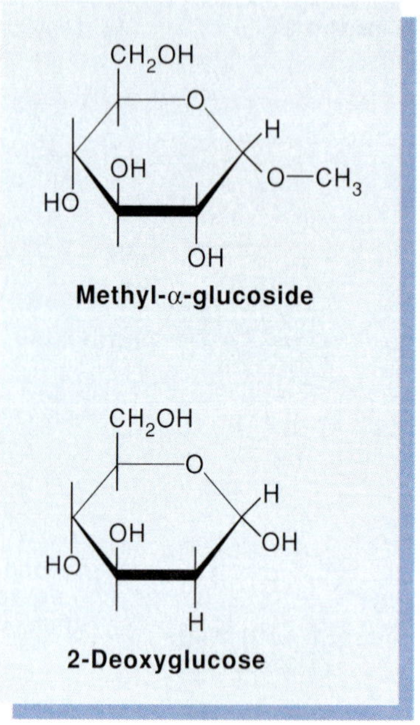

Figure 32.9

Electron micrograph of membrane vesicles obtained from *E. coli* spheroplasts subjected to osmotic lysis. These closed, spherical structures have been shown to transport a variety of compounds normally accumulated by whole cells as long as any required energy source is provided. They also have the advantage of lacking cytoplasmic components that could affect the transport rate (77,000X). (From H. R. Kaback, "Transport studies in bacterial membrane vesicles," *Science* 186:882, 1974. Copyright © 1974 by the AAAS.)

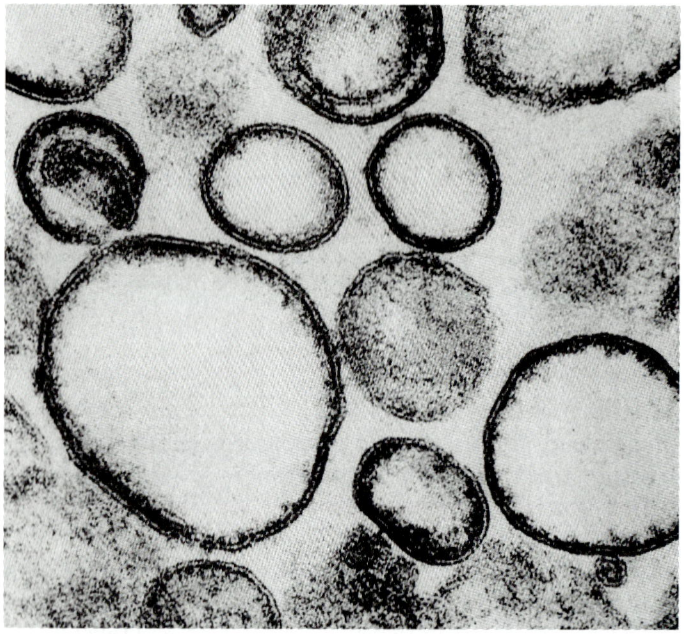

Figure 32.10

(*Left*) Hydropathy plot of the glucose transporter amino acid sequence. Hydropathy values are plotted with respect to position along the amino acid sequence. The numbers refer to putative membrane-spanning domains. (*Right*) Proposed model for the orientation of the glucose transporter in the membrane. The twelve putative membrane-spanning domains are numbered and shown as rectangles. The relative positions of acidic (Glu, Asp) and basic (Lys, Arg) amino acid residues are indicated by circled plus and minus signs, respectively. Uncharged polar residues within the membrane-spanning domains are indicated by their single-letter abbreviations: S, serine; T, threonine; H, histidine; N, asparagine; Q, glutamine. The predicted position of the N-linked oligosaccharide at Asn 45 is shown. The arrows point to the positions of known tryptic cleavage sites in the native, membrane-bound, erythrocyte glucose transporter. (Source: M. Mueckler et al., "Sequence and structure of a human glucose transporter," in *Science* 229:941–945, 1985. Copyright © 1985 American Association for the Advancement of Science, Washington, D.C.)

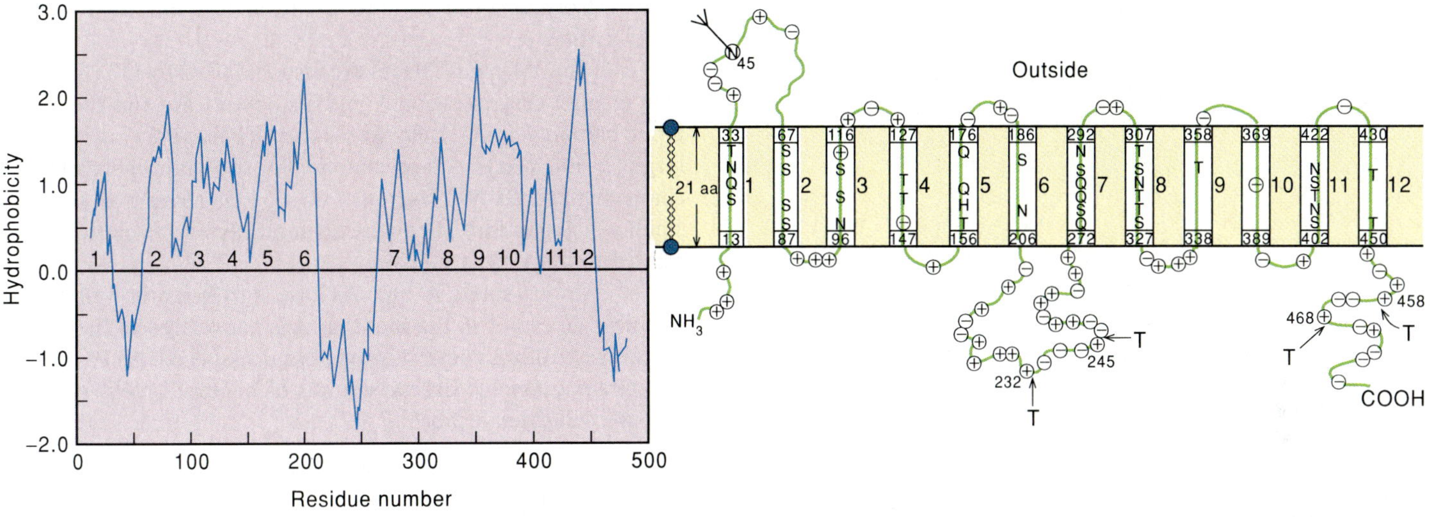

cell membrane. Kaback and others have shown that many substances transported by *E. coli* are also taken up by these vesicles as long as any required energy sources are provided. These vesicles have the advantage of lacking most cytoplasmic enzymes, so that the metabolism of the transport substrate is severely limited. Furthermore, they are especially suited for the study of energy-coupling mechanisms in active transport because the investigator can easily introduce various energy sources into the vesicle preparations and test their effects on the transport system of interest. Such experiments are sometimes difficult to interpret with whole cells, again because of the complications of multiple metabolic events occurring in the cytoplasm.

Examples of Facilitated Diffusion

Transport systems that appear to operate only by facilitated diffusion have been identified both in bacterial and animal cells. Glycerol, which can act as the sole carbon and energy source for many bacterial cells, has been shown to be transported in several bacteria by nonconcentrative systems. Studies using *E. coli* by E. Lin and collaborators have provided evidence for a "facilitator protein" that allows glycerol permeation in this organism. The facilitator increases the rate at which glycerol or any small, straight-chain, neutral molecule crosses the phospholipid bilayer but cannot accumulate these compounds within the cell.

The diameter of the pore formed by the glycerol facilitator is about 0.4 nm. Recently, the DNA sequence of the *glpF* gene encoding the glycerol facilitator was determined, and computer searches revealed that it is very similar in structure to two other potential pore-forming proteins, one from animal cells and one from plant cells. One of these proteins is called the major intrinsic protein (MIP) of the lens cells of the mammalian eye. It apparently facilitates the flow of electrolytes, small solutes, and fluid from the extracellular environment between two lens fiber cells into the cell cytoplasm. The second protein similar to the glycerol facilitator is called nodulin-26 and is found in the membranes of soybean nodules, which house bacteria and function in nitrogen fixation. It therefore appears that as we gain a clearer understanding of the mechanism of the *E. coli* glycerol facilitator, we may be better able to conceptualize the complex processes involved in cell–cell communication in higher organisms.

In animal cells, glucose, a common energy source, appears to be transported by a variety of mechanisms, depending on cell type. Both active (see the following discussion) and facilitated glucose transport systems have been characterized, sometimes both in the same type of cell. In human erythrocytes and hepatocytes, glucose is apparently transported only by facilitated diffusion. The carrier for this sugar has been isolated, and its gene has been cloned and sequenced. The purified glucose carrier has been reintroduced into phospholipid vesicles and was shown to rapidly equilibrate glucose across the bilayer, but no evidence was found consistent with any ability to carry out active transport. The primary amino acid sequence of the hepatocyte glucose facilitator, deduced from the nucleotide sequence of its gene, has allowed construction of models for how this glycoprotein may be folded in the membrane. One such model is shown in figure 32.10. This sequence is also very similar

Figure 32.11

Exchange diffusion of anions carried out by the human erythrocyte anion-channel (band 3) protein. This exchange is important in lung capillaries for the removal of CO_2 generated by cellular metabolism from the red cell cytoplasm.

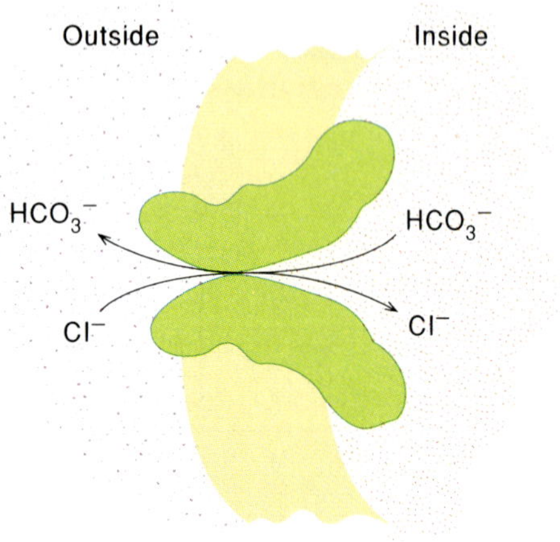

in certain aspects to those of transport proteins from bacteria and yeast that participate in secondary active transport of various carbohydrates, as we will later see.

A third example of facilitated diffusion in biological membranes is the process carried out by the anion-transport (band 3) protein of erythrocytes, whose structure we considered in chapter 7 (see fig. 7.33). The anion transporter allows CO_2 accumulated in the erythrocyte from respiring tissues to diffuse rapidly out of these cells in lung capillaries in the form of bicarbonate anion HCO_3^-. This process has been shown to be accompanied by an influx of Cl^- into the erythrocyte. Much work has established that the anion transporter catalyzes the one-for-one exchange of HCO_3^- for Cl^-. These two molecules always flow in a fashion determined by the sums of their concentration gradients, and thus this protein catalyzes a form of facilitated diffusion. Transport of HCO_3^- out of the cell is obligatorily coupled to the influx of a monovalent anion such as Cl^-, as shown in figure 32.11. This process, called exchange diffusion, is an example of a more general transport process termed antiport. Antiport involves the coupled transport of two different molecules that move in opposite directions across a membrane. "Antiporters" may carry out facilitated diffusion, as does the anion transport protein, or they can catalyze active transport, as we will discuss later.

Ion-Translocating ATPases Carry Out One Type of Primary Active Transport

Transmembrane, ATP-hydrolyzing enzymes use the energy in the terminal γ-phosphate anhydride bond of ATP to drive the active uptake or extrusion of an ion (or ions) into or out of a cell or organelle. Four types of these enzymes are now recognized: the P-type (plasma-membrane-localized), cation-translocating ATPases; the F-type (F_1/F_o), ion-translocating ATPases; the V-type (vacuolar), H^+-translocating ATPases; and the A-type, anion-translocating ATPases.

The P-type ATPases are sensitive to vanadate (a phosphate analog) and form an acyl phosphate intermediate during ATP hydrolysis. They are usually composed of monomeric or multimeric complexes in which a single polypeptide chain of about 100 kDa serves as the catalytic unit.

The F-type ATPases are inhibited by dicyclohexylcarbodiimide (DCCD; see chapter 15) and azide and do not form a phosphorylated enzyme derivative. They consist of multisubunit F_1 and F_o moieties in which the F_1 moiety functions in ATP hydrolysis/synthesis, and the F_o moiety functions in ion translocation.

The V-type ATPases are also inhibited by DCCD (but only at high concentrations), and also by nitrate and the antibiotic bafilomycin A_1. Like the F-type ATPases they are usually large (>400 kDa) and are composed of two major polypeptide components of 70–89 kDa and of 60–65 kDa along with several smaller polypeptides. They are structurally and evolutionarily related to the F-type ATPases.

Only a single A-type ATPase is sufficiently well characterized to establish the fact that it is not related to the other three classes of ATPases; it is the oxyanion-specific ATPase that pumps toxic arsenic derivatives out of bacterial cells, a topic that we will discuss shortly.

The Plasma Membrane Na^+-K^+ ATPase of Eukaryotes

One of the first biological transport systems to be studied in detail was that responsible for maintaining the normally high concentrations of K^+ and low levels of Na^+ inside animal cells. In order for this to occur, both cations must be transported essentially unidirectionally against their concentration gradients. A breakthrough in explaining how this might be accomplished came in the 1950s, when J. C. Skou identified an enzyme from crab nerves that hydrolyzes ATP to ADP only in the presence of both Na^+ and K^+. This sodium-potassium ATPase has subsequently been shown to be a transmembrane protein that pumps Na^+ out of the cell and K^+ into the cell, accompanied by ATP hydrolysis. This coupling of chemical energy to concentrative transport is an example of primary active transport, as well as of an antiport system. The same protein catalyzes both ion translocations and the hydrolysis of ATP, as shown schematically in figure 32.12. Thus the energy released by ATP hydrolysis is "translated" by the Na^+-K^+ ATPase into potential energy in the form of Na^+ and K^+ electrochemical gradients across the plasma membrane. The Na^+-K^+ ATPase is an example of a P-type ATPase, of which we will discuss several other examples shortly.

An important tool in investigating the mechanism of the Na^+-K^+ ATPase has been the use of the specific inhibitor ouabain (fig. 32.13), a compound isolated from foxglove (*Digitalis*) and related plants. Ouabain binds to the outer surface of the enzyme and blocks both ion translocation and ATP hydrolysis by the Na^+-K^+ ATPase. Furthermore, it has been demonstrated that K^+ has a much higher affinity for the outside surface of the protein, while Na^+ binds much more tightly to the cytoplasmic domain. Finally, the site of ATP hydrolysis has

Figure 32.12

Primary active transport of Na^+ and K^+ by the sodium-potassium ATPase found in many animal cells. This enzyme maintains the normally high K^+ and low Na^+ concentrations within the cell at the expense of chemical energy stored in ATP. Each cycle of the enzyme results in the extrusion of three molecules of Na^+ out of the cell, the transport of two molecules of K^+ into the cell, and the hydrolysis of one molecule of ATP. The subunit structure and stoichiometry of this enzyme have been established. One αβ heterodimer is the active catalytic unit. The α subunit (M_r = 112,200) is associated with all known catalytic properties of the enzyme complex, while the β subunit (M_r = 57,000) is a glycoprotein of unknown function.

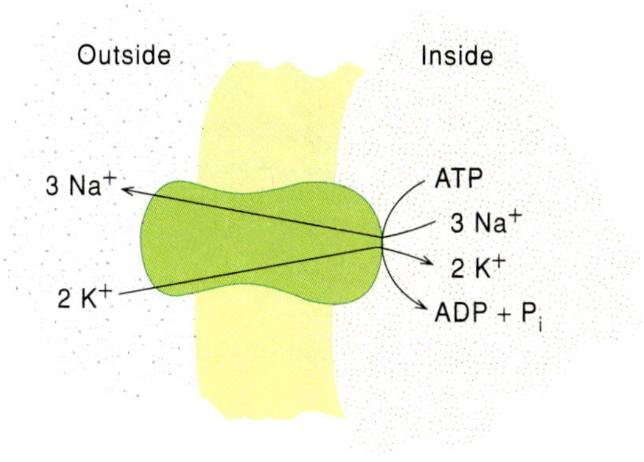

Figure 32.13

The structure of ouabain, a specific inhibitor of the Na^+-K^+ ATPase. Ouabain is a member of a class of compounds called *cardiotonic steroids. Digitalis,* the general name for a family of cardiotonic steroids that inhibit the Na^+-K^+ ATPase, indirectly causes an increase in intracellular Ca^{2+}, which is due to increased cytoplasmic Na^+ that arises because the cardiac sarcolemma contains a Na^+/Ca^{2+} antiport system. The increase in the intracellular Ca^{2+} concentration appears to directly stimulate contraction in cardiac muscle cells. This explains the use of digitalis in treating such conditions as congestive heart failure.

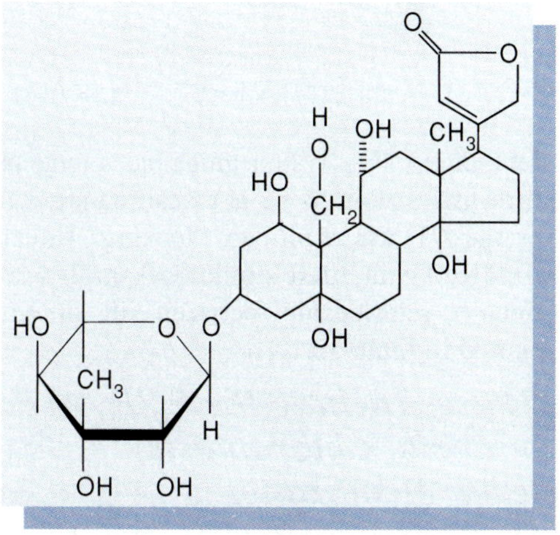

been localized to the inner surface of the cytoplasmic membrane. These observations support the schematic illustration shown in figure 32.12 and demonstrate that the Na^+-K^+ ATPase must be asymmetrically oriented in the plasma membrane.

The Na^+-K^+ ATPase exists in the membrane as a heterodimer consisting of 110-kDa α and 55-kDa β subunits. It has long been debated whether the functional unit is the heterotetramer ($\alpha_2\beta_2$) or the heterodimer ($\alpha\beta$). Recent evidence appears to favor the latter possibility. In the process of active Na^+ and K^+ transport, the cations become "trapped" or "occluded" in the structure of the protein. In this state, they exchange with the cations in the medium only very slowly. It has been found that the detergent-solubilized unit can occlude Rb^+ (a K^+ analog) or Na^+ with the same stoichiometry as the membrane-bound ATPase. These observations suggest that the cations are occluded in a cavity inside the $\alpha\beta$ unit, and that this unit therefore constitutes the minimal protein unit required for active Na^+ and K^+ transport.

Because the inhibitor ouabain binds on the outside of the Na^+-K^+ ATPase and yet inhibits ATPase activity, which resides on the cytoplasmic surface, information must be conducted through the protein by means of conformational changes in the enzyme. Studies with the purified Na^+-K^+ ATPase have shown further that the protein is covalently phosphorylated during its catalytic cycle, the site of phosphorylation being an aspartic acid residue. Phosphoryl Na^+-K^+ ATPase is formed in the presence of Na^+, ATP, and Mg^{2+}, while the dephosphorylation step is K^+-dependent. Ouabain binds tightly to the enzyme-phosphate intermediate and prevents dephosphorylation in the presence of K^+.

These observations and others have led to the formulation of the model depicted in figure 32.14 for the functioning of the Na^+-K^+ ATPase. Transport of Na^+ out of the cell and K^+ into the cytoplasm is thought to be mediated by conformational changes in the enzyme triggered by phosphorylation, dephosphorylation, and the binding of ligands at both surfaces. While figure 32.14 illustrates a reasonable mechanism for the Na^+-K^+ ATPase, it says little about how conformational changes in the protein can lead to unidirectional transport of ions through the membrane. This question has yet to be completely answered in any system, but, as we will see, a number of models have emerged for molecular mechanisms of such translocation events.

Other P-Type ATPases Exist for Ca^{2+} and H^+

Muscle cells contain a specialized endoplasmic reticular membrane system called the sarcoplasmic reticulum. This organelle acts as a repository for Ca^{2+} in resting muscle, and electrical stimulation of muscle cells causes rapid release of this ion from the sarcoplasmic reticulum into the sarcoplasm, where it triggers contraction of the myofibrils. Reuptake of Ca^{2+} into the reticular lumen is effected by a calcium ATPase, which catalyzes the following reaction:

$$2\ Ca^{2+}\ (out) + ATP\ (out) \xrightarrow{Mg^{2+}} 2\ Ca^{2+}\ (in) + ADP\ (out) + P_i\ (out)$$

In this reaction sequence, (out) refers to the outside of the sarcoplasmic reticulum (sarcoplasm), and (in) corresponds to the lumen of the system.

The Ca^{2+} ATPase has been extensively characterized, and like the Na^+-K^+ enzyme, it spans the membrane in an asymmetric fashion. In the presence of Ca^{2+} and ATP, formation of a covalent phosphoryl enzyme can also be demonstrated in this system. Thus many of the aspects of Ca^{2+} transport in the sarcoplasmic reticulum parallel those of Na^+ and K^+ translocation in the cytoplasmic membranes of animal cells. In fact,

Figure 32.14

Postulated mechanism of the functioning of the Na^+-K^+ ATPase. The yellow boundary separates events in which cation association or dissociation occurs on the inside and outside surfaces of the membrane, while the arrows relating E_1 and E_2 denote conformational changes in the enzyme, but are not meant to imply its physical translocation across the membrane. The enzyme is hypothesized to have at least two conformations, E_1 and E_2, both of which can exist in a phosphorylated state. E_1~P is a high-energy phosphoryl bond, in equilibrium with the ß-γ pyrophosphoryl bond of ATP. By contrast, E_2—P is of low energy, in equilibrium with inorganic phosphate. The changes in bond energies are presumed to be due to the free energies of Na^+, K^+, and Mg^{2+} binding. It should be noted that because ion binding may induce secondary conformational changes, both E_1 and E_2 may exist in both "empty" and "full" conformations. The "full" conformations, $E_2(K^+)$ and E_1~$P(Na^+)$, contain occluded (2 K^+) and (3 Na^+) portions, respectively. E_1 and E_2—P have their transport sites facing in and out, respectively. Only E_1 possesses high affinity for Na^+, while only E_2 possesses high affinity for K^+. Phosphorylation of E_1 drives the cyclical process in a unidirectional fashion, causing Na^+ to be translocated from the inside to the outside. Conversion of E_1~P to E_2—P decreases the affinity of the enzyme for Na^+ and increases its affinity for K^+. Dephosphorylation of E_2 drives translocation of K^+ from the outside to the inside. Conversion of E_2 to E_1 decreases the affinity of the enzyme for K^+ and enhances its affinity for Na^+. Release of K^+ from E_2 is greatly accelerated by ATP binding. Translocation of Na^+ out of the cell and K^+ into the cell is the net consequence of this series of conformational changes in the protein. Ouabain inhibits by binding to E_2—P, preventing dephosphorylation in the presence of K^+.

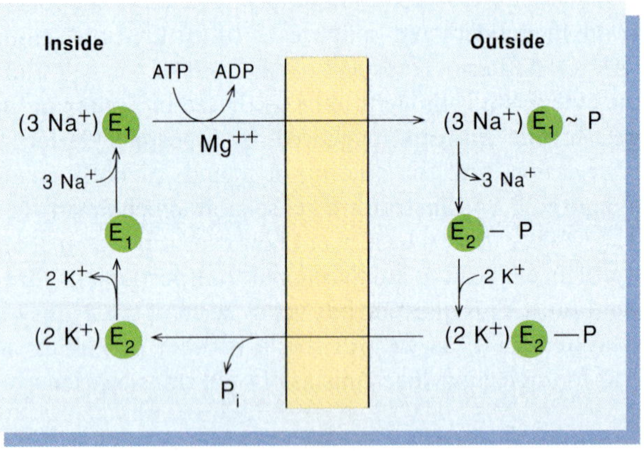

Figure 32.15

The F_1/F_0 ATP synthase or ATPase of bacterial, mitochondrial, and chloroplast membranes. (*a*) A proton gradient generated by electron transport can be used for ATP synthesis by means of chemiosmotic coupling. (*b*) Alternatively, in bacteria growing anaerobically, ATP hydrolysis is used to *generate* a proton gradient in the reverse reaction. This gradient can then be used to energize secondary active transport systems as well as for other processes in bacteria that use chemiosmotic energy, such as flagellar motility.

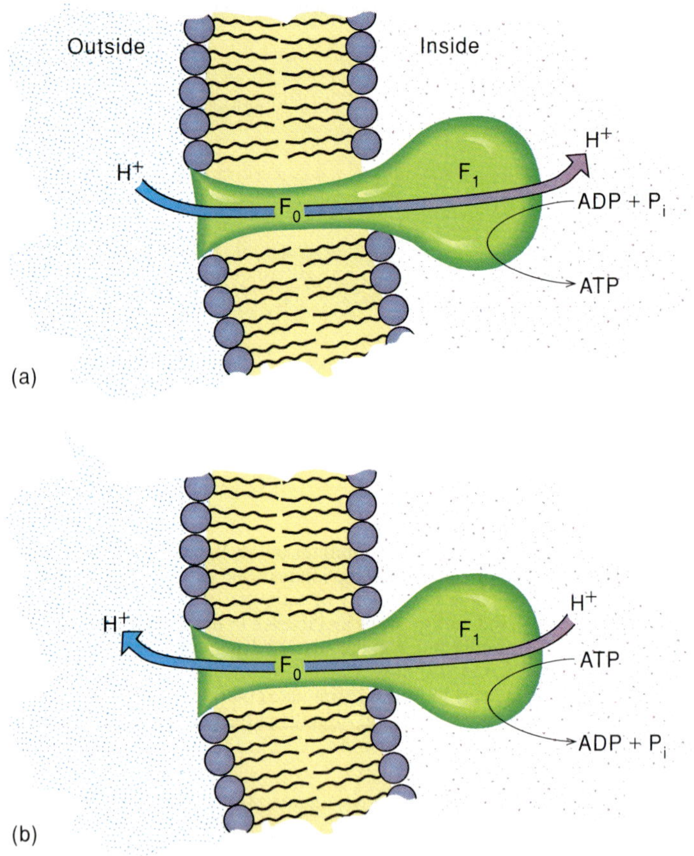

many animal cells also have a Ca^{2+} ATPase in their cytoplasmic membranes that is very similar in structure and mechanism to the sarcoplasmic reticular enzyme. The kinetic mechanism of the Ca^{2+} ATPase is thought to be analogous to that for the Na^+-K^+ enzyme shown in figure 32.14, with 2 Ca^{2+} replacing 3 Na^+, the polarity, however, being reversed. Again, at least two conformations of the protein are involved, and their interconversion is believed to be triggered by bound ligands and the state of phosphorylation of the enzyme. These structural and mechanistic similarities are expected since these proteins both belong to a single family of transport ATPases and presumably share a common evolutionary ancestor.

A third P-type ATPase is the H^+-translocating enzyme found in the plasma membrane of epithelial cells lining the stomach. This enzyme is responsible for acidification of the lumen and is structurally related to the Na^+-K^+ and Ca^{2+} ATPases. Each of these enzymes has a large catalytic polypeptide chain of about 100 kDa that is transiently phosphorylated during transport, and each is inhibited by the phosphate analog, pentavalent vanadate (VO_4^-). This anion has a trigonal bipyramidal structure and seems to be a transition-state analog of phosphate for the ATPase reactions, "locking" the enzyme in the transition state. Again, these similarities imply a reasonably close evolutionary relationship between all of the P-type ATPases, as noted in table 32.1.

F-Type ATPases Include the F_1/F_0 ATPases of Mitochondria, Chloroplasts, and Bacterial Membranes

In chapter 15 we discussed the second class of well-characterized ion-translocating ATPases (F-type). The F_1/F_0 ATPases (or ATP synthases) of mitochondria, chloroplasts, and bacteria appear to have a similar function in all three systems. Proton gradients formed as a result of electron transport are used to synthesize ATP from ADP and P_i as a consequence of H^+ flow down its electrochemical gradient through the F_1/F_0 ATPase (fig. 32.15*a*). It is also possible, however, for the F_1/F_0 ATPase

Table 32.2
Subunit Compositions of F_1 ATPases

Source	Molecular Weight (× 10³)	Subunit Sizes (× 10³)[a]				
		α	β	γ	δ	ϵ
E. coli	380	56	52	32	22	11
Thermophilic bacterium (PS3)	380	56	53	32	15	11
Beef heart mitochondrion	360	54	50	33	17	11
Chloroplast	325	59	56	37	17	13

Source: D. B. Wilson and J. B. Smith, "Bacterial transport proteins" in *Bacterial Transport,* edited by B. P. Rosen. Copyright © 1978 Dekker, New York, N.Y.
[a]The stoichiometry of subunits in the F_1 ATPase is $\alpha_3\beta_3\gamma\delta\epsilon$ for the bacterial enzymes.

to work in the reverse direction, i.e., to hydrolyze ATP and translocate protons against their electrochemical gradient (fig. 32.15*b*). This has been established in some bacteria under conditions where respiratory electron transport does not occur (e.g., anaerobiosis or the absence of a functional electron-transport chain in some strict anaerobes). Thus the F_1/F_0 ATPase also can be classified as a protein complex carrying out primary active transport. Unlike the P-type ATPases, however, there is no evidence that the F_1/F_0 ATPases are covalently phosphorylated during proton translocation.

We described the structure of the mitochondrial F_1/F_0 ATPase in chapter 15; the chloroplast and bacterial enzymes have very similar structures. The F_1 portion, which possesses the ATPase activity, consists of five different subunits (named α through ϵ), each of which has a similar molecular weight in each of these three sources (table 32.2). The F_0 portion, which spans the membrane, has been shown to be a proton channel. It consists of three different polypeptides (named *a, b,* and *c*) with molecular weights of 28,000, 19,000, and 8,500 respectively, in a subunit ratio of approximately 1:2:ca. 9. While subunit *b* is believed to function in binding F_0 to F_1, the *a* and/or *c* subunits function as the proton channel. Removal of F_1 from membranes results in an increased permeability of the bilayer to protons, a fact suggesting that F_1 acts as a "cap" that couples proton translocation to ATP synthesis or hydrolysis. The F_0 portion, which is rich in hydrophobic amino acids, has been shown to cause proton conductivity when incorporated into phospholipid vesicles.

Some elegant reconstitution studies by Y. Kagawa and co-workers first showed how the various subunits of the F_1/F_0 ATPase may be arranged on the membrane. As a result of information gained by purifying each subunit of the F_1 portion from the thermophilic bacterium PS3 and recombining them in various ways, they proposed the structure depicted in figure 32.16*a*. In this model, the complex of α and β subunits ($\alpha_3\beta_3$), which possesses ATPase activity by itself, is bound to the F_0 portion through subunits δ and ϵ. Subunit γ is believed to act as the "cap" or gate through which protons flow. Support for this model came from experiments showing that membranes containing F_0 and the δ plus ϵ subunits of F_1 were permeable to protons, but addition of the γ subunit blocked proton conductance. Other workers have proposed models for the arrangement of the subunits of the F_0 portion of the ATPase, one of which is shown in figure 32.16*b*. Because the subunit structures and functions of the F_1/F_0 ATPases from a number of sources are very similar, it seems reasonable to expect that the functions and topographies of the individual subunits will be related as well.

Interestingly, in the strictly anaerobic bacterium *Propionigenium modestum* an F_1/F_0 ATPase exists that normally couples Na^+ instead of H^+ translocation to ATP synthesis. The Na^+ translocation site is in the F_0 portion of this protein, and a hybrid enzyme composed of F_0 from *P. modestum* and F_1 from *E. coli* is a functional Na^+ pump. These results have important implications for the mechanism of ATP synthesis and hydrolysis by the F-type ATPases. All models in which protons play a role that can be performed only by protons must be incorrect. One such model that now appears unlikely is proton conduction through a "proton wire" consisting of a series of hydrogen-bonded residues within F_0 (see the later discussion on bacteriorhodopsin). Rather, these observations are more in accord with a mechanism in which the binding of cations to sites on the external surface of F_0 triggers a conformational change that exposes the cations to the inner membrane surface and simultaneously effects ATP synthesis by the F_1 moiety of the enzyme.

Vacuolar ATPases Pump Protons into Intracellular Vesicles and Vacuoles

The vacuolar ATPases (V-type), now recognized as a third distinct class of ATPases, are less well characterized than either the P-type or F-type enzymes. The V-type ATPases are present in the dynamic intracellular membrane networks of eukaryotic cells, which include the endoplasmic reticulum, Golgi, endosomes, coated vesicles, and lysosomes. They are also found in

Figure 32.16

Subunit structure of the F_1/F_0 ATPase. (*a*) A model for the arrangement of the subunits of the F_1 portion of the enzyme. The five subunits α, ß, γ, δ, and ε form a dissociable unit, which attaches to the integral membrane F_0 part of the complex. (Source: M. Yoshida et al., "Reconstitution of thermostable ATPase capable of energy coupling from its purified subunits" in *Proceedings. National Academy of Sciences USA* 74:936, 1977. Copyright © 1977 National Academy of Sciences, Washington, D.C.) (*b*) A schematic model of the F_0 part of the F_1F_0 ATPase of *E. coli,* shown both in side view (*top*) and top view (*bottom*). Subunits a, b, and c are indicated in their approximately correct stoichiometries. The essential carboxyl group (Asp 61, the DCCD binding site) is in contact with the lipid phase as shown. Note: Conformation is not indicated, as the "a" subunit probably traverses the membrane five times, while the "b" and "c" subunits pass through it one and two times, respectively. The F_0 part of the ATPase is more complex in eukaryotes and consists of four to six different subunits. (Reproduced from E. Schneider and K. Altendorf, "Bacterial adenosine 5′-triphosphate synthase (F_1F_0): Purification and reconstitution of F_0 complexes and biochemical and functional characterization of their subunits," in *Microbiology Reviews* 51:477–497, 1987.)

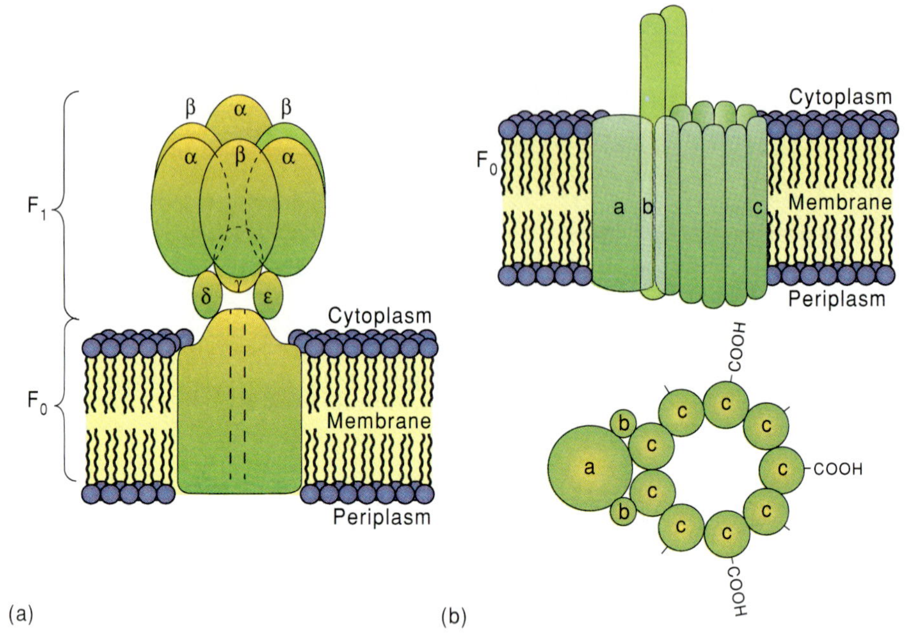

plant tonoplasts and the vacuoles of yeast and fungi. In all examples studied, these ATPases utilize cytoplasmic ATP to pump protons into the intracytoplasmic compartment. The vacuolar ATPases, of three or more distinct types of subunits, may be distantly related to the F_1/F_0 ATPases, as indicated by some of their biochemical properties. However, the vacuolar ATPases are not readily separable into F_1 and F_0 moieties, do not show a knoblike appearance by electron microscopy, and are solubilized only upon treatment with detergents. It therefore appears that this third class of ATPases and the F-type ATPases may have evolved from a common ancestor but diverged considerably. The recent demonstration of V-type ATPases in *Sulfolobus,* an archaebacterium, has led to the proposal that vacuolar ATPases of eukaryotes may have arisen by the internalization of plasma membrane V-type ATPases of an archaebacterial-like ancestral cell.

An Oxyanion-Specific ATPase Pumps Toxic Arsenic Derivatives Out of Bacterial Cells

The first member of a fourth class of ion-translocating ATPases has been discovered and characterized. This A-type ATPase is a plasmid-encoded enzyme complex that actively extrudes toxic oxyanions such as arsenite, antimonite, and arsenate from the cytoplasm of bacteria. Because the genes for the synthesis of this enzyme are encoded on a small, transferable plasmid, properties associated with the possession of this enzyme can be easily transferred from one population of bacteria to another. Since the plasmid confers on its host bacterium resistance to a variety of bacteriocidal agents, it serves as one kind of antibiotic resistance plasmid.

The arsenical resistance pump consists of three polypeptides, ArsA (63 kDa), ArsB (46 kDa), and ArsC (16 kDa). ArsB is an integral membrane protein that presumably serves as the oxyanion channel. ArsA is a peripheral membrane protein that is associated with ArsB and possesses the ATPase activity of the pump. The ArsC protein is also peripherally associated with the membrane, and it appears to function as a specificity protein. Surprisingly, sequence analyses suggest that the ArsA and ArsB proteins are related to enzymes involved in nitrogen fixation, which may have no transport function. It will be interesting to learn what functions any other proteins within this class of ATPases may possess.

Some Primary Active Transport Systems Are Driven by Electron Transport or Light

Membrane-bound electron-transport chains have as a primary function the extrusion of protons out of mitochondria and bacteria and into chloroplasts of plants and algae (chapter 16). These gradients can then be used for ATP synthesis as described earlier, or for secondary active transport processes linked to ion translocation (see the following discussion). The exact mechanisms of proton translocation during electron transport are still somewhat unclear, but both unidirectional proton channels and coenzyme Q are likely to be involved (chapter 16). In any case, energy derived from the oxidation-reduction reactions that take place within the membrane is more or less directly transformed into a chemical gradient of protons. For this reason, electron transport is generally thought of as a primary active transport event.

Light energy can likewise be used to drive active transport in a number of systems. For example, in photosynthesis, light energy absorbed by chlorophyll molecules is converted into a proton gradient at least partly by means of the electron-transport reactions that subsequently take place. The three-dimensional structure of the membrane-bound protein complex

Figure 32.17

Bacteriorhodopsin (BR). (*a*) Freeze-fracture electron micrograph of the membrane of *H. halobium*. Fine-grain regions are patches of purple membrane containing bacteriorhodopsin. (Magnification 48,000X.) (Courtesy of Walther Stoeckenius.) (*b*) Bacteriorhodopsin contains a covalently bound molecule of retinal, a derivative of vitamin A, attached by means of a Schiff base linkage to a lysine residue of the protein. Both the all-*trans* isomer and the 13-*cis* form can be isolated from the protein. Retinal is also the light-gathering chromophore of the vertebrate retina, where it exists as the 11-*cis* isomer attached by means of a Schiff base to the protein opsin (chapter 36). (*c*) The photochemical cycle of bacteriorhodopsin. Light absorption by the retinal moiety quickly leads to isomerization of the chromophore to the 13-*cis* isomer, followed by deprotonation of the Schiff base to the form absorbing maximally at 412 nm (BR_{412}). Two intermediate forms can be detected spectroscopically between the isomerization and deprotonation steps (BR_{625} and BR_{550}). Reprotonation and isomerization back to the all-*trans* form are slow steps involving two other intermediates (BR_{520} and BR_{640}). (Source: H. G. Khorana, "Bacteriorhodopsin, a membrane protein that uses light to translocate protons," in *Journal of Biological Chemistry* 263:7439–7442, 1988. Copyright © 1988 American Society for Biochemistry and Molecular Biology Inc., Bethesda, Md.)

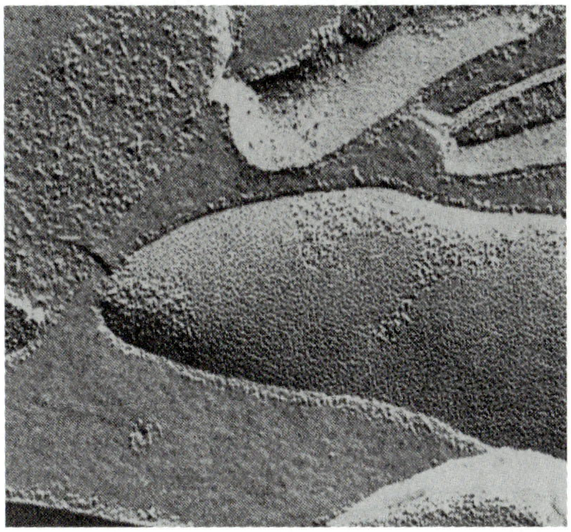

(a)

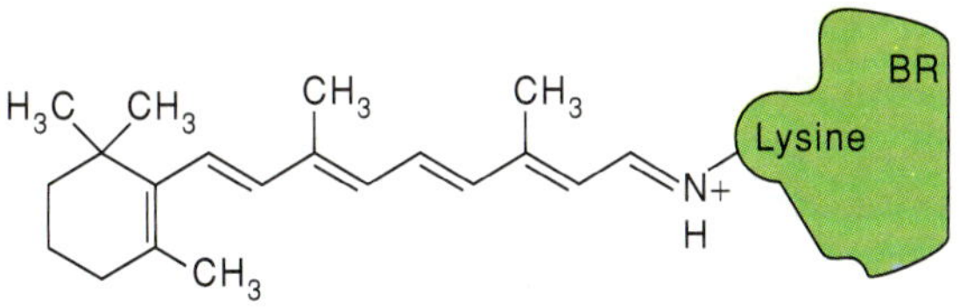

(b)

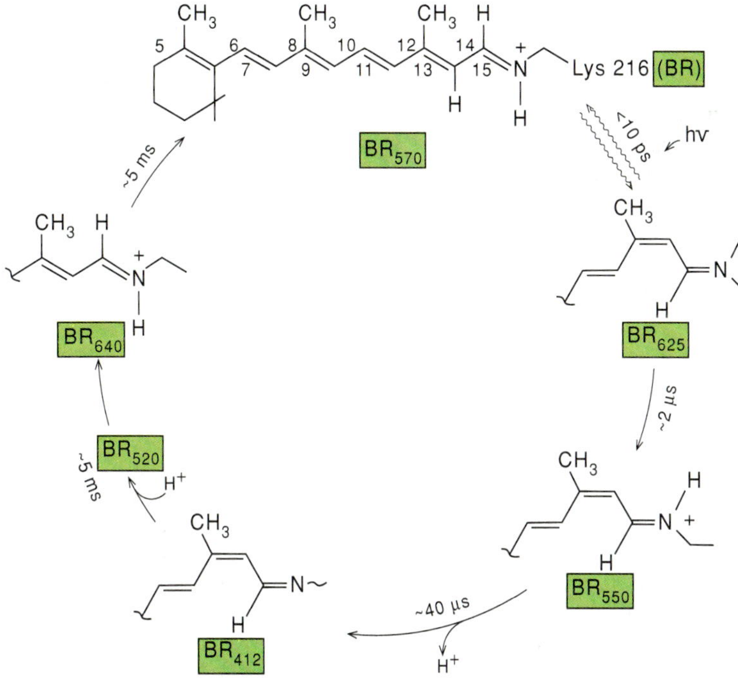

(c)

that initiates photosynthesis in the bacterium *Rhodopseudomonas viridis* was recently solved in Germany by J. Deisenhofer and H. Michel, a feat for which they were awarded the Nobel Prize. We described this structure in chapters 7 and 16, and it undoubtedly will be invaluable in determining the molecular mechanism involved in the transduction of light energy into a proton gradient in this photosynthetic system.

In halophilic bacteria, such as *Halobacterium halobium,* a remarkable light-driven proton pump, bacteriorhodopsin, has evolved in part to allow this organism to synthesize ATP under conditions of oxygen limitation. The structure of bacteriorhodopsin, which we considered in chapter 7, in some manner allows it to actively transport protons from inside the cell to the outside in response to light absorption. In 1971, Walther Stoeckenius and his co-workers showed that if O_2 was limited in cultures of *Halobacterium,* a purple membrane appeared on the surface of the cell in addition to its normally red, carotenoid-containing membrane (fig. 32.17*a*). The purple membrane is 75% protein by weight, and bacteriorhodopsin is the sole protein component. The purple color (λ_{max} = 570 nm) is due to the covalent association of one molecule of all-*trans* retinal, a derivative of vitamin A, to lysine 216 of bacteriorhodopsin by means of a Schiff base linkage (fig. 32.17*b*). It is the retinal moiety that absorbs photons and is responsible, along with aminoacyl residues of the protein, for the conversion of this form of energy into a proton gradient that can be used for ATP synthesis.

Although the photochemical reactions of bacteriorhodopsin are extremely complex, it is known that light absorption by retinal leads to deprotonation of the Schiff base to one of several intermediate forms, which absorbs light maximally at 412 nm. Associated with this process is the isomerization of the all-*trans* retinal (fig. 32.17*b*) to the 13-*cis* isomer. Reassociation of a proton with the retinal moiety then takes place more slowly, and the bacteriorhodopsin protein cycles back to the 570-nm (and all-*trans*) form through several other intermediate conformations that can be detected spectrophotometrically (fig. 32.17*c*). It is highly probable that this protonation-deprotonation

Figure 32.18

A helical-wheel model of bacteriorhodopsin, showing the relationships between the aspartic acid mutations that affect proton pumping and the amino acids that are believed to interact with the chromophore. The amino acids postulated to interact with retinal are shown in bold ovals. The aspartic residues in helices C and G whose substitution affects H^+ pumping are shown in red (D85, D96, and D212). (Source: H. G. Khorana, "Bacteriorhodopsin, a membrane protein that uses light to translocate proteins," in *Journal of Biological Chemistry* 263:7439–7442, 1988. Copyright © 1988 American Society for Biochemistry and Molecular Biology Inc., Bethesda, Md.)

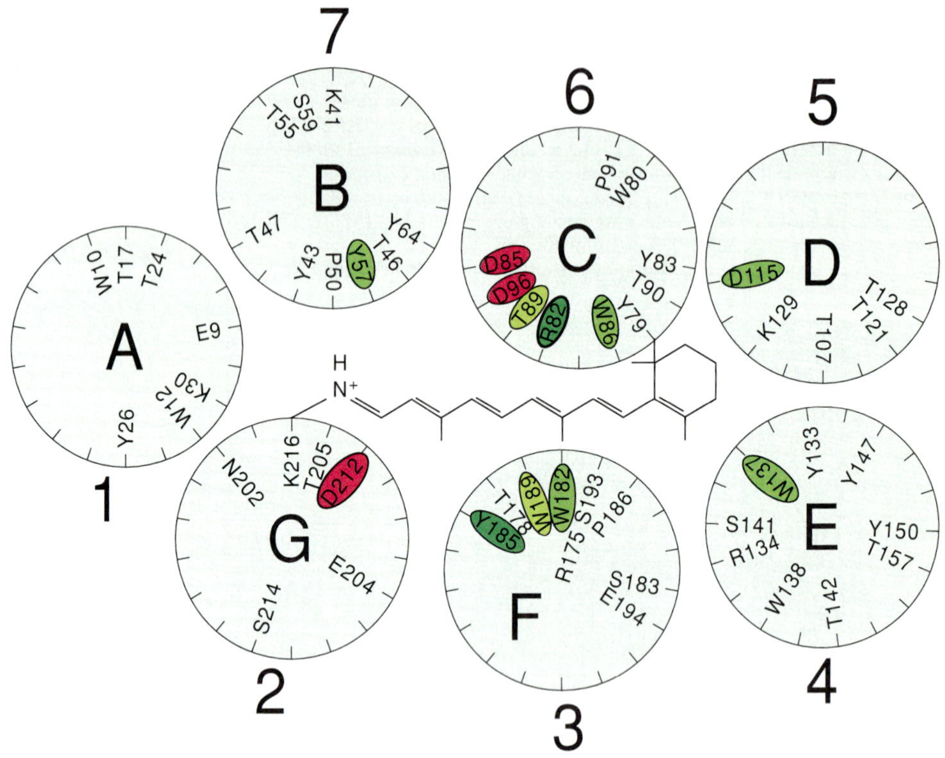

cycle involving multiple retinal and protein conformational states is the key to the mechanism of proton translocation and thus energy transformation carried out by bacteriorhodopsin.

Bacteriorhodopsin is one of several transport proteins that have been studied extensively in recent years by means of molecular genetic techniques. Through site-directed mutagenesis of specific amino acid residues, H. G. Khorana and coworkers have identified a number of residues that appear to be implicated in the proton-pumping mechanism. Several residues, which can reversibly accept and donate protons to form a proton path or "wire" through the interior of the molecule, have been tentatively identified by use of this technique, including Asp 85, Asp 96, Asp 212, and Tyr 185 (see fig. 7.35). The way in which these residues are situated with respect to the helical packing of bacteriorhodopsin and the retinal chromophore is shown in figure 32.18, which also shows residues that are believed to interact with the chromophore itself. The light-induced isomerization of retinal is therefore thought to initiate each cycle of proton conduction by allowing donation of the Schiff base proton to Asp 85 in the proton "wire," which ultimately leads to the release of a proton on the extracellular surface of the membrane. Reprotonation of the Schiff base is accomplished by Asp 96, which is protonated (directly or indirectly) via uptake of a proton from the cytoplasm. In this way, light energy can be used to drive the endergonic pumping of protons against their electrochemical gradient via a proton current involving amino acid residues of the protein, the Schiff base chromophore, and multiple conformational states of bacteriorhodopsin.

Some Bacterial Transport Systems Use Specific Binding Proteins

The last primary active transport mechanisms that we will consider are the so-called ATP-binding cassette (ABC-type) transport systems of the Gram negative bacteria. As we discussed in chapter 7, proteins that bind specific sugars, amino acids, and inorganic ions are found in the periplasmic spaces of Gram negative bacteria, and corresponding membrane-anchored proteins are found in Gram positive bacteria. Some of the compounds for which such binding proteins have been demonstrated are listed in table 32.3. Cold osmotic shock of Gram negative bacterial cells or the transformation of the cells to spheroplasts releases these proteins, and a concomitant decrease in the ability of these cells to transport binding-protein-linked substrates is observed. Thus these periplasmic proteins apparently are important either in concentrating their substrates at the cell surface or in interacting with the transmembrane permeases (or both). In all cases that have been studied, the binding proteins are not themselves responsible for the transmembrane transport of the molecules that they recognize. Rather, specific integral membrane proteins exist for this purpose (fig. 32.19).

The integral membrane proteins that, in part, comprise the binding-protein-dependent systems have been characterized in several instances. One such system is the maltose transport system in *E. coli.* It is known that four distinct proteins are required for the translocation of maltose across the cytoplasmic membrane. These are illustrated in figure 32.19. In addition to the maltose-binding protein in the periplasm (the E protein),

Table 32.3
Binding-Protein-Dependent or ATP-Binding Cassette (ABC-Type) Transport Systems

A. Binding-Protein-Dependent Uptake Systems in Enteric Bacteria

- I. **Oxyanions**
  - A. Phosphate
  - B. Sulfate
- II. **Amino Acids**
  - A. Glutamine
  - B. Lysine/arginine/ornithine
  - C. Histidine
- III. **Sugars**
  - A. Arabinose
  - B. Maltose
  - C. Galactose/glucose
  - D. Ribose
- IV. **Other Compounds**
  - A. Oligopeptides
  - B. Dicarboxylic Acids
  - C. Tricarboxylic Acids
  - D. Vitamin B_{12}
  - E. Iron and Iron Complexes

B. Related Binding-Protein-Independent Export Systems in Bacteria and Eukaryotes

- I. **Drugs**
  - A. Multidrug resistance (Mdr) in Animals
  - B. Drug resistance (Drr) in *Streptomyces pencetins*
- II. **Polysaccharides**
  - A. Polyribosylribitol phosphate (Bex) in *Hemophilus influenzae*
  - B. Polyneuraminic acid (Kps) in *E. coli*
- III. **Proteins and Peptides**
  - A. α-Hemolysin (Hly) in *E. coli*
  - B. a-type Mating Pheromone (Ste6) in Yeast
  - C. Antigenic peptides of the mammalian immune system (MHCII)
- IV. **Other Compounds**
  - A. Chloride (the Cystic Fibrosis Transmembrane regulator (CFTR) in animals)
  - B. Opines in *Agrobacterium tumefaciens*
  - C. Arsenicals (arsenate, arsenite, antimonite) in bacteria

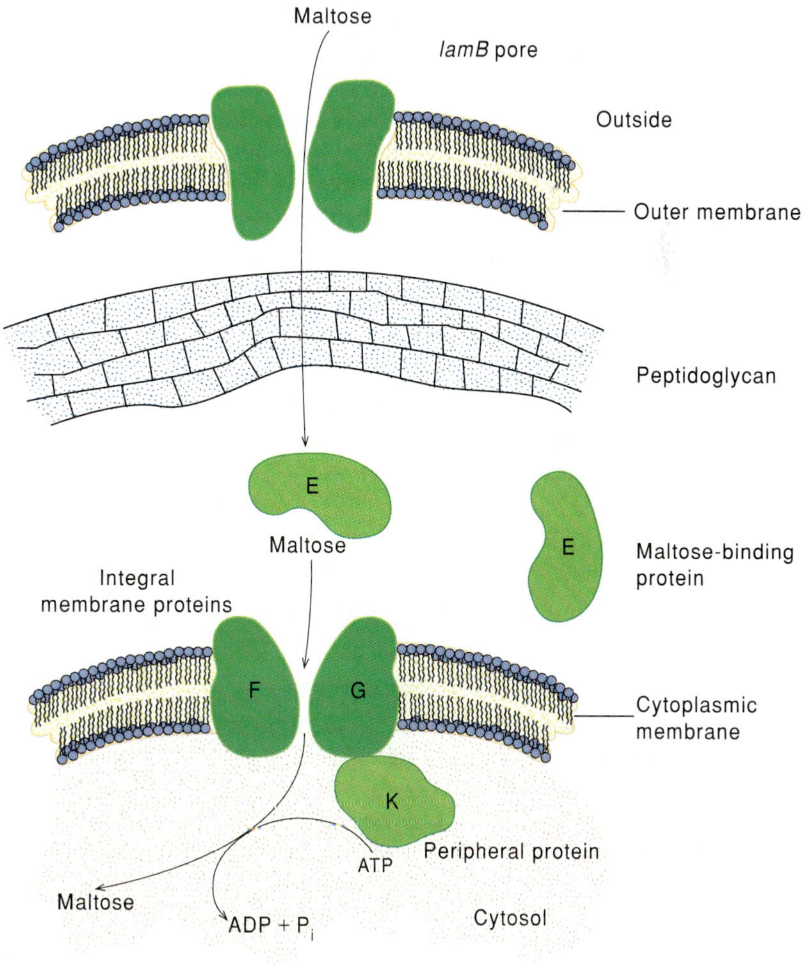

Figure 32.19

Schematic illustration of solute transport by the binding-protein systems of Gram negative bacteria using the maltose system as an example. Maltose passes through a pore in the outer membrane (the *lamB* gene product, which is also the receptor for bacteriophage λ), interacts with its periplasmic binding protein (E) and is then transported by inner membrane permeases (F and G) in a process dependent on ATP. The K protein is a peripheral membrane protein that is necessary for maltose transport, and is probably an ATPase (see text). The localizations and interactions of these proteins have been determined by biochemical and genetic analyses. The malE binding protein has been shown to interact both with the lamB protein pore in the outer membrane and with the maltose permease complex in the cytoplasmic membrane. It may act to facilitate and provide stereospecificity to the transport processes across both membranes.

two integral membrane proteins (F and G) span the membrane. While the E protein associates with the transmembrane transport components on the external surface, the K protein is localized to the inner membrane surface, probably owing to a specific association with the G protein (see fig. 32.19). The K protein is believed to function by energizing transport and also in the regulation of transport activity.

Energy for the active accumulation of binding-protein substrates is provided by ATP. Thus the peripheral membrane components of these transport systems that are present on the cytoplasmic side of the membrane (e.g., the MalK protein in figure 32.19) possess ATP-binding sites, and the energization of transport of both maltose and histidine has been shown to be fully provided by ATP in reconstituted systems *in vitro.* The mechanism whereby ATP hydrolysis allows the active accumulation of binding-protein substrates has yet to be determined, but again, protein conformational changes are undoubtedly involved.

A large number of homologous ATP-binding proteins are essential components of the periplasmic binding-protein-dependent permeases of bacteria. These structurally and functionally similar permeases are specific for such diverse substrates as peptides, amino acids, sugars, various intermediary metabolites, anions, and vitamins (as noted in table 32.3). Several of these systems have been characterized in some detail and, regardless of their specificities, they all have similar protein components, are similarly organized, and probably function by a common mechanism (see fig. 32.19).

Interestingly, ATP-binding proteins that are homologous to proteins comprising bacterial binding-protein systems have also been found in eukaryotic cells. The best characterized of these is the multidrug-resistance protein, or P-glycoprotein, of mammalian tumor cells. This protein functions to pump certain drugs out of tumor cells, and therefore the pump exhibits an opposite polarity for active transport as compared with the binding-protein systems. Other homologous proteins include the protein that is defective in patients suffering from cystic fibrosis (a genetic disease that is characterized by defective cellular Cl^- transport) and an export protein of yeast that mediates secretion of the mating pheromone, *a*-factor, from the cell. It is probable that all of these transport systems have a common evolutionary ancestor from which this class of transport proteins has diversified over many millions of years to accommodate many related functions involving primary active transport.

Secondary Active Transport Involves the Uptake of One Solute Coupled to the Transport of Another Solute

In chapter 15 we learned that a gradient of protons across a biological membrane is a form of potential energy that can be used to drive the synthesis of ATP from ADP and P_i. Expulsion of protons out of mitochondria or bacterial cells during electron transport results in both a difference in H^+ concentration (ΔpH) and in electrical charge ($\Delta\psi$; interior negative) across the membrane. Recall that both these components contribute to the so-called proton-motive force (pmf or Δp; see chapter 15), which can be expressed as

$$\Delta p = \Delta\psi - \frac{2.3RT}{F}\Delta\text{pH} \qquad \textbf{(9)}$$

The pmf, or Δp, is related to the free energy change experienced by protons, which tends to drive them across the cellular or organellar membrane. Since lipid bilayers are relatively impermeable to protons, such a flow must be mediated by membrane proteins. In the example of the F_1/F_0 ATPase, the exergonic flow of H^+ through the enzyme, driven by Δp, is used to drive the endergonic synthesis of ATP. It should also be apparent that Δp could be used to drive active transport if the inward flow of H^+ were coupled to the uptake of a particular solute. Examples of such proton cotransport or symport mechanisms are found in mitochondria, bacteria, and probably all eukaryotic cells. Because the primary active transport of H^+ is in part responsible for Δp, active transport systems depending on Δp for energization are said to carry out secondary active transport by means of chemiosmotic coupling.

The best-characterized H^+ symport system is the lactose transport system of *E. coli.* This system has been extensively studied by P. Mitchell, T. H. Wilson, and H. R. Kaback, among many others. Evidence that the active accumulation of this sugar by the *lacY* gene product, the lactose permease, involves the cotransport of H^+ can be summarized as follows:

1. Agents such as 2,4-dinitrophenol (see fig. 15.22), which collapse proton gradients across membranes, inhibit lactose transport in whole cells and membrane vesicles.
2. Uptake of lactose, or the nonmetabolizable analog thiomethyl-β-D-galactoside (TMG) is stimulated by reducing the extracellular pH of energy-depleted cells.
3. A 1:1 ratio of entry of TMG and H^+ has been demonstrated in energy-depleted cells.
4. Lactose transport in *E. coli* membrane vesicles, which cannot synthesize ATP in the absence of a source of ADP, is greatly stimulated by compounds such as D-lactate, which are oxidized and donate electrons to the electron-transport chain in this system.
5. Mutations in the *lacY* gene have been isolated in which lactose uptake is not tightly coupled to H^+ entry. These mutant proteins are severely impaired in the active transport of lactose, but have an increased ability to facilitate its diffusion across the membrane compared to the normal permease.
6. The lactose permease has been purified and the transport function of the pure protein has been reconstituted in a vesicular phospholipid membrane. In this system, the active accumulation of lactose in response to an artificially imposed Δp has been demonstrated. In fact, the highly purified, reconstituted permease was found to exhibit essentially the same kinetic properties as the transport system in membrane vesicles from *E. coli.* Thus a single protein, the lactose permease, is both necessary and sufficient for catalysis of lactose:proton symport.

Figure 32.20

Transport events carried out by the bacterial lactose permeases. (*a*) In response to a proton gradient, lactose is actively accumulated in the cell. (*b*) In the presence of a lactose gradient, protons can be concentrated in the cytoplasm. (*c*) In the presence of the $\Delta\psi$, both lactose and protons can be actively accumulated. The physiological situation in respiring cells corresponds to (*a*), in which a $\Delta\psi$ is usually also present.

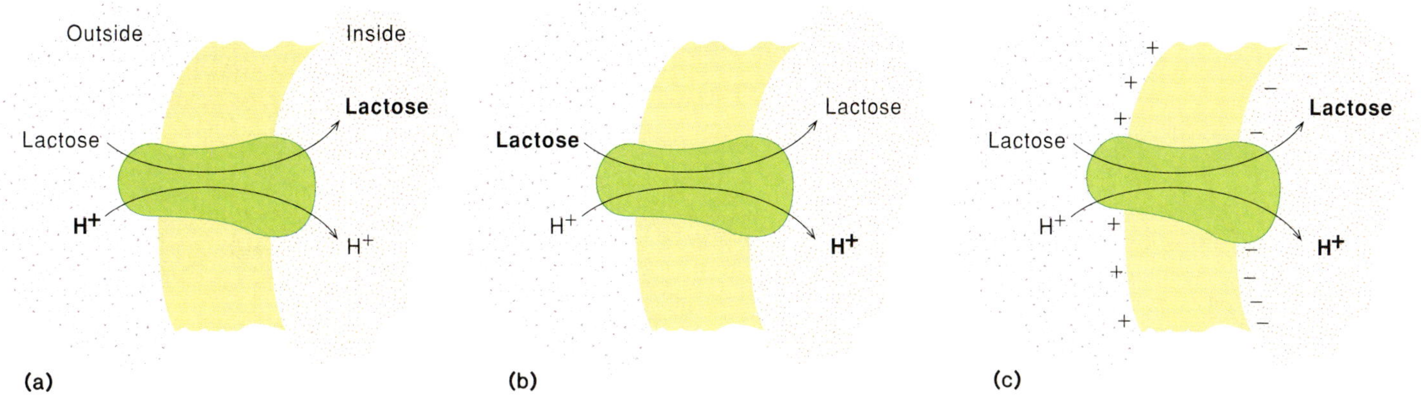

The preceding observations establish that the active accumulation of lactose in *E. coli* is obligatorily coupled to the simultaneous entry of H^+ flowing down its concentration and/or electrical gradients (fig. 32.20*a*). If this is true, then a gradient of lactose (or TMG) should produce a Δp in cells in which this value is initially zero. Indeed, it has been shown that addition of TMG to the outside of energy-depleted cells causes acidification of the cytoplasm as a result of H^+-TMG symport (fig. 32.20*b*).

The structure of the lactose permease is fairly well understood. The most significant advance leading to the elucidation of its structure resulted from the cloning and sequencing of the *lacY* gene, allowing deduction of the amino acid sequence of the lactose permease. Hydropathy analyses and other methods for studying membrane protein topography (see chapter 7) suggested that this protein consists of twelve extended hydrophobic segments, with a mean length of about 24 residues per segment. Both the C and N terminals appear to be localized to the cytoplasmic surface of the membrane. On the basis of circular dichroic measurements on the purified lactose permease protein, approximately 85% of the amino acid residues are probably arranged in helical secondary structure. Since about 70% of the 417 amino acid residues were found to be included within the twelve hydrophobic segments, it was predicted that the embedded segments are largely α-helical, with each helix extending the entire thickness of the membrane, as shown in figure 32.21.

As the result of both biochemical and molecular genetic approaches, the structure of the active site of the lactose permease is now coming to light. The galactoside-binding site resides within a segment of the protein that is embedded within the phospholipid bilayer. This conclusion was reached by the isolation and DNA sequence analysis of mutants of the *lacY* gene that encoded permease proteins with altered substrate specificities (e.g., that could recognize other disaccharides, such as maltose and sucrose, with a higher affinity than the wild-type protein). These studies revealed that Ala 177, Tyr 236, Thr 266, Ser 306, and Ala 389 of the native permease were likely to be at or near the substrate-binding site. Mutations in several other residues, including Arg 302, His 322, and Glu 325, have provided evidence that the side chains of these amino acids could possibly form a "charge-relay" system near the galactoside-binding site that could be responsible for the coupling of H^+ and lactose transport. This type of mechanism is reminiscent of the "proton wire" hypothesized for bacteriorhodopsin. It is reasonable to propose that the residues involved in sugar recognition/transport and in H^+ translocation might form a hydrophilic channel through the membrane. One such arrangement of six of the twelve transmembrane helices containing residues important for lactose permease function is shown in figure 32.22. Any such channel would have to be "gated" in the sense that lactose and H^+ translocation could normally occur only when both ligands are bound to the permease at the same time, presumably via a conformational change in the protein.

Transport systems such as the *E. coli* lactose permease, in which electrical charges are carried across the membrane without simultaneous compensation of the electrical potential so generated (for example, by the movement of other charged species across the membrane) are termed electrogenic. Electrogenic transport systems are always affected by the electrical potential $\Delta\psi$ across a biological membrane, even if the transported ion is not H^+. In the lactose transport system, a $\Delta\psi$ (interior negative) is sufficient to drive active accumulation of this sugar in the absence of a ΔpH because Δp has both $\Delta\psi$ and ΔpH components [equation (9)], as shown in fig. 31.19*c*. In animal cells and in some bacteria, such as *Halobacterium*, that live in high concentrations of Na^+, Na^+-solute symport systems are common means for the accumulation of salts, carbohydrates, and amino acids. Because these systems are also electrogenic, they are sensitive to $\Delta\psi$ and to $\Delta[Na^+]$, but not to ΔpH. For these systems

Figure 32.21

Secondary structure model of the lactose permease, based on the hydropathy profile of the protein. Hydrophobic segments are shown in boxes as transmembrane, α-helical domains connected by hydrophilic segments. The carboxyl terminus and hydrophilic segments 5 and 7 (with the amino terminus as hydrophilic segment 1) have been shown to be on the cytoplasmic surface of the membrane. Amino acids are represented by their single-letter codes. (Source: H. R. Kaback, "Use of site-directed mutagenesis to study the mechanism of a membrane transport protein," in *Biochemistry* 26:2071–2076, 1987. Copyright © 1987 American Chemical Society, Washington, D.C.)

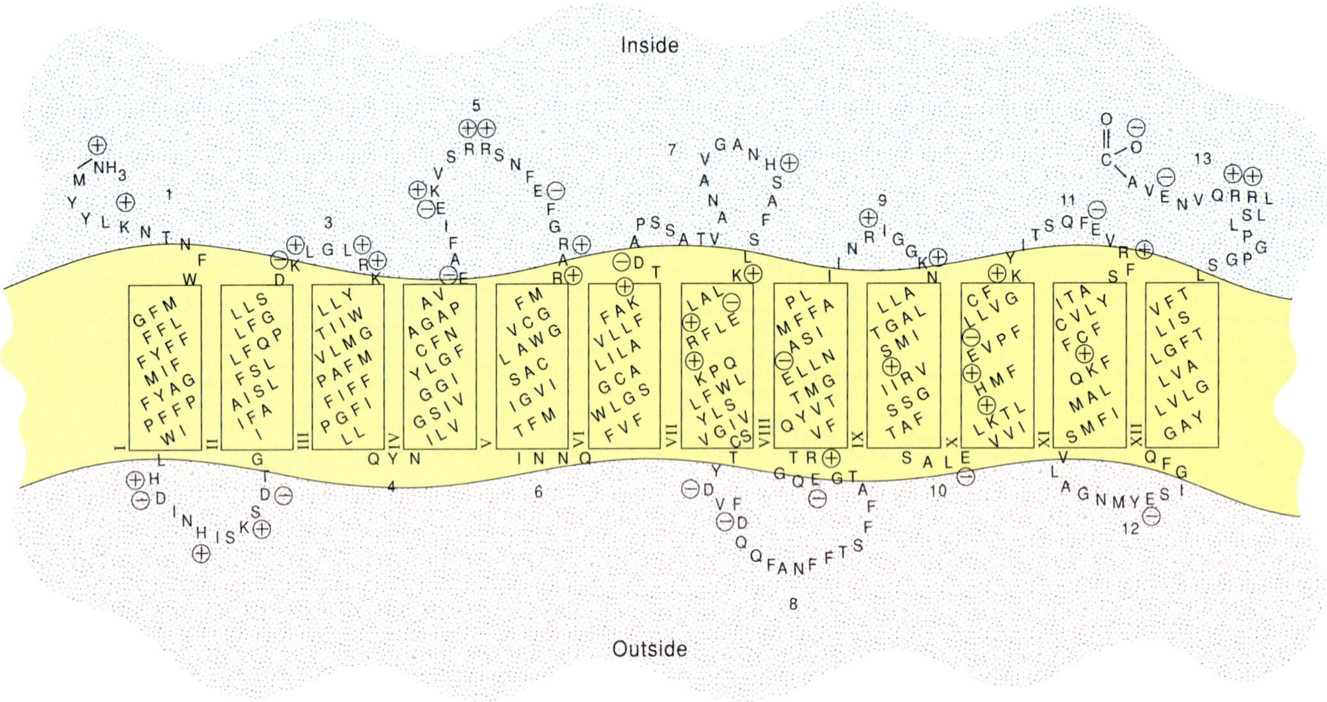

in animal cells, the Na^+-K^+ ATPase ensures that Na^+ flowing in through these Na^+-solute "symporters" is rapidly pumped back out to maintain the Na^+ gradient (fig. 32.23).

A list of some of the better-characterized secondary active transport systems is given in table 32.4. The amino acid sequences of many of the symporters listed are known. Some of these proteins show no homology to other known transport proteins or to each other (e.g., the *E. coli* lactose and melibiose permeases). On the other hand, the *E. coli* proton symporters for xylose and arabinose show some degree of homology to those for glucose, galactose, and maltose from yeast. Even more surprisingly, representatives of this latter group of permeases are homologous to parts of the glucose facilitators of human hepatocytes and rat brain cells. These observations suggest that the molecular mechanism of carrier-mediated facilitated diffusion may be similar to that of solute:cation symport except for the number of molecular species translocated in a single step (one versus two).

The Mitochondrial ATP/ADP Exchanger Is an Electrogenic Antiporter

Because ATP is synthesized within the mitochondria of eukaryotic cells, while metabolic processes that utilize ATP are largely confined to the cytoplasm or nucleus, a mechanism must exist for transport of this molecule across the inner mitochondrial membrane. The protein responsible for this function is an ATP/ADP exchanger that carries out exchange of intramitochondrial ATP for ADP formed from metabolic reactions in the cytoplasm. This antiport is electrogenic because, at pH 7, ATP molecules average about one more negative charge than do ADP molecules (fig. 32.24).

Despite the fact that the [ATP]/[ADP] ratio is usually higher in the cytoplasm than in the mitochondrion in actively respiring cells, the ATP/ADP exchanger still preferentially expels ATP from the organelle with the concomitant inward movement of ADP. This can be explained by the presence of a membrane potential $\Delta\psi$ (inside negative), resulting, in part, from the Δp formed across the mitochondrial inner membrane during respiration. This potential favors the outward transport of ATP and inward transport of ADP because this process results in the net translocation of one negative charge from inside to outside (see fig. 32.24).

The ATP/ADP exchanger, purified by M. Klingenberg and his associates, is a dimer of identical 30-kDa polypeptides. Each subunit is believed to span the membrane six times, and both the C and N terminals are localized to the cytoplasmic side of the membrane. Because the protein loses its binding affinity for nucleotides if it is dissociated into monomers, this dimer may be necessary for carrying out the exchange process. Biochem-

Figure 32.22

Hypothetical model for the tertiary structure of lactose permease. The model depicts twelve cylindrical transmembrane segments, in an α-helical conformation, connected by hydrophilic segments. Six of these helices are shown to form a hydrophilic channel through which protons and lactose might flow. Residues important for substrate binding and lactose: proton symport are shown projecting into the channel. Some of these include: Arg 302, His 322, and Glu 325, which have been proposed to be part of a charge-relay system for H^+ transport, and Ala 177, Tyr 236, Thr 266, Ser 306, and Ala 389, which are likely to be near the substrate-binding site. (Source: J. C. Collins et al., "Isolation and characterization of lactose permease mutants with an enhanced recognition of maltose and diminished recognition of cellobiose," in *Journal of Biological Chemistry* 264:14698–14703, 1989. Copyright © 1989 American Society for Biochemistry and Molecular Biology Inc., Bethesda, Md.)

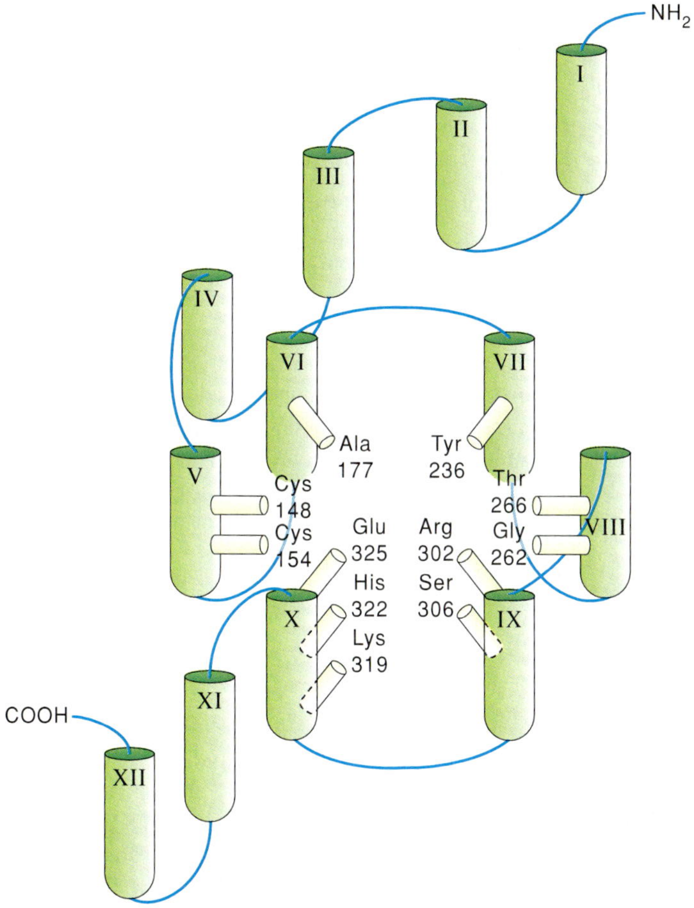

Figure 32.23

Na^+ symport is a common means of actively transporting sugars and amino acids in animal cells. For example, glucose is actively accumulated by this mechanism in intestinal and renal epithelial cells. The Na^+-K^+ ATPase maintains the Na^+ gradient essential to these secondary active transport systems.

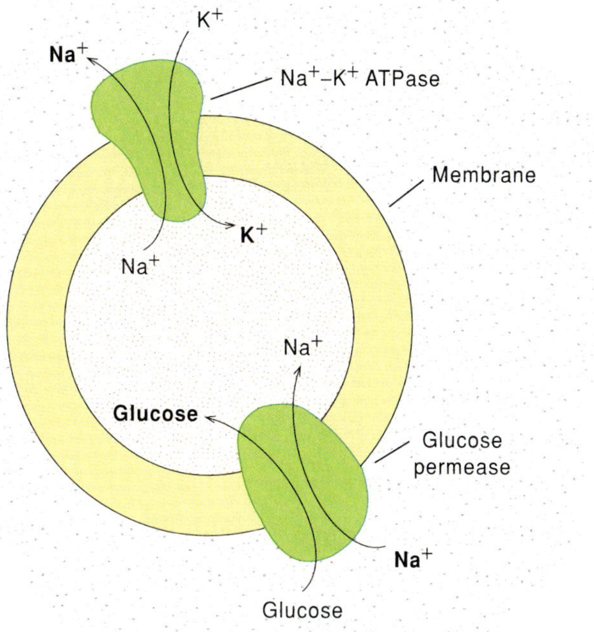

ical and kinetic studies on the ATP/ADP exchanger support a two-state "gated-pore" mechanism for this protein. The evidence for this mechanism can be summarized as follows:

1. There appears to be only one nucleotide-binding site per dimeric ATP/ADP exchanger.
2. The nucleotide-binding site has a higher affinity for ADP on the outer surface of the membrane than on the inner surface, while the opposite is true for ATP.
3. This site was shown never to be on both sides of the membrane simultaneously.
4. The respiratory poison atractylic acid binds to the form preferring ADP, while the antibiotic bongkrekic acid (see chapter 15) binds to the form that shows specificity for ATP.

Table 32.4
Some Secondary Active Transport Systems

Substrate	Cotransported Ion	Organism/Tissue
Neutral amino acids	Na^+	Eukaryotic cells
Glucose	Na^+	Some animal cells (intestine, kidney, choroid plexus)
Lactose	H^+	*E. coli* and some other bacteria
Dicarboxylic acids	H^+	*E. coli*
Proline	Na^+	*E. coli*
Glutamate	Na^+	*E. coli*
Melibiose	Na^+ or H^+	*E. coli*
Xylose	H^+	*E. coli*
Arabinose	H^+	*E. coli*
Glucose	H^+	Yeast
Maltose	H^+	Yeast
Galactose	H^+	Yeast

Figure 32.24

The ATP/ADP exchanger of the mitochondrial inner membrane. This dimeric protein carries out the exchange of intramitochondrial ATP for ADP formed in the cytoplasm by metabolic reactions. A membrane potential (interior negative) favors this exchange because of its electrogenic nature at pH values near neutrality.

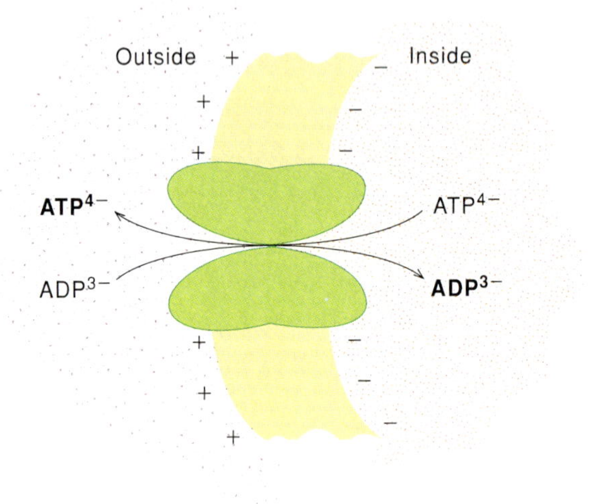

Figure 32.25

Molecular model for the functioning of the ATP/ADP exchanger of the mitochondrial inner membrane. The dimeric protein appears to have only one binding site for adenine nucleotides, perhaps involving amino acid residues from both subunits. When facing the outer surface of the membrane, this site has a high affinity for ADP (pentagon), and a low affinity for ATP (rectangle). The opposite specificities are observed when the site faces the lumen of the organelle. These states are interconvertible, even in the purified carrier, demonstrating that they arise from different conformations of the same protein. This model is analogous to that shown in figure 32.35 except that energy is apparently not required for the conformational interconversion. (Source: M. Klingenberg, "Membrane protein oligomeric structure and transport function," in *Nature* 290:449, 1981. Copyright © 1981 Macmillan Magazines Ltd., London, England.)

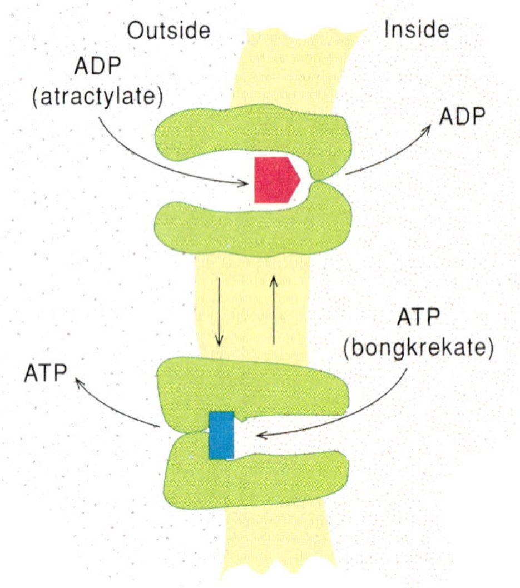

These results strongly support the transport model presented in figure 32.25. A single nucleotide-binding site per dimer is involved in transport by this protein. The configuration of this site, and thus its binding specificity, depends on the side of the membrane that it faces, and these two states are interconvertible through protein conformational changes that also result in translocation of any nucleotide bound to the protein. A conformational change can be triggered by binding of either ADP at the outside or ATP at the inside of the membrane surface. Once transported through the membrane, the nucleotide is now in a binding site that has little affinity for it, and it dissociates from the protein. The two forms of the adenine nucleotide exchanger appear to have quite different conformations indeed. Antibodies prepared against the atractylic acid-protein complex do not cross-react with the form that binds bongkrekic acid, and vice versa!

The Bacterial PEP:Sugar Phosphotransferase System Exemplifies Group Translocation

The final type of biological transport process that we will consider is group translocation. In this mechanism, chemical modification of the transported solute is tightly coupled to the translocation step (see fig. 32.7). The best-characterized example of group translocation is the phosphoenolpyruvate (PEP)-dependent sugar phosphotransferase system (PTS) in bacteria, which was discoverd by Saul Roseman and co-workers in 1964. The PTS both transports and phosphorylates sugars as they pass through the cytoplasmic membrane with PEP as the phosphoryl donor. It consists of several soluble and membrane-bound proteins in *E. coli* that catalyze the following set of reactions:

$$\text{PEP} + \text{enzyme I} \rightleftharpoons \text{enzyme I}{\sim}\text{P} + \text{pyruvate} \qquad \textbf{(10)}$$

$$\text{enzyme I}{\sim}\text{P} + \text{HPr} \rightleftharpoons \text{HPr}{\sim}\text{P} + \text{enzyme I} \qquad \textbf{(11)}$$

$$\text{HPr}{\sim}\text{P} + \text{sugar}^{a}_{(\text{out})} \xrightarrow[\text{(Enzyme III}^{a}\text{)}]{\text{Enzyme II}^{a}} \text{HPr} + \text{sugar}^{a}\text{-P}_{(\text{in})} \qquad \textbf{(12)}$$

In this reaction scheme, enzyme I and HPr are soluble, cytoplasmic proteins that participate in phosphoryl transfer reactions common to all sugars transported by the PTS in *E. coli*. In contrast, enzyme II is an integral membrane protein that is usually specific for only one sugar (as indicated by the superscript) and acts as the permease as well as the phosphoryl transfer enzyme. Enzyme III, which is required for the transport of some but not all sugars, also is sugar-specific. The spatial arrangement of the reactions that occur during PTS sugar transport is depicted in figure 32.26. The PTS allows for unidirectional sugar transport because the sugar-phosphate product is very impermeable to the lipid bilayer and is "trapped" once it is transported into the cell. The cost to the cell is one ATP equivalent (in the form of PEP) for each sugar molecule accumulated by means of the PTS.

Figure 32.26

Schematic representation of the reactions carried out by the bacterial sugar phosphotransferase system (PTS). A phosphoryl moiety from PEP is sequentially transferred to enzyme I, HPr, and enzyme III^{glc} before its ultimate transfer to glucose by an integral membrane protein specific for this sugar (enzyme II^{glc}). Although the general PTS enzymes, enzyme I and HPr, are soluble proteins, there is some evidence that they are associated peripherally with the membrane as shown. In *E. coli* there are at least seven different enzymes II, each specific for the sugars recognized by the PTS in this organism: glucose, mannose, fructose, *N*-acetylglucosamine, mannitol, glucitol, and galactitol. HPr has been shown to be a monomer (M_r = 9,500), enzyme I is a dimer (subunit M_r = 70,000), and at least some of the enzymes II are dimers. The mannitol-specific enzyme II (or mannitol permease) has been shown to consist of a single kind of polypeptide chain (M_r = 68,000).

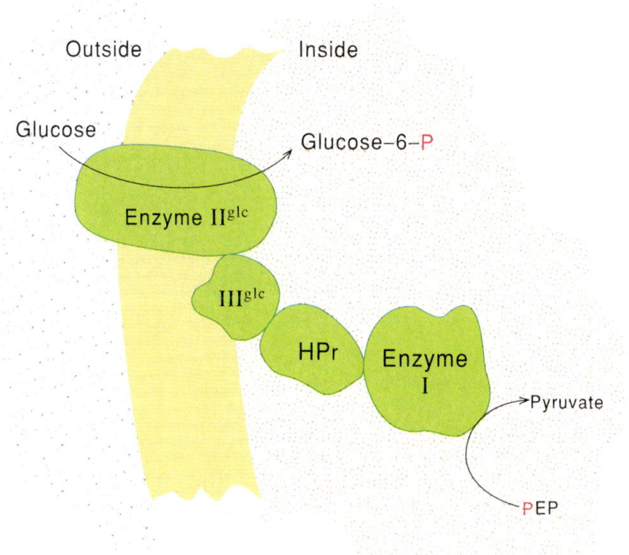

The phosphotransferase systems of *E. coli* and *Salmonella typhimurium* have been extensively studied. All the general proteins have been purified and shown to be transiently phosphorylated, as shown in reactions (10) through (12). The phosphate is attached to a single histidine residue in each of the soluble components (enzyme I, HPr, and enzyme III), as well as to one or two residues in each of the sugar-specific enzymes II, before being transferred to the incoming sugar. Such enzyme phosphorylation reactions may provide the driving force for transport by means of conformational changes in the enzymes II, as it apparently does for the ion-translocating ATPases.

One of the enzymes II, that specific for the sugar-alcohol mannitol, has been solubilized from the *E. coli* membrane, purified to homogeneity, and reconstituted in an artificial membrane. Additionally, the gene encoding the protein has been cloned and sequenced, so that like the lactose permease, the mannitol enzyme II is well understood structurally and functionally. The N-terminal half of the protein (M_r = 68,000) is strongly hydrophobic, traversing the membrane at least six times. In contrast, the C-terminal half of this protein exhibits the hydrophilic properties of a typical water-soluble protein and is localized to the cytoplasmic (inner) surface of the membrane. This particular protein, which functions without the aid of an enzyme III, is phosphorylated twice during catalytic turnover. A phosphoryl group is first transferred from phospho HPr to the N-3 position of a histidiyl residue in an enzyme III-like domain in the C-terminal part of the protein; next, the phosphoryl group is transferred to a cysteinyl residue also in the C-terminal half of the molecule; and finally, dephosphorylation of this residue is accompanied by the coupled phosphorylation and transport of the sugar across the membrane.

The DNA sequences of over a dozen genes encoding PTS permeases (enzymes II of enzyme II–III pairs) have been determined, and consequently the aminoacyl sequences of the encoded proteins are known. Computer analyses have revealed that most of these permeases are evolutionarily related, and show features that appear to be distinctive for the PTS permeases. For example, their N terminals can potentially form amphiphilic α helices with hydrophobic residues on one side of the helix and hydrophilic residues on the other side, a motif that may be important for insertion of these proteins into the membrane. In addition, site-specific mutagenesis studies and deletion analyses of some of these proteins have provided evidence for the involvement of specific residues in specific functions. Phosphoryl transfer between the individual PTS proteins/domains involves distinct catalytic residues, and as expected, the sugar-binding and translocation site is within the hydrophobic, transmembrane domains. These latter domains have been shown to exhibit similarities with some of the symporters and antiporters discussed previously in this chapter. Thus it is possible that the group-translocating permeases and the facilitators may share some common structural and mechanistic features.

Molecular Models of Transport Mechanisms

In the last section we examined the wide variety of mechanisms used by cells to accumulate and expel nutrients and ions. In order to fully understand transport at the molecular level, however, it is necessary to determine how the structures of transport permeases allow the often unidirectional translocation of hydrophilic molecules through the hydrophobic phospholipid bilayer. It is reasonable to assume that such proteins contain specific binding sites for their substrates by analogy with enzymes, and indeed this has been established for many transport permeases. Unlike most soluble enzymes, however, permeases also must be able to carry out a vectorial process: "picking up" a substrate on one side of the membrane and depositing it on the other. In no instance do we know the exact molecular details of how this process is accomplished. However, as more transport proteins are purified and their structures determined, clues are emerging that will undoubtedly allow this question to be answered for at least a few systems in the near future. In this section we will examine some of these clues and how they pertain to various models for the function of transport permeases.

Figure 32.27

In theory, a solute could be transported across a membrane by a mobile carrier that either rotates through or traverses the plane of the membrane (*a*), or by a more or less fixed channel or pore that may be controlled by a gate (*b*). Examples of both have been found among the ion-translocating antibiotics.

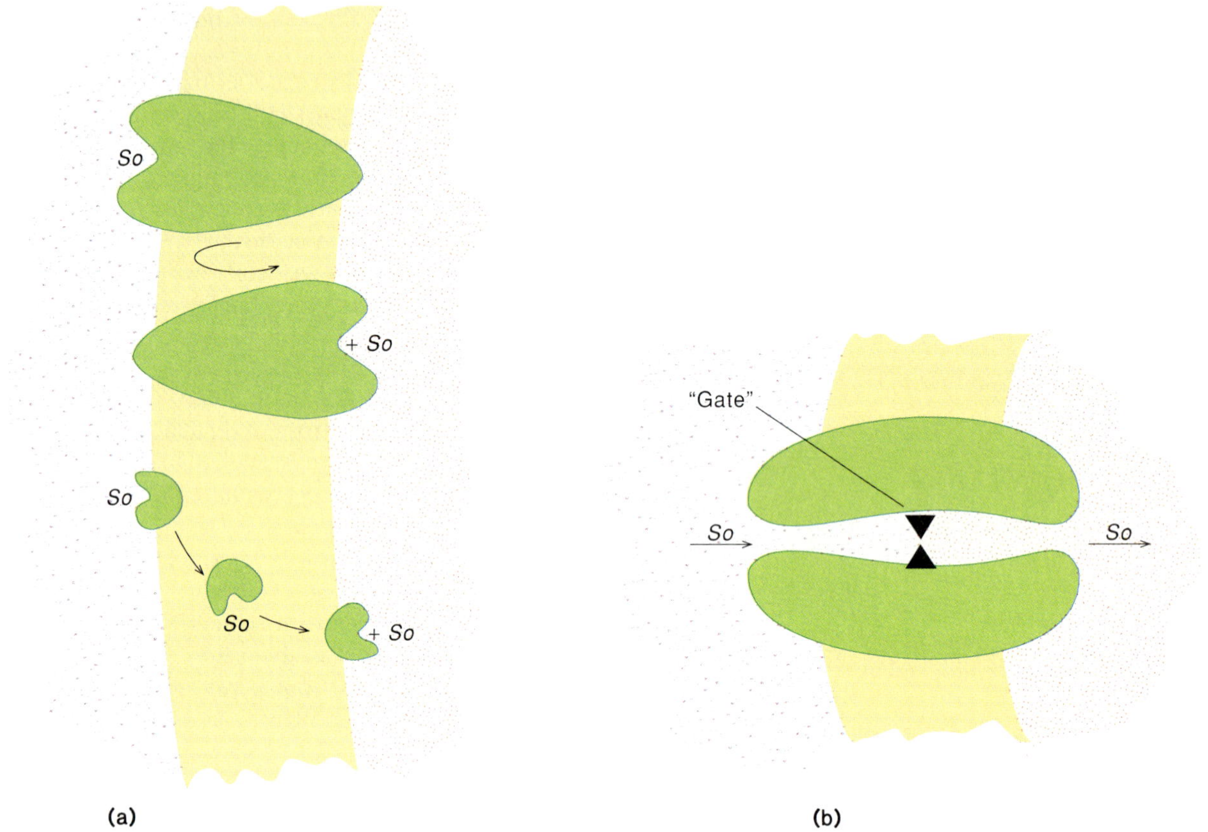

Mobile Carriers or Pores Could Explain Transport

At least two conceptual models could explain the transport of a substance through a biological membrane. In the mobile-carrier mechanism, the transporter or a solute-binding moiety of the permease is assumed to physically change its orientation in the membrane during translocation, either by shuttling back and forth across the bilayer or by rotation through the plane of the membrane (fig. 32.27*a*). On the other hand, the pore model assumes that the protein is more or less fixed in its intramembrane orientation and forms a hydrophilic pore that is stereospecific for its transport substrate. This pore is usually thought of as being gated in the sense that it opens only transiently in response to proper solute recognition (fig. 32.27*b*).

As model systems for studying biological transport, the ion-translocating antibiotics have provided evidence for both of the mechanisms illustrated in figure 32.27. Valinomycin, an antibiotic isolated from a species of the bacterium *Streptomyces,* is a cyclic peptide containing the sequence D-valine, L-lactate, L-valine, and D-hydroxyisovalerate repeated three times (see fig. 15.28). It is extremely selective for the binding of K^+ and makes both artificial bilayers and biological membranes permeable to this cation. This result can be explained by the fact that the valinomycin-K^+ complex effectively shields the hydrophilic groups of valinomycin on the interior of the molecule, while the hydrophobic side chains are exposed to solvent, as shown in figure 15.28. For this reason, the valinomycin-K^+ complex is soluble in lipid bilayers, while free K^+ is highly water-soluble. A second well-studied antibiotic is gramicidin A, also isolated from a bacterium (*Bacillus brevis*). It is a 15-amino-acid linear polypeptide that is a cation-specific ionophore with much less specificity than valinomycin for any particular cation (fig. 32.28).

The mechanisms of ion translocation carried out by both valinomycin and gramicidin A have been extensively studied. In one experiment, the conductance of an artificial phospholipid bilayer ($T_m = 41°$ C) to K^+ was measured as a function of temperature in the presence and absence of these antibiotics. The results showed that K^+ conductance was high in this model system throughout the temperature range studied with gramicidin A. In contrast, the K^+ permeability of the bilayer in the presence of valinomycin was low below 41° C, but sharply increased when the temperature was increased above this value.

Figure 32.28

The structure of gramicidin A, a linear antibiotic that facilitates the transmembrane transport of many cations.

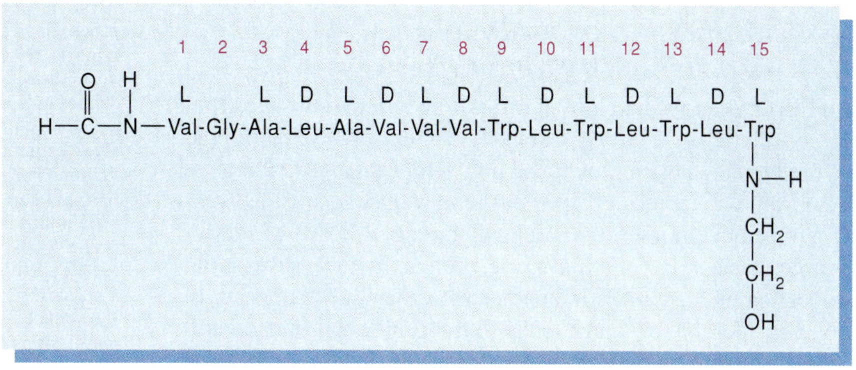

Figure 32.29

Mechanism of ion translocation by valinomycin. The carbonyl oxygens chelating K^+ are shown as "spokes" on the circular antibiotic molecule. In the aqueous phase, these oxygens are more freely available to the solvent; i.e., valinomycin undergoes a conformational change upon going from a lipid to an aqueous environment, as illustrated in the top of the figure. The lipid-soluble form is the one that complexes K^+ and traverses the membrane by diffusion, as shown in the lower part of the figure. (Source: Y. A. Ovchinnikov, "Physico-chemical basis of ion transport through biological membranes: Ionophores and ion channels," in *European Journal of Biochemistry* 94:321, 1979. Copyright © 1979 Springer-Verlag, Berlin, Germany.)

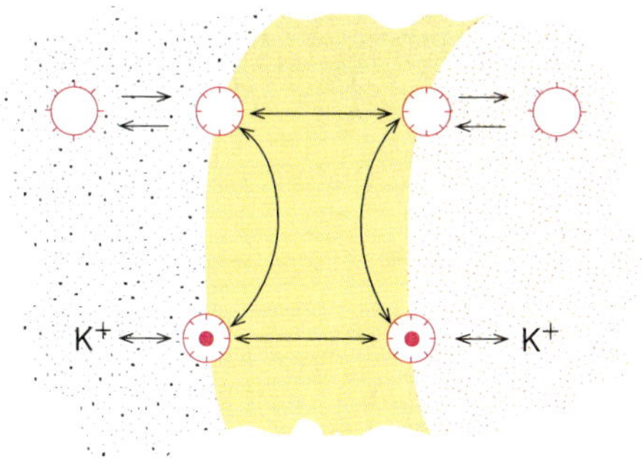

Hence ion translocation by valinomycin apparently depends on the physical state of the phospholipid bilayer, while the permeability induced by gramicidin A is insensitive to this parameter. This and many other experiments have led to the conclusion that valinomycin is a mobile carrier that can rapidly diffuse through the bilayer, while gramicidin A forms a static pore through the bilayer.

The mechanisms of ion translocation by valinomycin and gramicidin A as they are believed to occur are illustrated in figures 32.29 and 32.30. A single valinomycin-K^+ complex traverses the bilayer by simple diffusion. In contrast, physical measurements have shown that gramicidin A forms a transmembrane pore by means of head-to-head dimerization of helical monomers, each of which has a hydrophilic aqueous channel through the axis of the helix. The structure of each monomer is not an α helix because of the alternating D- and L-amino acid isomers in the chain. Rather, it is believed to be a so-called β helix as depicted in figure 32.30. Thus, at least in these model systems, both mobile-carrier and pore mechanisms of solute translocation can be readily demonstrated.

Figure 32.30

Model for the gramicidin A transmembrane channel. Gramicidin A molecules in a head-to-head dimer form a left-handed helix with a hydrogen-bonding pattern similar to that found in a parallel ß-pleated sheet. This hydrogen-bonding pattern is possible because of the alternating D and L configurations of the amino acid residues. The hollow, 4-Å core of the helix created by this structure is big enough to permit transport of alkali metal cations. (From Donald Voet and Judith G. Voet, *Biochemistry*. Copyright © 1991 John Wiley & Sons, Inc., New York. Reprinted by permission.)

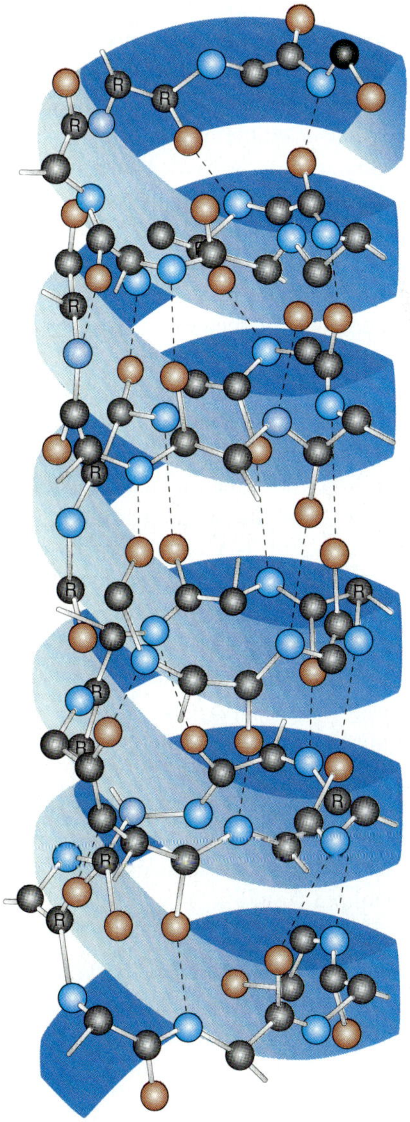

Most Biological Transport Systems Use Pores, Although Elements of the Mobile-Carrier Model May Also Apply in Some Cases

Studies on the topographies of integral membrane proteins that we considered in chapter 7 suggest a more or less static and asymmetric disposition of these polypeptides across biological membranes. Furthermore, any large movements, such as rotations through the membrane plane (see fig. 32.27*a*) are unfavorable, especially for membrane proteins with a substantial proportion of their mass in contact with the aqueous phase. In fact, considerable evidence leads to the conclusion that several classes of proteins that carry out transport do so by means of transmembrane pores or channels. These structures may or may not be static and/or gated. Nonspecific transport, where solute recognition is minimal, is presumably mediated by relatively static pore structures. By contrast, stereospecific solute permeation may involve conformational changes in the permease that transiently "open" the channel in response to binding of the solute recognized by the transport system, by energy-coupling mechanisms, or by both.

In most cases for which detailed structure-function information is available, the evidence is in favor of a pore-type mechanism for biological transport proteins, although elements of the mobile-carrier mechanism may also apply in some cases. In several instances, presumptive transmembrane pores formed by an integral membrane protein have been visualized by electron microscopy. The outer membrane of *E. coli* has been shown to be permeable to molecules with molecular weights below about 600. This permeability is conferred by a class of proteins called porins, which are integral constituents of the outer membrane of Gram negative bacteria, as well as of the outer membranes of mitochondria and chloroplasts.

In *E. coli,* porin molecules are arranged in a hexagonal lattice of trimers in the outer membrane, and optical filtration and computer processing of many negatively stained preparations viewed in the electron microscope give images that have been interpreted as representing transmembrane pores (fig. 32.31). Very recently, the three-dimensional structure of a similar porin molecule from the outer membrane of *Rhodobacter capsulatus* was solved at 3-Å resolution by x-ray crystallography. This structure gives definitive evidence for a pore through the center of each porin monomer as shown in fig. 32.32. Both the *E. coli* and *R. capsulatus* porins contain pores with a diameter of about 1 nm, which is consistent with the molecular weight exclusion limit given earlier. Unlike most transmembrane proteins studied to date, the membrane-spanning regions of porins consist almost entirely of β-type structures (β sheet and β barrel) as also shown in figure 32.32. That these molecules function as relatively nonspecific static pores has been shown by reconstitution experiments. When incorporated into artificial phospholipid vesicles, porin molecules allow rapid diffusion of most small hydrophilic molecules through the bilayer, and thus solute recognition by these proteins is minimal.

Figure 32.31

Arrangement of porin "pores" in the outer membrane of *E. coli.* Computer processing of many electron micrographs gives the image shown in (*a*), while a technique involving optical diffraction and filtering of micrographs gives the image shown in (*b*). A schematic interpretation of these images, shown in (*c*), suggests "triplet indentations" that are penetrated by the negative stain (dark circles) and which are the presumptive pores. A, B, and C are centers of local threefold symmetry and *L*, the lattice constant, is 7.7 nm. The center-to-center spacing *D* of the pores is 3 nm, and each is partially surrounded by a kidney-shaped region that excludes the negative stain and is probably part of a single molecule of porin. The view is perpendicular to the plane of the outer membrane. (From A. C. Steven, B. ten Heggeler, R. Müller, J. Kistler, and J. P. Rosenbusch, "Ultrastructure of a periodic protein layer in the outer membrane of *Escherichia coli*." Reproduced from the *Journal of Cell Biology*, 1977, 72:292, by copyright permission of the Rockefeller University Press.)

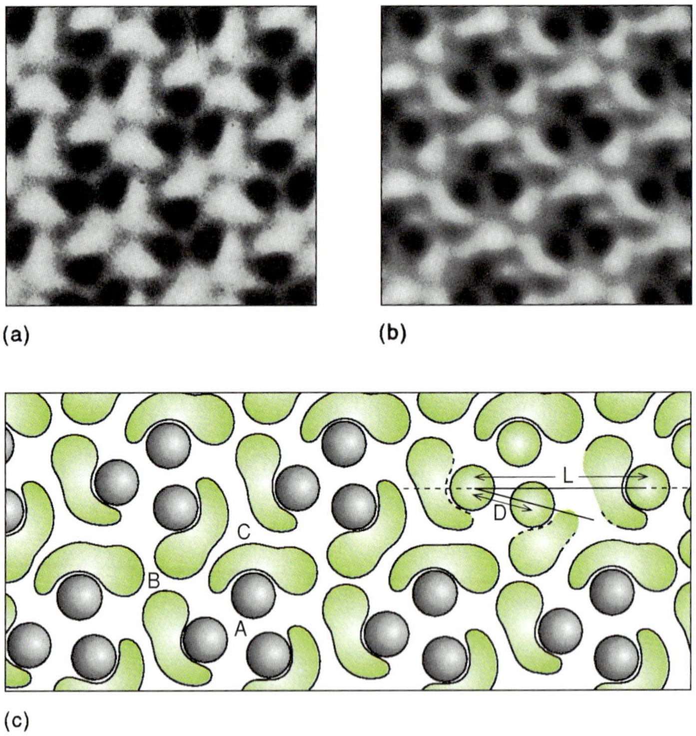

Nonspecific transmembrane pores also have been demonstrated in the plasma membranes of eukaryotic cells. Well-studied examples of these pores are the so-called gap junctions, which connect neighboring cells of similar types in tissues such as liver, intestine, kidney, brain, and cardiac muscle. In the region of the gap junction, the plasma membranes of two different cells are closely apposed and appear to have channels connecting them (fig. 32.33*a*). In the surface view of gap junctions, the region appears as a hexagonal lattice of protein hexamers, each of which seems to form a hydrophilic pore (fig. 32.33*b*). Ultrastructural studies have produced detailed models of the structure of the gap junction, one of which is shown in figure 32.34. The protein hexamers in each membrane are thought to span the bilayer, and apposition of two hexamers forms an aqueous channel between the cells.

Figure 32.32

Structure of the polypeptide backbone of the outer membrane porin of *Rhodobacter capsulatus*. (*a*) Side view of one subunit of the porin trimer. Note the extensive ß structures running perpendicular to (arrow) and at slight angles to the plane of the membrane. (*b*) Porin trimer viewed from the top (outside) of the membrane, showing how the ß structures form a type of ß barrel within which each pore resides. (From M. S. Weiss, T. Wacher, J. Weckesser, W. Welte, and G. E. Schultz, The three-dimensional structure of porin from *Rhodobacter capsulatus* at 3-Å resolution, *FEBS Lett.* 267:268–272, 1990.)

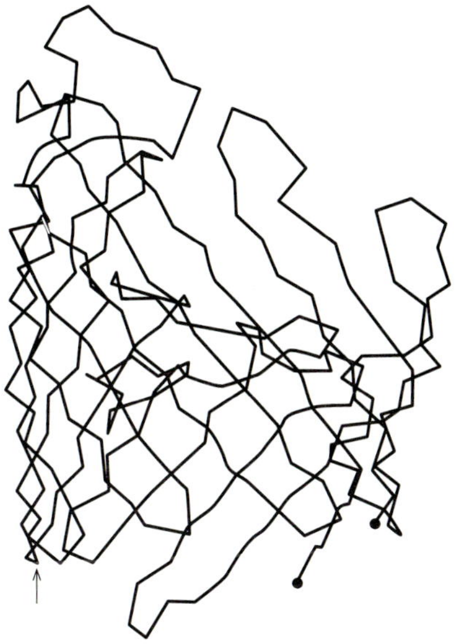

(a)

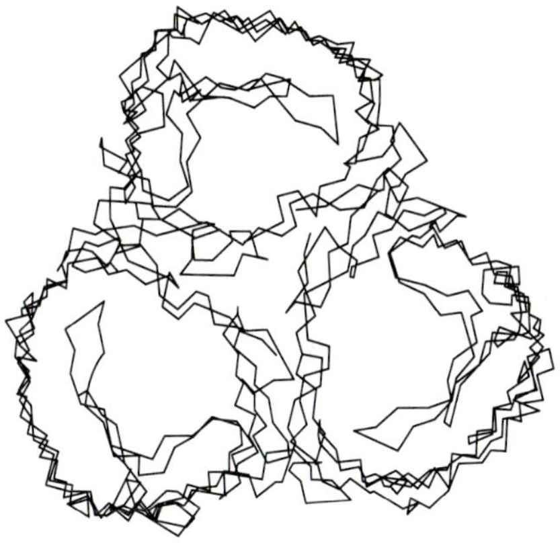

(b)

Figure 32.33

Isolated gap-junction sheets viewed in the electron microscope (*a*) parallel and (*b*) perpendicular to the planes of the two apposed membranes. The view in (*a*), part of which is outlined within a rectangle, arose from a curling up of a sheet on its edge and shows links (gaps or pores) penetrated by negative stain bridging the space between the two cell membranes. The perpendicular view in (*b*) shows annuli (rings), each composed of a protein hexamer surrounding a pore that has been filled with negative stain. The specimens were stained with uranyl acetate (300,000×). (From P. N. T. Unwin and G. Zampighi, "Structure of the junction between communicating cells." Reprinted by permission from *Nature* 283, 545; © 1980 Macmillan Magazines Limited.)

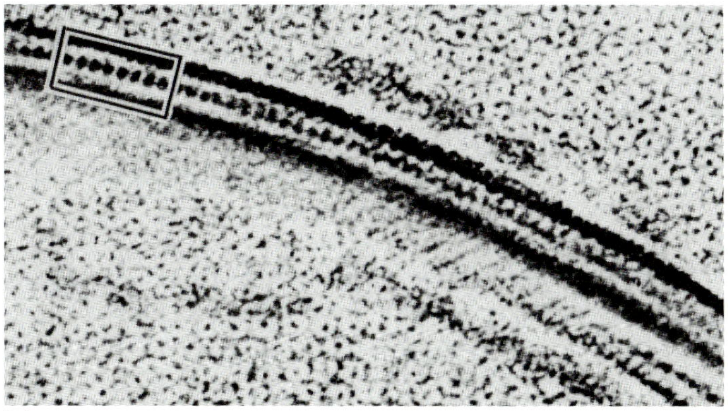

(a)

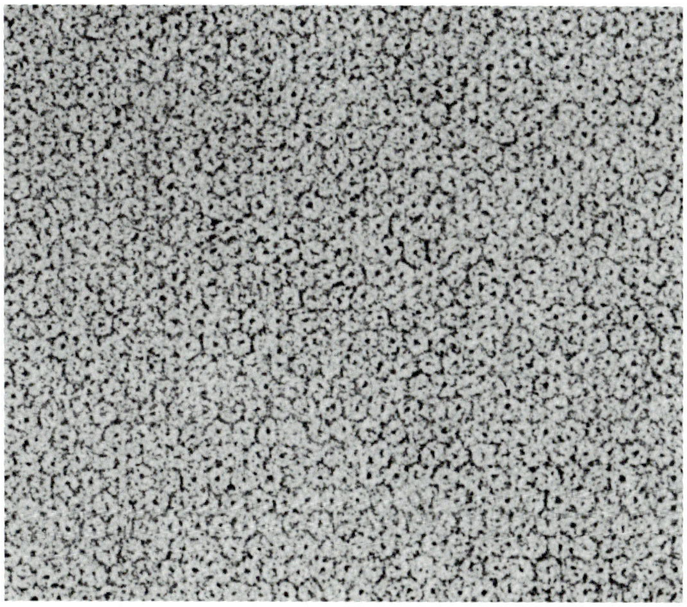

(b)

Figure 32.34

Molecular model of gap-junction pores as inferred from chemical, electron microscopic, and x-ray diffraction studies. A cross section, nearly perpendicular to the two membrane planes, is shown. The arrows, representing aqueous channels, are drawn through cross sections of two pores, formed by apposition of two hexamers (or *connexons*), each spanning its own lipid bilayer membrane. The end-on views of each connexon at the right and bottom of the illustration show that they protrude beyond the lipid bilayer on both sides of the plasma membrane. (Source: L. Makowski et al., "Gap junction structures. II. Analysis of the x-ray diffraction data," in *Journal of Cell Biology* 74:629, 1977. Copyright © 1977 Rockefeller University Press, New York, N.Y.)

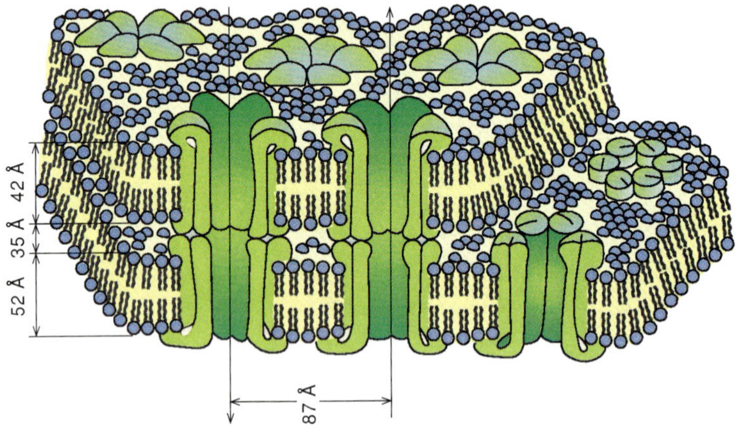

Figure 32.35

Generalized molecular model for transport involving a conformational rearrangement of subunits of an oligomeric carrier protein. In active transport, energy (such as that from the hydrolysis of ATP) could be used to effect this conformational change as shown. Alternatively, binding of the transport substrate could be enough to trigger this rearrangement in systems carrying out facilitated diffusion. (Source: S. J. Singer, "The molecular organization of membranes," in *Annual Review of Biochemistry* 43:805, 1974. Copyright © 1974 Annual Reviews Inc., Palo Alto, Calif.)

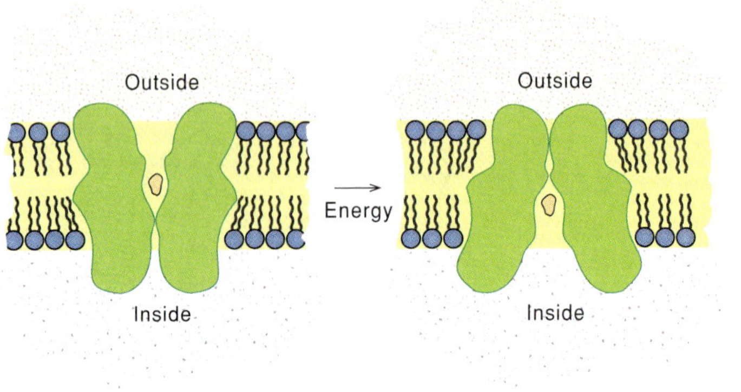

Gap junction pores have been shown to be permeable to hydrophilic molecules with masses as large as 1,000 to 2,000 daltons, depending on the cell type. They are believed to play roles in conducting electric impulses through particular cell types (nerve, muscle, and epithelia), as well as in providing avenues for the free flow of small metabolites between cells. The permeability of gap junction pores is regulated by the cytoplasmic concentration of Ca^{2+}; low concentrations ($<10^{-7}$ M) lead to open channels, while higher concentrations tend to close the channels in a graded manner. This sensitivity to $[Ca^{2+}]$ may play a role in the regulation of intracellular communication between cells connected by gap junctions.

Although porins and gap junctions are relatively nonspecific aqueous channels in biological membranes, it is reasonable to imagine that solute-specific pores can be formed by permease proteins as a result of their binding specificities. A number of workers have suggested similar models for how more specific permeases recognize and translocate their substrates. Aqueous channels between or within transmembrane subunits of transport proteins may respond to solute binding by undergoing conformational changes that alter the relative position of the solute-binding site with respect to the membrane and thereby, at least transiently, open the channels (fig. 32.35). This conformational change might be triggered solely by substrate binding, in which case facilitated diffusion would be the consequence. Alternatively, metabolic energy might expedite this process in one direction or the other, in which case active transport could result. Although conformational changes resulting in solute translocation are most easily envisioned between subunits of an oligomeric protein as depicted in figure 32.35, there is also evidence that at least some permeases may accomplish this process in a monomeric form.

A Unifying Model for Transmembrane Transport

Although, as we have seen, there exists a wide variety of seemingly divergent transport mechanisms and permeases, there appear to be a few general principles which can be used to develop a unifying concept for transmembrane transport:

1. The translocation domain of a transport protein is a hydrophilic channel, or pore, at least in part formed by apposition of a number of transmembrane α helices (usually), or in the case of porins by transmembrane β structures.
2. This channel is usually gated for permeases recognizing specific substrates, such that the bound transport substrate is in equilibrium with the aqueous phase on only one side of the membrane at any given moment.
3. Translocation of the transport substrate involves a conformational change in the protein, exposing the binding site to the other membrane surface.
4. In facilitated diffusion, the conformational change is induced by substrate binding and serves only to equilibrate the concentrations of the transport substrate on both sides of the membrane. In this case, the binding affinity of the substrate to the permease would be approximately the same for binding sites exposed on either side of the membrane.

5. In active transport or group translocation into the cell, the binding affinity of the transport substrate is higher when the binding site is exposed extracellularly than when it is exposed intracellularly. This difference allows for binding of the substrate on the extracellular surface (and subsequently transport), even when there is a much higher concentration of the transport substrate inside the cell or organelle. Translocation of the "loaded" binding site to the inside surface therefore requires energy to disrupt the favorable interactions of substrate and permease as the binding site changes from a high-affinity to a low-affinity form. This energy may be provided by light (bacteriorhodopsin), ATP hydrolysis (ion-translocating ATPases; bacterial binding-protein transport systems), or cotransport of an ion down its electrochemical gradient (lactose permease).

These principles, and the unifying model derived from them, are depicted in figure 32.36. This model is consistent with most of the available evidence concerning the structures and mechanisms of the best-studied transport permeases and is also consistent with the evidence for evolutionary relatedness among even seemingly diverse transport systems. Although the detailed molecular mechanism will differ for each type of permease, the overall mechanisms are likely to be similar. For example, for bacteriorhodopsin the channel need only be very narrow (the proton "wire") and is probably gated by a rather subtle conformational change initiated by the light-induced isomerization of retinal. In this case, it would be predicted that proton pumping by bacteriorhodopsin would be rather insensitive to the physical state (gel or liquid crystalline) of the membrane, and this fact has been confirmed experimentally. On the other hand, conformational changes concomitant with translocation of larger solutes, such as nucleotides or sugars, would probably be more substantial to move the larger binding site within the channel to the other side of the membrane. This has been observed directly with the ATP/ADP exchanger as discussed earlier, and is consistent with the fact that the transport activity of the lactose permease is much more sensitive to temperature than is that of bacteriorhodopsin. Thus, although the pore model best describes most features of transport permeases, at least the substrate-binding sites of some of these proteins exhibit mobility through the permeability barrier of the membrane through protein conformational changes.

Perspectives for the Future

The advent of powerful techniques in molecular genetics has greatly increased our understanding of the probable structures of transport permeases during the past decade. These approaches, coupled with more classical biochemical and biophysical techniques, will undoubtedly continue to provide valuable insights into structure-function relationships in these proteins. Since methods have now been developed to crystallize membrane proteins, three-dimensional structures of transport permeases at atomic resolution will begin to emerge in the near future. The prospects are therefore bright that before the end of the century a combination of all of these techniques will allow us to determine, at a molecular level, the transport mechanisms of some of the better-characterized transport permeases described in this chapter. The recent recognition that most of these proteins are probably arranged similarly in membranes from a wide variety of sources will greatly simplify this task. The goal of understanding transmembrane processes vital to all cells and organisms, including nutrient transport, energy interconversion, photosynthesis, and intercellular communication, is clearly within reach.

Figure 32.36

Models for facilitated diffusion (*a*) and active transport (*b*), incorporating many of the general features of most transport permeases. The substrate-binding site is comprised of transmembrane α helices (for clarity, only two are shown) with hydrophilic side chains projecting into the channel. These side chains interact with the substrate (filled triangle) when it binds to the outside of the membrane system (red lines). In facilitated diffusion, the affinity of the permease for the substrate is approximately the same on both sides of the membrane (*a*). In active transport, energy is used to drive the permease into a conformation in which the binding site "moves" across the membrane and simultaneously adopts a conformation with only a low affinity for the substrate (*b*). The situation depicted in (*b*) can lead to substrate accumulation.

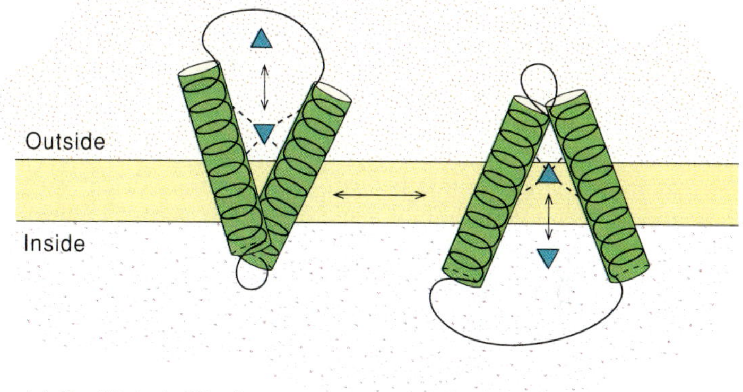

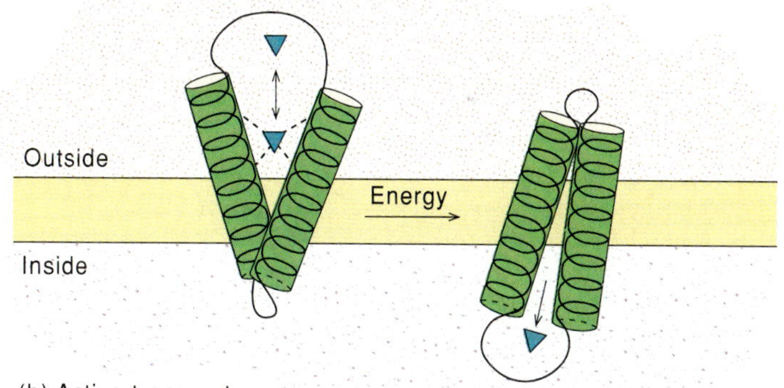

Summary

In this chapter we have considered the theory, energetics, and mechanisms of transport in cells. The chief points that we discussed are as follows.

1. Cells and organelles must be capable of rapidly transporting hydrophilic solutes across the hydrophobic membrane barrier that bounds them. This process can occur by simple diffusion or can be mediated by a hydrophilic transmembrane pore or carrier. In simple diffusion, the solute must be appreciably soluble in the domain of the bilayer. The kinetics of its movement across the membrane obey Fick's law, and the net transport rate in either direction is directly proportional to the concentration gradient of the solute. In carrier-mediated transport, the transport rate is saturable with respect to solute concentration because only a finite number of permeases exist in the membrane. The kinetics of this process are described by the Michaelis-Menten equation when initial transport rates are measured. Thus kinetic measurements frequently distinguish between these two mechanisms, although in many cases cells may use more than a single system to transport a given solute.
2. Net transport of a molecule across a biological membrane in a given direction will occur spontaneously only if ΔG for the process is negative. For an uncharged molecule, this means that in the absence of energy input, the compound will reach equal concentrations on both sides of the membrane barrier at equilibrium. Simple and facilitated diffusion, therefore, can lead only to equilibration of an uncharged solute across the membrane. However, if the molecule being transported has a net charge, unequal concentrations of it can occur at equilibrium without the input of additional energy if there is an independently maintained electric potential across the membrane. Both simple and facilitated diffusion are therefore examples of energy-independent biological transport mechanisms.
3. Many cellular transport systems, however, can carry out the active accumulation of metabolites and ions. Because ΔG for solute accumulation is positive, the process must be coupled to an exergonic event in which ΔG is negative. Primary active transport systems may use the chemical energy of ATP, light energy, or electron flow to drive the uphill concentration of the solute. In secondary active transport, ion gradients, which themselves are maintained by primary active transport processes, are used to drive the active accumulation of a second solute. Finally, in group translocation, the transported substrate is chemically modified by the membrane permease during the transport event. This modification may result in the expenditure of metabolic energy, as in the case of many bacterial sugar transport systems, and may provide the driving force for solute accumulation.
4. Primary active transport systems that use ATP as a source of energy have been identified in both eukaryotic and prokaryotic cells. The Na^+-K^+ ATPase of animal cells is an asymmetrically oriented transmembrane protein that maintains high intracellular levels of K^+ by carrying out the active accumulation of this ion and the concomitant extrusion of Na^+. A second, well-studied ion-translocating ATPase, the Ca^{2+} ATPase of sarcoplasmic reticulum, catalyzes the active accumulation of Ca^{2+} into the lumen of this organelle by a similar mechanism. The F_1/F_o proton-translocating ATPase found in bacteria and eukaryotic organelles couples H^+ translocation with ATP synthesis or hydrolysis, thereby allowing the interconversion of chemiosmotic and chemical energy. At least four evolutionarily distinct classes of ion-translocating ATPases are now believed to exist. Other primary active transport systems include the ATP-binding cassette (ABC-type) solute transporters, the H^+-translocating carriers of electron-transport chains, and bacteriorhodopsin.
5. Na^+ and proton gradients across biological membranes are forms of potential energy that can be used to drive secondary active transport. In this type of transport, the flow of one of these cations down its electrochemical gradient can be coupled by cotransport, or symport, with the transport of another molecule against its concentration gradient. An example of this type of mechanism is the H^+-lactose symport system of *E. coli* and other bacteria.
6. The ATP/ADP exchanger found in the inner mitochondrial membrane represents a case in which membrane potential can drive the transport of a negatively charged molecule. The phosphoenolpyruvate (PEP)-dependent sugar phosphotransferase system (PTS) in bacteria is the best-studied group translocation system. From both sequence comparisons and structural analyses, it appears that many of these seemingly diverse transport permeases are structurally and mechanistically related.
7. A membrane permease could function mechanistically as a mobile carrier or a relatively static pore. Examples of both types of transport processes are found among the ion-translocating antibiotics. Examples of nonspecific pores in cells include the porins of Gram negative bacteria and gap junctions in animal cells. Specific transport permeases probably function as gated pores or channels in which substrate binding, and energy in active transport, serve to "move" the substrate-binding site from one side of the membrane to the other. These proteins therefore exhibit certain characteristics of both the pore and mobile-carrier models. From the information now available, it appears that a unifying model may explain transmembrane transport regardless of the many differences in detail of biological transport systems. The future appears bright for an understanding of biological transport mechanisms at the molecular level.

Selected Readings

Ames, G. F.-L., C. S. Mimura, S. R. Holbrook, and V. Shyamala, Traffic ATPases: A superfamily of transport proteins operating from *E. coli* to humans. *Adv. Enzymol.* 65:1–47, 1992. Excellent review of binding-protein-dependent transport systems in bacteria and related eukaryotic transport proteins that are also ATPases.

Boyer, P. D., A perspective of the binding change mechanism for ATP synthesis. *FASEB J.* 3:2164–2178, 1989. Reviews evidence for the mechanism of H^+-coupled ATP synthesis by the F_1/F_o ATPase.

Bronner, F., and A. Kleinzeller (eds.), *Current Topics in Membranes and Transport.* New York: Academic Press. Continuing series reviewing some of the current problems in biological transport.

Dharmavaram, R. M., and J. Konisky, Characterization of a P-type ATPase of the archaebacterium *Methanococcus voltae. J. Biol. Chem.* 264:14085–14089, 1989. Recent discovery of a P-type ATPase in an archaebacterium, which may have significant evolutionary implications.

Friedlander, M., and M. Mueckler (eds.). Molecular biology of receptors and transporters, *Int. Rev. Cyto.* 137A, 1992. This recent volume contains 8 review articles on various types of transport systems in both eucaryotic and procaryotic cells.

Gennis, R. B., *Biomembranes: Molecular Structure and Function.* N.Y.: Springer-Verlag, 1989. Excellent book dealing with both membrane protein structure and function.

Higgins, C. F., M. P. Gallagher, M. L. Mimmack, and S. R. Pearce, A family of closely related ATP-binding subunits from prokaryotic and eukaryotic cells. *Bioessays* 8:111–118, 1988. An interesting summary of recent evidence for the structural and functional relatedness of seeming divergent proteins that bind ATP.

Inesi, G., D. Lewis, D. Nikic, A. Hussain, and M. Kirtley, Long-range intramolecular linked functions in the calcium transport ATPase. *Adv. Enzymol.* 65:182–215. Recent review of the structure and mechanism of the Ca^{2+}-ATPase, with an emphasis on how Ca^{2+} transport and ATP hydrolysis may be linked.

Kaback, H. R., E. Bibi, and P. D. Roepe, β-Galactoside transport in *E. coli:* a functional dissection of lac permease. *Trends. Biochem. Sci.* 15:309–314, 1990. Structure-function relationships in the lactose permease.

Meadow, N. O., D. K. Fox, and S. Roseman, The bacterial phosphoenolpyruvate:glycose phosphotransferase system. *Ann. Rev. Biochem.* 59:497–542, 1990. Recent review of this carbohydrate transport system that is commonly found in anaerobic and facultatively anaerobic bacteria.

Saier, M. H., Jr., *Mechanisms and Regulation of Carbohydrate Transport in Bacteria.* New York: Academic Press, 1985. Reviews the various mechanisms of carbohydrate transport in bacteria, as well as transport regulation.

Silverman, M. Structure and function of hexose transporters. *Ann. Rev. Biochem.* 60:757–794, 1991. Excellent review of recent work on sugar transporters in eukaryotic cells.

Spudich, J. L., and R. A. Bogomolni, Sensory rhodopsins of halobacteria. *Ann. Rev. Biophys. Biophys. Chem.* 17:193–215, 1988. Review of structure and function of bacteriorhodopsin and other related proteins of halobacteria.

Problems

1. (a) Using Fick's law, show that the diffusion coefficient D has the dimensions of area per unit time.
   (b) The diameter of a pore channel is about 10^{-9} m and its length is 4×10^{-9} m. In planar membrane bilayers, glucose traverses this channel at the rate of about 50 molecules per channel per second at room temperature when the concentration of glucose is 3×10^{-6} M on one side of the membrane. Calculate the diffusion coefficient for glucose through the pore channels under these conditions.
2. The Nernst equation relates the electric potential $\Delta\psi$ resulting from an unequal distribution of a charged solute across a membrane permeable to that solute to the ratio between the concentration of solute on one side and on the other:

$$m\,\Delta\psi = \frac{-2.3RT}{F}\log\frac{[So]_1}{[So]_2}$$

   where m is the charge on the solute, $2.3RT/F$ has a value of about 60 mV at 37° C, and $[So]_1$ and $[So]_2$ refer to the concentrations of solute on either side of the membrane. Consider a planar phospholipid bilayer separating two compartments of equal volume. Side 1 contains 50-mM KCl and 50-mM NaCl, while side 2 contains 100-mM KCl.
   (a) If the membrane is made permeable only to K^+, e.g., by addition of valinomycin, what will be the magnitude of $\Delta\psi$?
   (b) If the membrane is made permeable to H^+ and K^+, in which direction will H^+ initially flow?
   (c) If the membrane could be made selectively permeable to both K^+ and Cl^-, what would be the value of $\Delta\psi$ and the ion concentrations on both sides of the membrane at equilibrium? (Hint: Initially, K^+ would diffuse down its concentration gradient accompanied by an equivalent amount of Cl^-. Equilibrium would be established when the potentials due to K^+ and Cl^- were equal to each other and to the overall membrane potential.)
3. Membrane vesicles of *E. coli* that possess the lactose permease are preloaded with KCl and are suspended in an equal concentration of NaCl. It is observed that these vesicles actively, although transiently, accumulate lactose if valinomycin is added to the vesicle suspension. No such active uptake is observed if KCl replaces NaCl in the suspending medium. Explain these results in light of what you know about the mechanism of lactose transport and the properties of valinomycin.

4. Intracellular vacuoles in the yeast *Saccharomyces cerevisiae* are membrane-bounded organelles that are known to concentrate within them a variety of basic amino acids, including arginine (net charge = +1). Vesicles prepared from these vacuoles lack an electron-transport chain, and arginine uptake into them is dependent on extravesicular ATP. A membrane potential $\Delta\psi$ has no effect on ATP-dependent arginine uptake in the absence of a proton gradient, while proton ionophores and dicyclohexylcarbodiimide (a known inhibitor of the F_1/F_o ATPase) greatly inhibit accumulation of arginine by this system. Upon addition of ATP in the absence of arginine, the intravesicular pH of these vesicles drops. Describe a mechanism for the energization of arginine transport in this system, taking into account all these observations.
5. In *E. coli,* lactose is taken up by means of proton symport, maltose by means of a binding-protein-dependent (ABC-type) system, melibiose by means of Na^+ symport, and glucose by means of the phosphotransferase system (PTS). Although this bacterium normally does not transport sucrose, suppose that you have isolated a strain that does. How would you determine whether one of the four mechanisms just listed is responsible for sucrose transport in this mutant strain?
6. In some instances, the efflux of a radioactively labeled transport substrate out of preloaded cells or vesicles is transiently stimulated by addition of the same nonradioactive transport substrate to the outside. This phenomenon is known as trans-stimulation and occurs with transport systems that are reversible (i.e., can operate in either direction). Can you think of an explanation for trans-stimulation in view of what is known about the molecular mechanisms of transmembrane transport?
7. Outline a molecular mechanism by which, and the conditions under which, an H^+ symport system (such as the *E. coli* lactose permease system) might operate to actively accumulate a metabolite such as lactose.
8. Predict the effects of the following on the initial rate of glucose transport into vesicles derived from animal cells that accumulate this sugar by means of Na^+ symport. Assume that initially $\Delta\psi = 0$, $\Delta pH = 0$ (pH = 7), and the outside medium contains 0.2-M Na^+, while the vesicle interior contains an equivalent amount of K^+.
   (a) Valinomycin.
   (b) Gramicidin A.
   (c) Nigericin.
   (d) Preparing the membrane vesicles at pH 5 (in 0.2-M KCl), resuspending them at pH 7 (in 0.2-M NaCl), and adding 2,4-dinitrophenol.
9. You are growing some mammalian cells in culture and measure the uptake of D-glucose and L-glucose (see data below). What type of transport is observed with these sugars? (Hint: plot *V* versus [sugar] and 1/*V* versus 1/[sugar].) Explain the significance of these data.

	V (mM cm s^{-1}) × 10^7	
[Sugar] (mM)	**D-glucose**	**L-glucose**
0.100	166	4.8
0.167	252	8.0
0.333	408	16
1.000	717	50

10. If the cells in problem 9 are treated with $HgCl_2$ (mercury reacts with —SH groups in proteins), the rate of transport of D-glucose is the same as L-glucose. What is indicated about D-glucose transport?
11. The translocation of K^+ was studied using an artificial membrane system (this membrane system had a phase transition T_m of 41° C) with valinomycin and gramicidin. The results showed that K^+ translocation with gramicidin was high over a broad temperature range, while valinomycin translocated K^+ at a high rate only above 41° C. What models of translocation are suggested by these data for each of these polypeptide antibiotics?
12. While there are many different and diverse types of transport systems in cells, these systems seem to be evolutionarily related and can be accommodated within a unified model. Outline the overall mechanism that is similar in these transport systems.

33

CHAPTER

Immunobiology

When vertebrates are invaded by foreign agents, they can mobilize a versatile set of adaptive processes to form specifically reactive cells and proteins. These immune responses constitute the principal means of defense against pathogenic microorganisms and viruses, and probably also against host cells that undergo transformation into cancer cells.

Throughout most of this century, the subject of immunology has attracted some of the keenest minds in biology. As a result, the intricacies of this fascinating subject have been largely unraveled, and its understanding is providing a strong bridge between the fields of biochemistry and physiology. In this chapter we will give a brief introduction to some of the major findings in immunobiology and the experiments that have led to our current level of understanding.

Overview of the Immune System

The immune system was first studied in humans, but mice became a popular subject for immune system studies when researchers began to appreciate how close the mouse and human systems were in their organization and action. Currently both systems are under study in many different laboratories. Although we will focus here on mice, we will often refer to parallel observations on humans.

The immune system is an example of a developmental process that takes place in the mature organism. In the new cells that are constantly being generated, changes arise in the genes of the immune system. This variability results from DNA splicing and point mutations. It benefits the organism by providing the cells of the immune system with the widest possible range of specificities. As soon as the organism is invaded by foreign agents, usually viruses or bacteria, the immune system is activated. Those immune-system cells that carry the specific immune receptors for interacting with the foreign agent are stimulated to proliferate. In a matter of days, clones of immune-system cells with the appropriate specificity have been produced and the organism fends off the invader with the specific immunologic tools provided by those cells. The clones of cells tend to persist for some time, usually months to years, which accounts for the fact that the second time the same foreign invader makes its presence known, it is usually rejected promptly and without crisis.

There are two different classes of white cells or lymphocytes associated with the immune response; these are the B cells and the T cells. In mammals, B lymphocytes mature in the bone marrow and then migrate to secondary lymphoid organs (fig. 33.1). Upon exposure to foreign substances known as antigens, they proliferate and produce immunoglobulin proteins known as antibodies, which they secrete into the bloodstream.

Figure 33.1

B cells and T cells follow different pathways of development. In mammals, B lymphocytes mature in the bone marrow and then migrate to secondary lymphoid organs. Upon exposure to foreign substances known as antigens, they proliferate to produce immunoglobulins. T lymphocytes mature in the thymus gland. They also can be stimulated to proliferate by exposure to an appropriate antigen.

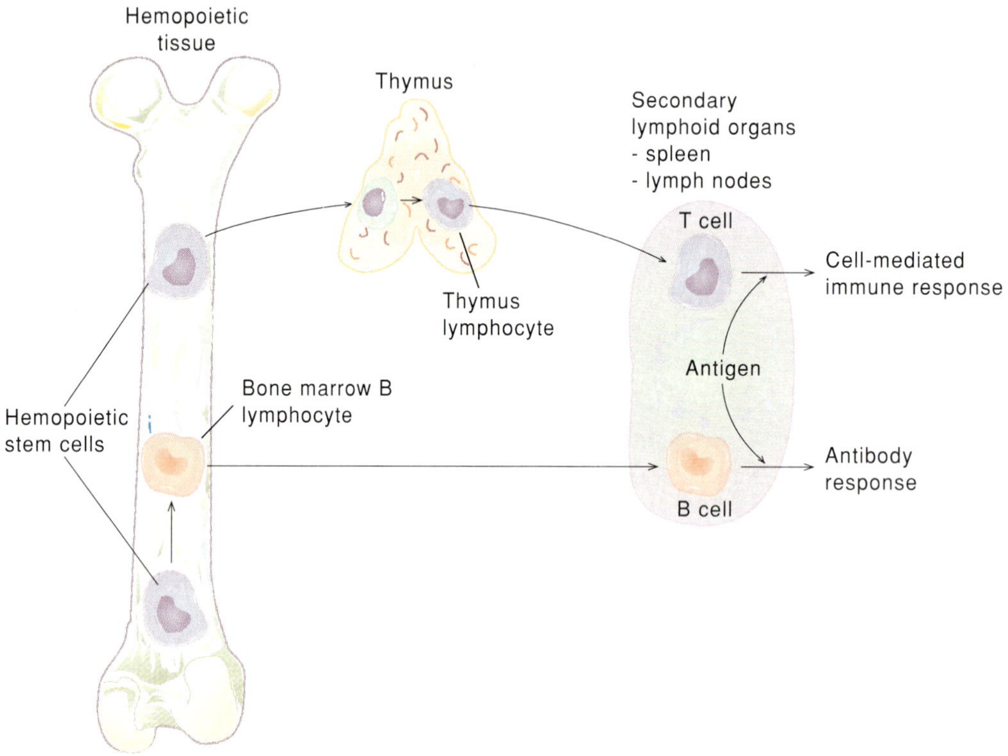

T lymphocytes, by contrast, mature in the thymus gland before migrating to the secondary lymphoid organs. They also can be stimulated to proliferate by exposure to an appropriate antigen, but their specific effector molecules remain firmly bound to the cellular membrane, as opposed to being secreted. Three main types of T cells have been recognized; T killer cells, which specifically destroy target cells; T helper cells, which promote the maturation of antigen-stimulated B and T cells; and T suppressor cells, which block the effects of T helper cells.

The Humoral Response: B Cells and T Cells Working Together

There are two basically different types of immune response. The first, called the humoral or antibody response, involves the concerted action of both B cells and T cells, and the active agents are the antibody or immunoglobulin proteins secreted by the B cells into the bloodstream. The combined B-T immune response is characteristic of most vertebrates. The second type of immune response, called the cell-mediated response, involves only T cells, and the active agent is the circulating T cell itself, which attacks the foreign agent. This type of immune response is limited to certain groups of vertebrates, including mammals. We will first discuss the B-T cell response mediated by antibodies.

Immunoglobulins Are Extremely Varied in Their Specificities

A detailed investigation of immunoglobulin structure provided the first leads as to how immunoglobulins of such a wide variety are synthesized. We have already provided a description of immunoglobulin structure (see fig. 5.25). For the purposes of this discussion, however, we need to describe certain features of the structure in greater detail. The most common type of immunoglobulin is the 7S molecule, known as immunoglobulin G (IgG). This immunoglobulin has a molecular weight of about 150,00 daltons. It is composed of four polypeptide chains, two heavy (H) chains with molecular weights of about 50,000 and two light (L) chains with molecular weights of about 25,000 (fig. 33.2).

To obtain a detailed understanding of immunoglobulin structure, it was necessary to isolate pure antibody proteins for sequencing. Fortunately, it was discovered that certain plasma cell tumors known as myelomas produce enormous amounts of pure immunoglobulins, which have the same gross structure as the mixed immunoglobulins isolated from serum. Each myeloma appears to originate from a single cell turned cancerous. As a result, each myeloma serves as the source of a pure antibody protein whose sequence and other properties can be determined. In some cases myelomas synthesize both heavy and

Figure 33.2

Structure of immunoglobulin G (IgG). The light (L) and heavy (H) chains have repeating domains, each with about 110 amino acid residues and an approximately 60-membered S—S bonded loop. Domains with variable sequences (V_L, V_H) are represented by blue bars; those with constant sequences (C_L, C_H) are represented by pink bars.

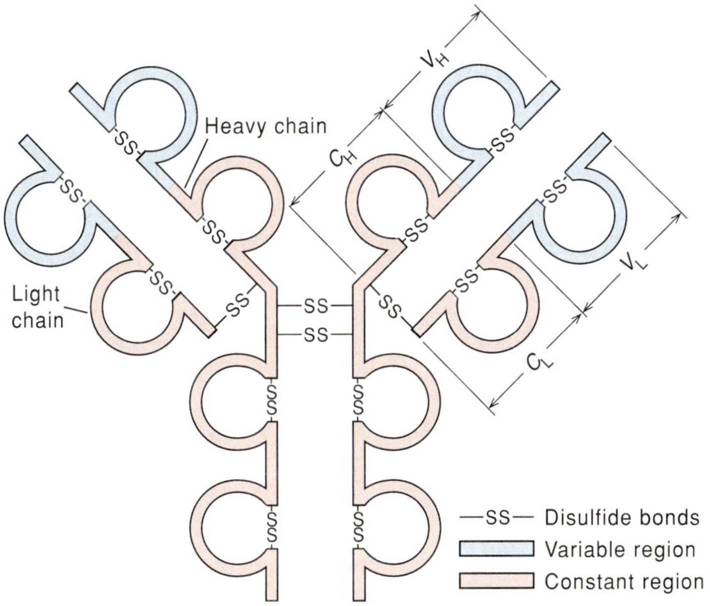

light chains, like normal antibody-forming cells, but in other cases they synthesize only one or the other type of chain. Comparison of the sequences from a number of different immunoglobins derived from myelomas has shown that both the heavy and the light chains for immunoglobulins of the same class are divided into regions of relatively constant sequence (the *C* segments) and regions of relatively variable sequence (the *V* segments). In the intact immunoglobulins the antigen-binding domain is composed exclusively of *V* segments originating from the H and L chains (see fig. 33.2), each tetramer containing two equivalent sites for the binding of a specific antigen.

As information on antibody sequences has accumulated, it has become increasingly clear that within the *V* regions of both the heavy and light chains there are three segments that account for most of the variability (fig. 33.3). These regions are called the hypervariable regions, and they are believed to be the parts of the antibody molecule in most direct contact with the antigen in the antigen-antibody complex.

Immunoglobulin G (IgG), the tetrameric species we have been discussing thus far, is not the only type of immunoglobulin found (table 33.1). Most of the other known antibodies—IgM, IgA, IgD, and IgE—also involve a closely related tetrameric structure, sometimes forming larger aggregates and always associated with different functions. For instance, IgM is a 19S antibody accounting for 5 to 10% of the serum Ig. It is an aggregate of five tetramers that is formed early during the immune reaction, soon to be diminished in quantity and overshadowed by large amounts of IgG. IgA occurs in various polymeric forms. It normally accounts for about 15% of the total

Figure 33.3

Approximate locations of the hypervariable regions in the heavy and light chains of IgG. Each hypervariable segment is believed to make up part of the site that binds to antigen, known as the complementarity-determining region (CDR). The CDR regions are located in the loop regions of the variable domains where they can make close contact with antigen (see fig. 5.25).

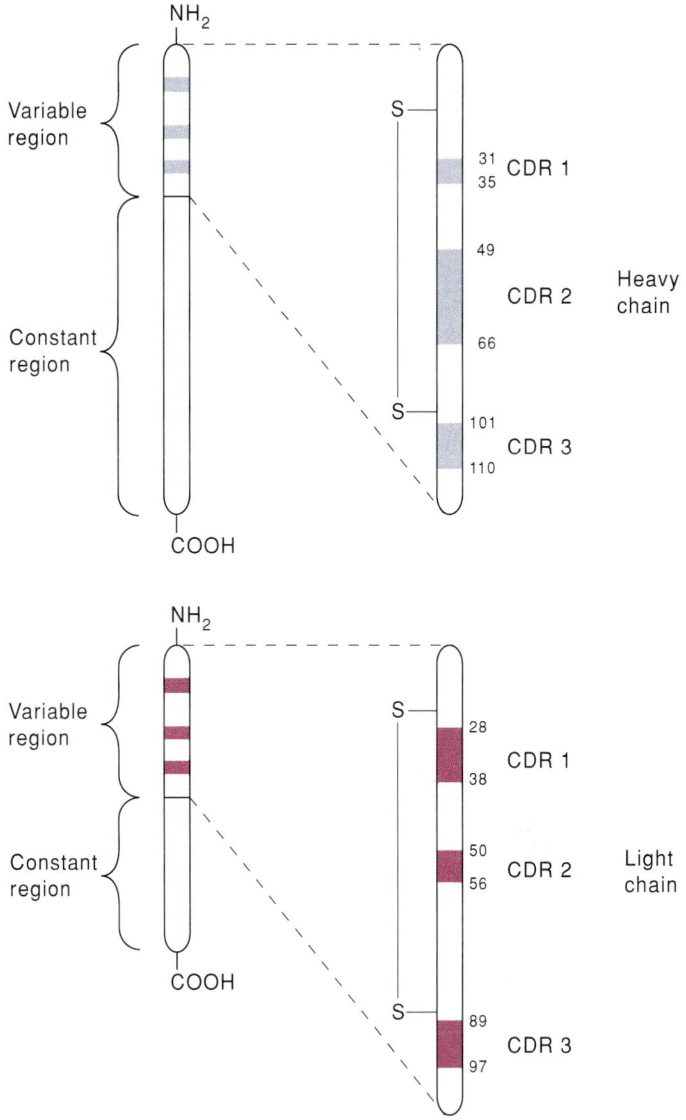

immunoglobulin found in serum, and in addition it is the principal immunoglobulin in exocrine secretions. Each of the different classes of immunoglobulin possesses distinct heavy chains. Indeed, even within the human IgG class, antisera tests reveal four different types of IgG (IgG1 through IgG4), each with a distinctive H chain, comprising about 70, 19, 8, and 3%, respectively, of the total IgG proteins. The light chains, of which there are two types, are common to all classes of immunoglobulins.

Immunoglobulins that are produced in response to antigens are themselves highly immunogenic (that is, they stimulate antibody formation) when injected into genetically nonidentical organisms. The serological responses induced by

Table 33.1
Different Isotypes Found in Humans

Class	Heavy Chain	Subclasses	Light Chain	Molecular Formula	Molecular Weight (Daltons)
IgG	γ	$\gamma1$, $\gamma2$ $\gamma3$, $\gamma4$ $\alpha1$, $\alpha2$	κ or λ	$\gamma_2\kappa_2$ $\gamma_2\lambda_2$	150,000
IgA	α	None	κ or λ	$(\alpha_2\kappa_2)^{a}_{a}$ $(\alpha_2\lambda_2)^{a}_{a}$	160,000 320,000
IgM	μ	None	κ or λ	$(\mu_2\kappa_2)_5$ $(\mu_2\lambda_2)_5$	900,000
IgD	δ	None	κ or λ	$(\delta_2\kappa_2)$ $(\delta_2\lambda_2)$	185,000
IgE	ϵ	None	κ or λ	$(\epsilon_2\kappa_2)$ $(\epsilon_2\lambda_2)$	200,000

using immunoglobulins as antigens or immunogens have been useful in classifying them according to their antigenic determinants.

So-called isotypic determinants are shared by all immunoglobulin molecules of a given class. For example, the human IgG molecules are classified into four isotypes because they are recognized by heterologous antisera produced in one species against the immunoglobulins of another species. Within a given isotype the different immunoglobulins can usually be separated into sets, called allotypes, that are distinguished by minor antigenic differences. Allotypic differences usually reflect alternative amino acid substitutions within otherwise quite similar amino acid sequences of the *C* region of the antibody. Allotypes show typical Mendelian inheritance patterns. Antisera useful for discriminating between allotypes are usually made by injecting an individual who lacks a specific allotype with immunoglobulins from an individual who carries the allotype. Finally, idiotype refers to the specific antigenic determinant of the antibody. Some idiotypic determinants are limited to a single immunoglobulin; others are shared by a small number of immunoglobulins. Whereas isotypic and allotypic differences usually result from differences in the constant portion or *C* segments of the immunoglobulin polypeptide chain, idiotypic differences usually result from differences in the variable portion or *V* segment. As we know already, this is the region that contains the binding site for the foreign agent or antigen.

Antibody Diversity Is Augmented by Unique Genetic Mechanisms

Antibodies have been studied most extensively in the mouse, an ideal vertebrate for both genetic and biochemical manipulations. It is believed that mice can synthesize more than a million antibodies with different antigenic specificities. This enormous diversity of proteins is generated from a limited amount of genetic information with the help of two mechanisms: somatic recombination and somatic mutation.

DNA Splicing Brings Different Parts of the Antibody Gene Together The involvement of somatic recombination was first demonstrated by examining the structure of a specific antibody gene in embryonic and adult immunoglobulin-forming tissue. The adult tissues favored for many studies are myelomas. These tumorous tissues, which we mentioned earlier, produce a homogeneous population of polypeptide chains. They have served as a convenient source of pure immunoglobulin polypeptide chains as well as their mRNAs.

Recall that the generalized structure of the predominant serum antibody consists of two identical heavy chains and two identical light chains (see fig. 33.2). In the mouse there are two classes of light chains, which differ appreciably in the constant regions of the polypeptide chains. These are referred to as the kappa (κ) and the lambda (λ) light chains. Using highly inbred, genetically identical (isogenic) strains of mice, Tonegawa and his co-workers isolated the DNA from the embryo and from two different myeloma tumor cells, one that produces homogeneous λ light chains (strain H2020) and one that produces κ light chains (strain MOPC321). These DNAs were exhaustively digested with the *Eco*R1 restriction enzyme and electrophoresed on an agarose gel. The gels contained an enormous variety of restriction fragments representing total nuclear DNA, and consequently no discrete pattern of bands could be seen with a strain for nucleic acid. However, when the electrophoresed gels containing the DNA were denatured and hybridized with ^{32}P-labeled DNA containing the sequences found in the RNA for the λ chain, a specific pattern of bands showed up in autoradiographs (fig. 33.4). The R1 digest of DNA from

Figure 33.4

Analysis of DNA fragments containing λ_1 gene sequences from mouse embryo and myeloma cells. High-molecular-weight DNAs extracted from myeloma H2020, a λ-chain producer (A), from a 13-day-old BALB/c embryo (B), and from myeloma MOPC321, a κ-chain producer (C) were digested to completion with *Eco*R1, electrophoresed on agarose gel, and transferred to nitrocellulose membrane filters and hybridized with a nick-translated *Hha*I fragment of the plasmid B1 DNA. (Source: C. Brack et al., "A complete immunoglobulin gene is created by somatic recombination," in *Cell* 15:1–14, 1978. Copyright © 1978 Cell Press, Cambridge, Mass.)

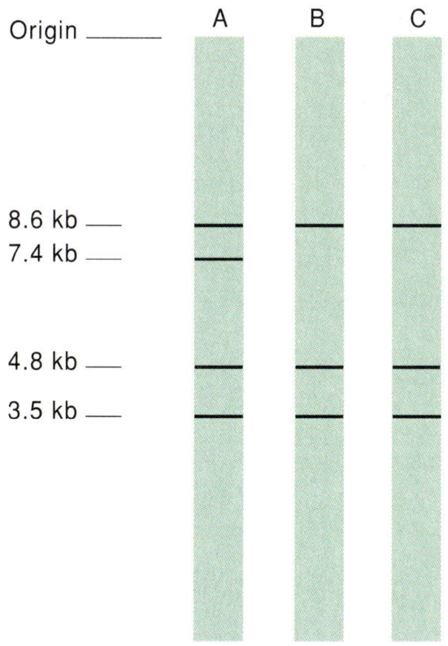

the λ-containing myeloma (H2020) showed four bands; the DNAs of the embryo and the κ-containing myeloma (MOPC321) showed three bands in common with the first DNA but were missing the fourth band.

These results strongly suggested that at some point during development from the embryonic state, a rearrangement had taken place in the H2020 myeloma cells on one of the homologous pairs of chromosomes carrying the λ-chain gene. Further analyses with more specific radioactive probes, containing sequences from either C_λ (the constant region of the λ chain) or V_λ (the variable region of the λ chain), were done. Only the 7.4-kb fragment originating from the λ myeloma cell hybridized to both probes. Thus the *Eco*R1 fragment must contain both V_λ and C_λ DNA. The 8.6-kb fragment hybridized to the C_λ but not the V_λ probe. Therefore this fragment must contain C_λ sequences but no V_λ sequences. Both the 3.5-kb fragment and the 4.8-kb fragment hybridized exclusively to the V_λ probe, so they must contain only V_λ sequences. Still further analyses indicated that the 7.4-kb fragment arose from a recombinational event between the 3.5-kb fragment and the 8.6-kb fragment (fig. 33.5). The 4.8-kb fragment originates from another V_λ sequence located in the same chromosome. The mouse has only two V_λ sequences in the embryo. As a rule, only one of the pairs of homologous chromosomes recombines to yield a productive antibody gene. This explains why the 3.5-kb and the 8.6-kb bands are still visible in the H2020 myeloma DNA preparation. The κ myeloma cell producer shows the same pattern as the embryonic cell. Had it been probed with a DNA carrying the V_κ antibody sequence, it would have shown differences from the embryonic pattern.

Figure 33.5

Arrangement of mouse λ_1 gene sequences in embryos and λ_1 chain-producing plasma cells. The vertical arrows point to *Eco*R1 restriction sites. The colored dashed diagonal lines point to hypothesized splice points that will explain the difference in structure of the region in the two cell types. The boxed regions represent coding regions, and I_1 and I_2 indicate first and second introns in the spliced gene.

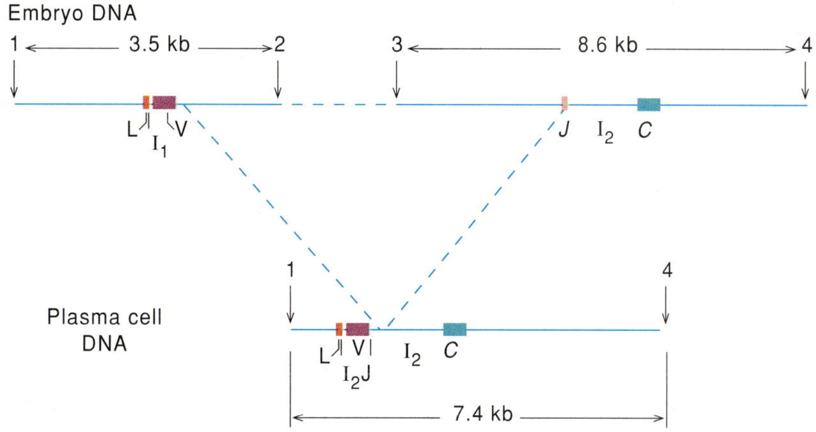

The success achieved by Tonegawa and others in this type of sequence analysis of the immunoglobulin genes led to a massive research effort and a general picture of antibody gene organization. All antibody polypeptide chains are derived from split genes. Antibody genes are encoded by three unlinked gene families located on different chromosomes. In the human, chromosomes 2, 22, and 14 encode those gene clusters associated with the kappa (κ) light chains, the lambda (λ) light chains, and the heavy chains, respectively.

The light chain is encoded by three distinct DNA elements—a variable (V_L), a joining (J_L), and a constant (C_L) element (fig. 33.6). During the differentiation of a B cell, a V_L element (encoding the first 95 amino-terminal residues) and a J_L segment (encoding about 13 amino acids) are joined by DNA splicing. A complete light-chain gene transcript consists of three exons and two introns, arranged in the following order; 5′L, I_1, *VJ*, I_2, *C*, 3′. Here L refers to sequences in the leader or signal peptide, which is removed in the mature immunoglobulin, and I_1 and I_2 refer to the first and second introns, respectively, which are removed by RNA splicing. The heavy chain is encoded by four distinct DNA elements—a V_H, a *D*, a J_H, and a C_H element. In heavy-chain variable region formation, a V_H element (encoding about 99 amino acids), a *D* element (encoding 1 to 15 amino acids), and a J_H element (encoding about 15 amino acids) are joined in two DNA-splicing steps. The heavy-chain gene family has several closely linked C_H genes, which determine the immunoglobulin class.

Class switching, involving an additional DNA-splicing operation, frequently occurs with heavy-chain genes. Thus an immature B cell, which initially expresses a μ chain, results in IgM antibodies. Subsequently the same cell can be induced to differentiate further so that the same V_H region of the genome becomes relocated next to a C_γ gene, causing the cell to produce an IgG antibody with the same V_H region associated with a different C_H region. In cases of class switching, the second DNA splicing involves removal of the intervening genes. Thus in the case of C_μ to C_γ switching, the region containing C_δ must be removed by the second splice reaction (fig. 33.7). In the case of C_μ to C_α class switching, all of the *C* genes between C_μ and C_α must be removed by the second splice reaction.

Class switching of the heavy-chain genes illustrates a temporally regulated process associated with the DNA rearrangements that can occur in the antibody-forming genes. Since any given cell normally synthesizes only one type of antibody, there must be a special type of regulatory mechanism that responds in a systematic way to antigenic stimulation. In B-cell development it is known that the V_H-D-J_H joining occurs first, leading to the formation of μ heavy chains. Subsequently V_L-J_L joining occurs, with the production of functional light chains.

As a rule, the DNA-splicing reactions leading to the joining process occur in only one of the alleles for each of these chains. The mechanism for this phenomenon, known by the name allelic exclusion, is not understood. Another type of exclusion process results in the production of only κ or λ light chains in a given B cell. This process is called isotypic exclusion. Indirect evidence that the κ gene family is expressed before the λ gene family comes from the examination of human B-cell lines that are active κ- or λ-chain producers. In all B cells producing λ chains, the C_κ genes are either deleted or rearranged. By contrast, in all cells producing κ chains, the C_λ genes are in the germ-line configuration. The obvious inference is that a B cell becomes a κ producer first and then becomes a λ producer. Direct proof of this order is still lacking. Other evidence exists for regulatory signals that may control isotypic exclusion. When a pre-B cell is fused with a myeloma cell that has been producing a functional heavy chain but no functional light chain, the hybrid cell is capable of expressing a new light chain as a result of a *V-J* joining of gene segments in the pre-B cell. If, however, the myeloma cell was already producing a functional light chain, no new light chains or DNA rearrangements result from the cell fusion process.

Figure 33.6

Organization and DNA splicing of the mouse immunoglobulin genes. The organization is depicted before and after somatic cell rearrangement. For each class only one example of a rearranged chromosome is shown. A light chain is encoded by three distinct DNA elements, a variable (*V*), a joining (*J*), and a constant (*C*) element. Boxes indicate exons; lines connecting boxes indicate introns. Somatic cell rearrangement involves the joining of a *V* and a *J* segment on the same chromosome. The heavy chain is encoded by four distinct types of DNA elements, a *V*, a *D*, a *J*, and a *C* element. In heavy-chain splicing a *V* element, a *D* element, and a *J* element are joined in two DNA splicing steps. The splice between *D* and *J* elements occurs first. Boxes that are filled are preserved after the DNA splicings. The splices shown provide only one example of many that might occur in this system.

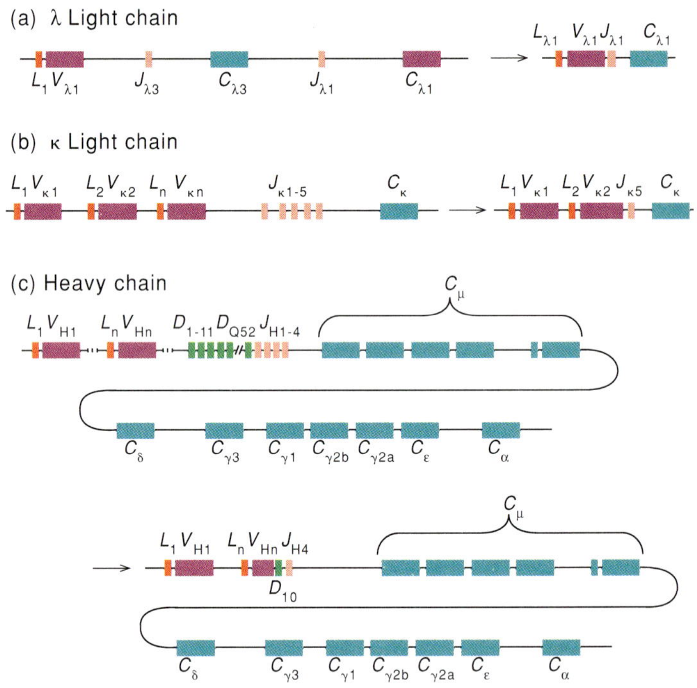

Alternative Pathways Exist for RNA Splicing of Heavy-chain mRNAs The heavy-chain transcript contains numerous exons and introns. In addition to the introns between the L_H and V_H exons and between the V_H-J_H and C_H exons, each C_H region

Figure 33.7

Class switching of heavy-chain genes involves additional DNA splicing. Switching from C_μ to $C_{\gamma 1}$ is illustrated. The additional splicing involves removal of the continuous segment carrying the C_μ through the $C_{\gamma 3}$ genes.

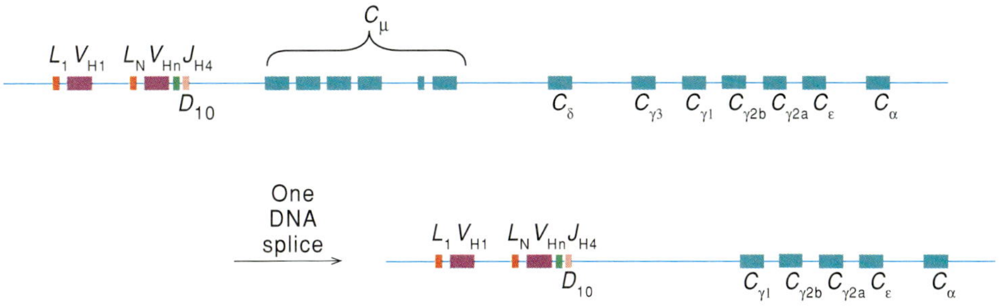

contains several introns. The hinge regions of γ chains (see fig. 5.25) and the intramembrane and cytoplasmic portions of all C_H chains are encoded by independent exons. Most of the RNA-splicing operations are presumed to occur by standard mechanisms similar to those described in chapter 28. In the case of the heavy-chain genes, it has been shown that alternative RNA-splicing patterns involving the C_H region can yield molecules that will function as membrane-bound or secreted forms of the antibody. Thus there are two membrane exons that have been detected in the 3′ flanking regions of both the C_μ and C_γ genes. Alternative pathways of RNA splicing give rise to two mRNAs, which encode separately the membrane-bound and the secreted forms of the heavy chains. It seems likely that the alternative splicing pattern is temporally regulated, since the membrane-bound forms are favored before and the secreted forms are favored after antigenic stimulation. Amino acid as well as nucleotide sequencing of the μ_M chain (μ_M stands for the membrane-bound form of the μ chain) has revealed a 26-residue hydrophobic segment at the COOH terminus, which anchors in the lipid bilayer of the membrane.

Table 33.2
Estimated Number of Germ-Line Gene Segments for Different Mouse Antibody Genes

	Light Chains		Heavy Chains
	κ	λ	*H*
C	1	4	8
V	90–300	2	100–200
J	5	4	4
D	—	—	12

From Geoffrey Zubay, *Genetics*. Copyright © 1987 Benjamin/Cummings Publishing Company Inc., Menlo Park, Calif. Reprinted by permission of the author.

Somatic Mutation Contributes to Antibody Diversity Various estimates, some exceeding a million, have been put forward for the number of different antibodies an organism can make. A good deal of this diversity results from the different combinations of heavy and light chains that can be generated by making alternative splicings between the different gene segments that make up the light and heavy chains (table 33.2). Superimposed on the variation from this source, there is additional diversity provided by somatic mutation within the coding regions. Several subsets of κ and heavy chains and their germ-line *V* segments have been analyzed by cloning and sequencing. The results confirm that somatic mutations amplify the diversity encoded in the germ-line genome.

A $V\kappa$ probe prepared for the κ genes expressed in a particular myeloma, MOPC167, detected only one major band in the Southern gel blot analysis of total cellular DNA. This unusual situation permitted a relatively simple sequence comparison to be made between this segment, found in the germ-line, and its rearranged counterparts in two myeloma lines, MOPC167 and MOPC511. Four and five nucleotide differences were found, respectively, in the $V\kappa$ regions of these two lines. The changes were completely different in both lines. This investigation and other similarly directed investigations show that the regions in which mutations are introduced during somatic differentiation are highly restricted to the V_L-J_L or V_H-D-J_H segments. The density of mutations observed is about three times higher in the complementarity-determining regions (CDRs) of the variable segments than in the connecting framework regions (FRs) of the variable segments. (See figure 33.3 for the approximate location of the CDR regions.) The complementarity-determining regions are the regions in the immunoglobulin that play a critical role in the interaction with antigen. Mutations produced in these regions must occur either during or after DNA splicing, as they are not found in unrearranged (unspliced) DNA. Somatic mutation clearly increases the number of possible antibodies almost without limit. Most importantly, it leads to selectable changes in just those regions of the immunoglobulins that are responsible for antigen recognition. Several mechanisms for hypermutation have been proposed; none have been verified.

Figure 33.8

DNA splicing of an antibody gene brings an enhancer element (E) close to the promoter region (P) of the gene. This step is depicted for a heavy-chain gene after the initial two splices (see fig. 33.6).

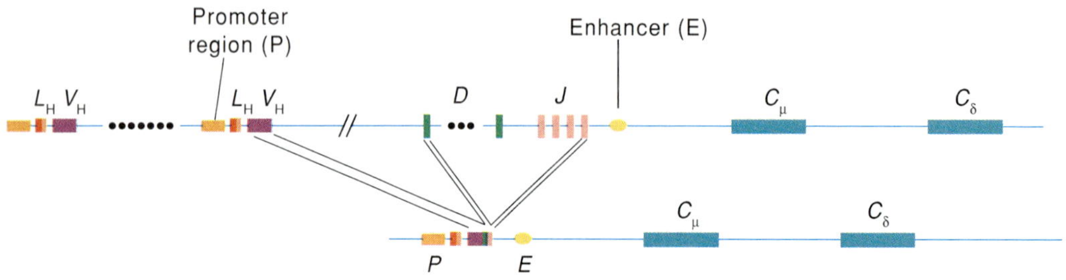

DNA Splicing Brings an Enhancer Element Close to the Promoter DNA sequences derived from the germ-line J_H-C_μ region are required for accurate and efficient transcription from a functionally rearranged V_H promoter (fig. 33.8). Similar to viral transcriptional enhancer elements (see chapters 28 and 34), these cellular sequences stimulate transcription from either the homologous V_H gene segment promoter or a heterologous SV40 promoter. They are active when placed on either the 5′ or the 3′ side of the rearranged V_H gene segment and they function when their orientation is reversed. However, unlike viral enhancers, the immunoglobulin gene enhancer appears to act in a tissue-specific manner, since it is active in mouse B cells but not in mouse fibroblasts.

The discovery of this enhancer suggests another benefit that results from splicing; it brings an enhancer in close proximity to the promoter just when that promoter activity becomes an important part of the cell's function.

Interaction of B Cells and T Cells Is Required for Antibody Formation

Thus far, our discussion of antibody formation has centered on those reactions that occur at the gene level and on the structure of the antibody. However, since the immune response *in vivo* involves reactions between whole cells, we must also consider what happens at the level of cell-to-cell interactions. Some of the pluripotent stem cells of the bone marrow replicate in an undifferentiated state, maintaining their pluripotency, while others replicate and differentiate at the same time. The latter cells can differentiate along various pathways (fig. 33.9). Some of those that become committed to the lymphocyte pathway differentiate further into B-cell and T-cell lineages. T cells mature in the thymus, whereas B cells mature in the marrow in most vertebrates. Subsequently both cell types relocate in secondary lymphatic organs, the spleen and the lymph nodes, and also more diffusely in all tissues. Here they remain dormant until they encounter specific antigens that trigger them to complete their differentiation into fully mature effector B cells and effector and regulator T cells. Each B cell contains immunoglobulins of a specific idiotype bound to its surface, usually of the IgM type and to a lesser extent of the IgD and IgG types.

Figure 33.9

Pluripotent stem cells in the marrow give rise to more stem cells and to various progenitors. The lymphocyte progenitors give rise to B-cell and T-cell lineages. The final differentiation step in both B-cell and T-cell development requires antigenic stimulation.

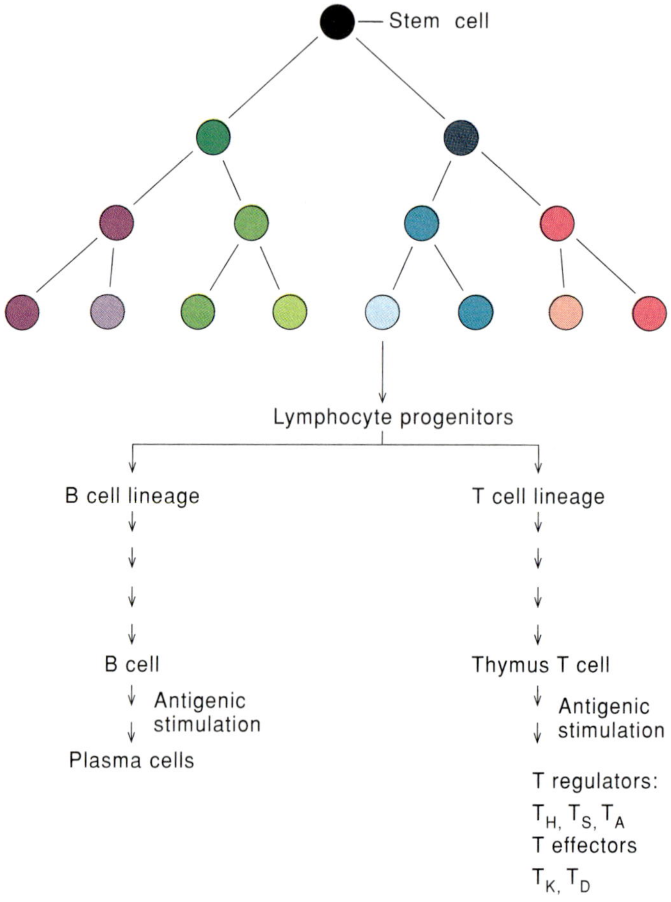

Antigens Stimulate the Formation of B-cell Clones The first time a foreign antigen is injected into the bloodstream, there is a long delay before specific antibody appears, and the response is fairly weak. The second time the same antigen is injected, the response is both more rapid and more intense (fig. 33.10). A

Figure 33.10

Kinetics of IgM and IgG appearance in the serum following a first and a second exposure to the same antigen. Upon first exposure there is a delay of several days before any antibody appears in the serum. IgM appears before IgG. After many days the serum level falls. If the system is exposed to a second equivalent dose of antigen, the appearance of IgM and IgG is much faster and the maximum level of IgG is much higher.

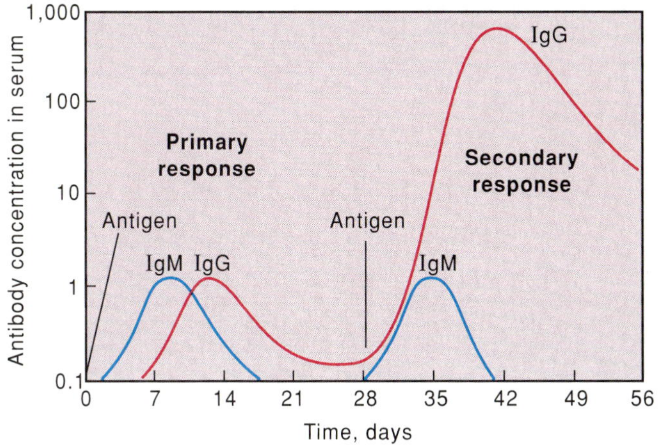

Figure 33.11

An immature B cell (*a*) and a fully developed antibody-producing B cell (*b*). (Electron micrographs courtesy of D. Zucker-Franklin, New York University Medical Center.)

(a)
(b)

great deal of research effort has gone into seeking an explanation for this difference between primary and secondary responses. Specific labeling studies show that secondary lymphatic organs contain a very heterogeneous mixture of B cells, carrying membrane-bound antibodies of different idiotypes. When sufficient amounts of antigen bind to the surface antibody, the B cell becomes triggered to divide and make more cells of the same type. Some of these continue to divide, making clones of memory cells in the lymphatic organs, while others terminally differentiate into plasma cells richly laced with endoplasmic reticulum that is geared to the production of the antibody of the same idiotype as that which is membrane-bound (fig. 33.11). Whereas the idiotype and consequently the V_L and V_H parts of the antibody are the same as in the original, unstimulated B cell, the *C* regions change. The first secreted antibody is of the secretory IgM type. This is gradually replaced by the secretory IgG type, which dominates in the secondary response. As we have noted (see fig. 33.7), the replacement of IgM by IgG is indicative of an additional DNA splicing.

Helper T Cells Trigger B Cell Division and Differentiation

The importance of T cells in mediating the B-cell response to most antigens is demonstrated by the fact that thymectomized animals, which are depleted in T cells, usually show a greatly attenuated reaction to foreign antigen. A normal response can be restored by transplantation of thymus tissue from a genetically identical animal. Activated T helper cells interact with B cells whose antibodies can recognize the same antigen that stimulated the helper T cell. The complex between the T and B cell results in the clustering of a large number of transmembrane proteins causing a process known as CAP formation. CAP formation by some unknown means triggers the proliferation and differentiation of the B cells involved.

Figure 33.12

Stimulation of antibody-forming cells involves three types of cells. Typically antigen is phagocytosed by a macrophage. The phagocyte partially digests the antigen and "presents" the processed antigen on its outer plasma membrane. A specific helper T cell binds the antigen to a specific receptor bound to its outer plasma membrane. The T cell usually has many of the same types of receptors and binds many copies of the antigen in a similar way. This stimulates T-cell proliferation. These T cells interact with B cells that display similar processed antigen. Several B-cell receptors involved in a similar way become focused in a local region of the outer plasma membrane, a process known as CAP formation. CAP formation triggers the proliferation and differentiation of the B cells involved.

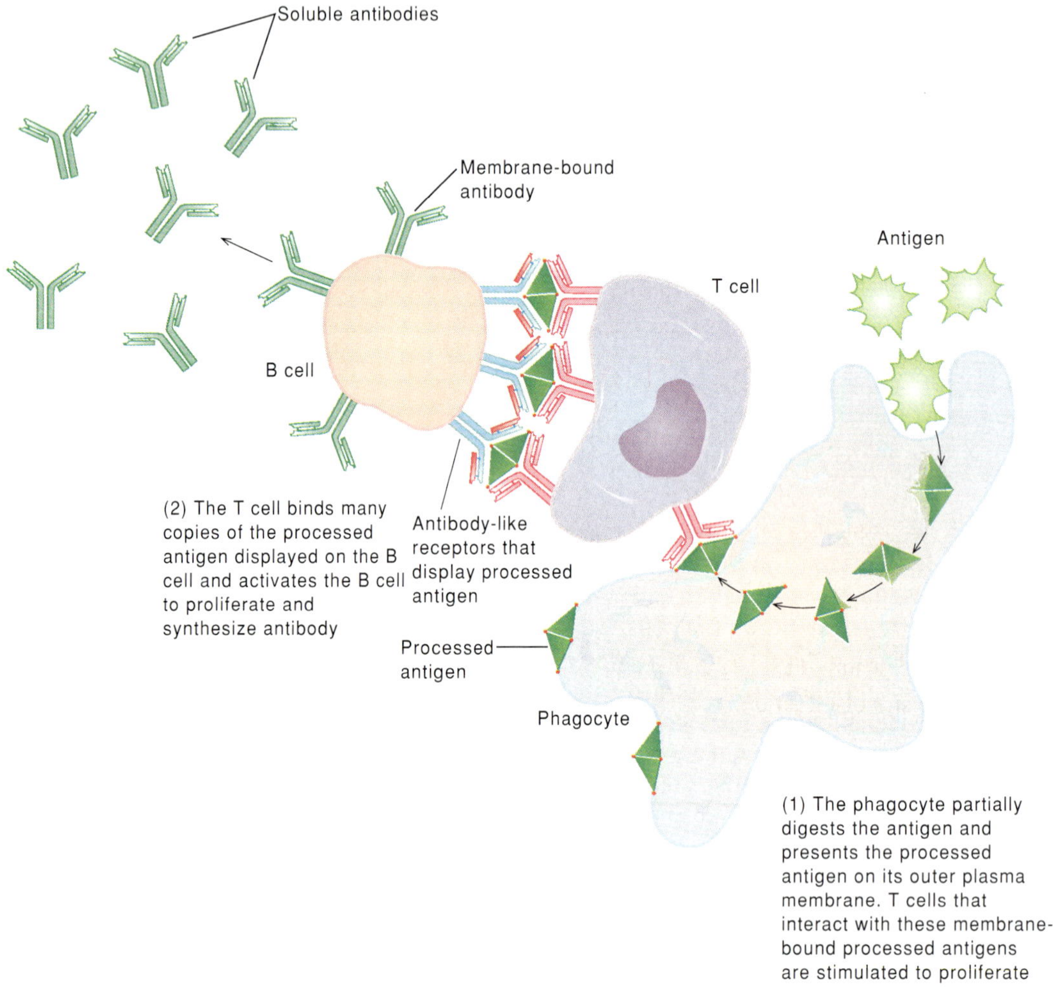

Support for the role of T helper cells in focusing the antigen comes from the observation that T-independent stimulation of B cells is frequently elicited by polymeric antigens. Such polymeric antigens constitute another way of presenting a large amount of antigen at one location on the B-cell membrane and thereby focusing the B-cell receptors.

Before they can help other lymphocytes respond to antigen, helper T cells must be activated themselves. This activation occurs when a helper T cell recognizes a foreign antigen bound on the surface of a specialized antigen-presenting cell. The latter cells are found in most tissues; they are derived from bone marrow and comprise a heterogeneous group, including dendritic cells in lymphoid organs and certain types of macrophages. Together with B cells, which can present antigen to helper T cells, these are the main cell types that react with helper T cells. The sequence of events involving antigen-processing cell, T helper cell, and antibody-forming B cell is depicted in figure 33.12.

Note that the mechanism proposed for helper T-cell function requires that T cells have surface receptors that will recognize antigen-processing cells, antigen itself, and the appropriate B cell. We will shortly discuss the nature of T-cell receptors, as well as the surface structures they recognize, in a broader context.

The Complement System Facilitates Removal of Microorganisms and Antigen-Antibody Complexes

Many antigen-antibody complexes are eliminated by phagocytosis. Others are attacked by complement, a group of serum proteins that aids in the defense against microorganisms. Individuals with a deficiency in their complement system are subject to repeated bacterial infections, just as are individuals deficient in antibodies themselves.

Figure 33.13

The principal stages in complement activation. Complement activation occurs exclusively on the microbial cell membrane, where it is triggered by bound antibody or microbial envelope polysaccharides, both of which activate early complement components. There are two sets of early components belonging to two distinct pathways of complement activation. Activation of each complement system involves a cascade of proteolytic reactions. Each component of the complement system is a proenzyme that is activated by the preceding component of the chain by a limited proteolytic cleavage. The ultimate result of this chain reaction is the development of a complex that attacks the cell membrane.

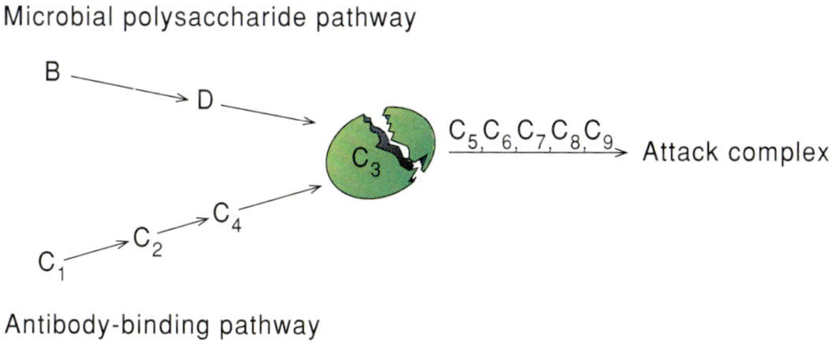

There are about twenty different proteins in the complement system—proteins C1–C9, factors B and D, and a series of regulatory proteins. All these proteins are made in the liver, and they circulate freely in the blood and extracellular fluid. Activation of the complement system involves a cascade of proteolytic reactions. In addition to forming membrane attack complexes, the proteolytic fragments released during the activation process promote dilation of blood vessels and the accumulation of phagocytes at the site of infection.

The activation of complement occurs exclusively on the microbial cell membrane, where it is triggered either by bound antibody or microbial envelope polysaccharides, both of which activate early complement components. There are two sets of early components belonging to two distinct pathways of complement activation: C1, C2, and C4 belong to the pathway that is triggered by antibody binding; factors B and D belong to the alternative pathway that is triggered by microbial polysaccharides (fig. 33.13). The early components of both pathways ultimately act on C3. The early components and C3 are proenzymes that are activated sequentially by limited proteolytic cleavage. As each proenzyme in the sequence is cleaved, it is activated to generate a serine protease, which cleaves the next proenzyme in the sequence. Many of these cleavages liberate small peptide fragments and expose a membrane-binding site on the larger fragment. The larger fragment binds tightly to the target cell membrane by its newly exposed membrane-binding site, and helps to carry out the next reaction in the sequence. In this way complement activation is confined largely to the cell surface, where it began.

The components of the complement attack complex have a very short half-life if released from the complex. This limitation confines their destructive action to the point of assembly, the surface membrane of the microorganism.

The Cell-Mediated Response: A Separate Response by T Cells

Most B cells are fixed in the secondary lymphatic organs, the spleen and the lymph nodes. They are stimulated by antigen only if the antigen is transported to the lymph nodes. In many cases, however, antigens are poorly transported. To rid the organism of antigens that are not readily transported, a second type of immune system exists that is independent of the B cells. This second type of immunity is not found in all vertebrates; it is restricted to mammals, birds, certain amphibians, and bony fishes. Long before the distinction between T and B cells was appreciated, it was known that certain types of immunity, which manifest a delayed-type hypersensitivity response, could be transferred from immunized animals to nonimmunized animals with leukocytes, but not with antibody-containing serum. Subsequently other forms of leukocyte-mediated immunity were discovered that involved increased phagocytic and bacteriocidal activity of macrophages, lysis of virus-infected cells and tumor cells, and the rejection of skin grafts from genetically dissimilar organisms.

There are two types of cell-mediated immunity, involving basically different types of T cells. In the delayed-type hypersensitivity response, the T cell, T_D, that reacts specifically with antigens secretes lymphokines (box 33A) that attract and activate macrophages or other leukocytes, thereby causing a slowly developing inflammatory response. A second type of T cell, known as the T killer cell, T_K, reacts specifically with antigen that is bound to target cells, causing their lysis.

Tolerance Prevents the Immune System from Attacking Self-Antigens

A thorough understanding of the immune system requires an explanation for how the immune system is able to discriminate between foreign antigens and its own antigens (self-antigens). The ability to recognize self-antigens and thus to avoid making an antagonistic response to them is called tolerance. Some serious diseases are believed to be due to a breakdown in the self-tolerance mechanism. Conversely, there are conditions under which a foreign antigen is able to establish itself so that it becomes tolerated, for either a short or a long time. Tolerance and rejection are so closely related that it seems likely that a full understanding of one will entail a full understanding of the other.

Several conditions favor the establishment of tolerance: (1) Tolerance is much easier to establish in the fetus or the newborn than in the adult. Thus if an organism is exposed as a fetus to a soluble foreign antigen or a foreign tissue graft, it may become indefinitely tolerant to the antigen. This situation is illustrated for two genetically nonidentical strains of mice in figure 33.14. Cells from the Y mouse are injected into a newborn X mouse. The same X mouse as an adult is the recipient of a skin graft (transplant) from a Y-type mouse. Normally such a graft would be rapidly rejected. However, because of the early exposure to the Y cells, the graft is tolerated indefinitely. This

33A
BOX

The Role of Lymphokines (Interleukins)

The interactions between immune and inflammatory cells are mediated in large part by certain proteins, called lymphokines or interleukins (IL), that are able to promote cell growth, differentiation, and functional activation. Interleukins resemble hormones, which also function as intercellular messengers. In contrast to hormones, however, they are secreted by isolated cells rather than discrete glands. Several interleukins have been described; each has unique biological activities as well as some that overlap with the others. For example, macrophages produce IL1 and IL6, whereas T cells produce IL2–IL6 and bone marrow stromal cells produce IL7. IL1 and IL6 not only play important roles in immune cell function, they also stimulate a spectrum of inflammatory cell types and induce fever. IL2 is a potent proliferative signal for T cells. IL1, IL3, IL4, and IL7 enhance the development of a variety of hematopoietic precursors. IL4–IL6 also serve to enhance B-cell proliferation and antibody production.

Figure 33.14

Establishment of tolerance to tissue grafts. If a newborn X mouse is injected with Y cells when it is young, it is tolerant to Y tissue transplants when it becomes an adult. If the X mouse is not exposed to Y cells at an early age, it will readily reject the tissue graft.

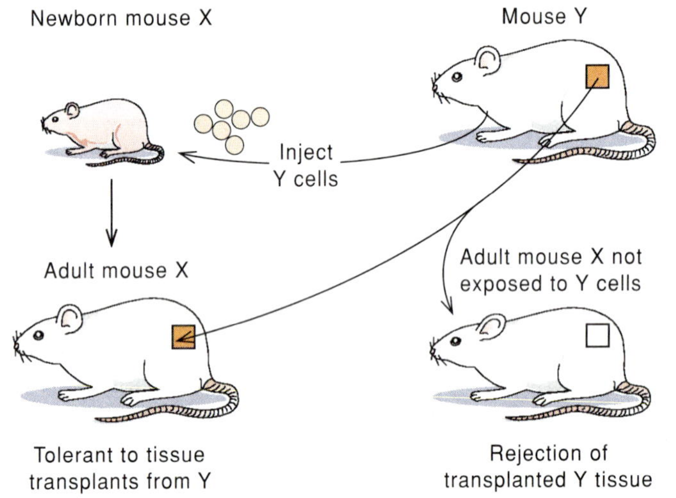

result indicates that the mouse that was exposed to the Y cells just after birth recognizes Y cells later in life as self-antigens. (2) Very high levels of antigen frequently favor the development of tolerance, whereas intermediate levels of antigen favor the development of an immune response. (3) Frequently the route of antigen administration is critical in the development of tolerance. An intravenously injected antigen is more likely to promote a tolerant condition than an antigen injected into the tissue.

Two cellular mechanisms may account for most forms of tolerance. In one of these the clones of T or B cells that could otherwise be activated by the antigen are destroyed or inactivated. In the other mechanism, the clones in question are still present but do not respond because they are blocked by T_S cells.

In general, B cells are more difficult to make tolerant than T cells. This difference is evident from observations that longer and higher doses of antigen are required to establish tolerance to the B-dependent system. Moreover, the period of tolerance is generally shorter when B cells are involved.

Adult mice may be made tolerant to foreign antigens by destroying their immune system with whole-body irradiation and then supplying them with transplants of bone marrow and thymus from a foreign donor, thus refurbishing their system with competent B and T cells, respectively. With the new immune system the animal is now tolerant to any tissue from a mouse that is isogenic with the mouse used to supply the marrow and the thymus.

The irradiated host organism also supplies a useful means of testing the state of B and T cells from a nonirradiated isogenic mouse. B cells may be effectively transferred by bone marrow transplants, and T cells may be transferred by thymus tissue transplants. The donor cells may come from a normal, untreated isogenic strain. In this event the irradiated recipient will show normal immunologic behavior. Alternatively, the donor cells may come from a donor that has been made either sensitive or tolerant to a particular antigen. By doing varous combinations of experiments using different isogenic donors, it is possible to determine the relative importance of B-cell and T-cell involvement with different antigens.

T Cells Recognize a Combination of Self and Nonself

All T cells, whether they fight infection directly or regulate the activity of other effector cells, recognize antigen only in combination with a cell membrane surface. Moreover, the antigen must be associated with a cell from an organism that is genetically similar or identical to that of the T cell. This requirement was first demonstrated by an experiment in which T cells were removed from a mouse that had been immunized with a virus. Immunization meant that the extracted T cells had been activated to direct themselves against cells infected with this

Figure 33.15

Activated T cells will bind and kill only those cells that display a foreign antigen and a familiar cell-surface antigen.

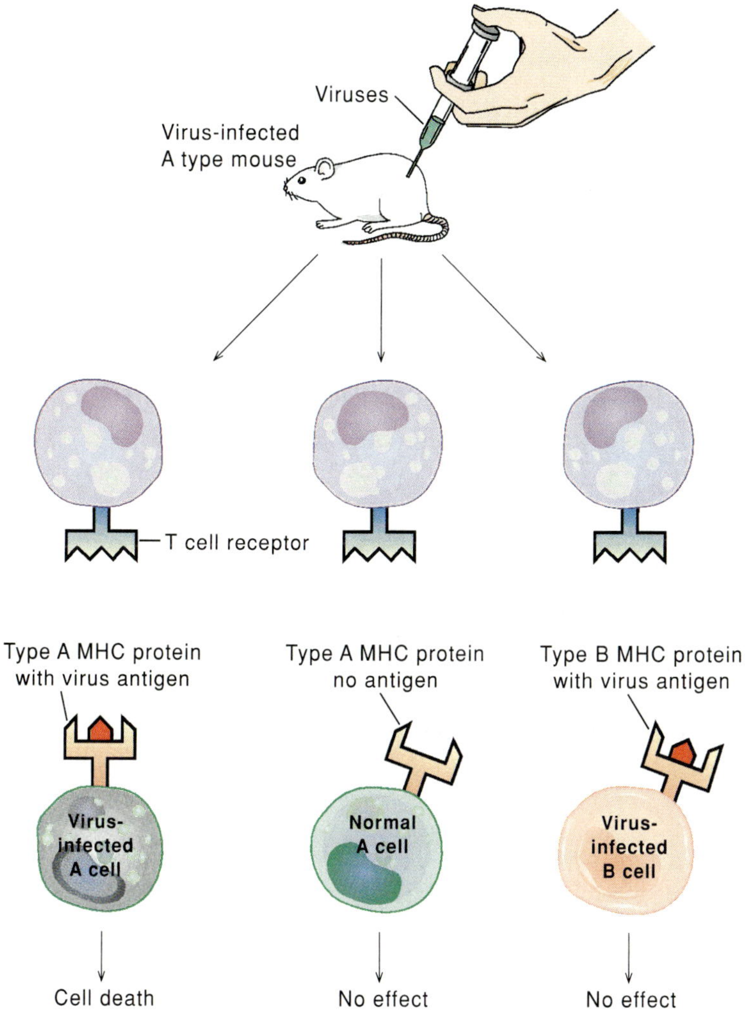

particular virus. Thus these T cells could kill fibroblasts infected with the virus in tissue culture. However, they could not kill fibroblasts infected with the virus if the fibroblasts came from a mouse of a different genetic background. Thus killer T cells do not recognize just any foreign antigens on any cell membrane surface; they recognize foreign antigens only in combination with determinants present on their host's own cells (fig. 33.15). These common determinants are a subset of the proteins encoded by the so-called major histocompatibility complex (MHC).

MHC Molecules Account for Graft Rejection

MHC molecules were recognized long before their major biological function was appreciated. They were initially defined as the main target antigens in transplantation reactions, and because of this they became known as major histocompatibility antigens. Experiments on mice demonstrated that graft rejection is an immune response to the surface antigens of the grafted cells. Subsequently it was shown that these reactions are mediated by T cells and that they are directed against a family of glycoproteins encoded by a group of genes called the major histocompatibility complex (MHC). MHC molecules are transmembrane proteins found on the cell membranes of all higher vertebrates.

A vertebrate does not normally need to be protected against invasion by foreign vertebrate cells, so the antagonistic reaction of its T cells to foreign MHC molecules was both an obstacle to transplant operations and an enigma to immunologists. The enigma was solved when it was discovered that MHC molecules contributed to the binding of T lymphocytes on those host cells that have foreign antigen on their surface, as in the case of the virus-infected cells we have just described (see fig. 33.15).

There Are Two Major Types of MHC Proteins, Class I and Class II

There are two major classes of MHC proteins which have similar types of structures (fig. 33.16). Class I molecules contain three external domains, each about 90 residues in length, a transmembrane region, and a cytoplasmic domain. The third external domain is noncovalently associated with a small polypeptide known as the β_2 microglobulin. Class II molecules are composed of two noncovalently associated polypeptide chains, α and β, whose overall structure resembles that of the class I complex.

There are several genes that encode each type of MHC protein, and there are several alleles that represent different MHC proteins for each of these genes. This gives rise to a tremendous variety of MHC proteins, with the consequence that it is extremely unlikely that any two members of the same species will possess the same assortment of MHC proteins. Thus two humans are not likely to carry the same MHC proteins on their cell membranes unless they are genetically identical twins. This is the reason why transplants between two individuals are almost always rejected—it is usually just a matter of time. Highly inbred strains of mice, which are genetically identical, will display the same MHC proteins and accordingly will take grafts from each other without any rejection.

The two classes of MHC proteins are displayed on different cell types. Class I MHC proteins are found on almost all nucleated cells including killer T cells. Class II MHC proteins are found mainly on cells involved in the immune response including antigen-presenting cells, B cells and T helper cells, but not T killer cells.

T-Cell Receptors (TCRs) Resemble Membrane-Bound Antibodies

The specific receptors found on the membranes of T cells resemble the highly specific antibodies found on the membranes of B cells. Specific T-cell receptors (TCRs) are composed of two disulfide-linked peptide chains (called α and β). Each of these chains share with antibodies the distinctive property of a variable amino-terminal region and a constant carboxyl-terminal region.

Figure 33.16

Structures of class I and class II molecules as they would appear in the lipid bilayers of the cell membrane. Models show striking similarities between the two types of molecules. S—S indicates disulfide bridges in the class I and class II molecules and the β_2 microglobulin.

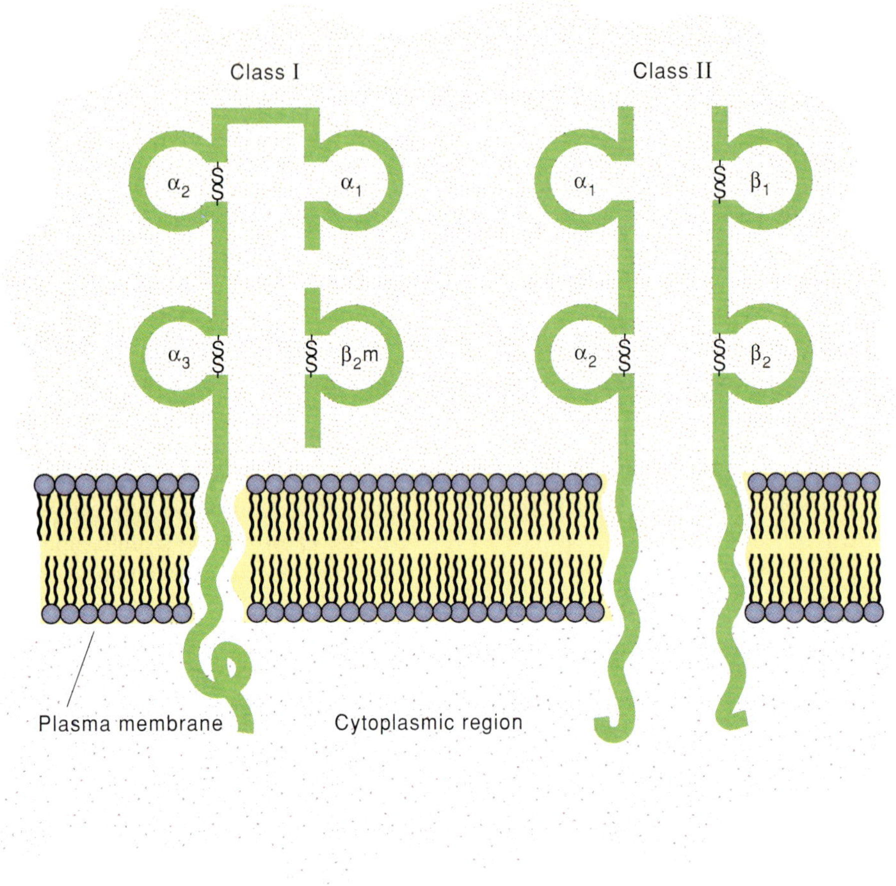

With one exception, all the mechanisms used by B cells to generate antibody diversity are also used by T cells to generate T-cell receptor diversity. The one mechanism that does not appear to operate in T-cell receptor diversification is somatic hypermutation. This is presumably because mutation would be likely to generate killer T cells that would wantonly attack self-molecules. This is much less of a problem for B cells, since most self-reactive B cells could not be activated without the aid of self-reactive helper T cells.

Additional Cell Adhesion Proteins Are Required to Mediate the Immune Response

In addition to membrane-bound antibodies, TCR proteins and MHC proteins, there are other transmembrane proteins required to obtain the specific cell–cell interactions necessary to mediate the immune response. These additional proteins are also important in a number of nonimmune reactions where cell–cell adhesion is important to the organism. We will only consider a limited number of these additional proteins because of the enormous complexity of this subject. Three transmembrane proteins that are directly involved in the T-cell reaction with other cells are members of a family of cell–cell adhesion proteins (the CD family of proteins). The CD3 protein is found on both killer and helper T cells; it forms a complex with the membrane-bound TCR proteins. The CD8 and CD4 proteins interact with this complex but unlike the CD3 proteins, CD8 is only found on killer T cells, while CD4 is only found on helper T cells. The presence of CD4 and CD8 proteins limits the range of binary complexes formed by T cells with other cells. This is because CD4 and CD8 have selective affinities for MHC class II and MHC class I proteins respectively. As a result killer T cells only form complexes with cells that display a processed foreign antigen in a complex with the MHC I protein; this includes most nucleated cells other than those that belong to the immune system. Similarly helper T cells are limited to reactions with cells, which display the appropriate processed antigen in a complex with the MHC II protein; this includes B cells or phagocytes (fig. 33.17).

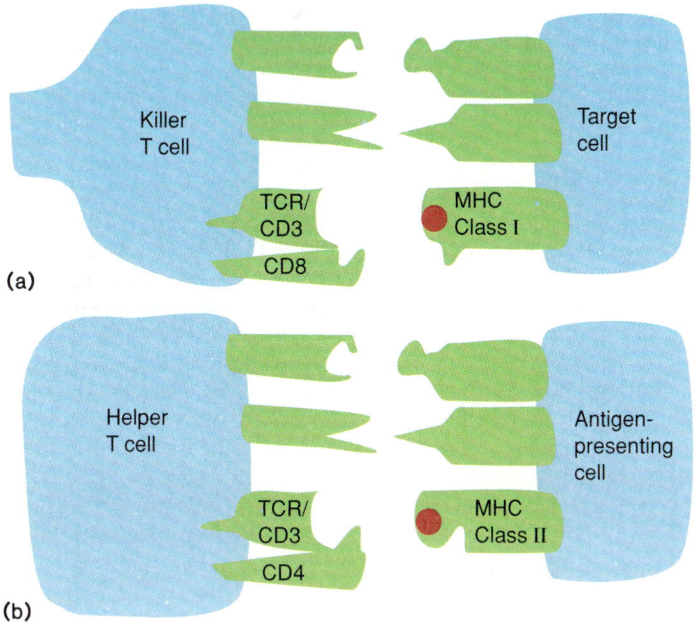

Figure 33.17

Antigen-specific reactions involving T cells. (*a*) Killer T cells interact and destroy target cells that display complexes of processed antigen with an MHC I protein. Effective interaction requires that the TCR protein interact specifically with the processed antigen. All of the remaining reactions between the two cells are general reactions that occur between other killer T cells and other target cells. (*b*) Helper T cells interact specifically with antigen-presenting B cells, provided a specific complex is made between the TCR protein and the processed antigen complexed to the MHC II protein and the processed antigen complexed to the MHC II protein on the B cell. The unlabeled membrane proteins in this figure represent other nonspecific protein–protein interactions that strengthen the binary cell reaction. Processed antigen is indicated in red.

From this discussion we can see that a productive interaction between T cells and other cells requires a highly specific reaction between a TCR protein and a processed antigen that is supplemented by less specific interactions between other transmembrane proteins on the two cell types.

Left out of this discussion is an explanation for the source of the processed antigen on the antigen-presenting B cell. Immature B cells display a membane-bound antibody, which will interact with circulating complementary antigen. Once bound to the antibody the antigen is internalized, processed and ultimately displayed by the MHC II protein on the cell surface of the same B cell. It is the interaction between such a B cell and the appropriate helper T cell that triggers the B cell to divide and differentiate.

Immune Recognition Molecules Are Evolutionarily Related

There is a striking structural similarity between immune recognition proteins that almost certainly reflects their evolution from a common ancestor. All of them contain one or more immunoglobulinlike domains, each with about 90 amino acids stabilized by a conserved disulfide linkage. The simplest member of this family of genes is β_2 microglobulin. Possibly this or a similar protein was the ancestor for the remaining immune recognition proteins.

Summary

In this chapter we have considered the major findings that have led to our understanding of the immune system in vertebrates, expecially mice and humans. The following points are the highlights of our discussion.

1. There are two different classes of white cells (lymphocytes) associated with the immune response, the B cells and the T cells. B cells produce antibodies that are secreted and aggregate foreign substances known as antigens. T cells are of different types; some of them assist B cells to become antibody-forming cells, while others mount an attack on antigens of their own.
2. The most common type of antibody is a tetramer composed of two identical (H) chains and two identical light (L) chains. Each chain is divided into a relatively constant sequence (the *C* segments) and a relatively variable sequence (the *V* segments). Within the variable regions there are subsegments known as the hypervariable regions. These subsegments are the regions that specifically bind different antigens.
3. According to some estimates, each organism is capable of making more than a million antibodies from a limited amount of genetic information. This feat is accomplished with the help of somatic recombination and somatic mutation. Somatic recombination involves splicing at the DNA level. It brings specific *C* and *V* regions together to make a unit that can be transcribed. Somatic mutation is focused on those parts of the gene that encode the hypervariable regions.
4. T helper cells focus the antigen on an immature B cell. In response, some B cells turn into antibody-forming cells and some B cells turn into memory cells for production of antibodies at a later date. The first time an organism is exposed to a specific antigen it takes longer to mount an immunological response. That is because of the lack of memory cells for making a specific antibody.
5. Complement is a group of serum proteins that aids in the defense against microorganisms and removal of antibody-antigen complexes.
6. T cells give rise to an immune response of their own. There are two types of cell-mediated immunity, involving basically different types of T cells. In the delayed-type response, the T cell reacts specifically with antigens and secretes lymphokines. These are substances that attract macrophages or other leukocytes, thus producing a slowly developing inflammatory response. A second type of T cell reacts specifically with antigen bound to target cells and causes their lysis.
7. Tolerance prevents the immune system from attacking self-antigens. To understand tolerance we must appreciate how T cells work. T cells recognize a combination of self and nonself. The cell-surfake antigens that are recognized by T cells are known as the major histocompatibility complex (MHC). If two organisms carry the same histocompatibility antigens, the tissues from the two organisms are completely compatible. The histocompatibility antigens resemble antibodies in structure.

Selected Readings

Ada, G. L., and G. Nossal, The clonal selection theory. *Sci. Am.* 257(2):62–69, 1987.

Honjo, T., and S. Habu, Origin of immune diversity: Genetic variation and selection. *Ann. Rev. Biochem.* 54:803–830, 1985.

Hood, L., M. Kronenberg, and T. Hunkapiller, T cell antigen receptors and the immunoglobulin super gene family. *Cell* 40:225–229, 1985.

Hunkapiller, T., and L. Hood, The growing immunoglobulin gene superfamily. *Nature* 323:15–17, 1986.

Kaappes, D., and J. L. Strominger, Human class II major histocompatibility genes and proteins. *Ann. Rev. Biochem.* 57: 991–1028, 1988.

Klein, J., *Immunology.* Blackwell Scientific Publications, 1990.

Leder, P., The genetics of antibody diversity. *Sci. Am.* 246(5):102–115, 1982.

Lieber, M. R., Site-specific recombination in the immune system. *FASEB J.* 5:2934–2944, 1991.

Milstein, C., Monoclonal antibodies. *Sci. Am.* 243(4):66–74, 1980. An exciting, tissue-culture technique for obtaining pure antibodies specific for an antigen of choice.

Mizel, S. B., The interleukins. *FASEB J.* 3:2379–2388, 1989.

Nowak, M. A., R. M. Anderson, A. R. McLean, T. F. W. Wolks, J. Goudsmit, and R. M. May, Antigenic diversity thresholds and the development of AIDS. *Science* 254:963–969, 1991.

Singer, S. J., Intercellular communication and cell–cell adhesion. *Science* 255:1671–1677, 1992. A comprehensive review of the major cell–cell interactions important in immune reactions.

Springer, T. A., Adhesion receptors of the immune system. *Nature* 346:425–434, 1990. An excellent review.

Tonegawa, S., Somatic generation of antibody diversity. *Nature* 302: 801–803, 1983. A classic paper.

Problems

1. Where do T cells and B cells mature in the body? Where do the mature T and B cells reside in the body?
2. Why does humoral immunity require both B and T cells?
3. What are the two types of cell-mediated immunity?
4. Explain the role of DNA recombination in generating antibody diversity. How does somatic mutation contribute to diversity?
5. How can DNA splicing help stimulate a higher level of transcription of an immunoglobulin gene?
6. What role does RNA splicing play in antibody production?
7. You have two strains of genetically nonidentical mice (strain A and strain B). How could an adult mouse from strain A be made to tolerate a transplant from strain B?
8. How could all the immune recognition proteins have evolved from a single common ancestor protein?
9. If an antibody and its antigen are mixed in a test tube at some optimum concentration, a precipitate is formed. If either the antibody or the antigen is in great excess, no precipitate is formed. Explain these results from what you know about antibody structure.
10. Recently antibodies with catalytic activity have been produced. How would you develop a catalytic antibody for a specific reaction? (Hint: How does an enzyme work?)
11. If you inject a pure protein into a rabbit and raise antibodies to a high titer, the antibodies produced will be a mixed population with various binding constants. Why would a single protein produce many different antibodies?
12. If you added a thiol compound such as 2-mercaptoethanol at a high concentration (1 mM) to an antibody and antigen mixture, no precipitation complex would form even if the concentration of antibody and antigen were optimal. Explain this observation. (Hint: How do thiol compounds work?)

34

CHAPTER

Carcinogenesis and Oncogenes

In the complex developmental process that leads to a mature human being from a fertilized egg, each cell takes up a role in a specialized organ or tissue. Each group of cells proliferates to a certain point and then stops. In the case of a solid organ such as the liver, we see cells of several different types that grow until the organ takes on a fixed size and shape and then cease to grow. Normally, further growth will occur only at the very slow rate required for replacement of aging cells. However, if part of the liver is amputated, a rapid regenerative process ensues so that the liver regrows to approximately the same size it had before the amputation. Other organs or tissues may show more or less growth than the liver, but as a rule, new cells will appear only as old or damaged cells are replaced. Furthermore, the cells of normal tissues do not as a rule mix.

Normally you do not find liver cells mixing with intestinal cells or any other types of cells, suggesting that cells have surface properties that favor interaction of like cells. This was demonstrated in the 1940s by experiments with embryonic tissues. If embryonic tissues are dissociated into single cells and allowed to reassociate in the presence of other cell types, they reassociate with cells of a similar type. It is now known that there are adhesion molecules located on cell surfaces that encourage such associations. This was alluded to in chapter 33 when we were describing the interaction of T cells with other cell types of the immune system. Cell adhesion molecules that favor the association of like cells are of at least two types: some require Ca^{2+} for adhesion, the cadherins, and others do not, the cell adhesion molecules (CAMs). When these molecules are not present, as in culture transformed cell lines, like cells adhere poorly to each other or to other cultured cells. It seems likely that, when tumor cells become invasive, the character of their cell adhesion molecules has been altered. We mention this in passing here as it is probably an important property of malignant cells, but we will have little more to say about it in this chapter as this aspect of malignancy is not well characterized.

Cancer cells are relatively easy to recognize when they occur in groups in the organism (fig. 34.1). A cluster of cancer cells is more rapidly dividing, and it usually clashes with the architecture of normal cells in its immediate surroundings. From cytological inspection alone, we might speculate that cancer cells are breaking the rules for normal cell growth and associations. It appears that the processes that regulate cells are disrupted; cell growth is uncontrolled, and patches of morphologically distinct cells show by their invasive character that they do not recognize normal territorial boundaries between cells of different types. Cancer cells are deadly because ultimately they invade and disrupt a vital area. In this chapter we present a brief overview of some aspects of carcinogenesis and oncogenes.

Figure 34.1

A benign glandular adenoma (*a*) and a malignant glandular adenocarcinoma (*b*). In both cases the tumor cells differ morphologically from the surrounding tissue and show a pattern of disorganized growth.

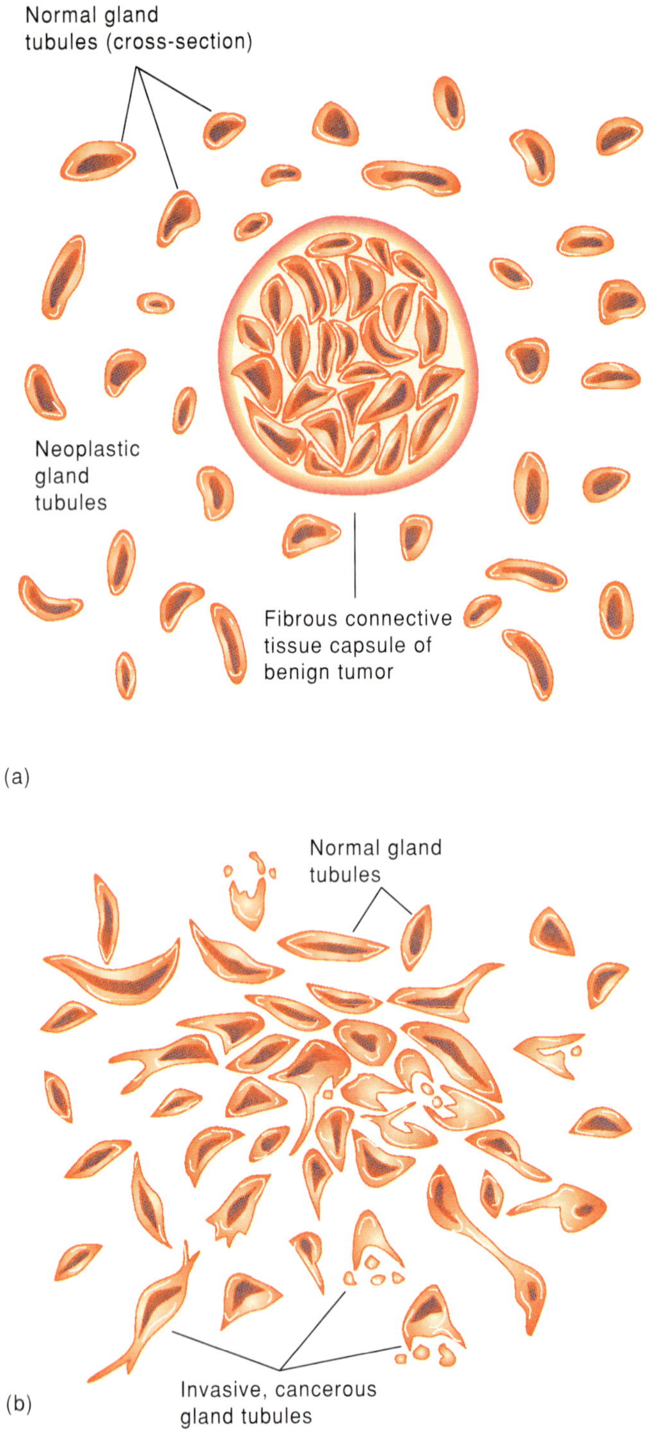

Cancers: Cells Out of Control

Cancer is not a stage in the orderly evolution of a complex organism; it is an organismic catastrophe. Perhaps it is only reasonable to expect that multicellular eukaryotes, with their many dividing cells, will occasionally produce a mutant cell that gets out of control and destroys the system. The notion that cancer cells are out of control was proposed more than half a century ago. In the intervening time we have learned a great deal about the biochemistry of growth regulation. This knowledge has helped us to understand the molecular basis of cancer in many cases. Conversely, the study of cancer has added to our knowledge of normal growth regulation.

Transformed Cells Are Closely Related to Cancer Cells

In the 1950s, systematic methods were developed for taking cells directly from the tissues of whole animals and inducing them to grow as single cells in liquid culture.

An outgrowth of this technology was the development of pure cell lines—cells that divide indefinitely. Primary cultures of dispersed single cells from tissue fragments usually die out after 50 or so duplications. However, during the growth of a primary culture, occasional altered cells appear that possess a somewhat different morphology; they frequently take over the culture, because they are capable of initiating new cultures with far fewer cells. A clone that is derived from one of these indefinitely dividing variants—which has in effect become immortal—is designated a cell line. The transition of primary culture cells to cells that can grow indefinitely is considered to be a step in the direction of becoming cancerous.

Typically, continuous cell lines will grow in culture only on a solid support or "anchor" (the surface of a Petri dish, for example) and in the presence of relatively high concentrations of nutrients. Even then, they will divide only as long as the culture is sparse. When the cell density increases beyond a critical point the growth rate decreases sharply; in fact, the cells of some continuous lines stop dividing altogether once they have formed a confluent monolayer.

At the density that normally halts growth, rare transformed cells continue to multiply and may reach cell densities up to twenty times higher than those of untransformed cells. By picking and subculturing such transformed cells, it is possible to establish clonal lines of transformants and to ask in what ways such cells differ from their untransformed progenitors.

The differences between transformed and untransformed cells can be classified in three ways: changes in cell growth, changes in cell-surface properties, and genetic alterations. These changes are too great in number and too diverse in quality to be caused directly by one or a few mutations. Clearly, then, most of the observed alterations must be secondary and tertiary effects that have occurred as a consequence of some primary event. It is the aim of much current work with tumor-causing agents, especially certain viruses, to discover the molecular nature of this primary event.

Figure 34.2

AIDS virion structure. Numbers associated with proteins are kilodalton masses (e.g., gp120 is a 120-kilodalton protein).

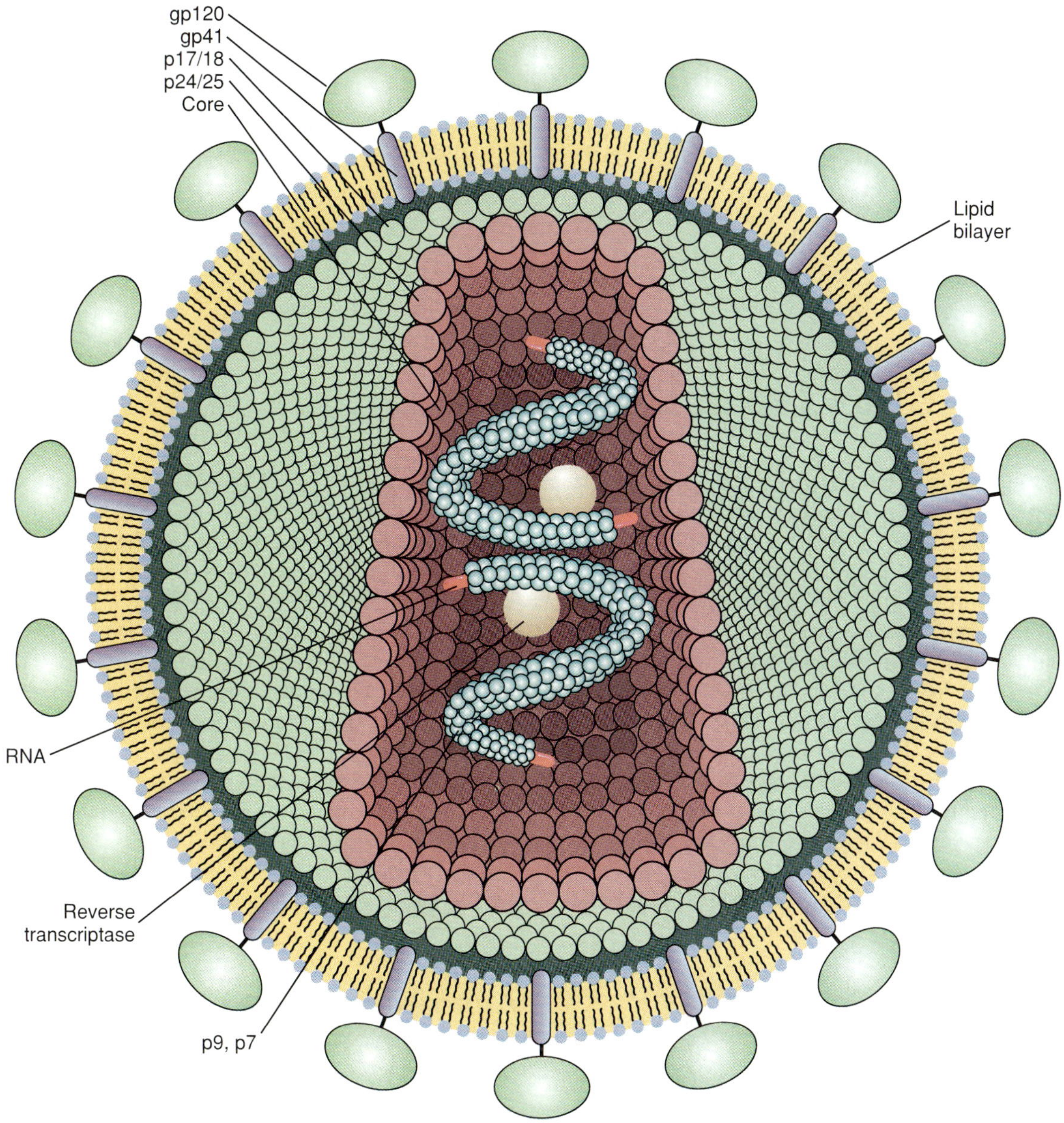

Transformed cells sometimes, but not always, give rise to tumors when injected into an isogenic animal. There may be numerous reasons for the failure to produce tumors on all occasions. One of the best-known reasons for lack of tumor production is immunological rejection. The transformed cell resulting from exposure to a tumor virus frequently displays viral antigens that are recognized as foreign to the animal host. This recognition excites a positive immune response that can lead to rejection of the transformed cells.

Just as transformed cells frequently do not develop into tumors when injected into whole animals, so is it that tumor cells do not always grow well when dispersed into tissue culture. Nevertheless, tumor cells usually adapt more readily to growth in tissue culture than normal cells taken from the whole animal.

Environmental Factors Influence the Incidence of Cancers

In seems very unlikely that we will find cures for many typcs of cancer in the foreseeable future. Therefore it would be a good idea to put considerable effort into cancer prevention, especially since there are reasons for believing that the vast majority of cancers are preventable. The frequency of different kinds of

Table 34.1
Variation in the Incidence of Cancers by Country

Type of Cancer	Country of High Incidence	Country of Low Incidence	Ratio (High/Low)
Skin	Australia	India	200
Esophagus	Iran	Nigeria	200
Lung	England	Nigeria	35
Prostate	United States	Japan	15
Liver	Mozambique	England	40
Breast	Canada	Israel	7
Rectum	Denmark	Nigeria	20
Colon	United States	Nigeria	10
Ovary	Denmark	Japan	6
Stomach	Japan	Uganda	25

cancer varies enormously in different human populations (table 34.1). For example, breast and colon cancer are much higher in the United States than in Japan, whereas stomach cancer is much higher in Japan. In these cases, diet is most likely the culprit. Liver cancer is highest in Third World countries. Once again, diet seems the most likely cause. Lung cancer is more prevalent in countries where smoking is popular, and indeed, it is well documented that heavy smokers are much more likely to get lung cancer than nonsmokers. Overall, it is clear that much of the variation in cancer incidence is environmental rather than genetic, since these differences tend to disappear from one generation to the next in migrant populations. For example, the high incidence of stomach cancer in Japan is not matched by a similar figure among Japanese-Americans in the United States.

There is a growing conviction that the carcinogenic agents in the environment are active as cancer-causing agents because they produce mutations. The hunt for carcinogens has been facilitated by Bruce Ames, who developed a test for carcinogens based on the mutagenic action of a compound on bacteria.

Ames tests have shown that many carcinogens originate from food and chemical pollutants. In addition to chemicals, ionizing radiations are carcinogens. For example, there are strong indications that the ionizing radiation we get from the sun is the major cause of skin cancer. Thus the incidence of skin cancer in the United States is much higher in the South than in the North. High dosages of radioactive materials or x-rays are also strongly correlated with the incidence of numerous forms of cancer.

In many cases, exposure to certain viruses is correlated with specific types of cancers. For example, liver cancer is 300 times more common in individuals who have previously been infected with hepatitis B virus. Although we can only speculate on the precise reasons for this, such numbers indicate that hepatitis resulting from hepatitus B virus infection is almost certainly the major cause of human liver cancer.

A cancer of the B lymphocytes known as Burkitt's lymphoma is strongly correlated with infection by the Epstein-Barr virus. Once again, we do not know the precise chain of events that lead to cancer. Hepatitis B and Epstein-Barr are DNA viruses. Certain RNA viruses have also been shown to be correlated with cancer. The best known is the human immunodeficiency virus (HIV-1, the AIDS virus; fig. 34.2). The AIDS virus is believed to cause cancer indirectly by incapacitating the helper T cells that are required for the humoral immune reaction (see chapter 33). Various types of cancer follow infection by the AIDS virus, a fact testifying to the key role that the immune system plays in protecting the body from cancer. The RNA virus known as human T-cell leukemia virus type I (HTLV-I) causes T-cell leukemia. It is believed that this virus carries a cancer-causing gene (an oncogene) whose continuous expression upsets the normal metabolism, causing infected cells to become transformed to cancer cells.

Cancerous Cells Are Genetically Abnormal

In the previous section we stated that there was a strong correlation between the mutagenic properties of a chemical and its potency as a carcinogen. This is consistent with other evidence that mutation often plays a causal role in carcinogenesis.

The most graphic illustrations of the correlation between genetic abnormality and cancer come from the frequent association of certain types of tumors with highly specific chromosomal translocations. Chromosomal translocations involve the movement of a segment of chromosome from one location to another. Frequently translocations occur between different chromosomes in a reciprocal fashion. This is the case for translocations between human chromosome 8 and the chromosomes that carry the antibody genes, 14, 2, and 22.

A reciprocal exchange of segments of chromosome from the ends of chromosomes 8 and 14 is frequently found in the genome of cancer cells of the Burkitt's lymphoma (fig. 34.3). Burkitt's lymphoma is a tumor of B lymphocytes; for unknown

Figure 34.3

Translocation associated with Burkitt's lymphoma. The protooncogene *c-myc,* on the end of chromosome 8q, is translocated to the end of chromosome 14q, adjacent to the immunoglobulin constant gene, *Ig-Cμ*.

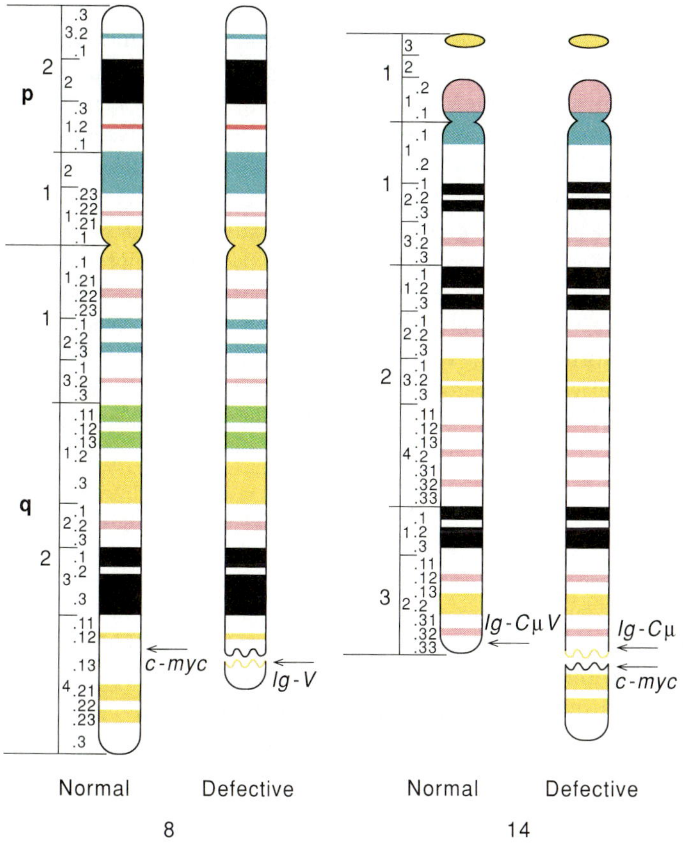

Figure 34.4

The Philadelphia chromosome arises by a reciprocal translocation between chromosomes 9 and 22. Note that the *abl* protooncogene is moved in the translocation process.

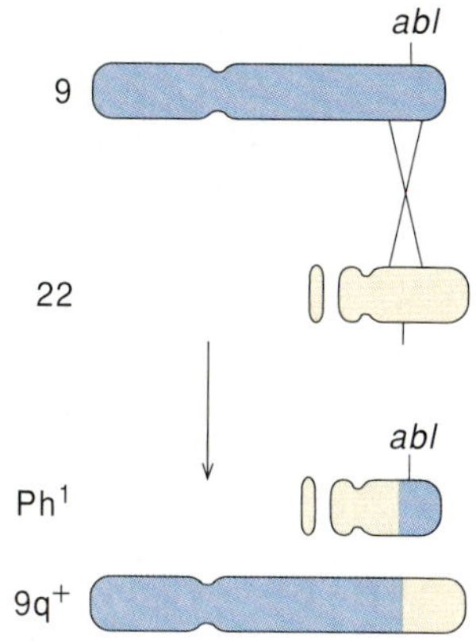

reasons, its occurrence is most common in Africa. The resulting tumor cells usually secrete antibodies. The fact that antibody genes, which are so active in B cells, map to the same chromosomal regions involved in the specific translocations in Burkitt's tumors led to the suggestion that a specific oncogene on chromosome 8 might be activated if it were placed near one of these antibody genes. This suggestion proved to be correct; the oncogene involved is *c-myc.* We will have more to say about *c-myc* later.

The first chromosome abnormality found to be associated with a cancer was the Philadelphia chromosome, which arises as the result of a reciprocal translocation between chromosomes 9 and 22 (fig. 34.4). The tumor associated with this translocation is chronic myelogenous leukemia and results from the activation of the *c-abl* oncogene, which is normally located on chromosome 9.

Whereas it is a common belief that tumor formation involves mutation, certain properties associated with the etiology of teratocarcinoma do not fit with this concept. Teratocarcinoma is a unique type of tumor in which the tumor is associated with a cluster of cells of different types including stem cells and differentiated cells of various types. Only the stem cells are oncogenic. These stem cells may be replicated *in vitro.* Upon subcutaneous injection these cells produce tumors, but upon injection into a blastocyst embryo these stem cells revert to normal cells, which can be identified in the adult animal by the genetic markers they carry. If these stem cells bear a mutation that accounts for their oncogenicity, then it is hard to see how this mutation could revert to wild type by placing it in the cytoplasmic environment of the embryo. Possibly the teratoma is an exceptional situation where the tumor is caused by abnormal development not necessitating a mutation.

Some Tumors Arise from Genetically Recessive Mutations

The oncogenes *c-myc* and *c-abl* are dominant to their wild-type alleles. Many tumors arise from oncogenes that show recessive behavior when compared with their wild-type alleles. Such is the case with the genetic locus implicated in the malady known as bilateral retinoblastoma, in which tumors develop in the retinas of both eyes. Human subjects with this condition usually inherit a genetically recessive mutation on chromosome 13. Sometimes the mutation is microscopically visible as a small deletion. In general, recessive mutations are not apparent unless they are present in the homozygous state. However, in the course of the very extensive cell duplication that ensues in retinal development, there is a high probability that at least one cell in each retina will mutate in the wild-type allele of the genetic locus in question. This change will result in a homozygous state and in the conversion of the normal cell to a tumorous cell that will proliferate at a rapid rate. As a result of this somatic mutation, both eyes usually develop tumors. It seems likely that genes that give rise to tumors when they are defective or lacking, as in retinoblastoma, are of the regulatory type. Some other tumors that result from recessive genetic lesions are listed in table 34.2.

Table 34.2
Recessive Genetic Lesions in Human Tumors

Tumor	Chromosomal Locus
Retinoblastoma and osteosarcoma	13 (q14)
Wilm's tumor (mephroblastoma)	11 (p13)
Embryonal tumor of the kidneys, muscle, liver, and adrenal glands	11p —
Bladder carcinoma	11p
Acoustic neuroma or meningioma	22

Many Tumors Arise by Mutational Events in Cellular Protooncogenes

A growing number of cellular genes have been implicated in a wide variety of cancers. These genes do not lead to cancers in their normal state, but only after a mutational event has somehow resulted in their altered expression. For that reason such genes are called protooncogenes in their normal state and oncogenes in their mutated, cancer-causing state. Most protooncogenes, including *c-abl* and *c-myc,* were not known until very similar genes were found as a result of studies of tumor-causing viruses.

Oncogenes Are Frequently Associated with Tumor-Causing Viruses

The first virus linked to tumor production was discovered in 1908 by Ellerman and Bang. Their finding was followed in 1911 by the better known discovery by Peyton Rous of a virus that produces sarcomas in chickens (now known as Rous sarcoma virus, or RSV). Later (1932) Shope showed that a papilloma virus produced cutaneous fibroma tumors in rabbits, and Lucke (1934) showed that adenoviruses produced renal adenocarcinoma in the frog. These and other discoveries made in the 40s and 50s indicated that exposure to certain viruses could result in tumor production. However, viral oncogenes had not yet been identified, and skeptics still classified cancer-causing viruses along with other carcinogens.

The development of tissue culture techniques in the 1950s permitted a systematic investigation of the causal link between tumor viruses and cellular transformation in culture. In many ways the properties of cells transformed in culture resembled those of tumor cells *in vivo,* and so the tissue culture system was considered to be a convenient way to explore the oncogenic properties of viruses. It was noted with some tumor viruses that the response of cells of different species was quite different. On cells that were normally permissive for virus replication, no tumors arose. This was hardly surprising, in view of the fact that the virus replication cycle, particularly for the DNA viruses, usually results in cell death. What was surprising was the finding that the same virus on another cell type, in which the virus could not replicate, would sometimes result in cell transformation. Observations of this sort led Renato Dulbecco to the hypothesis that there was a parallel between lysogeny of *E. coli* by λ and transformation of animal cells by oncogenic viruses. In λ, the viral genome can exist either as an independent entity, leading to virus replication and cell death, or as an integrated passive state, leading to a lysogenic cell (see chapter 30). With λ the two states—lytic or active, and lysogenic or passive—are attainable in the same cell type. Tumor-causing viruses are capable of existing in two states also. However, in this case the two states, actively replicating and passively integrated, are not normally attainable in the same cell type. Usually two cell types of different species are required. In one cell type the viral genome invariably exists in the actively replicating form, causing immediate cell death; in another cell type the viral genome can exist only in a passive integrated form. Dulbecco suggested that in the integrated state the viral genome expresses a limited number of gene products that lead to cellular transformation.

So far so good. Dulbecco's suggestion appeared to constitute a reasonable hypothesis for explaining the oncogenicity of certain DNA viruses. But it was also known that certain RNA viruses cause tumors. Indeed, one of the first known cancer-causing viruses, the Rous sarcoma virus, was an RNA virus. To explain the oncogenicity of RNA viruses, Dulbecco proposed that cancer-causing RNA viruses form a DNA copy of their genome that can be integrated into the host cellular genome. This bold proposal—that an RNA genome could be converted into a DNA copy—was confirmed by experiments in both David Baltimore's and Howard Temin's laboratories. Both groups found an enzyme, reverse transcriptase, uniquely associated with RNA tumor-causing viruses that could synthesize DNA from an RNA template.

The finding that many viruses cause cancer only after inserting their genome into the host genome has spurred new efforts in cancer research. More and more attention has been paid to viruses. With the help of genetic manipulations, the cancer-causing genes within the viruses have been identified and their gene products have been studied. The results of these studies are providing a broad platform of understanding about the kinds of biochemical processes that result in the conversion of normal cells to cancer cells.

The Role of DNA Viral Genes in Transformation Reflects Their Role in the Permissive Infectious Cycle

Usually tumor-causing DNA viruses possess oncogenes that bear no sequence relationships to any host genes. We can probably best understand the role of DNA viral oncogenes in cellular transformation by appreciating the role these genes play in the virus lytic cycle (table 34.3). First of all, the oncogenes we have found in the adenovirus, polyoma virus, and SV40 virus are all expressed early in virus infection. These genes also trigger a transient transformation process in most cells they infect, where

Table 34.3
Properties of DNA Virus Oncogenes

Viral Gene and/or Gene Product	Function in Lytic Infection	Function in Transformation	Other Properties
Adenovirus			
E1A 9 kDa, 26 kDA, 32 kDa	Required for expression of other early genes	Immortalizes cells; partial transformation	General transcription activator
E1B 9 kDa, 55 kDa	Required for lytic infection	Required for full transformation and tumorigenic cells Either 19 kDa or 55 kDa will suffice	Associates with p53[a] 55-kDa protein found primarily in nucleus; 19-kDa protein in endoplasmic reticulum and in nucleus
SV40			
Big T 90 kDa	Binds to specific sites; necessary, for DNA synthesis; inhibition of early RNA synthesis, and activation of late RNA synthesis	Immortalizes and transforms	Stimulates host DNA synthesis on microinjection; associates with p53[a]
Small t	None (?)	Required for full transformation on some cell lines	—
Polyoma			
Big T 100 kDa	Binds to DNA; necessary for DNA synthesis and regulation of early and late RNA synthesis	Immortalizes cells; only amino terminal 40% fragment is required	Associates with p53[a]; ATPase activity; gyrase activity
Middle T 56 kDa	None (?)	Transforms cells and makes them tumorigenic	Associates with $pp60^{c\text{-}src}$ [b]
Small t	None (?)	Transformed cells adhere less firmly when small t is present	Found in nucleus

[a]p53 is a cellular protein with unknown function. It is found in high concentrations in some tumors and it may be an oncogene.
[b]$pp60^{c\text{-}src}$ is a cellular protooncogene.

a permissive virus replication cycle is not possible. Thus it seems likely that these genes are primarily regulatory. In SV40 and polyoma the so-called big-T genes are clearly regulatory in productive infection, as they are involved in regulating both viral DNA and RNA synthesis. In adenovirus there is no comparable gene, but the early genes *E1A* and *E1B* may play a similar, albeit indirect role. *E1A* functions as a transcription activator.

Another lead as to how a DNA virus transforms cells is to look at the cellular proteins they influence. There is a striking parallel here between the functions of adenovirus *E1B,* SV40 big T, and polyoma big T in this regard. These three genes are required for immortalization or transformation, and they all form a tight association with the cellular protein p53, which is considered to be a good candidate for a protooncogene-related protein.* Finally, the polyoma middle-T protein, which is required for transformation by polyoma virus, forms a strong complex with the cellular protein encoded by the protooncogene *c-src.* It seems likely that these associations between viral oncogenes and host proteins result in changes in regulatory functions, and that if we knew what regulatory functions were being affected we would be on the threshold of understanding the biochemical mechanism(s) involved in transformation. We will have more to say about this after we have considered the retroviruses.

*In its normal protooncogenic form, p53 inhibits SV40 replication; in its mutated oncogenic state, it does not inhibit SV40 replication, nor does it bind to T antigen.

p53 Is the Most Common Gene Associated with Human Cancers

The p53 protein acts as a negative regulator of cell growth; it was first detected through its association with the SV40 big-T oncoprotein in virus-transformed cells. Viral oncoproteins like SV40 T antigen and adenovirus *E1B* sequester p53 protein in inactive complexes. Whereas this results in an increase in the steady-state concentration of p53, the p53 is no longer able to reach its normal site of action in the nucleus.

Many mutant *p53* alleles favor growth and transformation in cells that continue to carry intact, wild-type *p53* gene copies; accordingly, such mutant *p53* alleles act in a dominant fashion. The dominant behavior of mutant *p53* could be explained by the normal active form of the protein. In its active form p53 protein assembles into homotetramers and higher order homo-oligomeric structures. Defective subunits of such an oligomerizing protein (for example, mutant p53 molecules) may participate in forming a multi-subunit complex together with wild-type monomers and, in so doing, poison the function of the complex as a whole.

Figure 34.5

Hypothesis for how a retrovirus picks up a cellular protooncogene. The RNA-DNA cycle for retroviruses is shown in figure 26.28. The proviral DNA is believed to recombine the cellular DNA in such a way that it picks up a region of the cellular DNA containing a protooncogene.

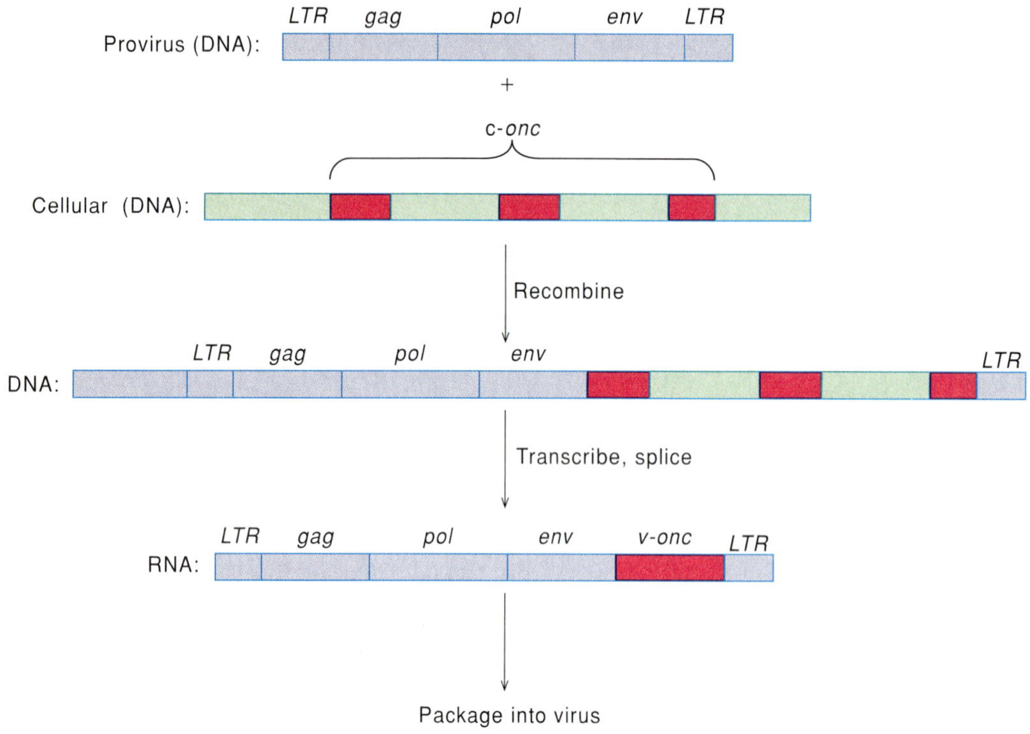

Normal p53 protein binds DNA in a sequence-specific manner and thus most likely regulates gene transcription. Cotransfection experiments show that wild-type p53 activates the expression of genes adjacent to a p53 DNA-binding site. Cells bearing oncogenic forms of p53 have lost this activity.

The best demonstration that the loss of active normal p53 explains the oncogenic behavior of mutant p53 comes from studies in which a null mutation was introduced into the gene by homologous recombination in murine embryonic stem cells. Mice homozygous for the null allele appear normal but are prone to the development of a variety of neoplasms by 6 months of age. These observations indicate that a normal *p53* gene is dispensable for embryonic development, but that its absence predisposes the animal to neoplastic disease. They also indicate that the oncogenicity of the mutant form of *p53* is most likely due to its negative effect on the normal protein.

Somatic mutations of *p53* have been implicated as causal events in the formation of a large and ever-increasing number of common tumors, including those involving the hematopoietic organs, bladder, liver, brain, breast, lung, and colon. In view of the fact that there seem to be so many other oncogenes, we must ask why *p53* is so often implicated in commonly occurring tumors. This appears to stem from its genetic and biochemical traits. Point mutations create carcinogenic *p53,* and such simple genetic changes occur readily. Unlike *ras,* for example, where point mutations productive for cancer are limited to specific changes in two or three codons, the cancer-favoring missense mutations of *p53* can occur in at least 30 distinct codons in its reading frame. In addition, these point mutations often create dominant alleles that produce shifts in cell phenotype even without a reduction to homozygosity.

Retroviral-Associated Oncogenes That Are Involved in Growth Regulation

The first intensively studied oncogene associated with a retrovirus was the RSV *src* gene. It is clear that the *src* gene found in RSV, *v-src,* is structurally quite similar to a normal host gene found in the chicken genome called *c-src.* Subsequent work has shown that at least 30 other animal retroviruses exist that have acquired a cellular oncogene during their evolution. This uncanny association of retroviruses with cancer-causing genes has made retroviruses useful devices for scanning the cellular genome for the presence of protooncogenes (fig. 34.5).

Table 34.4
Retroviral Oncogenes and Their Cellular Homologs

Oncogene	Viral Origin[a]	Viral Gene Product	Cellular Homolog	Activity	Subcellular Location
sis	Simian sarcoma virus	$p28^{sis}$	PDGF B-chain	PDGF agonist	Cytoplasm
src	Rous sarcoma virus	$pp60^{v\text{-}src}$	$pp60^{c\text{-}src}$	Tyrosine kinase	Plasma membrane
fps[b]	Fujinami sarcoma virus	$p140^{gag\text{-}fps}$			
abl	Abelson murine virus	$P120^{gag\text{-}abl}$	$p150^{c\text{-}abl}$	Tyrosine kinase	Plasma membrane
erbB	Avian erythroblastosis virus	$gp65^{erbB}$	Truncated EGF receptor	Tyrosine kinase	Plasma membrane
myc	Avian myelocytomatosis virus MC29	$P110^{gag\text{-}myc}$	$p58^{c\text{-}myc}$	Binds DNA	Nucleus
H-ras	Harvey murine sarcoma virus	$p21^{v\text{-}Hras}$	$p21^{c\text{-}H\text{-}ras}$	Threonine kinase binds GDP or GTP	Plasma membrane
K-ras	Kirsten murine sarcoma virus	$p21^{v\text{-}Kras}$	$p21^{v\text{-}Kras}$		
N-ras	See footnote *c*.		$p21^{c\text{-}N\text{-}ras}$		

[a]Only one example of a virus strain is given for each oncogene.
[b]*fps* and *fes* are homologous genes from chicken and cat, respectively.
[c]The transforming gene product, $p21^{N\text{-}ras}$, identified by transfection experiments.

A listing of some of the better-characterized retrovirus-associated cellular oncogenes appears in table 34.4. In this table the name of the oncogene is indicated in the leftmost column. Proceeding to the right in the table we have indicated: (1) the virus of origin, (2) the viral gene product, (3) the cellular homolog of the viral product; (4) the activity associated with the viral gene product, and (5) the subcellular location of the viral gene product, which in most cases is similar to the subcellular location of the cellular homolog.

The src *Gene Product*

The protein encoded by the *src* gene has a molecular weight of 60 kDa. Like many other oncogenic proteins, it is bound to the inner plasma membrane and it possesses a tyrosine phosphokinase activity; that is, it catalyzes the addition of a phosphate group to the tyrosine hydroxyl groups of various proteins. It is considered highly likely that the kinase activity of the src protein is associated with its transforming activity, since its loss in mutant *src* genes leads to loss of transforming activity. However, since many different proteins are phosphorylated by the src kinase, it is not clear which phosphorylations are crucial to the complex physiological response that triggers transformation.

The sis *Gene Product*

Many oncogenes are associated with tumors produced from a restricted class of differentiated cells. Such is the case with the *sis* oncogene, found in the simian sarcoma virus. This oncogene is believed to be closely related to the gene for platelet-derived growth factor, PDGF. PDGF is a small protein, synthesized in platelets, that stimulates the growth and division of target cells that carry a membrane-bound specific receptor for PDGF—in particular, the mesenchymal cells involved in wound healing. Tumors originate when the target cells carrying the PDGF receptor mutate to cells capable of synthesizing their own PDGF. This situation is believed to lead to uncontrolled growth because the factor continuously stimulates cell proliferation in the very cells in which it is synthesized (fig. 34.6).

The PDGF protein contains two polypeptide chains, A and B, with molecular weights of 28,000 and 32,000, respectively. The amino acid sequence of the PDGF protein is very similar to that of the predicted sequence of the transforming p23 sis protein of the SSV virus, indicating a common ancestral origin. Furthermore, it is believed that the SSV virus causes tumors by synthesizing large amounts of the viral p23 sis protein, which stimulates the unregulated proliferation of target cells that carry the PDGF receptor.

The erbB *Gene Product*

Epidermal growth factor (EGF) is a small mitogenic protein that stimulates the proliferation of cells carrying specific membrane-associated EGF receptors. The EGF receptor has a strong amino acid sequence homology with $gp65^{erbB}$, the transforming protein of avian erythroblastosis virus (AEV). The EGF receptor has tyrosine kinase activity, like the src protein. In addition, the EGF receptor contains an extracellular domain that binds the EGF growth factor.

The *v-erbB* oncogene acts to expand a pool of highly mitotic, undifferentiated erythroid precursor cells, but these are poorly tumorigenic, because they differentiate at high rates into post-mitotic, end-stage red cells. Another potential oncogene *v-erbA* on its own acts to block differentiation of these erythroid precursors, but creates no tumors because it is unable to provide the mitogenic impetus needed to expand the pool of stem cells.

Figure 34.6

Mechanism of mitogenesis in normal and transformed cells. (*a*) Schematic representation of the growth-factor-related mitogenic pathway in normal cells. Here, (1) represents the growth factor, (2) the growth factor receptor, and (3) the intracellular messenger system that transmits the mitogenic signal from the receptor to the nucleus. (*b*) Schematic representation of a possible perturbation of the growth-factor-related mitogenic pathway in transformed cells. Here, (1) represents endogenous production of growth factor that may stimulate the cell. The endogenously produced factor may be secreted and interact with growth factor receptors at the cell surface (as shown) or, alternatively, activate the receptor in an intracellular compartment.

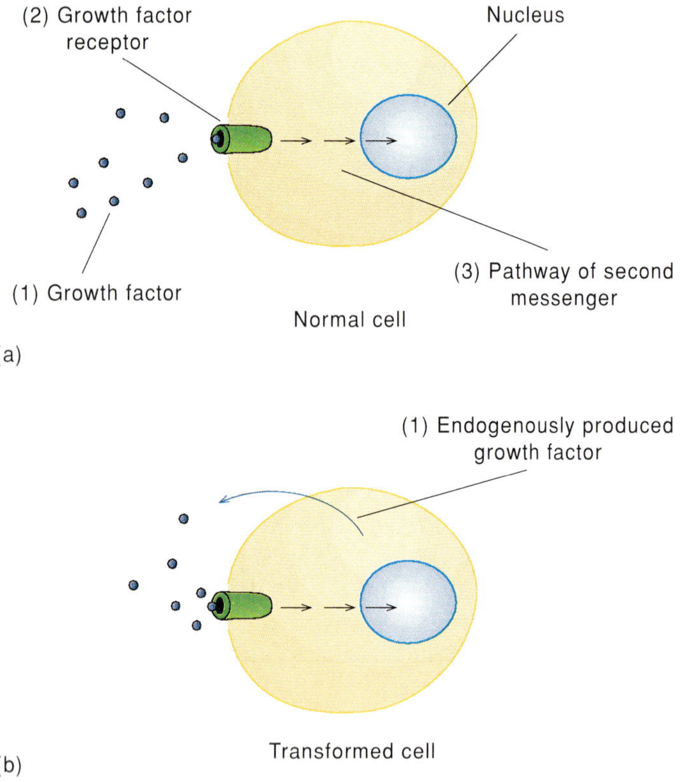

Figure 34.7

Interaction between the *ras* protooncogene and GDP. The protein is shown as a green ribbon, interacting with GDP (dark purple, yellow, and orange). The dark purple rectangle represents the guanosine; the dark yellow pentagon, the ribose; and the orange circles, the phosphates of GDP. The domains of the protein interacting with each of these parts of GDP are represented by sleeves on the protein ribbon, color-coded to match the corresponding part of GDP. The P domain of the protein contains glycine 12; this amino acid is in a critical position next to the phosphates of the nucleotide. The N and C termini of the protein are labeled.

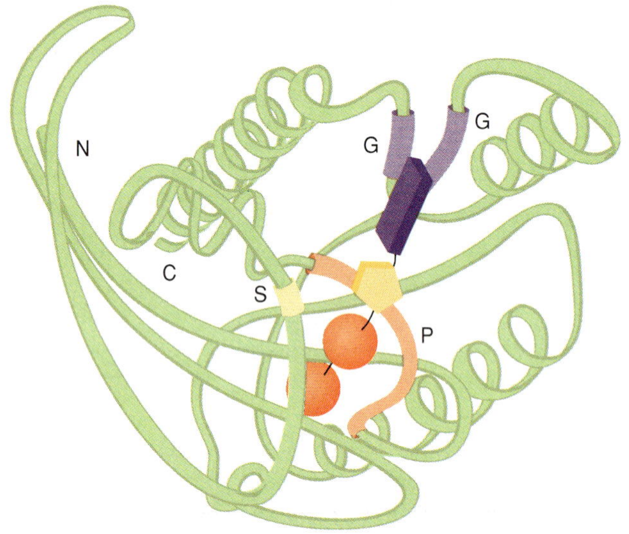

The two genes, ***erbA*** and ***erbB***, carried into erythroid precursors by avian erythroblastosis virus, act in concert to create an aggressive erythroleukemia; ***v-erbB*** drives expansion of the pool of undifferentiated precursor cells, while *v-erbA* blocks their conversion by the differentiation pathway. The *v-erbA* allele that participates in formation of chicken erythroleukemias is a mutant version of a transcriptional regulatory protein, the chicken thyroid hormone (triiodothyronine) receptor. Function of the wild-type receptor protein is blocked in the presence of *v-erbA*, because the latter occupies critical DNA-binding sites in a way that precludes association by the wild-type receptor protein and inhibits transcription.

The ras *Gene Product*

One of the best understood protooncogenes is *ras*. The *ras* gene found in normal mammalian cells is closely related to the *ras* oncogenes of Harvey and Kirsten murine sarcoma viruses, *H-ras* and *K-ras*, respectively. *Ras* oncogenes have also been isolated with the help of DNA transfection techniques. Activated *ras* genes have been isolated from a wide variety of dissimilar neoplasms, including carcinomas, sarcomas, neuroblastomas, lymphomas, and leukemias. Some of these *ras* genes are associated with the murine sarcoma viruses (e.g., *N-ras* in table 34.4). All members of the *ras* gene family encode closely related proteins of approximately 21 kDa, which have been designated p21. Unlike the oncogenes associated with many retroviruses, the level of p21 expression is similar in normal cells and in many different human tumor-cell lines. Nucleotide sequence analysis of the *H-ras* transforming gene isolated from a human bladder carcinoma cell line has indicated that the transforming activity of this gene is the consequence of a point mutation altering amino acid 12 of wild-type p21 from glycine to valine (fig. 34.7).

Subcellular fractionation and immunofluorescence of normal cellular and viral ras proteins have indicated that p21 is localized on the inner face of the plasma membrane. Both the normal and the mutant ras proteins bind GTP specifically and strongly, but only the normal protein has GTPase activity.

We understand normal *ras* gene function better than that of most other protooncogenes because the yeast *Saccharomyces cerevisiae* carries *ras* genes, and many researchers have taken advantage of the powerful genetic tools available in yeast to study them. *S. cerevisiae* contains two closely related but distinct genes, *RAS1* and *RAS2*, that encode proteins that are highly homologous to the mammalian ras proteins. While neither *RAS1* or *RAS2* by itself is an essential gene, *RAS* function is required for the continued growth and viability of yeast cells. Thus *ras1⁻*, *ras2⁻* double mutants are nonviable unless they carry a suppressor mutation *bcyL*. The *bcyL* mutation suppresses lethality in adenylate-cyclase-deficient yeast. We will see the significance of this fact presently. Yeast strains that are $ras2^-$ are noticeably depressed in cAMP levels. In $ras1^-$, $ras2^-$, *bcyL* triple mutants the levels of cAMP are virtually undetectable. But cells containing *RAS2 val19*, a *RAS2* allele with a missense mutation analogous to the one that activates the transforming potential of mammalian *ras* genes, have cAMP levels significantly higher than those observed in wild-type cells. Membranes from $ras1^-$, $ras2^-$, *bcyL* triple mutants lack the GTP-stimulated adenylate cyclase activity present in membranes from wild-type cells, and membranes from the *RAS2 val19* yeast strain have elevated levels of a GTP-independent adenylate cyclase activity. Mixing membranes from $ras1^-$, $ras2^-$ yeast with membranes from an adenylate-cyclase-deficient yeast leads to the reconstitution of a GTP-dependent adenylate cyclase. It appears, then, that *RAS* encodes a protein or proteins that somehow regulate activity of membrane-bound adenylate cyclase in a GTP-dependent manner. In the yeast mutant *RAS2 val19* and in the homologous mammalian mutant, the regulatory properties of this protein appear to be disrupted so that the adenylate cyclase is overactive and no longer requires GTP for activation. Further experiments have been done to show that the *ras* genes from yeast cells can function in mammalian cells and vice versa. Thus it has been shown that yeast carrying $ras1^-$, $ras2^-$ mutations remain viable if they carry the mammalian *H-ras* gene. Conversely, a mutant yeast *RAS* gene has been isolated that can bring about tumorigenic transformation in mammalian cells.

From what has been said, you might expect that ras proteins would be involved in activating adenylate cyclase in mammalian cells. However, there is no indication that this is the case. In mammals, other heterotrimeric GTP-binding proteins are involved in the adenylate cyclase pathway. Apparently, the *ras* system is a useful signaling module that has been adapted to various uses in different organisms; it could be likened to an electronic switching device, which could serve different functions depending upon the devices it is attached to. In this connection we note that a structural analysis indicates a similarity between $p21^{ras}$ structure and the G-domain of the protein synthesis elongation factor Tu(EF-Tu) from *Escherichia coli*.

The myc *Gene Product*

The *c-myc* gene, identified originally as the cellular homolog of the transforming determinant carried by avian myelocytomatosis virus, is altered in association with a broad spectrum of neoplasms. Consistent with the observation that altered *c-myc* is associated with tumors of diverse origin, it has been observed that normal *c-myc* is expressed in a variety of tissues. Thus *c-myc* appears to encode a function associated with a ubiquitous metabolic pathway.

The *c-myc* gene encodes a 49-kDa protein that is highly concentrated in the nucleus of the cell. The concentration of c-myc protein normally varies appreciably with the metabolic state of the cell, increasing by more than an order of magnitude in cells just prior to chromosome duplication and cell division.

In most human tumors associated with the *c-myc* gene, the concentration of c-myc protein is greatly amplified. The Burkitt lymphoma, which we discussed earlier, is a notable exception. In this case the c-myc transcript is marginally, and in some cases not at all, increased by comparison with control lymphoblastoid cell lines. Recall that in the Burkitt lymphoma, reciprocal chromosomal translocations are found that involve a chromosome carrying *c-myc* protooncogene (chromosome 8) and one of the chromosome segments (usually chromosome 14) bearing immunoglobulin genes.

In at least one Burkitt lymphoma cell line, the structure of the amino-acid-coding portion of the translocated *c-myc* gene, and hence its predicted protein product, has not been altered, a fact suggesting that activation of *c-myc* must be mediated via a regulatory disturbance. Normally, c-myc protein synthesis is strongly regulated with respect to the cell cycle and tightly repressed in quiescent cells. The translocated *c-myc* gene appears to be somewhat deregulated with respect to its expression during various phases of the cell cycle.

Members of the *myc* gene family have been implicated in the control of normal cell proliferation as well as in neoplasia. A more direct role for *myc* genes in transformation is indicated by their ability to transform primary rat embryo fibroblasts in association with the *c-H-ras* oncogene.

An important clue to *c-myc* function was the discovery in the conserved carboxy-terminal region of three structural motifs, the leucine zipper (LZ), helix-loop-helix (HLH) and basic region (B). These motifs were originally defined in a number of other sequence-specific DNA-binding proteins but had not previously been found within a single protein.

The comparatively weak homo-oligomerization efficiency of the c-myc HLH/LZ suggested that a partner protein(s) might exist that hetero-oligomerizes with c-myc to form a specific DNA-binding complex. A protein termed max (for myc-associated "X" factor), was shown to interact with c-myc in a manner that required the integrity of the c-myc HLH/LZ motif. Max was also demonstrated to associate with N- and L-myc proteins, but not with the HLH/LZ proteins USF or AP-4, nor with several other HLH or leucine zipper proteins. In DNA-binding assays, c-myc-max specifically recognized a c-myc DNA-binding site (CACGTG) in a manner that required both an intact max basic region and the HLH/LZ motifs.

What is the molecular function of the c-myc-max complex *in vivo?* At present the characteristics of max do not suggest obvious new hints to myc function other than confirming a role for the B/HLH/LZ regions in DNA binding.

It has been reported that the myc amino-terminal and central regions can specifically interact with the retinoblastoma (RB) tumor suppressor protein *in vitro.* This finding prompts the suggestion that RB may directly regulate myc function, but it is not yet known whether the observed interaction reflects a physiologically relevant association *in vivo.*

The jun *and* fos *Gene Products*

The *jun* oncogene was discovered as a 0.93-kb insert in the genome of a replication-defective retrovirus, avian sarcoma virus 17 (ASV17), isolated from a chicken sarcoma. ASV17 causes fibrosarcomas in chickens and oncogenic transformation in cultured avian embryonic fibroblasts. The oncogenic potential of ASV17 is due to the presence of the *jun* gene, which is derived from the cell genome.

A great deal of excitement was generated by the discovery that the jun protein can substitute for the yeast transcription activator GCN4. Comparative analysis of the sequences of the two proteins showed that there was a strong homology over about one-third of their length in the DNA-binding parts. Subsequently it was discovered that the human transcription factor AP-1 also had very similar properties to both jun and GCN4. All three of these proteins bind to the same consensus sequence, TGACTCA.

The oncogene of the FBJ murine osteosarcoma virus (*fos*) also codes for a related nuclear protein that participates in transcriptional regulation. In human fibroblasts the fos protein is mostly associated with a cellular protein p39 with a molecular weight of 39 kDa. The fos-p39 complex binds specifically to DNA containing the AP-1 binding site. Since fos alone does not show specific DNA binding, it is believed that p39 is responsible for this affinity. Sequence studies showed that p39 and jun are identical proteins. Although jun, like GCN4, can form homodimers that bind to DNA, the heterodimers formed between fos and jun show a greater affinity for the AP-1 binding site. The heterodimers are also more effective in transcription activation; therefore the heterodimer is probably the functionally relevant state of the jun and fos proteins.

Structural analysis indicates that the fos and jun proteins belong to a class of DNA-binding proteins that share the conserved structural motif known as the leucine zipper (see fig. 31.20). Thus the dimerization of these two proteins is mediated by hydrophobic interaction between the leucine side chains of two leucine-zipper domains.

In its normal context as a cellular gene product, jun is not oncogenic. By contrast, as we have already mentioned, the viral *jun* gene of ASV17 is an effective carcinogen both in the animal and in cell culture. This difference appears to be due to the constitutively elevated expression of *jun* when the *jun* gene is in the virus.

The jun-fos protein complex interacts with regulatory regions of numerous genes. We have yet to find out which of these target genes are involved in aberrant cellular growth.

The Transition from Protooncogene to Oncogene

All of the oncogenes thus far discovered are associated with a cellular homolog that is required for normal growth. The transition of the protooncogene to an oncogene is accompanied by abnormal expression of the gene products. In many cases, e.g., *src, jun,* or *sis,* excessive amounts of the gene product are synthesized. In some other cases, e.g., *H-ras* or *K-ras,* normal amounts of the oncogenic product are synthesized but the protein encoded by the oncogene is altered so that it behaves differently. In still other cases, e.g., *myc,* the oncogene product is similar to the cellular homolog in amount as well as structure, but the time during the cell cycle when it is produced is altered.

Oncogene products assume specific locations within the cell. Usually they are associated with one of two locations: the plasma membrane or the nucleus. This specificity is consistent with the hypothesis that oncogene products are associated with normal cellular metabolism relating to regulation of cell proliferation; it seems likely that elements regulating cell proliferation would be found at the membrane and nucleus of the cell.

It is not surprising to find oncogenes with varying specificity for producing tumors. Some components of the cell proliferation regulatory apparatus are probably quite general. An oncogene like *myc* is probably associated with one of these and is therefore associated with tumors of widely varying origins. By contrast, an oncogene like *sis* is specifically associated with cells that are designed to be triggered into proliferation by PDGF. Hence tumors associated with this oncogene are limited to cell types possessing the PDGF membrane receptor.

Carcinogenesis Is a Multistep Process

A wide range of observations indicate that tumorigenesis is a multistep process involving several mutations, each of which results in discrete changes in the cellular metabolism. If this is the case, then we might expect that any particular oncogene would have the capacity for affecting only one step in the overall process. We have seen that a multistep biochemical pathway ordinarily requires a separate enzyme for each step. Enzymes that function in the different steps of a pathway are sometimes said to complement one another. The question is, in carcinogenesis do different oncogenes show a similar complementation? For example, can purely cellular oncogenes such as *N-ras* complement one or the other of the polyoma oncogenes? In fact, middle-T antigen and *N-ras* oncogene produced no new phenotypes when they were cotransfected into rat embryo fibroblasts (REFs), but large-T antigen and *ras* together achieved dramatic results, producing rapidly expanding foci. This study shows that the conversion of a normal cell to a tumor cell can be achieved by the complementary action of two distinct oncogenes; in this case, one is cellular and one is viral.

Similar studies have shown that the *H-ras* and the *myc* oncogenes can cooperate to produce dense foci of morphologically transformed cells from REF cultures. Thus two cellularly derived oncogenes have been shown to complement one another to produce a fully transformed phenotype. Experiments of this sort suggest that at a minimum, two different oncogene functions are required to convert a normal cell into a tumorigenic one, but it is too early to say that only two are required in general. Nevertheless, these results are very exciting because they are the beginning of genetic complementation assays that should help us to classify oncogenes into different complementary functions.

One further comment on the multistep nature of carcinogenesis: In our discussion here, we have focused on some of the early steps in the process that lead to uncontrolled growth, and have said nothing about those transitions that convert transformed cells into invasive cells. We have also concentrated on oncogenes related to cancer-causing viruses, because these are most likely to be the first understood in terms of their biochemical function.

Summary

From the time of their discovery, cancer cells have always appeared to be unruly. We have reached that stage in our understanding of cancer cells where we can point to specific aberrations in the genome. In this chapter we have taken the view that an understanding of cancer can be achieved by analyzing the regulatory pathways involved in growth control, because most cancers appear to originate from mutations in specific genes involved in growth regulation. The following points are the highlights of our discussion.

1. Many properties of transformed cells grown in tissue culture resemble cancer cells. The factors that lead to uncontrolled growth *in vivo* can be studied *in vitro* by the effects they have on tissue culture cells.
2. To judge by the frequency of occurrence of cancers in different countries, environmental factors have more influence on the incidence of cancers than genetic factors do. This conclusion is reinforced by studies of migrant populations.
3. Chromosomal translocations are frequently associated with specific types of cancer. This is direct evidence that genetic abnormalities can lead to cancer.
4. A number of tumors arise from recessive mutations in which the mutations appear to be in growth-control genes.
5. Growth-control genes that lead to cancers when they are altered in some way are referred to as protooncogenes. They become oncogenes, that is, cancer-causing genes, by mutation. Host protooncogenes are frequently very similar in structure to oncogenes carried by tumor-causing viruses.
6. Dulbecco proposed that cancer-causing viruses insert their oncogenes into the host genome. It appears that cancer-causing viruses are associated with a limited number of DNA viruses and RNA viruses known as retroviruses, which replicate through a DNA intermediate. It is this DNA intermediate that usually gets inserted into the host genome.
7. Abnormal expression accompanies the transition from protooncogene to oncogene. Three types of abnormalities occur: (1) excessive production of the gene product, (2) altered behavior of the gene product, such as a change in its regulatory properties, and (3) expression at a time during the cell cycle when the gene is not normally expressed.
8. A fully developed cancer appears to arise in steps, each step showing a breakdown in normal regulation.

Selected Readings

Aaronson, S. A., Growth factors and cancer. *Science* 254:1146–1152, 1991.

Donehower, L. A., M. Harvey, B. L. Slagle, M. J. McArthur, C. A. Montgomery, J. S. Butel, and A. Bradley, Mice deficient for p53 are developmentally normal but susceptible to spontaneous tumours. *Nature* 356:215–221, 1992. A remarkable new technique for obtaining null mutations demonstrates that embryogenesis is normal in the absence of p53. However, animals develop a variety of neoplasms in the first 6 months when p53 is lacking.

Downward, J., The ras superfamily of small GTP-binding proteins. *Trends Biochem. Sci.* 15:469–472, 1990.

Gallo, R. C., The AIDS virus. *Sci. Am.* 256(1):46–56, 1987.

Halauska, F. G., Y. Tsujimoto, and C. M. Croce, Oncogene activation by chromosome translocation in human malignancy. *Ann. Rev. Genet.* 21:321–345, 1987.

Hausen, H., Viruses in human cancers. *Science* 254:1167–1173, 1991.

Henderson, B. E., R. K. Ross, and M. C. Pike, Toward the primary prevention of cancer. *Science* 254:1131–1144, 1991.

Jacks, T., A. Fazeli, E. M. Schmitt, R. T. Bronson, M. A. Goodell, and R. A. Weinberg, Effects of an Rb mutation in the mouse. *Nature* 359:295–300, 1992.

Levine, A. J., The p53 protein and its interactions with the oncogene products of the small DNA tumor viruses. *Virology* 177:419–426, 1990.

Linzer, D. I. H., The marriage of oncogenes and anti-oncogenes. *Trends Genet.* 4:245–247, 1988.

Liotta, L. A., Cancer cell invasion and metastasis. *Sci. Am.* 266:54–63, 1992. A most important aspect of carcinogenesis that we have not dealt with in our short chapter.

Malkin, D., F. P. Li, L. C. Strong, J. F. Fraumeni, C. E. Nelson, D. H. Kim, J. Kassel, M. A. Gryka, F. Z. Bischoff, M. A. Tainsky, and S. H. Friend, Germ line p53 mutations in a familial syndrome of breast cancer, sarcomas, and other neoplasms. *Science* 250:1233–1238, 1990.

Milburn, M. V., L. Tong, A. M. deVos, A. Brunger, Z. Yamaizumi, S. Nishimura, and S. H. Kim, Molecular switch for signal transduction: structural differences between active and inactive forms of protooncogenic ras proteins. *Science* 247:939–945, 1990.

Solomon, E., J. Borrow, and A. D. Goddard, Chromosome aberrations and cancer. *Science* 254:1153–1160, 1991.

Varmus, H. E., Reverse transcription. *Sci. Am.* 257(3):56–64, 1987.

Weinberg, R. A., Finding the anti-oncogene. *Sci. Am.* 259(3):44–51, 1988.

Weinberg, R. A., The retinoblastoma gene and cell growth control. *Trends Biochem. Sci.* 15:199–202, 1990.

Weinberg, R. A., Tumor suppressor genes. *Science* 254:1138–1146, 1991. Provides an update on the mechanisms of action of a number of better known oncogenes.

Problems

1. What are some of the differences between a cancer cell and a normal cell?
2. Consider the following statements: "The vast majority of cancers are preventable" and "Cancer is a disease of aging, i.e., predominantly older people get cancer." Do you agree with either statement? Explain your answer.
3. The methionine analog L-ethionine (which has an ethyl instead of a methyl group attached to the sulfur) causes liver cancer in rats. This analog is a nonmutagenic carcinogen that inhibits cellular methylation. How might this analog cause cancer? (Hint: What role does DNA methylation play in gene expression?)
4. You have just gotten your first job in Houston, Texas, and are presented with the following observations:
   (a) West of Houston there is a very high incidence of a variety of skin cancers.
   (b) South of Houston the incidence of skin cancer is low.
   (c) The region west of Houston was settled by immigrants from Germany who are mostly farmers, while the region south of Houston was settled, in large part, by Hispanics, who also do mostly outdoor work.

   The Texas Health Department asks you to explain why there is such a high incidence of skin cancer west of Houston. (Give a biochemical explanation.)
5. Renato Dulbecco suggested that DNA viruses that cause cancer did so by a mechanism similar to lysogeny of *E. coli* by λ virus, that is, by integration of the viral genome into the host cell DNA. How can RNA viruses cause cancer?
6. Speculate on where and how retroviral oncogenes originated.
7. Why is an oncogene like *myc* found associated with many different kinds of cancer, while the oncogene *sis* is found only in cancers with the PDGF cell-surface receptor?
8. What are some of the ways a cellular protooncogene can be converted to an active oncogene?
9. How does overexpression of the *src* gene cause cancer?
10. The AIDS virus (HIV) can cause many types of cancers (Kaposi's sarcoma, B-cell lymphoma, cancers of the rectum and tongue, etc.), yet HIV does not have a viral oncogene. Explain how HIV can cause cancer.

35
CHAPTER

Neurotransmission

We considered the structures of biological membranes, as well as the mechanisms by which ions and metabolites are transported across them, in chapters 7 and 32. Clearly, however, cells must do much more than simply assimilate nutrients and expel waste materials. Unicellular, free-living organisms such as bacteria and protozoa must be able to navigate and successfully compete in environments that often contain many perils to their livelihood. Multicellular organisms have the additional problem of coordination and signaling among many diverse, differentiated cell types to ensure the efficient functioning of the organism as a whole. The necessary interactions of a cell with its environment and with other cells must obviously involve processes associated with the cytoplasmic membrane. In many cases, transmembrane transport is integrally involved in the signaling processes between cells and their environment.

In chapter 24 we dealt with one type of communication mechanism in animals. Hormones released in one part of the body interact with receptors, which are sometimes membrane-bound, to affect processes occurring in other cells that are often quite distant from the cells releasing the signal. This mechanism allows for efficient coordination of the wide variety of metabolic events continually taking place in higher organisms. Another, quite different phenomenon takes place in the cells of sensory organs—for example, in the retina cells of the eye of vertebrate animals. Light energy is converted by these cells into nerve impulses, which are translated by the brain into an image of our surroundings. This interconversion involves transmembrane movements of ions and will be considered in chapter 36.

In this chapter we will examine the mechanism by which nerve impulses are propagated, to illustrate how excitable tissues function and interact in higher organisms. As we will show, the fundamental mechanisms revealed by studies of membrane transport in both prokaryotic and eukaryotic cells (chapter 32) can account in large part for the seemingly more complex phenomena of nerve-impulse propagation and transmission.

Nerve-Impulse Propagation

As early as the late eighteenth century, experiments by Galvani and Volta suggested that the transmission of nerve impulses and muscular contraction involved electrical signals. In 1902, J. Bernstein first proposed that the unequal distribution of K^+ across the nerve-cell membrane and the selective permeability of this membrane for the potassium ion were responsible for a resting potential known to exist in nerve and muscle fibers. He further believed that excitation of a nerve cell involved a transient collapse in this selective permeability such that other ions were able to penetrate the nerve-cell membrane and abolish the resting potential. If these changes in ion permeability could move down the axon of a nerve cell and be transmitted to other cells, they could provide the basis for propagation of nerve impulses.

In the 1930s, the isolated squid giant nerve axon became available for experimentation, and its size made it especially amenable to electrophysiological measurements. Experiments pioneered by A. L. Hodgkin and A. F. Huxley soon established the essential ionic movements associated with impulse propagation, and the resultant local changes in membrane potential could be measured. It became clear that the transmembrane transport of ions was important in nerve cells for their signaling function and for the actual signal conductance itself. As we will see, two basic types of transmembrane transport proteins are intimately involved in these processes: ion channels, which participate in the propagation of the impulse down the axon, and neurotransmitter receptors (also ion channels), which are involved in the transmission of signals between nerve cells and their target cells.

An Unequal Distribution of Ionic Species Results in a Resting Transmembrane Potential

To understand how nerve impulses are generated along the axon of a nerve cell (fig. 35.1), we must first understand the basis for the electric potentials that exist across the neuronal membrane. An unequal distribution of ionic species across a biological membrane that is permeable to these molecules can result in a transmembrane electric potential, $\Delta\Psi$ (chapter 32). For a membrane system permeable to several ionic species, the numerical value of $\Delta\Psi$ can be approximated by the Goldman equation, derived by D. E. Goldman in 1943:

$$\Delta\Psi \text{ (``in'' relative to ``out'')} = \frac{2.3RT}{F}\log_{10}\left(\frac{\Sigma P_c[\text{C}]_{\text{out}} + \Sigma P_a[\text{A}]_{\text{in}}}{\Sigma P_c[\text{C}]_{\text{in}} + \Sigma P_a[\text{A}]_{\text{out}}}\right) \quad (1)$$

where C and A are univalent cations and anions, respectively, and P_c and P_a refer to their permeability coefficients* across the membrane of interest. Since multivalent ions are generally not quantitatively significant in contributing to $\Delta\Psi$ in resting neuronal membranes, we usually ignore them in calculating $\Delta\Psi$. If the membrane is selectively permeable to one ion only, for example C, equation (1) reduces to the familiar Nernst equation:

$$\Delta\Psi = E_c = \frac{2.3RT}{F}\log_{10}\frac{[\text{C}]_{\text{out}}}{[\text{C}]_{\text{in}}} \quad (2)$$

where E_c refers to the equilibrium electric potential of C.

It is the unequal distribution of protons and other cations that gives rise to transmembrane potentials that can drive ATP synthesis and secondary active transport in many cells (chapter 32). In nerve cells, a similar situation exists as summarized in table 35.1. Thus the external environment of the nerve cell contains a high concentration of sodium ions and a low concentration of K^+, while the reverse is true for the cytoplasm of the nerve cell, called the axoplasm. Furthermore, the extracellular concentration of Cl^- is 5 to 10 times that of the axoplasm.

*The permeability coefficient is equal to the diffusion coefficient D divided by the width of the membrane l.

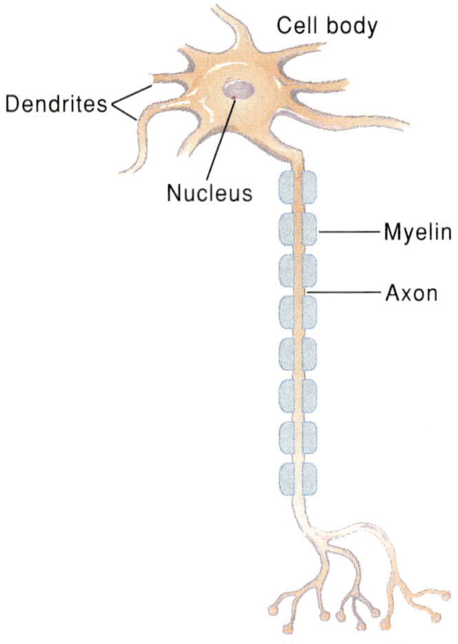

Figure 35.1

Schematic diagram of a typical motor neuron (a nerve cell conducting impulses to muscle cells).

A number of other impermeant anions—largely organic molecules, proteins, and nucleic acids—maintain an approximate charge neutrality in the axoplasm. Resting nerve cells are much more permeable to K^+ than to Na^+, and it is this selective permeability that allows an electric potential to develop across the membrane in the presence of a K^+ concentration gradient. It is, of course, the role of the Na^+-K^+ ATPase to maintain these unequal distributions of cations across the neuronal membrane, as we discussed in chapter 32.

To calculate the resting membrane potential $\Delta\Psi$ across the axonal membrane from equation (1), we can use the values in table 35.1 and the permeabilities of Na^+ and Cl^- relative to K^+ (0.04 and 0.45, respectively), assuming an intracellular Cl^- concentration of 50 mM:

$$\Delta\Psi = 60 \log\left(\frac{20 + 0.04(440) + 0.45(50)}{400 + 0.04(50) + 0.45(560)}\right) = -62 \text{ mV}$$

This value is close to the experimentally measured resting potential across a squid axonal membrane.

An Action Potential Is the Transient Change in Membrane Potential Occurring during Nerve Stimulation

The use of giant axons from squid nerve cells in electrophysiological experiments has greatly aided our understanding of the electrical events that take place during nerve stimulation. An experimental apparatus for measuring changes in the potential across the membrane of such an axon, which has a diameter of approximately 0.5 mm, is schematically illustrated in figure 35.2.

Table 35.1
Ionic Concentrations Inside (Axoplasm) and Outside (Blood) the Squid Giant Axon

Ion	Inside (mM)	Outside (mM)
Na^+	50	440
K^+	400	20
Cl^-	40–150	560

Source: S. W. Kuffler and J. G. Nicholls, *From Neuron to Brain.* © 1976 Sinauer Associates, Sunderland, Mass.

Figure 35.2

A device for eliciting and recording action potentials along the squid giant nerve axon. Brief closure of the switch connected to the stimulating electrode causes a current pulse into the axoplasm. If an impulse is generated, resultant potential changes can be detected by the recording electrode, which is connected to an oscilloscope or other recording device.

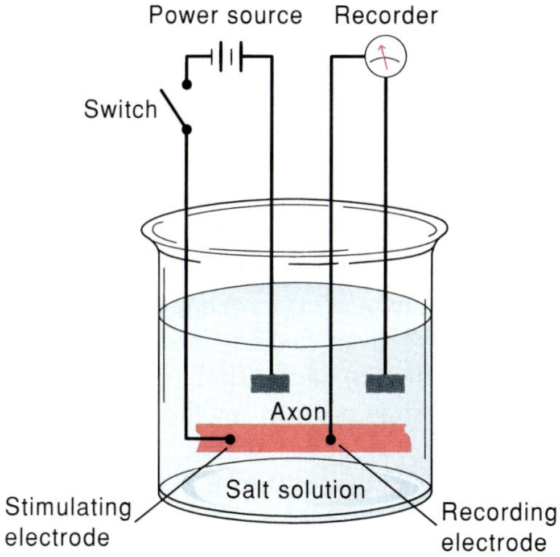

The apparatus consists of a pair of stimulating electrodes connected to a current source and a second pair of recording electrodes located slightly farther down the axonal segment. The latter electrodes are connected to a sensitive recording device, such as an oscilloscope, and can be used to measure time-dependent changes in the membrane potential.

Initially, the potential measured in the system shown in figure 35.2 is about −60 mV; i.e., the resting membrane potential (fig. 35.3). If a brief current is applied at the stimulating electrodes, a time-dependent change in the membrane potential may be recorded on the oscilloscope, as shown in figure 35.3*a*. This so-called action potential occurs only if the stimulus is sufficient to depolarize the membrane by about 20 mV (i.e., to about −40 mV). Weaker stimuli give small local potential changes, while current pulses greater than this threshold value give a curve similar in shape and height to that shown in figure 35.3,

Figure 35.3

A typical action potential V (blue) that might be recorded by the instrument in figure 35.2 if the stimulating current is sufficient to depolarize the membrane by at least 20 mV. The membrane potential returns to its resting value of about −60 mV within about 5 ms. Accompanying changes in relative Na^+ permeability, g_{Na^+} (red), and K^+ permeability, g_{K^+} (green), are also shown. Depolarization is correlated with an increase in g_{Na^+}, while repolarization is accompanied by a decrease in g_{Na^+} and a transient increase in g_{K^+}. Note that the membrane potential transiently becomes more negative than the resting value (hyperpolarizes) until g_{K^+} returns to its normal value. The time of appearance of the action potential after stimulation at $T = 0$ depends on the distance between the recording and stimulating electrodes. (Source: S. W. Kuffler and J. G. Nicholls, *From Neuron to Brain.* Copyright © 1976 Sinauer Associates, Sunderland, Mass.)

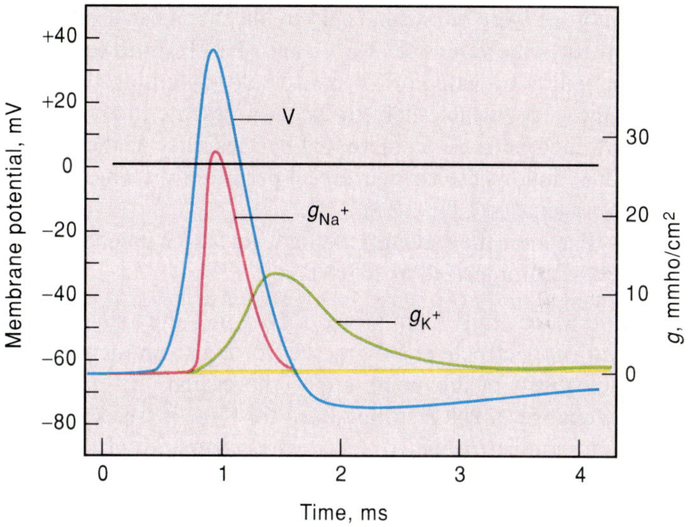

independent of the magnitude of the stimulus. During the development of the action potential, the value of $\Delta\Psi$ across the axonal membrane rises in about 1 ms to nearly +40 mV. This is followed by a somewhat slower return to the resting potential, during which time the membrane potential drops transiently below the resting value, to about −75 mV. This value, which is close to the one predicted by the Nernst equation if the membrane is permeable only to K^+ (the potassium equilibrium potential), marks the state referred to as hyperpolarization.

Classic experiments by A. L. Hodgkin and A. F. Huxley established that the changes in membrane potential occurring during nerve stimulation are due to transient changes in the permeability of the membrane to Na^+ and K^+ ions. As illustrated in figure 35.3, the rapid rise in $\Delta\Psi$ to a positive value is accompanied by a large increase in the relative permeability of Na^+, while the return to the resting potential is correlated with the inactivation of Na^+ permeability and a transient increase in K^+ permeability. An important conclusion from these observations is that the permeabilities of the axonal membrane to Na^+ and K^+ depend on the membrane potential. Thus depolarization above the threshold, leading to a more positive $\Delta\Psi$, first leads to an increased permeability of Na^+ followed by inactivation of this phenomenon and an increase in the membrane permeability of K^+. The latter events tend to hyperpolarize the membrane, increasing the negative $\Delta\Psi$ and decreasing the Na^+ permeability.

As we mentioned in the preceding section, the membrane potential depends on the relative permeabilities and concentration gradients of electrolytes across the membrane [equation (1)]. The resting potential is largely dependent on the K^+ gradient, since unstimulated nerve membranes have a high permeability only to this cation. At the height of depolarization, however, the membrane is much more permeable to Na^+ than to K^+. Because the Na^+ gradient is opposite to that of K^+, Na^+ influx rapidly causes the membrane potential to become positive, approaching but never reaching the value it would have if the membrane were permeable only to Na^+. You can see this if you substitute the values of $[Na^+]_{in}$ and $[Na^+]_{out}$ into equation (2). You will get a value of 57 mV for $\Delta\Psi$, assuming that the membrane is permeable only to Na^+. In reality, this value is never attained because an increased permeability of the membrane to K^+ follows the change in Na^+ permeability, which itself is only transient (see fig. 35.3).

The ionic movements leading to an action potential can therefore be summarized as follows:

1. Stimulation leads to an influx of Na^+ into the axoplasm, down its electrochemical gradient, owing to an increased permeability of the membrane to this cation.
2. The change in $\Delta\Psi$ resulting from this flow increases the membrane permeability of K^+, which flows out of the cell and down its electrochemical gradient; the result is the reestablishment of a negative $\Delta\Psi$, accompanied by an inactivation of the influx of Na^+.
3. The membrane potential eventually becomes sufficiently negative to return both K^+ and Na^+ permeabilities to their normal values, and $\Delta\Psi$ resumes its resting level.

The events shown in figure 35.3 record fluctuations in $\Delta\Psi$ and ionic currents at one point on the axonal membrane as a function of time during passage of an action potential through this point. The action potential, however, is conducted down the axon as a wave of depolarization-repolarization events through the following mechanism: Depolarization of a given area of the membrane causes current to flow in the axoplasm from the more positive (depolarized) region to neighboring regions. This flow, in turn, triggers an action potential across the neighboring section of the membrane, and so forth (fig. 35.4). Thus the action potential provides a mechanism whereby the transmitted signal is constantly amplified to maintain a constant amplitude. In a regular cable without amplification, the propagated pulse decreases with distance as a result of resistance and leakage. Without the action potential, a current pulse would therefore be reduced to an insignificant level after traveling a very short length along the axon. In nerve cells of invertebrate animals, as well as in many cells in vertebrates, nerve impulses are therefore conducted along axons and dendrites by these local currents and action potentials.

Figure 35.4

Nerve impulses in unmyelinated nerves are conducted through local current movements that propagate the action potential. A portion of the resting axonal membrane is shown at the top. Arrival of an action potential causes local depolarization of the membrane (*middle*), which is propagated from left to right by local currents shown by the arrows. (Source: A. L. Hodgkin, *Proceedings. Royal Society of London. Series B. Biological Sciences* 148:1, 1957. Copyright © 1957 Royal Society of London, London, England.)

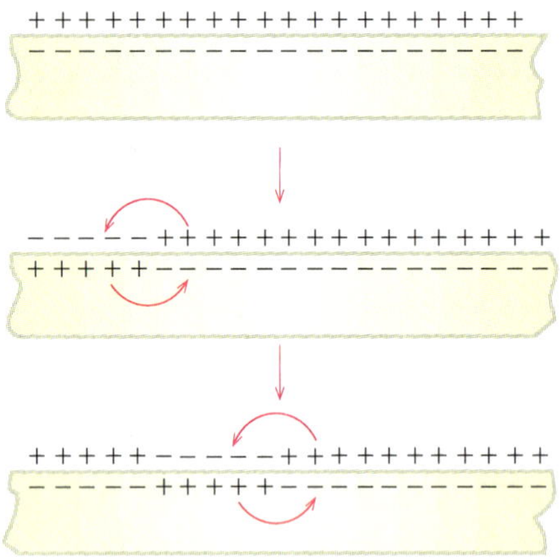

The axons of many nerve cells of higher animals, however, are also surrounded by a multilayered myelin sheath, each layer consisting of a typical lipid bilayer membrane (fig. 35.5). At intervals, the spacing of which depends on the fiber diameter, this myelin insulation is interrupted by the so-called nodes of Ranvier (fig. 35.6). Nerve impulses in these types of nerve fibers are conducted in a saltatory manner, that is, with action potentials "jumping" from node to node, where the axonal membranes are in direct contact with the extracellular fluid (see fig. 35.6). The insulation provided by the myelin sheath allows for efficient current conduction within the axoplasm by preventing signal loss, and as a result, this type of impulse propagation can be much more rapid than that observed in unmyelinated fibers of similar diameter. Indeed, impulse propagation velocities of over 100 m/s have been recorded in myelinated nerve fibers.

Gated Ion Channels

By a slight modification of the experimental setup shown in figure 35.2, it is possible to hold the membrane potential across an axonal membrane constant at any predetermined value. This is accomplished by connecting the recording electrodes to a device called a feedback amplifier. When any potential change is sensed by the electrodes, the feedback amplifier compensates for it by applying a current to keep the voltage constant. In such a voltage-clamped situation, ionic movements can be inferred from the amount of current necessary to hold $\Delta\Psi$ at its predetermined value. It was the use of a voltage clamp by Hodgkin and Huxley that allowed them to deduce the movements of Na^+ and K^+ that accompany the appearance of the action potential.

Figure 35.5

Cross section of a myelinated nerve axon from the superior cervical ganglion of the rabbit. Note the multilayered membrane of the myelin sheath that serves as an electric insulator. (Micrograph courtesy of Dr. T. Lentz.)

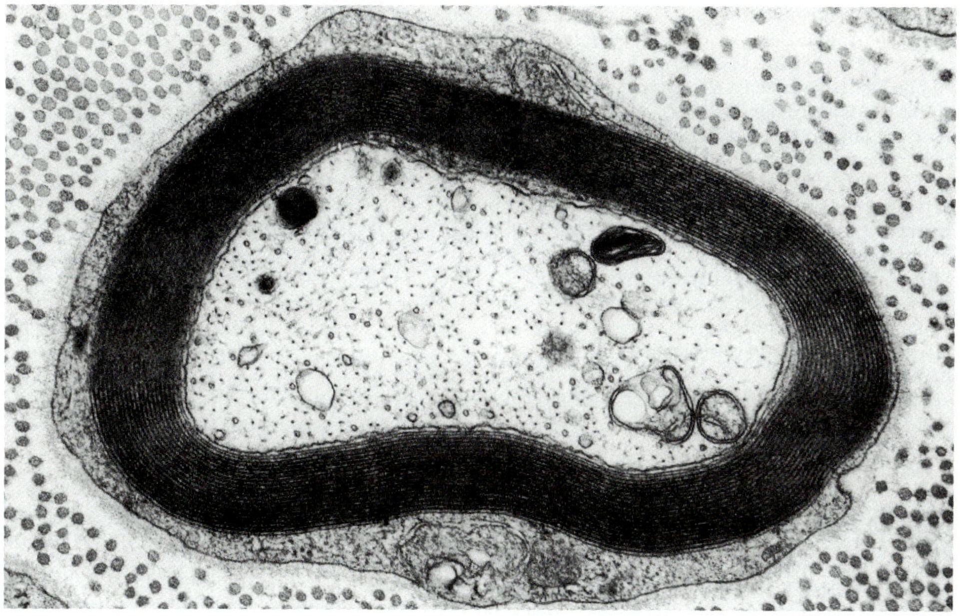

Figure 35.6

Schematic illustration of a longitudinal cross section of a myelinated axon. The sheath is interrupted at various intervals by the nodes of Ranvier, where the cytoplasmic membrane is exposed to the surrounding fluid. Nerve impulses are conducted in a saltatory manner in myelinated axons, as illustrated in this schematic diagram. Impulse propagation is from left to right and the arrows show the accompanying local current movements.

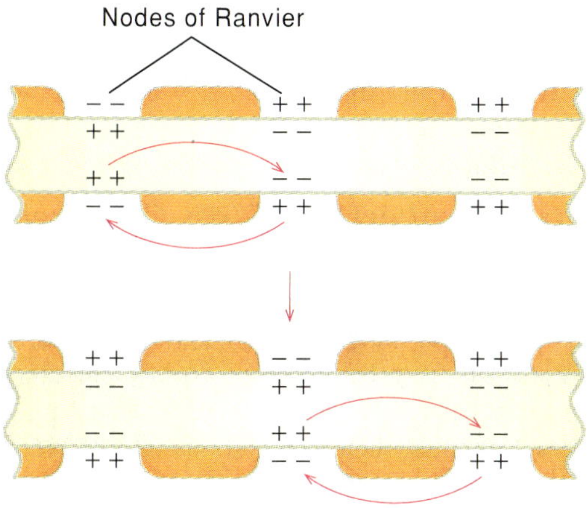

Because the changes in membrane permeability to Na^+ and K^+ were not superimposable in time, they tentatively concluded that two different "channels" were involved in the transmembrane movements of these ions.

Separate Channels for Na^+ and K^+ Have Been Found in Excitable Cell Membranes

Compounds that specifically block the conductance of the nerve membrane to either Na^+ or K^+ provided the first direct evidence for separate Na^+ and K^+ channels in the nerve-cell membrane. Tetrodotoxin and saxitoxin (fig. 35.7) are both potent nerve poisons that specifically inhibit the transmembrane movement of Na^+ by binding to the outside surface of nerve-cell membranes without affecting K^+ permeability, as measured in voltage-clamp experiments. Similarly, tetraethylammonium ions (fig. 35.8) have been shown to bind to the axoplasmic membrane surface and to specifically block the outward flow of K^+.

A large body of evidence has now accumulated that shows that K^+ and Na^+ movements through axonal membranes are mediated by separate channels that act as gated pores (chapter 32). During a typical action potential lasting about 5 ms, it has been calculated that at least 60 Na^+ ions pass through a single Na^+-specific channel (i.e., about 12,000 per second). Since the neuronal Na^+-K^+ ATPase normally pumps less than 200 Na^+ ions per second out of the cell, this pump cannot be directly involved in the development of the action potential. Instead, a process involving passive diffusion through a transmembrane pore best explains these rapid Na^+ fluxes. Furthermore, voltage-clamp studies by B. Hille established that molecules such as the K^+-monohydrate complex, with dimensions larger than about 0.4 nm^2 in cross section (the size of a monohydrated Na^+ ion), do not pass readily through the Na^+ channel. However, smaller molecules, such as monohydrated Li^+ ions, and other small cations such as hydroxylamine and hydrazine, do. This mechanism, that is, exclusion on the basis of size rather than chemical structure, is more characteristic of a pore than of a specific membrane permease (chapter 32).

Additional properties of the Na^+ channel have been inferred from many other experiments. For example, the pH dependence of Na^+ permeation, as well as other experimental observations, suggests that a negatively charged carboxylate

Figure 35.7

(*a*) The structure of tetrodotoxin, a compound that specifically blocks Na^+ channels in nerve cell membranes. This extremely toxic compound is found in the liver and ovaries of the Japanese puffer fish (*Spheroides rubripes*). (*b*) The structure of saxitoxin, a compound found in certain marine dinoflagellates ("plankton"), which are constituents of the so-called red tide. Mussels and clams that have fed upon these organisms are therefore extremely poisonous, and commercial shell fishing is banned in areas where these dinoflagellates appear. Saxitoxin is also a Na^+-channel blocking agent, and both compounds most probably interact with the channel through their positively charged guanidino groups.

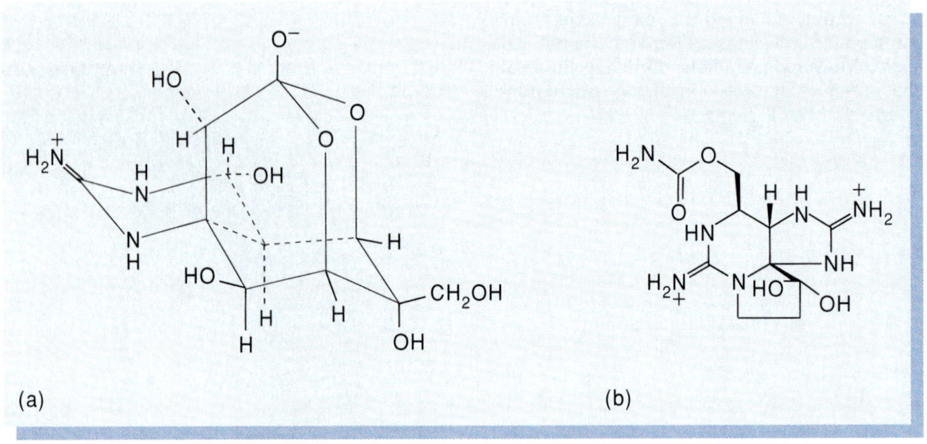

Figure 35.8

Tetraethylammonium ion, a compound that specifically inhibits K^+ channels in nerve cell membranes. The ethyl groups presumably sterically hinder passage of this cation through the channel, "plugging" the channel and therefore blocking the transport of K^+.

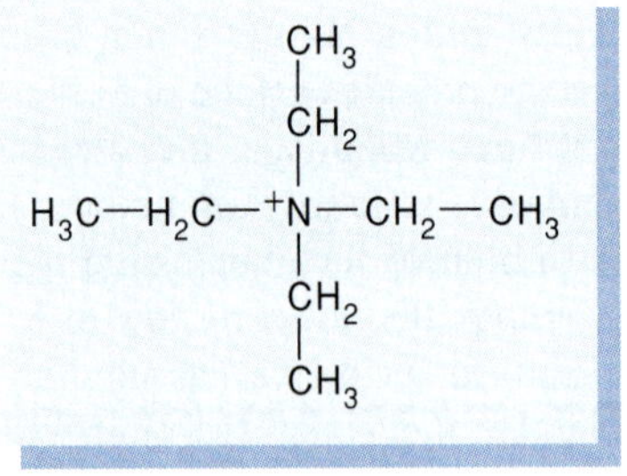

group ($-COO^-$) is present in the pore and is involved in interacting with the cationic molecules that can traverse the Na^+ channel. It is presumably this interaction that allows the positively charged guanidinium groups of tetrodotoxin and saxitoxin (see fig. 35.7) to bind to the channel, but the sizes of these poisons prevent their passage through the membrane, with the result that Na^+ conduction is blocked.

Estimates of the number of Na^+ channels in a variety of nerve types have been obtained by the use of radioactively labeled tetrodotoxin or saxitoxin molecules. From these studies, it is apparent that in unmyelinated nerve fibers, the density of Na^+ channels in the membrane is quite low. For example, the olfactory nerve of the garfish has only about 30 to 40 Na^+ channels per square micrometer of membrane, which corresponds to only about 0.2% of the total surface area of the phospholipid bilayer. In the squid giant axon, which has a fiber diameter some 2,500 times that of the garfish olfactory nerve, this value is increased only to several hundred Na^+ channels per square micrometer, or about 2% of the surface area. The situation is quite different, however, in myelinated nerves of vertebrate animals. In this case, Na^+ channels are found in significant numbers only at the nodes of Ranvier, as we might anticipate from the mechanism of impulse propagation in these types of nerve cells (see fig. 35.6). Here, the channel density is on the order of 10^4 channels per square micrometer, or about 60% of the total membrane surface area. Indeed, Na^+ fluxes at the nodes of Ranvier have been estimated to be 10 to 100 times larger than those in unmyelinated axons during propagation of the action potential.

In contrast to the situation with Na^+ channels, there are many fewer K^+ channels in the axonal membrane (about one-tenth the number in the squid giant axon). Nevertheless, it has been demonstrated that the K^+ pore is also quite specific, barring the passage of cations both smaller (Na^+) and larger (Cs^+; tetramethylammonium ions) than K^+. It seems likely, therefore, that both the diameter of the channel and specific interactions of channel components with the K^+-H_2O complex confer cation selectivity on this channel as well.

The Gating Properties of Ion Channels

How do channels selective for Na^+ and K^+ ions "sense" the membrane potential and react by opening or closing during various stages of action-potential propagation? Although this question is still far from being answered in molecular detail, a number of clues have emerged from work on the gating properties of these channels, their purification and reconstitution into artificial membrane systems, and the recent cloning and sequencing of genes encoding Na^+ and K^+ channels. The results of these investigations show that ion channels in nerve and muscle cell membranes can assume multiple conformations, one of which is more permeable to the ion in question than the others.

An important observation that bears on this question is the discovery of gating currents, first predicted by Hodgkin and Huxley. Under appropriate conditions, a small current opposite in direction to that carried by Na^+ during the opening of the Na^+ channels can be detected just before the development of the action potential. This current is short in duration (0.1 ms) and precedes the inward movement of Na^+ (fig. 35.9). One widely accepted proposal is that the Na^+ channel has a "built-in" voltage sensor that has a large dipole moment. Depolarization of the membrane, which elicits the action potential, would displace or rearrange the dipole in response to the change in the electric field, and this process would be detected as the gating current. The change in the orientation of the dipole could then be the trigger by which the Na^+ channel is opened, presumably by means of a conformational change in this protein. Indeed, a candidate for such a voltage sensor in both Na^+ and K^+ channels has recently been recognized from the primary structures of these proteins (see the following discussion).

Additional evidence for voltage-dependent conformational changes that allow the gating of the Na^+ channel has been obtained using nerve poisons from North African scorpions and the sea anemone. These toxins, which are basic polypeptides, bind to sites on the Na^+ channel that are distinct from those that bind the channel-blocking agents tetrodotoxin and saxitoxin. They appear to exert their physiological effects by slowing inactivation of the Na^+ channel after its initial opening during the action potential. Binding of scorpion toxin to Na^+ channels in nerve and muscle-cell membranes has been shown to be voltage-dependent, a fact suggesting that this toxin recognizes a conformation of the channel that also depends on the membrane potential (the "open" conformation). Toxin binding would thus reflect the conformational state of the channel, which, in turn, depends on $\Delta\Psi$. These and other experiments have led to the conclusion that both the Na^+ and K^+ channels can exist in at least three conformations: closed (resting $\Delta\Psi$), open, and inactive. The latter two states are responsible for the transient increase and then decrease in Na^+ and K^+ permeabilities, in sequence, observed during the action potential.

Further Evidence Concerning the Structure and Function of Ion Channels

One major approach to understanding structure-function relationships in the ion channels of nerve-cell membranes has been to purify these proteins and to study their properties in artificial membrane systems, as we described for transport permeases in chapter 32. The ability of Na^+ channels to tightly bind specific neurotoxins has been used as an assay in such purifications. Na^+ channels have been purified from the electric organ of the electric eel (*Electrophorus electricus*), which is rich in such channels, as well as from a number of higher vertebrate sources. A major component of all these preparations was a protein with a polypeptide-chain molecular weight of about 260,000, while variable amounts of a second polypeptide (M_r = ca. 37,000) were also found in some preparations. These two subunits have been referred to as the α and β subunits of the Na^+ channel, respectively.

Several highly purified preparations of the Na^+ channel have been reconstituted into proteoliposomes. In these experiments, reconstituted Na^+ channels were activated by channel activators and blocked by channel inhibitors such as saxitoxin and tetrodotoxin. Moreover, the cation specificity of reconstituted Na^+ channels was the same as that for the native channel:

Figure 35.9

Illustration of the gating current that precedes the inward-directed Na^+ current associated with depolarization during the action potential. The gating current is believed to be related to the voltage-dependent opening of the Na^+ channels and perhaps may reflect rearrangement of a dipolar "voltage sensor" associated with the channel gate. (Source: C. M. Armstrong and F. Bezanilla, "Charge movement associated with the opening and closing of the activation gates of the Na channels," in *Journal of General Physiology* 63:533, 1974. Copyright © 1974 Rockefeller University Press, New York, N.Y.)

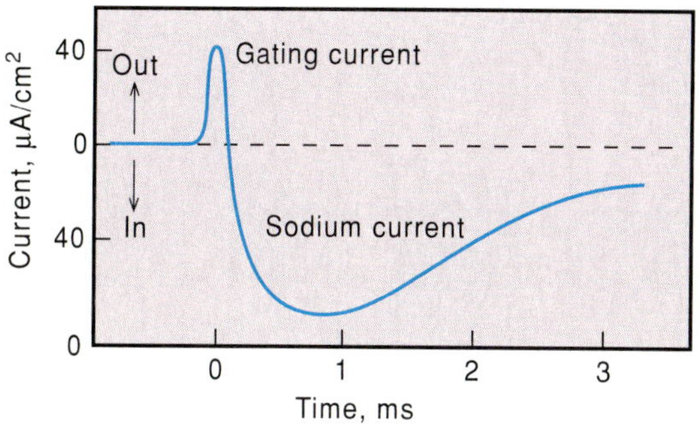

$Li^+ = Na^+ > K^+ > Rb^+ > Cs^+$. Further, in their purified form these channels retained a voltage dependency similar to that observed in the cells from which they were obtained. Since most of these properties could be attributed to the α subunit in many of these systems, it appears that this polypeptide alone is sufficient for most of the biological functions attributed to the Na^+ channel in nerve and muscle cells.

A major breakthrough in attempts to understand the structure and function of ion channels of excitable cells came with the cloning and sequencing of the cDNA that encodes the α subunit of the Na^+ channel from *Electrophorus electricus* by S. Numa and collaborators in 1984. The deduced amino acid sequence showed that this protein consists of 1,820 amino acid residues and has several remarkable features. First, the protein consists of four highly homologous domains, each of which may span the membrane six times, with both the N- and C-terminal regions on the cytoplasmic surface of the membrane (fig. 35.10). The membrane-spanning regions include both potential hydrophobic α helices (regions 1, 2, 5, and 6 in figure 35.10) and amphipathic α helices (regions 3 and 4). Strikingly, potential membrane-spanning region 4 (S4) of all four domains contains a stretch in which every third aminoacyl residue is basic (either lysine or arginine). A number of workers have proposed that this region may be the positively charged voltage sensor that moves outward in response to depolarization of the nerve-cell membrane, thus producing the gating current. In contrast, proposed membrane-spanning region 3 (S3) can form a highly amphipathic helix with a net negative charge. Synthetic peptides containing the amino acid sequence of S3 have been shown to form ion-selective channels in phospholipid bilayer membranes. Moreover, computer-assisted three-dimensional structural predictions have shown that four S3 peptides could form a tetrameric transmembrane pore with hydrophilic side chains lining this channel (fig. 35.11). The dimensions of the channel are strikingly similar to those deduced for the native Na^+ channel

Figure 35.10

A functional map of the Na^+ channel α subunit. The transmembrane folding model of the α subunit is depicted with experimentally demonstrated sites of cAMP-dependent phosphorylation (P), interaction of site-directed antibodies that define transmembrane orientation (≻), covalent attachment of α-scorpion toxins (ScTx), glycosylation (ΨΨ), and modulation of channel inactivation (h). (Source: W.A. Catterall, "Structure and function of voltage-sensitive ion channels," in *Science* 242:50, 1988.)

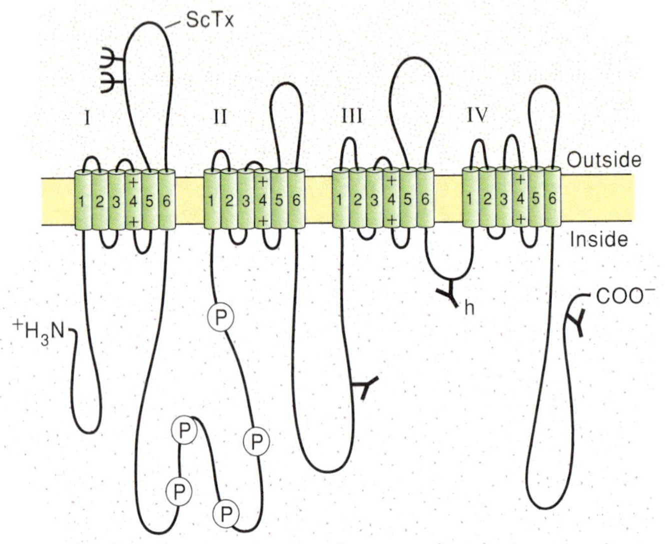

Figure 35.11

End view of energy-optimized parallel tetramer of the sodium channel S3 segment. Colors: light blue, α-carbon backbone; red, acidic residues; blue, basic residues; yellow, polar and neutral residues; and purple, lipophilic residues. This view is from the intracellular face of the membrane, at which the N terminus of each segment is predicted to be located.

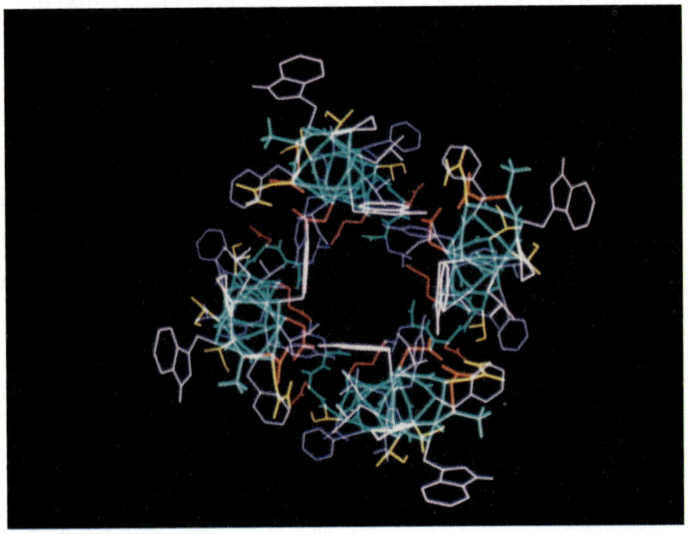

on the basis of its size selectivity. The negatively charged, acidic residues inside and at the entrances to the pore could account for its cation selectivity and also for the results of the pH-dependence studies of Na^+-channel function mentioned earlier. Additionally, the loop between S5 and S6 has recently been shown to have a role in channel function and may therefore comprise part of the pore. However, whether each of the four homologous domains of the Na^+ channel, or a complex including, for example, the S3 region and the S5/S6 loop from each of these domains, forms the functional channel remains to be determined. Finally, a region between homologous domains I and II that is exposed in the cytoplasm has been found to be phosphorylated in a cyclic-AMP-dependent manner (see fig. 35.10). Phosphorylation of Na^+ channels inhibits Na^+ influx, and therefore may have a regulatory role in excitable cells.

Genes encoding Na^+ channels from higher organisms have also been cloned and sequenced, and they have a structure that is highly homologous to the one we have just described. More recently, the gene corresponding to the *Shaker* locus of *Drosophila* has also been cloned and sequenced. True to its name, a mutation in this locus causes severe neurological malfunction in the affected fruit fly, which is due to a defect in one type of K^+ channel. The deduced amino acid sequence of this K^+ channel revealed a 616-residue polypeptide that showed significant primary and secondary structural homology to each of the four domains of the Na^+ channel. Furthermore, models of the intramembrane structure of this protein were very similar to that of a single Na^+ channel domain, complete with approximately six membrane-spanning regions, an S4-like voltage sensor domain, and an S3-like domain with an enrichment of negatively charged residues. These properties of the K^+ channel from *Drosophila,* and the fact that even unicellular eukaryotes, such as *Paramecium,* and bacteria, such as *E. coli,* have voltage-sensitive K^+-selective channels, have led to the suggestion that all such channels may have had a common evolutionary ancestor that was similar in structure to the present-day K^+ channels. Presumably, tandem duplication events involving such an ancestral gene could have given rise to the present-day Na^+ channel gene, consisting of four tandem segments that are highly homologous to each other and to the K^+-channel gene.

Very recently, the powerful tool of site-directed mutagenesis has been used to study structure-function relationships in ion channels. Such studies have been aided by the development of techniques to study specific mutant proteins expressed in a model system, oocytes from South American frogs of the genus *Xenopus.* These relatively large cells are easily manipulable, and mRNA transcribed from normal or mutant ion channel genes can be injected into the oocytes and expressed. Voltage-clamp studies of *Xenopus* oocytes, into which mRNAs encoding various ion channels have been injected, have shown that the channel proteins are expressed in a functional form in the oocyte membrane. Thus far, use of this technique has yielded its greatest benefits in the area of understanding the mechanism of ion channel inactivation.

Recent experiments by R. W. Aldrich and co-workers have provided strong evidence that ion channel inactivation may operate by a relatively simple "ball-and-chain" mechanism (fig. 35.12). These workers studied a number of site-directed mutants of the *Drosophila Shaker* gene for their inactivation properties after injecting the corresponding mRNA for the gene into *Xenopus* oocytes. Previous experiments had suggested that the extreme N terminus of the *Shaker* gene product (K^+ channel) may be involved in inactivation, so the focus was on mutations in this region. The results of the experiments can be summarized as follows:

1. Mutations within the first eleven amino acid residues of this protein that resulted in hydrophobic to polar substitutions gave a K^+ channel in which inactivation was slowed relative to the wild-type protein. Furthermore, mutations that neutralized or eliminated positively charged residues at positions 14, 16, 17, 18, and/or 19, immediately following the hydrophobic N terminus, also slowed the inactivation rate of the channel.
2. A region immediately C terminal to the positively charged cluster of amino acids (residues 23–37) was also studied. This region is rich in hydrophilic amino acids, and when deletion mutations were made in this region, inactivation was faster than in the wild-type protein. Conversely, insertion mutations in this region slowed inactivation of the channel.
3. Finally, a deletion mutant that lacked most of the first twenty aminoacyl residues failed to inactivate, as we would expect from the previous results. However, inactivation could be restored by adding back a synthetic peptide that had the same sequence as the amino-terminal twenty amino acid residues of the native *Shaker* channel. The degree of restoration of inactivation was directly proportional to the amount of peptide added.

The simplest interpretation of these results is that the first twenty amino acid residues of this K^+ channel comprise a "ball," with the positively charged residues on the surface and the hydrophobic residues forming the core. It is this structure that is responsible for K^+ channel inactivation by virtue of its binding to the inside surface of the pore, presumably at least partially by interacting with negatively charged residues near the pore opening. Mutations that disrupt this structure disrupt inactivation. The next twenty residues or so are hypothesized to be a "chain" that can "swing" the ball into or out of the pore opening. Shortening the chain by deletion mutation brings the ball closer to the opening and thus speeds up inactivation, while the reverse is true for insertion mutations. However, the free ball can inactivate a mutant protein lacking an attached ball, as long as it is provided in a high enough concentration. Perhaps the blocking effect of the triethylammonium cation, which we mentioned earlier, is simply due to its mimicking the positively charged region of the inactivation "ball." Figure 35.12 is thus a reasonable schematic model for how inactivation may work, at least in the case of one ion channel. Application of similar molecular biological techniques to the cloned genes of other ion channels will no doubt cast further light on structure-function relationships in these ubiquitous proteins.

Figure 35.12

"Ball-and-chain" model for inactivation of the K^+ channel. (*a*) Open conformation, (*b*) inactive conformation. (Source: "Research News," in *Science* 250:507, 1990.)

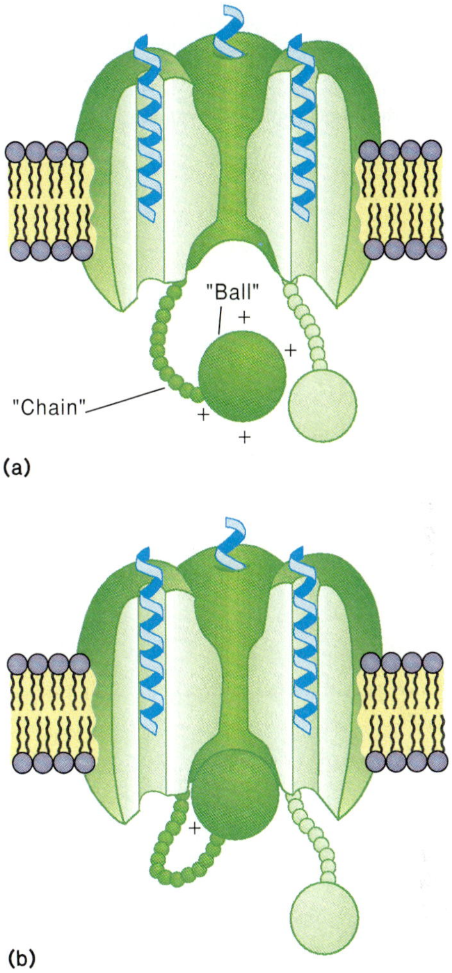

Synaptic Transmission: A Chemical Mechanism for Communication between Nerve Cell and Target Cell

Nerve impulses, propagated along the axon by mechanisms outlined in the preceding sections, must be transmitted between nerve cells or from nerve cells to muscle or glandular tissues in order for their effects (e.g., contraction or secretion) to take place. This type of intercellular communication can occur either by an electrical mechanism of coupling or by chemical transmission across specific connections, called synapses, between

Figure 35.13

Schematic diagram of a synaptic junction in which acetylcholine is the chemical transmitter. Arrival of an action potential at the terminus of the presynaptic cell (*top*) stimulates Ca^{2+} uptake, which triggers release of acetylcholine (ACh) from vesicles near the terminus of the presynaptic cell. Release is accomplished by vesicle fusion with the plasma membrane, and the interaction of acetylcholine with its receptors in the postsynaptic membrane triggers depolarization, thus propagating the action potential in the postsynaptic cell (*bottom*).

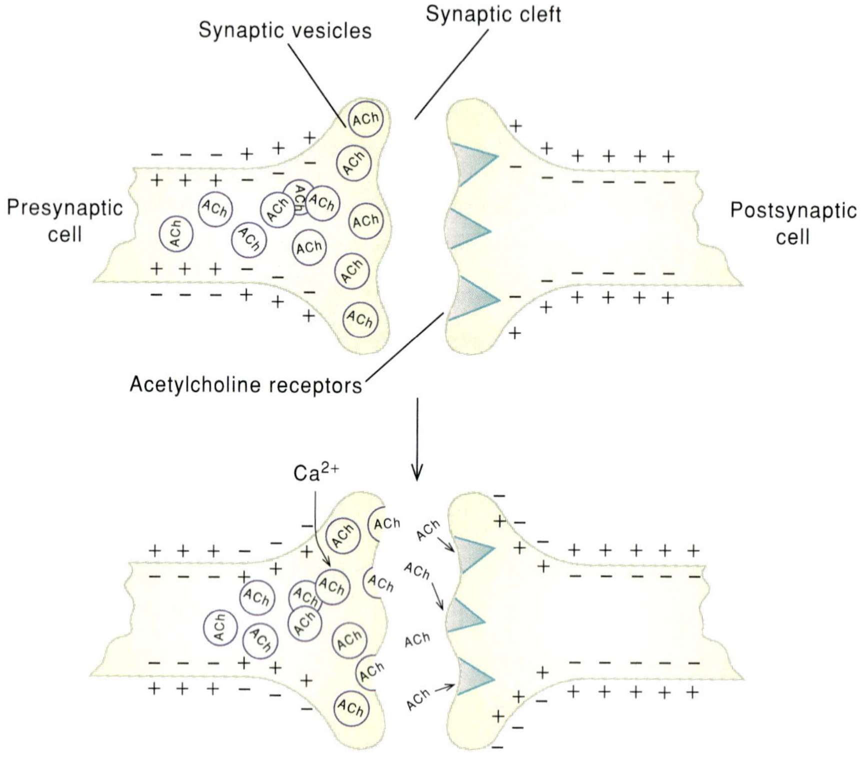

Figure 35.14

Acetylcholine is synthesized by cholineacetyltransferase, which esterifies choline with an acetyl group from acetyl-CoA.

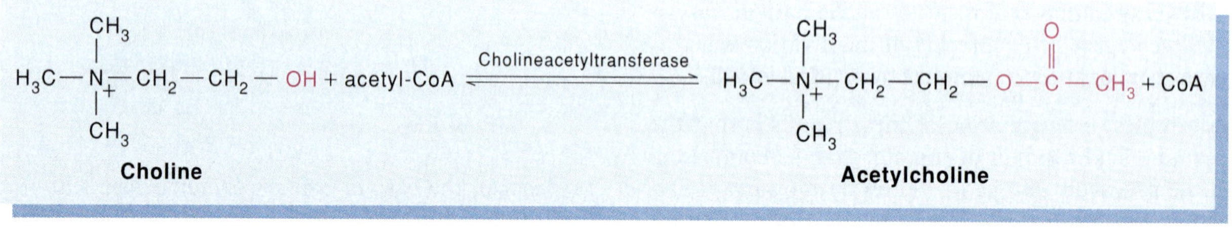

nerve cells and target cells. An example of electrical transmission was given in chapter 32. Direct cell-to-cell contact by means of gap junctions can allow action potentials to be transmitted between cells by ionic mechanisms. In another mechanism, specific chemicals called neurotransmitters are released at the presynaptic membrane of one cell in response to membrane depolarization. These chemicals then diffuse to receptors in the postsynaptic membrane of the recipient cell, where a new action potential may be generated.

Acetylcholine Is a Common Chemical Neurotransmitter

The process of synaptic transmission is depicted schematically in figure 35.13 for acetylcholine, the excitatory neurotransmitter at vertebrate neuromuscular junctions (motor end plates).

Acetylcholine is synthesized in nerve cells by the enzyme cholineacetyltransferase (fig. 35.14) and is packaged in units of 10^3 to 10^4 molecules within synaptic vesicles, which are abundant near the cytoplasmic membrane of the presynaptic axon. The arrival of an action potential triggers a large increase in the permeability of the presynaptic membrane to Ca^{2+}, and this ion flows into the axoplasm down its chemical gradient. Fusion of the synaptic vesicles with the plasma membrane and the concomitant release of acetylcholine into the synaptic cleft are promoted by this increase in intracellular Ca^{2+}. In a typical neuromuscular synaptic junction, several hundred synaptic vesicles empty their contents into the synaptic cleft by this mechanism in response to a single action potential. The resultant large increase in the local concentration of acetylcholine is "sensed" by a protein, the acetylcholine receptor, located in the cytoplasmic membrane of the postsynaptic cell. Binding of the neurotransmitter to many receptor molecules triggers an action potential in the recipient cell. The acetylcholine is then rapidly degraded into acetate and choline by an enzyme in the synaptic cleft called acetylcholinesterase, and the resting potential of the postsynaptic membrane is soon restored.

How does the binding of a neurotransmitter to its receptor promote depolarization of the postsynaptic membrane? One of the first clues came from studies by B. Katz and collaborators, who showed that acetylcholine increases the ionic permeability of the postsynaptic membrane. Subsequent studies with radioactive tracers and by the voltage-clamp technique established that the membrane permeabilities to both Na^+ and K^+ were increased simultaneously by the action of acetylcholine. Because in the resting cell K^+ permeability is already quite high relative to Na^+ permeability, the major effect of acetylcholine binding and receptor channel opening is a substantial influx of Na^+ down its concentration gradient. The inward Na^+ current tends to collapse (or depolarize) the negative resting potential toward zero. This local perturbation in $\Delta\Psi$ is enough to initiate a new action potential in the recipient neuronal or muscular membrane as long as a sufficient number of receptor molecules bind the neurotransmitter. In fact, as we will see in a later section, the receptor itself contains the ion channel through which both Na^+ and K^+ can flow. The number of occupied receptors at a given moment thus dictates the magnitude of the inward flow of Na^+ and the resulting magnitude of the change in the membrane potential. Rapid hydrolysis of acetylcholine by acetylcholinesterase bound to the postsynaptic membrane then quickly reduces the number of transmitter-receptor complexes and repolarizes the membrane until a new action potential triggers the release of more neurotransmitter from the presynaptic membrane.

To elucidate the steps just outlined, researchers have made use of specific inhibitors of both acetylcholinesterase and the acetylcholine receptor. Compounds such as eserine (fig. 35.15*a*) block the hydrolytic enzyme and thus can be used to study the effects of acetylcholine in cholinergic systems (those using acetylcholine as a transmitter) under conditions where it cannot be hydrolyzed. Likewise, certain organophosphorus compounds efficiently inhibit acetylcholinesterase by forming stable covalent intermediates with an active-site serine in the enzyme. The widely used insecticides parathion and malathion are examples of this class of nerve poisons (fig. 35.15*b*). Neuromuscular junctions exposed to acetylcholinesterase inhibitors are paralyzed because the persistent presence of acetylcholine prevents repolarization of the postsynaptic membrane to restore its excitability. In fact, in such a situation the acetylcholine receptor eventually becomes desensitized, i.e., it remains closed to ion flow for long intervals even in the presence of the neurotransmitter.

Specific blocking agents of the acetylcholine receptor include *d*-tubocurarine, an active component of the neurotoxin curare (fig. 35.16), and the snake venom poisons α-bungarotoxin (from snakes of the genus *Bungarus*) and cobratoxin. The latter are small basic proteins with masses of around 7,000 daltons. All three of these substances interact noncovalently with the receptor and interfere with acetylcholine binding, thus blocking depolarization of the postsynaptic membrane. They are referred to as antagonists of the cholinergic systems. Another type of acetylcholine receptor inhibitor is exemplified by the divalent cation decamethonium (fig. 35.17), which "locks" the ion

Figure 35.15

Acetylcholinesterase inhibitors. (*a*) Eserine, or *physostigmine*, is an alkaloid that forms a relatively stable covalent carbamoyl intermediate with an active-site serine on the enzyme that is hydrolyzed only very slowly. (*b*) Parathion and malathion are organophosphorus compounds that also form stable covalent complexes with the active-site serine of acetylcholinesterase. They are widely used agriculturally as insecticides.

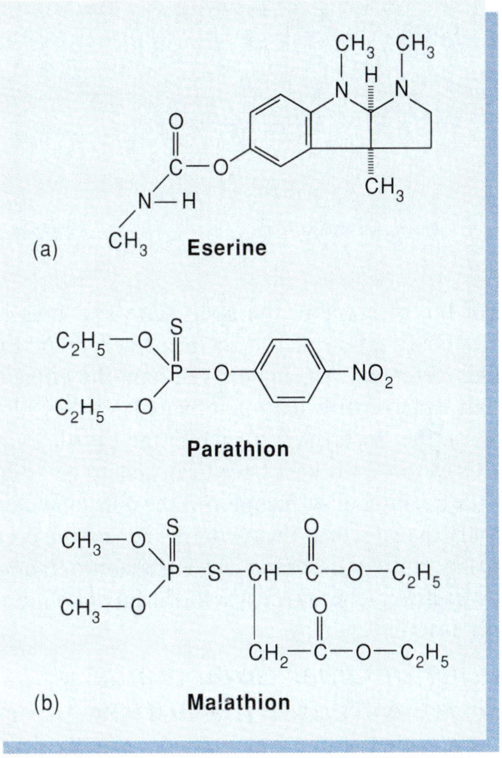

Figure 35.16

The structure of *d*(+)-tubocurarine, an active component of the neurotoxin curare. This compound binds to the acetylcholine-binding site on its receptor, preventing synaptic transmission and subsequent depolarization of the postsynaptic cell membrane.

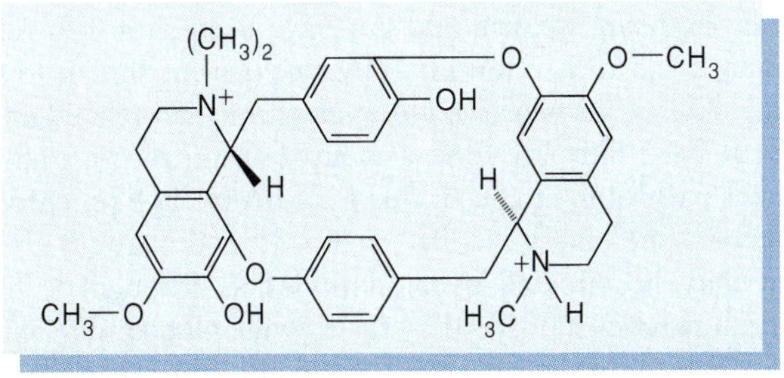

Figure 35.17

Decamethonium ion, an agonist of cholinergic systems. This compound binds to the acetylcholine receptor, but because it cannot be degraded, it causes persistent depolarization of the postsynaptic membrane.

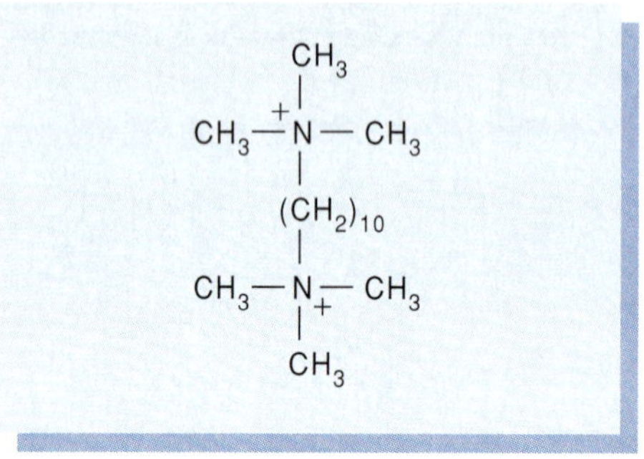

channel of the receptor in the open state and thus leads to a constant depolarization of the recipient cell membrane. Such compounds, referred to as agonists, mimic the effect of acetylcholine but cannot be rapidly inactivated, so they block resensitization of the postsynaptic membrane. With agonists and antagonists as tools, workers have been able to investigate properties of the acetylcholine receptor in the open and closed states and to study the effects of these compounds on the permeability of the postsynaptic membrane. Agonists and antagonists have also been useful in the purification of acetylcholine receptors, as we will describe shortly.

A Number of Other Compounds Also Serve as Neurotransmitters

The best-documented examples of chemical neuromessengers other than acetylcholine are the catecholamines, certain amino acids (and derivatives), and a variety of peptides. The catecholamines are all derived biosynthetically from L-tyrosine (fig. 35.18) and include the hormones norepinephrine and epinephrine (adrenaline). Because these compounds are also synthesized in the adrenal gland, neurons that use these substances as chemical transmitters are said to be adrenergic. Sympathetic nerve fibers that innervate smooth-muscle cells in internal organs such as the heart, spleen, and gut have been shown to release norepinephrine at their terminals by mechanisms similar to those used by cholinergic neurons. Norepinephrine and related amines also have been shown to serve as neurotransmitters in a number of nerve pathways in the brain. Like acetylcholine, catecholamines may be inactivated by chemical modifications. Inactivation may be effected by a methylation reaction or by an oxidation reaction catalyzed by the enzyme monoamine oxidase (see fig. 35.18). In some cases, they can also be resorbed through the presynaptic membrane after their release, providing an additional mechanism for removal from the synaptic cleft.

The hydroxylation of tyrosine, which is the first unique step in the biosynthesis of catecholamines, is catalyzed by tyrosine hydroxylase and yields the compound 3,4-dihydroxyphenylalanine (L-DOPA). The neurological disorder Parkinson's disease is associated with an underproduction in the human brain of the catecholamine transmitter dopamine, which is derived from L-DOPA by a decarboxylation reaction (see fig. 35.18). L-DOPA has therefore been found to be an effective drug in some instances in the treatment of Parkinson's disease. Interestingly, overproduction of dopamine in the brain also occasionally occurs and appears to be associated with psychological disorders such as schizophrenia. In these cases, dopamine-receptor blocking drugs, such as chlorpromazine (fig. 35.19), have been found to be useful therapeutic agents.

Amino acids that are believed to have roles as neurotransmitters include glutamic acid and glycine. The amino acid derivatives histamine (synthesized by the decarboxylation of histidine), 5-hydroxytryptamine (or serotonin, derived from tryptophan), and gamma-aminobutyric acid (or GABA, a decarboxylation product of glutamic acid) have all been shown to be transmitters in various systems as well. The reactions involved in the biosynthesis of these compounds are diagrammed in figure 35.20. GABA is used most often as an inhibitory transmitter. Interaction of this compound with its receptor on many postsynaptic membranes results in a large increase in the membrane permeability to Cl^- and/or K^+ ions. This change inhibits the postsynaptic cell, often by hyperpolarizing its membrane (recall, for example, that the equilibrium potential of K^+ is more negative than the resting potential). Most target tissues, in fact, are innervated by more than one type of nerve fiber, each using a different neurotransmitter. Thus different signals can be relayed to the recipient cells, some stimulatory and others inhibitory. Given the complexity of the nervous systems of higher animals, additional neurotransmitters are continually being discovered. Some compounds currently thought to be chemical transmitters are listed in table 35.2.

The Acetylcholine Receptor Is the Best-Understood Neurotransmitter Receptor

The structure and properties of the acetylcholine receptor are by far the best understood among all neurotransmitter receptors. One reason for this is their abundance in postsynaptic membranes found in the electric organs of the electric eel (*Electrophorus*) and the electric ray (*Torpedo*). These organs contain stacks of cells (electroplaxes), each cell of which receives nerve endings on one side but not on the other. Release of acetylcholine by the presynaptic nerve endings thus depolarizes the postsynaptic membrane by virtue of the binding of the neurotransmitter to the membrane-bound receptors, while the other face of the cell remains at its resting potential. In this way each cell, when stimulated, can attain a potential difference of over 100 mV between its two faces, and thousands of such stacked cells, present in the electric organ, can consequently emit an electric discharge of several hundred volts.

Acetylcholine-receptor-rich membranes can easily be isolated from the electric organs of *Electrophorus* and *Torpedo*. These membranes contain from 10 to 50 $\times$ 10^4 receptor sites per square micrometer, as measured by the binding of radioactively labeled antagonists such as α-bungarotoxin. Electron microscopy and x-ray diffraction have further revealed that such membrane fragments have a similar density of particles, each

Figure 35.18

Biosynthesis from L-tyrosine and inactivation by monoamine oxidase of catecholamine neurotransmitters. Norepinephrine, epinephrine, and dopamine are confirmed neurotransmitters in various systems, and L-DOPA is a probable neurochemical messenger (see table 35.2).

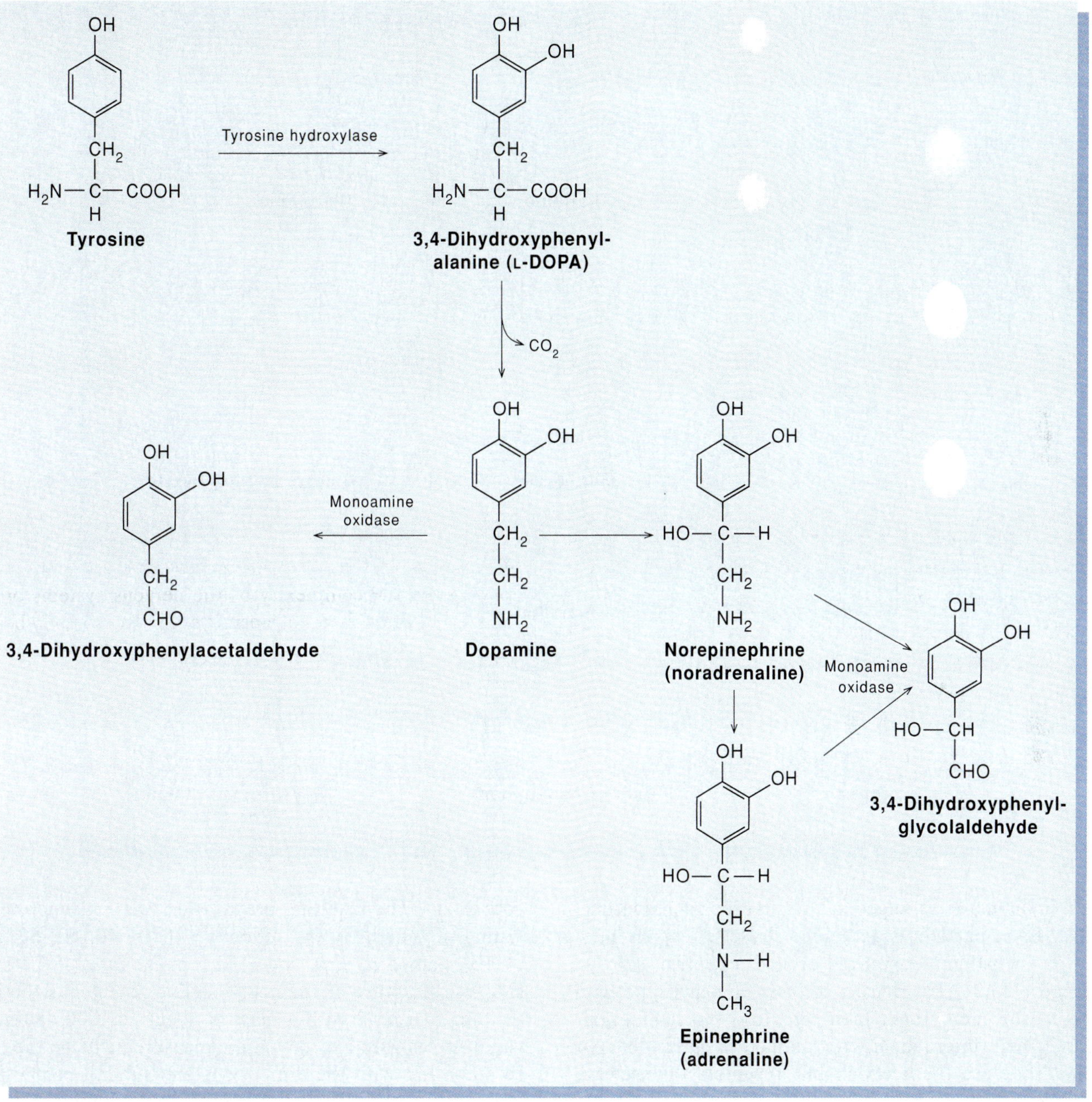

Figure 35.19

The structure of chlorpromazine, a drug that blocks dopamine receptors and has been used in the treatment of psychological disorders such as schizophrenia.

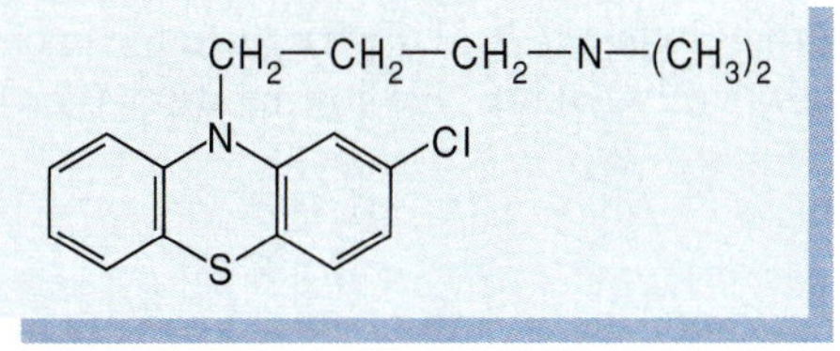

Figure 35.20

Biosynthesis of the neurotransmitters histamine, gamma-aminobutyrate (GABA), and serotonin (5-hydroxytryptamine).

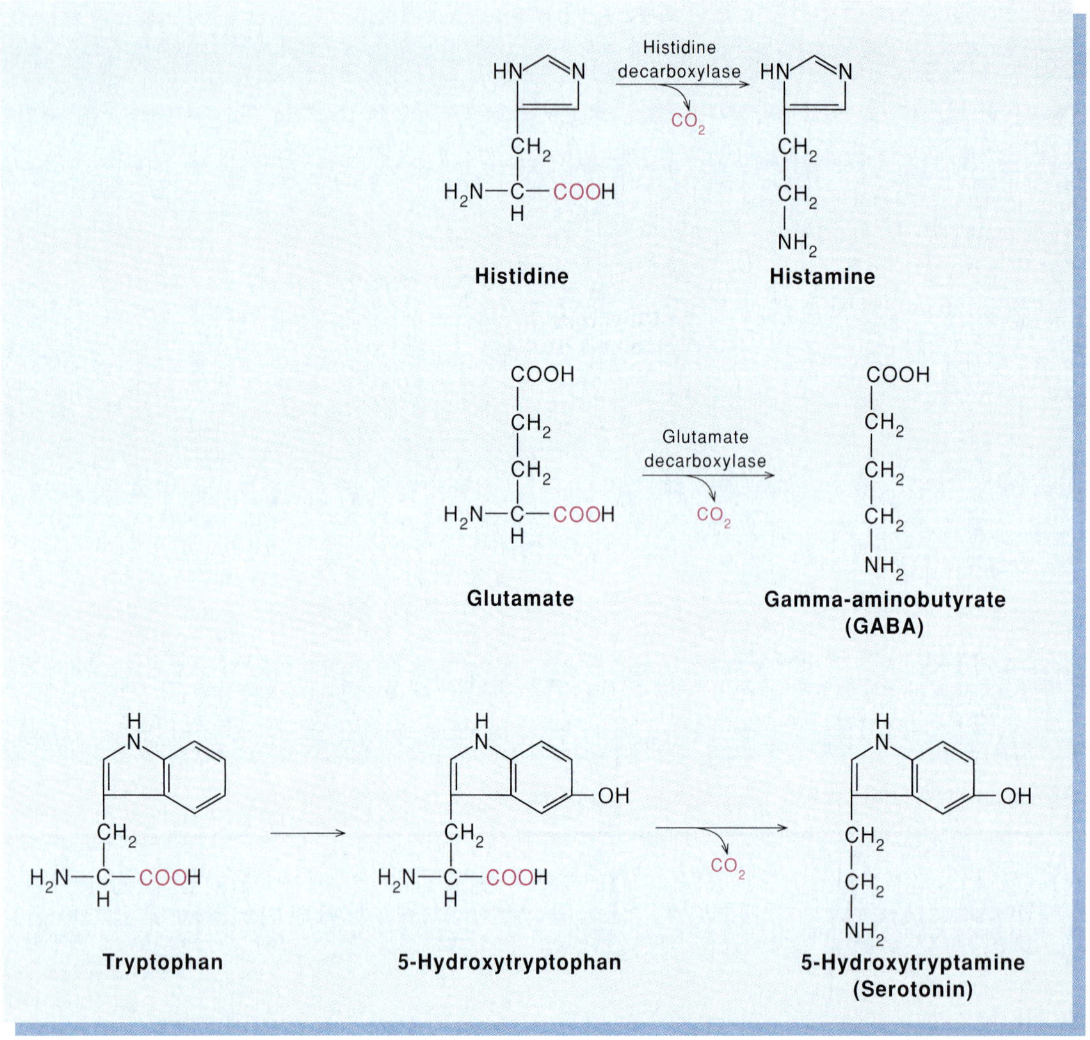

about 8 nm in diameter, arranged in a regular hexagonal lattice (fig. 35.21). Each particle is actually a "rosette," apparently consisting of five subunits arranged around a central axis, as shown in figure 35.21. The density of these oligomeric protein molecules on the membrane, their size (see the next paragraph), and their interaction with antibodies have demonstrated that they are the acetylcholine receptors themselves. Furthermore, closed membrane vesicles derived from such membrane fragments can be made permeable to Na^+ by adding agonists of the acetylcholine receptor, an effect that is blocked by antagonists such as the snake venom toxins.

The acetylcholine receptor was purified to homogeneity for the first time by solubilizing electroplax membranes of *Torpedo* with nonionic detergents and subjecting them to affinity chromatography on columns containing covalently bound cobratoxin. The resulting preparation was a glycoprotein consisting of four different subunits (M_r = 40, 50, 60, and 65 $\times 10^3$, named α, β, γ, and δ) in a molar ratio of 2:1:1:1. The simplest structure of the acetylcholine receptor is therefore a pentamer ($\alpha_2\beta\gamma\delta$) with a mass of about 255,000 daltons. This value agrees with hydrodynamic measurements on the purified receptor and with the dimensions and subunit composition of the particle seen by electron microscopy (see fig. 35.21). Interestingly, the four different subunits of this protein, all of which have been fully sequenced, exhibit extensive sequence homology, an indication that they all arose by duplication and divergence of a single ancestral gene. Computer analyses of these amino acid sequences suggest a common motif of secondary structure for each subunit inside the lipid bilayer (see the following discussion).

Table 35.2
Neurochemical Messengers

Compounds	Status[a]
Acetylcholine	C
Catecholamines	
Norepinephrine (noradrenaline)	C
Epinephrine (adrenaline)	C
L-DOPA	P
Dopamine	C
Octopamine	C
Amino acids (and derivatives)	
Glutamate	C
Aspartate	P
Glycine	C
Proline	Pos
Gamma-aminobutyrate (GABA)	C
Tyrosine	Pos
Taurine	P
Alanine	Pos
Cystathione	Pos
Histamine	C
Serotonin (5-hydroxytryptamine)	C
Peptides	
Substance P	P
Cholecystokinin	P
Neurotensin	P
Enkephalins	P
Somatostatin	P

[a]C = confirmed neurotransmitter; P = probable; Pos = possible.

Reactive affinity labels that covalently bind to the acetylcholine receptor have been prepared by chemically modifying known agonists and antagonists of cholinergic systems. In most cases examined, the α subunit was labeled by these treatments. These and other studies have shown that the α subunit contains the binding site for acetylcholine, and the residues most important for this interaction appear to be localized between aminoacyl residues 180 and 200 of this subunit. This region of the protein is believed to be part of an extracellular, hydrophilic domain of the receptor, as we might expect from its apparent function. The purified receptor from *Torpedo* has been functionally reconstituted into phospholipid vesicles. These proteoliposomes became permeable to Na^+ in the presence of an acetylcholine analog, while α-bungarotoxin blocked Na^+ permeability, as we would expect if the complex as purified comprised the authentic receptor molecule.

Figure 35.21

A view of the arrangement of acetylcholine receptors on electroplax membranes from *Torpedo californica*. (*a*) Membrane tubes formed spontaneously from membrane vesicles showing a crystalline surface lattice of acetylcholine receptors. Stained with uranyl acetate. (*b*) Computer filtration of micrographs similar to that in (*a*), showing the arrangement of protein density in a single receptor around the central depression (presumed to be the ion channel). (From J. Kistler, R. M. Stroud, M. W. Klymkowsky, R. A. Lalancette, and R. H. Fairclough. "Structure and function of an acetylcholine receptor." Reproduced from the *Biophysical Journal*, 1982, 37:371, by © permission of the *Biophysical* Society. Micrograph courtesy of Dr. R. Stroud.)

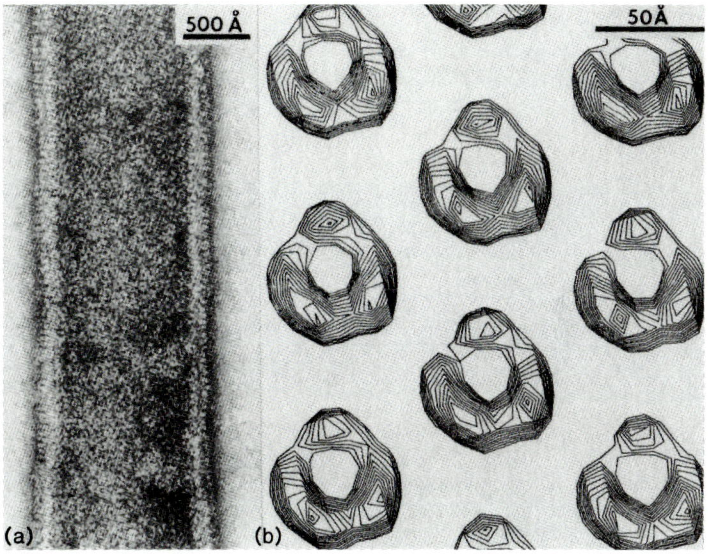

Because all five subunits of the acetylcholine receptor span the postsynaptic membrane, it was proposed that the channel corresponds to the central "hole" in the rosette seen by electron microscopy of electroplax membranes (see fig. 35.21). In fact, the five membrane-spanning subunits of the acetylcholine receptor have been recently resolved in three dimensions by electron microscopy of crystallized postsynaptic membranes of *Torpedo*. The complex consists of five rod-shaped structures, of about equal cross section, which lie largely perpendicular to the membrane. The five subunits are contained within a pentagonally symmetrical shell, about 140 Å long and 80 Å in diameter, which delineates a water-filled channel along the central axis of the cylinder. This cylinder extends about 40 Å into the cytoplasm and nearly 70 Å into the synaptic cleft. The channel opening to the synaptic cleft is wide, but narrows as it extends through the membrane into the interior of the cell (fig. 35.22). Treatment with an acetylcholine analog, carbamylcholine, produces a visible conformational change in this complex (see fig. 35.22). Thus, for the first time, we have a view of the changes in quaternary conformation that are probably the basis for opening and closing of the receptor ion channel.

The genes for all four of the acetylcholine receptor subunits have been cloned and sequenced. As we mentioned earlier, there is a great deal of amino acid sequence homology among

Figure 35.22

Conformational changes induced in the acetylcholine receptor upon binding of an agonist. *Top*: A view toward the face of the δ subunit before (N) and after (C) exposure to carbamylcholine. Note the change in tilt of the δ subunit (dotted line) and the outward movement of the γ subunit upon agonist binding. *Bottom*: Views of the unliganded and liganded receptor toward the face of the γ subunit.

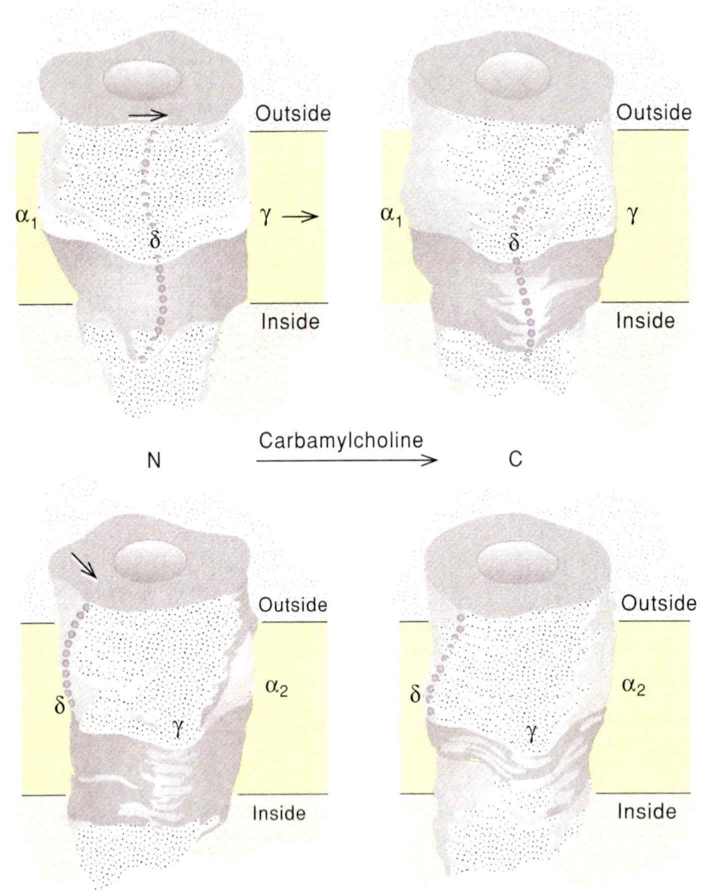

them. In fact, some workers have proposed that in all four subunits the polypeptides each span the membrane four times in a similar fashion as illustrated in figure 35.23. The potential membrane-spanning regions of each subunit (named M1 through M4 in figure 35.23) all contain polar residues, especially serine and threonine. With the *Xenopus* oocyte expression system described earlier for the Na^+ channel, it has been possible recently to use site-directed mutagenesis to explore regions important for channel function. The results have implicated membrane-spanning region M2, in particular, in channel formation and function. When serine and threonine residues within M2 were replaced with alanine residues, the ion conductance of the channel decreased, and when negatively charged or glutamine residues flanking the M2 regions of all four subunits were eliminated (see fig. 35.23), a similar effect was noted. Therefore it has been proposed that rings of negatively charged residues, contributed by all five M2 regions in the acetylcholine receptor complex, flank the cation-specific pore on both sides of the membrane, while hydrophilic residues within the pore provide the polar channel through which the ions can travel. In fact, a synthetic peptide corresponding in sequence to the M2 region of the δ subunit has been shown to form cation-selective channels in artificial phospholipid bilayers, and computer modeling studies have shown that a pentameric structure is the most likely arrangement of these helices in the membrane.

Thus considerable evidence points to a gated-pore-type mechanism for ion permeability conferred by the acetylcholine receptor, as is the case for the Na^+ and K^+ channels. At least two conformations of the protein, corresponding to open and closed states, have been recognized in the presence of agonists and antagonists, respectively. Furthermore, voltage-clamp studies of *Xenopus* oocytes, in which the acetylcholine receptor subunits have been expressed, have revealed current pulses having a square shape and a constant amplitude in the presence of cholinergic agonists. These properties are similar to those displayed by Na^+ and K^+ channels, as well as by bacterial porins and certain ionophore antibiotics as we discussed in chapter 32. It has been calculated that about 10^4 molecules of Na^+ flow through the receptor channel *in vivo* in the millisecond or so that it is open. A pore model allowing for more or less free diffusion of ions through the channel in its open state, rather than a slower, carrier-mediated mechanism, is therefore likely for the acetylcholine receptor.

Excitability Is Found in Many Different Cell Types

Although nerve cells are among the most extensively studied excitable cells, excitability has been demonstrated not only in other cell types in higher animals but also in organisms as simple as unicellular protozoans and in algae. In ciliated protozoa, for example, membrane depolarization events have been shown to regulate swimming behavior, while in certain algae localized transmembrane movements of ions appear to play a role in early development. Membrane excitability thus appears to be a common phenomenon in many different types of cells and to play a role in a variety of physiological processes. In the following chapter we will examine one such process, vision. In this case photons—impinging upon specialized nerve cells containing photoreceptors—initiate nerve impulses that are sent to the brain, which translates these signals into an image of our environment.

Figure 35.23

A model for the arrangement of the subunits of the acetylcholine receptor in the membrane. The amino acid sequences of each of the subunits (α, ß, γ, δ, respectively, from left to right) at each position are shown for the proposed transmembrane (M1–M4) and flanking regions. Solid circles denote polar, uncharged residues; boxes indicate positively charged residues; and dotted boxes show negatively charged residues. The acetylcholine binding region in the α subunit is toward the N terminus of the transmembrane regions and is not shown.

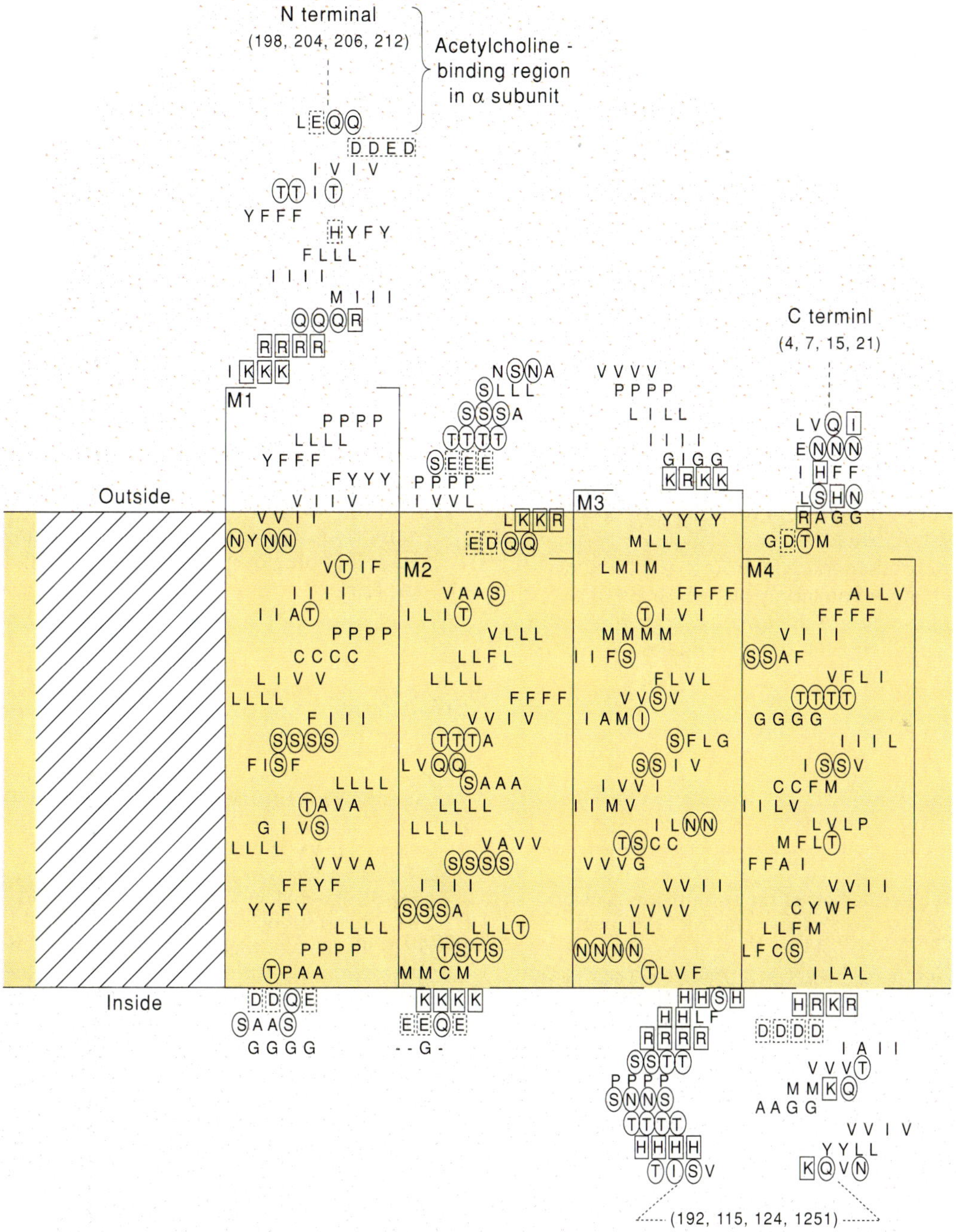

Summary

In this chapter we have examined the way in which nerve impulses are propagated along the neurons and are transmitted to receptor cells. The following points were central to our discussion.

1. Nerve cells are highly specialized for the reception of external signals and the transmission of those signals to other cells in the organism. Signal transmission within a neuron involves waves of membrane depolarization-repolarization events that travel the length of the axon to the nerve endings. The arrival of such a wave at a single region of the axon can be recorded as an action potential, which is generated by a transient influx of Na^+, followed by a transient efflux of K^+ and then a return of the membrane potential to its resting state.
2. Separate, voltage-gated channels for Na^+ and K^+ have been demonstrated in many types of excitable cells, and have been shown to be responsible for the ionic movements and changes in membrane potential associated with the action potential. Na^+ channels from a number of sources consist of an α subunit ($M_r = 260{,}000$) and variable amounts of a β subunit ($M_r = 37{,}000$). The best-characterized K^+ channel is that encoded by the *Drosophila Shaker* locus. It is a 616-residue polypeptide that has significant primary and secondary structural homology to each of the four domains of the Na^+ channel.
3. Transmission of the action potential from a neuron to another cell occurs most commonly by the release of chemical neurotransmitters at the synapse, triggered by the arrival of the action potential at the terminus of the axon. Neurotransmitter compounds include acetylcholine, the catecholamines such as epinephrine and norepinephrine, certain amino acids, and a variety of neuropeptides. Neurotransmitters bind to receptors in the postsynaptic membrane, initiating a response in the target cell that is, again, triggered by the influx of cations.
4. The best-studied neurotransmitter receptor is that for acetylcholine. It consists of four different polypeptide subunits (α, β, γ, and δ) in a molar ratio of 2:1:1:1 with a total mass of about 255,000 daltons. Biochemical and electron-microscopic techniques have established that all five subunits span the postsynaptic membrane. These subunits comprise a pentagonally symmetrical shell, which delineates a water-filled channel that is undoubtedly important in the translocation of cations initiated by ligand binding.
5. Recently, molecular biological techniques and modeling studies have shown how both the ion channel proteins and neurotransmitter receptors may be arranged in the membrane, and have identified regions in these proteins that are important for ion translocation and ligand binding. Site-directed mutagenesis studies of a *Drosophila* K^+ channel also provide evidence as to how the opening and closing of such channels may occur. Further application of these techniques will probably continue to increase our understanding of structure-function relationships in these proteins.

Selected Readings

Brisson, A., and P. N. T. Unwin, Quaternary structure of the acetylcholine receptor. *Nature* 315:457, 1985. This article describes the three-dimensional electron image analysis of tubular crystals of the acetylcholine receptor grown from native membrane vesicles.

Catterall, W., Structure and function of voltage-sensitive ion channels. *Science* 242:50, 1988. Excellent review of the subject through 1987.

Galzi, J-L., A. Devillers-Thiery, N. Hussy, S. Bertrand, J-P. Changeux, and D. Bertrand. Mutations in the channel domain of a neuronal nicotinic receptor convert ion selectivity from cationic to anionic. *Nature* 359:500–505, 1992. In this article, site-directed mutagenesis is used to probe amino acid residues in the acetylcholine receptor that may be important for ion selectivity.

Hille, B., *Ionic Channels in Excitable Membranes,* 2nd edition. Sunderland, MA:Sinauer, 1991. This book covers recent research on a variety of nerve and muscle ion channels. Theory of ion channel function and techniques of analysis are also reviewed.

Hoshi, T., W. N. Zagotta, and R. W. Aldrich, Biophysical and molecular mechanisms of *Shaker* potassium channel inactivation. *Science* 250:533, 1990. Evidence for the "ball-and-chain" mechanism of ion channel inactivation is presented in this article.

Katz, B., *Nerve, Muscle and Synapse.* New York: McGraw-Hill, 1966. Excellent overview of early and classical work on neurotransmission.

Kuffler, S. W., and J. G. Nicholls, *From Neuron to Brain: A Cellular Approach to the Function of the Nervous System.* Sunderland, Mass.: Sinauer Associates, 1984. This book covers aspects of neurotransmission and neurophysiology at the cellular level in an easily readable form.

Rehm, H., and B. L. Temple, Voltage-gated K^+ channels of the mammalian brain. *FASEB J.* 5:164, 1991. Excellent synopsis of recent work on structure-function relationships in the K^+ channel, including molecular genetic approaches.

Problems

1. (a) Calculate the membrane potential $\Delta\Psi$ across a resting nerve-cell membrane in the presence of tetrodotoxin. (Assume that the resting permeability to Na^+ is due to a small, steady-state level of "open" Na^+ channels.)
   (b) What would the value of $\Delta\Psi$ be if the resting axonal membrane were permeable only to Cl^-? (Assume an axoplasmic Cl^- concentration of 50 mM.)
2. Explain why a nerve-cell membrane exhibits an "all or none" response (i.e., action potential) independent of the magnitude of an electric or chemical stimulus (above a threshold value).
3. List the criteria for demonstrating that a particular compound acts as a neurotransmitter in a given system, assuming that the mechanism of synaptic transmission in most instances is analogous to that found in cholinergic systems.
4. In reconstituted transport systems, it is often important to demonstrate that the rate of transport is similar to that observed *in vitro,* that is, that the transport protein is fully functional in the reconstituted state. For systems in which *in vivo* fluxes are very rapid (e.g., cation flux through the acetylcholine receptor), it is often difficult to measure these rates directly in reconstituted vesicles. Thallous ion (Tl^+) is known to pass readily through the "open" state of the acetylcholine receptor. It also very efficiently quenches the fluorescence emission of the fluorophore 8-aminonaphthalene-1,3,6-trisulfonate (ANTS), which is relatively impermeable to phospholipid bilayers.
   (a) Using this information, outline a series of experiments to measure Tl^+ fluxes in reconstituted proteoliposomes containing purified acetylcholine receptors.
   (b) Actual Tl^+ fluxes into proteoliposomes containing an average of two acetylcholine receptor channels per vesicle in the presence of agonists have been measured to be 200 moles/(liter · s). If the average inner diameter of such vesicles is 400 Å, what is the number of Tl^+ ions transported per second by each activated acetylcholine receptor channel? How does this value compare with the rate of Na^+ flux measured *in vivo?*
5. Describe in your own words, and critique, the evidence that voltage-sensitive K^+ channels may be regulated by a "ball-and-chain" mechanism.
6. Patients with Parkinson's disease are treated with DOPA and inhibitors of dopa decarboxylase. This treatment will raise levels of dopamine in certain areas of the brain. However, dopamine given as a drug has no effect! How does this treatment of Parkinson's disease work? (Hint: Think about the blood-brain barrier and membrane permeability.)
7. Draw a chemical reaction mechanism for the reaction of the nerve poison parathion with acetylcholinesterase (see fig. 35.15*b*). How do these types of inhibitors prevent repolarization of the postsynaptic membrane?
8. What are the data which support a gated-pore-type model for ion permeability conferred by the acetylcholine receptor, as opposed to a carrier-mediated mechanism?
9. Earlier in this text, and in this chapter as well, we have cited evidence for protein structure-function relationships that has come by applying the powerful tools of molecular genetics. From what you have learned so far, what do you think are both the advantages and limitations of this kind of approach? (Use proteins described in this chapter as examples.)

36

CHAPTER

Vision

The photochemistry of vision is different from that of photosynthesis. Perhaps this is not surprising. Animals that can see use light to obtain information; photosynthetic organisms use light as a source of energy. However, vision and photosynthesis both start with the excitation of an electron from one molecular orbital to another orbital of higher energy. The excited molecule must then undergo a transformation to a metastable product. Since chlorophyll can undergo such a transformation with a quantum yield near 1.0, it is not hard to imagine an eye that uses electron-transfer reactions like those that work so well in photosynthesis. In fact, there is a synthetic eye that works very much that way—a silicon-diode-array television camera tube. But the eyes of multicellular animals are different. Instead of chlorophyll, their light-sensitive cells contain a complex called rhodopsin, which consists of a protein, opsin, and a linear polyene, 11-*cis*-retinal. Instead of undergoing oxidation when it is excited with light, the retinal isomerizes.

In this chapter we will examine the biochemistry involved in this reaction to light. We will focus primarily on vertebrates, with only occasional references to invertebrate photochemistry, but we will conclude with a short section on the light-reactive substance found in one species of bacterium.

The Visual Pigments Found in Rod and Cone Cells

The eyes of vertebrates are marvelously complex organs, with many different types of specialized cells (fig. 36.1). Light rays entering the eye are refracted by the cornea, the clear tissue at the front of the eye. The light traverses an aqueous chamber and reaches the lens, which is densely packed with proteins called crystallins. Adjustments in the shape of the lens focus a sharp optical image onto the retina, a thin layer of tissue that lines the back of the eye. The retina is a neural tissue with several different layers of cells. Some of these cells, the rod and cone cells, contain the visual pigments. Other cells make synaptic connections to the rods or cones and to additional neural cells that carry impulses to the brain.

In both rods and cones, the light-sensitive molecules are collected in a layered system of membranes at one end of the cell (figs. 36.2 and 36.3). The membranes form by invaginations of the cytoplasmic (plasma) membrane near the middle of the cell. In cone cells, the membranes remain contiguous with the cytoplasmic membrane. In rods, they pinch off to form a stack of autonomous flattened vesicles, or disks, in the outer segment of the cell. The outer segment of each rod contains from 500 to 2,000 of these disks. This region of the cell is connected by a thin cilium to the inner segment, which is packed with mitochondria and ribosomes. The basal part of the cell ends in a synaptic junction with another neural cell called a bipolar cell. In a living rod cell, disks are constantly forming at the base of the outer segment and moving in a file to the tip of the outer segment, where they are sloughed off and phagocytosed by the underlying epithelial cells. It takes about ten days for a disk to make this journey. Exactly why the disks need to be replaced so rapidly is unclear.

Figure 36.1

Structure of the human eye.

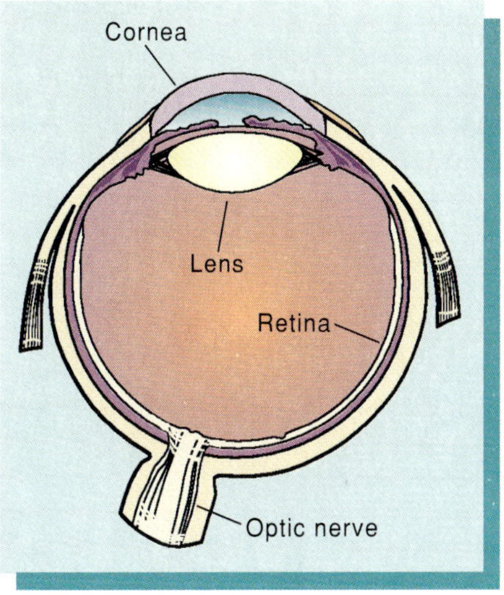

Figure 36.2

Thin-section electron micrograph (59,200X) of a portion of a rod cell in a rabbit retina. Part of the outer segment is shown at the top, and part of the inner segment at the bottom. (D = disks; M = mitochondrion; C = cilium.) (Courtesy of Dr. Ann H. Milam.)

Figure 36.3

Schematic diagram of a rod cell. The orientation of the cell is the same as that in figure 36.2. In the retina, many rod cells are stacked side by side, with the outer segments all pointing out to the periphery of the eye. Light enters the cells end-on through the inner segment, after passing through several layers of other neural cells. In cone cells, the outer segments are shorter, and are conical rather than cylindrical in shape.

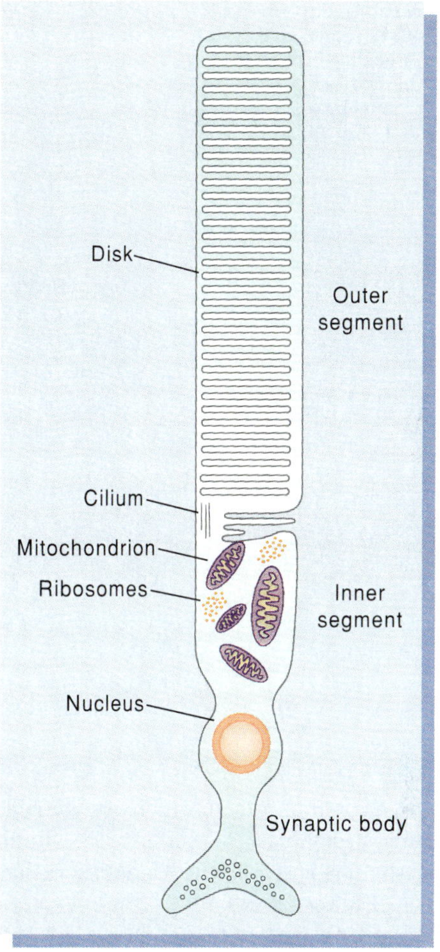

Rod cells are specially adapted for vision in dim light. Cones provide visual acuity in bright light, and also serve for perception of color. Animals such as owls, which have very high visual sensitivity in dim light but cannot distinguish colors, have only rod cells. Some animals, such as pigeons, have only cones and are skilled at distinguishing colors in bright light but inept at night vision. Primates have both rods and cones, with rods considerably outnumbering cones in all but the central portion of the retina.

*Rhodopsin Consists of 11-*cis*-Retinal Bound to a Protein, Opsin*

Figure 36.4 shows the structures of 11-*cis*-retinal and its more stable isomer all-*trans*-retinal. The retinals are related to the alcohol retinol, or vitamin A_1. These compounds cannot be synthesized *de novo* by mammals, but they can be formed from

Figure 36.4

Structures of retinals, retinols, and ß-carotene. The structure of 11-*cis*-retinal (*top*) indicates the numbering system used for the carbons. For simplicity, the molecules are shown here as being planar, although the most stable conformer of 11-*cis*-retinal actually is twisted about the C-6—C-7 and C-12—C-13 single bonds to reduce steric crowding of the methyl groups. In rhodopsin, 11-*cis*-retinal is bound by a protonated Schiff's base linkage to a lysine of opsin.

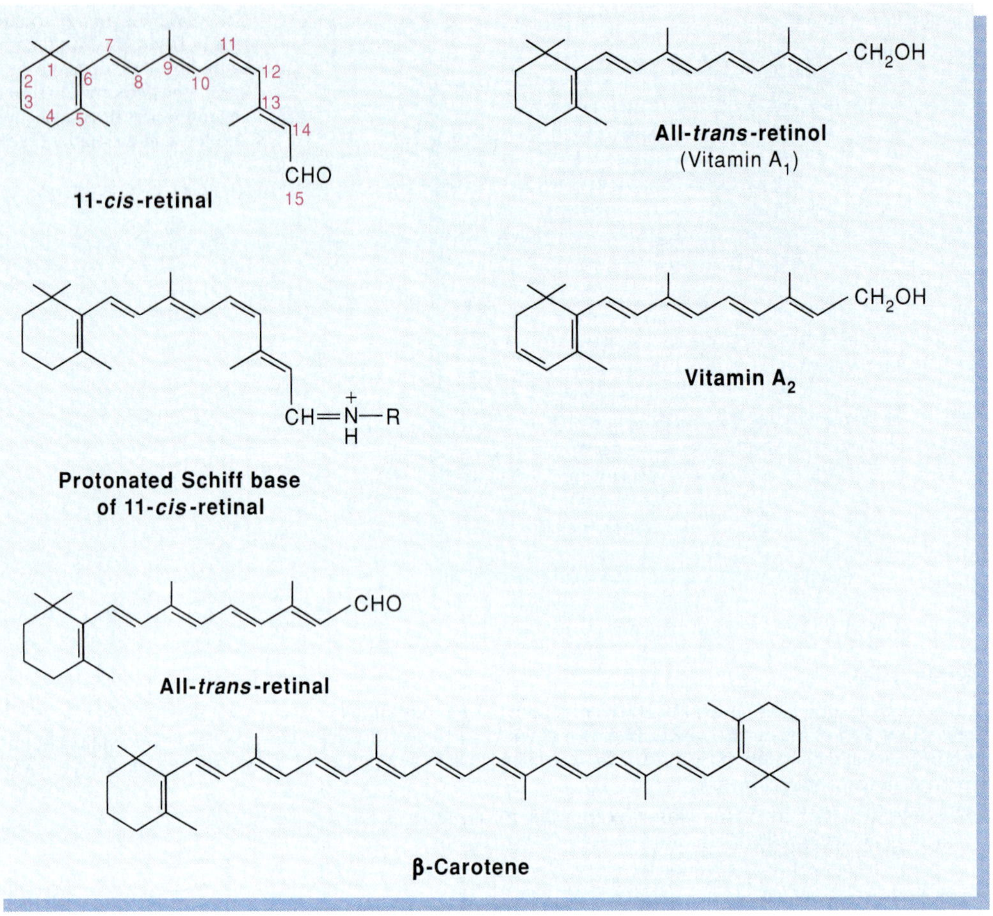

carotenoids, such as β-carotene, which are abundant in carrots and some other vegetables. A deficiency of vitamin A causes night blindness, along with a serious deterioration of the eyes and a number of other tissues.

There are animals with eyes in several different phyla, including Mollusca, Arthropoda, and Annelida, in addition to Chordata. The eyes of arthropods differ substantially from those of molluscs and chordates anatomically, and they apparently originated independently, after the phyla had separated in evolution. Some animals, such as sea turtles and amphibians in certain stages of development, use a retinal that has an additional double bond in the ring. The parent molecule in this case is vitamin A_2, or 3,4-dehydroretinal (see fig. 36.4). In all cases, however, the photochemically active protein complex is made from the 11-*cis* isomer of the aldehyde. 11-*cis*-Retinals must somehow be particularly fit for the task of responding to light. (Some unicellular organisms, such as *Euglena,* have light-sensitive organelles that contain a carotenoid instead of a retinal, but little is known of the biochemistry of these primitive receptors.)

Opsin is an intrinsic membrane protein with a molecular weight of about 38,000. It accounts for about 95% of the protein in the disk membranes. Labeling studies of isolated disks indicate that the carboxyl-terminal end of rhodopsin is exposed to the cytoplasmic space outside the disk, while the amino end is exposed to the space inside the disk. Seven hydrophobic α-helical stretches of the protein probably lace back and forth across the phospholipid bilayer. The retinal is bound to a lysyl-NH_2 group by a Schiff base, or aldimine linkage (see fig. 36.4), in a region that is buried in the bilayer. Rhodopsin also has several carbohydrate groups bound near the amino-terminal end.

In solution, 11-*cis*-retinal absorbs maximally near 380 nm, but in rhodopsin the peak is at 500 nm (fig. 36.5). The absorption spectrum of rhodopsin is essentially identical to the spectrum of the sensitivity of the rod cells (after correction for absorption in the cornea and lens). The perception of color by cones depends on the fact that there are three different types of cones with different absorption spectra. One of these absorbs blue light (440 nm) maximally, the second absorbs green (530 nm), and the third absorbs yellow (570 nm). In some species of birds, the third component absorbs maximally at 630 nm. However, all cones contain 11-*cis*-retinal bound to proteins that are similar to opsin. (Some fish use another strategy altogether for color vision. Their eyes have carotenoid-containing oil droplets of three colors, and these act as filters in front of the receptor cells.)

Figure 36.5

Absorption spectra of rhodopsin and of 11-*cis*-retinal in hexane solution. The absorption of rhodopsin in the 280-nm region is due mainly to the opsin.

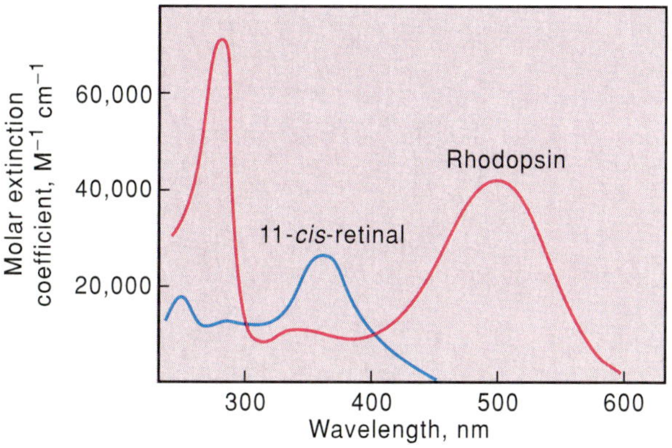

*Light Isomerizes the Retinal of Rhodopsin to All-*trans

The discovery that rhodopsin contains the 11-*cis* isomer of retinal came as a surprise. In the early 1950s, Ruth Hubbard, George Wald, and their co-workers found that rhodopsin decomposes into retinal and the apoprotein opsin when the retina is exposed to light. The retinal that was released was the all-*trans* isomer. When opsin was mixed with a crude preparation of retinal, rhodopsin was regenerated. Crystalline all-*trans*-retinal, however, did not bind to opsin. The regeneration achieved with the crude preparation proved to be due to a small amount of the 11-*cis* isomer in the mixture. Since rhodopsin binds 11-*cis*-retinal in the dark, but releases all-*trans*-retinal after illumination, Hubbard and Kropf concluded that the action of light in vision is to isomerize the chromophore about its C-11=C-12 double bond. This change in structure must somehow be translated into an electrophysiological signal that can be transmitted to the brain.

Transformations of Rhodopsin Can Be Detected by Changes in Its Absorption Spectrum

The release of all-*trans*-retinal from opsin takes several minutes and is too slow to be an obligatory step in visual perception. It appears to be a step in the regeneration of the active form of rhodopsin. To explore the changes in rhodopsin that precede the release of all-*trans*-retinal from the protein, Toru Yoshizawa and Wald measured the optical absorbance changes that occurred when they illuminated rhodopsin at low temperatures. Illumination at liquid N_2 temperature (77 K) caused the absorption band of the rhodopsin to shift from 500 to 543 nm. The product of this transformation is now called bathorhodopsin. Bathorhodopsin is stable indefinitely in the dark at 77 K, but if it is warmed above about 130 K it decays spontaneously to a species that absorbs maximally at 497 nm. This species is called lumirhodopsin. If the sample is warmed further, to about

Figure 36.6

Photochemical transformations of rhodopsin. The numbers in parentheses indicate the optical absorption maxima of the intermediates. The numbers in color on the right are approximate half-times for the conversions in rod outer segment membranes near 37° C. With isolated rhodopsin, some of the reactions are slower and are multiphasic. The steps following the formation of bathorhodopsin have progressively higher thermal activation energies. They can be blocked by lowering the temperature below the temperatures indicated on the left.

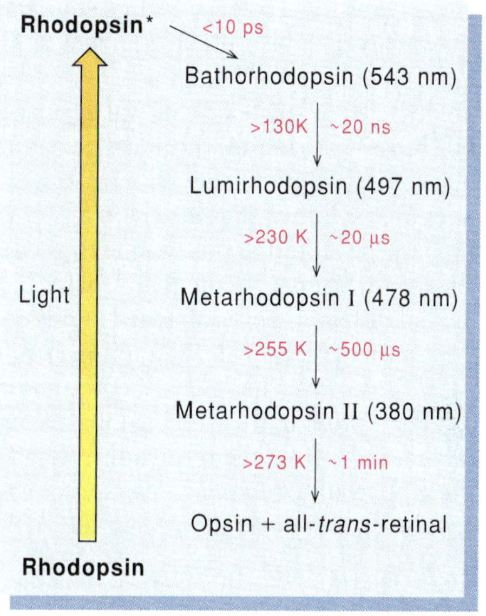

230 K, lumirhodopsin decays to metarhodopsin I, which absorbs at 478 nm. Above about 255 K, metarhodopsin I decays to metarhodopsin II, which absorbs at 380 nm. These transformations are outlined in figure 36.6.

Subsequent kinetic measurements by other investigators have shown that rhodopsin passes through the same series of states if it is excited with a short flash at physiological temperatures. The numbers on the right side of figure 36.6 give approximate half-times for the transformations at 37° C. Agreement between the kinetic studies and low-temperature experiments may seem routine, but it might not have turned out this way at all. The trapping of a metastable state at low temperature can depend on the thermodynamic properties of the state, as well as on its position in the kinetic sequence. For example, side products that normally are unimportant could accumulate when the normal pathway is blocked by lowering the temperature.

How can the absorption spectrum of rhodopsin go through such wild changes from 500 to 543 nm, and eventually to 380 nm? In thinking about this, we are reminded that rhodopsin's initial absorption spectrum is already shifted by 120 nm compared with the spectrum of free 11-*cis*-retinal (fig. 36.5). Further, the absorption maxima of the cone pigments vary from 450 to 630 nm, in spite of the fact that these pigments all contain 11-*cis*-retinal. The explanation for these spectral differences depends partly on the fact that the nitrogen atom of the retinylidine Schiff base linkage in rhodopsin is protonated and is therefore positively charged (fig. 36.7). The protonation state of the nitrogen has been demonstrated clearly by NMR and resonance Raman spectroscopy. When the retinal absorbs light and is raised to an excited state, the electron density on the nitrogen increases, and the positive charge moves to the opposite end of the molecule, as represented roughly in figure 36.7. Because of the redistribution of charge, the relative energies of the excited and ground states are extremely sensitive to the positions of other charged, dipolar, or polarizable groups nearby. An arrangement of charged groups that stabilizes the excited state relative to the ground state will shift the absorption spectrum to longer wavelengths.

There must be at least one charged group near the retinylidine Schiff base in rhodopsin, the anionic counterion that is needed to balance the positive charge on the nitrogen. The anion is probably the carboxylate of the side chain of a glutamate residue, Glu 113, because mutating this residue to glutamine causes a dramatic change in the pK_a of the Schiff base. In the mutant rhodopsin, the Schiff base is not protonated at physiological pH, and the absorption maximum is at 380 nm instead of 500 nm. The structure of the protein thus can affect the absorption spectrum of the pigment profoundly. Moving the carboxylate group of Glu 113 farther away from the retinylidine Schiff base would destabilize the ground state relative to the excited state, shifting the absorption spectrum to longer wavelengths.

Figure 36.7

Excitation of the protonated Schiff's base causes a movement of positive charge from the nitrogen toward the ring. The C-11═C-12 bond loses much of its double-bond character. The valence bond diagrams shown here should not be taken too literally, but they give a good qualitative picture of the redistribution of electrons that occurs when the molecule is excited.

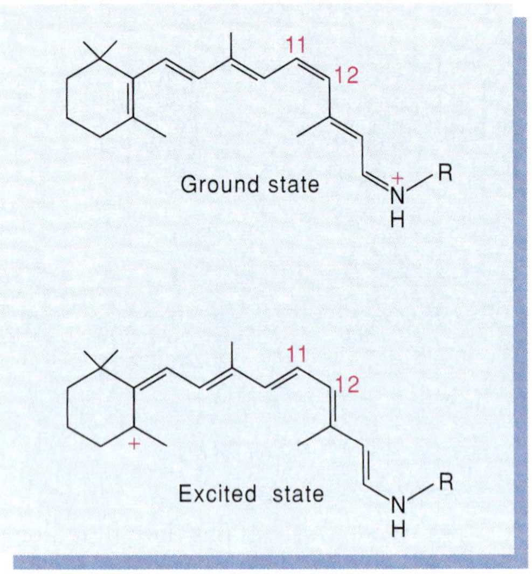

Isomerization of the Retinal Causes Other Structural Changes in the Protein

Let us now consider bathorhodopsin, which appears to be the first metastable product of the photochemical reaction. If bathorhodopsin is excited with long-wavelength light at 77 K, it can be converted back to rhodopsin. Resonance Raman measurements support the view that the retinal in bathorhodopsin has isomerized to the all-*trans* form, but that it continues to be held as a protonated Schiff base. Additional evidence that bathorhodopsin contains all-*trans*-retinal has been obtained by studying a modified form of rhodopsin, isorhodopsin, which contains the 9-*cis* isomer of retinal. When isorhodopsin is illuminated, it gives rise to bathorhodopsin with 10^{-11} s, just as rhodopsin (11-*cis*) does. The only common product that could form directly from both the 9-*cis* and 11-*cis* isomers is the all-*trans* isomer.

Isomerization of the retinal Schiff base can occur when the molecule is excited with light, because the C-11═C-12 bond loses much of its double-bond character in the excited state. The valence bond diagrams of figure 36.7 illustrate this point qualitatively. In the ground state of rhodopsin, the potential energy barrier to rotation about the C-11═C-12 bond is probably on the order of 30 kcal/mole. This barrier essentially vanishes in the excited state. In fact, molecular orbital calculations suggest that the energy of the excited molecule is minimal when the C-11═C-12 bond is twisted by about 90° (fig. 36.8). The 11-*cis* and all-*trans* molecules thus have a common excited state. When the molecule decays from the excited state to a ground state, it can end up in either of these isomeric forms. It turns out that the excited state decays to bathorhodopsin (all-*trans*) about 67% of the time and to rhodopsin (11-*cis*) about 33% of the time. Isorhodopsin (9-*cis*) and other isomers are formed only in small amounts (about 1%). Similar ratios of products hold no matter whether we start by exciting rhodopsin or bathorhodopsin.

The high energy barrier to the isomerization of rhodopsin in the dark is physiologically important because it limits the "noise" in our perception of light. If the barrier were low enough to be overcome thermally, the discrimination of light from dark would be more difficult.

Because opsin binds 11-*cis*-retinal, but not all-*trans*-retinal, the all-*trans* molecule that is created in bathorhodopsin must find itself initially in a binding site that is tailored for the 11-*cis* structure. In solution, 11-*cis*-retinal is less stable than all-*trans*-retinal, because of steric repulsion between the methyl group on C-13 and the hydrogen atom on C-10 (see fig. 36.4). On the protein, there must be compensating interactions that decrease the free energy of the 11-*cis* complex relative to that of the all-*trans* complex. These interactions will be disrupted when the pigment is isomerized. The pronounced shift of the absorption spectrum to longer wavelengths in bathorhodopsin could be explained if the isomerization results in a movement of the Schiff base nitrogen, increasing the distance between the positively charged nitrogen and its counterion (fig. 36.9). Such a movement would destabilize the ground state of bathorhodopsin relative to the excited state, as we discussed earlier for rhodopsin. A repositioning of the Schiff base also is likely to lead to changes in the structure of the protein surrounding the binding site. If the positively charged nitrogen moves closer to a second anionic group, as illustrated in figure 36.9, a new hydrogen bond could form between this group and the Schiff base nitrogen, and we would expect the protein to respond by adjusting the positions of other nuclei nearby.

Figure 36.8

Potential energy surfaces of ground and excited states in rhodopsin, as functions of the angle of rotation of the bond between carbons 11 and 12. A rotational angle of 0° means that the retinyl group is 11-*cis* (rhodopsin); 180° means that it is all-*trans* (bathorhodopsin). In the ground state (lower curve), rhodopsin is stabilized with respect to bathorhodopsin, possibly because the positive charge of the protonated Schiff's base interacts more favorably with a negatively charged group of the protein (fig. 36.9). The energy of the excited state (upper curve) is less sensitive to the position of the Schiff's base than the energy of the ground state is, because the positive charge has moved to the opposite end of the retinal molecule (fig. 36.7). The energy of the excited state is minimal when the rotation angle is about 90°. Rhodopsin is raised from the ground state to the excited state by light (vertical arrow). It relaxes back to the ground state along a path like that indicated by the dashed arrow, ending up as bathorhodopsin.

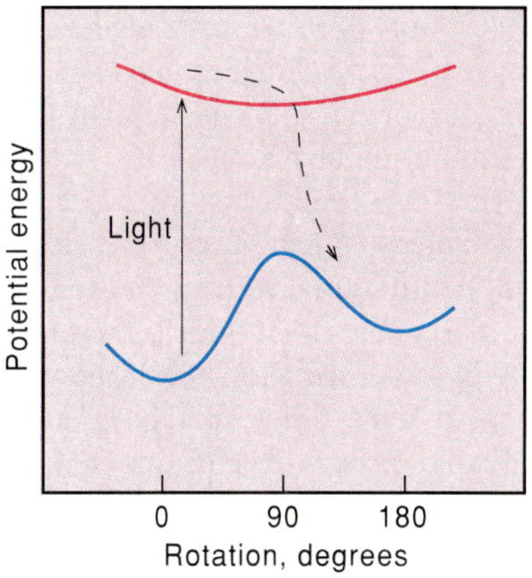

Figure 36.9

A model for the photochemical conversion of rhodopsin (*top*) to bathorhodopsin (*bottom*). In rhodopsin, the proton of the Schiff's base nitrogen is hydrogen bonded to a counterion, A_1^-. The retinal is twisted about the C-12—C-13 and C-6—C-7 single bonds. (As noted in figure 36.4, free retinal is twisted in this way, but the amount of twisting in rhodopsin is still uncertain.) When rhodopsin is converted to bathorhodopsin (*bottom*), the ring end of the retinal is presumed to be locked in position. Isomerization about the C-11═C-12 double bond flips the protonated nitrogen and its positive charge away from the counterion A_1^-. The separation of electrical charge could account for the shift of the absorption spectrum to longer wavelengths. The isomerization also can lead to proton movements. The proton of the Schiff's base could be passed to another basic group (A_2^-), and A_1^- might pick up a proton from another acidic group (A_3H). Other groups of the protein, including the lysine attached to the retinal, also must adjust to the new geometry of the retinyl Schiff's base. (For additional discussion, see B. Honig et al., An external point-charge model for wavelength regulation in visual pigments, *J. Am. Chem. Soc.* 101:7084, 1979, and Photoisomerization, energy storage, and charge separation: A model for light-energy transduction in visual pigments and bacteriorhodopsin, *Proc. Natl. Acad. Sci. USA* 76:2503, 1979.)

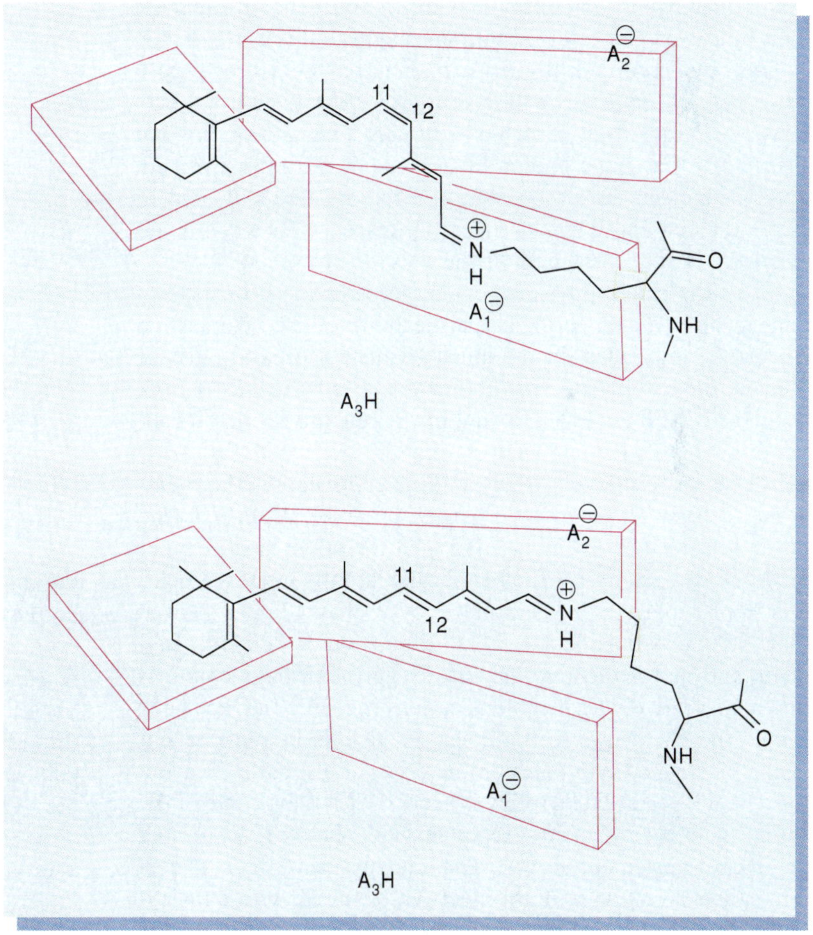

The reorganization of rhodopsin that is set in motion by isomerization of the retinylidine Schiff base continues as the system relaxes through the states lumirhodopsin, metarhodopsin I, and metarhodopsin II. Physical measurements indicate that isolated rhodopsin undergoes particularly substantial changes in protein conformation during the transition from metarhodopsin I to metarhodopsin II. The extent of these changes and the kinetics of formation of metarhodopsin II are sensitive to the type and number of phospholipids surrounding the protein. Conformation changes also undoubtedly occur in this step when the rhodopsin is in place in the disk membrane, but they are subtler and may be confined to small regions of the protein. The nitrogen of the retinyl Schiff base loses its proton as metarhodopsin II is formed, and a second, unidentified group takes up a proton from the solution. The loss of the positive charge on the nitrogen accounts for the shift of the absorption maximum to 380 nm, where unprotonated Schiff bases of all-*trans*-retinal absorb. The Schiff base linkage also becomes accessible to reagents in the aqueous solution during or shortly after the formation of metarhodopsin II.

The formation of metarhodopsin II is fast enough to be an obligatory step in visual transduction. It clearly is associated with changes in the interactions between rhodopsin and its surroundings. A reasonable hypothesis, therefore, is that the changes in protein structure allow metarhodopsin II to initiate an interaction with some other component of the disk membrane. We will explore the nature of this component in the following sections.

The Conductivity Change That Results from Absorption of a Photon

The human eye is amazingly sensitive. After being in the dark for a time, we can perceive continuous light that is so weak that an individual rod cell absorbs a photon, on the average, only once every 38 minutes. A flash of light is detectable if approximately six rods each absorb one photon. This means that the absorption of a single photon by any one of the approximately 3×10^7 molecules of rhodopsin in a rod must be sufficient to excite the cell and trigger a neuronal response. The combined responses of six rods can elicit the sensation of seeing.

The electrophysiological response of a rod or cone to light involves a change in the permeability of the cytoplasmic membrane to cations. In the inner segment and basal parts of the cell, the cytoplasmic membrane contains a Na^+-K^+ pump, which uses ATP to move Na^+ out of the cell and K^+ in. (Recall our discussion of ion pumps in chapter 32.) K^+ can diffuse back out of the cell relatively freely, and its efflux causes the membrane to become negatively charged by about 20 mV on the inside relative to the outside. Na^+, on the other hand, cannot readily pass back across the membrane into the inner segment. Electrophysiological measurements by William Hagins showed that Na^+ flows outside the cell to the region of the outer segment, and that it reenters the outer segment through channels that are selectively permeable to cations (fig. 36.10). Once back inside, the Na^+ flows to the inner segment to be pumped out again. The round trip takes on the order of a minute. When a rod cell of a vertebrate's retina is excited with light, some of the cation channels in the outer segment suddenly close, decreasing the inward movement of Na^+. The interruption of the Na^+ current causes an increase in the electrical potential across the cytoplasmic membrane. This hyperpolarization causes a change in the movement of an unidentified chemical transmitter to the bipolar cell that makes a synaptic junction with the rod. The bipolar cell then sends a signal to a ganglion cell, the third component in the hierarchy of retinal neurons.

The hyperpolarization of the rod and the response of the bipolar cell are graded effects. Absorption of a single photon causes the membrane potential to increase by about 1 mV, with the effect peaking about 1 s after the excitation. Up to about 100 photons per rod, the more light the cell absorbs, the larger the amplitude of the hyperpolarization and the faster the hyperpolarization occurs. The reaction of the ganglion cell, however, is all-or-none. When the ganglion cell is triggered, it responds with an action potential that proceeds to the brain. Sensitivity to weak light is enhanced (with a sacrifice in spatial resolution) by having multiple rod cells connected to each bipolar cell and multiple bipolar cells connected to each ganglion cell. The output of cone cells is not summed in this way. This difference partly explains why the cones provide better visual acuity but lower sensitivity than the rods.

Figure 36.10

Na^+ is pumped out of the inner segment (solid arrows) and diffuses back into the cell through channels in the outer segment (dashed arrows). In rods of vertebrates, the Na^+ channels are held open in the dark, and they close in the light. In invertebrates, the channels open in the light.

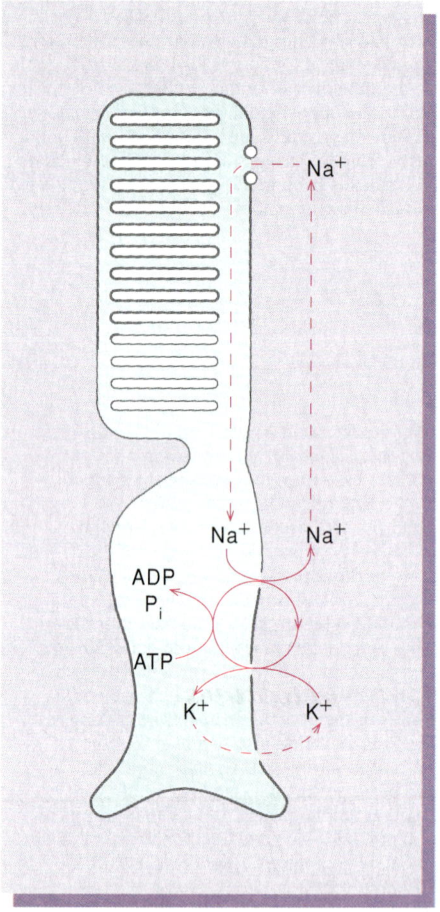

Analogous events occur in the eyes of invertebrates, except that light causes an increase in the cation permeability of the receptor cell rather than a decrease. In the rods of vertebrates, the absorption of a single photon decreases the current of Na^+ into the outer segment by about 3%. The inflow of approximately 10^6 Na^+ ions is transiently prevented. Exactly how many Na^+ channels have to close in order to achieve this effect is uncertain, but estimates have ranged from 25 to 1,000. Since only one molecule of rhodopsin is excited, the cell must have a mechanism for amplifying the effect of light by a substantial factor. Recall also that the cytoplasmic membrane of the rod

Figure 36.11

cGMP increases the electrical conductivity of the rod-cell cytoplasmic membrane. The conductivity was measured with a patch of membrane held at the tip of a micropipette. By varying the ions in the solution, it is possible to show that cGMP affects the movement of Na^+ and other cations across the membrane. Note the sigmoidal concentration dependence: the Hill coefficient is 1.6. (For more details, see E. E. Fesenko, S. S. Kolesnikov, and A. L. Lymbarsky, Induction by cyclic GMP of cationic conductance in plasma membrane of retinal rod outer segment, *Nature* 313:310, 1985.)

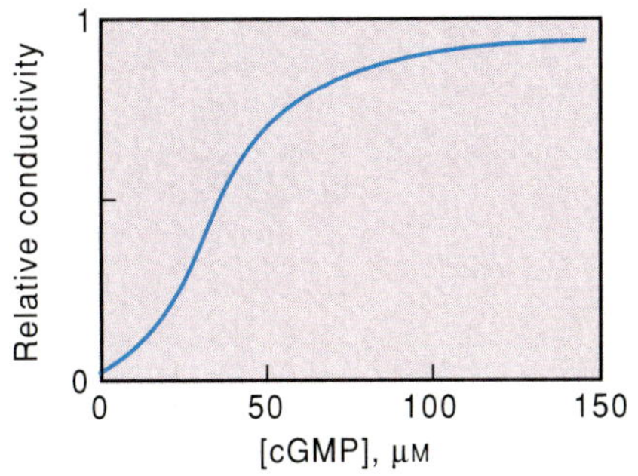

outer segment is not contiguous with the disk membranes that hold the rhodopsin (see fig. 36.3). This fact suggests that the excitation of rhodopsin causes a change in the concentration of a diffusible transmitter, which moves from the disk to the cytoplasmic membrane.

The Effect of Light Is Mediated by Guanine Nucleotides

There is strong evidence that the diffusible component that moves between the disk and the cytoplasmic membrane is 3′,5′-cyclic-GMP (cGMP). Electrophysiological measurements have shown that cGMP causes an increase in the permeability of the cytoplasmic membrane to Na^+ (fig. 36.11). This appears to be a direct effect of cGMP on the Na^+ channels, rather than an indirect effect mediated by a kinase, because it can be seen in the absence of ATP.

The disk membranes of the rod outer segment contain a phosphodiesterase, which hydrolyzes 3′,5′-cyclic-GMP (cGMP) to 5′-GMP. If the disk membranes are kept in the dark, the phosphodiesterase remains in a relatively inactive state. When the disks are illuminated and rhodopsin is converted to metarhodopsin II, the activity of the phosphodiesterase increases. For each molecule of rhodopsin that is excited, on the order of 500 molecules of the phosphodiesterase are activated. Illumination thus results in a decrease in the cGMP content of the cell. A drop in the cGMP concentration is in the right direction to cause a decrease in Na^+ permeability of the cell membrane (see fig. 36.11), and thus to cause a hyperpolarization of the membrane. In agreement with this scheme, injecting extra cGMP into rod outer segments temporarily inhibits the hyperpolarization caused by light.

Unlike rhodopsin, the phosphodiesterase is a peripheral membrane protein that can be readily solubilized. Its activation by light requires the presence of GTP and is associated with the binding of GTP to a second peripheral membrane protein called transducin. Transducin is a member of the family of G proteins that participate in the activation or inhibition of adenylate cyclase by hormones in other tissues (see chapter 24). Like other G proteins, transducin consists of three subunits, α, β, and γ (fig. 36.12). In the resting state, the α subunit contains a molecule of bound GDP. When rhodopsin is transformed to metarhodopsin II by light, it interacts with transducin, causing GTP to displace the bound GDP. Once GTP is attached, the α subunit separates from the β and γ subunits and binds to an inhibitory subunit of the phosphodiesterase. The removal of the inhibitory component activates the phosphodiesterase.

Each molecule of metarhodopsin II that is generated by light appears to be able to trigger about 500 molecules of transducin to bind GTP in place of GDP. This number agrees with the number of phosphodiesterase molecules that are activated, and accounts for much of the amplification in the response to light. Additional amplification results from the enzymatic action of each phosphodiesterase on many molecules of cGMP. These effects can be demonstrated with purified preparations of rhodopsin, transducin, and the phosphodiesterase. Purified, illuminated rhodopsin that has been incorporated into phospholipid vesicles is capable of activating transducin, and the complex of transducin's α subunit with GTP is capable of activating the phosphodiesterase in the absence of rhodopsin.

In order for the eye to respond rapidly to changing light intensities, the activation of the phosphodiesterase must be a transient response that quickly switches off again. The activation of transducin's α subunit is reversed by hydrolysis of the bound GTP to form bound GDP and free inorganic phosphate (see fig. 36.12). When this happens, the α subunit separates from the inhibitory polypeptide of the phosphodiesterase and recombines with the β and γ subunits of transducin. The inhibitory polypeptide then recombines with the phosphodiesterase, returning the phosphodiesterase to its resting, inactive state. We can prolong activation of the phosphodiesterase indefinitely if, instead of adding GTP, we add a nonhydrolyzable analog of GTP.

To switch off the response to light, two additional things must happen: Metarhodopsin II must decay to a form that is incapable of activating additional molecules of transducin, and the cGMP that was hydrolyzed must be resynthesized in order to reopen the cation channels in the plasma membrane. The inactivation of metarhodopsin appears to involve phosphorylation of the protein and binding of the phosphorylated rhodopsin to another protein called the 48K protein, or arrestin. The regeneration of the cGMP is carried out by a guanylate cyclase, which catalyzes the synthesis of cGMP from GTP. This enzyme is stimulated by still another regulatory protein, but is strongly inhibited by Ca^{2+}. Ca^{2+} enters the rod outer segment through the same cation channels that admit Na^+, and it is exported by a transport protein that exchanges Ca^{2+} for Na^+. The Ca^{2+} concentration in the cell thus decreases upon illumination, when the cation channels close. The drop in the Ca^{2+} concentration results in an activation of the guanylate cyclase. Changes in the intracellular Ca^{2+} concentration also appear to underlie the ability of the eye to adapt to large changes in illumination intensity.

Figure 36.12

The large arrows show the main steps in the activation of the cGMP phosphodiesterase by light. Light converts rhodopsin (R) to a metastable form (M_{II}, probably metarhodopsin II). M_{II} reacts catalytically with transducin, which contains GDP bound to one of its three subunits (α). The interaction with M_{II} changes the specificity of the nucleotide binding site so that GDP is released and GTP is bound. This causes the α subunit to dissociate from the other two subunits of transducin (ß and γ). In the dark, the phosphodiesterase (PDE) is inactive, because of the presence of an inhibitory polypeptide (I). The I polypeptide is removed by binding to the complex of the transducin α subunit and GTP. The active phosphodiesterase then can hydrolyze cGMP to 5′-GMP. M_{II} can go on to activate several hundred additional molecules of transducin, before slower enzymatic reactions (thin arrows) gradually convert it to an inactive form. The inactivation involves conversion of the rhodopsin to a phosphorylated form (R-P) and an interaction with another protein (48K). The GTP bound to transducin's α subunit eventually is hydrolyzed to GDP. When this happens, I dissociates and returns to the phosphodiesterase, the three subunits of transducin reassemble, and the system returns to its resting state. (For details, see M. Chabre, Trigger and amplification mechanisms in visual phototransduction, *Ann. Rev. Biophys. Biophys. Chem.* 14:331, 1985.)

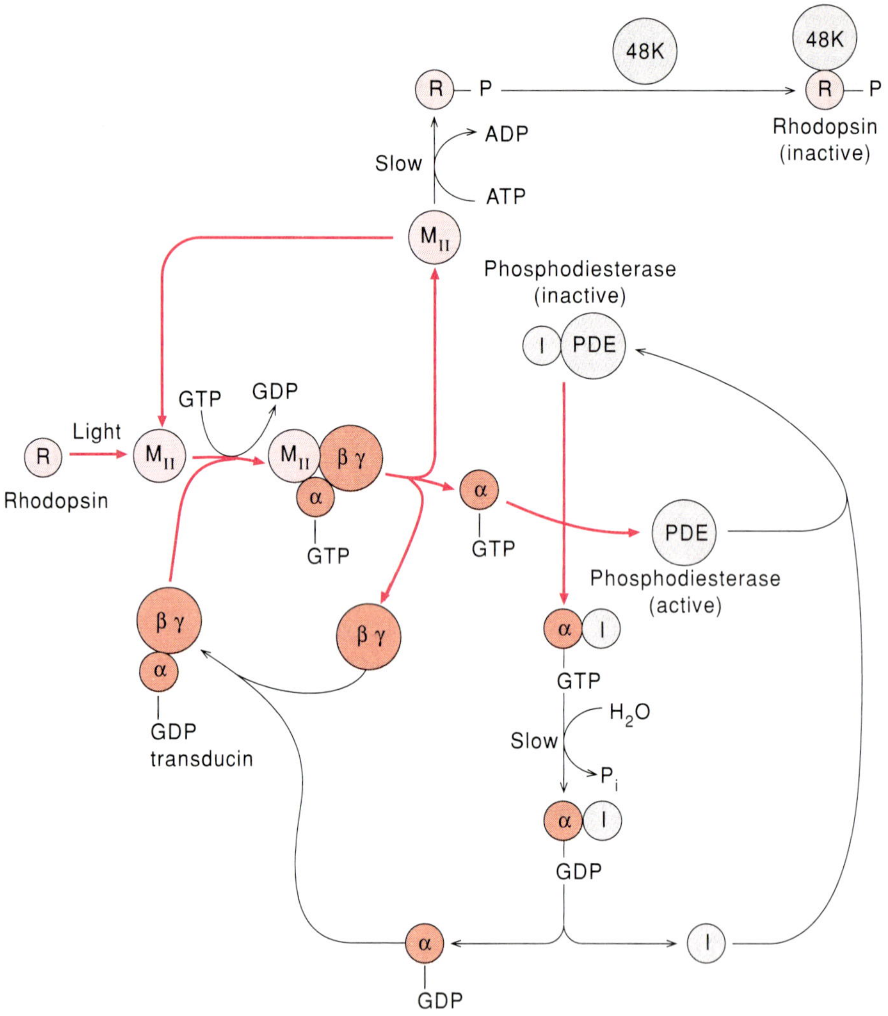

Regeneration of 11-*cis*-Retinal by Way of a Retinyl Ester

The transformations of rhodopsin begin with 11-*cis*-retinal bound to opsin and culminate in the release of all-*trans*-retinal from opsin. How is the 11-*cis*-isomer regenerated? The regeneration turns out to be more complex than we might have expected (fig. 36.13). The all-*trans*-retinal first is reduced to all-*trans*-retinol by NADH. The all-*trans*-retinol then leaves the retina and is taken up by the pigment epithelium, a separate tissue at the back of the eye, where it is esterified by an acyl-transferase. This enzyme transfers a fatty acid from phosphatidylcholine to the retinol, forming an all-*trans*-retinyl ester. The retinyl ester then breaks down in an unusual, concerted reaction that couples isomerization of the retinol and the release of the free fatty acid. 11-*cis*-Retinal is regenerated by the oxidation of the 11-*cis*-retinol in the pigment epithelium. It then returns to the retina and is taken up again by opsin. The free energy that is needed to drive the conversion of retinol from all-*trans* to the less stable 11-*cis* isomer thus is provided by the breakdown of a phospholipid.

Rhodopsin Movement in the Disk Membrane

Each molecule of rhodopsin that is converted to metarhodopsin II is capable of causing some 500 molecules of transducin to bind GTP. Since this happens within about 0.5 s after the absorption of a photon, either rhodopsin or transducin, or both, must move about rapidly enough to encounter thousands of reaction partners per second. Rhodopsin is firmly embedded in the disk membrane and transducin is attached to the surface of the membrane, so they need to diffuse only in two dimensions in the plane of the membrane. But is it likely that they could move this rapidly? Actually, evidence that rhodopsin can diffuse rapidly within the membrane was obtained prior to the discovery of transducin and the phosphodiesterase.

The initial indications that rhodopsin moves around in the membrane came from studies of linear dichroism of rod outer segment disks. A material is said to exhibit linear dichroism if the strength of its optical absorption, when measured with polarized light, depends on the orientation of the polarizer. The polarization of a light beam is defined by the orientation of the

Figure 36.13

11-*cis*-retinal is regenerated from all-*trans*-retinal in a series of reactions that involve the formation of a retinyl ester as an intermediate. Most of these reactions occur in the pigment epithelium. (R_1 and R_2 = fatty acid side chains; R_3 = choline.)

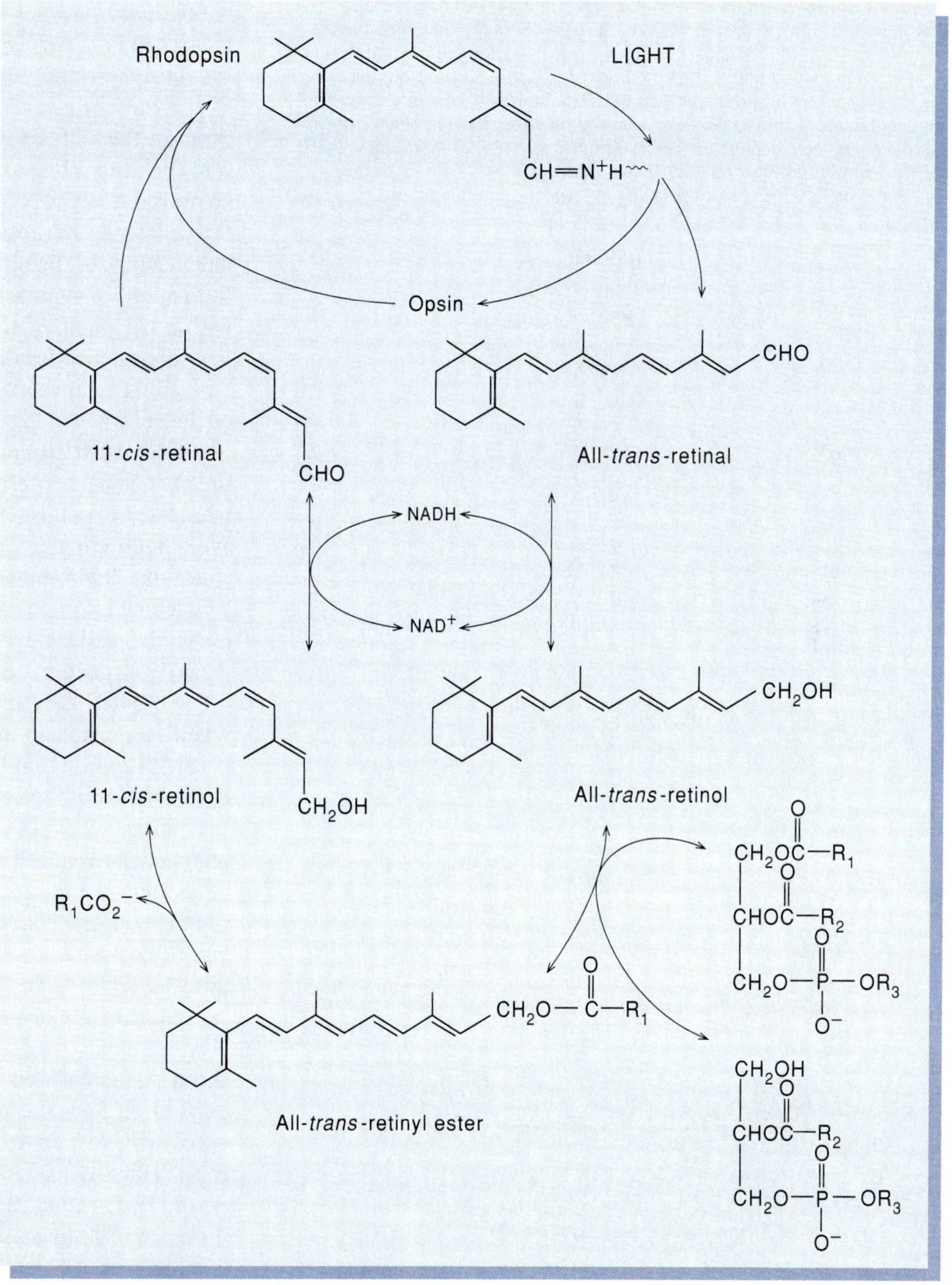

light's electric field, as indicated in figure 16.6. Molecules in solution usually do not exhibit linear dichroism because they are free to tumble about. At any given time, the solution contains molecules with all possible orientations relative to the polarizer, so the absorbance of the solution will not change if we rotate the polarizer. But if all the molecules are fixed in position with their molecular axes aligned, the system generally will exhibit linear dichroism. In the case of retinal or its Schiff base, light is absorbed best when the polarization makes the electric vector of the light parallel to the long axis of the molecule. You can get a feeling for this by examining the diagrams in figure 36.7. In order to convert the molecule from the ground state to the excited state, the electric field of the light must move electron density away from the ring end of the molecule and in the direction of the nitrogen.

The eyes of vertebrates are well suited for measurements of linear dichroism, because the rod cells are neatly aligned with their long axes perpendicular to the plane of the tissue. Suppose that we send a weak beam of polarized light through the rods from the side (fig. 36.14*a* and *b*). The polarization of the beam then could be made either parallel or perpendicular to the long axis of the cells. The planes of the disk membranes in the rod outer segments are aligned perpendicular to the cell axis, so the polarization of the light will be either

Figure 36.14

When rod cells are illuminated from the side, the light can be polarized either perpendicular (*a*) or parallel (*b*) to the planes of the disk membranes. The absorbance of the rhodopsin is much greater for parallel polarization than it is for perpendicular. When the cells are illuminated end-on, the light's electric field has to be parallel to the membranes, no matter how the light is polarized (*c* and *d*). With end-on illumination, the absorbance of the rhodopsin is independent of the polarization. These observations show that the 11-*cis*-retinyl groups (small arrows in disks) are held parallel to the plane of the membrane but can point in any direction in this plane.

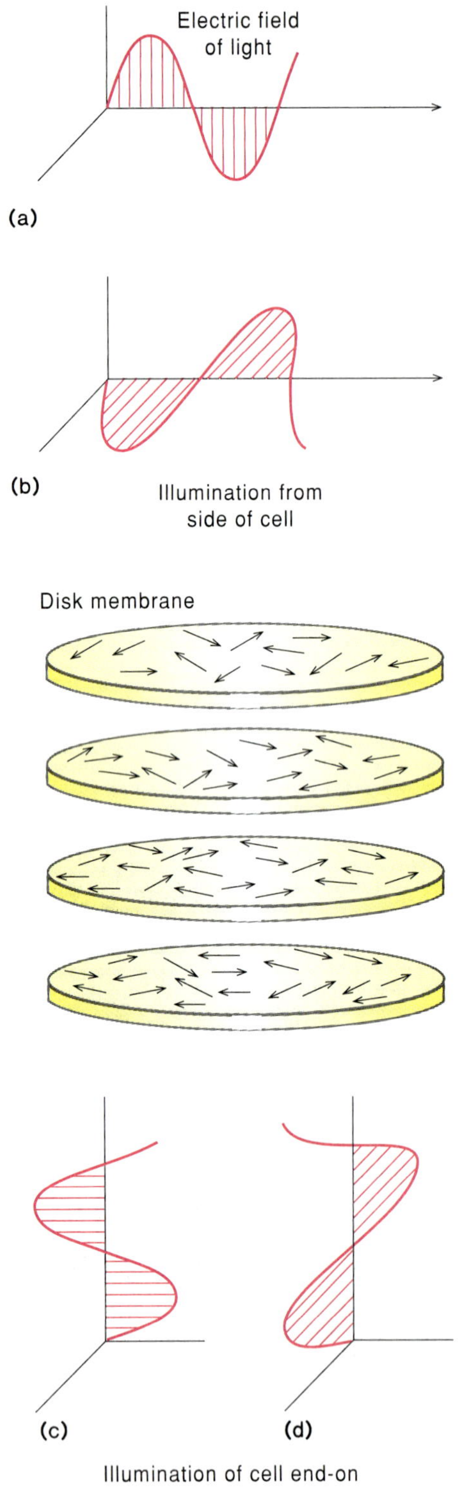

perpendicular or parallel to the membrane surface. It turns out that the rhodopsin absorbs much more strongly when the light is polarized parallel to the membranes than it does when the polarization is perpendicular. This means that the long axes of the retinyl groups must be held more or less parallel to the plane of the membrane. Now suppose that we send the light beam through the cells end-on (see fig. 36.14*c* and *d*). The polarization then must be parallel to the plane of the disk membranes, no matter how the polarizer is turned. With light coming through the cells end-on, the absorbance of the rhodopsin is found to be independent of the direction of the polarization. This means that the retinyl groups are not aligned in any particular direction within the plane of the membrane, but instead can take on all possible orientations in this plane.

It is interesting to note in passing that the orientation of the retinyl groups in the retinas of vertebrates optimizes our visual sensitivity for unpolarized light. Most of the light reaching the retinas passes through the rods end-on and is absorbed well, whatever its polarization (see fig. 36.14*c* and *d*). Insects, however, can distinguish between horizontally and vertically polarized light. This ability can be advantageous, because light that is reflected by smooth surfaces is partially polarized. Insects may make this distinction by having separate sets of cells with differently aligned retinyl groups.

How can measurements of linear dichroism reveal whether rhodopsin molecules move about within the membrane? Suppose that the retina is exposed to a flash of polarized light that enters the rods end-on. Some of the retinyl groups in the disk membranes will be oriented parallel to the polarization, and some will not. The light will *selectively* excite those molecules that are parallel to the polarization, because these have the highest absorbance. Most of the molecules that are excited will be converted to bathorhodopsin and the series of states that descend from it, and their absorption spectrum will change. If we measure the absorbance changes a short time after the excitation flash, using polarized measuring light that passes through the cells end-on, we see that the absorbance changes depend on how the polarization of the measuring light is oriented with respect to the excitation polarization. The polarized measuring light selectively measures molecules that have a particular orientation. Before the excitation flash, there is no linear dichroism for measurements made end-on; after the flash, there is. However, if the rhodopsin molecules can rotate in the plane of the membrane, the orientations of the molecules that were excited will in time become randomized, and the linear dichroism that is induced by the excitation will disappear. The kinetics of this disappearance will depend on how rapidly rhodopsin rotates.

Experiments of this sort were done by Richard Cone. Rhodopsin proved to rotate surprisingly rapidly. The time required for a rotation is about 2×10^{-5} s. This means that the environment of rhodopsin in the membrane must be highly fluid.

To determine whether rhodopsin can diffuse laterally in the membrane, Cone and Paul Leibman independently used a microspectrophotometric technique. They excited individual rod outer segments end-on near one edge of the disks through a microscope. Initially, the excitation caused absorbance changes only in the region that was illuminated, but within a few seconds after the excitation, the absorbance changes became uniformly distributed across the disk. This means that rhodopsin must be able to diffuse rapidly in the plane of the membrane. The diffusion coefficient is calculated to be about 5×10^{-9} cm^2/s. At this speed, rhodopsin molecules would collide with each other 10^5 to 10^6 times a second, and it would take only about a second for a single rhodopsin to collide with most of the molecules of transducin on the surface of the disk.

Bacteriorhodopsin: A Bacterial Pigment-Protein Complex That Resembles Rhodopsin

Halobacterium halobium is a red-colored, halophilic (salt-loving) bacterium that thrives in tidal salt flats. Its red color comes from two components of its cytoplasmic membrane: carotenoids and a purple pigment-protein complex that bears a striking resemblance to rhodopsin. The purple complex, bacteriorhodopsin, was first described by Walther Stoeckenius and his colleagues. Bacteriorhodopsin is not distributed randomly throughout the cell membrane, but rather is collected in patches that are held together by strong interactions among the bacteriorhodopsins. The patches, or purple membranes, can be isolated relatively simply, if the cells are disrupted by being placed in distilled water. Bacteriorhodopsin is the only protein in the purple membrane. It forms a highly ordered two-dimensional array that is almost crystalline in nature. Electron diffraction studies have shown that the individual proteins are folded into seven α-helical stretches extending from one side of the membrane to the other (see chapter 7).

Like rhodopsin, bacteriorhodopsin contains retinal bound to a lysine as a protonated Schiff base. The retinal isomerizes when it is excited with light. However, bacteriorhodopsin differs from rhodopsin in that the retinal starts out in the all-*trans* form rather than as the 11-*cis* isomer. The main product of the isomerization is the 13-*cis* isomer. Like rhodopsin, the excited bacteriorhodopsin progresses through a series of metastable states with different absorption spectra. The first of these states resembles bathorhodopsin in having an absorption spectrum that is shifted markedly to longer wavelengths. A later state resembles metarhodopsin II in that the retinyl Schiff base has lost its proton. A major difference, however, is that the transformations of bacteriorhodopsin are cyclic. Rather than dissociating from the protein, the retinal returns to the all-*trans* form, and the complex relaxes back to its original state and is ready to operate again.

The role of bacteriorhodopsin in *H. halobium* is to pump protons across the cytoplasmic membrane. We discussed some aspects of this pumping in chapter 32. As the excited bacteriorhodopsin undergoes its cyclic series of transformations, a proton is taken up from the solution inside the cell and released on the extracellular side of the membrane. This probably is the proton that is initially on the nitrogen atom of the Schiff base, because the Schiff base is deprotonated and then reprotonated in the course of the cycle. The isomerization of the retinal could move the proton with the nitrogen from a position where it equilibrates with the cytosol to a position where it equilibrates with the extracellular solution. Figure 36.9 will serve to illustrate this idea. The figure was drawn to show the isomerization of 11-*cis*-retinal to all-*trans*-retinal that occurs in rhodopsin, but the principles here are conceptually the same. The isomerization of the retinyl chain pushes the proton on the Schiff base nitrogen from the region of one functional group (A_1^- in the figure) to the region of another (A_2^-). From A_2H, a proton could be conducted to the solution on the extracellular side of the membrane. When the retinal then returns to its original state, the proton lost by the nitrogen could be replaced by one from a third group (A_3H) on the cytoplasmic side of the membrane.

As in other types of photosynthetic bacteria, the protons that are pumped across the membrane by bacteriorhodopsin in the light can reenter the cell by way of a proton-conducting ATP-synthase, generating ATP. *H. halobium* also contains a transport system that moves Na^+ ions out of the cell in exchange for protons coming in. Removal of Na^+ is a major chore for an organism that lives in water containing 4-M NaCl. The cytoplasmic membrane also has two other pigment-protein complexes that execute photochemical cycles similar to the transformations of bacteriorhodopsin. One of these, halorhodopsin, acts as a light-driven pump for Cl^-. The other, sensory rhodopsin, is involved in phototaxis, the tendency of the cells to swim in the direction of a source of light.

Summary

In this chapter we have described the biochemical mechanisms involved in vision. The main points of our discussion are as follows.

1. Light rays entering the eye of a vertebrate are refracted by the cornea and focused to form an image on the retina. The retina contains several layers of cells. Some of these cells, the rods and cones, contain the visual pigments that are responsible for the initial response to light; other cells make synaptic connections to rods or cones and to additional neural cells that carry impulses to the brain.
2. The light-sensitive protein complex in the rods, rhodopsin, consists of 11-*cis*-retinal bound as a Schiff base to a protein, opsin. Rhodopsin is an integral constituent of membranes that form a stack of disks at one end of the cell. Cone cells, which are responsible for the perception of color, contain similar complexes in infoldings of the plasma membrane.
3. When rhodopsin absorbs light, the retinal isomerizes to the all-*trans*-isomer. This change initiates a series of transformations of the pigment-protein complex that result in an interaction with another protein, transducin. The

interaction with rhodopsin causes one of transducin's subunits to take up a molecule of GTP in exchange for bound GDP, to dissociate from the other subunits, and to react with a phosphodiesterase that hydrolyzes cGMP. This reaction activates the phosphodiesterase. The resulting drop in cGMP concentration leads to a decrease in the Na^+ permeability of the cytoplasmic membrane. A hyperpolarization of the membrane then triggers the transmission of a signal at the rod's synaptic junction to the adjacent neural cell.

4. Bacteriorhodopsin, a bacterial pigment-protein complex, resembles rhodopsin (it contains all-*trans*-retinal), but serves a different function. Like rhodopsin, bacteriorhodopsin progresses through a series of metastable states when it is excited with light. In the course of these transformations, protons are taken up from the solution inside the cell and are released on the extracellular side of the membrane, generating an electrochemical potential gradient for protons across the membrane. Protons reenter the cell by way of an ATP-synthase, energizing the formation of ATP.

Selected Readings

Chabre, M., Trigger and amplification mechanisms in visual phototransduction. *Ann. Rev. Biophys. Biophys. Chem.* 14:331, 1985. A review covering the structure of rhodopsin and the roles of cGMP, GTP, the phosphodiesterase, and transducin in the response of the rod cell to light.

Cone, R., Rotational diffusion of rhodopsin in visual photoreceptor membrane. *Nature* 236:39, 1972. Excitation of the retina with polarized light creates a transient linear dichroism. The decay of the dichroism is explained by rotation of rhodopsin.

Fesenko, E. E., S. S. Kolesnikov, and A. L. Lymbarsky, Induction by cyclic GMP of cation conductance in plasma membrane of retinal rod outer segment. *Nature* 313:310, 1985. cGMP increases the Na^+ conductance of the cytoplasmic membrane.

Nathans, J., Rhodopsin: Structure, function, and genetics. *Biochem.* 31:4923–4932, 1992. A review of recent work on the structure and reactions of rhodopsin.

Stoeckenius, W., Purple membrane of halobacteria: A new light-energy converter. *Acc. Chem. Res.* 13:337, 1980. A review of the structure and function of bacteriorhodopsin and the purple membrane.

Wald, G., The molecular basis of visual excitation. *Nature* 219:800, 1968. Wald's Nobel Prize address, describing early work on the isomerization of retinal in rhodopsin.

Problems

1. What chemical characteristics would you look for in a compound that absorbs visible light?
2. The spectrophotometers in biochemistry laboratories are very sensitive because they use a photomultiplier tube as a detector and convert photons of light into an electrical signal. The signal subsequently is multiplied to produce a strong signal. How do our eyes multiply the signal derived from light to increase sensitivity?
3. Draw traces showing how the optical absorbance of a fresh suspension of rod outer segment disks might change as a function of time when the suspension is excited with a short flash of light at 37° C. Show the absorbance at (a) 545 nm and (b) 480 nm. Select the time scale for each trace judiciously, so that the traces illustrate the kinetics of the major absorbance changes that occur at the two wavelengths. (You may need to use two traces with different time scales at each wavelength to show both the initial absorbance change and its decay.)
4. If rhodopsin is illuminated at 500 nm at 77 K, the absorbance of the sample at 500 nm decreases. If the sample is then illuminated at 550 nm (still at 77 K), the absorbance at 500 nm increases again. Explain.
5. Why is the absorption spectrum of metarhodopsin II so different from that of metarhodopsin I?
6. Draw traces showing how the membrane potential of a rod changes with time when the cell is excited with (a) one photon, and (b) two photons. Assume that each photon converts one molecule of rhodopsin to bathorhodopsin and one to metarhodopsin II. (The probability of conversion actually is only about 0.67.)
7. Provide evidence and a model for the involvement of cGMP as a mediator in visual transduction.
8. A frog retina is excited with a flash of polarized light that passes through the rods end-on. The flash causes optical absorbance changes at 500 nm, which are measured with polarized light that also passes through the rods end-on.
   (a) Draw traces showing the kinetics of the absorbance changes measured with light polarized parallel to the excitation polarization and with light polarized perpendicular to the excitation polarization. The traces should cover the time period up to 100 μs after the excitation.
   (b) Similar experiments are done with a retina that has been treated with glutaraldehyde, which causes cross-linking of the rhodopsins in the membrane. Draw traces showing the kinetics of the absorbance changes expected in this case. Assume that the cross-linking immobilizes the rhodopsin molecules but does not otherwise affect their photochemical transformations.
9. In solution, 11-*cis*-retinal maximally absorbs light near 380 nm, but in rhodopsin the peak is at 500 nm. Explain the spectral shift in the retinal bound to the protein.
10. How do our eyes detect color? Why do we not see color in dim light?

A Student's Guide to Methods of Biochemical Analysis

Biochemistry is an experimental science, in which a great variety of techniques are used. We provide examples of the use of various methods throughout this text, usually in connection with the accounts of specific experiments. This approach has the advantage of allowing you to appreciate immediately the relevance of a particular technique. It does mean, however, that the examples are scattered throughout the text. Our aim here is to supply a guide to the locations of those examples within the framework of a unifying outline, so that you can see at a glance which methods of analysis are most useful for certain purposes and so that you can easily find the places in the text where those methods are described. In the outline, you are referred to the figures or boxes that contain such descriptions; you will sometimes find additional information in the adjoining text.

Our descriptions of techniques are brief. There are a number of excellent printed sources where you can find more complete descriptions, as well as comprehensive surveys of available techniques. For example, Academic Press has published more than two hundred books in their series *Methods in Enzymology,* and Oxford Press has published more than eighty books in their series *Biochemistry: A Practical Approach.* In addition, you will find descriptions of the procedures for manipulating DNA in the remarkable three-volume set *Molecular Cloning: A Laboratory Manual,* by J. Sambrook, E. F. Fritsch, and T. Maniatis, 2d ed. (Cold Spring Harbor, N.Y.: Cold Spring Harbor Laboratory, 1989).

I. Separation of major cellular components
- Comparison of differential centrifugation with isopycnic centrifigution (fig. 7.20).
- Separation of periplasmic proteins and inner and outer membranes of a Gram negative bacterium (fig. 7.23).

II. Subfractioning of major cellular components into their constituent molecules
- Methods of protein purification are described in the latter part of chapter 5, which covers differential precipitation with $(NH_4)_2SO_4$ (box 5B), column fraction procedures (box 5C), gel electrophoresis (box 5D), isoelectric focusing (box 5D), and differential centrifugation (box 5E). The following, related material appears in other chapters.
- Analysis of protein mixtures by polyacrylamide gel electrophoresis (fig. 7.31).
- Analysis of ribosomal proteins by two-dimensional gel electrophoresis (box 28A).
- Method for lipid fractionation (box 7A).
- Detergent solubilization of biological membranes (fig. 7.29).
- Separation of N-linked carbohydrates by lectin-affinity chromatography (fig. 6.26).

III. Structural characterization of individual cellular components

A. **Amino acids and protein primary structure**
- Determination of p*K* for amino acids (fig. 3.2, 3.3, and 3.4).
- pH titration curve for a protein (fig. 3.6).
- Column method for analysis of amino acid mixtures (fig. 3.14).
- Determination of amino acid composition of a protein (figs. 3.15 and 3.16).
- Dansyl chloride method for N-terminal amino acid determination (box 3B).
- Determination of the amino acid sequence of the polypeptide chain(s) of a protein (figs. 3.17 through 3.22).

Separation of amino acid derivatives by thin-layer chromatography (fig. 3.23).
Measurement of ultraviolet absorption in solution (box 3A).
Determination of essential amino acids in rats and humans (table 18.2).

B. **Properties of intact proteins**
Most methods of determining protein conformation are described in chapter 4. These include predicting protein tertiary structure from the primary structure (fig. 4.37), X-ray diffraction methods for determining the structure of fibrous proteins (box 4A) and of globular proteins (box 4B), different ways of visualizing molecular structures of proteins (box 4C), and radiation techniques for examining protein structure (box 4D). The following, related material appears in other chapters.
Measuring the binding of inducer to repressor by equilibrium dialysis (box 30B).
Domain-swap experiment demonstrating bifunctional nature of a DNA-binding protein (fig. 31.5).
Electron microscopic examination of striated muscle indicating the structural changes that accompany muscular contraction (fig. 5.17).
Determination of the size of protein molecules in solution by sedimentation and diffusion (box 4E).

C. **Lipids and membranes**
Isolation and analysis of phospholipids (box 7A).
Hydrophobic interaction chromatography (fig. 7.30).
Hydropathy plot to determine segments of a polypeptide chain most likely to be membrane-bound (figs. 7.37 and 32.10).
Differential scanning calorimetry for estimating stability of a membrane structure (fig. 7.39).
Determination of the rate of diffusion of a molecule across a semipermeable membrane (figs. 32.2, 32.3 and 32.4).

D. **Carbohydrates**
Use of glycosidases for structural analysis of oligosaccharides (fig. 6.27).
Polarimetry (box 6A).
Nuclear magnetic resonance spectroscopy for structural analysis of oligosaccharides (box 6B).

E. **Nucleic acids**
Use of ultraviolet absorbance to study DNA denaturation and renaturation (fig. 25.22).
Use of reassociation kinetics to study sequence complexity of DNA (figs. 25.25 and 25.26).
Use of enzyme probes to investigate bacterial chromosome structure (fig. 25.28).
Assessment of supercoiling by ultracentrifugation (fig. 25.19) or electrophoresis (fig. 25.21).
X-ray diffraction analysis of DNA (box 25A).
Techniques using nucleic acid renaturation (box 25D).
Equilibrium density-gradient centrifugation for analysis of DNAs of different densities (box 25B).
Chromosome analysis by gel electrophoresis (box 25C).
Southern blot analysis to characterize DNA sequences in spliced immunoglobulin genes (fig. 33.4; also see fig. 27.27).
Hybridization followed by CsCl density-gradient sedimentation for structural analysis of phage transcripts (fig. 28.2).
Determination of DNA protein-binding sites by chemical protection (fig. 28.15).
Determination of intron size and location by selective annealing and electron microscopy (fig. 28.25).
Footprinting technique for determining location of site of protein binding to DNA (box 28B).
Techniques for manipulating DNA (see listing under part VI of this outline).

IV. **Characterization of individual reactions *in vitro***
Chapter 8 is concerned with kinetic methods for analyzing enzyme-catalyzed reactions. These include determination of the Michaelis constant, the turnover number, and the specificity constant, special kinetic considerations for reactions involving two substrates, and analysis of the effects of temperature, pH, and various types of inhibitors. Chapters 9 and 10 are concerned with methods for analyzing mechanisms of enzyme-catalyzed reactions. Chapter 10 deals exclusively with the unique properties of regulatory enzymes. The following, related material appears in other chapters.
Calculation of the equilibrium constant, K_{eq}, when the numbers of reactants and products are not equal (box 13A).
Calculation of the equilibrium constant for a reaction from the thermodynamic data (chapter 2).
Apparatus for measuring the difference between the standard redox potentials, E^0, of two redox couples (fig. 15.10).
Titration curves for redox couples (fig. 15.11).
Assay for the activity of fatty acid synthase (box 17A).
In vitro studies to determine properties of the recA enzyme (fig. 26.34).
Instrument for recording action potentials (figs. 35.2 and 35.3).
Isotope labeling technique for determining stereospecificity of an enzyme-catalyzed reaction (fig. 11.7 and associated text).

V. ***In vitro* or *in vivo* characterization of reactions in pathways**
Methods for pathway analysis are reviewed at the end of chapter 12. The following, related material appears in other chapters.
Difference spectra method for examining the redox states of respiratory carriers in intact mitochondria (fig. 15.13).

Inhibitors used to analyze electron transport in mitochondria (fig. 15.14).
Determination of the P/O ratio in respiring mitochondria (fig. 15.21).
Use of uncouplers to study the mechanism of ATP synthesis in respiring mitochondria (fig. 15.23).
Use of enzyme inhibitors to study ATP synthesis in mitochondria (fig. 15.24).
In vitro methods for studying the mechanism of ATP synthesis in respiring mitochondria (figs. 15.26, 15.27, 15.29, 15.32, and 15.34).
Analysis of O_2 production by illumination of chloroplasts (figs. 16.16, 16.20, and 16.23).
Spectral methods for analysis of photosystems I and II (fig. 16.21).
Manipulation of pH to study ATP synthesis of chloroplasts (fig. 16.26).
Spectral methods for the analysis of phytochrome behavior (fig. 16.23).
Use of equilibrium density-gradient centrifugation to determine overall mode of DNA replication (figs. 26.3 and 26.6).
Use of autoradiography to determine mode of DNA replication in eukaryotic chromosomes (figs. 26.4 and 26.6).
Use of pulse-chase labeling to follow precursor-product relationship in DNA synthesis (fig. 26.9).
Use of synthetic polynucleotides for determining genetic code assignments (fig. 29.7).
In vitro assay for translation release factors (fig. 29.20).
Device for recording action potentials (figs. 35.2 and 35.3).
Use of polarized light to determine orientation of optical pigments in rod cells (fig. 36.14).
Tissue transplantation technique used to establish the mechanism for immunologic tolerance (fig. 33.14).
Pathway analysis by genetic complementation (box 18A).
Use of isotopes as tracers to delineate the valine and isoleucine pathways (box 18C).
Use of isotopes and mutants to demonstrate that indole is not a true intermediate in the tryptophan pathway (box 18D).
Precursors of purine ring determined by isotope labeling (fig. 20.12).
Use of antibiotics to determine pathway for cell-wall biosynthesis in bacteria (figs. 21.16, 21.17, and 21.19).
Use of inhibitors to determine stages in N-linked carbohydrate synthesis (table 21.3).
Use of mutants to determine stages in N-linked and O-linked carbohydrate synthesis (table 21.4).
Assay for transit of glycoproteins between Golgi membranes (box 21D).
Use of substrates and inhibitors for demonstrating the circular nature of the Krebs cycle (fig. 14.3).
Demonstration of the prochiral nature of citrate in the Krebs cycle (fig. 14.9).

VI. Genetic methods

Chapter 27 is concerned with techniques of DNA manipulation. These include the Maxam-Gilbert and Sanger methods for sequencing DNAs and RNAs, the synthesis of oligonucleotides, the amplification of sequences by the polymerase chain reaction, the use of restriction enzymes to cut DNA into well-defined fragments, procedures for cloning DNA into *E. coli,* yeast, mammals, and plants, site-directed mutagenesis, and the use of recombinant DNA techniques for mapping the globin gene family and the cystic fibrosis gene. The following, related material appears in other chapters.
DNA transformation (fig. 25.2).
Demonstration that nucleic acids carry the genetic information in viruses (figs. 25.3 and 25.4).
Pathway analysis by genetic complementation (box 18A).
Use of mutants to determine stages in N-linked and O-linked carbohydrate synthesis (table 21.4).
Procedure for constructing a congenic strain (fig. 33.16).
Tissue transplantation technique to establish the mechanism for immunologic tolerance (fig. 33.14).
Genetic analysis of *lac* operon expression by mutant studies on haploid and diploid cells (table 30.1).
Mutant studies on expression of the tryptophan operon of *E. coli* (table 30.2).
Nuclear transplantation assay for totipotent nucleus (fig. 31.12).
Genetic analysis of the interaction between regulatory proteins involved in early *Drosophila* development (fig. 31.31).
Mutant analysis of genes responsible for characteristics of specific body segments (fig. 31.33).
Genetic analysis of genes regulating galactose metabolism in yeast (table 31.1).
Determination of the location of developmental gene expression in the *Drosophila* embryo by *in situ* labeling with specific probes followed by autoradiography (box 31A).
Analysis of key events in early *Drosophila* development through detection of specific transcripts at various times and locations (fig. 31.29).
Use of mutants to determine the significance of DNA polymerase I of *E. coli* (chapter 26).
Determination of the crucial amino acid residues in the hemoglobin molecule by observations on the invariant residues found in all hemoglobins (fig. 5.13) and the amino acid residues that lead to pathologic conditions (fig. 5.14).

Common Abbreviations in Biochemistry

A	adenine
ACh	acetylcholine
ACTH	adrenocorticotropic hormone
ADP	adenosine-5′-diphosphate
AIDS	acquired immune deficiency syndrome
Ala (A)	alanine
AMP	adenosine-5′-monophosphate
Asn (N)	asparagine
Asp (D)	aspartic acid
ATP	adenosine-5′-triphosphate
BChl	bacteriochlorophyll
bp	base pair
BPG	D-2,3-bisphosphoglycerate
C	cytosine
Cys (C)	cysteine
CAP	catabolite gene activator protein
CDP	cytidine-5′-diphosphate
CMP	cytidine-5′-monophosphate
CTP	cytidine-5′-triphosphate
CoA or CoASH	coenzyme A
CoQ	coenzyme Q (ubiquinone)
cAMP	adenosine 3′, 5′-cyclic monophosphate
cGMP	guanosine 3′, 5′-cyclic monophosphate
cyt	cytochrome
d	2′-deoxy-
DHAP	dihydroxyacetone phosphate
DHF	dihydrofolate
DHFR	dihydrofolate reductase
DMS	dimethyl sulfate
DNA	deoxyribonucleic acid
cDNA	complementary DNA
DNase	deoxyribonuclease
DNP	2,4-dinitrophenol
ER	endoplasmic reticulum
FAD	flavin adenine dinucleotide (oxidized form)
$FADH_2$	flavin adenine dinucleotide (reduced form)
FBP	fructose-1,6-bisphosphate
fMet	*N*-formylmethionine
FMN	flavin mononucleotide (oxidized form)
$FMNH_2$	flavin mononucleotide (reduced form)
F1P	fructose-1-phosphate
F6P	fructose-6-phosphate
G	guanine
G protein	guanine-nucleotide binding protein
GDP	guanosine-5′-diphosphate
Gly (G)	glycine
GMP	guanosine-5′-monophosphate
Gln (Q)	glutamine
Glu (E)	glutamic acid
GSH	glutathione
GSSG	glutathionine disulfide
GTP	guanosine-5′-triphosphate
Hb	hemoglobin
HDL	high-density lipoprotein
HETPP	hydroxyethylthiamine pyrophosphate
HGPRT	hypoxanthine-guanosine phosphoribosyl transferase
His (H)	histidine
HIV	human immunodeficiency virus
HMG-CoA	β-hydroxy-β-methylglutaryl-CoA
HPLC	high-performance liquid chromatography
IDL	intermediate-density lipoprotein
IF	initiation factor
IgG	immunoglobulin G
Ile (I)	isoleucine
IMP	inosine-5′-monophosphate
IP_1	inositol-1-phosphate
IP_3	inositol-1,4,5-triphosphate
IPTG	isopropylthiogalactoside
K_m	Michaelis constant
kb	kilobase pair
kDa	kilodaltons

LDL	low-density lipoprotein
Leu (L)	leucine
Lys (K)	lysine
Man	mannose
MHC	major histocompatability complex
Met (M)	methionine
NAD^+	nicotinamide-adenine dinucleotide (oxidized form)
NADH	nicotinamide-adenine dinucleotide (reduced form)
$NADP^+$	nicotinamide-adenine dinucleotide phosphate (oxidized form)
NADPH	nicotinamide-adenine dinucleotide phosphate (reduced form)
NDP	nucleoside-5′-diphosphate
NAM	*N*-acetylmuramic acid
NMR	nuclear magnetic resonance
NTP	nucleoside-5′-triphosphate
Phe (F)	phenylalanine
P_i	inorganic orthophosphate
PEP	phosphoenolpyruvate
PFK	phosphofructokinase
PG	prostaglandin
2PG	2-phosphoglycerate
3PG	3-phosphoglycerate
PIP_2	phosphatidylinositol-4,5-bisphosphate
PK	pyruvate kinase
PLP	pyridoxal-5-phosphate
PP_i	inorganic pyrophosphate
Pro (P)	proline
PRPP	phosphoribosylpyrophosphate
PS	photosystem
Q	ubiquinone or plastoquinone
QH_2	ubiquinol or plastoquinol
RER	rough endoplasmic reticulum
RF	release factor or replicative form
RFLP	restriction-fragment length polymorphism
RNA	ribonucleic acid
hnRNA	heterogeneous nuclear RNA
mRNA	messenger RNA
rRNA	ribosomal RNA
snRNA	small nuclear RNA
tRNA	transfer RNA
snRNP	small ribonucleoprotein
RNase	ribonuclease
Ru1,5P	ribulose-1,5-bisphosphate
Ru5P	ribulose-5-phosphate
R5P	ribose-5′-phosphate
RSV	Rous sarcoma virus
s	Svedberg constant
SAM	*S*-adenosylmethionine
SDS	sodium dodecyl sulfate
Ser (S)	serine
S7P	sedoheptulose-7-phosphate
SRP	signal recognition particle
T	thymine
THF	tetrahydrofolate
Thr (T)	threonine
TLC	thin-layer chromatography
TMV	tobacco mosaic virus
TPP	thiamine pyrophosphate
Trp (W)	tryptophan
TTP	thymidine-5′-triphosphate
Tyr (Y)	tyrosine
U	uracil
UDP	uridine-5′-diphosphate
UDPG	UDP-glucose
UMP	uridine-5′-monophosphate
UQ	ubiquinone
Val (V)	valine
VLDL	very-low-density lipoprotein
XMP	xanthosine-5′-monophosphate
Xu5P	xylulose-5′-phosphate

Appendix A
Some Major Discoveries in Biochemistry

In this appendix we list, in chronological order, some of the most important discoveries made in biochemistry in the past two centuries. It is impossible, for reasons of space, to give credit to every worker who has made a significant contribution, but it is possible to identify certain events as milestones, and thus to show how progress in this field has accelerated with the passage of time.

1770–1774
Priestly showed that oxygen is produced by plants and consumed by animals.

1773
Rouelle isolated urea from urine.

1828
Wohler synthesized the first organic compound, urea, from inorganic components.

1838
Schleiden and Schwann proposed that all living things are composed of cells.

1854–1864
Pasteur proved that fermentation is caused by microorganisms.

1864
Hoppe-Seyler crystallized hemoglobin.

1866
Mendel demonstrated the segregation and independent assortment of alleles in pea plants.

1893
Ostwald showed that enzymes are catalysts.

1905
Knoop deduced the β oxidation mechanism for fatty acid degradation.

1907
Fletcher and Hopkins showed that lactic acid is formed quantitatively from glucose during anaerobic muscle contraction.

1910
Morgan discovered sex-limited inheritance in *Drosophila.*

1912
Warburg postulated a respiratory enzyme for the activation of oxygen.

1913
Michaelis and Menten developed a kinetic theory of enzyme action.

1922
McCollum showed that lack of vitamin D causes rickets.

1926
Sumner crystallized the first enzyme, urease.

1926
Jansen and Donath isolated vitamin B_1 (thiamine) from rice polishings.

1926–1930
Svedberg invented the ultracentrifuge and determined the molecular weights of macromolecules.

1928
Levene showed that nucleotides are the building blocks of nucleic acids.

1928
Szent-Gyorgyi isolated ascorbic acid (Vitamin C).

1928–1933
Warburg deduced the iron-prophyrin presence in the respiratory enzyme.

1931
Engelhardt discovered that phosphorylation is coupled to respiration.

1932
Warburg and Christian discovered the "yellow enzyme," a flavoprotein.

1933
Krebs and Henseleit discovered the urea cycle.

1933
Embden and Meyerhof demonstrated the intermediates in the glycolytic pathway.

1935
Schoenheimer and Rittenberg first used isotopes as tracers in the study of intermediary metabolism.

1935
Stanley first crystallized a virus, tobacco mosaic virus.

1937
Krebs discovered the citric acid cycle.

1937
Warburg showed how ATP formation is coupled to the dehydrogenation of glyceraldehyde-3-phosphate.

1938
Hill found that cell-free suspensions of chloroplasts yield oxygen when illuminated in the presence of an electron acceptor.

1939
C. Cori and G. Cori demonstrated the reversible action of glycogen phosphorylase.

1939
Lipmann postulated the central role of ATP in the energy-transfer cycle.

1939–1946
Szent-Gyorgyi discovered actin and actin-myosin complex.

1940
Beadle and Tatum deduced the one gene–one enzyme relationship.

1942
Bloch and Rittenberg discovered that acetate is the precursor of cholesterol. Subsequently Woodward and Bloch determined the complete pathway for cholesterol biosynthesis.

1943
Chance applied spectrophotometric methods to the study of enzyme–substrate interactions.

1943
Martin and Synge developed partition chromatography.

1944
Avery, MacLeod, and McCarty demonstrated that bacterial transformation is caused by DNA.

1947–1950
Lipmann and Kaplan isolated and characterized coenzyme A.

1948
Leloir discovered the role of uridine nucleotides in carbohydrate metabolism.

1948
Hogeboom, Schneider, and Palade refined the differential centrifugation method for fractionation of cell parts.

1948
Kennedy and Lehninger discovered that the tricarboxylic acid cycle, fatty acid oxidation, and oxidative phosphorylation all take place in mitochondria.

1950–1953
Chargaff discovered the base equivalences in DNA.

1951
Pauling and Corey proposed the α-helix structure for α-keratins.

1951
Lynen postulated the role of coenzyme A in fatty acid oxidation.

1952
Palade, Porter, and Sjostrand perfected thin sectioning and fixation methods for electron microscopy of intracellular structures.

1952–1954
Zamecnik and his colleagues developed the first cell-free systems for the study of protein synthesis.

1953
Sanger and Thompson determined the complete amino acid sequence of insulin.

1953
Horecker, Dickens, and Racker elucidated the 6-phosphogluconate pathway of glucose catabolism.

1953
Watson and Crick and Wilkins determined the double-helix structure of DNA.

1955
Ochoa and Grunberg-Manago discovered polynucleotide phosphorylase.

1956
Kornberg discovered the first DNA polymerase.

1956
Umbarger reported that the end product isoleucine inhibits the first enzyme in its biosynthesis from threonine.

1956
Ingram showed that normal and sickle-cell hemoglobin differ in a single amino acid residue.

1956
Anfinsen and White concluded that the three-dimensional conformation of proteins is specified by their amino acid sequence.

1957
Hoagland, Zamecnik, and Stephenson isolated tRNA and determined its function.

1957
Sutherland discovered cyclic AMP.

1958
Weiss, Hurwitz, and Stevens discovered DNA-directed RNA polymerase.

1958
Meselson and Stahl demonstrated that DNA is replicated by a semiconservative mechanism.

1960
Kendrew reported the x-ray analysis of the structure of myoglobin.

1961
Jacob and Monod proposed the operon hypothesis.

1961
Jacob, Monod, and Changeux proposed a theory of the function and action of allosteric enzymes.

1961
Mitchell postulated the chemiosmotic hypothesis for the mechanism of oxidative phosphorylation.

1961
Nirenberg and Matthaei reported that polyuridylic acid codes for polyphenylalanine.

1961
Marmur and Doty discovered DNA renaturation.

1962
Racker isolated F_1 ATPase from mitochondria and reconstituted oxidative phosphorylation in submitochondrial vesicles.

1966
Maizel introduced the use of sodium dodecylsulfate (SDS) for high-resolution electrophoresis of protein mixtures.

1966
Gilbert and Muller-Hill isolated the lac repressor.

1968
Meselson and Yuan discovered the first DNA restriction enzyme. Shortly thereafter Smith and Wilcox discovered the first restriction enzyme that cuts DNA at a specific sequence.

1969
Zubay and Lederman developed the first cell-free system for studying the regulation of gene expression.

1973
Cohen, Chang, Boyer, and Helling reported the first DNA cloning experiments.

1975
Sanger and Barrell developed rapid DNA-sequencing methods.

1977
Starlinger discovered the first DNA insertion element.

1977
Splicing of RNA simultaneously discovered in Broker's and Sharp's laboratories.

1978
Shortles and Nathans did the first experiments in directed mutagenesis.

1978
Tonegawa demonstrated DNA splicing for an immunoglobulin gene.

1981
Cech discovered RNA self-splicing.

1981
Steitz determined the structure of CAP protein.

1981–1982
Palmiter and Brinster produced transgenic mice.

1983
Mullis amplified DNA by the polymerase chain reaction (PCR) method.

1984
Schwartz and Cantor developed pulsed field gel electrophoresis for the separation of very large DNA molecules.

Appendix B
Answers to Selected Problems

Chapter 2

1. Reactions within the cell occur under a much more restricted range of temperature, pressure, pH, and reactant concentration than is possible *in vitro.*
3. Intensive thermodynamic parameters are independent of the amount of material in the state (e.g., temperature, density), whereas extensive parameters depend on the amount of material (energy, mass).
5. Free energy changes are state functions independent of pathway. A reaction mechanism defines a specific reaction path. You can predict free energy changes on the basis of a reaction mechanism but not vice versa.
7. The reaction will proceed toward oxaloacetate formation in the cell if the concentration of products is kept low. Oxidation of NADH by the mitochondrial electron-transport system and utilization of oxaloacetate in the formation of citrate shifts the malate–oxaloacetate reaction toward oxaloacetate production.
9. **(a)** For reaction (P1), $\Delta G^{\circ\prime} = -2.4$ kcal/mole; **(b)** $\Delta G^{\circ\prime}$ of ATP hydrolysis is -7.9 kcal/mole.
11. A minimum of 2.7 moles of protons is required per mole of ADP phosphorylated.
13. **(a)** If the ratio of NADH/NAD^+ is 1:10, the ratio of lipoamide/dihydrolipoamide is 0.98. **(b)** If the NADH/NAD^+ ratio is 10:1, the ratio is 9.8×10^{-3}.
15. K'_{eq} is 1.5×10^{-2} with pyrophosphatase absent; 4.8×10^3 with pyrophosphatase present.

Chapter 3

1. **(a)** $K_a = 1.11 \times 10^{-4}$; $pK_a = 3.95$; **(b)** pH = 3.47.
3. 1 mmole of NaOH
5. pH 2: Arg, His, Lys; pH 7: Arg, Asp, Glu, Lys; pH 12: Arg, Asp, Glu, Tyr, Cys
7. The two principal forms differ in degree of protonation of the R group (ϵ-amino group). Each will have a fully deprotonated carboxylate (net negative charge) and an unprotonated α-amino group (neutral). The equilibrium mixture will contain 0.028 mmoles protonated and 0.072 mmoles unprotonated ϵ-amino group.
9. Polyhistidine will bear little positive charge at pH 7.5 and even less at pH 10, hence the polymer will be insoluble. At pH 5.5, the fraction of imidazole groups protonated will be greater and the polymer more soluble.
11. At pH 6.1 the negatively charged glutamate will bind to the basic column, arginine will bind to the acidic column, and alanine will pass unretarded through both columns.
13. NH_2-Ala-Ala-Lys-Ala-Ala-Phe-Ala
15. NH_2-Ala-Ser-Lys-Phe-Gly-Lys-Tyr-Asp

Chapter 4

1. Consider the entropic effect of decreasing water organization by moving the hydrophobic residue side chains from an aqueous to a nonaqueous environment.
3. The α helix is a rodlike element that cannot easily change direction. Loops, β bends, and "random" structure break the helical structure and allow these directional changes.
5. An α helix broken at the Pro-Asn-Ala region with the hydrophobic residues on the exterior should insert into the membrane.
7. The metal at the active site must be aligned in reasonably precise geometry. Conservation of residues around the metal ligands preserves the geometry of metal binding.
9. The two proteins in question may share common epitopes and cross-react. Alternatively, the "pure" protein may have been contaminated with a trace of the other protein.
11. Dimer of 100,000-M_r units. Each 100,000-M_r unit is a dimer of a 25,000-M_r and a 75,000-M_r protein joined covalently by disulfide bonds.
13. Consider right-hand twist of extended α helices, right-hand crossovers, and right-hand twist of the β-sheet structure.
15. A_6, A_5B, A_4B_2, A_3B_3, A_2B_4, A_5B, B_6 held together by ionic or hydrophobic forces or by covalent (disulfide) bonds.

Chapter 5

1. Step	Specific Activity (U/mg)	Yield (%)	Purification (*n*-fold)
Cell extract	0.039	100	1
$((NH_4)_2SO_4)$ fractionation	0.091	85	2.3
Heat treatment	0.12	73	3.1
DEAE chromatography	4.3	62	110
CM-cellulose	29	50	740
Bio-Gel A	32	41	820

The DEAE chromatography (36-fold purification by this step) yields greatest purification. The enzyme was 0.12% of the protein in the initial extract.

3. The buffer pH must be between above pH 6 but below pH 9 to assure that the protein is negatively charged and the diethylaminoethyl groups are positively charged.
5. The volume following the heat treatment step is too large to allow efficient chromatography by gel exclusion (a column containing approximately 60 liters of resin would be required).
7. Contaminant B can be separated from both contaminant A and the ribonuclease by gel exclusion chromatography. Contaminant A and the ribonuclease can be separated by ion-exchange chromatography.
9. The characteristic absorbance features at 280 nm and 288 nm are consistent with the presence of tryptophan as a component of the amino acid composition of protein A, whereas protein B has little if any tryptophan.
11. The molecular weight is approximately 39,000.
13. Increasing the net positive charge is accomplished either by (**a**) substitution of a neutral amino acid for an acidic amino acid or (**b**) substitution of a basic amino acid for a neutral amino acid.
15. The quaternary structure is a tetramer of approximately 16,000-M_r subunits associated through ionic or electrostatic interactions.
17. [Hb] = 4.8 mM and is approximately equal to the concentration of GBP.

Chapter 6

1. D-mannose is the 2-epimer of D-glucose; D-galactose is the 4-epimer of D-glucose.
3. Carbohydrates as branched structures with multiple linkage sites conceivably could have attained the catalytic function now provided by proteins. However, there may be functional groups provided by amino acids (arginine, isoleucine, for example) that would be difficult to attain with carbohydrates.
5. Pure α or β anomers of the sugars are thermodynamically less stable than the mixture and mutarotate via opening and reclosing the sugar ring.
7. Cellulose is a linear polymer of $\beta(1,4)$-linked Glc residues. Polymer strands H bond with each other to form stable, insoluble fibrils suited for structural roles. Starch and glycogen are polymers of glucose in $\alpha(1,4)$ linkages with $\alpha(1,6)$ branches. The secondary structure is a hydrated helical coil with multiple nonreducing termini. These termini in glycogen provide the "rabid" release of stored glucose upon metabolic demand.
9. 2,3,4,6-tetra-*O*-methyl-D-mannose; 2,4,6-tri-*O*-methyl-D-glucose; 2,3,4-tri-*O*-methyl-D-galactose
11. D-Man $\alpha(1,6)$ ⟩ D-Gal (1,4) D-Glc
D-Man $\beta(1,4)$
13. Carbohydrates are linked to the amide N of Asn in the sequence Asn-X-Ser(Thr). Other Asn residues not in that configuration will be exceedingly poor candidates for glycosylation.

Chapter 7

1. Plants provide a source of linolenic and linoleic acids, some of which is present in meat and milk of herbivores.
3. Triacylglycerols are neutral lipids, readily soluble in organic but not aqueous media. Phospholipids are amphipathic because the polar phosphoglyceryl portion is soluble in aqueous media, while the fatty acids esterified to the glyceryl portion are hydrophobic.
5. Review Haworth representations of Gal and Glc. The linkage to ceramide is a β-glycosidic linkage.
7. Triton X-100 is present above the critical micellar concentration and will probably not pass through the dialysis membrane. It is neutral and could be used in the ion-exchange chromatography. Sodium deoxycholate is present below CMC, should dialyze easily, and could be used for the gel exclusion experiment. The ionic nature of NaDOC would limit its use in ion exchange.
9. The hydrophobic amino acid side chains on the exterior of the integral membrane protein interact with the hydrophobic lipid of the membrane exterior and are stable in the nonaqueous environment. These residues pack in the interior, hydrophobic environment of globular proteins.
11. (**a**) Temperature above phase transition, greater lipid mobility. (**b**) Unsaturated fatty acid will decrease T_m. (**c**) Short-chain fatty acids will decrease T_m. (**d**) Integral proteins will increase and broaden temperature range of the phase change.
13. Glycosyl residues are hydrophilic and will not readily pass through the hydrophobic interior of a membrane.
15. Removal of the detergent has caused aggregation of the detergent-solubilized protein.

Chapter 8

1. Reaction order is the power to which a reactant concentration is raised in defining the rate equation. The example is first order in A and B, second order overall, second order in C.
3. K_m is $(k_2 + k_3)/k_1$, whereas K_S is k_2/k_1. K_m is equal to the [ES] dissociation constant K_S when the value of k_3 is much smaller than the value of k_2.
5. Nonlinearity develops because substrate concentration is less than saturating and velocity is proportional to substrate concentration. Substrate depletion causes decreased velocity; product concentration increases and reverse reaction becomes significant.
7. When [S] is 10^{-3} M, $v_0 = V_{max}$ (10^{-7} M min^{-1}).
When [S] is 10^{-5} M, $v_0 = 9.0 \times 10^{-8}$ M min^{-1} (90% V_{max}).
When [S] is 10^{-6} M, $v_0 = 5.0 \times 10^{-8}$ M min^{-1} (50% V_{max}).
When [S] is 2×10^{-2} M, $v_0 = V_{max}$, thus no increase in rate.

9. (a) $K_m = 10^{-4}$ M; (b) $V_{max} = 10^{-8}$ M s^{-1}; (c) $k_{cat} = 100$ s^{-1}; (d) $v_0 = 1.7 \times 10^{-9}$ M s^{-1}
11. Intersecting initial velocity plots are consistent with sequential (ordered or random) but not Ping-Pong mechanism.
13. $V_{max} = 3.3 \times 10^{-7}$ M s^{-1}; $K_m = 2.4 \times 10^{-4}$ M; $k_{cat} = 3.3 \times 10^4$ s^{-1}; specificity constant $= 1.4 \times 10^8$ M^{-1} s^{-1}
15. Irreversible inactivating reagents decrease the amount of active enzyme without altering the kinetic constants of the remaining active enzyme. Lineweaver-Burk analysis of the data reveals an apparent slope and *y*-intercept effect that may be confused with noncompetitive inhibition. Enzyme reversibly inhibited is reactivated upon dilution, while irreversibly inactivated enzyme is not.

Chapter 9

1. Each enzyme uses the "catalytic triad" (Asp-His-Ser) at the active site and an enzyme-bound intermediate is formed. The binding site for the side chain of the residue contributing the carboxyl to the bond that is cleaved differs in each case.
3. RNase activity depends on acid–base reaction catalyzed by two histidyl residues at the active site. Removal of the C′-2 OH proton is best accomplished by an unprotonated imidazole (pH > p*K*) and the addition of a proton to the C-5′ (—O$^-$) is accomplished by a protonated imidazole (pH < p*K*). The best compromise is pH at p*K* of imidazole (pH 6), where each of the histidines exists with the maximal fraction of each imidazole in the correct form. correct form.
5. Release of water during the conformational change initiated by binding the substrate provides a hydrophobic environment to stabilize the nonpolar side chain of the substrate.
7. Renaturation of denatured protein is dictated by the primary structure of the protein. The trypsin family of enzymes and carboxypeptidase A are synthesized as proenzymes that are proteolytically activated. The proteolyzed, active enzymes have primary structures different from the gene product and are not active upon renaturation. In addition, zinc is a cofactor required for carboxypeptidase A activity.
9. The structure of the transition-state analog is complementary to the structure of the active site. The analog thus binds tightly to the active site.
11. For the iron-dependent form, as an example,

$$Fe^{3+} + O_2^- \rightarrow Fe^{2+} + O_2$$

$$Fe^{2+} + O_2^- + 2\,H^+ \rightarrow Fe^{3+} + H_2O_2$$

13. (a) Substitution of Asp for Lys 86 markedly decreases activity and several explanations are possible: The lysine is a critical residue either at or near the active site or is essential for maintaining a catalytically competent conformation of the enzyme. (b) Lysines 21 and 101 are probably outside the catalytic site and may not be evolutionarily conserved. Their replacement with aspartate yielded no great change in enzymatic activity. (c) Lysine 86 is essential for enzymatic activity and would be conserved.
15. An intermediate covalently bound to the active site is an essential part of the mechanism in covalent catalysis. In noncovalent catalysis, the intermediates are bound through ionic, hydrophobic, or hydrogen bonds.

Chapter 10

1. Consider the reactions

$$\text{Protein(Ser)} + \text{ATP} \rightarrow \text{Protein(Ser-P}_i) + \text{ADP}$$

$$\text{Protein(Ser-P}_i) + \text{HOH} \rightarrow \text{Protein(Ser)} + \text{P}_i$$

$$\text{Sum: ATP} + \text{HOH} \rightarrow \text{ADP} + \text{P}_i$$

3. If G, H, and J are essential end products, each would be likely to inhibit its own production without inhibiting production of the others. The committed step (enzyme 1) may be cumulatively inhibited by each of the products. G and H might cumulatively inhibit enzyme 4. J could inhibit enzymes 7 and 8.
5. ATP binds at the active site and is a substrate, and also binds at an allosteric site, inhibiting the enzyme.
7. The substrate concentration required for half-maximal activity ($S_{0.5}$) of an allosterically regulated enzyme will depend on the cumulative effects of allosteric activators and/or inhibitors also present. Hence $S_{0.5}$ may be decreased with allosteric activators and may be increased with allosteric inhibitors.
9. Cooperativity in the tetrameric hemoglobin depends on the interaction of the two α and two β subunits. Separation of the tetramer into two α-β dimers would destroy the cooperativity exhibited by the tetramer. There should be more O_2 bound at lower O_2 tensions. 2,3-Bisphosphoglycerate should not stabilize the deoxyform of the dimer. The O_2-binding curve for the dimer should resemble that of myoglobin.
11. Multisubunit complexes respond cooperatively by transmitting the conformational change at one subunit to an adjacent subunit with subsequent changes in substrate binding. There may be local conformational changes in a subunit, as with LADH, that do not alter substrate binding to adjacent subunits.
13. The sulfhydryl groups on the activated enzymes must be reoxidized to the disulfide form. Whether the disulfide is formed by an enzymatically catalyzed process or occurs spontaneously via low-molecular-weight oxidants is not known. Irreversibly modified enzymes must in principle be degraded to allow recycling of the amino acids. Although the cell can reclaim the amino acids, there is still a large metabolic expense in *de novo* synthesis of the protein.
15. Free calmodulin would respond to Ca^{2+} signal by forming a Ca^{2+}-calmodulin complex, which then must collide with and bind to a target protein. This scenario requires the binding of three to four Ca^{2+} atoms per molecule of calmodulin, then a bimolecular reaction of Ca^{2+}-calmodulin with the target protein. Calmodulin complexed to the target protein is a single entity that must collide with Ca^{2+} atoms. The latter reaction should be kinetically less complex and would occur with a higher probability.

Chapter 11

1. The carboxyl group on the acyl substituent of biotin and lipoic acid forms an amide bond with the ϵ-amino groups of lysine in the cofactor-containing subunit. The cofactors are bound to flexible extended structures that allow the chemically reactive portion of the cofactor to move between several catalytic centers present on different subunits.

3. The two pyridine nucleotides (NADH, NADPH), although functionally similar, are specific coenzymes for dehydrogenases. The dehydrogenase-dependent specificity provides a mechanism frequently used to differentiate anabolic and catabolic processes. Pathways may be regulated by either $NADH/NAD^+$ or $NADPH/NADP^+$ ratios.

5. (a) 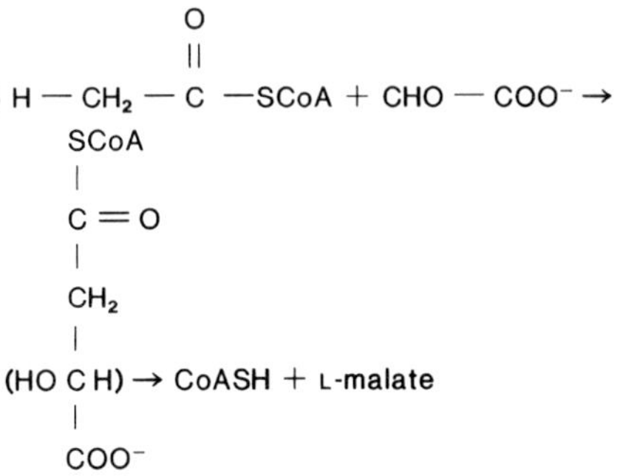

(b) The thioester activates the carboxyl group of acetate and stabilizes the enolate formed upon enzyme-catalyzed abstraction of the α proton. Abstraction of an α proton from acetate, forming the dianion, is improbable. **(c)** Coenzyme A is a good leaving group and hydrolysis of the thioester is thermodynamically favorable. The products malate and CoASH are more stable when compared with the malylthioester.

7. Flavins are fully reduced upon accepting two electrons but may transfer electrons to one-electron acceptors (iron-sulfur proteins and hemoproteins). The free radical remaining after transfer of a single electron from FADH is resonance-stabilized by the isoalloxazine ring system.

9. (a) Reduced flavins in solution rapidly reduce O_2 to superoxide and H_2O_2, metabolites that are toxic to the cell. Enzyme-bound flavins are usually shielded from rapid oxidation by O_2. **(b)** Tightly bound NAD(P) is an advantage to enzymes catalyzing rapid $H{:}^-$ removal and readdition in a stereospecific fashion. Freely diffusing NADH is an advantage in transferring reducing equivalents among various enzyme-catalyzed reactions.

11. The products of the oxidation, NH_4^+ and H_2O_2, are toxic to most cells. In PLP-dependent transamination the amino group is retained in organic molecules of low toxicity.

13. (a) Flavin; **(b)** PLP; **(c)** biotin; **(d)** thiamine pyrophosphate.

15. NAD^+ is cleaved at the glycosidic bond between nicotinamide and the ribose ring. Adenosine diphosphate ribose is transferred to the acceptor and nicotinamide is released.

Chapter 12

1. Thermodynamics dictates the feasibility of a reaction and the direction in which a reaction proceeds to reach equilibrium, but does not specify the rate of the reaction or the reaction pathway. Kinetics describes that rate at which a reaction approaches equilibrium. The most thermodynamically favorable reaction is of no biological value if it is not adequately catalyzed.

3. Using different high-energy intermediates, or different portions of the same high energy intermediate, and different forms of reductant allows the cell ample sites to control catabolism and anabolism differentially.

5. The "committed step" in a reaction sequence steers the metabolite to a sequence of reactions whose intermediates have no other function in the cell. Control of the committed step prevents wasteful accumulation of these single-purpose intermediates and obviates the necessity of controlling each enzyme in a pathway.

Chapter 13

1. **(a)** Bacteria are adept at fermenting numerous sugars. **(b)** The NADH formed by glyceraldehyde-3-phosphate dehydrogenase is used to reduce pyruvate to lactate. There is no net oxidation, but the products are of lower free energy than substrates. Energy difference between products and substrates is available to drive the phosphorylation of ADP. **(c)** The culture could also be decarboxylating pyruvate to acetaldehyde and CO_2. The acetaldehyde is subsequently reduced to ethanol, using the NADH from the glyceraldehyde-3-phosphate dehydrogenase reaction. **(d)** Fluoride forms an inhibitor (fluorophosphate) of enolase. Inhibition of enolase would increase the amount of 2-phosphoglycerate, but the amount of PEP would decline as a result of pyruvate kinase activity.

3. **(a)** Hydrolysis rather than phosphorolysis of the thioester bound to glyceraldehyde-3-phosphate dehydrogenase releases 3-phosphoglycerate. **(b)** There would be no net yield of ATP if glycolysis began with glucose. **(c)** The obligate aerobe could oxidize the pyruvate, with the formation of sufficient ATP for growth.

5. Arsenate substitutes for phosphate at the glyceraldehyde-3-phosphate dehydrogenase reaction and forms a transient arsenocompound that spontaneously hydrolyzes, regenerating the arsenate. Arsenate acts catalytically rather than stoichiometrically in dispelling the energy of the thioester bond. Arsenate is ineffective at the pyruvate kinase step because the active phosphate was derived initially from ATP, not orthophosphate.

7. If the $\Delta G^{\circ\prime}$ for the hexokinase reaction is -5.0 kcal/mole, the K'_{eq} is 4,800. The glucose concentration in equilibrium with 1 mM glucose-6-phosphate must be at least 42 nM, a concentration easily attainable in the cell. However, the K_m of hexokinase for glucose is about 100 μM. The glucose concentration in actively metabolizing cells must be several orders of magnitude greater than 40 nM to have reasonable hexokinase activity.

9. **(a)** Alcohol dehydrogenase reduces acetaldehyde to ethanol to regenerate NAD^+ from the NADH produced in glycolysis. **(b)** Alcohol dehydrogenase oxidizes ethanol to acetaldehyde. Acetaldehyde is subsequently oxidized to acetyl CoA_2.

11. Consider enolase, aldolase, and phosphohexoisomerase.

13. **(a)** Review pyruvate carboxylase and PEP carboxykinase reactions. **(b)** $\Delta G^{\circ\prime}$ overall is app. -1.0 kcal/mole if the hydrolysis of PEP releases 14 kcal/mole and if hydrolysis of ATP (GTP) releases 7.5 kcal/mole. If the overall free

energy change for the coupled reactions is as calculated, the formation of PEP from pyruvate is metabolically feasible.

15. (a) PFK-2 will be active when the insulin/glucagon ratio is high (for example, when the animal is fed a carbohydrate-rich meal), or when adrenalin is released. FBPase-2 will be active when the insulin/glucagon ratio is low (blood glucose is low). **(b)** If fructose-2,6-bisphosphate were still present although not being actively synthesized, it would continue to stimulate PFK-1 and inhibit FBPase-1. **(c)** Activation of FBPase-2 causes the hydrolysis of fructose-2,6-bisphosphate, alleviates inhibition of FBPase-1, and prevents activation of PFK-1. **(d)** Insulin/glucagon ratio decreases. Thus FBPase-2, FBPase-1, pyruvate carboxylase, and PEPCK activities should increase. PFK-2 and PFK-1 activities will decline. **(e)** The administered insulin will restore the insulin/glucagon ratio. The cAMP level in the liver cell will decrease, as will the activities of FBPase-2, FBPase-1, and PEP carboxykinase.

Chapter 14

1. (a) False. Lipoamide transacetylase catalyzes reduction of the disulfide on lipoamide concomitantly with oxidation and transfer of the hydroxyethyl group from thiamine pyrophosphate. Dihydrolipoamide dehydrogenase catalyzes oxidation of dihydrolipoamide and reduction of NAD^+. (b) False. Hydrolysis of acetyl-CoA thioester should yield as much free energy as succinyl-CoA hydrolysis, a process coupled to ADP phosphorylation (succinate thiokinase). (c) False. The methyl group of acetyl-CoA could be derived from pyruvate, from β oxidation of long-chain fatty acids, or from amino acid metabolism. **(d)** False. If aconitase failed to discriminate between the ($-CH_2-COO^-$) groups, half the CO_2 would arise from oxaloacetate and half from the acetate carboxylate. The two CO_2 molecules released by oxidative decarboxylation of isocitrate and α-ketoglutarate are derived from the carboxyl groups of oxaloacetate with which the acetyl-CoA was condensed. (e) False. Malate can easily be dehydrated to fumarate by reversal of the fumarase reaction.
3. (a) Glyceraldehyde-3-phosphate, C-3; **(b)** acetyl-CoA, methyl of acetate; **(c)** citrate, C-2 (derived from methyl of acetyl-CoA); **(d)** α-ketoglutarate, C-4; **(e)** succinate, C-2, C-3; **(f)** malate, both C-2 and C-3.
5. (a) Hydroxypyruvate → B → D → C → A → pyruvate (b) Hydroxypyruvate + NADH + H^+ → pyruvate + NAD^+ + HOH (c) A = phosphoenolpyruvate; B = glycerate; C = 2-phosphoglycerate; D = 3-phosphoglycerate. A, C, and D are glycolytic intermediates. **(d)** ATP used to phosphorylate glycerate is regenerated from PEP.
7. Lipoic acid bound to the acyltransferase in the multienzyme complexes (a) provides a greater coenzyme concentration at the active site than may be seen if the cofactor must diffuse into and away from the complex; (b) decreases the diffusion path required for a product of one active site to move to the next catalytic center.
9. If the complexes were deficient in dihydrolipoamide dehydrogenase, dihydrolipoamide would be reoxidized to lipoamide more slowly than necessary to maintain activity of the complexes. ATP production would be lowered as a consequence of reduced TCA cycle activity.
11. In bacteria, acetate is used as a source of energy via the TCA cycle and oxidative phosphorylation and as a carbon source for synthesis of intermediates for cell growth and division. Competition between ICDH and isocitrate lyase for isocitrate is shifted to anabolism by phosphorylation of ICDH and shifted to catabolism upon dephosphorylating (reactivating) ICDH. The fraction of ICDH in the active form dictates the level of isocitrate available to isocitrate lyase. The isocitrate lyase is apparently not regulated at the metabolite level.
13. In plants, the TCA cycle and glyoxalate bypass are located in different subcellular organelles. The TCA cycle is located in mitochondria and the glyoxalate bypass is located in glyoxosomes. Communication between the compartments includes flow of succinate to the mitochondria for oxidation to malate.
15. In mammalian liver, acetyl-CoA provides reducing equivalents by its oxidation in the TCA cycle and subsequent oxidative phosphorylation to provide ATP. Acetyl-CoA is an obligatory activator of the pyruvate carboxylase, whose activity provides oxaloacetate for gluconeogenesis.

Chapter 15

1. Iron-sulfur protein: one electron, no H^+ transferred, no semiquinone; flavoprotein: isoalloxazine ring, two electrons, one H^+/electron, semiquinone formed; quinone: two electrons, one H^+/electron, semiquinone formed.
3. (a) $\Delta E^{\circ\prime} = +110$ mV, $\Delta G^{\circ\prime} = -2.5$ kcal/mole
   (b) $\Delta E^{\circ\prime} = +580$ mV, $\Delta G^{\circ\prime} = -53$ kcal/mole O_2 reduced.
   (c) $\Delta E^{\circ\prime} = +270$ mV, $\Delta G^{\circ\prime} = -12$ kcal/mole
   (d) $\Delta E^{\circ\prime} = -430$ mV, $\Delta G^{\circ\prime} = +20$ kcal/mole
5. At +300 mV, the cytochrome *c* couple will be 91% oxidized and 9% reduced.
7. (a) Data are consistent with diffusion of cytochrome *c* between complex III and the cytochrome oxidase. Electrons are probably donated and accepted from the same area on the cytochrome *c* structure. **(b)** The lysines surrounding the heme crevice on cytochrome *c* should interact electrostatically with anionic groups (Glu, Asp) spatially located on complex III and cytochrome oxidase to facilitate interaction with the cytochrome.
9. (a) P/O = 2.3; (b) P/O = 0.2; (c) P/O = 2.7; (d) P/O = 0.5.
11. (a) Tightly coupled mitochondria exhibit an obligatory codependence of electron transfer and phosphorylation of ADP. In "perfectly" coupled mitochondria, no electron transport to O_2 should occur in the absence of ADP and phosphate. **(b)** Respiratory control is a function of coupled mitochondria. Oxidation of NADH and succinate is blocked when ATP levels are high; glycolysis and TCA cycle are inhibited. When ATP demand is high, ADP levels will stimulate the oxidation of succinate and NADH. Low ATP levels and high NAD^+/NADH ratio will stimulate glycolysis and the TCA cycle. **(c)** Brown-fat mitochondria contain the uncoupler thermogenin and the electron-transport system is partially uncoupled. The oxidation of fat leads to nonshivering thermogenesis, a process important to animals aroused from hibernation.

13. (a) Membrane potential is app. 180 mV, negative on matrix side of membrane. **(b)** An oxidative phosphorylation uncoupler allows H^+ equilibration across the membrane, collapsing both membrane potential and H^+ gradient.
15. (a) Transport of H^+ is in the same orientation in these SMPs as in the mitochondria, but the cytoplasmic aspect of the inner mitochondrial membrane is facing the interior of the vesicle. **(b)** Mechanical or chemical uncoupling allows free equilibration of H^+ across the SMP membrane. Succinate is oxidized without vectorial accumulation of H^+. **(c)** Atractyloside inhibits ADP transport into mitochondria. The SMP have the F_o-F_1 complex exposed to bulk solvent and do not have a transport barrier to ADP. The ADP/ATP transporter is not required, so atractyloside would have no effect in this system. SMP oxidize extravesicular NADH but intact mitochondria do not.

Chapter 16

1. **(a)** 33 kcal/Ein; **(b)** excited state reduction potential is −1.0 volt.
3. If oxidation of compound A requires operation of both photosystem I and photosystem II, oxidation of compound A will diminish if chloroplasts are illuminated with red light (>680 nm) and will be inhibited by DCMU. If compound A is oxidized by photosystem I, DCMU will not inhibit the oxidation, nor will the action spectrum of compound A oxidation exhibit this phenomenon of "red drop."
5. Light energy absorbed by antenna chlorophyll molecules is transferred to a photocenter trap (P_{680} or P_{700}) where an electron is transferred initially to pheophytin, then to bound plastoquinone. If the potential is decreased to 0 mV, the quinone pool will be reduced, electron transfer from pheophytin will diminish, and the steady-state level of activated chlorophyll will increase. The alternative energy dissipation pathway, fluorescence, increases.
7. The quantum yield of O_2 evolution from plants as a function of light absorbed parallels the absorption spectrum until the wavelength impinged is greater than approximately 700 nm. Although the chloroplasts absorb wavelengths greater than 700 nm, O_2 evolution declines because the oxygenic photosystem II is no longer activated at these wavelengths.
9. Photophosphorylation is the process of ATP generation driven by the proton gradient generated during the light-driven electron transfer (Z scheme). Noncyclic photophosphorylation requires net electron transfer from a donor (HOH) to an acceptor ($NADP^+$) with vectorial H^+-translocation. Cyclic photophosphorylation is driven by a proton gradient generated by electron transfer from photosystem I through the quinone pool and the electron-transfer system, back to photosystem I.
11. CO_2 and O_2 each bind to the catalytic site of the ribulose bisphosphate carboxylase/oxygenase (RubisCO) and react with the ribulose bisphosphate. The enzyme uses either CO_2 or O_2 as (alternative) substrate, and O_2 competes with CO_2 for the bound ribulose bisphosphate.
13. CO_2 is released by oxidative decarboxylation of malate in the bundle sheath cells of C_4 plants, causing the CO_2/HCO_3 concentration to exceed the CO_2/HCO_3^- concentration in the mesophyll cell. The larger concentration of CO_2 increases the fraction of ribulose bisphosphate that is carboxylated rather than oxygenated. Net photorespiration is lower in C_4 plants as compared with C_3 plants.
15. **(a)** $NADP^+$ is reduced to NADPH by the noncyclic electron flow in the chloroplast to $NADP^+$-ferredoxin oxidoreductase. **(b)** 6 moles of ATP are used to phosphorylate 6 moles of glycerate-3-phosphate to glycerate-1,3-bisphosphate. An additional 3 moles of ATP are consumed during phosphorylation of 3 moles of ribulose-5-phosphate to 3 moles of ribulose-1,5-bisphosphate. **(c)** The ATP consumed by the Calvin cycle remains unaltered, but 2 additional moles of ATP are used to regenerate PEP from pyruvate catalyzed by the pyruvate, phosphate dikinase. **(d)** NADPH is used in the mesophyll cell to reduce oxaloacetate to malate, but the NADPH is regenerated upon oxidative decarboxylation of malate to CO_2 and pyruvate.

Chapter 17

1. **(a)** Oxidation of 1 mole of glucose yields 32 moles of ATP. 350 kcal of energy are stored as ATP or 1.9 kcal/gm. **(b)** Oxidation of 1 mole of palmitate yields 106 moles of ATP. 1200 kcal of energy are stored as ATP, or 4.7 kcal/g. **(c)** Lipids are more highly reduced than are carbohydrates and supply more reducing equivalents to the electron-transport system than do carbohydrates. **(d)** Lipids have approximately 2.5 times greater energy storage per gram than do carbohydrates and are stored as compact, hydrophobic globules. Storage of an equivalent energy as carbohydrate would require at least 2.5 times the mass, not considering the water of hydration that would accompany the carbohydrate.
3. Carnitine acyltransferases (CAT) catalyze the reversible formation of fatty acyl carnitine esters from the acyl-CoA thioesters. The carnitine esters, but not the CoA derivatives, are transported across the inner mitochondrial membrane.
5. Propionyl-CoA, a product of β oxidation of odd-chain-length fatty acids and of pristanic acid, is metabolized to L-methylmalonyl-CoA in two steps. The L-methylmalonyl-CoA is substrate for the vitamin-B_{12}-dependent mutase. The product, succinyl-CoA, is metabolized in the TCA. A deficiency of vitamin B_{12} could limit the mutase reaction.
7. **(a)** Ketone body formation in liver supplies an easily transported, water-soluble, energy-rich metabolite that can be used in lieu of glucose in many nonhepatic tissues. **(b)** β-hydroxybutyrate supplies an additional hydride (2 reducing equivalents) compared with acetoacetate. **(c)** Consider the reactions catalyzed by β-hydroxybutyrate dehydrogenase, 3-ketoacyl-CoA transferase, and thiolase.
9. The NADPH-dependent dienoyl-CoA reductase reduces the Δ^2-*trans* $\Delta^{4,7}$-*cis* decatrienoyl-CoA produced after removal of the first 4 acetyl-CoA units from linolenyl-CoA in β oxidation. The product, Δ^3-*trans*, Δ^7-*cis* decadienoyl-CoA, is isomerized to the Δ^2-*trans*, Δ^7-*cis* decadienoyl-CoA by the enoyl-CoA isomerase.
11. **(a)** The ^{14}C-labeled methyl group of acetyl-CoA will be C-16 of palmitate. **(b)** Only one deuterium atom from each labeled malonyl-CoA will remain in the reduced lipid chain. Carbons 2, 4, 6, 8, 10, 12, and 14 of palmitic acid will each have one deuterium label. **(c)** The ^{14}C label will be lost by decarboxylation and no label will remain in the palmitate.

13. Glucose-6-phosphate dehydrogenase, 6-phosphogluconate dehydrogenase, and the $NADP^+$-malic enzyme are sources of the 14 moles of NADPH required for biosynthesis of palmitate.
15. The thioesterase activity of the fatty acid synthase prefers the palmitoyl acyl carrier protein thioester as substrate.

Chapter 18

1. Glutamine, the product of glutamine synthase, is the source of nitrogen required in the synthesis of a number of diverse, structurally unrelated compounds synthesized by different pathways. Total inhibition of the glutamine synthase by a single product would in turn inhibit the synthesis of all compounds requiring glutamine.
3. In the absence of serine hydroxymethyltransferase, the mutant would be unable to synthesize glycine. Glycine supplementation would be required. Metabolism of glycine contributes to the one-carbon pool, as does methionine. Addition of methionine should decrease the demand for glycine contribution to the one-carbon pool and stimulate growth of the mutant.
5. Pyruvate, from glycolysis of glucose, is carboxylated to oxaloacetate or oxidized to acetyl-CoA. These metabolites enter the Krebs cycle, are metabolized to α-ketoglutarate and oxaloacetate, then transaminated to aspartate or glutamate. Asn, Gln, and Pro are synthesized from Asp or Glu. The cycle replenishes intermediates via the anaplerotic reactions (e.g., carboxylation of pyruvate to form oxaloacetate).
7. Serine carbon: α carboxyl: C-3 or C-4 of glucose; α carbon: C-2 or C-5 of glucose; β carbon: C-1 or C-6 of glucose.
9. Glutamate is phosphorylated and reduced to L-glutamyl-γ-semialdehyde. The aldehyde condenses with the α-amino group on the same molecule to form Δ^1-pyrroline-5-carboxylate. Reduction of the latter compound yields proline. Transamination implies that the α-keto acceptor and the α-amino donor are separate molecules.
11. Tryptophan synthase catalyzes the cleavage of indole glycerol phosphate, with retention of indole as an enzyme-bound intermediate and release of glyceraldehyde-3-phosphate. The enzyme-bound indole replaces the hydroxyl group of serine to form tryptophan.
13. Hydroxyproline is formed by a posttranslational modification of proline residues in the protein. The ^{14}C-labeled hydroxyproline is not incorporated directly into the collagen because there is no genetic codon to specify the incorporation of hydroxyproline.

Chapter 19

1. Pyridoxal phosphate deficiency would curtail the transamination of α-amino group amino acids to recipient α-ketoacids. Disposition of the amino group via the urea cycle and subsequent catabolism of the carbon skeleton would be diminished.
3. Glutaminase catalyzes the hydrolytic cleavage of the amide group from glutamine, leaving glutamate. Glutamate dehydrogenase catalyzes the NAD^+-dependent oxidative deamination of glutamine to ammonium ion, α-ketoglutarate and NADH.
5. The α-ketoacid pyruvate, a product of glycolysis, is transaminated to form alanine and as such transports the α-amino groups, from amino acids catabolized in muscle, to the liver. In the liver, transamination of alanine with α-ketoglutarate yields pyruvate, a gluconeogenic substrate, and glutamate, an amino donor in urea biosynthesis.
7. Glutamate is oxidatively deaminated to α-ketoglutarate, a TCA cycle intermediate subsequently oxidized to oxaloacetate. Aspartate is transaminated directly to oxaloacetate. The TCA cycle regenerates oxaloacetate, preventing the complete oxidation of oxaloacetate directly. Oxaloacetate is reduced to L-malate, transported to the cytoplasm, and oxidatively decarboxylated by the $NADP^+$-dependent malic enzyme to pyruvate and NADPH. Pyruvate is completely oxidized in the mitochondria.
9. The amidino group of arginine is required for creatine biosynthesis. Creatine phosphate, a phosphoramide, is a "high-energy" phosphate reservoir for the muscle.
11. Each of these amino acids is catabolized to succinyl-CoA, a TCA cycle intermediate. Succinate is oxidized to L-malate, shuttled to the cytoplasm, and oxidized to oxaloacetate. Oxaloacetate is a gluconeogenic substrate.
13. Phenylalanine and tyrosine are catabolized to 4-hydroxyphenylpyruvate, whose further metabolism is blocked by the dioxygenase deficiency. A diet containing only maintenance levels of phenylalanine and deficient in tyrosine is recommended to prevent accumulation of the 4-hydroxyphenylpyruvate.
15. (a) L-glutathione is synthesized in successive steps catalyzed by γ-glutamyl cysteine synthase and glutathione synthase. Glutathione synthesis is directed by the substrate specificity of these enzymes. (b) Decreased glutathione synthesis would increase the probability of oxidative damage to the cell.

Chapter 20

3. Humans have two carbamoyl syntheses, one in the mitochondria (CPase I) and one in the cytosol (CPase II). The mitochondrial enzyme is used in urea and arginine synthesis, while the cytosolic enzyme (part of the multienzyme complex) is dedicated to the synthesis of pyrimidines and is the first enzyme in that pathway. Bacteria have only one CPase and the regulation is at the first committed enzyme in pyrimidine biosynthesis, ATCase—a form of regulation that makes biochemical sense.
5. Large amounts of dATP would accumulate and inhibit ribonucleotide reductase. The lymphocytes could not make the large amounts of dNTPs needed for DNA synthesis and cell division.
7. Because Lesch-Nyhan patients lack the enzyme hypoxanthine-guanine phosphoribosyltransferase, they accumulate high levels of PRPP, which stimulates purine biosynthesis to high levels, leading to large amounts of uric acid being made. The brain may not have high levels of *de novo* purine biosynthesis and probably relies on the salvage pathway enzymes for its purine nucleotides.

9. Antifolates indirectly inhibit the synthesis of thymidylate and prevent DNA synthesis; therefore cells that are growing rapidly may be killed by methotrexate. Normal cells that are actively dividing (in the cell cycle) and replicating their DNA are also sensitive to the toxic effects of antifolates. This is why we see harmful side effects of anticancer drugs.
11. GTP requires the hydrolysis of eight high-energy bonds, while CTP requires the hydrolysis of five high-energy bonds.
13. Allopurinol inhibits xanthine oxidase, preventing the oxidation of 6-mercaptopurine so that it is not as rapidly degraded.

Chapter 21

1. Lactose synthase is a galactosyltransferase using UDP-galactose and glucose to make the disaccharide lactose. The enzyme is composed of two peptides, the galactosyltransferase and α-lactalbumin. The galactosyltransferase is found in all tissues in the body, but the α-lactalbumin (which has no catalytic activity) is found only in milk. α-Lactalbumin modifies substrate specificity of the galactosyltransferase to use glucose so that lactose can be produced.
4. Sugars have large numbers of hydroxyl groups that form glycosidic bonds with the anomeric carbons of other sugars and can generate a large number of structures from a limited number of monosaccharides (over 80 different glycosidic linkages are known).
7. In mutants that lack Dol-P-Man synthase, the glycosylphosphatidylinositol (GPI) anchor is not made and the glycoproteins that would be on the plasma membrane are secreted outside the cell (not attached).
10. Patients with I-cell disease do not phosphorylate the mannose residues on the glycoproteins that are lysosome-bound. These "lysosomal hydrolases" are therefore secreted.

Chapter 22

1. Phospholipids are essential to living organisms, including humans. Any defects would be lethal in the early stages of development and would never be observed.
3. The lung surfactant dipalmitoylphosphatidylcholine is synthesized by removing the fatty acid in the C-2 position and then reacylating with palmitoyl-CoA. This process is called remodeling.
5. Some eukaryotic membranes contain significant amounts of ether-linked glycerophospholipids (for example, 50% of all phospholipids in heart tissue are ether-linked in the C-1 position). These phospholipids are called plasmalogens when they have a hydrocarbon chain linked to the SN-1 carbon by a vinyl ether linkage and are called alkylacylglycerophospholipids when linked by an ether linkage. Archaebacteria have ether-linked lipids in their cell membranes.
7. This person could still make PGE_2 because linoleic acid (which is considered an essential fatty acid in animals) can be desaturated and elongated to form arachidonic acid.
9. Arachidonic acid is stored in membranes as phospholipids with C_{20} polyunsaturated fatty acids in the SN-2 position. Phospholipase A_2 releases arachidonic acid, which is then used to synthesize prostaglandins, which induce inflammation.

Chapter 23

1. The major regulation occurs with HMG-CoA reductase. This enzyme is regulated at three levels: gene expression, rate of degradation of HMG-CoA reductase, and phosphorylation/dephosphorylation of HMG-CoA reductase. High cholesterol levels reduce HMG-CoA reductase mRNA, which reduces synthesis and increases degradation, and the enzyme already present is phosphorylated to reduce its activity. LDL receptors will be reduced in number.
3. Any of the enzymes in the synthesis of acetoacetate (acetoacetyl-CoA thiolase, HMG-CoA synthase, and HMG-CoA lyase) might be defective. Different isozymes are used for ketone bodies and cholesterol biosynthesis; therefore normal cholesterol biosynthesis would not be a clue for the deficient enzyme.
5. The patient could be treated with lovastatin and a bile-acid-binding resin, which together could lower serum cholesterol levels about 50%. Lovastatin inhibits HMG-CoA reductase, lowering cholesterol biosynthesis. Bile-acid-binding resins increase the excretion of bile acids, which stimulates the liver to convert more cholesterol to bile acids.
9. HDL and other lipoproteins would have a reduced level of cholesterol esters, but serum levels of free cholesterol would be elevated.

Chapter 24

2. Nothing will happen! Normally you would expect a person to go into insulin shock as a result of a drop in blood glucose levels, but a starving person's brain is using ketone bodies for energy and the blood glucose is already low.
4. Progestin and estrogen inhibit secretion of FSH and LH by the pituitary and thus prevent follicular growth and ovulation.
6. The patient most probably has a primary defect.
9. Vitamin D can be considered both a hormone and a vitamin. Its mode of action is like that of other steroid hormones and it is synthesized in the body. It can be given in the diet, for example, in supplemented milk, and would then be called a vitamin.

Chapter 25

2. Investigators used two different strains of TMV that had identifiable proteins. They made different reconstituted viruses and found that the progeny viral proteins were always determined by the type of RNA in the reconstituted virus.
4. At the pH extremes, some of the bases have either positive charges (low pH) or negative charges (high pH), which are disruptive of duplex structure.
6. 7-M urea, 90% formamide, pure water, and the T4 gene-32-encoded protein lower the T_m of the DNA. Increased salt raises the T_m of the DNA duplex. Urea and formamide disrupt hydrophobic forces that hold the duplex together (it was earlier thought that these compounds disrupted hydrogen bonding, but water is a better hydrogen-bonding reagent). Higher salt shields the anionic phosphate groups from each other, limiting the electrostatic repulsion of the DNA strands. The T4 protein binds to single-strand regions and pulls the equilibrium to the denatured state.

8. These proteins would bind to the Z DNA and pull the equilibrium to the Z structure. Therefore, even if the B form predominated in pure DNA at equilibrium, the binding of protein to the Z DNA would cause this region of the DNA to be in the Z configuration.
10. A is a DNA-RNA hybrid duplex.
12. T4 DNA will renature 25 times faster than *E. coli* DNA. Unsheared T4 DNA will anneal about 20 times faster than sheared DNA.

Chapter 26

2. The location of the origin is close to gene *c*, which has the highest number of gene copies. The pattern observed agrees with a bidirectional mode of replication.
4. Both ϕX174 and *E. coli* DNA replication are inhibited by rifampicin, which inhibits RNA polymerase. ϕX174 is inhibited during the replication of the RF since it is dependent on the synthesis of a viral protein A, which generates the 3′-OH by cleavage of the DNA. The synthesis of this protein is dependent on the transcription of gene *A* by RNA polymerase. The initiation of replication in *E. coli* is dependent on RNA polymerase to synthesize the RNA primer at *oriC*.
6. It would take about 33 min to replicate the chromosome, since you have a bidirectional mode of replication with two replication forks. If you had multifork replication (one round of replication starting before the other finished), then a 20-min division time would be possible.
9. This experiment by Cairns and DeLucia showed that DNA Pol I was not the primary DNA polymerase responsible for replicating DNA in *E. coli* and suggested that another enzyme was involved.
12. Reverse transcriptase is the unique viral protein that converts the viral RNA into DNA. A specific tRNA serves as a primer (all template-dependent DNA polymerases require a primer, even if they have a RNA template).
14. Cytosine spontaneously deaminates to uracil, which would then lead to a point mutation if not removed. The thymine base allows the organism to identify uracil produced by spontaneous deamination, which would not be possible if you had A-U base pairs in DNA.

Chapter 27

2. The indirect methods of sequencing would not detect many of the posttranscriptional modifications found in RNA, such as 2′-O-methylation, various base methylations, etc.
6. Digest the circular viral DNA with *Eco*R1 and ligate the structure with DNA ligase. This digestion would generate a hybrid virus with only one 72-bp sequence.
9. Human serum albumin is synthesized in the liver; therefore the levels of mRNA for serum albumin would be high. The cDNA made from liver would have many copies of albumin cDNA, so that the job of cloning this sequence would be easier. All human tissues should contain the albumin gene in one copy per haploid amount of DNA, so any tissue would serve to establish a genomic library.
11. If a fragment of DNA of interest has linked to it the thymidine kinase gene (*tk*), then if a cell has taken up this DNA it will also contain the *tk* gene. If the cell line used for cloning contains a mutant (tk^-) that does not allow it to grow in a selection media (HAT, or medium containing hypoxanthine, amethopterin, and thymidine), then only cells taking up the DNA with the *tk* marker will grow.

Chapter 28

1. The base composition of the newly synthesized RNA would be: C = 20%, G = 25%, U = 15%, and A = 40%.
3. mRNA has a short half-life (a couple of minutes), while tRNA and rRNA are very stable (half-life measured in hours) and accumulate to make up most of the RNA in the cell (95%).
5. The D and T loops in tRNA interact with each other to form the tertiary structure, leaving only the anticodon with a single-stranded loop able to be cleaved by RNase.
11. The 5′ terminus is capped (G^mpppX—), generating a guanosine nucleoside with 2′ and 3′ -OH groups and a 5′-5′ pyrophosphate linkage. The other end of the mRNA has poly(A) added, so the 3′ end is adenosine. Therefore most of the mRNA in the cell has a guanosine and an adenosine 2′ and 3′ -OH and no typical triphosphate ending (pppN—).
15. If an intron or part of an intron containing a stop signal for translation was not removed, this mRNA would be longer but would yield a shorter polypeptide. Alternative splicing would remove the intron or use an alternative splice site, generating a shorter mRNA and a longer polypeptide.

Chapter 29

2. The large size of tRNAs (larger than a three-nucleotide anticodon) results from the requirements of specific charging of the tRNA by its aminoacyl tRNA synthase. The enzyme must recognize its tRNA and attach the correct amino acid (a correspondence sometimes called the second genetic code). Also, the structure of the ribosome requires some unique spacing between the codon–anticodon interaction and the site of peptide bond formation.
4. If tRNAs were not charged correctly, the genetic code would have a high error rate. The correct charging of the tRNA with the correct amino acid is just as important as the codon–anticodon interaction that defines the genetic code.
12. PEST sequences are regions rich in proline, serine, and acetic amino acids that are common to short-lived eukaryotic proteins. Regulatory proteins generally have short half-lives, while structural proteins are more stable, as is required by their function in the cell.
16. The incorporation of the valine analog prevents the formation of the active tetramer structure of hemoglobin. This unassembled globin would be a preferred target for reaction with ubiquitin, which would lead to rapid degradation.

Chapter 30

2. **(a)** Both the wild type and z^- would express a low level of acetylase with no treatment. **(b)** When lactose is added, only the wild type would produce a lot of the acetylase. **(c)** IPTG would induce a high level of acetylase in both strains.
6. IPTG does not have to be metabolized by galactosidase to form an active inducer. It is stable and builds up to high levels in the cell.

8. At the enzyme level, the end product of the histidine pathway (histidine) inhibits the first enzyme in the pathway by simple linear feedback inhibition. Transcription is regulated by attenuation; no feedback repression is observed with this operon.
12. Bacterial genes are regulated in a reversible manner, depending on metabolic needs, while viral genes are turned on only once, in a predesignated sequence.
14. The expression of the *crp* gene is negatively autoregulated by cAMP-CAP. A divergent promoter is activated by cAMP-CAP, and the RNA produced binds to the 5′ end of the *crp* mRNA, producing a rho-independent terminator. The binding of this antisense RNA stops the transcription of the *crp* gene.

Chapter 31

2. In eukaryotes, transcription and processing of the mRNA occurs in the nucleus while translation occurs in the cytosol. Attenuation control is unlikely because you can't form a transcriptional-translational complex.
6. Hypomethylation of the 5′ flanking sequences is correlated with the expression of some genes in eukaryotes. 5-Azacytidine can be incorporated into DNA and generate hypomethylated regions in the DNA, leading to gene activation.
9. Histones bind to DNA by electrostatic interaction of basic amino acids with the negative charge on the phosphate in the deoxyribose phosphate backbone of the DNA duplex. While many regulatory proteins initially interact (nonspecific binding) with DNA by charge interactions, high-affinity binding (at specific binding sites) occurs by hydrophobic interactions, which are stabilized by high salt.
12. The gene for dihydrofolate reductase can be amplified to a high copy number, giving a large gene dose. The large amount of dihydrofolate reductase produced binds the methotrexate, leaving some active enzyme to meet the cell's need for tetrahydrofolate. A DNA probe for dihydrofolate reductase gene could be used to measure the amount of specific DNA amplified (the "dot blot" procedure with genomic DNA could be used to compare normal and resistant cells).
14. The enhancer sequence in the DNA can bind to tissue specific protein(s). The DNA can then loop to allow the enhancer proteins to come in close contact with the promoter of the gene being stimulated, even at great distances (5,000 bp).

Chapter 32

1. **(b)** $D = 5.7 \times 10^{-6}$ cm²/sec.
2. **(a)** $\Delta\Psi = +18$ mV; **(b)** from side 1 to side 2; **(c)** Side 1: 64.3 mM K^+, 50 mM Na^+, 144.3 mM Cl^-. Side 2: 85.7 mM K^+, 85.7 mM Cl^-. $\Delta\Psi = 7.5$ mV.
5. Sucrose uptake should be inhibited by a proton ionophore if uptake is by a proton symport. If a protein-binding system was operational, then membrane vesicles or cells subjected to osmotic shock would be defective in uptake. If a Na^+ symport was involved, then uptake would be dependent on extracellular Na^+. If a PTS was operational, then sucrose phosphorylation would be dependent on PEP and not ATP in a crude cell extract.
10. If Hg^{2+} inhibits the uptake of D-glucose, this suggests that the uptake is dependent on a protein permease (facilitated diffusion) that has — SH groups present.

Chapter 33

1. B cells mature in the bone marrow and then migrate to secondary lymphoid organs, while T cells mature in the thymus and then migrate to different areas of the body.
3. There is the delayed-type hypersensitivity response, in which T cells react with antigens and secrete lymphokines. A second type of T cell, known as a "killer cell," reacts with antigens bound to target cells and causes the death of the latter.
7. Irradiate the mouse from strain A with whole-body radiation to destroy the immune system and then transplant bone marrow and thymus from strain B. This new immune system is now tolerant to transplants from strain B mice.
10. Develop antibodies to the transition-state intermediate or a transition-state intermediate analog.
12. The thiol reagent would disrupt disulfide bonds and allow the peptides that form the antibody structure (four peptide chains) to dissociate.

Chapter 34

3. Expression of some genes is induced by hypomethylation of DNA on the 5′ side of the gene. Therefore some genes, not usually active in adult tissue, may be expressed. For example, some of the genes important in development can become oncogenes when expressed at the "wrong" time (oncofetal genes).
5. RNA viruses such as retroviruses can cause cancer by a reverse transcription mechanism, using reverse transcriptase to produce a DNA copy of their genome that can then integrate into the host-cell genome.
10. HIV can cause different cancers by destroying the immune system (T4 cells) and preventing immune surveillance.

Chapter 35

1. **(a)** $\Delta\Psi = -71$ mV; **(b)** $\Delta\Psi = -63$ mV.
3. Criteria are (i) excitation of proper postsynaptic cell, (ii) presence of the compound in presynaptic terminal, (iii) release upon stimulation, and (iv) receptors on postsynaptic membrane.
6. L-DOPA can pass the blood-brain barrier but dopamine cannot. Inhibitors of dopa carboxylase cannot pass the blood-brain barrier; therefore the DOPA is not converted into dopamine until it enters the brain.
8. The high rate of Na^+ flux can be accounted for only by a pore model, as opposed to the slower carrier-mediated mechanism.

Chapter 36

1. Compounds that absorb visible light have many alternating double bonds, i.e., conjugated double bonds.
4. Rhodopsin is converted to bathorhodopsin and then to rhodopsin.
5. The Schiff's base linkage between the retinal and opsin is protonated in metarhodopsin I but not in metarhodopsin II.
10. Cone cells contain 11-*cis*-retinal bound to proteins like opsin. There are three types of cones, which absorb blue, green, and yellow light, thereby allowing us to see color. Rod cells are connected to each bipolar cell and have multiple bipolar cells connected to each ganglion cell, thereby summing the signals from the rods. The cone cells are not connected in this manner, and therefore are not sensitive to dim light.

Glossary

A

A form. A duplex DNA structure with right-handed twisting in which the planes of the base pairs are tilted about 70° with respect to the helix axis.

Acetal. The product formed by the successive condensation of two alcohols with a single aldehyde. It contains two ether-linked oxygens attached to a central carbon atom.

Acetyl-CoA. Acetyl-coenzyme A, a high-energy ester of acetic acid that is important both in the tricarboxylic acid cycle and in fatty acid biosynthesis.

Actin. A protein found in combination with myosin in muscle and also found as filaments constituting an important part of the cytoskeleton in many eukaryotic cells.

Actinomycin D. An antibiotic that binds to DNA and inhibits RNA chain elongation.

Activated complex. The highest free energy state of a complex in going from reactants to products.

Active site. The region of an enzyme molecule that contains the substrate binding site and the catalytic site for converting the substrate(s) into product(s).

Active transport. The energy-dependent transport of a substance across a membrane.

Adenine. A purine base found in DNA or RNA.

Adenosine. A purine nucleoside found in DNA, RNA, and many cofactors.

Adenosine diphosphate (ADP). The nucleotide formed by adding a pyrophosphate group to the 5′-OH group of adenosine.

Adenosine triphosphate (ATP). The nucleotide formed by adding yet another phosphate group to the pyrophosphate group on ADP.

Adenylate cyclase. The enzyme that catalyzes the formation of cyclic 3′,5′-adenosine monophosphate (cAMP) from ATP.

Adipocyte. A specialized cell that functions as a storage depot for lipid.

Aerobe. An organism that utilizes oxygen for growth.

Affinity chromatography. A column chromatographic technique that employs attached functional groups that have a specific affinity for sites on particular proteins.

Alcohol. A molecule with a hydroxyl group attached to a carbon atom.

Aldehyde. A molecule containing a doubly bonded oxygen and a hydrogen attached to the same carbon atom.

Alleles. Alternative forms of a gene.

Allosteric enzyme. An enzyme whose active site can be altered by the binding of a small molecule at a nonoverlapping site.

Angstrom (Å). A unit of length equal to 10^{-8} cm.

Anomers. The sugar isomers that differ in configuration about the carbonyl carbon atom. This carbon atom is called the anomeric carbon atom of the sugar.

Antibiotic. A natural product that inhibits bacterial growth (is bacteriostatic) and sometimes results in bacterial death (is bacteriocidal).

Antibody. A specific protein that interacts with a foreign substance (antigen) in a specific way.

Anticodon. A sequence of three bases on the transfer RNA that pair with the bases in the corresponding codon on the messenger RNA.

Antigen. A foreign substance that triggers antibody formation and is bound by the corresponding antibody.

Antiparallel β-pleated sheet (β sheet). A hydrogen-bonded secondary structure formed between two or more extended polypeptide chains.

Apoactivator. A regulatory protein that stimulates transcription from one or more genes in the presence of a coactivator molecule.

Asexual reproduction. Growth and cell duplication that does not involve the union of nuclei from cells of opposite mating types.

Asymmetric carbon. A carbon that is covalently bonded to four different groups.

Attenuator. A provisional transcription stop signal.

Autoradiography. The technique of exposing film in the presence of disintegrating radioactive particles. Used to obtain information on the distribution of radioactivity in a gel or a thin cell section.

Autoregulation. The process in which a gene regulates its own expression.

Autotroph. An organism that can form its organic constituents from CO_2.

Auxin. A plant growth hormone usually concentrated in the apical bud.

Auxotroph. A mutant that cannot grow on the minimal medium on which a wild-type member of the same species can grow.

Avogadro's number. The number of molecules in a gram molecular weight of any compound (6.023×10^{23}).

B

B cell. One of the major types of cells in the immune system. B cells can differentiate to form memory cells or antibody-forming cells.

B form. The most common form of duplex DNA, containing a right-handed helix and about 10 (10.5 exactly) base pairs per turn of the helix axis.

β bend. A characteristic way of turning an extended polypeptide chain in a different direction, involving the minimum number of residues.

β oxidation. Oxidative degradation of fatty acids that occurs by the successive oxidation of the β-carbon atom.

β sheet. A sheetlike structure formed by the interaction between two or more extended polypeptide chains.

Base analog. A compound, usually a purine or a pyrimidine, that differs somewhat from a normal nucleic acid base.

Base stacking. The close packing of the planes of base pairs, commonly found in DNA and RNA structures.

Bidirectional replication. Replication in both directions away from the origin, as opposed to replication in one direction only (unidirectional replication).

Bilayer. A double layer of lipid molecules with the hydrophilic ends oriented outward, in contact with water, and the hydrophobic parts oriented inward toward each other.

Bile salts. Derivatives of cholesterol with detergent properties that aid in the solubilization of lipid molecules in the digestive tract.

Biochemical pathway. A series of enzyme-catalyzed reactions that results in the conversion of a precursor molecule into a product molecule.

Bioluminescence. The production of light by a biochemical system.

Blastoderm. The stage in embryogenesis when a unicellular layer at the surface surrounds the yolk mass.

Bond energy. The energy required to break a bond.

Branchpoint. An intermediate in a biochemical pathway that can follow more than one route in subsequent steps.

Buffer. A conjugate acid-base pair that is capable of resisting changes in pH when acid or base is added to the system. This tendency will be maximal when the conjugate forms are present in equal amounts.

C

cAMP. 3′,5′ cyclic adenosine monophosphate. The cAMP molecule plays a key role in metabolic regulation.

CAP. The catabolite gene activator protein, sometimes incorrectly referred to as the CRP protein. The latter term, in small letters (*crp*), should be used to refer to the gene but not to the protein.

Capping. Covalent modification involving the addition of a modified guanidine group in a 5′-5″ linkage. It occurs only in eukaryotes, primarily on mRNA molecules.

Carbohydrate. A polyhydroxy aldehyde or ketone.

Carboxylic acid. A molecule containing a carbon atom attached to a hydroxyl group and to an oxygen atom by a double bond.

Carcinogen. A chemical that can cause cancer.

Carotenoids. Lipid-soluble pigments that are made from isoprene units.

Catabolism. That part of metabolism that is concerned with degradation reactions.

Catabolite repression. The general repression of transcription of genes associated with catabolism that is seen in the presence of glucose.

Catalyst. A compound that lowers the activation energy of a reaction without itself being consumed.

Catalytic site. The site of the enzyme involved in the catalytic process.

Catenane. An interlocked pair of circular structures, such as covalently closed DNA molecules.

Catenation. The linking of molecules without any direct covalent bonding between them, as when two circular DNA molecules interlock like the links in a chain.

cDNA. Complementary DNA, made *in vitro* from the mRNA by the enzyme reverse transcriptase and deoxyribonucleotide triphosphates.

Cell commitment. That stage in a cell's life when it becomes committed to a certain line of development.

Cell cycle. All of those stages that a cell passes through from one cell generation to the next.

Cell line. An established clone originally derived from a whole organism through a long process of cultivation.

Cell lineage. The pedigree of cells resulting from binary fission.

Cell wall. A tough outer coating found in many plant, fungal, and bacterial cells that accounts for their ability to withstand mechanical stress or abrupt changes in osmotic pressure. Cell walls always contain a carbohydrate component and frequently also a peptide and a lipid component.

Chelate. A molecule that contains more than one binding site and frequently binds to another molecule through more than one binding site at the same time.

Chemiosmotic coupling. The coupling of ATP synthesis to an electrochemical potential gradient across a membrane.

Chimeric DNA. Recombinant DNA whose components originate from two or more different sources.

Chiral compound. A compound that can exist in two forms that are nonsuperimposable images of one another.

Chlorophyll. A green photosynthetic pigment that is made of a magnesium dihydroporphyrin complex.

Chloroplast. A chlorophyll-containing photosynthetic organelle, found in eukaryotic cells, that can harness light energy.

Chromatin. The nucleoprotein fibers of eukaryotic chromosomes.

Chromatography. A procedure for separating chemically similar molecules. Segregation is usually carried out on paper or in glass or metal columns with the help of different solvents. The paper or glass columns contain porous solids with functional groups that have limited affinities for the molecules being separated.

Chromosome. A threadlike structure, visible in the cell nucleus during metaphase, that carries the hereditary information.

Chromosome puff. A swollen region of a giant chromosome; the swelling reflects a high degree of transcription activity.

***Cis* dominance.** Property of a sequence or a gene that exerts a dominant effect on a gene to which it is linked.

Cistron. A genetic unit that encodes a single polypeptide chain.

Citric acid cycle. *See* tricarboxylic acid (TCA) cycle.

Clone. One of a group of genetically identical cells or organisms derived from a common ancestor.

Cloning vector. A self-replicating entity to which foreign DNA can be covalently attached for purposes of amplification in host cells.

Coactivator. A molecule that functions in conjunction with a protein apoactivator. For example, cAMP is a coactivator of the CAP protein.

Codon. In a messenger RNA molecule, a sequence of three bases that represents a particular amino acid.

Coenzyme. An organic molecule that associates with enzymes and affects their activity.

Cofactor. A small molecule required for enzyme activity. It could be organic in nature, like a coenzyme, or inorganic in nature, like a metallic cation.

Complementary base sequence. For a given sequence of nucleic acids, the nucleic acids that are related to them by the rules of base pairing.

Configuration. The spatial arrangement in which atoms are covalently linked in a molecule.

Conformation. The three-dimensional arrangement adopted by a molecule, usually a complex macromolecule. Molecules with the same configuration can have more than one conformation.

Consensus sequence. In nucleic acids, the "average" sequence that signals a certain type of action by a specific protein. The sequences actually observed usually vary around this average.

Constitutive enzymes. Enzymes synthesized in fixed amounts, regardless of growth conditions.

Cooperative binding. A situation in which the binding of one substituent to a macromolecule favors the binding of another. For example, DNA cooperatively binds histone molecules, and hemoglobin cooperatively binds oxygen molecules.

Coordinate induction. The simultaneous expression of two or more genes.

Cosmid. A DNA molecule with *cos* ends from λ bacteriophage that can be packaged *in vitro* into a virus for infection purposes.

Cot curve. A curve that indicates the rate of DNA-DNA annealing as a function of DNA concentration and time.

Cytidine. A pyrimidine nucleoside found in DNA and RNA.

Cytochromes. Heme-containing proteins that function as electron carriers in oxidative phosphorylation and photosynthesis.

Cytokinin. A plant hormone produced in root tissue.

Cytoplasm. The contents enclosed by the plasma membrane, excluding the nucleus.

Cytosine. A pyrimidine base found in DNA and RNA.

Cytoskeleton. The filamentous skeleton, formed in the cytoplasm, that is largely responsible for controlling cell shape.

Cytosol. The liquid portion of the cytoplasm, including the macromolecules but not including the larger structures, such as subcellular organelles or cytoskeleton.

D

D loop. An extended loop of single-stranded DNA displaced from a duplex structure by an oligonucleotide.

Dalton. A unit of mass equivalent to the mass of a hydrogen atom (1.66×10^{-24} g).

Dark reactions. Reactions that can occur in the dark, in a process that is usually associated with light, such as the dark reactions of photosynthesis.

***De novo* pathway.** A biochemical pathway that starts from elementary substrates and ends in the synthesis of a biochemical.

Deamination. The enzymatic removal of an amine group, as in the deamination of an amino acid to an α-keto acid.

Dehydrogenase. An enzyme that catalyzes the removal of a pair of electrons (and usually one or two protons) from a substrate molecule.

Denaturation. The disruption of the native folded structure of a nucleic acid or protein molecule; may be due to heat, chemical treatment, or change in pH.

Density-gradient centrifugation. The separation, by centrifugation, of molecules according to their density, in a gradient varying in solute concentration.

Dialysis. Removal of small molecules from a macromolecule preparation by allowing them to pass across a semipermeable membrane.

Diauxic growth. Biphasic growth on a mixture of two carbon sources in which one carbon source is used up before the other one is mobilized. For example, in the presence of glucose and lactose, *E. coli* will utilize the glucose before the lactose.

Difference spectra. Display comparing the absorption spectra of a molecule or an assembly of molecules in different states, for example, those of mitochondria under oxidizing or reducing conditions.

Differential centrifugation. Separation of molecules and/or organelles by sedimentation rate.

Differentiation. A change in the form and pattern of a cell and the genes it expresses as a result of growth and replication, usually during development of a multicellular organism. Also occurs in microorganisms (e.g., in sporulation).

Diploid cell. A cell that contains two chromosomes ($2N$) of each type.

Dipole. A separation of charge within a single molecule.

Directed mutagenesis. In a DNA sequence, an intentional alteration that can be genetically inherited.

Dissociation constant. An equilibrium constant for the dissociation of a molecule into two parts (e.g., dissociation of acetic acid into acetate anion and proton).

Disulfide bridge. A covalent linkage formed by oxidation between two SH groups either in the same polypeptide chain or in different polypeptide chains.

DNA. Deoxyribonucleic acid. A polydeoxyribonucleotide in which the sugar is deoxyribose; the main repository of genetic information in all cells and most viruses.

DNA cloning. The propagation of individual segments of DNA as clones.

DNA library. A mixture of clones, each containing a cloning vector and a segment of DNA from a source of interest.

DNA polymerase. An enzyme that catalyzes the formation of 3′-5′ phosphodiester bonds from deoxyribonucleotide triphosphates.

Domain. A segment of a folded protein structure showing conformational integrity. A domain could comprise the entire protein or just a fraction of the protein. Some proteins, such as antibodies, contain many structural domains.

Dominant. Describing an allele whose phenotype is expressed regardless of whether the organism is homozygous or heterozygous for that allele.

Double helix. A structure in which two helically twisted polynucleotide strands are held together by hydrogen bonding and base stacking.

Duplex. Synonymous with double helix.

Dyad symmetry. Property of a structure that can be rotated by 180° to produce the same structure.

E

Ecdysone. A hormone that stimulates the molting process in insects.

Edman degradation. A systematic method of sequencing proteins, proceeding by stepwise removal of single amino acids from the amino terminal of a polypeptide chain.

Eicosanoid. Any fatty acid with twenty carbons.

Electrophoresis. The movement of particles in an electrical field. A commonly used technique for analysis of mixtures of molecules in solution according to their electrophoretic mobilities.

Elongation factors. Protein factors uniquely required during the elongation phase of protein synthesis. Elongation factor G (EF-G) brings about the movement of the peptidyl-tRNA from the A site to the P site of the ribosome.

Eluate. The effluent from a chromatographic column.

Embryo. Plant or animal at an early stage of development.

Enantiomorphs. Isomers that are mirror images of one another.

Endergonic reaction. A reaction with a positive free energy change.

End-product (feedback) inhibition. The inhibition of the first enzyme in a pathway by the end product of that pathway.

Endocrine glands. Specialized tissues whose function is to synthesize and secrete hormones.

Endonuclease. An enzyme that breaks a phosphodiester linkage at some point within a polynucleotide chain.

Endopeptidase. An enzyme that breaks a polypeptide chain at an internal peptide linkage.

Endoplasmic reticulum. A system of double membranes in the cytoplasm that is involved in the synthesis of transported proteins. The rough endoplasmic reticulum has ribosomes associated with it. The smooth endoplasmic reticulum does not.

Energy charge. The fractional degree to which the AMP-ADP-ATP system is filled with high-energy phosphates (phosphoryl groups).

Enhancer. A DNA sequence that can stimulate transcription at an appreciable distance from the site where it is located. It acts in either orientation and either upstream or downstream from the promoter.

Entropy. The randomness of a system.

Enzyme. A protein that contains a catalytic site for a biochemical reaction.

Epimers. Two stereoisomers with more than one chiral center that differ in configuration at one of their chiral centers.

Equilibrium. In chemistry the point at which the concentrations of two compounds are such that the interconversion of one compound into the other compound does not result in any change in free energy.

Escherichia coli (E. coli). A Gram negative bacterium commonly found in the vertebrate intestine. It is the bacterium most frequently used in the study of biochemistry and genetics.

Established cell line. A group of cultured cells derived from a single origin and capable of stable growth for many generations.

Ether. A molecule containing two carbons linked by an oxygen atom.

Eukaryote. A cell or organism that has a membrane-bounded nucleus.

Excision repair. DNA repair in which a damaged region is replaced.

Excited state. An energy-rich state of an atom or a molecule, produced by the absorption of radiant energy.

Exergonic reaction. A chemical reaction that takes place with a negative change in free energy.

Exon. A segment within a gene that carries part of the coding information for a protein.

Exonuclease. An enzyme that breaks a phosphodiester linkage at one or the other end of a polynucleotide chain so as to release single or small nucleotide residues.

F

F factor. A large bacterial plasmid, known as the sex-factor plasmid because it permits mating between F^+ and F^- bacteria.

Facultative aerobe. An organism that can use molecular oxygen in its metabolism but that also can live anaerobically.

Fatty acid. A long-chain hydrocarbon containing a carboxyl group at one end. Saturated fatty acids have completely saturated hydrocarbon chains. Unsaturated fatty acids have one or more carbon–carbon double bonds in their hydrocarbon chains.

Feedback inhibition. *See* end-product inhibition.

Fermentation. The energy-generating breakdown of glucose or related molecules by a process that does not require molecular oxygen.

Fingerprinting. The characteristic two-dimensional paper chromatogram obtained from the partial hydrolysis of a protein or a nucleic acid.

Fluorescence. The emission of light by an excited molecule in the process of making the transition from the excited state to the ground state.

Frameshift mutations. Insertions or deletions of genetic material that lead to a shift in the translation of the reading frame. The mutation usually leads to nonfunctional proteins.

Free energy. That part of the energy of a system that is available to do useful work.

Furanose. A sugar that contains a five-membered ring as a result of intramolecular hemiacetal formation.

Futile cycle. *See* pseudocycle.

G

G_1 phase. That period of the cell cycle in which preparations are being made for chromosome duplication, which takes place in the S phase.

G_2 phase. That period of the cell cycle between S phase and mitosis (M phase).

Gametes. The ova and the sperm, haploid cells that unite during fertilization to generate a diploid zygote.

Gel exclusion chromatography. A technique that makes use of certain polymers that can form porous beads with varying pore sizes. In columns made from such beads, it is possible to separate molecules, which cannot penetrate beads of a given pore size, from small molecules that can.

Gene. A segment of the genome that codes for a functional product.

Gene amplification. The duplication of a particular gene within a chromosome two or more times.

Gene splicing. The cutting and rejoining of DNA sequences.

General recombination. Recombination that occurs between homologous chromosomes at homologous sites.

Generation time. The time it takes for a cell to double its mass under specified conditions.

Genetic map. The arrangement of genes or other identifiable sequences on a chromosome.

Genome. The total genetic content of a cell or a virus.

Genotype. The genetic characteristics of an organism (distinguished from its observable characteristics, or phenotype).

Globular protein. A folded protein that adopts an approximately globular shape.

Goldman equation. An equation expressing the quantitative relationship between the concentrations of charged species on either side of a membrane and the resting transmembrane potential.

Golgi apparatus. A complex series of double-membrane structures that interact with the endoplasmic reticulum and that serve as a transfer point for proteins destined for other organelles, the plasma membrane, or extracellular transport.

Gluconeogenesis. The production of sugars from nonsugar precursors such as lactate or amino acids. Applies more specifically to the production of free glucose by vertebrate livers.

Glycogen. A polymer of glucose residues in 1,4 linkage and 1,6 linkage at branchpoints.

Glycogenic. Describing amino acids whose metabolism may lead to gluconeogenesis.

Glycolipid. A lipid containing a carbohydrate group.

Glycolysis. The catabolic conversion of glucose to pyruvate with the production of ATP.

Glycoprotein. A protein linked to an oligosaccharide or a polysaccharide.

Glycosaminoglycans. Long, unbranched polysaccharide chains composed of repeating disaccharide subunits in which one of the two sugars is either *N*-acetylglucosamine or *N*-acetylgalactosamine.

Glycosidic bond. The bond between a sugar and an alcohol. Also the bond that links two sugars in disaccharides, oligosaccharides, and polysaccharides.

Glyoxylate cycle. A pathway that uses some of the enzymes of the TCA cycle and some enzymes whereby acetate can be converted into succinate and carbohydrates.

Glyoxysome. An organelle containing key enzymes of the glyoxylate cycle.

Gram molecular weight. For a given compound, the weight in grams that is numerically equal to its molecular weight.

Ground state. The lowest electronic energy state of an atom or a molecule.

Growth factor. A substance that must be present in the growth medium to permit cell proliferation.

Growth fork. The region on a DNA duplex molecule where synthesis is taking place. It resembles a fork in shape, since it consists of a region of duplex DNA connected to a region of unwound single strands.

Guanine. A purine base found in DNA or RNA.

Guanosine. A purine nucleoside found in DNA and RNA.

H

Hairpin loop. A single-stranded complementary region that folds back on itself and base-pairs into a double helix.

Half-life. The time required for the disappearance of one half of a substance.

Haploid cell. A cell containing only one chromosome of each type.

Heavy isotopes. Forms of atoms that contain greater numbers of neutrons (e.g., ^{15}N, ^{13}C).

Helix. A spiral structure with a repeating pattern.

Heme. An iron-porphyrin complex found in hemoglobin and cytochromes.

Hemiacetal. The product formed by the condensation of an aldehyde with an alcohol; it contains one oxygen linked to a central carbon in a hydroxyl fashion and one oxygen linked to the same central carbon by an ether linkage.

Henderson-Hasselbalch equation. An equation that relates the pK_a to the pH and the ratio of the proton acceptor (A^-) and the proton donor (HA) species of a conjugate acid-base pair.

Heterochromatin. Highly condensed regions of chromosomes that are not usually transcriptionally active.

Heteroduplex. An annealed duplex structure between two DNA strands that do not show perfect complementarity. Can arise by mutation, recombination, or the annealing of complementary single-stranded DNAs.

Heteropolymer. A polymer containing more than one type of monomeric unit.

Heterotroph. An organism that requires preformed organic compounds for growth.

Heterozygous. Describing an organism (a heterozygote) that carries two different alleles for a given gene.

Hexose. A sugar with a six-carbon backbone.

High-energy compound. A compound that undergoes hydrolysis with a high negative standard free energy change.

Histones. The family of basic proteins that is normally associated with DNA in most cells of eukaryotic organisms.

Holoenzyme. An intact enzyme containing all of its subunits with full enzymatic activity.

Homologous chromosomes. Chromosomes that carry the same pattern of genes, but not necessarily the same alleles.

Homopolymer. A polymer composed of only one type of monomeric building block.

Homozygous. Describing an organism (a homozygote) that carries two identical alleles for a given gene.

Hormone. A chemical substance made in one cell and secreted so as to influence the metabolic activity of a select group of cells located at other sites in the organism.

Hormone receptor. A protein that is located on the cell membrane or inside the responsive cell and that interacts specifically with the hormone.

Host cell. A cell used for growth and reproduction of a virus.

Hybrid (or chimeric) plasmid. A plasmid that contains DNA from two different organisms.

Hydrogen bond. A weak attractive force between one electronegative atom and a hydrogen atom that is covalently linked to a second electronegative atom.

Hydrolysis. The cleavage of a molecule by the addition of water.

Hydrophilic. Preferring to be in contact with water.

Hydrophobic. Preferring not to be in contact with water, as is the case with the hydrocarbon portion of a fatty acid or phospholipid chain.

Hydrophobic bonding. The association of nonpolar groups with each other in aqueous solution.

Hydroxyapatite. A calcium phosphate gel used, in the case of nucleic acids, to selectively absorb duplex DNA-RNA from a mixture of single-stranded and duplex nucleic acids.

I

Icosahedral symmetry. The symmetry displayed by a regular polyhedron that is composed of twenty equilateral triangular faces with twelve corners.

Imine. A molecule containing a nitrogen atom attached to a carbon atom by a double bond. The nitrogen is also covalently linked to a hydrogen.

Immunofluorescence. A cytological technique in which a specific fluorescent antibody is used to label an antigen. Frequently used to determine the location of an antigen in a tissue or a cell.

Immunoglobulin. A protein made in a B plasma cell and usually secreted; it interacts specifically with a foreign agent. Synonymous with antibody. It is composed of two heavy and two light chains linked by disulfide bonds. Immunoglobulins can be divided into five classes (IgG, IgM, IgA, IgD, and IgE) based on their heavy-chain component.

In vitro. Literally, "in glass," describing whatever happens in a test tube or other receptacle, as opposed to what happens in whole cells of the whole organism (*in vivo*).

Induced fit. A change in the shape of an enzyme that results from the binding of substrate.

Initiation factors. Those protein factors that are specifically required during the initiation phase of protein synthesis.

Intercalating agent. A chemical, usually containing aromatic rings, that can sandwich in between adjacent base pairs in a DNA duplex. The intercalation leads to an adjustment in the DNA secondary structure, as adjacent base pairs are usually close-packed.

Interferon. One of a family of proteins that are liberated by special host cells in the mammal in response to viral infection. The interferons attach to an infected cell, where they stimulate antiviral protein synthesis.

Intervening sequence. *See* intron.

Intron. A segment of the nascent transcript that is removed by splicing. Also refers to the corresponding region in the DNA. Synonymous with intervening sequence.

Inverted repeat. A chromosome segment that is identical to another segment on the same chromosome except that it is oriented in the opposite direction.

Ion-exchange resin. A polymeric resinous substance, usually in bead form, that contains fixed groups with positive or negative charge. A cation exchange resin has negatively charged groups and is therefore useful in exchanging the cationic groups in a test sample. The resin is usually used in the form of a column, as in other column chromatographic systems.

Isoelectric pH. The pH at which a protein has no net charge.

Isomerase. An enzyme that catalyzes an intramolecular rearrangement.

Isomerization. Rearrangement of atomic groups within the same molecule without any loss or gain of atoms.

Isozymes. Multiple forms of an enzyme that differ from one another in one or more of the properties.

K

K_m. *See* Michaelis constant.

Ketogenic. Describing amino acids that are metabolized to acetoacetate and acetate.

Ketone. A functional group of an organic compound in which a carbon atom is double-bonded to an oxygen. Neither of the other substituents attached to the carbon is a hydrogen. Otherwise the group would be called an aldehyde.

Ketone bodies. Refers to acetoacetate, acetone, and β-hydroxybutyrate made from acetyl-CoA in the liver and used for energy in nonhepatic tissue.

Ketosis. A condition in which the concentration of ketone bodies in the blood or urine is unusually high.

Kilobase. One thousand bases in a DNA molecule.

Kinase. An enzyme catalyzing phosphorylation of an acceptor molecule, usually with ATP serving as the phosphate (phosphoryl) donor.

Kinetochore. A structure that attaches laterally to the centromere of a chromosome; it is the site of chromosome tubule attachment.

Krebs cycle. *See* tricarboxylic acid (TCA) cycle.

L

Lampbrush chromosome. Giant diplotene chromosome found in the oocyte nucleus. The loops that are observed are the sites of extensive gene expression.

Law of mass action. The finding that the rate of a chemical reaction is a function of the product of the concentrations of the reacting species.

Leader region. The region of an mRNA between the 5′ end and the initiation codon for translation of the first polypeptide chain.

Lectins. Agglutinating proteins usually extracted from plants.

Ligase. An enzyme that catalyzes the joining of two molecules together. In DNA it joins 3′-OH to 5′ phosphates.

Linkers. Short oligonucleotides that can be ligated to larger DNA fragments, then cleaved to yield overlapping cohesive ends, suitable for ligation to other DNAs that contain comparable cohesive ends.

Linking number. The net number of times one polynucleotide chain crosses over another polynucleotide chain. By convention, right-handed crossovers are given a plus designation.

Lipid. A biological molecule that is soluble in organic solvents. Lipids include steroids, fatty acids, prostaglandins, terpenes, and waxes.

Lipid bilayer (*see* Bilayer). Model for the structure of the cell membrane based on the hydrophobic interaction between phospholipids.

Lipopolysaccharide. Usually refers to a unique glycolipid found in Gram negative bacteria.

Lyase. An enzyme that catalyzes the removal of a group to form a double bond, or the reverse reaction.

Lysogenic virus. A virus that can adopt an inactive (lysogenic) state, in which it maintains its genome within a cell instead of entering the lytic cycle. The circumstances that determine whether a lysogenic (temperate) virus will adopt an inactive state or an active lytic state are often subtle and depend on the physiological state of the infected cell.

Lysosome. An organelle that contains hydrolytic enzymes designed to break down proteins that are targeted to that organelle.

Lytic infection. A virus infection that leads to the lysis of the host cell, yielding progeny virus particles.

M

M phase. That period of the cell cycle when mitosis takes place.

Meiosis. Process in which diploid cells undergo division to form haploid sex cells.

Membrane transport. The facilitated transport of a molecule across a membrane.

Merodiploid. An organism that is diploid for some but not all of its genes.

Mesosome. An invagination of the bacterial cell membrane.

Messenger RNA (mRNA). The template RNA carrying the message for protein synthesis.

Metabolic turnover. A measure of the rate at which already existing molecules of the given species are replaced by newly synthesized molecules of the same type. Usually isotopic labeling is required to measure turnover.

Metabolism. The sum total of the enzyme-catalyzed reactions that occur in a living organism.

Metamorphosis. A change of form, especially the conversion of a larval form to an adult form.

Metaphase. That stage in mitosis or meiosis when all of the chromosomes are lined up on the equator (i.e., an imaginary line that bisects the cell).

Micelle. An aggregate of lipids in which the polar head groups face outward and the hydrophobic tails face inward; no solvent is trapped in the center.

Michaelis constant (K_m). The substrate concentration at which an enzyme-catalyzed reaction proceeds at one-half maximum velocity.

Michaelis-Menten equation (also known as the Henri-Michaelis-Menten equation). An equation relating the reaction velocity to the substrate concentration of an enzyme.

Microtubules. Thin tubules, made from globular proteins, that serve multiple purposes in eukaryotic cells.

Mismatch repair. The replacement of a base in a heteroduplex structure by one that forms a Watson-Crick base pair.

Missense mutation. A change in which a codon for one amino acid is replaced by a codon for another amino acid.

Mitochondrion. An organelle, found in eukaryotic cells, in which oxidative phosphorylation takes place. It contains its own genome and unique ribosomes to carry out protein synthesis of only a fraction of the proteins located in this organelle.

Mitosis. The process whereby replicated chromosomes segregate equally toward opposite poles prior to cell division.

Mobile genetic element. A segment of the genome that can move as a unit from one location on the genome to another, without any requirement for sequence homology.

Molecularity of a reaction. The number of molecules involved in a specific reaction step.

Monolayer. A single layer of oriented lipid molecules.

Mutagen. An agent that can bring about a heritable change (mutation) in an organism.

Mutagenesis. A process that leads to a change in the genetic material that is inherited in subsequent generations.

Mutant. An organism that carries an altered gene or change in its genome.

Mutarotation. The change in optical rotation of a sugar that is observed immediately after it is dissolved in aqueous solution, as the result of the slow approach of equilibrium of a pyranose or a furanose in its α and β forms.

Mutation. The genetically inheritable alteration of a gene or group of genes.

Myofibril. A unit of thick and thin filaments in a muscle fiber.

Myosin. The main protein of the thick filaments in a muscle myofibril. It is composed of two coiled subunits (M_r about 220,000) that can aggregate to form a thick filament, which is globular at each end.

N

Nascent RNA. The initial transcripts of RNA, before any modification or processing.

Negative control. Regulation of the activity by an inhibitory mechanism.

Nernst equation. An equation that relates the redox potential to the standard redox potential and the concentrations of the oxidized and reduced form of the couple.

Nitrogen cycle. The passage of nitrogen through various valence states, as the result of reactions carried out by a wide variety of different organisms.

Nitrogen fixation. Conversion of atmospheric nitrogen into a form that can be converted by biochemical reactions to an organic form. This reaction is carried out by a very limited number of microorganisms.

Nitrogenous base. An aromatic nitrogen-containing molecule with basic properties. Such bases include purines and pyrimidines.

Noncompetitive inhibitor. An inhibitor of enzyme activity whose effect is not reversed by increasing the concentration of substrate molecule.

Nonsense mutation. A change in the base sequence that converts a sense codon (one that specifies an amino acid) to one that specifies a stop (a nonsense codon). There are three nonsense codons.

Northern blotting. *See* Southern blotting.

Nuclease. An enzyme that cleaves phosphodiester bonds of nucleic acids.

Nucleic acids. Polymers of the ribonucleotides or deoxyribonucleotides.

Nucleohistone. A complex of DNA and histone.

Nucleolus. A spherical structure visible in the nucleus during interphase. The nucleolus is associated with a site on the chromosome that is involved in ribosomal RNA synthesis.

Nucleophilic group. An electron-rich group that tends to attack an electron-deficient nucleus.

Nucleosome. A complex of DNA and an octamer of histone proteins in which a small stretch of the duplex is wrapped around a molecular bead of histone.

Nucleotide. An organic molecule containing a purine or pyrimidine base, a five-carbon sugar (ribose or deoxyribose), and one or more phosphate groups.

Nucleus. In eukaryotic cells, the centrally located organelle that encloses most of the chromosomes. Minor amounts of chromosomal substance are found in some other organelles, most notably the mitochondria and the chloroplasts.

O

Okazaki fragment. A short segment of single-stranded DNA that is an intermediate in DNA synthesis. In bacteria, Okazaki fragments are 1,000–2,000 bases in length; in eukaryotes, 100–200 bases in length.

Oligonucleotide. A polynucleotide containing a small number of nucleotides. The linkages are the same as in a polynucleotide; the only distinguishing feature is the small size.

Oligosaccharide. A molecule containing a small number of sugar residues joined in a linear or a branched structure by glycosidic bonds.

Oncogene. A gene of cellular or viral origin that is responsible for rapid, unruly growth of animal cells.

Operon. A group of contiguous genes that are coordinately regulated by two *cis*-acting elements, a promoter and an operator. Found only in prokaryotic cells.

Optical activity. The property of a molecule that leads to rotation of the plane of polarization of plane-polarized light when the latter is transmitted through the substance. Chirality is a necessary and sufficient property for optical activity.

Organelle. A subcellular membrane-bounded body with a well-defined function.

Osmotic pressure. The pressure generated by the mass flow of water to that side of a membrane-bounded structure that contains the higher concentration of solute molecules. A stable osmotic pressure is seen in systems in which the membrane is not permeable to some of the solute molecules.

Oxidation. The loss of electrons from a compound.

Oxidative phosphorylation. The formation of ATP as the result of the transfer of electrons to oxygen.

Oxido-reductase. An enzyme that catalyzes oxidation-reduction reactions.

P

Palindrome. A sequence of bases that reads the same in both directions on opposite strands of the DNA duplex (e.g., GAATTC).

Pentose. A sugar with five carbon atoms.

Pentose phosphate pathway. The pathway involving the oxidation of glucose-6-phosphate to pentose phosphates and further reactions of pentose phosphates.

Peptide. An organic molecule in which a covalent amide bond is formed between the α-amino group of one amino acid and the α-carboxyl group of another amino acid, with the elimination of a water molecule.

Peptide mapping. Same as fingerprinting.

Peptidoglycan. The main component of the bacterial cell wall, consisting of a two-dimensional network of heteropolysaccharides running in one direction, cross-linked with polypeptides running in the perpendicular direction.

Periplasm. The space between the inner and outer membranes of a bacterium.

Permease. A protein that catalyzes the transport of a specific small molecule across a membrane.

Peroxisomes. Subcellular organelles that contain flavin-requiring oxidases and that regenerate oxidized flavin by reaction with oxygen.

Phenotype. The observable trait(s) that result from the genotype in cooperation with the environment.

Phenylketonuria. A human disease caused by a genetic deficiency in the enzyme that converts phenylalanine to tyrosine. The immediate cause of the disease is an excess of phenylalanine, which can be alleviated by a diet low in phenylalanine.

Pheromone. A hormonelike substance that acts as an attractant.

Phosphodiester. A molecule containing two alcohols esterified to a single molecule of phosphate. For example, the backbone of nucleic acids is connected by 5′-3′ phosphodiester linkages between the adjacent individual nucleotide residues.

Phosphogluconate pathway. Another name for the pentose phosphate pathway. This name derives from the fact that 6-phosphogluconate is an intermediate in the formation of pentoses from glucose.

Phospholipid. A lipid containing charged hydrophilic phosphate groups; a component of cell membranes.

Phosphorylation. The formation of a phosphate derivative of a biomolecule.

Photoreactivation. DNA repair in which the damaged region is repaired with the help of light and an enzyme. The lesion is repaired without excision from the DNA.

Photosynthesis. The biosynthesis that directly harnesses the chemical energy resulting from the absorption of light. Frequently used to refer to the formation of carbohydrates from CO_2 that occurs in the chloroplasts of plants or the plastids of photosynthetic microorganisms.

Pitch length (or pitch). The number of base pairs per turn of a duplex helix.

Plaque. A circular clearing on a lawn of bacterial or cultured cells, resulting from cell lysis and production of phage or animal virus progeny.

Plasma membrane. The membrane that surrounds the cytoplasm.

Plasmid. A circular DNA duplex that replicates autonomously in bacteria. Plasmids that integrate into the host genome are called episomes. Plasmids differ from viruses in that they never form infectious nucleoprotein particles.

Polar group. A hydrophilic (water-loving) group.

Polar mutation. A mutation in one gene that reduces the expression of a gene or genes distal to the promoter in the same operon.

Polarimeter. An instrument for determining the rotation of polarization of light as the light passes through a solution containing an optically active substance.

Polyamine. A hydrocarbon containing more than two amino groups.

Polycistronic messenger RNA. In prokaryotes, an RNA that contains two or more cistrons; note that only in prokaryotic mRNAs can more than one cistron be utilized by the translation system to generate individual proteins.

Polymerase. An enzyme that catalyzes the synthesis of a polymer from monomers.

Polynucleotide. A chain structure containing nucleotides linked together by phosphodiester (5′-3′) bonds. The polynucleotide chain has a directional sense with a 5′ and a 3′ end.

Polynucleotide phosphorylase. An enzyme that polymerizes ribonucleotide diphosphates. No template is required.

Polypeptide. A linear polymer of amino acids held together by peptide linkages. The polypeptide has a directional sense, with an amino and a carboxyl-terminal end.

Polyribosome (polysome). A complex of an mRNA and two or more ribosomes actively engaged in protein synthesis.

Polysaccharide. A linear or branched chain structure containing many sugar molecules linked by glycosidic bonds.

Porphyrin. A complex planar structure containing four substituted pyrroles covalently joined in a ring and frequently containing a central metal atom. For example, heme is a porphyrin with a central iron atom.

Positive control. A system that is turned on by the presence of a regulatory protein.

Posttranslational modification. The covalent bond changes that occur in a polypeptide chain after it leaves the ribosome and before it becomes a mature protein.

Primary structure. In a polymer, the sequence of monomers and the covalent bonds.

Primer. A structure that serves as a growing point for polymerization.

Primosome. A multiprotein complex that catalyzes synthesis of RNA primer at various points along the DNA template.

Prochiral molecule. A nonchiral molecule that may react with an enzyme so that two groups that have a mirror-image relationship to each other are treated differently.

Prokaryote. A unicellular organism that contains a single chromosome, no nucleus, no membrane-bound organelles, and has characteristic ribosomes and biochemistry.

Promoter. That region of the gene that signals RNA polymerase binding and the initiation of transcription.

Prophage. The silent phage genome. Some prophages integrate into the host genome; others replicate autonomously. The prophage state is maintained by a phage-encoded repressor.

Prophase. The stage in meiosis or mitosis when chromosomes condense and become visible as refractile bodies.

Proprotein. A protein that is made in an inactive form, so that it requires processing to become functional.

Prostaglandin. An oxygenated eicosanoid that has a hormonal function. Prostaglandins are unusual hormones in that they usually have effects only in that region of the organism where they are synthesized.

Prosthetic group. Synonymous with coenzyme except that a prosthetic group is usually more firmly attached to the enzyme it serves.

Protamines. Highly basic, arginine-rich proteins found complexed to DNA in the sperm of many invertebrates and fish.

Protein subunit. One of the components of a complex multicomponent protein.

Proteoglycan. A protein-linked heteropolysaccharide in which the heteropolysaccharide is usually the major component.

Protist. A relatively undifferentiated organism that can survive as a single cell.

Proton acceptor. A functional group capable of accepting a proton from a proton donor molecule.

Proton motive force (Δ*p*). The thermodynamic driving force for proton translocation. Expressed quantitatively as $\Delta G_{H+}/F$ in units of volts.

Protooncogene. A cellular gene that can undergo modification to a cancer-causing gene (oncogene).

Pseudocycle. A sequence of reactions that can be arranged in a cycle but that usually do not function simultaneously in both directions. Also called a futile cycle, since the net result of simultaneous functioning in both directions would be the expenditure of energy without accomplishing any useful work.

Pulse-chase. An experiment in which a short labeling period is followed by the addition of an excess of the same, unlabeled compound to dilute out the labeled material.

Purine. A heterocyclic ring structure with varying functional groups. The purines adenine and guanine are found in both DNA and RNA.

Puromycin. An antibiotic that inhibits polypeptide synthesis by competing with aminoacyl-tRNA for the ribosomal binding site A.

Pyranose. A simple sugar containing the six-membered pyran ring.

Pyrimidine. A heterocyclic six-membered ring structure. Cytosine and uracil are the main pyrimidines found in RNA, and cytosine and thymine are the main pyrimidines found in DNA.

Pyrophosphate. A molecule formed by two phosphates in anhydride linkage.

Q

Quaternary structure. In a protein, the way in which the different folded subunits interact to form the multisubunit protein.

R

R group. The distinctive side chain of an amino acid.

R loop. A triple-stranded structure in which RNA displaces a DNA strand by DNA-RNA hybrid formation in a region of the DNA.

Rapid-start complex. The complex that RNA polymerase forms at the promoter site just before initiation.

Recombination. The transfer to offspring of genes not found together in either of the parents.

Redox couple. An electron donor and its corresponding oxidized form.

Redox potential (*E*). The relative tendency of a pair of molecules to release or accept an electron. The standard redox potential ($E°$) is the redox potential of a solution containing the oxidant and reductant of the couple at standard concentrations.

Regulatory enzyme. An enzyme in which the active site is subject to regulation by factors other than the enzyme substrate. The enzyme frequently contains a nonoverlapping site for binding the regulatory factor that affects the activity of the active site.

Regulatory gene. A gene whose principal product is a protein designed to regulate the synthesis of other genes.

Renaturation. The process of returning a denatured structure to its original native structure, as when two single strands of DNA are reunited to form a regular duplex, or an unfolded polypeptide chain is returned to its normal folded three-dimensional structure.

Repair synthesis. DNA synthesis following excision of damaged DNA.

Repetitive DNA. A DNA sequence that is present in many copies per genome.

Replica plating. A technique in which an impression of a culture is taken from a master plate and transferred to a fresh plate. The impression can be of bacterial clones or phage plaques.

Replication fork. The Y-shaped region of DNA at the site of DNA synthesis; also called a growth fork.

Replicon. A genetic element that behaves as an autonomous replicating unit. It can be a plasmid, phage, or bacterial chromosome.

Repressor. A regulatory protein that inhibits transcription from one or more genes. It can combine with an inducer (resulting in specific enzyme induction) or with an operator element (resulting in repression).

Resonance hybrid. A molecular structure that is a hybrid of two structures that differ in the locations of some of the electrons. For example, the benzene ring can be drawn in two ways, with double bonds in different positions. The actual structure of benzene is in between these two equivalent structures.

Restriction-modification system. A pair of enzymes found in most bacteria (but not eukaryotic cells). The restriction enzyme recognizes a certain sequence in duplex DNA and makes one cut in each unmodified DNA strand at or near the recognition sequence. The modification enzyme methylates (or modifies) the recognition sequence, thus protecting it from the action of the restriction enzyme.

Reverse transcriptase. An enzyme that synthesizes DNA from an RNA template, using deoxyribonucleotide triphosphates.

Rho factor. A protein involved in the termination of transcription of some messenger RNAs.

Ribose. The five-carbon sugar found in RNA.

Ribosomal RNA (rRNA). The RNA parts of the ribosome.

Ribosomes. Small cellular particles made up of ribosomal RNA and protein. They are the site, together with mRNA, of protein synthesis.

RNA (ribonucleic acid). A polynucleotide in which the sugar is ribose.

RNA polymerase. An enzyme that catalyzes the formation of RNA from ribonucleotide triphosphates, using DNA as a template.

RNA splicing. The excision of a segment of RNA, followed by a rejoining of the remaining fragments.

Rolling-circle replication. A mechanism for the replication of circular DNA. A nick in one strand allows the 3′ end to be extended, displacing the strand with the 5′ end, which is also replicated, to generate a double-stranded tail that can become larger than the unit size of the circular DNA.

S

S phase. The period during the cell cycle when the chromosome is replicated.

Salting in. The increase in solubility that is displayed by typical globular proteins upon the addition of small amounts of certain salts, such as ammonium sulfate.

Salting out. The decrease in protein solubility that occurs when salts such as ammonium sulfate are present at high concentrations.

Salvage pathway. A family of reactions that permits, for instance, nucleosides as well as purine and pyrimidine bases resulting from the partial breakdown of nucleic acids to be reutilized in nucleic acid synthesis.

Satellite DNA. A DNA fraction whose base composition differs from that of the main component of DNA, as revealed by the fact that it bands at a different density in a CsCl gradient. Usually repetitive DNA or organelle DNA.

Scissile. Capable of being cut smoothly or split easily.

Second messenger. A diffusible small molecule, such as cAMP, that is formed at the inner surface of the plasma membrane in response to a hormonal signal.

Secondary structure. In a protein or a nucleic acid, any repetitive folded pattern that results from the interaction of the corresponding polymeric chains.

Semiconservative replication. Duplication of DNA in which the daughter duplex carries one old strand and one new strand.

Sigma factor. A subunit of RNA polymerase that recognizes specific sites on DNA for initiation of RNA synthesis.

Single-copy DNA. A region of the genome whose sequence is present only once per haploid complement.

Somatic cell. Any cell of an organism that cannot contribute its genes to a subsequent generation.

SOS system. A set of DNA repair enzymes and regulatory proteins that regulate their synthesis so that maximum synthesis occurs when the DNA is damaged.

Southern blotting. A method for detecting a specific DNA restriction fragment, developed by Edward Southern. DNA from a gel electrophoresis pattern is blotted onto nitrocellulose paper; then the DNA is denatured and fixed on the paper. Subsequently the pattern of specific sequences in the Southern blot can be determined by hybridization to a suitable probe and autoradiography. A Northern blot is similar, except that RNA is blotted instead onto the nitrocellulose paper.

Splicing. *See* RNA splicing.

Sporulation. Formation from vegetative cells of metabolically inactive cells that can resist extreme environmental conditions.

Stacking energy. The energy of interaction that favors the face-to-face packing of purine and pyrimidine base pairs.

Steady state. In enzyme-kinetic analysis, the time interval when the rate of reaction is approximately constant with time. The term is also used to describe the state of a living cell where the concentrations of many molecules are approximately constant because of a balancing between their rate of synthesis and breakdown.

Stem cell. A cell from which other cells stem or arise by differentiation.

Stereoisomers. Isomers that are nonsuperimposable mirror images of each other.

Steroids. Compounds that are derivatives of a tetracyclic structure composed of a cyclopentane ring fused to a substituted phenanthrene nucleus.

Structural domain. An element of protein tertiary structure that recurs in many structures.

Structural gene. A gene encoding the amino acid sequence of a polypeptide chain.

Structural protein. A protein that serves a structural function.

Subunit. Individual polypeptide chains in a protein.

Supercoiled DNA. Supertwisted, covalently closed duplex DNA.

Suppressor gene. A gene that can reverse the phenotype of a mutation in another gene.

Suppressor mutation. A mutation that restores a function lost by an initial mutation and that is located at a site different from the initial mutation.

Svedberg unit (S). The unit used to express the sedimentation constant *s*: $1S = 10^{-13}$ s. The sedimentation constant *s* is proportional to the rate of sedimentation of a molecule in a given centrifugal field and is related to the size and shape of the molecule.

Synapse. The chemical connection for communication between two nerve cells or between a nerve cell and a target cell such as a muscle cell.

Synapsis. The pairing of homologous chromosomes, seen during the first meiotic prophase.

T

Tandem duplication. A duplication in which the repeated regions are immediately adjacent to one another.

TCA cycle. *See* tricarboxylic acid cycle.

Template. A polynucleotide chain that serves as a surface for the absorption of monomers of a growing polymer and thereby dictates the sequence of the monomers in the growing chain.

Termination factors. Proteins that are exclusively involved in the termination reactions of protein synthesis on the ribosome.

Terpenes. A diverse group of lipids made from isoprene precursors.

Tertiary structure. In a protein or nucleic acid, the final folded form of the polymer chain.

Tetramer. Structure resulting from the association of four subunits.

Thioester. An ester of a carboxylic acid with a thiol or mercaptan.

Thymidine. One of the four nucleosides found in DNA.

Thymine. A pyrimidine base found in DNA.

Topoisomerase. An enzyme that changes the extent of supercoiling of a DNA duplex.

Transamination. Enzymatic transfer of an amino group from an α-amino acid to an α-keto acid.

Transcription. RNA synthesis that occurs on a DNA template.

Transduction. Genetic exchange in bacteria that is mediated via phage.

Transfection. An artificial process of infecting cells with naked viral DNA.

Transfer RNA (tRNA). Any of a family of low-molecular-weight RNAs that transfer amino acids from the cytoplasm to the template for protein synthesis on the ribosome.

Transferase. An enzyme that catalyzes the transfer of a molecular group from one molecule to another.

Transformation. Genetic exchange in bacteria that is mediated via purified DNA. In somatic cell genetics the term is also used to indicate the conversion of a normal cell to one that grows like a cancer cell.

Transgenic. Describing an organism that contains transfected DNA in the germ line.

Transition state. The activated state in which a molecule is best suited to undergoing a chemical reaction.

Translation. The process of reading a messenger RNA sequence for the specified amino acid sequence it contains.

Transport protein. A protein whose primary function is to transport a substance from one part of the cell to another, from one cell to another, or from one tissue to another.

Tricarboxylic acid (TCA) cycle. The cyclical process whereby acetate is completely oxidized to CO_2 and water, and electrons are transferred to NAD^+ and FAD. The TCA cycle is localized to the mitochondria in eukaryotic cells and to the plasma membrane in prokaryotic cells. Also called the Krebs cycle.

Trypsin. A proteolytic enzyme that cleaves peptide chains next to the basic amino acids arginine and lysine.

Tryptic peptide mapping. The technique of generating a chromatographic profile characteristic of the fragments resulting from trypsin enzyme cleavage of the protein.

Tumorigenesis. The mechanism of tumor formation.

Turnover number. The maximum number of molecules of substrate that can be converted to product per active site per unit time.

U

Ultracentrifuge. A high-speed centrifuge that can attain speeds up to 60,000 rpm and centrifugal fields of 500,000 times gravity. Useful for characterizing and/or separating macromolecules.

Unidirectional replication. *See* bidirectional replication.

Unwinding proteins. Proteins that help to unwind double-stranded DNA during DNA replication.

Urea cycle. A metabolic pathway in the liver that leads to the synthesis of urea from amino groups and CO_2. The function of the pathway is to convert the ammonia resulting from catabolism to a nontoxic form, which is subsequently secreted.

UV irradiation. Electromagnetic radiation with a wavelength shorter than that of visible light (200–390 nm). Causes damage to DNA (mainly pyrimidine dimers).

V

van der Waals forces. Refers to two types of interactions, one attractive and one repulsive. The attractive forces are due to favorable interactions among the induced instantaneous dipole moments that arise from fluctuations in the electron charge densities of neighboring nonbonded atoms. Repulsive forces arise when noncovalently bonded atoms come too close together.

Viroids. Pathogenic agents, mostly of plants, that consist of short (usually circular) RNA molecules.

Virus. A nucleic-acid-protein complex that can infect and replicate inside a specific host cell to make more virus particles.

Vitamin. A trace organic substance required in the diet of some species. Many vitamins are precursors of coenzymes.

W

Watson-Crick base pairs. The type of hydrogen-bonded base pairs found in DNA, or comparable base pairs found in RNA. The base pairs are A-T, G-C, and A-U.

Wild-type gene. The form of a gene (allele) normally found in nature.

Wobble. A proposed explanation for base-pairing that is not of the Watson-Crick type and that often occurs between the 3′ base in the codon and the 5′ base in the anticodon.

X

X-ray crystallography. A technique for determining the structure of molecules from the x-ray diffraction patterns that are produced by crystalline arrays of the molecules.

Y

Ylid. A compound in which adjacent, covalently bonded atoms, both having an electronic octet, have opposite charges.

Z

Z form. A duplex DNA structure in which there is the usual type of hydrogen bonding between the base pairs but in which the helix formed by the two polynucleotide chains is left-handed rather than right-handed.

Zwitterion. A dipolar ion with spatially separated positive and negative charges. For example, most amino acids are zwitterions, having a positive charge on the α-amino group and a negative charge on the α-carboxyl group but no net charge on the overall molecule.

Zygote. A cell that results from the union of haploid male and female sex cells. Zygotes are diploid.

Zymogen. An inactive precursor of an enzyme. For example, trypsin exists in the inactive form trypsinogen before it is converted to its active form, trypsin.

Credits

Portions of this text have been adopted from Geoffrey Zubay, *Genetics*. To be published.

Line Art

Chapter 1

1.1: From Sylvia S. Mader, *Inquiry Into Life,* 6th ed. Copyright © 1991 Wm. C. Brown Communications, Inc., Dubuque, Iowa. All Rights Reserved. Reprinted by permission. **1.10, 1.12, 1.13, and 1.22:** Illustration copyright by Irving Geis. **1.15:** Illustration copyright by Irving Geis. Coordinates courtesy of D. C. Phillips, Oxford.

Chapter 3

3.1, 3.9, and 3.10: Illustration copyright by Irving Geis.

Chapter 4

4.1, 4.2, 4.4*a*, 4.5, 4.8, and 4.9*a*, *b*: Illustration copyright by Irving Geis. **4.11:** Line art courtesy of Karl A. Piez. **4.12–4.14, 4.18*c*, and 4.20:** Illustration copyright by Irving Geis. **4.21, 4.23, 4.24, and 4.28–4.34:** From Jane S. Richardson. **4.39*a* and 4.42:** From Lansing M. Prescott, et al., *Microbiology.* Copyright © 1990 Wm. C. Brown Communications, Inc. Dubuque, Iowa. All Rights Reserved. Reprinted by permission. **Box 4C: Figures 1–4:** Courtesy of Irving Geis.

Chapter 5

5.2, 5.7–5.11, 5.13 and 5.14: Illustration copyright by Irving Geis. **5.16:** From John W. Hole, Jr., *Human Anatomy and Physiology,* 5th ed. Copyright © 1990 Wm. C. Brown Communications, Inc., Dubuque, Iowa. All Rights Reserved. Reprinted by permission. **5.23 and 5.24:** From Jane S. Richardson. **5.25:** Illustration copyright by Irving Geis.

Chapter 6

6.16: Illustration copyright by Irving Geis. **6.21** (*top*): From B. Alberts, *Molecular Biology of the Cell.* Copyright © 1983 Garland Publishing Inc., New York, N.Y. Reprinted by permission.

Chapter 7

7.43*a*: Reprinted by permission from Dr. Vincent Marchesi. **7.43*b*:** From S. B. Shohet and S. E. Lux, *Hospital Practice 19* (Nov. 1984):90. Redrawn based on a drawing by Robert Margulies. Reprinted by permission.

Chapter 9

9.4, 9.5, 9.19, and 9.30: Illustration copyright by Irving Geis.

Chapter 15

15.27: Illustration copyright by Irving Geis.

Chapter 25

25.1 and 25.4: From Robert Tamarin, *Principles of Genetics,* 3d ed. Copyright © 1991 Wm. C. Brown Communications, Inc., Dubuque, Iowa. All Rights Reserved. Reprinted by permission. **25.9*a*, *b*:** Reprinted by permission from *Nature,* vol. 171, p. 737. Copyright © 1953 Macmillan Magazines Ltd. **25.13:** Illustration copyright by Irving Geis. **25.14:** From A. Wang et al., "Left-handed Double Helical DNA: Variations in the Backbone Conformation" in *Science,* 211. Copyright © 1981 by the American Association for the Advancement of Science. Reprinted by permission. **25.28:** From Geoffrey Zubay, *Genetics.* Copyright © 1987 Benjamin/Cummings Publishing Company Inc. Reprinted by permission of the author. **25.30*b*:** From Robert F. Weaver and Philip W. Hedrick, *Basic Genetics.* Copyright © 1991 Wm. C. Brown Communications, Inc., Dubuque, Iowa. All Rights Reserved. Reprinted by permission.

Chapter 26

26.12: From D. L. Ollis et al., "Structure of large fragment of *Escherichia coli* DNA polymerase I complexed with dTMP." Reprinted by permission from *Nature,* Vol. 313, pp. 762–766. Copyright © 1985 Macmillan Magazines Ltd. **26.13:** From C. M. Joyce and T. A. Steitz, "DNA polymerase I: From crystal structure to function via genetics" in *Trends in Biochemical Sciences,* 12:288, 1987. Copyright © 1987 Elsevier Trends Journals, Cambridge, England. Reprinted with permission. **26.27:** From J. Tooze, *The Molecular Biology of DNA Tumor Viruses.* Copyright © 1980 Cold Spring Harbor Press, New York. Reprinted by permission. **26.29:** From Robert F. Weaver and Philip W. Hedrick, *Basic Genetics.* Copyright © 1991 Wm. C. Brown Communications, Inc., Dubuque, Iowa. All Rights Reserved. Reprinted by permission. **26.33:** From Geoffrey Zubay, *Genetics.* Copyright © 1987 Benjamin/Cummings Publishing Co. Inc. Reprinted by permission of the author.

Chapter 27

27.11: From J. L. Marx, "Multiplying genes by leaps and bounds" in *Science,* 240:1408–1410, June 10, 1988. Copyright © 1988 by the American Association for the Advancement of Science. Reprinted by permission. **27.13:** From Robert F. Weaver and Philip W. Hedrick, *Basic Genetics.* Copyright © 1991 Wm. C. Brown Communications, Inc., Dubuque, Iowa. All Rights Reserved. Reprinted by permission. **27.34:** From J. R. Riordan et al., "Identification of the cystic fibrosis gene: Cloning and characterization of complementary DNA" in *Science,* 245:1066, Sept. 8, 1989. Copyright © 1989 by the American Association for the Advancement of Science. Reprinted by permission.

Chapter 28

28.2: From Geoffrey Zubay, *Genetics.* Copyright © 1987 Benjamin/Cummings Publishing Company Inc., Menlo Park, CA. Reprinted by permission of the author. **28.3*b*:** Illustration copyright by Irving Geis. **28.6:** From Robert F. Weaver and Philip W. Hedrick, *Genetics.* Copyright © 1989 Wm. C. Brown Communications, Inc., Dubuque, Iowa. All Rights Reserved. Reprinted by permission. **28.7:** Source: R. R. Gutell et al., "Comparative anatomy of 16S-like ribosomal RNA" in *Progress in Nucleic Acid Research and Molecular Biology,* 32:155–216, 1985. Copyright © 1985 Academic Press Inc., San Diego, CA. **Box 28A:** Courtesy of Bishwajit Nag and Robert R. Traut. **28.8 and 28.9:** From Robert F. Weaver and Philip W. Hedrick, *Genetics.* Copyright © 1989 Wm. C. Brown Communications, Inc., Dubuque, Iowa. All Rights Reserved. Reprinted by permission. **28.17*b*:** From Robert Tamarin, *Principles of Genetics,* 3d ed. Copyright © 1991 Wm. C. Brown Communications, Inc., Dubuque, Iowa. All Rights Reserved. Reprinted by permission.

Chapter 29

29.5*b*: From Quigley and Rich, *Science,* 185:436, 1974. Copyright © 1974 by the American Association for the Advancement of Science, Washington, DC. Reprinted by permission. **29.6:** From C. Bernabau and J. A. Lake, *Proceedings. National Academy of Sciences USA,* 79:3111–3115, 1982. Copyright © 1982 National Academy of Sciences, Washington, DC. Reprinted by permission of the author. **29.8:** From Robert F. Weaver and Philip W. Hedrick, *Genetics.* Copyright © 1989 Wm. C. Brown Communications, Inc., Dubuque, Iowa.

All Rights Reserved. Reprinted by permission. **29.11:** From L. H. Schulman and J. Abelson, "Recent excitement in understanding transfer RNA identity" in *Science,* 240:1591, June 17, 1988. Copyright © 1988 by the American Association for the Advancement of Science, Washington, DC. Reprinted by permission. **29.23:** From *Molecular Cell Biology,* 2nd edition. By James Darnell, Harvey Lodish, and David Baltimore. Copyright © 1990 by Scientific American Books. Reprinted with permission of W. H. Freeman and Company.

Chapter 31

31.16: From Geoffrey Zubay, *Genetics.* Copyright © 1987 Benjamin/Cummings Publishing Company Inc., Menlo Park, CA. Reprinted by permission of the author.

Chapter 32

32.16*b:* From E. Schneider and K. Altendorf, "Bacterial adenosine 5′-triphosphate synthase (F1FO): Purification and reconstitution of Fo complexes and biochemical and functional characterization of their subunits" in *Microbiological Reviews,* 51:477–497, 1987. Copyright © 1987 American Society for Microbiology, Washington, DC. Reprinted by permission. **32.31*c:*** From A. C. Steven et al., "Ultrastructure of a periodic protein layer in the outer membrane of *Escherichia coli.*" Reproduced from the *Journal of Cell Biology,* 1977, vol. 72, p. 292 by copyright permission of the Rockefeller University Press. **32.32:** Source: M. S. Weiss et al., "The three-dimensional structure of porin from *Rhodobacter capsulatus* at 3 A resolution" in *FEBS Letters* 267:268–272, 1990. Copyright © 1990 Elsevier Science Publishers B.V., Amsterdam, Netherlands. **32.32:** Source: M. S. Weiss et al, "The three-dimensional structure of porin from *Rhodobacter capsulatus* at 3 A resolution" on *FEBS Letters* 267:268–272, 1990. © 1990 Elsevier Science Publishers B.V., Amsterdam, Netherlands.

Chapter 33

33.2, 33.6, 33.7, 33.10, and 33.16–33.19: From Geoffrey Zubay, *Genetics.* Copyright © 1987 Benjamin/Cummings Publishing Company Inc., Menlo Park, CA. Reprinted by permission of the author.

Chapter 34

34.1: From B. Alberts et al., *Molecular Biology of the Cell,* 2d ed. Copyright © 1989 Garland Publishing Inc., New York, NY. Reprinted by permission. **34.2:** © Scientific American, Inc., George V. Kelvin. **34.3:** From J. Yunis, "The chromosomal basis of human neoplasia" in *Science,* 221:227–236, Jan. 1, 1983. Copyright © 1983 by the American Association for the Advancement of Science. Reprinted by permission. **34.5:** From Robert F. Weaver and Philip W. Weaver, *Genetics.* Copyright © 1989 Wm. C. Brown Communications, Inc., Dubuque, Iowa. All Rights Reserved. Reprinted by permission. **34.6:** From Geoffrey Zubay, *Genetics.* Copyright © 1987 Benjamin/Cummings Publishing Company Inc., Menlo Park, CA. Reprinted by permission of the author. **34.7:** From L. Tong, et al., "Structure of the *ras* protein" in *Science,* 245:243, July 21, 1989. Copyright © 1989 by the American Association for the Advancement of Science. Reprinted by permission.

Chapter 35

35.10: From W. A. Catterall, "Structure and function of voltage-sensitive ion channels" in *Science* 242:50, Oct. 7, 1988. Copyright © 1988 American Association for the Advancement of Science. Reprinted by permission. **35.12:** Reprinted by permission from Suzanne Black. **35.22:** From N. Unwin et al., "Arrangement of the acetylcholine receptor subunits in the resting and desensitized states, determined by cryoelectron microscopy of crystallized *Torpedo* postsynaptic membranes." Reproduced from the *Journal of Cell Biology,* 1988, vol. 107, p. 1123 by copyright permission of the Rockefeller University Press. **35.23:** From M. Noda et al., "Structural homology of *Torpedo California* acetylcholine receptor subunits." Reprinted by permission from *Nature,* vol. 302, p. 528. Copyright © 1983 Macmillan Magazines Ltd.

Index

Page numbers followed by an "F" refer to illustrations; "T" indicates material in tables.

A

B

D

E

F

I

J

K

L

M

N

O

P

T

U

X

Y

Z